Pascal Lorenz (Ed.)

Networking – ICN 2001

First International Conference on Networking
Colmar, France, July 9-13, 2001
Proceedings, Part II

Springer

Series Editors

Gerhard Goos, Karlsruhe University, Germany
Juris Hartmanis, Cornell University, NY, USA
Jan van Leeuwen, Utrecht University, The Netherlands

Volume Editor

Pascal Lorenz
University of Haute Alsace
IUT de Colmar – Department GTR
34 rue du Grillenbreit
68008 Colmar, France
E-mail: lorenz@ieee.org

Cataloging-in-Publication Data applied for

Die Deutsche Bibliothek - CIP-Einheitsaufnahme

Networking : first international conference ; proceedings / ICN 2001,
Colmar, France, July 9 - 13, 2001. Pascal Lorenz (ed.). - Berlin ;
Heidelberg ; New York ; Barcelona ; Hong Kong ; London ; Milan ; Paris ;
Singapore ; Tokyo : Springer
 Pt. 2 . - (2001)
 (Lecture notes in computer science ; Vol. 2094)
 ISBN 3-540-42303-6

CR Subject Classification (1998): C.2, K.4.4, H.4.3, H.5.1, H.3, K.6.4-5

ISSN 0302-9743
ISBN 3-540-42303-6 Springer-Verlag Berlin Heidelberg New York

This work is subject to copyright. All rights are reserved, whether the whole or part of the material is concerned, specifically the rights of translation, reprinting, re-use of illustrations, recitation, broadcasting, reproduction on microfilms or in any other way, and storage in data banks. Duplication of this publication or parts thereof is permitted only under the provisions of the German Copyright Law of September 9, 1965, in its current version, and permission for use must always be obtained from Springer-Verlag. Violations are liable for prosecution under the German Copyright Law.

Springer-Verlag Berlin Heidelberg New York
a member of BertelsmannSpringer Science+Business Media GmbH

http://www.springer.de

© Springer-Verlag Berlin Heidelberg 2001
Printed in Germany

Typesetting: Camera-ready by author, data conversion by Steingräber Satztechnik GmbH, Heidelberg
Printed on acid-free paper   SPIN 10839419   06/3142   5 4 3 2 1 0

Preface

The International Conference on Networking (ICN01) is the first conference in its series aimed at stimulating technical exchange in the emerging and important field of networking. On behalf of the International Advisory Committee, it is our great pleasure to welcome you to the International Conference on Networking. Integration of fixed and portable wireless access into IP and ATM networks presents a cost effective and efficient way to provide seamless end-to-end connectivity and ubiquitous access in a market where demands on Mobile and Cellular Networks have grown rapidly and predicted to generate billions of dollars in revenue. The deployment of broadband IP - based technologies over Dense Wavelength Division Multiplexing (DWDM) and integration of IP with broadband wireless access networks (BWANs) are becoming increasingly important. In addition, fixed core IP/ATM networks are constructed with recent move to IP/MPLS over DWDM. More over, mobility introduces further challenges in the area that have neither been fully understood nor resolved in the preceding network generation. This first Conference ICN01 has been very well perceived by the International networking community. A total of 300 papers from 39 countries were submitted, from which 168 have been accepted. Each paper has been reviewed by several members of the scientific Program Committee.

The program covers a variety of research topics which are of current interest, such as mobile and wireless networks, Internet, traffic control, QoS, switching techniques, Voice over IP (VoIP), optical networks, Differentiated and Integrated services, IP and ATM networks, routing techniques, multicasting and performance evaluation, testing and simulation and modeling. Together with four tutorials and four Keynote Speeches, these technical presentations will address the latest research results from the international industries and academia and reports on findings from mobile, satellite and personal communications on 3rd and 4th generation research projects and standardization.

We would like to thank the scientific program committee members and the referees. Without their support, the program organization of this conference would not have been possible. We are also indebted to many individuals and organizations that made this conference possible (Association "Colmar-Liberty", GdR CNRS ARP, Ministère de la Recherche, Université de Haute Alsace, Ville de Colmar, France Telecom, IEEE, IEE, IST, WSES). In particular, we thank the members of the Organizing Committee for their help in all aspects of the organization of this conference.

We wish that you will enjoy this International Conference on Networking at Colmar, France and that you will find it a useful forum for the exchange of ideas and results and recent findings. We also hope that you will be able to spend some times to visit Colmar, with its beautiful countryside and its major cultural attractions.

General Chair
Pascal LORENZ
University of Haute Alsace
IUT de Colmar, France
lorenz@ieee.org

Technical Program Chair
Guy OMIDYAR
Center for Wireless Communications
National University of Singapore, SG
and Computer Sciences Corporation, USA
gomidyar@cwc.nus.edu.sg

International Scientific Committee

H. Abouaissa (France) - University of Haute Alsace
R. Addie (Australia) - University of Southern Queensland
K. Begain (UK) - University of Bradford
A. Benslimane (France) - University of Belfort-Montbeliard
B. Bing (Singapore) - Ngee Ann Polytechnic
A. Brandwajn (USA) - University of California Santa Cruz
J.P. Coudreuse (France) - Mitsubishi
J. Crowcroft (UK) - University College London
S. Fdida (France) - LIP6
E. Fulp (USA) - Wake Forest University
B. Gavish (USA) - Vanderbilt University
H. Guyennet (France) - University of Franche-Comte
Z. Hulicki (Poland) - University of Cracow
R. Israel (France) - IEEE
A. Jajszczyk (Poland) - University of Mining & Metallurgy
A. Jamalipour (Australia) - University of Sydney
S. Jiang (Singapore) - National University of Singapore
S. Karnouskos (Germany) - GMD FOKUS
G. Kesidis (USA) – Pennsylvania State University
D. Khotimsky (USA) - Lucent Bell Labs
D. Kofman (France) - ENST Paris
S. Kota (USA) - Lockeed Martin
D. Kouvatsos (UK) - University of Bradford
S. Kumar (USA) - Ericsson
G.S. Kuo (Taiwan) - National Central University
F. Le Faucheur (France) - Cisco
M. Lee (Korea) - Dongshin University
P. Lorenz (France) - University of Haute Alsace
Z. Mammeri (France) - University of Toulouse
N. Mastorakis (Greece) - Military Institutions of University Education
H. Mouftah (Canada) - Queen's University
G. Omidyar (USA) - Computer Sciences Corp. (Program Chair)
J.J. Pansiot (France) - University of Strasbourg
M. Potts (Switzerland) - Martel
G. Pujolle (France) - University of Versailles-Saint-Quentin
S. Rao (Switzerland) - Ascom
A. Reid (UK) - British Telecom
S. Ritzenthaler (France) - Alcatel
P. Rolin (France) - ENST Bretagne
D. Sadok (Brazil) - Federal University of Pernambuco
R. Saracco (Italy) - CSELT
G. Swallow (USA) - Cisco
H. Tobiet (France) - Clemessy
M. Trehel (France) - University of Franche-Comte

V.A. Villagra (Spain) - University of Madrid
E. Vazquez Gallo (Spain) - University of Madrid
O. Yang (Canada) - University of Ottawa

Table of Contents   Part II

Mobility Management

Dynamic Resource Management Scheme for Multimedia Services
in Wireless Communication Networks 1
*D.-E. Lee, Chungwoon University, Korea ; B.-J. Lee, J.-W. Ko, Y.-C. Kim,
Chonbuk National University, Korea*

An Adaptive Handover-Supporting Routing Method
for ATM Based Mobile Networks, and Its Adaptation to IP Scenarios 11
*S. Szabo, S. Imre, Budapest University of Technology and Economics,
Hungary*

Location Stamps for Digital Signatures:
A New Service for Mobile Telephone Networks 20
*M. Kabatnik, University of Stuttgart, Germany ; A. Zugenmaier, University of
Freiburg, Germany*

A New Method for Scalable and Reliable Multicast System
for Mobile Networks 31
M. Hayashi, Hitachi Europe, France ; C. Bonnet, Institute Eurecom, France

An Adaptive Mobility Management Scheme to Support Internet Host Mobility   41
M. Woo, Sejong University, Korea

TCP Analysis

Modeling and Analysis of TCP Enhancement over Heterogeneous Links 51
M. Liu, N. Ehsan, University of Michigan, USA

TCP Throughput Guarantee Using Packet Buffering 61
S. Choi, C. Kim, Seoul National University, Korea

Modular TCP Handoff Design in STREAMS-Based TCP/IP Implementation ... 71
*W. Tang, L. Cherkasova, L. Russell, Hewlett Packard Labs., USA ;
M.W. Mutka, Michigan State University, USA*

An Efficient TCP Flow Control and Fast Recovery Scheme
for Lossy Networks 82
H.Y. Liao, Y.C. Chen, C.L. Lee, National Chiao Tung University, Taiwan

Bandwidth Tradeoff between TCP and Link-Level FEC 97
C. Barakat, E. Altman, INRIA, France

QoS (I)

Supporting QoS for Legacy Applications 108
C. Tsetsekas, S. Maniatis, I.S. Venieris, National Technical University of Athens, Greece

An Open Architecture for Evaluating Arbitrary Quality of Service Mechanisms in Software Routers 117
K. Wehrle, University of Karlsruhe, Germany

Measurement-Based IP Transport Resource Manager Demonstrator 127
V. Räisänen, Nokia Research Center, Finland

Packet-Size Based Queuing Algorithm for QoS Support 137
M.C. Choi, H.L. Owen, Georgia Tech Research Institute, USA ; J. Sokol, Siemens AG, Germany

Backbone Network Design with QoS Requirements 148
H.-H. Yen, F.Y.-S. Lin, National Taiwan University, Taiwan

Ad Hoc Networks

A Multi-path QoS Routing Protocol in a Wireless Mobile ad Hoc Network 158
W.-H. Liao, National Central University, Taiwan ; Y.-C. Tseng, National Chiao-Tung University, Taiwan ; S.-L. Wang, J.-P. Sheu, National Central University, Taiwan

Study of a Unicast Query Mechanism for Dynamic Source Routing in Mobile ad Hoc Networks 168
B.-C. Seet, B.-S. Lee, C.-T. Lau, Nanyang Technological University, Singapore

Ad-hoc Filesystem:
A Novel Network Filesystem for Ad-hoc Wireless Networks 177
K. Yasuda, T. Hagino, Keio University, Japan

A Review of Current On-demand Routing Protocols 186
M. Abolhasan, T. Wysocki, University of Wollongong, Australia ; E. Dutkiewicz, Motorola Australia Research Centre, Australia

Security

Construction of Data Dependent Chaotic Permutation Hashes to Ensure Communications Integrity 196
J. Scharinger, Johannes Kepler University, Austria

Secure Communication: A New Application for Active Networks 206
M. Günter, M. Brogle, T. Braun, University of Berne, Switzerland

Deployment of Public-Key Infrastructure in Wireless Data Networks 217
A.K. Singh, Infosys Technologies Limited, India

A Scalable Framework for Secure Group Communication 225
L.-C. Wuu, H.-C. Chen, National YunLin University, Taiwan

Authenticating Multicast Streams in Lossy Channels
Using Threshold Techniques 239
M. Al-Ibrahim, University of Wollongong, Australia ; J. Pieprzyk, Macquarie University, Australia

QoS (II)

Tuning of QoS Aware Load Balancing Algorithm (QoS-LB)
for Highly Loaded Server Clusters 250
*K. Kaario, Honeywell Industrial Automation & Control, Finland ;
T. Hämäläinen, J. Zhang, University of Jyväskylä, Finland*

The Incremental Deployability of Core-Stateless Fair Queuing 259
Y. Blanpain, H.-Y. Hsieh, R. Sivakumar, Georgia Institute of Technology, USA

A New IP Multicast QoS Model on IP Based Networks 268
H.S. Eissa, T. Kamel, Electronics Research Institute, Egypt

Integrated Management of QoS-Enable Networks Using QAME 277
*L. Zambenedetti Granville, L.M. Rockenbach Tarouco, M. Bartz Ceccon,
M.J. Bosquiroli Almeida, Federal University of Rio Grande do Sul, Brazil*

On Web Quality of Service: Approaches to Measurement
of End-to-End Response Time 291
M. Tsykin, Fujitsu Australia Limited, Australia

MPLS

Path Computation for Traffic Engineering in MPLS Networks 302
G. Banerjee, D. Sidhu, University of Maryland, USA

Minimum Regret Approach to Network Management under Uncertainty
with Applications to Connection Admission Control and Routing 309
V. Marbukh, National Institute of Standards and Technology, USA

MPLS Restoration Scheme Using Least-Cost Based Dynamic Backup Path 319
*G. Ahn, Electronics and Telecommunications Research Institute, Korea ;
W. Chun, Chungnam National University, Korea*

Connection Management in MPLS Networks Using Mobile Agents 329
S. Yucel, Marconi Communications, USA ; T. Saydam, University of Delaware, USA

General Connection Blocking Bounds and an Implication of Billing
for Provisioned Label-Switched Routes in an MPLS Internet Cloud 339
G. Kesidis, Pennsylvania State University, USA ; L. Tassiulas, University of Maryland, USA

Switches

FPCF Input-Queued Packet Switch for Variable-Size Packets 348
P. Homan, J. Bester, University of Ljubljana, Slovenia

A Cost-Effective Hardware Link Scheduling Algorithm
for the Multimedia Router (MMR) 358
*M.B. Caminero, C. Carrión, F.J. Quiles, University of Castilla-La Mancha, Spain ; J. Duato, Polytechnical University of Valencia, Spain ;
S. Yalamanchili, Georgia Institute of Technology, USA*

The Folded Hypercube ATM Switches 370
J.S. Park, N.J. Davis IV, Virginia Polytechnic Institute and State University, USA

Open Software Architecture for Multiservice Switching System 380
H.-J. Park, Y.-I. Choi, B.-S. Lee, K.-P. Jun, Electronics & Telecommunication Research Institute, Korea

A Multicast ATM Switch Based on PIPN 390
S.F. Oktug, Istanbul Technical University, Turkey

CORBA

Concurrent Access to Remote Instrumentation
in CORBA-Based Distributed Environment 399
A. Stranjak, Lucent Technologies, Ireland ; D. Kovačić, I. Čavrak, M. Žagar, University of Zagreb, Croatia

Design and Implementation
of CORBA-Based Integrated Network Management System 409
J.-H. Kwon, HyComm Incorporated, USA ; J.-T. Park, Kyungpook National University, Korea

Framework for Real-Time CORBA Development 422
Z. Mammeri, J. Rodriguez, University of Toulouse, France ;
P. Lorenz, IUT de Colmar, University of Haute Alsace, France

Development of Accounting Management Based Service Environment
in Tina, Java and Corba Architectures 438
A. Sekkaki, University Hassan II, Morocco ; L.M. Cáceres Alvarez,
W. Tatsuya Watanabe, C. Becker Westphall, Federal University of Santa
Catarina, Brazil

A QoS System for CaTV Networks 449
J. Leal, J.M. Fornés, University of Sevilla, Spain

Mobile Agents

Towards Manageable Mobile Agent Infrastructures 458
P. Simões, P. Marques, L. Silva, J. Silva, F. Boavida, University of Coimbra,
Portugal

Dynamic Agent Domains in Mobile Agent Based Network Management 468
R. Sugar, S. Imre, Technical University of Budapest, Hungary

Networking in a Service Platform Based on Mobile Agents 478
M. Palola, VTT Electronics, Finland

Realizing Distributed Intelligent Networks
Based on Distributed Object and Mobile Agent Technologies 488
M.K. Perdikeas, O.I. Pyrovolakis, A.E. Papadakis, I.S. Venieris, National
Technical University of Athens, Greece

ATM Networks (I)

A New Cut-Through Forwarding Mechanism
for ATM Multipoint-to-Point Connections 497
A. Papadopoulos, Computers Technology Institute, Greece ;
T. Antonakopoulos, V. Makios, University of Patras, Greece

Cell-by-Cell Round Robin Service Discipline for ATM Networks 507
H.M. Mokhtar, R. Pereira, M. Merabti, Liverpool John Moores University, UK

Delay and Departure Analysis of CBR Traffic in AAL MUX
with Bursty Background Traffic 517
C.G. Park, D.H. Han, Sunmoon University, Korea

Virtual Path Layout in ATM Path with Given Hop Count 527
S. Choplin, INRIA, France

Simulation-Based Stability of a Representative, Large-Scale ATM Network
for a Distributed Call Processing Architecture 538
R. Citro, Intel Corporation, USA ; S. Ghosh, Stevens Institute of Technology, USA

Voice over IP (I)

Proposed Architectures for the Integration of H.323 and QoS over IP Networks   549
R. Estepa, J. Leal, J.A. Ternero, J.M. Vozmediano, University of Sevilla, Spain

Third-Party Call Control in H.323 Networks – A Case Study 559
A. Miloslavski, V. Antonov, L. Yegoshin, S. Shkrabov, J. Boyle, G. Pogosyants, N. Anisimov, Genesys Telecommunication Labs, USA

Measurement-Based MMPP Modeling of Voice Traffic in Computer Networks
Using Moments of Packet Interarrival Times 570
N.S. Kambo, D.Z. Deniz, T. Iqbal, Eastern Mediterranean University, Turkey

Evaluation of End-to-End QoS Mechanisms in IP Networks 579
F.A. Shaikh, S. McClellan, University of Alabama at Birmingham, USA

Web-Enabled Voice over IP Call Center 590
S. Kuhlins, University of Mannheim, Germany ; D. Gutacker, OSI mbH, Germany

Active Networks

ANMP: Active Network Management Platform
for Telecommunications Applications 599
W.-K. Hong, M.-J. Jung, Korea Telecom, Korea

An Active Network Architecture: Distributed Computer or Transport Medium .   612
E. Hladká, Z. Salvet, Masaryk University, Czech Republic

An Active Network for Improving Performance of Traffic Flow
over Conventional ATM Service 620
E. Rashid, T. Araki, Hirosaki University, Japan ; T. Nakamura, Tohoku University, Japan

An Active Programmable Harness
for Measurement of Composite Network States 628
J.I. Khan, A.U. Haque, Kent State University, USA

Protocol Design of MPEG-4 Media Delivery with Active Networks 639
S. Go, J.W. Wong, University of Waterloo, Canada ; Z. Wu, Bond University, Australia

ATM Networks (II)

Prediction and Control of Short-Term Congestion in ATM Networks
Using Artificial Intelligence Techniques 648
G. Corral, A. Zaballos, J. Camps, J.M. Garrell, University Ramon Llull, Spain

Monitoring the Quality of Service on an ATM Network Carrying MMB
Traffic Using Weighted Significance Data 658
A.A.K. Mouharam, M.J. Tunnicliffe, Kingston University, UK

ATM Traffic Prediction
Using Artificial Neural Networks and Wavelet Transforms 668
*P. Solís Barreto, Catholic University of Goiás, Brazil ; R. Pinto Lemos,
Federal University of Goiás, Brazil*

Threshold-Based Connection Admission Control Scheme in ATM Networks:
A Simulation Study 677
*X. Yuan, North Carolina A & T State University, USA ; M. Ilyas, Florida
Atlantic University, USA*

ABR Congestion Control in ATM Networks Using Neural Networks 687
K. Dimyati, C.O. Chow, University of Malaya, Malaysia

Voice over IP (II)

An Architecture for the Transport of IP Telephony Services 697
S. Guerra, J. Vinyes, D. Fernández, Technical University of Madrid, Spain

Architectural Framework for Using Java Servlets in a SIP Environment 707
R. Glitho, R. Hamadi, R. Huie, Ericsson Research Canada, Canada

A Practical Solution for Delivering Voice over IP 717
*S. Milanovic, Serco Group plc, Italy; Z. Petrovic, University of Belgrade,
Yugoslavia*

QoS Guaranteed Voice Traffic Multiplexing Scheme over VoIP Network
Using DiffServ 726
E.-J. Ha, J.-H. Kwon, J.-T. Park, Kyungpook National University, Korea

VoIP over MPLS Networking Requirements 735
*J.-M. Chung, Oklahoma State University, USA ; E. Marroun, H. Sandhu, Cisco
Systems, USA ; S.-C. Kim, Oklahoma State University, USA*

Video Communications

A System Level Framework for Streaming 3-D Meshes over Packet Networks . 745
G. Al-Regib, Y. Altunbasak, Georgia Institute of Technology, USA

Techniques to Improve Quality-of-Service in Video Communications
via Best Effort Networks 754
B.E. Wolfinger, M. Zaddach, Hamburg University, Germany

Simulation of a Video Surveillance Network
Using Remote Intelligent Security Cameras 766
*J.R. Renno, M.J. Tunnicliffe, G.A. Jones, Kingston University, UK ;
D.J. Parish, Loughborough University, UK*

Cooperative Video Caching for Interactive and Scalable VoD Systems 776
E. Ishikawa, C. Amorim, Federal University of Rio de Janeiro, Brazil

Optimal Dynamic Rate Shaping for Compressed Video Streaming 786
M. Kim, Y. Altunbasak, Georgia Institute of Technology, USA

ATM Networks (III)

IP Stack Emulation over ATM 795
I.G. Goossens, I.M. Goossens, Free University of Brussels, Belgium

ATM Network Restoration
Using a Multiple Backup VPs Based Self-Healing Protocol 805
S.N. Ashraf, INT, France ; C. Lac, France Télécom R&D, France

A New Consolidation Algorithm for Point-to-Multipoint ABR Service
in ATM Networks 815
*M. Shamsuzzaman, A.K. Gupta, B.-S. Lee, Nanyang Technological University,
Singapore*

Experimental TCP Performance Evaluation
on DiffServ Assured Forwarding over ATM SBR Service 825
*S. Ano, T. Hasegawa, KDDI R&D Laboratories Inc, Japan ; N. Decre, ENST,
France*

PMS: A PVC Management System for ATM Networks 836
C. Yang, S. Phan, National Research Council of Canada, Canada

Modelization

Buffer-Size Approximation for the Geo/D/1/K Queue 845
P. Linwong, Tohoku University, Japan ; A. Fujii, Miyagi University, Japan ;
Y. Nemoto, Tohoku University, Japan

Client-Server Design Alternatives: Back to Pipes but with Threads 854
B. Roussev, Susquehanna University, USA ; J. Wu, Florida Atlantic University,
USA

Towards a Descriptive Approach
to Model Adaptable Communication Environments 867
A.T.A. Gomes, S. Colcher, L.F.G. Soares, PUC-Rio, Brazil

All-to-All Personalized Communication Algorithms
in Chordal Ring Networks.. 877
H. Masuyama, H. Taniguchi, T. Miyoshi, Tottori University, Japan

Formal and Practical Approach to Load Conditions in High Speed Networks ... 890
A. Pollak, University of the Armed Forces Munich, Germany

Author Index 895

Table of Contents   Part I

3rd to 4th Generation

Bandwidth Management for QoS Support in Mobile Networks 1
S.-H. Lee, D.-S. Jung, Electronics & Telecommunication Research Institute, Korea ; S.-W. Park, Hannam University, Korea

3G and Beyond & Enabled Adaptive Mobile Multimedia Communication 12
T. Kanter, Ericsson Radio Systems AB, Sweden ; T. Rindborg, D. Sahlin, Ericsson Utvecklings, Sweden

Creation of 3rd Generation Services
in the Context of Virtual Home Environment 27
J. Oliveira, University of Porto, Portugal ; R. Roque, Portugal Telekom, Portugal ; S. Sedillot, INRIA, France ; E. Carrapatoso, University of Porto, Portugal

Internet (I)

WLL Link Layer Protocol for QoS Support of Multi-service Internet 37
H. Pham, B. Lavery, James Cook University, Australia ; H.N. Nguyen, Vienna University of Technology, Austria

The Earliest Deadline First Scheduling with Active Buffer Management
for Real-Time Traffic in the Internet 45
X. Hei, D.H.K. Tsang, Hong Kong University of Science and Technology, Hong Kong

Pricing and Provisioning for Guaranteed Internet Services 55
Z. Fan, University of Birmingham, UK

Price Optimization of Contents Delivery Systems with Priority 65
K. Yamori, Y. Tanaka, Waseda University, Japan ; H. Akimaru, Asahi University, Japan

An Approach to Internet-Based Virtual Call Center Implementation 75
M. Popovic, V. Kovacevic, University of Novi Sad, Yugoslavia

Traffic Control

Implementation and Characterization of an Advanced Scheduler 85
F. Risso, Politecnico of Torino, Italy

A Performance Study of Explicit Congestion Notification (ECN)
with Heterogeneous TCP Flows 98
R. Kinicki, Z. Zheng, Worcester Polytechnic Institute, USA

Diffusion Model of RED Control Mechanism 107
R. Laalaoua, S. Jedruś, T. Atmaca, T. Czachórski, INT, France

A Simple Admission Control Algorithm for IP Networks 117
K. Kim, P. Mouchtaris, S. Samtani, R. Talpade, L. Wong, Telcordia
Technologies, USA

An Architecture for a Scalable Broadband IP Services Switch 124
M.V. Hegde, M. Naraghi-Poor, J. Bordes, C. Davis, O. Schmid, M. Maher,
Celox Networks Inc., USA

Mobile and Wireless IP

On Implementation of Logical Time in Distributed Systems Operating
over a Wireless IP Network 137
D.A. Khotimsky, Lucent Technologies, USA ; I.A. Zhuklinets, Mozhaysky
University, Russia

Performance of an Inter-segment Handover Protocol
in an IP-based Terrestrial/Satellite Mobile Communications Network 147
L. Fan, M.E. Woodward, J.G. Gardiner, University of Bradford, UK

Performance Evaluation of Voice-Data Integration
for Wireless Data Networking 157
M. Chatterjee, S.K. Das, The University of Texas at Arlington, USA ;
G.D. Mandyam, Nokia Research Center, USA

A Hard Handover Control Scheme Supporting IP Host Mobility 167
Y. Takahashi, N. Shinagawa, T. Kobayashi, YRP Mobile Telecommunications
Key Technology Research Laboratories, Japan

Internet (II)

A Flexible User Authentication Scheme for Multi-server Internet Services 174
W.-J. Tsaur, Da-Yeh University, Taiwan

NetLets: Measurement-Based Routing
for End-to-End Performance over the Internet 184
N.S.V. Rao, Oak Ridge National Laboratory, USA ; S. Radhakrishnan,
B.-Y. Choel, University of Oklahoma, USA

Using the Internet in Transport Logistics –
The Example of a Track & Trace System 194
K. Jakobs, C. Pils, M. Wallbaum, Technical University of Aachen, Germany

Distributed Management of High-Layer Protocols and Network Services
through a Programmable Agent-Based Architecture 204
*L.P. Gaspary, L.F. Balbinot, R. Storch, F. Wendt, L. Rockenbach Tarouco,
Federal University of Rio Grande do Sul, Brazil*

Differentiated Services

A Buffer-Management Scheme for Bandwidth and Delay Differentiation
Using a Virtual Scheduler 218
*R. Pletka, P. Droz, IBM Research, Switzerland ; B. Stiller, ETH Zürich,
Switzerland*

Enhanced End-System Support for Multimedia Transmission
over the Differentiated Services Network 235
*H. Wu, Tsinghua University, China ; H.-R. Shao, Microsoft Research, China ;
X. Li, Tsinghua University, China*

Investigations into the Per-Hop Behaviors of DiffServ Networks 245
Z. Di, H.T. Mouftah, Queen's University, Canada

An Adaptive Bandwidth Scheduling for Throughput and Delay Differentiation   256
*H.-T. Nguyen, H. Rzehak, University of Federal Armed Forces Munich,
Germany*

Evaluation of an Algorithm for Dynamic Resource Distribution
in a Differentiated Services Network 266
*E.G. Nikolouzou, P.D. Sampatakos, I.S. Venieris, National Technical
University of Athens, Greece*

GPRS and Cellular Networks

Quality of Service Management in GPRS Networks 276
P. Stuckmann, F. Müller, Aachen University of Technology, Germany

Case Studies and Results on the Introduction of GPRS
in Legacy Cellular Infrastructures 286
*C. Konstantinopoulou, K. Koutsopoulos, P. Demestichas, E. Matsikoudis,
M. Theologou, National Technical University of Athens, Greece*

Traffic Analysis of Multimedia Services in Broadband Cellular Networks 296
*P. Fazekas, S. Imre, Budapest University of Technology and Economics,
Hungary*

Scheduling Disciplines in Cellular Data Services
with Probabilistic Location Errors 307
J.-L. Chen, H.-C. Cheng, H.-C. Chao, National Dong Hwa University, Taiwan

WDM and Optical Networks

Restoration from Multiple Faults in WDM Networks
without Wavelength Conversion 317
C.-C. Sue, S.-Y. Kuo, National Taiwan University, Taiwan

An All-Optical WDM Packet-Switched Network Architecture
with Support for Group Communication 326
*M.R. Salvador, S.Heemstra de Groot, D. Dey, University of Twente,
The Netherlands*

Performance Consideration for Building
the Next Generation Multi-service Optical Communications Platforms 336
S. Dastangoo, Onex™ Communications Corporation, USA

Traffic Management in Multi-service Optical Network 348
H. Elbiaze, T. Atmaca, INT, France

Performance Comparison of Wavelength Routing Optical Networks
with Chordal Ring and Mesh-Torus Topologies 358
*M.M. Freire, University of Beira Interior, Portugal ; H.J.A. da Silva,
University of Coimbra, Portugal*

Differentiated and Integrated Services

Achieving End-to-End Throughput Guarantee for Individual TCP Flows
in a Differentiated Services Network 368
X. He, H. Che, The Pennsylvania State University, USA

Performance Evaluation of Integrated Services in Local Area Networks 378
J. Ehrensberger, Swiss Federal Institute of Technology, Switzerland

Integrating Differentiated Services with ATM 388
S. Manjanatha, R. Bartoš, University of New Hampshire, USA

Management and Realization of SLA for Providing Network QoS 398
*M. Hashmani, M. Yoshida, NS Solutions Corporation, Japan ; T. Ikenaga,
Y. Oie, Kyushu Institute of Technology, Japan*

Optimal Provisioning and Pricing of Internet Differentiated Services
in Hierarchical Markets 409
*E.W. Fulp, Wake Forest University, USA ; D.S. Reeves, N.C. State University,
USA*

Keynote Speech

Adding Interactive Services in a Video Broadcasting Network 419
R. Jäger, BetaResearch, Germany

Wireless ATM

A Discrete-Time Queuing Analysis of the Wireless ATM Multiplexing System   429
M.M. Ali, X. Zhang, J.F. Hayes, Concordia University, Canada

A QoS Based Distributed Method for Resource Allocation
in Unlicensed Wireless ATM Systems 439
G.F. Marias, L. Merakos, University of Athens, Greece

An Adaptive Error Control Mechanism for Wireless ATM 449
P.R. Denz, A.A. Nilsson, North Carolina State University, USA

A Review of Call Admission Control Schemes in Wireless ATM Networks 459
*D.D. Vergados, N.G. Protopsaltis, C. Anagnostopoulos, J. Anagnostopoulos,
M.E. Theologou, E.N. Protonotarios, National Technical University of Athens,
Greece*

Multicast (I)

Analysis and Evaluation of QoS-Sensitive Multicast Routing Policies 468
E. Pagani, G.P. Rossi, University of Milano, Italy

Multicast Performance of Multistage Interconnection Networks
with Shared Buffering 478
*D. Tutsch, International Computer Science Institute, USA ;
M. Hendler, G. Hommel, Technical University of Berlin, Germany*

Performance Evaluation of PIM-SM Recovery 488
*T. Čičić, S. Gjessing, University of Oslo, Norway ; Ø. Kure, Norwegian
University of Science and Technology, Norway*

Reducing Multicast Inter-receiver Delay Jitter – A Server Based Approach 498
J.-U. Klöcking, C. Maihöfer, K. Rothermel, University of Stuttgart, Germany

Multicast Routing and Wavelength Assignment in Multi-Hop Optical Networks   508
*R. Libeskind-Hadas, Harvey Mudd College, USA ; R. Melhem, University
of Pittsburgh, USA*

Real-Time Traffic

Improving the Timed Token Protocol 520
Y. Bouzida, R. Beghdad, EMP, Algeria

Design of a Specification Language and Real-Time APIs
for Easy Expression of Soft Real-Time Constraints with Java 530
K.-Y. Sung, Handong University, Korea

Feedback-Controlled Traffic Shaping for Multimedia Transmissions
in a Real-Time Client-Server System 540
G.-M. Muntean, L. Murphy, University College Dublin, Ireland

Bandwidth Reallocation Techniques
for Admitting High Priority Real-Time Calls in ATM Networks 549
L.K. Miller, University of Toledo, USA ; E.L. Leiss, University of Houston, USA

Temporal Control Specifications and Mechanisms
for Multimedia Multicast Communication Services 559
H.-Y. Kung, National Pingtung University of Science and Technology, Taiwan;
C.-M. Huang, National Cheng Kung University, Taiwan

Wireless (I)

Improving Fairness and Throughput in Multi-Hop Wireless Networks 569
H.-Y. Hsieh, R. Sivakumar, Georgia Institute of Technology, USA

Dynamic Allocation of Transmitter Power in a DS-CDMA Cellular System
Using Genetic Algorithms 579
J. Zhou, Y. Shiraishi, U. Yamamoto, Y. Onozato, Gunma University, Japan

The Coexistence of Multicast and Unicast over a GPS Capable Network 589
T. Asfour, A. Serrhrouchni, ENST, France

An Approach for QoS Scheduling on the Application Level
for Wireless Networks 599
D.-H. Hoang, D. Reschke, TU Ilmenau, Germany

Multicast (II)

A Host-Based Multicast (HBM) Solution for Group Communications 610
V. Roca, A. El-Sayed, INRIA Rhône-Alpes, France

An Evaluation of Shared Multicast Trees with Multiple Active Cores 620
D. Zappala, A. Fabbri, University of Oregon, USA

QoS Routing Protocol for the Generalized Multicast Routing Problem (GMRP) ... 630
H. Bettahar, A. Bouabdallah, Technical University of Compiègne, France

Feedback Scalability for Multicast Videoconferencing 640
H. Smith, California Polytechnic State University, USA ; M. Mutka, L. Yang, Michigan State University, USA

Group Communication and Multicast 649
J. Templemore-Finlayson, S. Budkowski, INT, France

Routing

Avoiding Counting to Infinity in Distance Vector Routing 657
A. Schmid, O. Kandel, C. Steigner, University of Koblenz-Landau, Germany

Proposal of an Inter-AS Policy Routing and a Flow Pricing Scheme
to Improve ASes' Profits 673
N. Ogino, M. Suzuki, KDD R&D Laboratories Inc., Japan

Stigmergic Techniques for Solving Multi-constraint Routing
for Packet Networks 687
T. Michalareas, L. Sacks, University College London, UK

Dynamic Capacity Resizing of Virtual Backbone Networks 698
S.H. Rhee, Kwangwoon University, Korea ; J. Yoon, H. Choi, I. Choi, ETRI/National Security Research Institute, Korea

Wireless (II)

Throughput Improvements Using the Random Leader Technique
for the Reliable Multicast Wireless LANs 709
H.-C. Chao, National Dong Hwa University, Taiwan ; S.W. Chang, Southern Information System Inc., Taiwan ; J.L. Chen, National Dong Hwa University, Taiwan

Performance Evaluation and Implementation of QoS Support
in an 802.11b Wireless LAN 720
V. Mirchandani, E. Dutkiewicz, Motorola Australian Research Centre, Australia

Increasing Throughput and QoS in a HIPERLAN/2 System
with Co-channel Interference 727
J. Rapp, Aachen University of Technology, Germany

An Intra-media Multimode Wireless Communication Terminal
for DSRC Service Networks 737
M. Umemoto, Yokosuka ITS Research Center, Japan

Traffic Analysis, Modeling and Simulation

Simulation of Traffic Engineering in IP-Based Networks 743
T.T.M. Hoang, W. Zorn, University of Karlsruhe, Germany

An Algorithm for Available Bandwidth Measurement 753
J. He, C.E. Chow, J. Yang, T. Chujo, Fujitsu Laboratories of America Inc., USA

Influence of Network Topology on Protocol Simulation 762
D. Magoni, J.-J. Pansiot, Université Louis Pasteur, France

An Adaptive Flow-Level Load Control Scheme for Multipath Forwarding 771
Y. Lee, Y. Choi, Seoul National University, Korea

Performance Evaluation of CANIT Algorithm
in Presence of Congestion Losses 780
H. Benaboud, N. Mikou, University of Bourgogne, France

User Applications

Project Driven Graduate Network Education 790
A. Van de Capelle, E. Van Lil, J. Theunis, J. Potemans, M. Teughels, K.U. Leuven, Belgium

Modelling User Interaction with E-commerce Servers 803
H. Graja, J. McManis, School of Electronic Engineering, Ireland

XML Smartcards 811
P. Urien, H. Saleh, A. Tizraoui, Bull, France

The Influence of Web Page Images on the Performance of Web Servers 821
C. Hava Muntean, J. McManis, J. Murphy, Dublin City University, Ireland

Network Resilience in Multilayer Networks:
A Critical Review and Open Issues 829
F. Touvet, D. Harle, University of Strathclyde, UK

Author Index 839

Dynamic Reource Management Scheme for Multimedia Services in Wireless Communication Networks

Dong-Eun Lee[1], Bong-Ju Lee[2], Jae-Wook Ko[3], and Young-Chon Kim[4]

[1] Dept. of Internet and Computer, Chungwoon Univ., Korea
delee@cwunet.ac.kr
[2] Dept. of Image Eng., Chonbuk National Univ., Korea
[3] Dept. of Infomation and Communication, Chonbuk National Univ., Korea
[4] Dept. of Computer Eng., Chonbuk National Univ., Korea
{bjlee, jwko, yckim}@networks.chonbuk.ac.kr
http://networks.chonbuk.ac.kr

Abstract. We present a dynamic wireless resource management scheme considering the call level QoS and mobility characteristics for multimedia services in wireless communication networks. The objectives of proposed scheme are to improve the utilization of wireless resources and to reduce the blocking probability of new call while guaranteeing the required QoS of handoff calls with various service classes.
The wireless resources are reserved preferentially for handoff calls in order to provide their required call level QoS. Those resources can also be dynamically allocated as much as request probability of each class for new calls. The request probability of each class is determined by using the mobility characteristics, channel occupancy, and correction factor of each classes to keep the new call blocking proability as low as possible.

1 Introduction

With the increasing demands for mobilie multimedia services, future wireless networks will adopt micro/pico-cellular architecture in order to provide the higher capacity needed to support broadband services under the limited radio spectrum.[1] Due to its flexible bandwidth allocation, efficient multiplexing of burst traffic, and provision of a wide range of wireless broadband services, the wireless asynchronous transfer mode networks is a promising solution for the next-generation wireless communication system. However, handoff will occur frequently in the networks because the cell size is much smaller to support higher capacity on the limited radio spectrum. As the handoff rate increase, bandwidth management and traffic control strategy (call admission control, handoff procedure, etc.) become more challenging problems in wireless networks.

Handoff is the action of switching a call in progress in order to maintain continuity and the required quality of service (QoS) of the call when a mobile terminal moves from one cell to another.[2] In the wireless environment, we need to consider two additional QoS parameters: dropping probability of handoff calls and blocking probability of new calls from the standpoint of call-level QoS. If the base station (BS) has no idle channels, it may drop the handoff request and cause forced termination of the call in progress. From the subscriber's point of view, forced termination due to handoff calls is less desirable than blocking of a new call. Therefore, the QoS of handoff calls must be guaranteed while allowing high utilization of wireless channels.

Up to the present, intensive research on resource management schemes such as guard channel scheme (GCS) and handoff queuing scheme has been progress to guarantee the QoS by assigning the handoff calls to a higher priority. In general, handoff prioritization schemes result in decrement of handoff failures and increment of call blocking, which, in turn, reduce total admitted traffic and were able to support only single service classes rather multi service classes.[2,3] Meanwhile another approach such as complete sharing scheme (CSS), upper limit scheme (ULS), guaranteed minimum scheme (GMS) is also studied in traditional ATM networks and some wireless ATM networks to support multi service classes.[4,5] However, these schemes do not consider peculial characteristics and handoff in wireless environments. Therefore efficient resource management scheme is required to integrate services with various QoS considering the mobility characteristics.

In [6], we propose a dynamic resource management scheme (DCRS) based on mobility. The objective of DCRS is to guarantee the required dropping probability of handoff calls while keeping the blocking probability as low as possible. In this paper, we propose an extended DCRS to support multimedia services with various QoS in wireless communication networks. The performance of the proposed scheme is evaluated by numerical analysis and simulation.

The remainder of this paper is organized as follows. Chapter 2 describes the proposed resource management scheme and chapter 3 performs numerical analysis. Chapter 4 analyzes the call level performance by comparing the traditional resource management scheme. Finally, we conclude in chapter 5.

2 Dynamic Resource Management Scheme for Multimedia

Figure 1 shows the wireless resource model of the proposed resource management scheme for multi service class. We assume that total number of wireless resources in a BS is C unit bandwidth. Both handoff and new calls without regard to

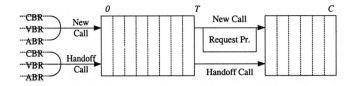

Fig. 1. Wireless resource model for multimedia

service classes share equally the normal bandwidth, which are radio bandwidth below the threshold. The guard bandwidth, the remaining bandwidth above the threshold, is reserved preferentially for handoff calls regardless of service classes. Those channels, however, can also be allocated as much as the request probability for new calls instead of immediate blocking unlike GCS.

The request probability reflects the possibility that the BS permits each classes of new call to allocate the wireless bandwidth among the guard bandwidth. For this purpose we consider following two circumstances.

Case 1: The arrival rate of handoff calls is larger than that of new calls. The request probability for new calls should be reduced to give more opportunity to handoff calls. Extremely, it must be selected closely to zero in the case that all the requested calls are handoff calls. With this, more wireless bandwidth can be allocated to handoff calls in the guard bandwidth and dropping probability of handoff can be guaranteed.

Case 2: The arrival rate of handoff calls is smaller than that of new calls. The request probability of new calls should be increased to give more opportunity to new calls. Request probability must be selected one for fair assignment between handoff and new calls in the case that all the requested calls are new call. With this, more wireless bandwidth can be allocated to new calls in the guard bandwidth and it has the advantage of reduction of new call blocking probability and increase of the channel utilization.

Considering the previous two circumstances, we define the request probability of class i new call in equation 1. It is dynamically determined considering the total number of bandwidth in BS (C), threshold between normal bandwidth and guard bandwidth (T), current number of used bandwidth ($n \cdot b$), mobility characteristics (α), and correction factor class i call (K_i).

$$P_{r,i}(n) = \begin{cases} 1 & , 0 \leq n \cdot b \leq T \\ \left(\frac{C - n \cdot b}{C_g}\right)^{\alpha K_i} & , T < n \cdot b \leq C - b_i \\ 0 & , C - b_i < n \cdot b \leq C \end{cases} \quad (1)$$

where C_g is the guard bandwidth above the threshold, mobility characteristics α is the ratio of arrival rate of handoff calls (λ_h) to arrival rate of new calls (λ_n), row vector $K = [K_1, K_2, \cdots, K_k]$ is the correction factor for assigning the channel allocation priority on each classes, and column vector $b^{-1} = [b_1, b_2, \cdots, b_k]$ is requiring bandwidth vector. When C_g is larger than b_i, α is positive real, and K_i is positive integer, it can be easily proved that $P_{r,i}(n)$ ranged from zero to one.

Figure 2 shows an example of request probability determined by the proposed formula under C of 125 unit bandwidths, threshold of 100 unit bandwidths, and various call mobility. We also assume that bandwidth requirements of class 1, 2, 3 is 1, 2, 6 unit bandwidth respectively. The request probability increases according to decreasing of mobility characteristics and decreases according to increasing of mobility characteristics under the same network situation. When the mobility characteristics equals zero, the request probability is set to one and

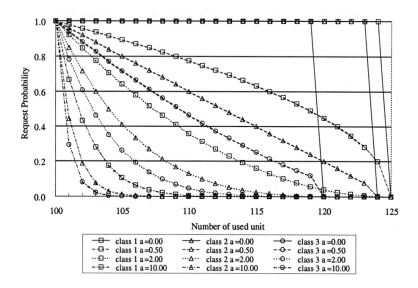

Fig. 2. Service probability of new call ($K_1 = 1, K_2 = 2, K_3 = 3$)

all the requests of new calls is admitted like CSS. When the mobility characteristic is larger than one, the request probability is abruptly decreased in order to provide more service opportunity to handoff calls. If the mobility characteristic is much larger than one, the request probability will be zero like GCS. Request probabilities of each class are different from each other under the same mobility characteristics according to priority level of each class (Class 1 > Class 2 > Class 3). Therefore the proposed scheme has the capability of flexible QoS control by using the mobility characteristics and correction factor.

3 Numerical Analysis

Let the requiring bandwidth of each k classes be column vector $b^{-1} = [b_1, b_2, \cdots, b_{i-1}, b_i, b_{i+1}, \cdots, b_k]$ where b_i is the request number of unit bandwidth of class i calls. The unit bandwidth corresponds to a data slot of MAC frame. We define that set of system states is row vector $n = [n_1, n_2, \cdots, n_{i-1}, n_i, n_{i+1}, \cdots, n_k]$ where n_i is the number of class i calls using the resource. If each of the C units bandwidth is viewed as a server, a class i call simultaneously requires b_i servers. At the time of call departure, all b_i servers must be simultaneously released. Accordingly, another set of system states can be defined as follows.

$$n^{\pm} = [n_1, n_2, \cdots, n_{i-1}, n_i \pm 1, n_{i+1}, \cdots, n_k]$$

We assume that the number of mobile users is much larger than the total number of bandwidth in a BS so that call arrivals may approximate to Poisson process. Following are the traffic parameters used to evaluate the performance of extended DCRS:

- C : the maximum number of available bandwidth in the BS
- C_g : guard bandwidth above the threshold
- b_i : requiring bandwidth of class i
- $\lambda_{h,i}$: arrival rate of class i handoff calls which has a Poisson property
- $\lambda_{n,i}$: arrival rate of class i new calls which has a Poisson property
- $\mu_{h,i}$: service rate of class i handoff calls, in which residence time has a exponential distribution of mean $1/\mu_{h,i}$
- $\mu_{n,i}$: service rate of class i new calls, in which residence time has a exponential distribution of mean $1/\mu_{n,i}$

From the above assumption, performance metrics are listed in Table 1.

Table 1. Analysis parameters

	$0 \leq n \cdot b \leq T$	$T < n \cdot b \leq C - b_i$	$C - b_i < n \cdot b \leq C$
Arrival rate	$P_{r,i}(n) = 1, \lambda_i(n)$	$P_{r,i}(n) = [0,1], \lambda_i(n)$	$P_{r,i}(n) = 0, \cdot$
Service rate	$n_i \mu_i$	$n_i \mu_i$	$n_i \mu_i$
Blocking probability	0	$(1 - P_{r,i}(n))P(n)$	$P(n)$
Dropping probability	0	0	$P(n)$

$\lambda_i(n) \equiv \lambda_{h,i} + P(n)\lambda_{n,i}\beta(n)$, $\mu_i \equiv \mu_{h,i} + \mu_{n,i}$
where $\beta(n) = 1$ for $P_{r,i} \geq \text{Random}(0,1)$, $\beta(n) = 0$ for $P_{r,i} < \text{Random}(0,1)$

The steady state probability distribution of extended DCRS is given by equation 2 by using the Markovian equilibrium balance equation.

$$P(n) = \prod_{i=1}^{k} \left\{ \frac{1}{n_i!} \prod_{i=1}^{n_i} \alpha_i(n) G^{-1}(\Omega_{DCRS}) \right\} \qquad (2)$$

where $\alpha_i(n) = \lambda_i(n_i^-)/\mu_i$, $G(\Omega)$ is normalizing constant of state space of Ω and Ω_{DCRS} is the set of admissible state space of extended DCRS given by:

$$\Omega_{DCRS} = \{n : 0 \leq n \cdot b \leq C\}.$$

The blocking probability experienced by a class i call is given by equation 3. It is the sum of state probability not included when new call request is issued.

$$P_{b,i} = \sum_{n \in B_i^+} (1 - P_{r,i}(n))P(n) \\ = \frac{G(B_i^+)}{G(\Omega_{DCRS})} \qquad (3)$$

where $B_i^+ = \{n : n \in \Omega_N, n \cdot b > T\}$.

The dropping probability experienced by a class i call is given by equation 4.

$$P_{d,i} = \sum_{n \in D_i^+} P(n)$$
$$= \frac{G(D_i^+)}{G(\Omega_{DCRS})} \qquad (4)$$

where $D_i^+ = \{n : n \in \Omega_H, n_i^+ \notin \Omega_H\}$.

The resource utilization efficiency is given by equation 5.

$$P_u = \frac{\sum_{n \in \Omega_{DCRS}} n \cdot bP(n)}{C} \qquad (5)$$

4 Numerical Results and Discussions

In this section, we present numerical examples to evaluate the performance of our extended dynamic channel reservation scheme (DCRS) and two traditional scheme, complete sharing scheme (CSS) and extended guard channel scheme (GCS). Comparisons are conducted in terms of the dropping probability of the handoff call, the blocking probability of new call, and utilization of wireless resources. System parameters used in the analysis are summarized in table 2.

Table 2. System parameters

Total bandwidth of base station	125 units
Number of guardbandwidth	25 units
Offered load ratio of each class	1:1:1
Required bandwidth of each class	1,2,6 unit
Correction factor of each class	1,1,1
Call holding time	2 minute

Dropping probability of handoff call, as a function of offered load is illustrated in figure 3. Dropping probability is increased in proportion to the increased of offered load regardless of resource management scheme and service classes. As show in figure, we can observe that both extended DCRS except class 3 and GCS meet the constraints (target dropping probability is one percent) when offered load is 0.8. Dropping probability of class 3 can be reduced by controlling the correction factor and sophisticated control of correction factor is another issue of extended DCRS. In the case of CSS, target dropping probability of class 2 and 3 is not guaranteed and can not be controlled.

On the contrary, figure 4 shows that blocking probability of CSS is lower than other schemes, because all wireless resources can be allocated equally by new and handoff calls. Blocking probability of extended DCRS is lower than GCS by providing much room for new calls as much as request probability.

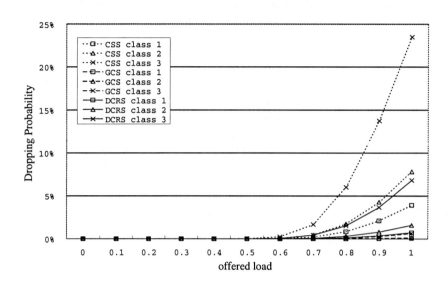

Fig. 3. Dropping probability ($\alpha = 1$)

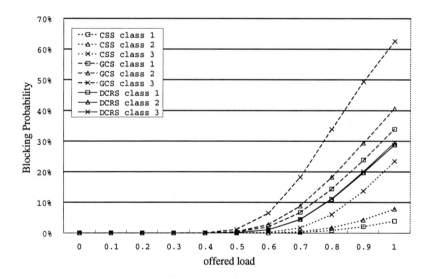

Fig. 4. Blocking probability ($\alpha = 1$)

the sequence of switches form the source to the destination. This route can guarantee the desired QoS parameters of the connection. When the call setup message travels via the DTL, every switch ensures that the required capacity is available at the moment. This process is called ACAC (Actual CAC). When the destination receives and accepts the call setup message, it sends back a receipt message. The receipt message travels backward along the DTL. After this call setup process, the path is ready for data communication.

Now we can examine the detailed operation of the new method. In the network, when a link is getting fully loaded, the switches at the end of the link decide to divert a connection. Equation (1) gives the probability of diverting a connection.

$$p(x) = \frac{e^{\frac{x}{a}} - 1}{b} \qquad (1)$$

where x is the load of the link and $p(x)$ refers to the non-linear transformation of x to the [0...1] range, therefore $p(x)$ likes a probability variable. The parameter a adjusts the slope of the curve, therefore adjusting the border of the light loaded region. In the light-loaded region, the algorithm does not detour connections very likely. By increasing this value, the algorithm becomes more "aggressive", so the detouring starts earlyer. Parameter b is used to normalize the function to the [0...1] range. The effect of the parameters can be seen in Fig. 1.

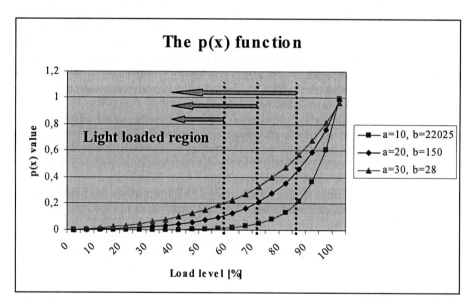

Fig. 1. The effect of the a and b parameters in equation 1. By modifying the parameters, the algorithm becomes more aggressive.

The reason for the non-linear transformation is the following: at low traffic (light loaded region), we do not need to divert any connections, but when the load approaches the link capacity, the probability of a diversion increases exponentially. At stated intervals every switch decides according to the value of $p(x)$, whether it should divert a connection. (The time interval is adaptive to the link load; at higher load level, it becomes shorter). The point of using function $p(x)$ is that the algorithm is more adaptive to link load, and can achieve better bandwidth utilization than using a given threshold limit.

Our new method requires the introduction of a few new signaling elements in addition to the ATM PNNI signaling system:
1. CONNECTION_LOCK, it ensures, that only one switch (at the end of the link) diverts a given connection at a time.
2. LINK_STATE_QUERY, in response to this message, the neighboring switches return the load level of their links. This is vital for a switch to choose the minimum loaded alternative path.
3. DIVERTING, this message modifies the routing tables in the affected switches. This message includes the ID of the initiating switch and the VPI/VCI number of the affected connection.

The algorithm uses the ATM traffic class descriptor to select the suitable connection for diverting. The following types of traffic are suitable for diverting: UBR (Undefined Bit Rate), ABR (Available Bit Rate), VBRnrt (Variable Bit Rate). These traffic types do not specify the CDV (Cell Delay Variation) and the CTD (Cell Transfer Delay) parameter, so these connections are appropriate for diverting to a longer path.

In case of heavy loaded neighbor links, the system could divert the same connection repeatedly from one link to another and inversely. To deal with this problem, the switch registers the path of the last diverted connection, and if further diversion is needed – because of the high load of the newly selected neighboring link – the new path of the connection can not be the same as the former one. Therefore the path of the connection soon evades the busy part of the network.

The new scheme does not require central call processing, it can be realized in a fully distributed manner. The information needed by the algorithm could be obtained from the standard ATM SETUP messages (i.e. ATM traffic descriptor, broadband capability, DTL) and by monitoring the load of the switches' links.

This routing method is also applicable for IP based mobile networks. The basic idea of IP flows (RFC 2460) is similar to the VC/VP concept of ATM [7]. An IP flow can be regarded as a sequence of packets sent from a source, using a particular host-to-host protocol for the transmission of data over the Internet, with a certain set of quality of service requirements. These flows can be manipulated by means of our algorithm in the same a manner as the ATM connection paths.

3. Simulation and Results

Simulations were performed to examine the efficiency of the new scheme in a small ATM based mobile environment (see Fig. 2). The core networks consists of regular ATM switches (referred as network nodes), the access network utilizes mobility enhanced ATM switches.

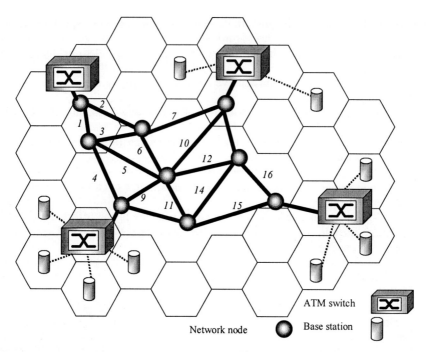

Fig. 2. The test network. The core networks consists of regular ATM switches, the access network utilizes mobility enhanced ATM switches connected to the base stations.

The simulation was written in C programming language. The program uses randomly moving users with 12 different traffic classes. The typical applications (e-mail, ftp, http session) are mapped to these traffic classes. The performance of the new scheme was compared to the standard operation of an ATM network in a mobile scenario.

Fig. 3. shows the flow chart of the algorithm. The first part of the algorithm monitors the link loads, and if the load is greater than parameter p_1, it searches a new path for one of the non-real time connections of the given link. A procedure named ROUTE provides the new path (or just a path segment) based on the input parameters (source and destination switch number, required traffic parameters). If this path setup is successful, the connection is diverted to the new path, and the old one is torn down. The second part of the algorithm is responsible for preventing the forming of non-optimal connection paths. It periodically calculates the shortest path between the source and the destination for every connection. If the current path is more than p_2 times longer than the shortest possible path, the algorithm tries to reroute the connection to the shortest available path.

The result show that the average load in the network has became equal. The maximum difference of the link loads - compared to the average link load - is less than 20 % in the case of the new algorithm. The standard ATM network shows bigger deviations with a maximum of 60 percentage. (See Fig. 4)

The number of successful call-setup and successful handover are increased by 5% compared to the traditional method. The ratio of the accepted real-time connections has risen compared to the original scenario. The time needed for signaling is also reduced. This is very advantageous from the view of real-time connections. The

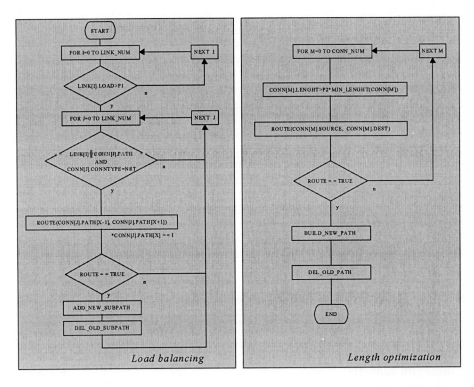

Fig. 3. Block scheme of the simulator. The load balancing and the length optimization processes are repeated periodically.

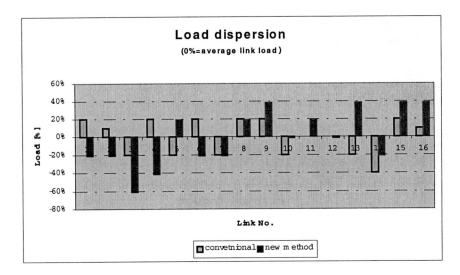

Fig. 4. Link load deviation in the network (0% is the mean load level). Every bar represents the load deviation of a link. The longer the bar, the bigger is the underload/overload of the given link (the link number can be seen under the bars).

signaling time decreases in the case of the new method due to the lack of actual CAC, because the general CAC method itself can support enough information on the call setup. The more successful handovers are resulting more active connections and bigger network load in case of the new method. The other reason for the average network load rising is that the longer (diverted) connection routes consume more resource in the network. An interesting question is the maximum improvement, which can be achieved by the algorithm. To give an absolute coefficient of improvement, the effect of the traffic type varying and the different user movements has to be eliminated, because the measured values are depending on the generator traffic matrix, and user movement properties. We have used random chosen traffic types, and random user movement patterns to minimize the effect of different traffic and movement properties. By running the simulation several times, the results are closing up to the real life scenario.

The simulations have shown that the algorithm's performance falls back at very high level (above ~95%) of network load. The reason for this effect is the additional load caused by the detouring method, but the detouring would not help either, it only causes some extra load. The solution to this problem is the modification of the $p(x)$ function, to adopt a fall-back characteristic at very high level of network load.

Table 1. Simulation results.

	No optimization	New algorithm	Gain
connection setup success	78 %	83 %	+ 5%
successful handoffs	66 %	71 %	+ 5%
Signaling time	11.5	10.3	- 1.2
Load/connection	70.7	82.7	+ 12
Average network load	63 %	76.79 %	+ 14%

4. Conclusions and Future Research

A new handover supporting routing scheme for mobile networks was presented. Our algorithm provides faster handover and more even network load. The growing number of successful handovers – achieved by our method – increases the users' content. We used large number of simulations to investigate the performance of the new method, and the results proved the effectiveness of our algorithm. An extension to mobile IP networks is also introduced.

We are planning to study the performance and effectiveness of the scheme applied to IP flows.

References

[1] D.Cox and D. Reudink, "Increasing channel occupancy in large-scale mobile radio systems: Dynamic channel reassignment," *IEEE Trans Commun.*, vol. COM-21, Nov 1973

[2] S. Chia, "Mixed cell architecture and handover," in *IEEEColloquium – Mobil Commun. in the year 2000*, London, U.K., June 1992

[3] C. Oliveira, Jaime Bae Kim, T. Suda, "An adaptive Bandwidth reservation scheme for high-speed multimedia wireless networks", *IEEE Journal on sel. areas in comm.*, vol. 16, NO. 6, August 1998

[4] Anthony S. Acampora, and Mahmoud Nagshineh, "An Architecture and Methodology for Mobile-Executed Handoff in cellurar ATM Networks," *IEEE,* 1994.

[5] Gopal Dommety, Malathi Veerarghavan *IEEE,* Mukesh Singhal, "A Route Optimization Algorithm and Its Application to Mobile Location Management in ATM Networks", *IEEE,* 1998.

[6] Private Network-Network Interface Specification Version 1.0 (PNNI 1.0). *The ATM forum Technical Committee (af-pnni-0055.000)*, 1996.

[7] RFC 2460, "Internet Protocol, Version 6 (IPv6) Specification", *www.ietf.org*, 2000.

Location Stamps for Digital Signatures: A New Service for Mobile Telephone Networks

Matthias Kabatnik[1] and Alf Zugenmaier[2]

[1] Institute of Communication Networks and Computer Engineering,
University of Stuttgart, Germany,
kabatnik@ind.uni-stuttgart.de

[2] Institute for Computer Science and Social Studies – Dept. of Telematics,
Albert-Ludwigs-University Freiburg, Germany,
zugenmai@iig.uni-freiburg.de

Abstract. Location aware services are expected to make up a large share of the mobile telephone market in the future. The services proposed so far make use of uncertified location information—information push services, guidance systems, positioning for emergency calls, etc. We propose a service that provides certified location information. Integrated with cryptographic digital signatures this service enables the determination of the current position of the signer in a provable way. Such certified location information—called location stamp—can be applied in electronic commerce, e.g. for determination of applicable laws or taxes.

1 Introduction

The benefit of modern cellular telephone networks is the ability of users to roam with their terminals being reachable at all times. The underlying system needs information about the user's location to be able to route calls to the mobile terminal. This information is provided by maintaining special registers that store information about the area and the cell where the user is located.

In cellular mobile communications networks like the GSM network the traffic volume is increasing steadily. However, the number of simultaneous calls per cell is limited. Therefore operators shrink cell sizes—at the hot spots like airports and train stations even down to picocells (several tens of meters in radius)—in order to keep up with the growing number of call attempts. Thus, location information of the mobile stations in the network becomes more accurate and enables services like location depended charging (low tariff in the so-called *home zone*). Other regulatory requirements (for example by the FCC) demand even more precise location information, e.g. to provide better emergency call response services. Therefore, mobile terminal tracking methods like triangulation with higher precision have been developed.

With the availability of precise location information new services are possible. One application of location information in daily business is signing of contracts. The location of signing may determine to which jurisdiction the contract is subject to. When manually signing a contract, it contains the location, the date, and the hand-written signature. The written location can be verified immediately since both contract partners are present; in an electronic commerce setting this location of the conclusion of a contract, i.e. where it is signed is not evident. How can one be sure where the partner is located?

We designed a new service that provides an approved location stamp bound to a digital signature in order to map the process of manual contract signing to modern communication systems. Additionally, we present a procedure that assures the correct mapping of the communications system's view on a mobile terminal to a user requesting a location stamp. In order to provide liability we make use of digital signatures based on asymmetric cryptographic algorithms.

The remainder of this paper is organized as follows: First we describe the service and the technical entities required to generate a location stamp that can be tied to a digital signature. In chapter three we show the protocol that can be used to perform this. In the next chapter we map this generic description to the functional and organizational entities of the GSM networks to show the feasibility of the service. Chapter five discusses security aspects of the protocol and shows its limits. The related work is presented in chapter six, and finally a summary is given.

2 Service Specification

The location stamp service (LSS) provides a location stamp which certifies at which position the subscriber is located when making a signature—or rather, at which position the subscriber's mobile station is visible to the network. This location stamp can be verified by any third party using the public key assigned to the LSS, and it can be used as a proof of the location information. In this scenario the service operator who signs the location is a trusted third party (i.e. the signer is neither the principal of the stamp nor the verifier of the stamp).

2.1 Location Measurement

There are two main types of location measurement: positioning and tracking. Positioning is a measurement which is performed within the mobile device (e.g. by an integrated GPS receiver). The calculation of the position is also done in the mobile device. The other type is tracking: Some external measurement infrastructure determines the position of the mobile device (e.g. by triangulation using the mobile's emitted signal).

Positioning has the advantage of greater accuracy, privacy of location information, and – for security critical applications like proving a location via a telecommunication channel – the disadvantage of requiring tamper proof hardware, and resistance against misleading the device with a fake environment like a GPS simulator. Tracking has the disadvantage of lower accuracy, possible loss of privacy against the network, but the advantages of being harder to deceive.

In our approach we have chosen tracking because it offers better security and because of its straightforward implementation: tracking information is available in a cellular telephone network since the location information is needed for routing and billing purposes.

2.2 Cryptographic Capabilities

The handwritten signature can be replaced by the digital signature in electronic commerce (e.g. [1]). A digital signature using asymmetric (public key) cryptography is created by calculating a hash of the document to be signed and encrypting this hash with the private key of the signer [2]. The signature can be verified by anybody who knows the public key of the signer. The private key has to be kept secret by the signer. To bind a public key to an identity, a certificate according to a standard like X.509 [3] is used. This certificate is signed by a certification authority that guarantees the fact that a certain public key belongs to the identity named in the certificate.

Next to the need of identification of users signing a contract there is the necessity to identify terminals within a network. This identification is needed to perform routing, addressing, and billing tasks. In mobile networks usually authentication mechanisms are provided to prove the terminal's claimed identity. Therefore, a mapping between the identity and some secret exists—usually a symmetric key—that must be known by the network.

We will use a generic identifier called MoID (Mobile IDentifier) and a secret key named K.

2.3 Binding Signature and Authentication

From the perspective of the network, only a terminal named by a non-ambiguous MoID can be located in the network. Because the task in our scenario is to certify that a certain person signs a document at a certain location, it is necessary to bind the capability to authenticate the user's identity to the same physical entity that is used to authenticate the identity of the mobile terminal. This can be done, e.g. by placing the private signature key and the key K corresponding to MoID on the same tamper resistant chip card. The chip card should also provide the capabilities for authentication and signing to prevent the keys from ever leaving the physical module. When both operations (proof of user ID with a signature and authentication of MoID) are performed at the same time, they must be performed at the same place. Thus, the location of the terminal is identical with the place of signature.

The physical binding of the two keys can be represented by a digital certificate issued together with the chip card.

3 Protocol

The location stamp protocol consists of six information flows which may consist of several messages each (e.g. in case of fragmentation or necessary acknowledgments). First the abbreviations are introduced, then the protocol itself is presented. The abbreviations used are:

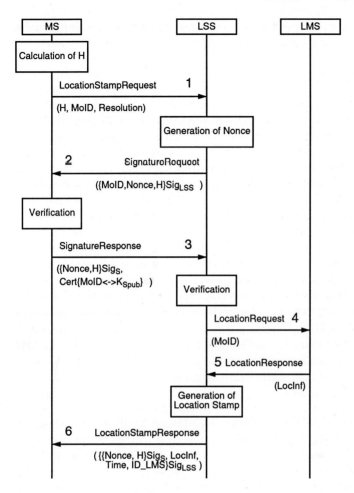

Fig. 1. Successful run of LSS protocol

MS	mobile system
LSS	location stamp service
LMS	location measurement system
H	hash value
MoID	mobile's identity
K_{Spub}	public key corresponding to the private key of the subscriber
{...}SigS	signature of the data in parenthesis with the private key of the subscriber
{...}SigLSS	signature of the data in parenthesis with the private key of the location stamp service
ID_LMS	identifier of the applied location measurement system (LMS)

An example of a successful protocol run is presented in Fig. 1.

The following explanations will refer to an information flow (IF) depicted by an arrow, and its preceding operation block (if any).

IF 1: The hash of the document is calculated. This can happen at any place at any time before the run of the protocol since the hash is simply another representation of the document that shall be signed. Thus, this calculation does not necessarily have to be performed within the mobile device. The mobile device initiates the protocol run by sending a service request to the service. The request contains the mobile's ID and the hash. The hash is included to provide some additional information for preparing the operation of IF 3. The communication between mobile and service can be settled on any kind of bearer service that is available. Additionally, the mobile supplies the desired resolution for the location stamp.

IF 2: The service generates a nonce value. A nonce is a kind of random value which may contain some time dependent information (e.g. 8 time related octets and 2 random octets). This value is used to prove recentness within the protocol [4]. If the mobile can generate a signature containing this nonce value, the service can derive that the counterpart has been alive after the reception of message 2. The message in IF 1 alone does not give this information since it could be a replay or a complete fake. The signature provided by the service is necessary to ensure that the nonce was produced by the service. Additionally the hash value within the message can provide the same effect like the nonce if it is unique (in cases when one document has to be signed several times an additional nonce value is necessary to protect against replay attacks). The signed nonce is sent to the mobile.

IF 3: First the mobile checks if the parameters H and MoID are the same as in IF 1. If this is not the case, no time consuming cryptographical operation has to be performed, the message can be discarded immediately, and the protocol terminates with an error. If the parameters fit, the signature of the service is validated. Now the mobile knows that the nonce was generated by the service and shall be used for signing the document represented by H. The mobile signs the concatenation of the document's hash H and the nonce, and sends the signature back to the service. Additionally, it may provide a certificate which can be used to prove the physical binding of the secret key belonging to its signature key K_{Spub} and the MoID. In order to reduce the amount of information that is transferred, this certificate may also be stored at the LSS.

IF 4: The LSS verifies the signature of the mobile to be sure of the origin of the value. Since the signature contains the nonce, it can be derived that it is fresh. The key used to produce this signature is located inseparably from the authentication mechanism of the MoID, therefore the service must locate the corresponding mobile station. The LSS prompts the location measurement system for the location of the mobile...

IF 5: ... and receives the location information.

IF 6: The LSS generates the location stamp by expressing the received location information with the accuracy requested by the mobile system in step

1 (resolution parameter). Additionally, the point of time is provided. The parameter ID_LMS is included to enable any verifier of the signature to apply his own estimation of the trustworthiness (cf. section 5.1). Finally, the mobile receives the requested location stamp.

4 Mapping onto the GSM Infrastructure

The GSM system, standardized by the ETSI in 13 series of standards [5], is used primarily in Europe and Asia. It has a share of about 65% of the world's digital cellular telephone networks. Two types of mobility are supported by the GSM system: mobility within a network and mobility across networks (roaming). The location information that handles the mobility inside a network is stored in the visitor location register VLR of the network the mobile station is currently booked into. The home location register HLR in the network of the provider is used to store information about which network the subscriber is currently using. It also stores personalization parameters of services of the subscriber.

GSM enables subscriber mobility in a sense that different subscribers may use the same terminal. To bind a device to an identity, a subscriber identity module (SIM) is used. It is generally placed in a chip card or module that can be plugged into different terminals. On the SIM card the international mobile subscriber identification (IMSI), a 15 digit number, is stored. To address this subscriber, a mobile station ISDN number (MSISDN) is used. More than one MSISDN may be assigned to a subscriber identified by the IMSI.

Location information in the GSM network arises naturally: it is necessary to route a call to its destination. This information has a resolution of the cell size. In the GSM standard [6],[7] a location service (LCS) is described that permits a more precise measurement. A location service client may request the location of a subscriber that is identified by its IMSI or MSISDN. The location service can communicate with the location measurement units through the mobile switching centre (MSC) to locate the terminal device and return the current position to the LCS client.

4.1 Location Stamp in GSM Networks

We propose the following mapping of the technical entities of the protocol as described above to the entities that are available within the GSM system:

The IMSI takes the role of the MoID. The public key of the subscriber (K_{Spub}) can be stored on the SIM card. This enables the network operator of the subscriber to issue a certificate that states the physical binding between the IMSI and K_{Spub}. The HLR has knowledge which network the mobile station is booked into. The MSC of this network can contact the location measurement units which can in turn locate the mobile. The location measurement itself depends on the desired resolutio n: it is either possible to just locate the cell the mobile is booked into—this can be performed with a lookup—or perform a measurement, e.g. based on time of arrival of a signal emitted from the mobile station

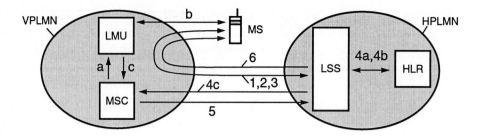

Fig. 2. Request for a location stamp by the mobile station. The figures and letters correspond to the information flows

at two or three base stations. The measurement itself is performed by location measurement units, the result is calculated by the location service (LCS) server.

As a transport protocol for the connection between the terminal and the LSS Unstructured Supplementary Service Data (USSD) or even Short Message Service (SMS) can be used. In Fig. 2 the entities and the information flows between them are depicted. In Fig. 3 a trace of a successful protocol run in the GSM network is shown. The information flows are:

IF 1, 2, 3: Same as in chapter 3. The communication between mobile and service can be settled on any kind of bearer service that is available (e.g. GPRS, USSD). The certificate binds the public key K_{Spub} to the IMSI, thus stating that the private key K_{Spub}^{-1} is stored on the SIM card corresponding to the IMSI. In the case of GSM the retrieval of location information (IF 4 of Fig. 1) maps to several IFs denoted by an appended letter (a, b, c).

IF 4a: The LSS verifies the signature of the mobile to be sure of the origin of the value. Since the signature contains the nonce, it can be derived that it is fresh. Since the key used to produce this signature is located inseparably in the SIM with the IMSI assigned, the service must locate the corresponding mobile station. The LSS prompts the HLR for the location of the mobile station ...

IF 4b: ... and the HLR provides the routing address of the MSC currently serving the mobile, which is indicated in the data base entry indexed by the IMSI.

IF 4c: The LSS assesses this MSC address. The VPLMN may not support location services, or the LSS provider may not consider it to be trustworthy either for its poor protection of secrecy and integrity (possible manipulation of location data) or simply for the provided accuracy of the location information. In the latter case the LSS aborts the request by sending an error message to the mobile. However, if the MSC supports location services and is considered to be trustworthy the location information request is sent to the MSC in the immediate mode (only minimum delay between request and response is acceptable) which in turn uses the infrastructure of the VPLMN to perform the necessary measurements. We assume that all necessary steps like encryption or physical separation of network links are taken to protect

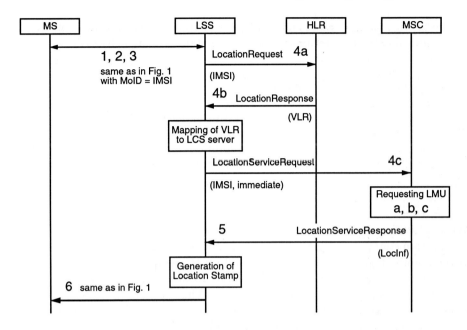

Fig. 3. LSS service in a GSM environment

the integrity of the communication between LSS and the infrastructure of the VPLMN.

IF a, b, c: The MSC determines the location involving one or more location measurement units (LMU)

IF 5: The MSC sends its answer to the LSS.

IF 6: The LSS generates the location stamp by expressing the received location information with the accuracy requested by the mobile system in step 1 (resolution parameter). Additionally, the point of time is provided. The parameter ID_LMS is included to enable any verifier of the signature to apply his own estimation of the trustworthiness of the LSS and the VPLMN. Finally, the mobile receives the requested location stamp.

In case the protocol cannot terminate successfully, an error is returned. There are two general types of errors: protocol internal errors and protocol external errors.

Protocol internal errors are related to the information flows 1-6. It is possible, that a message is not received in time or that some verification fails. In this case typical error messages can be generated and delivered to the participants of the protocol run.

Protocol external errors are related to the process of location measurement (IF a, b, c in the GSM example). These errors might be caused by either a negative judgement of the LMS with respect to the security qualities of the responsible LMS or simply by unavailability of location information with the desired accuracy. In cases low resolution information is available the LSS can

Enhancing the digitally signed documents with a location stamp is similar to adding a time stamp to it. Different standards exist, one of them is the PKIX draft [14].

7 Summary

We have presented a new type of service enhancing the capabilities of mobile communication terminals. Extending digital signatures with location stamps provided by a trusted third party can be used to improve electronic commerce applications. Additionally, the new service enables other applications like electronic travel logs or proof of service for field staff. We have proposed a service architecture which can be realized with existing telecommunication networks. Therefore, only limited effort has to be taken to bring the service into existence.

References

1. Directive 1999/93/EC of the European Parliament and of the Council: A Community Framework for Electronic Signatures, December 1999
2. D. W. Davies and W. L. Price: The Application of Digital Signatures Based on Public Key Cryptosystems, Proceedings of the Fifth International Computer Communications Conference, October 1980, pp. 525–530
3. ITU-T Recommendation X.509 Data Networks and Open System Communications – Directory – Information Technology – Open Systems Interconnection – The Directory: Authentication Framework, Geneva, June, 1997
4. Lowe, G.: A Hierarchy of Authentication Specification. In Proceedings of the 10th Computer Security Foundation Workshop. IEEE press, 1997
5. GSM 01.01: Digital cellular telecommunications system (Phase 2+); GSM Release 1999 Specifications, ETSI TS 101 855
6. GSM 02.71 Digital cellular communication system (phase 2+); location services; service description, stage 1, ETSI TS 101 723
7. GSM 03.71 Digital cellular communication system (phase 2+); location services; functional description, stage 2, ETS 101 724
8. Yahalom, R.; Klein, B.; Beth, Th.: Trust relationships in secure systems – a distributed authentication perspective. In Proceedings of the IEEE Conference on Research in Security and Privacy, pages 150–164, 1993
9. Shibuya, A.; Nakatsugawa, M.; Kubota, S.; Ogawa, T.: A high-accuracy pedestrian positioning information system using pico cell techniques; In: IEEE 51st Vehicular technology conference proceedings, Volume 1, pages 496–500, spring 2000
10. Hu, Yun-Chao: IMT-2000 mobility; In: Proceedings of the IEEE Intelligent Network workshop 2000, pp. 412–435, 2000
11. Tang, H.; Ruutu, J.; Loughney, J.: Problems and Requirements of Some IP Applications Based on Spatial Location Information, IETF draft ¡draft-tang-islf-req-00.txt¿, 2000
12. Denning, D.: Location-Based Authentication: Grounding Cyberspace for Better Security, Computer Fraud & Security, Elsevier Science, February 1996.
13. www.quova.com
14. Adams, C.; Cain, P.; Pinkas, D.; Zuccherato, R.:IETF draft <draft-ietf-pkix-time-stamp-12.txt>, December 2000

A New Method for Scalable and Reliable Multicast System for Mobile Networks

Masato Hayashi[1] and Christian Bonnet[2]

[1] Corporate Technology Group, Hitachi Europe, Ltd, c/o CICA,
2229, Route des Cretes, 06560 Valbonne, France
Tel. +33(0)4-9294-2569
Fax. +33(0)4-9294-2586
mhayashi@cica.fr
[2] Mobile Communications Department, Institute Eurecom
2229, Route des Cretes, B.P.193, 06904 Sophia Antipolis, France
Tel. +33(0)4-9300-2608
Fax. +33(0)4-9300-2627
christian.bonnet@eurecom.fr

Abstract. This paper proposes a new method aiming at realizing a scalable and reliable multicast system focused on mobile network. For reliability, retransmission scheme was adopted. The main idea of the proposed method is that network node does take care of error recovery by retransmission instead of a sender or receivers by which conventional methods have been done. Consequently terminal mobility can be dealt with more simple. This paper shows that the proposed method has much better performance compared with existing methods through simulation.

1 Introduction

Recent rapid advance of mobile communications technology allows users in mobile environment to enjoy not only voice service but also the Internet. And especially the enlarging capacity of wireless link makes it possible to accept the communication with large amount of data (file, picture and video, etc.) [1].

In these days, one of the main services is sure to be delivery-type service of, for example, music, game software and so on, to many mobile terminals (MT) utilizing the Internet. The purpose of our study is to develop a new method suitable for this kind of service over mobile network.

To realize delivery service, network must support some characteristics. First is an efficient data routing. Multicasting, which provides a very efficient routing way of one-to-multipoint communication, is suited to delivery service very well. Many kinds of multicasting protocols have been developed [2][3]. Data generated by the sender runs over the multicast tree that was organized in advance by the existing multicasting protocol to group members. The second is reliability that is to guarantee error-free in the sending data at all receivers. There are two classified techniques for reliability. One is feedback-less scheme, which adds to the multicast data, some redundant information so that receivers could correct the erroneous data using the redundant information. The other is feedback (reception state report) scheme; retransmission which sends again the same data to the erroneous receivers. Our concern is on feedback approach for complete (high) reliability. The third is high scalability, which is to be able to deal with extremely many mobile receivers with connection to networks, is essential to the service. The last is terminal mobility that is to guarantee the continuation of service session for moving receivers.

There are already several scalable and reliable multicast protocols with retransmission for reliability. They address the scheme to avoid implosion problem, which is caused by swamping on a sender with feedback messages from all receivers, in order to obtain scalability. SRM (Scalable Reliable Multicast) [4] proposes the random delay feedback scheme with multicast to reduce the implosion on the sender. A receiver suppresses its feedback after its random time passed if another receiver responds the same feedback. A successful receiver retransmits the data for erroneous MTs. PGM (Pretty Good Multicast) [5], LGMP (Local Group Multicast Protocol) [6], RMTP (Reliable Multicast Transport Protocol) [7] [8] and MESH protocol [9] employ local group (LG) concept to avoid the implosion and aim at shorter delay on recovery. In these methods a multicast group is divided into several LGs for local recovery. The main difference exists on the way to define LG and determine local controller (LC) of the group for the recovery. In PGM, error-free receiver volunteers to become LC. Hence LCs dynamically changes. On the other hand, LGMP, RMTP and MESH adopt fixed LC scheme. That is, LC is pre-defined depending on the multicast service type and network topology. MESH protocol [9] is targeted on the specific application of time-constrained data streams. Therefore, high reliability could not be gained since it protects delay-sensitive stream against errors.

As shown so far, the conventional methods can provide scalable and reliable multicast system. However, no consideration for mobile environment in these existing methods causes following the problems. For example, the low capacity of the wireless link between network and LC does not always lead to efficient recovery. The movement of LC could make local recovery complex. Consequently, these conventional methods do not have aspect to meet the described requirements.

This paper provides a new method of scalable and reliable multicast system not only for fixed networks, but also for mobile networks. Rest of the paper is organized as follows. Section II shows the conventional approach and discusses its characteristics. Section III presents the new method in detail. In the Section IV, the evaluation of the new method is shown through comparing to conventional methods by computer simulation. Finally, Section V describes the concluding remarks.

2   Conventional Approach and New Approach

Regarding the existing method, we categorized existing methods into two types of approach with respect to who does take care of recovery.

2.1 Conventional Approach

(1) Sender-Based approach (SB)
A sender performs retransmission with Nack (error report) or/and Ack (error-free report) feedback scheme. In SB approach, the sender repeats retransmission in multicast until all members get multicasting data correctly. This approach includes typical point-to-point reliable protocols (TCP, HDLC etc.) and early multicasting protocol. This approach causes implosion problem with the number of member increased. And the bandwidth for feedback is required over the whole network, i.e., resource consuming problem issues.

(2) Receiver-Based Approach (RB)
This approach employs local group concept. In this approach, recovery is carried out by a receiver which represents its LG instead of sender. LC (Local Controller) is predefined. It can avoid implosion problem thanks to local management of feedback and recovery. Recent most efforts, SRM, PGM, LGMP, RMTP and MESH that I cited in section1, belong to this approach.

As already explained, the problem for mobile network is that the low capacity and instable quality of the wireless link and the low performance of LC make local recovery inefficient. Other is that the movement of LC could cause terminal mobility difficult.

2.2 New Approach

(3) Network-Based Approach (NB)
The new approach also adopts local group concept. In this approach, it is a network node that takes care of recovery instead of a sender or receivers. The node forms LG dynamically to execute local recovery. Therefore, it is not difficult for the network to track the movement of MT with usage of terminal mobility function of mobile network. As a result, it is expected to solve the problems that conventional approach issues as explained previously, and to satisfy the four requirements that described in section1. The next section explains the new method including how to decide the network node as LC in the next section.

3 New Method

3.1 Network Architecture

We assume mobile network with IP-based which is constructed by router as network node. The network consists of several subnetworks in Figure1(a). Each subnetwork is connected by Border Router (BR) which keeps temporarily relaying data frames for recovery, or manages local group address and so on for management of subnetwork. Figure1(b) illustrates an example of subnetwork near MT. In this Figure, router3 and router5, which are LCs for local group1 and local group2, respectively, request and download recovery data from BR for its own group.

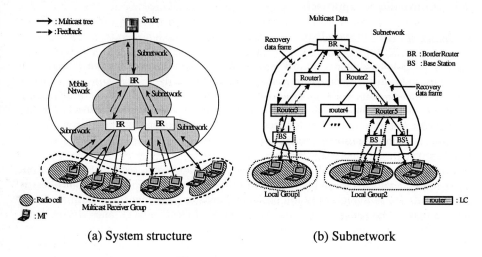

(a) System structure   (b) Subnetwork

Fig.1. System architecture

3.2 Algorithm for Decision of LC (Local Controller)

The policy of the algorithm is that it could reduce traffic generated in recovery and save the network resource as much as possible. Using Figure2, the algorithm is demonstrated as follows. After a given data frame is multicasted to all MTs of a multicast group, feedbacks from MTs come up to routers. The router that has received the feedbacks makes judgement if there is a coincidence with feedback for error from plural lower routers (Figure3). If not, the router orders the lower router

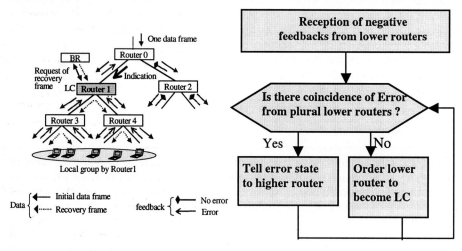

Fig.2. DSD algorithm   **Fig.3.** DSD : LC selection

with negative feedback to become a LC. If any, it sends negative feedback up to the higher router. In figure2, at router1, for instance, there is a coincidence with feedbacks with error from router3 and router4, and the router1 reports the coincidence to the higher router0. At the router0, the coincidence with the feedbacks breaks, and the router0 indicates the router1to become LC. Accordingly router1 as LC organizes its LG (Local Group) to carry out recovery in multicast; the router1 gets the recovery data and local multicast address from BR and defines the LG by a set of erroneous MTs under the routers of which the highest is router1, and then executes retransmission.

Thus, LC router is selected considering the geographical distribution of erroneous MT (we named the algorithm "DSD; Dynamic Selection based on the Distribution") and dynamically forms LG for recovery. In this way, we can obtain the characteristics that the more densely error happens, the higher router would become LC for covering erroneous MTs properly. As a result, efficient recovery and usage of resource can be expected.

3.3 Terminal Mobility

The new method, NB, could make realization of the mobility much easier. The process of mobility is able to make use of the existing mobile IP function of network because it is the network node that takes care of recovery in NB.

Here we just outline only the point of study for mobility since the objective of this paper is to prove that NB-DSD algorithm is effective for mobile networks.

The action sequence of mobility was considered on the timing of MT movement to the algorithm execution. In case of the movement of MT before NB-DSD algorithm execution, LC router in the destination of MT recovers it, while in the reverse case, the original LC router continues to recover it.

4 Evaluation

In this section, the evaluation of the new approach with NB-DSD method is described through comparison with conventional approaches (SB and RB).

4.1 Outline of Evaluation

The evaluation consists in verification of scalability, efficiency of reliability and effectiveness of dynamism of the new method. The mean transmission time per one data frame and the mean number of retransmissions were observed in the simulation. It is enough that the evaluation is done only on subnetwork with connection to MT in Figure1 since the highest LC router is BR.

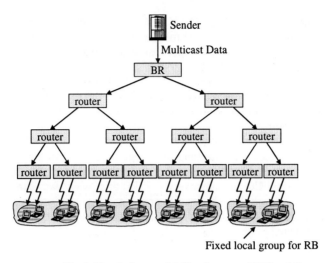

Fig.4. Simulation model (for the case of MTs =16)

Table 1. Notation

NM	The number of MT
Pr	Data frame error probability at wireless link
Cn	Capacity of the link inside network
Cr	Capacity of the receiver–network wireless link
Cs	Capacity of the sender–network link

4.2 Simulation Model and Assumptions

Figure4 is an example of simulation model for a subnetwork which consists of one sender, sixteen MTs and fifteen routers. There are some assumptions;
- Each router has two interfaces to lower router and one interface to upper router and the lowest router accommodates two MTs.
- Regarding the method RB, one local group is defined by every four MTs including LC for recovery.
- Error with data arises only on wireless link in accordance with probability of uniform random distribution.
- Multicast transmission is simulated for total file = 500[kB], data frame length = 2000[bit]. Table1 shows the parameter notation.

4.3 Results

(1) Scalability
For evaluation of scalability, the increase ratio of the mean transmission time per one frame was measured with a variation of the number of MTs that represents network scale as appeared in Figure5. The increase ratio was calculated using the standard

case of the transmission time for Nm = 4. In this Figure, only SB has much higher ratio with the number of MTs. It is because, in SB, the sender has to take care of retransmission for all of increased erroneous members, while, in RB and NB-DSD, each LC does it only for local erroneous members.

In Figure6 shows the number of retransmissions for the same conditions as that of Figure5. From this result, only on SB, the more the number of MTs increases, the more the sender repeats retransmission over the whole subnetwork, that causes wasting of network resource.

These results indicate that the local group scheme has much better scalability.

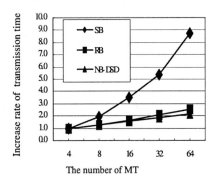 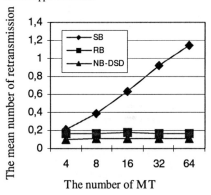

Fig.5. Increase ratio vs. Network scale

Fig.6. The number of retransmission vs. Network scale

(2) Efficiency of reliability
Figure7 illustrates the increase ratio of the mean transmission time with a variation of frame error probability of wireless link. This result also shows that SB has much worse performance in recovery than the local schemes. The fact suggests that SB cannot deal with recovery in efficient manner for deteriorated wireless link quality. And the result shows that NB-DSD has considerably (about by a half) lower ratio than SB.

Compared with RB and NB-DSD using the result of the number of retransmissions in Figure8 for the same conditions as that of Figure7, in which NB-DSD is lower than RB by about a half, we see that the quality of wireless uplink (MT • mobile network) impacts directly on recovery efficiency considering that the main difference between RB and NB-DSD is whether the wireless uplink is used for retransmission.

Moreover the uplink capacity also influences the efficiency in negative on RB as appears in Figure9. The influence is bigger particularly for the condition of low capacity of wireless uplink. Therefore, the conditions of wireless link exert worse influence on recovery efficiency of RB than that of NB-DSD.

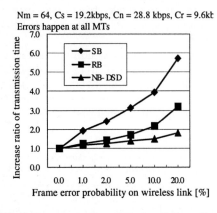
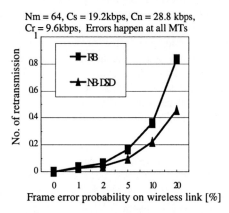

Fig.7. Increase ratio vs. Wireless link quality

Fig.8. The number of Retransmission vs. Wireless link quality

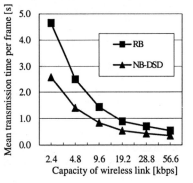
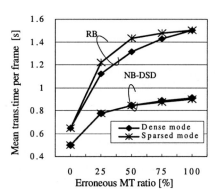

Fig.9. Mean time vs. Capacity of wireless link

Fig.10. Mean transmission time vs. Distribution of erroneous MTs

(3) Dynamism of NB-DSD algorithm
The effectiveness of the dynamism on NB-DSD is made clear through comparison with RB as fixed strategy. Figure10 shows the mean transmission time per one data frame with a variation of ratio of erroneous MT. The two kinds of the geographical distribution of erroneous MT were selected as parameter, i.e., sparse mode and dense mode. Sparse mode is the state that erroneous MTs lie scattered uniformly over the network. On the other hand, dense mode is that erroneous MTs lie together. In this result, it is evident that the transmission time is different depending on the distribution for RB method, while it is almost the same for NB-DSD method. The fact implies the

new algorithm can dynamically choose LC routers considering the state of erroneous MT distribution for covering erroneous MTs properly (i.e. the more densely error happens, the higher router would become LC for this area) so as to form several proper groups for local recovery towards the reduction of recovery traffic and network resource consumption.

5 Conclusion and Future Work

In this paper, we examined the problems of existing protocols for reliable multicast when applying them to mobile network, and proposed the new method. The new method (NB-DSD) was compared with two conventional generic approaches (SB and RB) with regard to scalability and efficiency of reliability and dynamism. Concerning terminal mobility, we can expect that the new method reduces greatly the difficulty of its realization.

Our conclusion by the evaluation is the following;
(1) The new method (NB-DSD) which employs dynamic local group concept is the most suitable for mobile network with regard to large scalability and efficiency for reliability and can be expected to realize a very large scale and high reliable multicast system.
(2) The dynamism of new method (NB-DSD) enables to absorb the volatile geographical distribution to reduce traffic generated in recovery and economize the network resource.

In the future, we will take additional metrics (e.g. link congestion etc.) into account for the decision LC algorithm in order to obtain more fitting algorithm for real network.

References

1. J.F.Huber, D.Weiler and H.Brand, "UMTS, the Mobile Multimedia Vision for IMT-2000: A focus on Standardization," IEEE Communication Magazine,Vol.38 No.9,2000
2. T.A.Maufer, "DEPLOYING IP MULTICAST IN THE ENTERPRISE," Prentice Hall PTR,1998.
3. C.K.Miller, "Multicast Networking and Applications," Addison Wesley,1998.
4. S.Floyd, V.Jacobson, S.McCanne,C.Liu and L.Zhang, "A Reliable Multicast Framework for Light-weight Sessions and Application Level Framing," in Proc. ACM SIGCOMM'95, Boston, MA, Aug. 1995, pp. 342-356.
5. T.Speakman, D.Farinarri, S.Lin, A.Tweedly, "Pretty Good Multicast (PGM) Transport Protocol Specification," Internet Draft, draft-spealman-pgm-spec-00.text, January 8, 1998.
6. M.Hofmann, "A Generic Concept for Large-Scale Multicast," in Proc. IZS'96, Feb. 1996, No.1044.
7. J.C.Lin and S.Paul, "RMTP: A Reliable Multicast Transport Protocol," in Proc. IEEE INFOCOM'96, March 1996, pp. 1414-1424.

8. S.Paul, K.K.Sabnani, J.C.Lin and S.Bhattacharyya, "Reliable Multicast Transport Protocol (RMTP)," IEEE Journal on Selected Areas in Com., Vol.15, No.3, April 1997, pp. 407-421.
9. M.T.Lucas, B.J.Dempsey and A.C.Weaver, "MESH: Distributed Error Recovery for Multimedia Streams in Wide-Area Multicast Networks," IEEE ICC'97, June 1997.

An Adaptive Mobility Management Scheme to Support Internet Host Mobility

Miae Woo

Department of Information and Communications Engineering
Sejong University
98 Kunja-Dong, Kwangjin-Ku, Seoul, Korea
`mawoo@sejong.ac.kr`

Abstract. To cope with the Internet host mobility in a cellular network environment, we proposed an adaptive mobility management scheme that can compensate drawbacks of Mobile IP. Our proposed scheme determines foreign agent care-of addresses adaptively according to user mobility. Consequently, it is different from other proposals for micro mobility, which statically assign the gateway in the domain as a foreign agent. Using such a scheme, it is possible to effectively meet the user's demands for different service qualities in the various environments considered in the cellular network and to reduce signaling overhead due to frequent handovers occurred in Mobile IP. The performance of the proposed scheme is examined by simulation. The results of simulation show that the proposed scheme can provide relatively stable points of attachment to the mobile node.

1 Introduction

This paper presents a new approach to Internet host mobility. Mobile IP[1] was designed to provide wide area mobility support. Adopting Mobile IP in the cellular network where a mobile host visits different subnets from time to time results in quite numbers of signaling messages to the home network and significant signaling delay. To overcome such drawbacks, several proposals[2-5] that separate local area mobility, say micro mobility, from wide area mobility have been made. These proposals choose a gateway that connects its domain to the global Internet as a representing foreign agent (FA) of its domain. During stay within a domain, the mobile node registers to its home agent with care-of address of the gateway only once, resulting in minimizing signaling overhead. On the other hand, QoS provision for the mobile node has to be sacrificed unless there is an additional QoS provision method in the visited domain.

In the cellular network, users expect to receive different services according to their mobility. For example, third-generation systems are expected to offer at least 144kb/s for very high-mobility users in vehicles, 384 kb/s for pedestrians with wide-area coverage, and 2 Mb/s for low-mobility users with local coverage [6]. In order to provide various service qualities that are appropriate to the state of user mobility, micro mobility scheme adopted in the network should use an adaptive scheme. In such a perspective, the micro mobility schemes proposed in the other literatures assign a gateway as a foreign agent statically, resulting in lack of adaptability. To effectively meet the user's demands for different service qualities in such various environments and to reduce signaling overhead due to handovers, we propose an adaptive foreign agent determination scheme that chooses the foreign agent care-of addresses adaptively according to user mobility.

We also propose a network architecture that can support such an adaptive foreign agent determination scheme. In the proposed architecture, it is assumed that the cellular network is all-IP network, since the wireless industry is evolving its core networks toward IP technology[7]. The base stations and the other nodes in the access network are all IP nodes, and any of them can be chosen as a foreign agent of a specific mobile node. The mobile nodes that are connected with the same base station may have different foreign agents. Choice for a foreign agent is based on the user mobility. By applying adaptive foreign agent determination, it is possible to provide the required QoS up to the base station that is not considered in the other proposals in micro mobility. In the extreme case, when foreign agent is fixed to be a base station, then our scheme becomes Mobile IP[1]. If foreign agent is fixed to be a gateway, then it corresponds to the other micro mobility proposals.

The proposed adaptive scheme has following advantages compared to the others.
- It can distribute loads throughout the IP nodes in the access network by locating foreign agents according to the mobility of mobile nodes. In other proposals for micro mobility, all loads for providing mobility service are concentrated on the gateway.
- It can provide a relatively stable point of attachment to the mobile node. Although overall signaling overhead due to handovers is larger than other micro mobility schemes, the effect of the overhead on the network performance can be regulated.
- If the base station is not chosen for the foreign agent, it introduces one more tunnel that is in between the foreign agent and the base station. So it uses intrinsic IP routing between the foreign agent and the serving base station. Consequently, it eliminates need for maintaining tables or caches in the nodes that are located in between the foreign agent and the serving base station en route.
- It can provide appropriate levels of QoS to the mobile nodes using network-intrinsic QoS provision mechanisms.

The rest of this paper is organized as follows: Section 2 presents an overall network architecture for the proposed scheme. Section 3 describes the proposed adaptive foreign agent determination scheme. Performance evaluation of the proposed scheme is presented in Section 4. Section 5 describes how the proposed scheme can provide an appropriate level of QoS. Section 6 concludes this paper.

2 An Overview of the Proposed Architecture

In this section, we present a network architecture for the adaptive mobility management. The proposed micro mobility architecture is depicted in Fig. 1. It is assumed that the proposed architecture supports three levels of network node hierarchy. The base stations belong to the lowest level. The intermediate nodes that are one hop away from the base stations towards the gateway are the elements of the second level. Finally, the gateway forms the third level. The network node hierarchy can be extended freely if required.

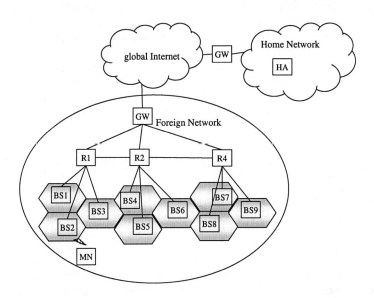

Fig. 1. Network architecture

A foreign network is consisted of a set of subnets, which correspond to the cells in the cellular network. In each subnet, there is a base station that provides an access point to the mobile nodes. The base station to that a mobile node is currently attached is called as a *serving base station* in the paper. In Fig. 1, BS2 is the serving base station to the mobile node MN. All base stations, intermediate nodes, and the gateway are assumed to be IP nodes and can be act as foreign agents if required.

In this paper, *FA level* is used to describe how many hops the candidate foreign agent away from the mobile node. In other words, FA level n means that the candidate foreign agent is n hops away from the mobile node towards the gateway. For each mobile node, its mobility is calculated at the handover time and an appropriate FA level is determined. The mobile node then uses the IP address of a node with the determined FA level as a foreign agent care-of address. The mobile node with FA level 1 uses the address of the serving base station as its foreign agent care-of address. The foreign agent care-of address of the mobile node with FA level 2 is the address of the intermediate node. The foreign agent care-of address of the mobile node with other FA level can be determined similarly.

Using the IP intrinsic reachability information, each base station can build and maintain a list of foreign agent care-of addresses. Each item in the list corresponds to the appropriate FA level. For example, the list of foreign agent care-of addresses maintained by the BS2 in the Fig. 1 is {(FA level 1 = IP address of BS2), (FA level 2 = IP address of R1), (FA level 3 = IP address of GW)}.

In the proposed architecture, we use the Mobile IP regional registration[2] with some modification for the micro mobility. Consequently, two types of registration, say home registration and regional registration, are considered. Home registration is a registration with the home agent in the home network of the mobile node. Home registration is executed when the mobile node needs to change the foreign agent. Regional registration is a registration with the foreign agent if the mobile node can continuously use the foreign agent of the previous subnet after changing the subnet. When a mobile node first arrives at a visited domain, it always performs a home registration. When a mobile node moves inside the domain, it needs to do either home registration or regional registration depending on the selected foreign agent in the new subnet.

In the following section, we propose a procedure to determine a foreign agent care-of address to be used by a moved mobile node and to determine the need for home registration or regional registration.

3 Adaptive Determination of a Foreign Agent

When a mobile node moves into an adjacent subnet, it should associate with the base station in the newly visiting subnet. Also, it is required to determine a foreign agent through which the mobile node should maintain connection with its home agent and the corresponding host.

An appropriate foreign agent can be chosen based on how frequently the mobile node changes subnets. In other words, user mobility can be modeled as handover frequency in the cellular network. In this paper, *mobility frequency* is used as the referring terminology. If mobility frequency is high, the mobile node moves fast. If mobility frequency is low, the mobile node is said to be in a stationary state.

In this section, we describe how to determine a foreign agent based on the mobility frequency of the mobile node.

3.1 Mobility State and Mobility Frequency

Mobility frequency is the quantitative term for how frequently a mobile node moves among subnets. Depending on mobility frequency, three mobility states, say stationary, slow-moving, and fast-moving states, are considered in this paper. Whenever a mobile node moves into an adjacent subnet, its mobility frequency and the corresponding mobility state are determined. According to the mobility state, an appropriate IP node is chosen for a foreign agent.

In general, mobility frequency $f_{t_n}(i)$ of a mobile node when it requires nth handover in the foreign network at time t_n can be calculated as follows:

$$f_{t_n}(i) = \frac{i}{t_n - t_{(n-i)}}$$

In the equation, i represents the window size of the moving average in calculating mobility frequency. When i is equal to 1, mobility frequency $f_{t_n}(1)$ is mobility rate from the last handover. Similarly, $f_{t_n}(i)$ is the average mobility frequency in the latest i handovers.

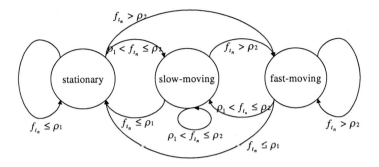

Fig. 2. Mobility state transition diagram

Stability of a point of attachment is a relative term in a mobile environment. If a mobile node maintains association with a foreign agent for a reasonable amount of time, then we could say that the mobile node is attached to a stable point. By analyzing the user's mobility pattern, one can determine the reasonable value for attachment time that is regarded as stable. Based on the determined value, threshold values among stationary state, slow-moving state, and fast-moving state can be determined. Let ρ_1 and ρ_2 be the determined threshold values for deciding mobility state of a mobile node as stationary state, slow-moving state, or fast-moving state. A mobility state is changed depending on mobility frequency as shown in Fig. 2. In the figure, f_{t_n} represents mobility frequency calculated. Selecting proper window size to calculate f_{t_n} is given in Section 3.2.

Each mobility state has the corresponding FA level. Mobility states of stationary, slow-moving, and fast-moving correspond to the FA level 1, 2, and 3 respectively. So, based on the determined mobility state, one can decide what the FA level is to be used and which node is used as a foreign agent for the mobile node.

In the next subsection, we develop a procedure for choosing an appropriate foreign agent care-of address.

3.2 Procedure for Choosing a Foreign Agent

To provide a stable point of attachment, it is desirable not to change a foreign agent unless it is necessary. For that purpose, we develop two provisions in determining a foreign agent. The first provision is to represent the trend of mobility pattern of a mobile node effectively. The second one is to maintain the same foreign agent if possible.

- **Provision 1:** The first provision is use of different window size values in calculating mobility frequency. Let the previous FA level denote the FA level of a mobile node in the previous subnet. If the previous FA level of the mobile node was 1, then the mobile node needs to register to the home agent with a new foreign agent care-of address. So in this case, it is better to represent the mobility of the mobile node as it is in the calculation of mobility frequency and to determine the FA level accordingly. Subsequently, it is desired to use $f_{t_n}(1)$ in this case. When the previous FA level was not 1, moving average of window size 2 is used to calculate mobility frequency. By using moving average, it is possible to average out mobility frequency and to represent the trend of mobility pattern of a mobile node.

- **Provision 2:** For the second provision, the foreign agent that is used in the previous subnet is checked whether it can be used continuously as a foreign agent in the new subnet. Let the newly determined FA level in the new subnet be 3 and the FA level in the previous subnet be 2. In such a case, if the foreign agent with level 3 in the new subnet can be reachable through the previous foreign agent with level 2, there is no need to change the foreign agent since the mobile node can continuously communicate through the foreign agent in the previous subnet. Such provision eliminates unnecessary signaling to the home network. In this case, the current serving base station of the mobile node is notified to the foreign agent using regional registration.

By incorporating the provision 1 and 2, we develop a procedure for determination of a foreign agent care-of address for a mobile node at its nth handover time. The proposed procedure is listed in below.

```
PROCEDURE FADetermination
// Provision 1
// Calculate mobility frequency
if (prevFAlevel == 1) {
            // use moving average of window size 1
      f_{t_n} = 1/(t_n - t_{n-1});
} else {   // use moving average of window size 2
      f_{t_n} = 2/(t_n - t_{n-2});
}

// Determine the mobility state and the new FA level
// depending on mobility frequency
if (f_{t_n} ≤ ρ_1)  // stationary state
      newFAlevel = 1;
elseif (f_{t_n} ≤ ρ_2)  // slow-moving state
      newFAlevel = 2;
else         // fast-moving state
      newFAlevel = 3;

Choose a newFAaddress with newFAlevel from the FA list
in the new serving base station;

// Provision 2
if ((newFAlevel >2) and (prevFAaddress is in the FA
list in the new base station)) {
      // use the previous FA in the newly visited subnet
      newFAlevel = prevFAlevel;
      newFAaddress = prevFAaddress;
}
if (newFAaddress is different from prevFAaddress)
      home registration is required;
```

```
else
    regional registration is required;
```

In the next section, we verify the performance of the proposed procedure through simulation study.

4 Performance Evaluation

We study how stable point of attachment the proposed procedure in Section 3.2 can provide by a simulation study. For the simulation, it is assumed that a mobile node stays in a subnet with an exponential distribution. The mean stay time of the mobile node in a subnet varies with its mobility state type. Let the mean stay time of a mobile node with stationary state be τ_1. We use τ_1 as a reference value for the mean stay time of a mobile node with other state. For the simulation, the mean stay times in the slow-moving and fast moving states are assumed to be $(1/8)\tau_1$ and $(1/40)\tau_1$ respectively. It is assumed that a mobile node maintains one mobility state during its stay in a subnet. When the mobile node moves to an adjacent subnet, new mobility state is determined based on the previous mobility state. The mobility state transition probability considered is as follows:

$$P = \begin{bmatrix} 0.7 & 0.2 & 0.1 \\ 0.25 & 0.5 & 0.25 \\ 0.1 & 0.2 & 0.7 \end{bmatrix}$$

In the simulation, we investigate the performance of five methods in terms of how many home registrations are required due to the change of a foreign agent, and compare their results to that of Mobile IP. Five methods used in the simulation are listed in below.

- Method 1: Mobility frequency is calculated with window size of 1. A FA level and the foreign agent care-of address are chosen according to the calculated mobility frequency.
- Method 2: Mobility frequency is calculated with window size of 2. A FA level and the foreign agent care-of address are chosen according to the calculated mobility frequency.
- Method 3: Mobility frequency is calculated with window size of 3. A FA level and the foreign agent care-of address are chosen according to the calculated mobility frequency.
- Method 4: Mobility frequency is calculated with the provision 1. A FA level and the foreign agent care-of address are chosen according to the calculated mobility frequency.
- Method 5: Mobility frequency is calculated with the provision 1. A FA level and the foreign agent care-of address are chosen using the provision 2. This is the proposed method in Section 3.2.

Method 1, 2 and 3 are used to see the effect of the window size values in calculating mobility frequency. Method 4 and 5 are used to see how our proposal produces better results.

In the simulation, we change the threshold values ρ_1 and ρ_2, and see their effect on the performance of each method. From the simulation result, it is observed the performance of each method depends on the threshold value ρ_1. However, the threshold value ρ_2 does not have much effect on the performance. Fig. 3 shows the effect of ρ_1 on the performance of the five methods. In the figure, the value of ρ_1 is represented as a ratio of the mean stay time of a mobile node in a stationary state τ_1. In the figure, the number of home registration required by Mobile IP is used as a reference value and the number of home registration required by each method is compared to it. As shown in the figure, Method 5, which is the proposed scheme in Section 3.2, requires the least number of home registrations. Also, the effect of chosen value for the threshold ρ_1 is well illustrated. When the ρ_1 value is less than $1/\tau_1$ ($\rho_1 = 0.4/\tau_1$ or $\rho_1 = 0.7/\tau_1$ in the figure), The number of home registration required by Method 5 is about 27% of Mobile IP. As the value for ρ_1 increases, the number of home registration required by Method 5 also increases.

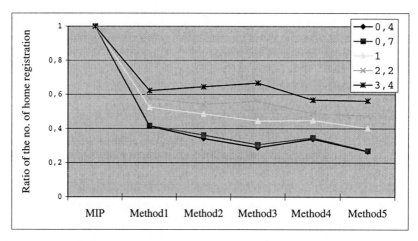

Fig. 3. The ratio of the number of home registration compared to Mobile IP in various ρ_1 values (in $1/\tau_1$ unit)

In the figure, it can be observed that the performance gets better as the window size increases when $\rho_1 \leq 1/\tau_1$. However, when we look at the results more closely, it is observed that there are lagging phenomena in determining the FA level as the window size increases especially when the mobility state changes from stationary to other states. For such phenomena, provision 1 in Section 3.2 seems to be effective.

We analyze how many times home registration is required in τ_1 time using the simulation result. The result is depicted in Fig. 4. When using Method 5, a mobile node requires to do home registration in every 1.5 τ_1 on average when $\rho_1 = 0.4/\tau_1$.

Such frequency of home registration is definitely larger than that of other micro-mobility schemes. However, as shown in Fig. 3 and Fig. 4, the number of home registrations in a unit time can be regulated by choosing the appropriate value for the threshold value ρ_1.

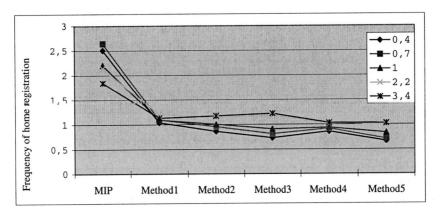

Fig. 4. Frequency of home registration in τ_1 in various ρ_1 values (in $1/\tau_1$ unit)

5 QoS Supports

In this section, we briefly discuss how our proposed scheme can provide the desired QoS to the mobile nodes. If any reservation protocol that can work through tunnels is employed, then end-to-end QoS provision can be achieved. For example, RSVP operation over IP tunnels[8] can work in such a situation.

If that is not the case, then we can consider two options for the macro mobility. One is Mobile IP[1], and the other one is Mobile IP with route optimization[9]. When Mobile IP is used, reservation from the corresponding node to the home network is only provided. When Mobile IP with route optimization is used, reservation from the corresponding node to the foreign agent can be established. As a result, the range of QoS supports is depend on the user mobility. For the stationary mobile node, QoS support from the corresponding node to the serving base station is possible. Subsequently, a certain level of guaranteed QoS can be provided to the mobile node. For the fast-moving mobile node, QoS support from the corresponding node to the gateway is possible. In this case, users are expected to get some important messages probably in a less bandwidth-consuming format, so degradation in QoS may be tolerable. By using such an adaptive scheme, the proposed method provides stable point of attachment as well as provides adequate QoS to the mobile user.

6 Conclusion

In this paper, we proposed an adaptive mobility management scheme to cope with the Internet host mobility in a cellular network environment. The first goal of the proposed scheme is to effectively meet the user's demands for different service qualities in the various environments considered in the wireless network. The second goal is to reduce signaling overhead due to frequent handovers. To achieve such goals, the proposed scheme determines the foreign agent care-of addresses adaptively according to user mobility. By adopting such an adaptive scheme, it can distribute loads for providing mobility service throughout the IP nodes in the access network while other proposals for micro mobility put all loads on the gateway.

The stability of the provided point of attachment is studied by simulation. The simulation results show that our scheme can provide a relatively stable point of attachment to the mobile node. Although overall signaling overhead due to handovers is larger than that of other micro mobility schemes, it can be regulated by adjusting the threshold values for determining mobility states so that network performs well enough.

References

1. C. Perkins, ed., "IP Mobility Support," IETF RFC 2002, Oct. 1996.
2. E. Gustafsson, A. Jonsson, and C.E. Perkins, "Mobile IP Regional Registration," Internet draft (work in progress) draft-ietf-mobileip-reg-tunnel-02.txt, Mar. 2000.
3. Campbell, et. al., "Cellular IP," Internet draft (Expired), <draft-ietf-mobileip-cellularip-00.txt>, Jan. 2000.
4. R. Ramjee, et. al., "IP Micro-Mobility Support using HAWAII," Internet draft, <draft-ietf-mobileip-hawaii-01.txt>, Jul. 2000.
5. S. Das, et. al., "TeleMIP: Telecommunications-Enhanced Mobile IP Architecture for Fast Intradomain Mobility," IEEE Personal Communications, Aug. 2000, pp. 50-58.
6. T. Ojanpera and R. Prasad, "An Overview of Third-Generation Wireless Personal Communications: A European Perspective," IEEE Personal Communications, Dec. 1998, pp. 59-65.
7. G. Patel and S. Dennett, "The 3GPP and 3GPP2 Movements Toward an All-IP Mobile Network," IEEE Personal Communications, Aug. 2000, pp. 62-64.
8. A. Terzis, et. al., "RSVP Operation Over IP Tunnels," RFC2746, Jan. 2000.

Modeling and Analysis of TCP Enhancement over Heterogeneous Links

Mingyan Liu and Navid Ehsan

University of Michigan, Ann Arbor, MI 48109-2122, USA,
{mingyan, nehsan}@eecs.umich.edu,

Abstract. In this paper we focus on one type of TCP enhancement commonly used to improve performance for connections over heterogeneous physical layer, namely, TCP connection splitting or split TCP, where the end-to-end TCP connection is split in segments so that each segment runs over a homogeneous environment and is optimized separately. This paper presents a simple model capturing some of the features of this scheme where split segments are essentially "coupled" TCP connections. We use simulation to validate our models and use our model to analyze situations where split TCP is preferred to the end-to-end scheme, and where the performance gain of splitting vanishes by investigating factors including initial window size, file size, number of on-going connections and loss.

1 Introduction

With the advances in wireless technology, wireless communication is becoming more and more common. It is well known that the original form of TCP does not do well over wireless links where packet losses due to link failure is more prevalent than losses due to congestion, since TCP takes any loss as a sign of congestion. TCP also does not do well over large bandwidth-delay-product links, e.g., a satellite link, since it takes a long time for the congestion window to fully utilize the channel, which is particularly inefficient for short connections.

Extensive research has been done over the past few years to improve the TCP performance over heterogeneous connections, see for example [1]. One type of enhancement completely segregates the wireless part of the connection from the wired part. This scheme is often called TCP splitting or spoofing [2] in satellite communication and I-TCP [3] in terrestrial wireless communication. In general, by this segregation, the end-to-end TCP semantics are inevitably broken or altered, e.g., MTCP [4]. Another type of enhancement maintains the TCP end-to-end semantics, but tries to reduce the impact of the wireless link. This involves differentiation between the wired part and the wireless part, although the impact of the wireless link cannot be completely singled out because of the end-to-end semantic [5–9]. There are also schemes that focus on the link layer characteristics and often involves link layer retransmission [10].

Among these solutions, split TCP was proposed, in view of TCP's inability to handle heterogeneous environment, with the idea that if a communication path consists of physical medium that are of very different characteristics, then the end-to-end performance is optimized by isolating one type of physical link from another and optimizing each separately. This violates the TCP end-to-end semantics, and will not work if the IP packet payload is encrypted [11], which has been a major criticism for this type of schemes. Nevertheless, split TCP is very commonly used in satellite based data services because there seems to be no other apparent solution to deal with the large bandwidth-delay-product of the satellite link without having to modify the end systems. On the other hand it has also been observed that when the satellite link becomes congested, the performance gain from TCP splitting reduces, and could even disappear [2].

This motivated us to do a quantitative study on split TCP as a solution to heterogeneous environment. This paper presents mathematical models to analyze the applicability of this scheme, and the amount of improvement under different situations by varying a number of parameters, including the initial window size, file size, and link propagation delay. This paper is organized as follows. In Section 2 we present the network model and give a brief introduction to the technique of connection splitting. In Section 3 we derive the model and use it to analyze the latency in file transfer. The accuracy of our model is discussed. Section 4 discusses our results and concludes the paper.

2 Split Connection and the Network Model

2.1 Network Model

Our analysis is based on a two link model with one end host on each side of the tandem of two links, and a proxy at the junction of the two links, as shown in Figure 1, where the two circles in the middle represent possibly coupled connection when split TCP is used. We assume one is a fixed terrestrial link, and the

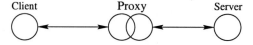

Fig. 1. Network Model

other is either a satellite or a terrestrial wireless link, which is represented by different propagation delay and error rate of the link. Without loss of generality, we assume that the client is located at the end of the wireless/satellite link and that the server is located at the end of the terrestrial link. File transfer is our main application of interest, and are considered to be from the server to the client.

When the end-to-end connection is used, the proxy functions as a normal router that forwards packets from the source to the destination. When split connection is used, the proxy typically acknowledges the packets received from the server on behalf of the client, and opens up a second connection between itself and the client, and sends packets received from the server to the client as if it were the server, thus the term "split connection". Same procedure is used for the other direction of the connection. The proxy is transparent to the end points. In general, the initial connection establishment (three-way handshake) and the final connection closing are done in exactly the same way as in an end-to-end connection. Thus the connection is only split in two during the data transfer stage.

The focus of this paper is to examine the response time or latency in file transfers, using end-to-end or split TCP. In our analysis we do not include the time for connection establishment (including file requests), since this duration is the same in both cases. Therefore our latency in file transfer is defined as the duration between when the server sends the first data packet of a file and when the client receives the last data packet of a file.

2.2 Assumptions and Parameters

We mainly consider the lossless case in this paper. When there is no loss, the congestion window grows exponentially during slow start, linearly during congestion avoidance and then stays at maximum window size. We assume that a file contains exactly M segments of maximum segment size (MSS). This is an approximation to an arbitrary file size whose last segment may be a fraction of MSS. However, in calculating latency, this does not affect our method of analysis, and also does not affect the comparison between using end-to-end TCP and using split TCP.

The $ssthresh$, W_{sst}, and the maximum window size W_{max} are assumed to be in number of segments rather than number of bytes to simplify our analysis. Again, these may not be the exact values, but they do not affect our study. The server, the proxy and the client each has a transmission rate of C_1, C_p, and C_2, respectively. Assuming packet length of $D = MSS + 40$ (TCP and IP headers), the time it takes for the server to transmit a packet is $\mu_1 = \frac{D}{C_1}$, and $\mu_p = \frac{D}{C_p}$, $\mu_2 = \frac{D}{C_2}$ for the proxy and the client, respectively. When splitting is used, we assume a per-packet processing delay of t_p at the proxy. All other processing delays are ignored. Through out our analysis, we assume the transmission time of an ACK to be negligible.

In addition, we assume that the transmission is only constrained by the congestion window and not the advertised receive window size. The one-way propagation delay on the server-proxy link and the proxy-client link are denoted by I_1 and I_2, respectively. We assume that the delay on each link is symmetric, although our analysis can be easily generalized to the asymmetric case. Most work in TCP analysis assumes an infinite source, e.g., [12–14]. However, when connection splitting is used, the window of the second connection (proxy-client) changes not only according to the window dynamics of TCP, but also according to the availability of packets (from the server-proxy connection), i.e., the first connection may not "catch up" with the second connection due to factors like initial window size, transmission rate, etc.. Thus the window of the second connection may be forced to grow at a slower rate. In this paper, we only consider the case where the window of the proxy-client connection is not constrained by the server-proxy connection. This, in the lossless case implies that the proxy-client connection uses an initial window size same as or bigger than the server-proxy connection, and that the proxy transmission rate is not significantly higher than that of the server. This also implies that in general $I_2 > I_1$, which is a reasonable assumption when the proxy-client link is a satellite link (for GEO satellite commonly used for data communication, the one-way propagation delay is 250 ms), but less accurate if it is a terrestrial wireless link. When loss is present, the implication is more complicated. The second connection can starve due to heavy congestion/loss on the server-proxy link regardless of the initial window size and propagation delay.

3 Latency Analysis

3.1 Lossless Links

Under lossless situation, latency is mostly due to transmission delay, propagation delay, processing delay, and the time the server spent in waiting for the window to ramp up to fully utilize the link capacity.

We first calculate the number of windows that is needed to cover a file of size M segment by extending the method introduced in [15]. Let w_0 denote the initial window size, and therefore using slow start, the first window has $w_0 2^0$ packets, the second window has $w_0 2^1$ packets, and so on. (We do not consider delayed ACK in our analysis, although our model can be easily modified to take delayed ACK into account. For example, as shown in [16] assuming one ACK is generated for every b packets received before the timer expires, then the rate of exponential growth of the congestion window is $r = 1 + \frac{1}{b}$, which equals 2 when no delayed ACK is used.) Let S be such that

$$w_0 2^{S-1} < W_{sst} \leq w_0 2^S, \tag{1}$$

if $M > \sum_{i=1}^{S} w_0 2^{i-1}$, i.e., the $ssthresh$ is achieved during the $(S+1)^{th}$ window given enough data. Therefore the $(S+1)^{th}$ window size is W_{sst} and the $(S+2)^{th}$ window size is $W_{sst} + 1$, and so on. Similarly, let M_x be such that

$$W_{sst} + M_x - S - 1 < W_{max} \leq W_{sst} + M_x - S, \tag{2}$$

i.e., the maximum window size is achieved during the $(M_x + 1)^{th}$ window if the file is big enough, and thus all subsequent windows have the same window size of W_{max}. Therefore the number of windows needed to transfer a file is given by the following:

$$K = \begin{cases} \min\{k : \sum_{i=1}^{k} w_0 2^{i-1} \geq M\} & k \leq S \\ \min\{k : \sum_{i=1}^{S} w_0 2^{i-1} + \sum_{i=S+1}^{k} (W_{sst} + i - S - 1) \geq M\} & S < k \leq M_x \\ \min\{k : \sum_{i=1}^{S} w_0 2^{i-1} + \sum_{i=S+1}^{M_x} (W_{sst} + i - S - 1) \\ \quad + \sum_{i=M_x+1}^{k} W_{max} \geq M\} & M_x < k \end{cases} \tag{3}$$

1. End-to-End Connection. After transmitting each window, the server will stall to wait for the first acknowledgment of that window if it takes longer for the ACK to arrive than it takes to transmit the entire window. The time it takes to transmit the kth window is a function of the packet transmission time at the sender (μ_1 in this case), which we denote by $t_k(\cdot)$, is given by

$$t_k(\mu_1) = \begin{cases} w_0 2^{k-1} \mu_1 & \text{if } k \leq S \\ (W_{sst} + k - S - 1)\mu_1 & \text{if } S < k \leq M_x \\ W_{max}\mu_1 & \text{if } M_x < k \end{cases}. \quad (4)$$

Therefore if $\mu_1 \geq \mu_p$, which indicates the case where the proxy transmits at least as fast as the server and thus packets will not experience queueing delay at the proxy, the total time it takes to transfer the file is

$$T_e(M) = M\mu_1 + \sum_{k=1}^{K-1} [\mu_1 + \mu_p + 2(I_1 + I_2) - t_k(\mu_1)]^+ + I_1 + I_2 + \mu_p, \quad (5)$$

where $[a]^+ = a$ for a positive and 0 otherwise. This latency reflects the total transmission time, the time that the server spends waiting for ACKs, and the time for the last window to reach the client.

When $\mu_1 < \mu_p$, packets could build up at the proxy waiting to be transmitted onto the slower link and therefore experience additional queueing delay at the proxy. In this case the ACKs of the same window arrive at the server approximately μ_p apart instead of μ_1, thus the server may need to wait for every ACK of the same window instead of stalling after sending out the entire window. The above stall time analysis therefore does not directly apply. We derive the latency by examining from the receiver (client) side. Since $\mu_p > \mu_1$, the client receives packets of the same window continuously at rate $1/\mu_p$. The time that the client is idle is therefore $[\mu_1 + \mu_p + 2(I_1 + I_2) - t_k(\mu_p)]^+$, where $t_k(\mu_p)$ is the time it takes the client to receive the kth window, and $t_k(\cdot)$ is given by (4). The latency is then

$$T_e(M) = M\mu_p + \sum_{k=1}^{K-1} [\mu_1 + \mu_p + 2(I_1 + I_2) - t_k(\mu_p)]^+ + I_1 + I_2 + \mu_1, \quad (6)$$

which reflects the time the client spends receiving the file, waiting for the next packet, and the time for the first window to reach the client. (5) and (6) can be combined and written as

$$T_e(M) = M \max(\mu_1, \mu_p) + \sum_{k=1}^{K-1} [\mu_1 + \mu_p + 2(I_1 + I_2) - t_k(\max(\mu_1, \mu_p))]^+$$
$$+ I_1 + I_2 + \min(\mu_p, \mu_1). \quad (7)$$

2. Connection Splitting. As we have pointed out earlier, we consider the situation where the proxy transmission is not constrained by the availability of packets from the server, i.e., $I_2 > I_1$, $\mu_p \geq \mu_1$, and that the server's initial window size is greater or equal to that of the proxy. Since a larger initial window size of the server-proxy connection does not change the end-to-end latency (additional packets will simply be queued at the proxy), we assume both connections use the same initial window size.

For initial window size w_0, we get S, M_x, and K from (1), (2) and (3) for both connections. The proxy receives the first packet from the server at time $\mu_1 + I_1$. Assuming there is t_p delay for TCP processing, the proxy starts sending this packet to the client at time $\mu_1 + I_1 + t_p$. From this point on, we only need to focus on the second connection since the latency is only determined by this connection. By following the previous analysis, the total latency is

$$T_s(M) = \mu_1 + I_1 + t_p + M\mu_p + \sum_{k=1}^{K-1} [\mu_p + 2I_2 - t_k(\mu_p)]^+ + I_2, \qquad (8)$$

where $t_k(\cdot)$ is given in (4). This latency reflects the initial delay for the first packet to arrive at the proxy, the total transmission time at the proxy, stall time and the time for the last packet to reach the client.

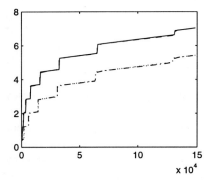 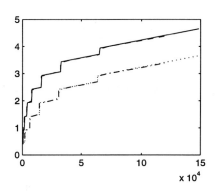

Fig. 2. Both graphs show latency (in seconds, vertical) vs. file sizes (in bytes, horizontal), by using end-to-end connections (graph on the left), and using split connections (graph on the right). There are four curves in each of the two graphs, two for initial window size of 1 (dashed for model, solid for simulation) and two for initial window size of 4 (dotted for model and dash-dotted for simulation). In both cases our model matches very well with the simulation

3. Validation of the Model. Figure 2 compares our numerical results with NS simulation [17], for both end-to-end TCP and split TCP. In this case, $C_1 = C_p = C_2 = 1$ Mbps. The initial window size is set to 1 and 4, respectively. Unless pointed out explicitly, our numerical results throughout this paper are based on the following parameters: $MSS = 512$ bytes, $I_1 = 150$ ms, $I_2 = 250$ ms, $ssthresh = 128$ segments, $W_{max} = 256$ segments. We see the curve from simulation and from our model almost overlap completely.

4. Results and Analysis. In this section we show a sequence of results from our model illustrating the performance (in terms of latency) of end-to-end connection and split connection under various situations.

Initial window size. Figure 3 shows the latency comparison between end-to-end connection and split connection when using different initial window size. The

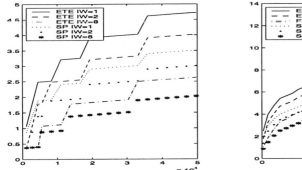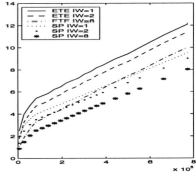

Fig. 3. Both graphs shows latency (in seconds, vertical) vs. file sizes (in bytes, horizontal). "ETE" for end-to-end connection; "SP" for connection splitting; "IW" for initial window size, $I_1 = 100$ ms, $I_2 = 250$ ms, $C_1 = C_2 = C_p = 1$ Mbps

gain from using split connection is obvious in this case, although using end-to-end connection with initial window size of 8 outperforms split connection with initial window size of 1 or 2, when the file size is below 200 Kbytes and 500 Kbytes, respectively. As the file becomes large enough to fully utilize the channel, the difference between Equations (7) and (8) stays constant, i.e., the summation term becomes fixed as M increases, and the rate of the growth is dominated by the term $M\mu_p$. This means in a lossless situation, for relatively small files (e.g., < 100 Kbytes), the same level of latency achieved by using split connection can be achieved using end-to-end connection by increasing the inital window size.

Delay ratio. Figure 4 compares the latency when using end-to-end connection and using split connection under different propagation delay of the server-proxy link (the delay of the proxy-client link is fixed at 250 ms). This result reflects the difference of the summation terms in equations (7) and (8).

Proxy transmission rate. Figure 5 shows the impact of the forwarding rate of the proxy has on the file transfer latency. When the proxy is as fast as the server, the gain from using split connection is shown in ealier results. However, when the proxy becomes the bottleneck, the gain diminishes quickly for both file sizes, i.e., μ_p dominates in the summation terms in equations (7) and (8). A slower forwarding rate can be an approximation to a more congested proxy (increased

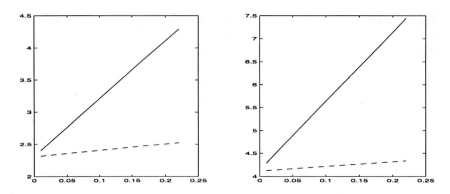

Fig. 4. Both graphs show latency (in seconds, vertical) vs. the propagation delay of the server-proxy link (in seconds, horizontal). File sizes fixed at 11 Kbytes and 110 Kbytes for the graphs on the left and right, respectively. Solid lines for end-to-end connection; dashed lines for connection splitting

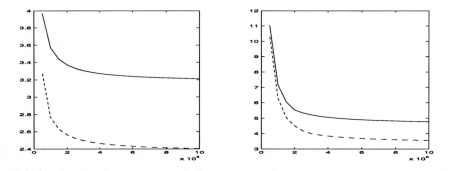

Fig. 5. Both graphs show latency (in seconds, vertical) vs. C_p, the transmission rate of the proxy (in bits per second, horizontal). File sizes are 11 Kbytes and 51 Kbytes, for the graph on the left and right, respectively. $C_1 = C_2 = 1$ Mbps, and $I_1 = 100$ ms, $I_2 = 250$ ms in both cases. Solid line is for end-to-end connection, and the dashed line is for split connection

queueing delay, etc.), thus we see as the proxy becomes more congested the percentage gain in reduced latency becomes marginal.

Persistent connection vs. non-persistent connection. We can also use our model to investigate the performance of different versions of HTTP. Using equations (7), (8) and adding the time spent in connection establishment ($3(I_1 + I_2)$ in both cases), we can obtain the latency in downloading webpages using either persistent connection (HTTP/1.1) or non-persistent connection (HTTP/1.0). Figure 6 shows the comparison. We see that only in very limited regions do non-persistent connections perform better even when using split TCP. This is mainly due to the overhead in establishing separate connections, and that each connection is not long enough to fully utilize the channel.

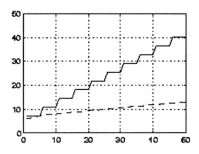 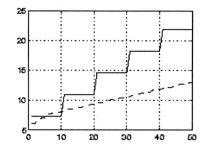

Fig. 6. Both graphs show latency (in seconds, vertical) vs. the number of objects in a base page. Each object is 10 Kbytes. The base page is also 10 Kbytes. Solid lines are for non-persistent connections using split TCP, and dashed lines are for persistent connection using end-to-end TCP. Graph on the left uses 5 paralell connections for non-persistent connection; graph on the right uses 10 paralell connections for non-persistent connection. $C_1 = C_2 = C_p = 1$ Mbps

3.2 Links with Random Losses

The model presented in the previous section can be combined with the TCP steady-state analysis, e.g., [12] and [13], to investigate cases where losses occur mainly on the proxy-client link. This may not be a very realistic assumption even if the proxy-client link is terrestrial wireless since the server-proxy link is always subject to congestion losses, but it does guarantee that the proxy or the second connection will not starve, so that it can be modeled using an infinite source. (Due to space limit we are not able to include our model and analysis. It will appear in a separate paper.)

4 Discussion and Conclusion

In conclusion, we presented a model for analyzing the latency of file transfer when using end-to-end TCP connection and when using split TCP connections over heterogeneous links. We investigated the performance comparison between the two by varying inital window size, file size, the proxy transmission rate and the propagation delays. In general split TCP results in lower latency from our analysis, however for large file sizes, slower proxy, and small ratio of propagation delay between the two links, the improvement is reduced.

As we pointed out earlier, when the proxy-client connection is constrained by the rate at which it receives packets from the server, the window will be forced to slow down and therefore yield larger file transfer latency. Characterizing the window dynamics under such constraint is a topic of our current research. On the other hand, when the server-proxy connection out-paces the proxy-client connection, packets will build up at the proxy. Characterizing the buffer dynamics at the proxy is another topic of our current research.

References

1. Pan, J.: Research: TCP over Something-Air Links - Wireless Links. http://bbcr.uwaterloo.ca/jpan/tcpair
2. Bharadwaj, V.: Improving TCP Performance over High-Bandwidth Geostationary Satellite Links. Technical Report MS 99-12, Institute for Systems Research, University of Maryland, College Park (1999) http://www.isr.umd.edu/TechReports/ISR/1999/
3. Bakre, A. V., Badrinath, B. R.: Implementation and Performance Evaluation of Indirect TCP. IEEE Trans. Computers **46**(3) (1997) 260–278
4. Brown, K., Singh, S.: A Network Architecture for Mobile Computing. IEEE INFOCOM (1996) 1388–1396
5. Schiller, J.: Mobile Communications. Addison-Wesley (2000)
6. Ratnam, K., Matta, I.: WTCP: An Efficient Mechanism for Improving TCP Performance Over Wireless Links. Proc. IEEE ISCC (1998) pages 74–78
7. Iftode, I., Cáceres, R.: Improving the Performance of Reliable Transport Protocol in Mobile Computing Environment. IEEE J-SAC, **13**(5) (1995) 850–857
8. Vaidya, N. H., Mehta, M., Perkins, C., Montenegro, G.: Delayed Duplicated Acknowledgments: A TCP-Unware Approach to Improve Performance of TCP over Wireless. Technical Report 99-003, TAMU (1999)
9. Balakrishnan, H., Padmanabhan, V. N., Seshan, S., Katz, R. H.: Comparison of Mechanisms for Improving TCP Performance over Wireless Links. IEEE/ACM Trans. Networking, **5**(6) (1997) 756–769
10. Parsa, C., Garcia-Luna-Aceves, J. J.: Improving TCP Performance Over Wireless Network at The Link Layer. ACM Mobile Networks & Applications Journal (1999)
11. Karir, M.: IPSEC and the Internet. Technical Report MS 99-14, Institute for Systems Research, University of Maryland, College Park (1999) http://http://www.isr.umd.edu/TechReports/ISR/1999/
12. Cardwell, N., Savage, S., Anderson, T.: Modeling TCP Latency. IEEE INFOCOM (2000)
13. Padhye, J., Firoiu, V., Towsley, D. F., Kurose, J. F.: Modeling TCP Reno Performance: A Simple Model and Its Empirical Validation. IEEE Trans. Networking **8**(2) (2000) 133–145
14. Lakshman, T. V., Madhow, U., Suter, B.: TCP Performance with Random Loss and Bidirectional Congestion. IEEE Trans. Networking **8**(5) (2000) 541–555
15. Kurose, J. F., Rose, K.: Computer Networking, A Top-Down Approach Featuring the Internet. Addison Wesley (2000)
16. Allman, M., Paxson, V.: On Estimating End-to-end Network Path Properties. SIGCOMM (1999)
17. UCB/LBNL Network Simulator. http://www.isi.edu/nsnam/ns/

TCP Throughput Guarantee Using Packet Buffering*

Sunwoong Choi and Chongkwon Kim

School of Computer Science and Engineering
Seoul National University, Korea
{schoi, ckim}@popeye.snu.ac.kr

Abstract. This paper deals with the TCP bandwidth guarantee problem in a Differentiated Services (Diffserv) network. Several researches proposed a RIO mechanism for the assured service. In RIO, IN packets experience lower drop probability than OUT packets to guarantee the promised bandwidth even under network congestion. However a token bucket marker fails to provide adequate performance to TCP connections because TCP generates bursty packets due to the unique TCP congestion control mechanism. We propose a new marker that uses a data buffer as well as a token buffer. The marker with a data buffer works well with RIO because it smooths TCP traffic. We show that the marker with a data buffer achieves the target throughput better than a marker with a token buffer only. We also show that the optimal data buffer size is proportional to reserved throughput and RTT.

1 Introduction

One of the most distinguishing features of the future Internet will be the QoS support capability. The traditional TCP/IP Internet provides best-effort packet delivery service. However, the future Internet will support diverse applications that require stringent or loose QoS guarantee. The mechanisms to support QoS in packet switched network have been actively studied in last 10 years and many clever methods and algorithms were proposed. In IETF, at least two approaches including Intserv and Diffserv were proposed for the QoS guarantee network architecture.

The Intserv (Integrated Services) architecture provides strict QoS guarantee using signaling and end-to-end resource reservation [1]. Before sending information, a subscriber must reserve buffer and bandwidth at each link on the path from the source to the destination using a signaling protocol such as RSVP [2]. Then the intermediate routers maintain the reservation states and handle packets after flow classification. The Intserv architecture may suffer from scalability problem [3], [4]. A flow is defined by a [source IP address, destination IP address, source port number, destination port number, protocol] tuple. More than

---

* This work is supported in part by the Ministry of Information & Communication of Korea and the Brain Korea 21 Project in 2000.

100,000 active flows may exist in a backbone router simultaneously. The overhead of maintaining of a separate queues, state information for each flow and per- packet scheduling may prevent the deployment of the Intserv architecture in large-scale public network [3], [4].

The Diffserv (Differentiated Services) architecture improves the scalability by removing the per-flow state management or packet processing burden at core routers. All functions that require per-flow treatment such as classification, marking, policing and shaping are performed at edge routers [3], [4], [5], [6]. Compared to a core router, an edge router has fewer active flows and packet arrivals. An edge router aggregates flow into a small number of QoS classes and monitors if each flow obeys the promised traffic generation pattern. An edge router marks a conformed packet as an IN packet and marks a non-conformed packet as OUT. The DS codepoint in the IP header records the QoS class and IN/OUT mark of each packet. A core router treats packets according to the DS codepoint and PHB (Per Hop Behavior).

We can implement service models on the general Diffserv QoS architecture. The service models that may be implementable are the PS (Premium Service) model and the AS (Assured Service) model. The PS emulates the leased line by transferring PS packets before other packets [5]. A core router may use priority queue to implement the PS. The AS provides loose bandwidth guarantee without explicit reservation. The RIO (RED with IN and OUT) mechanism was proposed to support the AS at a core router [6].

The concept of the AS model, providing throughput guarantee without explicit resource reservation, is rather new and many researchers studied the effective mechanism for the AS. The mechanism for the AS consists of two methods: One is buffer management methods and the other is packet marking methods at edge routers. RIO, which allocates different packet drop probabilities to IN and OUT packets, was the first proposed buffer management method and has received great attentions [6]. Later, several other buffer management mechanisms were proposed to enhance the performance of RIO. These mechanisms include ERED (Enhanced RED) [7], (r, RTT)-adaptive buffer management, and Dynamic RIO (DRIO) [8]. Compared to buffer management schemes, less research attention has been given to packet marking methods. Most prior work adopted simple packet regulating methods such as leaky bucket, and TSW (Time Sliding Window) [6]. Recently, Lin et al. proposed ETSW (Enhanced TSW) [8].

We address mechanisms to improve the performance of TCP connection in the AS architecture. Considerable efforts have been made to develop performance enhancement mechanism in last three years [7], [8], [9], [10]. Some of proposed methods were successful. However, many of them require non-trivial modification of the current TCP. Some of them rely on the interaction between network elements and the TCP protocol. These methods are harder to deploy than a method that works with the current TCP. We aim to develop a method that does not require TCP modification.

In this paper, we focus on the packet marking mechanism. We propose to use a data buffer in addition to a token buffer in the packet marker. To the best of our knowledge, no published work proposed to use a data buffer in the AS architecture. Note that the data buffer does not require TCP modification and works with the current TCP. The data buffer reduces the burstiness of the TCP packet generation process.

The data buffer is not without drawbacks. Buffering brings unnecessary delays. The buffering delay increases in proportion to the data buffer size. A large data buffer increases the RTT and may degrade TCP performance because TCP throughput decreases as the RTT increases [11]. Therefore, we should exercise care in determining the optimal data buffer size.

We examined the performance of the proposed method via computer simulation. Simulation results show that TCP connections achieve throughput closer to the target throughput with the new packet marker than with a token bucket based packet marker. We also investigated the optimal data buffer size varying the subscription level, the reserved bandwidth, and RTT. We observed that the benefit of the packet buffering increases as the subscription level grows. The optimal data buffer size is proportional to the reserved bandwidth and RTT.

The rest of the paper is organized as follows. In Section 2, we define the notion of the target bandwidth and explain the collision between TCP congestion control mechanism and the assured service mechanism. We explain the new packet marker in detail and discuss the effect of the packet buffering in Section 3. Section 4 contains extensive performance analysis. In addition to the performance comparison of the new method to existing method, we analyse the effect of data buffer size on the performance. We conclude the paper in Section 5.

2 Background

The AS architecture aims to provide throughput guarantee. To examine the issue of bandwidth sharing, we define target bandwidth. The computation of target bandwidth is quite straightforward as follows.

Let r_i be the reserved bandwidth of flow i, C be the capacity of a bottleneck link, and n be the number of flows. The bandwidth of $\sum_{i=1}^{n} r_i$ is reserved for IN packets. The surplus bandwidth at this link is then $(C - \sum_{i=1}^{n} r_i)$. This surplus bandwidth should be evenly shared by all the flows. Hence, the target bandwidth of flow i, t_i is given by

$$t_i = r_i + \frac{(C - \sum_{i=1}^{n} r_i)}{n} . \qquad (1)$$

2.1 Collision between TCP Congestion Control and the Current AS Mechanism

Current AS architecture interferes with the TCP congestion control mechanism and may not allocate promised bandwidth to TCP subscribers [7], [8], [9]. TCP congestion control is an AIMD model and the cwnd size fluctuates in a typical sawtooth pattern. When a TCP packet generation rate is smaller than the reserved bandwidth, the TCP connection cannot use the allocated bandwidth. If

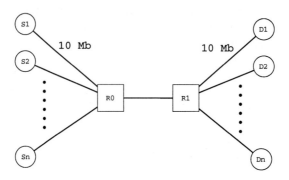

Fig. 1. Simulation Topology

a TCP connection increases its cwnd and generates more packets than reserved bandwidth, some packets are marked as OUT packets that will experience high drop probability. If one of them is discarded, then TCP reduces its cwnd size by half. As a result, a TCP connection may be not compensated for the loss of the reserved bandwidth during the under-utilization state.

The loss of the reserved bandwidth also results from the burstiness of TCP traffic. There are persistent gaps in the ack (acknowledgement) stream. Such gaps are commonly referred to as ack compression [12]. Since TCP uses ack to trigger transmissions, significant gaps in the acks cause the token buffer to overflow. The overflow of the token buffer results in a loss of reserved bandwidth. Thus the token bucket marker fails to allocate the promised rate to a bursty flow. In this paper, we focus on how to avoid the performance degradation due to the burstiness of TCP traffic.

It is not a good solution that we use a marker with a large token buffer. Even though large token buffer can prevent the token loss, it may generate many IN packets consecutively. IN packet burst may result in the loss of IN packets. Because the frequent loss of IN packet threatens the AS architecture, the size of the token buffer should be restricted.

3 Packet Buffering

In this section we address the packet buffering mechanism to improve the performance of TCP connection in the AS architecture and study its effect. We explain the problem of the token bucket marker and the rationale of the packet buffering. We evaluate the performance of the packet buffering using ns simulator [13].

We consider a simple topology with TCP Reno hosts. The network topology is shown in Figure 1, which is similar to that used in [8], [9]. There are two routers R0 and R1. All senders are connected to router R0 and all receivers are connected to router R1; all connections share one bottleneck link between R0 and R1. We use the token buffer whose size is the product of the reserved bandwidth and RTT. R0 and R1 maintain queues with size of 100 packets and

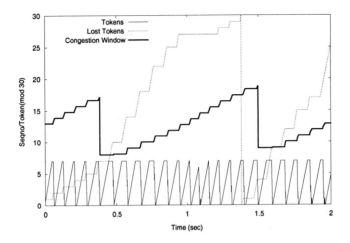

Fig. 2. The behavior of token bucket marker and TCP source

use RIO algorithm. The two thresholds and the dropping probability used for IN and OUT packets in RIO are 20/80/0.02 and 20/50/0.1, respectively. All senders are ftp sources, that is, they always have data to send.

3.1 Problem of the Token Bucket Marker

We examine the behavior of the token bucket marker and TCP source in the following experiment environment. There are six AS TCP flows and six best effort TCP flows. Each AS flow reserves 1Mbps bandwidth and the capacity of the bottleneck link is 10Mbps. The size of the token buffer is same as 7 packets.

Figure 2 shows the TCP congestion window of the sender, the number of lost tokens (given in packets modulo 30) for the connection and the number of tokens in the token buffer. We can observe that TCP flow suffers from the crucial token loss. The TCP congestion control mechanism is partly responsible for this phenomenon. Because TCP congestion control is an AIMD model, the cwnd size fluctuates in a typical sawtooth pattern. Thus, even when tokens exist, TCP sender may be throttled by the congestion window. As shown in Figure 2, many tokens are lost when cwnd size is small.

Another cause for token loss is the presence of persistent gaps in the ack stream. We can observe that TCP cwnd size increases in shape of stair in Figure 2. It is because sender receives acks at burst. Since TCP uses acks to trigger transmissions, such burstiness of ack stream causes a token loss.

3.2 Token Bucket Marker with Data Buffer

We propose to use a data buffer in additional to a token buffer in the packet marker to reduce the token loss. The marker with a data buffer stores a received packet when there are no tokens. If a new token is generated, the first packet in

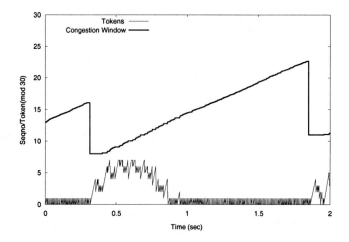

Fig. 3. The behavior of packet marker with a data buffer of size 3 packets and TCP source

the data buffer is transmitted as IN packet. If a data buffer overflows, the first packet in the data buffer is transmitted as OUT packet.

The data buffer reduces the burstiness of the TCP packet generation process. Figure 3 shows the behavior of the marker with a data buffer and cwnd size of TCP sender. We used a data buffer with size of 3 packets. We can observe that due to the smoothing effect of the data buffer, TCP cwnd size increases smoothly and few tokens are lost. Because the data buffer lessens the token losses, it provides flows with IN marking rate closer to the reserved bandwidth than the token bucket buffer.

In [14], Aggarwal studied the TCP pacing mechanism which transmits the packets evenly. A packet of paced TCP flow tends to be dropped due to the impact of other bursty flows. On the other hand, the packet of a bursty flow is dropped with the relatively small probability. From this reason, the paced TCP shows the degraded performance than the TCP Reno. However, the packet smoothing does not deteriorate the performance of TCP flow in the AS network. Because the IN packet is transmitted successfully for the most part regardless of the other flows, the packet buffering can improve the performance of AS TCP flow.

4 Evaluation Result

We study the performance of the packet buffering via computer simulation and exercise care in determining the optimal data buffer size. We investigate the optimal data buffer size varying the subscription level, the reserved bandwidth, and RTT.

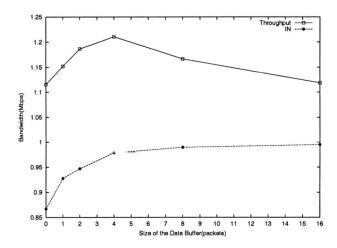

Fig. 4. The effect of the data buffer size

4.1 Effect of the Data Buffer Size

Simulation results show that TCP connections achieve throughput closer to the target throughput with the new packet marker than with a marker using a token buffer only. The performance of the new packet marker depends on the size of the data buffer. Figure 4 shows IN packet transmission rate and achieved throughput including both IN and OUT transmission rate of a TCP flow, which reserves 1 Mbps. There are six markers with a data buffer of a different size: 0, 1, 2, 4, 8, and 16. The marker with token buffer only is the case where the size of data buffer is zero.

Two important observations are visible in this figure. First, as the size of data buffer grows, IN packet transmission rate gets closer to the reserved bandwidth. It is because the data buffer reduces the burstiness of the TCP packet generation process and lessens the token losses. Note that the token buffer also can alleviate the problem of token loss. However, it is only when the token buffers are very large and the use of large token buffer allows large bursts of IN packets into the network that can result in loss of IN packets, thus defeating the service differentiation mechanism provided by RIO [7].

The second observation is that data buffer improves the performance of TCP flow, but too large data buffer may degrade TCP performance. The data buffer reduces token losses, but brings additional delays. Although the data buffer provides a TCP flow with IN packet transmission rate closer to the reserved bandwidth, a large data buffer increases the RTT and may degrade TCP performance. Therefore, it is important to determine the optimal data buffer size.

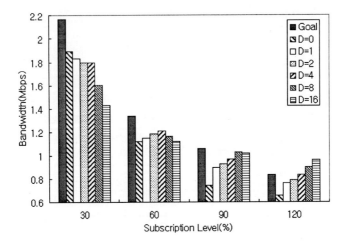

Fig. 5. The effect of the subscription level

4.2 Effect of the Subscription Level

We examine the impact of the subscription level on the performance of the proposed method. We performed a experiment varying the capacity of the bottleneck link into 20Mbps, 10Mbps, 6.67Mbps, and 5Mbps.

Since the sum of the reserved bandwidth is 6Mbps, the subscription level is 30%, 60%, 90%, and 120% respectively. We can see that the optimal data buffer size is larger as the subscription level grows in Figure 5. It relates to the drop probability of OUT packets. When the subscription level is low, OUT packets are dropped in a low probability as well as IN packets. Therefore the benefit of the packet buffering increases as the subscription level grows.

4.3 Effect of the Reserved Bandwidth

This experiment is designed to investigate the impact of the reserved bandwidth on the performance of the proposed method. For this study we ran the eight AS TCP flows with reservations of 2Mbps, 1Mbsp, 0.5Mbps, and 0.1Mbps, and two best effort TCP flows. The capacity of the bottleneck link is 12Mbps and the bandwidth of 7.2Mbps is reserved for IN packets.

Figure 6 shows the throughput according to the reserved bandwidth and the data buffer size. We can see that the token bucket marker provides too much bandwidth to the flow with the small reserved bandwidth and provides too small bandwidth to the flow with the large reserved bandwidth. Compared to the token bucket marker, the marker with a data buffer provides the throughput closer to the target bandwidth.

We can observe that the optimal data buffer size varies according to the reserved bandwidth. The optimal data buffer size is proportional to the reserved bandwidth. Because the reserved bandwidth is equal to the token generation

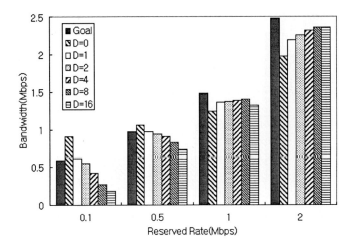

Fig. 6. The effect of the reserved bandwidth

rate, the packet buffering delay decreases as the reserved bandwidth increases. It is the reason that the optimal data buffer size is proportional to the reserved bandwidth.

4.4 Effect of the RTT

We also examine the impact of RTT on the performance of the proposed method. For this purpose, we performed the experiment where there are eight AS TCP flows and eight best effort TCP flow with the different RTT of 10, 20, and 40ms. Each AS TCP flow reserves 1Mpbs and the capacity of the bottleneck link is 14Mbps.

The result of the experiment is shown in Figure 7. We can see that the optimal data buffer size is proportional to the RTT of a flow. The reason is that TCP throughput decreases as the RTT increases [11]. If the packet buffering delay is constant, the performance degradation is severer as RTT grows.

5 Conclusions

In the paper, we have addressed mechanism to improve the performance of TCP flow in the AS architecture using the packet buffering. We proposed a new marker that uses a data buffer in addition to a token buffer. We presented that the marker with a data buffer achieved the target throughput better than a marker with a token buffer only. We also showed that the optimal data buffer size is proportional to the reserved bandwidth and RTT, and the benefit of the packet buffering increases as the subscription level grows.

Further study is required to infer the optimal data buffer size. We are now modeling the packet buffering mechanism to infer the optimal data buffer size. We expect to describe the effect of the packet buffering more accurately.

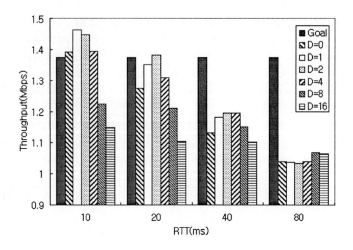

Fig. 7. The effect of RTT

References

1. R. Braden, D. Clark, S. Shenker, "Integrated Services in the Internet Architecture: An Overview," RFC 1633, Jun., 1994
2. L. Zhang, S. Deering, D. Estrin, S. Shenker, D. Zappala, "RSVP: A New Resource ReSerVation Protocol," *Transaction on Networking*, Sep., 1993
3. S. Blake, D. Black, M. Carlson, E. Davies, Z. Wang, W. Weiss, "An Architecture for Differentiated Services," RFC 2475, Dec., 1998
4. X. Xiao, L. Ni, "Internet QoS: the Big Picture," *IEEE Network*, Mar./Apr., 1999
5. K. Nichols, V. Jacobson, L. Zhang, "A Two-bit Differentiated Services Architecture for the Internet," RFC 2638, Jul., 1999
6. D. Clark, W. Fang, "Explicit Allocation of Best Effort Delivery Service," *Transactions on Networking*, Aug. 1998
7. W. Feng, D. Kandlur, D. Saha, K. Shin, "Understanding and Improving TCP Performance over Networks with Minimum Rate Guarantees," *Transactions on Networking*, Apr., 1999
8. W. Lin, R. Zheng, J. C. Hou, "How to Make Assured Services More Assured," *ICNP'99*
9. I. Yeom, A. Reddy, "Realizing throughput guarantees in a differentiated services network," *ICMCS'99*
10. W. Feng, D. Kandlur, D. Saha, K. Shin, "Adaptive Packet Marking for Providing Differentiated Services in the Internet," *ICNP '98*
11. J. Padhye, V. Firoiu, D. Towsley, J. Kurose, "Modeling TCP throughput: A simple model and its empirical validation," *SIGCOMM'98*
12. L. Zhang, S. Shenker, D. Clark, "Observations on the Dynamics of a Congestion Control Algorithm: The Effects of Two-Way Traffic," *SIGCOMM'91*
13. UCB, LBNL, VINT Network Simulator - ns http://www-mash.cs.berkeley.edu/ns/ns.html
14. A. Aggarwal, S. Savage, T. Anderson, "Understanding the Performance of TCP Pacing," *INFOCOM'2000*

Modular TCP Handoff Design in STREAMS-Based TCP/IP Implementation

Wenting Tang[1], Ludmila Cherkasova[1], Lance Russell[1], and Matt W. Mutka[2]

[1] Hewlett-Packard Labs, 1501 Page Mill Road,
Palo Alto, CA 94303, USA,
wenting,cherkasova,lrussell@hpl.hp.com
[2] Dept of Computer Science & Eng., Michigan State Universsity,
East Lansing, MI 48824, USA,
mutka@cse.msu.edu

Abstract. Content-aware request distribution is a technique which takes into account the content of the request when distributing the requests in a web server cluster. A handoff protocol and TCP handoff mechanism were introduced to support content-aware request distribution in a client-transparent manner. Content-aware request distribution mechanisms enable the intelligent routing inside the cluster to provide the quality of service requirements for different types of content and to improve overall cluster performance.
We propose a new modular TCP handoff design based on STREAMS-based TCP/IP implementation in HP-UX 11.0. We design the handoff functions as dynamically loadable modules. No changes are made to the existing TCP/IP code. The proposed plug-in module approach has the following advantages: *flexibility*-TCP handoff functions may be loaded and unloaded dynamically, without node function interruption; *modularity*- proposed design and implementation may be ported to other OSes with minimal effort.

1 Introduction

The web server cluster is the most popular configuration used to meet the growing traffic demands imposed by the World Wide Web. However, for clusters to be able to achieve scalable performance as the cluster size increases, it is important to employ the mechanisms and policies for a "balanced" request distribution.

The market now offers several hardware/software load-balancer solutions that can distribute incoming stream of requests among a group of web servers. Typically, the load-balancer sits as a front-end node on network and acts as a gateway for incoming connections (we often will call this entity a distributor). Incoming client requests are distributed more or less evenly to a pool of servers (back-end nodes).

Traditional load balancing solutions for a web server cluster try to distribute the requests among the nodes in the cluster based on some load information without regard to the requested content and therefore forwarding the client requests to a back-end node prior to establishing a connection with the client.

Content-aware request distribution takes into account the content (such as URL name, URL type, or cookies) when making a decision to which server the request is to be routed. Previous work on content-aware request distribution [4, 5, 1, 2] has shown that policies distributing the requests based on cache affinity lead to significant performance improvements compared to the strategies taking into account only load information.

HTTP protocol relies on TCP - a connection-oriented transport protocol. The front-end node (the request distributor) must establish a connection with the client to inspect the target content of a request prior to asssigning the connection to a back-end web server. A mechanism is needed to service the client request by the selected back-end node. Two methods were proposed to distribute and service the requests on the basis of the requested content in a client-transparent manner: the TCP handoff [4] and the TCP splicing. [3].

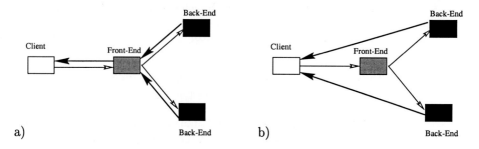

Fig. 1. Traffic flow with a) TCP splicing; b) TCP handoff.

TCP splicing is an optimization of the front-end relaying approach, with the traffic flow represented in Figure 1 a).

The TCP handoff mechanism was introduced in [4] to enable the forwarding of back-end responses directly to the clients without passing through the front-end, with traffic flow represented in Figure 1 b). The main idea behind the TCP handoff is to migrate the created TCP state from the distributor to the back-end node. The TCP implementation running on the front-end and back-ends needs a small amount of additional support for handoff. In particular, the protocol module needs to support an operation that allows the TCP handoff protocol to create a TCP connection at the back-end without going through the TCP three-way handshake with the client. Similar, an operation is required that retrieves the state of an established connection and destroys the connection state without going through the normal message handshake required to close a TCP connection. Once the connection is handed off to a back-end node, the front-end must forward packets from the client to the appropriate back-end node.

This difference in the response flow route allows substantially higher scalability of the TCP handoff mechanism than TCP splicing. In [1], authors compared performance of both mechanisms showing the benefits of the TCP handoff schema. Their comparison is based on the implementation of the TCP handoff mechanism in FreeBSD UNIX.

In this work, we consider a web cluster in which the content-aware distribution is performed by each node in a web cluster. Each server in a cluster may forward a request to another node based on the requested content (using TCP handoff mechanism).

STREAMS-based TCP/IP implementations, which are available in leading commercial operating systems, offers a framework to implement the TCP handoff mechanism as plug-in modules in the TCP/IP stack, and to achieve the flexibility and portability without too much performance penalty. As part of the effort to support a content-aware request distribution for web server clusters, we propose a new modular TCP handoff design. The proposed TCP handoff design is implemented as STREAMS modules. Such a design has the following advantages:

- *portability*: the STREAMS-based TCP/IP modules are relatively independent of the implementation internals. New TCP handoff modules are designed to satisfy the following requirements:
  - all the interactions between TCP handoff modules and the original TCP/IP modules are message-based, no direct function calls are made.
  - TCP handoff modules do not access and/or change any data structures or field values maintained by the original TCP/IP modules.

  This enables maximum portability, so that the TCP handoff modules may be ported to other STREAMS-based TCP/IP implementation very quickly.
- *flexibility*: TCP handoff modules may be dynamically loaded and unloaded as DLKM (Dynamically Loadable Kernel Module) modules without service interruption.
- *transparency*: no application modification is necessary to take advantage of the TCP handoff mechanism. This is a valuable feature for some applications where no source code is available.
- *efficiency*: the proposed TCP handoff modules are only peeking into the messages, with minimal functionality replicated from the original TCP/IP modules.

2 Cluster Architectures and Request Distribution Mechanisms

Different products have been introduced in the market for load balancing.

Popular Round-Robin DNS solutions [8] distribute the accesses among the nodes in the cluster in the following way: for a name resolution it returns the IP address list (for example, list of nodes in a cluster which can serve this content), placing a different address first in the list for each successive request. Round-Robin DNS is available as part of DNS which is already in use, i.e. there is no additional cost.

Other traditional load balancing solutions for a web server cluster try to **distribute the requests** among the nodes in the cluster without regard to the requested content and therefore forwarding client requests to a back-end node

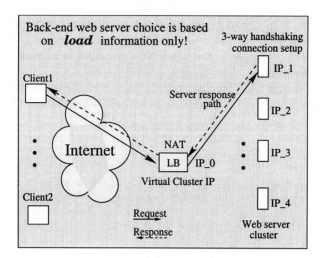

Fig. 2. Traditional load balancing solution (like Cisco's Local Director) in a web server cluster.

prior to establishing a connection with the client as shown in Figure 2. In this configuration, web server cluster appears as a single host to the clients. To the back-end web servers, the front-end load-balancer appears as a gateway. In essence, it intercepts the incoming web requests and determines which web server should get each one. Making that decision is the job of the proprietary algorithms implemented in these products. This code can take into account the number of servers available, the resources (CPU speed and memory) of each, and how many active TCP sessions are being serviced, etc. The balancing methods across different load-balancing servers vary, but in general, the idea is to forward the request to the least loaded server in a cluster.

Only the virtual address is advertised to the Internet community, so the load balancer also acts as a safety net. The IP addresses of the individual servers are never sent back to the web browser. The load-balancer rewrites the virtual cluster IP address to a particular web server IP address using *Network Address Translation* (NAT). Because of this IP address rewriting, both inbound requests and outbound responses pass through the load-balancer.

The **3-way handshaking and the connection set up** with original client is the responsibility of the chosen **back-end web server**. After the connection is established, the client sends to this server the HTTP request with specific URL to retrieve.

Content-aware request distribution intends to take into account the content (such as URL name, URL type, or cookies) when making a decision to which server the request has to be routed. The main technical difficulty of this approach is that it requires the establishment of a connection between the client and the request distributor. So the client will send the HTTP request to the distributor. The distributor can then make a decision to which back-end web server this request will be forwarded.

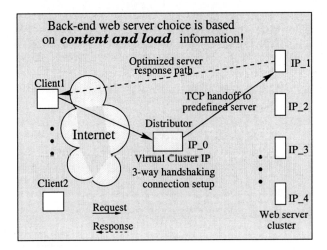

Fig. 3. Content-aware request distribution solution (front-end based configuration) in a web server cluster.

Thus, **3-way handshaking and the connection set up** between the client and **request distributor** happens first, as shown in Figure 3. After that, back-end web server is chosen based on the content of the HTTP request from the client. To be able to distribute the requests on the basis of requested content, the distributor component should implement either a form of TCP handoff [4] or the splicing mechanism [3]. Figure 3 shows request and response flow in case of TCP handoff mechanism.

In this configuration, the typical bottleneck is due to the front-end node which performs the functions of distributor. For realistic workloads, a front-end node, performing the TCP handoff, does not scale far beyond four cluster nodes [1]. Most of the overhead in this scenario is incurred by the distributor component.

Thus, another recent solution proposed in [1] is shown in Figure 4. It is based on alternative cluster design where the **distributor is co-located with the web server.** We will call this architecture CARD (Content-Aware Request Distribution).

For simplicity, we assume that the clients directly contact the distributor, for instance via RR-DNS. In this case, the typical client request is processed in the following way. 1) Client web browser uses TCP/IP protocol to connect to the chosen distributor; 2) the distributor component accepts the connection and parses the request, and decides on server assignment for this request; 3) the distributor hands off the connection using TCP handoff protocol to the chosen server; 4) the server application at the server node accepts the created connection; 5) the server sends the response directly to the client.

The results in [1] show good scalability properties of the CARD architecture when distributing requests with the LARD policy [4]. The main idea behind LARD is to partition the documents logically among the cluster nodes, aiming

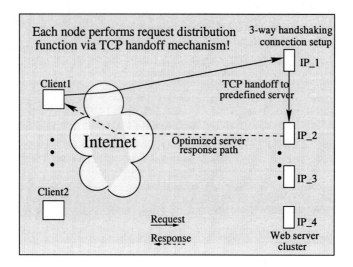

Fig. 4. Content-aware request distribution solution (cluster based, distributed configuration) in a web server cluster.

to optimize the usage of the overall cluster RAM. Thus, the requests to the same document will be served by the same cluster node that will most likely have the file in RAM.

Our TCP handoff modules are designed to support a content-aware request distribution for the CARD implementation shown in Figure 4.

3 STREAMS and STREAMS-Based TCP/IP Implementation

STREAMS is a modular framework for developing the communication services. Each stream has a *stream head*, a *driver* and multiple optional *modules* between the stream head and the driver (see Figure 5 a). Modules exchange the information by *messages*. Messages can flow in two directions: *downstream* or *upstream*. Each module has a pair of *queues*: *write queue* and *read queue*. When a message passes through a queue, the service routine for this queue may be called to process the message. The service routine may drop a message, pass a message, change the message header, and generate a new message.

The stream head is responsible for interacting with the user processes. It accepts the process request, translates it into appropriate messages, and sends messages downstream. It is also responsible for signaling to the process when new data arrives or some unexpected event happens.

The STREAMS modules for STREAMS-based TCP/IP implementation are shown in Figure 5 b). Transport Provider Interface (TPI) specification [7] defines the message interface between TCP and the upper module. Data Link Provider Interface (DLPI) specification [6] defines the message interface between driver

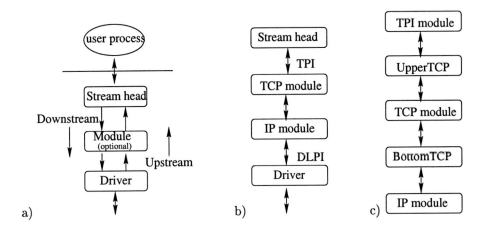

Fig. 5. a) STREAMS b) STREAMS-Based TCP/IP Implementation c) New Plug-in Modules for TCP Handoff in STREAMS-Based TCP/IP Implementation

and the IP module. These specifications define the message format, valid sequences of messages, and semantics of messages exchanged between these neighboring modules.

When the TCP module receives a SYN request for establishing the HTTP connection, the TCP module sends a T_CONN_IND message upstream. Under the TPI specification, TCP should not proceed until it gets the response from the application layer. However, in order to be compatible with BSD implementation-based applications, the TCP module continues the connection establishment procedure with the client. When the application decides to accept the connection, it sends the T_CONN_RES downstream. It also creates another stream to accept this new connection, and TCP module attaches the TCP connection state to this new stream. The data exchange continues on the accepted stream until either end closes the connection.

4 Modular TCP Handoff Design

The TCP handoff mechanism (shown in Figure 1 b) enables the response forwarding from the back-end web server nodes directly to the clients without passing through the distributing front-end.

In the CARD architecture, each node performs both front-end and back-end functionality: the distributor is co-located with the web server. We use the following denotations: the distributor-node accepting the original client connection request is referred to as FE (Front-End). In the case where the request has to be processed by a different node, thenode receiving the TCP handoff request is referred to as BE (Back-End).

Two new modules are introduced to implement the functionality of TCP handoff as shown in Figure 5 c). According to the relative position in the existing

TCP/IP stack, we refer to the module right on top of the TCP module in the stack as UTCP (UpperTCP), and the module right under the TCP module as BTCP (BottomTCP).

These two modules provide a wrapper around the current TCP module. In order to explain the proposed modular TCP handoff design and its implementation details, we consider typical client request processing. There are two basic cases:

remote request processing, i.e. when the front-end node accepting the request must handoff the request to a different back-end node assigned to process this request;

local request processing, i.e. when the front-end node accepting the request is the node which is assigned to process this request.

First, we consider the *remote request* processing. There are six logical steps to perform the TCP handoff of the HTTP request in the CARD architecture:

1) finish 3-way TCP handshaking (connection establishment), and get the requested URL; 2) make the routing decision: which back-end node is assigned to process the request; 3) initiate the TCP handoff process with the assigned BE node; 4) migrate the TCP state from FE to BE node; 5) forward the data packets; 6) terminate the forwarding mode and release the related resources on FE after the connection is closed.

Now, we describe in detail how these steps are implemented by the newly added UTCP and BTCP modules and original TCP/IP modules in the operating system.

– *3-way TCP handshake*

Before the requested URL is sent to make a routing decision, the connection has to be established between the client and the server. The proposed design depends on the original TCP/IP modules in the current operating system to finish the 3-way handshaking functionality. In this stage, $BTCP_{FE}$ allocates a connection structure corresponding to each connection request upon receiving a TCP SYN packet from the client. After that, $BTCP_{FE}$ sends the SYN packet upstream. Upon receiving a downstream TCP SYN/ACK packet from the TCP_{FE} module, $BTCP_{FE}$ records the initial sequence number associated with the connection, and sends the packet downstream. After $BTCP_{FE}$ receives an ACK packet from the client, it sends the packet upstream to TCP_{FE}. During this process, the $BTCP_{FE}$ emulates the TCP state transitions and changes its state accordingly.

In addition to monitoring the 3-way TCP handshaking, $BTCP_{FE}$ keeps a copy of the incoming packets for connection establishment (SYN packet, ACK to SYN/ACK packet sent by the client) and URL (Figure 6), for *TCP state migration* purpose, which is discussed later.

Also, because the TCP handoff should be transparent to server applications, the connection should not be exposed to the user level application before the routing decision is made. $UTCP_{FE}$ intercepts the T_CONN_IND message sent by TCP_{FE}. TCP_{FE} continues the 3-way handshaking without waiting for explicit messages from the modules on top of TCP.

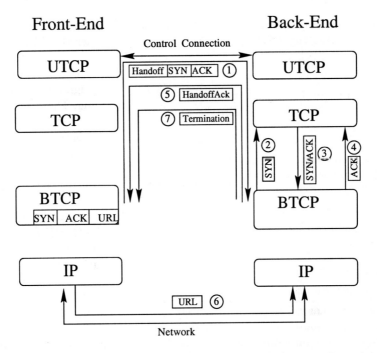

Fig. 6. Remote Request Processing Flow During TCP Handoff Procedure

- *URL parsing*

  $BTCP_{FE}$ parses the first data packet from the client, retrieves the URL and makes the distribution decision.

- *TCP handoff initiation*

  A special communication channel is needed to initiate the TCP handoff between FE and BE. A *Control Connection* is used for this purpose between two $UTCP_{FE}$ and $UTCP_{BE}$ as shown in Figure 6. This control connection is a pre-established persistent connection set up during the cluster initialization. Each node is connected to all other nodes in the cluster. The TCP handoff request is sent over the control connection to initiate the handoff process. Any communication between $BTCP_{FE}$ and $BTCP_{BE}$ modules goes through the control connection by sending the message to the $UTCP$ module first (see Figure 6). After $BTCP_{FE}$ decides to handoff the connection, it sends a handoff request to the $BTCP_{BE}$ (Figure 6, step 1). The SYN and ACK packets from the client and the TCP initial sequence number returned by TCP_{FE} are included in the message. $BTCP_{BE}$ uses the information in the handoff request to migrate the associated TCP state (steps 2-4 in Figure 6, which are discussed next). If $BTCP_{BE}$ successfully migrates the state, an acknowledgement is returned (Figure 6, step 5). $BTCP_{FE}$ frees the half-open TCP connection upon receiving the acknowledgement by sending a RST packet upstream to TCP_{FE} and enters forwarding mode. $UTCP_{FE}$

discards corresponding T_CONN_IND message when the T_DISCON_IND is received from the TCP_{FE}.

– *TCP state migration*

In the STREAMS environment it is not easy to get the current state of a connection at TCP_{FE}, to transfer it and to replicate this state at TCP_{BE}. First it is difficult to obtain the state out of the black box of the TCP module. Even if this could be done, it is difficult to replicate the state at BE. TPI does not support schemes by which a new half-open TCP connection with predefined state may be opened. In the proposed design, the half-open TCP connection is created by replaying the packets to the TCP_{BE} by the $BTCP_{BE}$. In this case, the $BTCP_{BE}$ acts as a client(Figure 6). $BTCP_{BE}$ uses the packets from $BTCP_{FE}$, updates the destination IP address of SYN packet to BE and sends it upstream (Figure 6, step 2). TCP_{BE} responds with SYN-ACK(Figure 6, step 3). $BTCP_{BE}$ records the initial sequence number of BE, discards SYN-ACK, updates the ACK packet header properly, and sends it upstream (Figure 6, step 4).

– *Data forwarding*

After the handoff is processed successfully, $BTCP_{FE}$ enters a forwarding mode. It forwards all the pending data in $BTCP_{FE}$, which includes the first data packet (containing the requested URL) (Figure 6, step 6). It continues to forward any packets on this connection until the forward session is closed.

During the data forwarding step, $BTCP_{FE}$ updates (corrects) the following fields in the packet: 1) the destination IP address to BE's IP address; 2) the sequence number of the TCP packet; 3) the TCP checksum.

For data packets that are sent directly from BE to the client, the $BTCP_{BE}$ module updates (corrects): 1) the source IP address to FE's IP address; 2) the sequence number; 3) TCP checksum. After that, $BTCP_{BE}$ sends the packet downstream.

– *Handoff connection termination*

The connection termination should free states at BE and FE. The data structures at BE is closed by the STREAMS mechanism. $BTCP_{BE}$ monitors the status of the handoffed connection and notifies the $BTCP_{FE}$ upon the close of the handoffed connection in TCP_{BE} (Figure 6, step 7). $BTCP_{FE}$ releases the resources related to the forwarding mechanism after receiving such a notification.

Local request processing is performed in the following way. After the $BTCP_{FE}$ finds out that the request should be served locally, the $BTCP_{FE}$ notifies $UTCP_{FE}$ to release the correct T_CONN_IND message to upper STREAMS modules, and sends the data packet (containing the requested URL) to the original TCP module (TCP_{FE}). $BTCP_{FE}$ discards all the packets kept for this connection and frees the data structures associated with this connection. After this, $BTCP_{FE}$ and $UTCP_{FE}$ send packets upstream as quickly as possible without any extra processing overhead.

5 Conclusion

Research on scalable web server clusters has received much attention from both industry and academia. A routing mechanism for distributing requests to individual servers in a cluster is at the heart of any server clustering technique. Content-aware request distribution (LARD, HACC, and FLEX strategies) [4, 5, 1, 2] has shown that policies distributing the requests based on cache affinity lead to significant performance improvements compared to the strategies taking into account only the load information.

Content-aware request distribution mechanisms enable intelligent routing inside the cluster to support additional quality of service requirements for different types of content and to improve overall cluster performance.

With content-aware distribution, based on TCP handoff mechanism, incoming requests must be handed off by distributor component to a back-end web server in a client-transparent way after the distributor has inspected the content of the request. The modular TCP handoff design proposed in this paper offers additional advantages: *portability, flexibility, transparency,* and *efficiency* to support scalable web server cluster design and smart request routing inside the cluster.

References

1. Mohit Aron, Darren Sanders, Peter Druschel and Willy Zwaenepoel. Scalable Content-Aware Request Distribution in Cluster-based Network Servers. In Proceedings of the USENIX 2000 Annual Technical Conference, San Diego, CA, June 2000.
2. L. Cherkasova. FLEX: Load Balancing and Management Strategy for Scalable Web Hosting Service. In Proceedings of the Fifth International Symposium on Computers and Communications (ISCC'00), Antibes, France, July 3-7, 2000, p.8-13.
3. A. Cohen, S. Rangarajan, and H. Slye. One the Performance of TCP Splicing for URL-Aware redirection. In Proceedings of the 2nd Usenix Symposium on Internet technologies and Systems, Boulder, CO, Oct, 1999.
4. V. Pai, M. Aron, G. Banga, M. Svendsen, P. Drushel, W. Zwaenepoel, E.Nahum: Locality-Aware Request Distribution in Cluster-Based Network Servers. In Proceedings of the 8th International Conference on Architectural Support for Programming Languages and Operating Systems (ASPLOS VIII), ACM SIGPLAN,1998, pp.205-216.
5. X. Zhang, M. Barrientos, J. Chen, M. Seltzer: HACC: An Architecture for Cluster-Based Web Servers. In Proceeding of the 3rd USENIX Windows NT Symposium, Seattle, WA, July, 1999.
6. Data Link Provider Interface (DLPI), UNIX International, OSI Work Group.
7. Transport Provider Interface (TPI), UNIX International, OSI Work Group.
8. T. Brisco: DNS Support for Load Balancing. RFC 1794, Rutgers University, April 1995.

An Efficient TCP Flow Control and Fast Recovery Scheme for Lossy Networks

H.Y. Liao, Y.C. Chen, and C.L. Lee

Department of Computer Science and Information Engineering
National Chiao Tung University
1001 Tahsueh Road, Hsinchu 300, Taiwan, ROC
{hyliao,ycchen,leecl}@csie.nctu.edu.tw

Abstract. The initial TCP Tahoe version uses the slow-start algorithm to deal with flow control and congestion avoidance. The later Reno version deploys both fast-retransmit and fast-recovery algorithms. Traditionally a segment loss is considered as owing to the network congestion. However, a packet loss may be caused by some other reason such as a transmission error in the wireless link. Due to this reason, we design a mechanism that subdivides the congestion control mechanism into two parts, the packet loss indication and the loss recovery. Regarding the former, we no longer treat the packet loss caused by the transmission error as an indication of network congestion. While for the latter, we proposed a modified scoreboard algorithm in TCP FACK to quickly recover the packet loss and prevent the retransmitted packet from being lost again.

1 Introduction

Nowadays, TCP (Transmission Control Protocol) [12][13] has become the most popular and dominant transport layer protocol. In the evolution of TCP, there are two representative implementation versions, TCP Reno [3] and Vegas [1]. The former refers to TCP with the earlier algorithms plus Fast Recovery. The main idea of TCP Reno is to provoke packet losses when the congestion window has grown large enough to surpass the current available link bandwidth. While in Vegas, the expected throughput can be achieved by adjusting the window size based on the calculated current throughput. It does not need to adjust the window size if the current throughput lies in between two predefined thresholds. Through the approach of Vegas, it is unnecessary to generate packet loss repeatedly, because the sending rate is already kept at an appropriate level. In the latter enhanced versions of TCP Reno, including TCP New-Reno [4][5], TCP SACK [10] and TCP FACK [9], some problems for TCP Reno are solved.

The above approaches of TCP implementations all assume that packet losses are due to the network congestion, and hence the overflow of buffers. However, packet loss may also be caused by unreliable transmission, such as a wireless channel or other lossy links. All current TCP implementation versions do not deal with the performance degradation under lossy networks. Whenever a packet loss occurs, the current TCP congestion control algorithm will reduce its congestion window to half, even if the loss is due to packet errors. This characteristic may degrade the TCP throughput.

In this work, we focus on the TCP flow control and fast recovery algorithm. We design a new algorithm based on Vegas-like flow control, which is delay-based. Through the proposed flow control algorithm, we can eliminate the impact caused by the packet error. Since the lost packet should be retransmitted in time, therefore in the fast recovery phase, we deploy TCP FACK's concept to obtain a more accurate number of outstanding packets, and give a robust scoreboard algorithm to obtain the status of retransmitted packets, so that it can solve the problem of timeout incurred by the loss of retransmitted packet. The scheme is derived from both Vegas and FACK, so we name it as Enhanced Vegas-Fack TCP, abbreviated as EVFack TCP.

In Section 2, we discuss the proposed EVFack scheme. We also give examples to show the detailed behavior of EVFack, and perform both analysis and comparisons with other TCP implementation versions. In Section 3, we perform the simulation to show that our proposed scheme improves the throughput on lossy links. Section 4 concludes the work.

2 Proposed Delay-Based Congestion Control and Fast Recovery Mechanism

2.1 Delay-Based Control

Among all TCP implementations, Vegas is the only one based on delay (RTT estimate) to probe the current available bandwidth. The goal of delay-based control is to maintain a stable buffer occupancy on the bottleneck link for a connection. Unlike other approaches, such as Reno, Sack and Fack, the congestion window size will be kept increasing until it overflows the bottleneck link buffer. Once the sender detects any segment loss, it will adjust the window size accordingly. The possible action is to reduce its congestion window to half. Figure 1 demonstrates the window variation of TCP Reno; this shows the window size changing to one half of the original value when a segment loss occurs.

From Figure 1, we see that the window increases and then decreases to half repeatedly, especially after 5.84 seconds because of the buffer overflow. With the delay-based version such as TCP Vegas, the window size variation will not be as large as other versions. Figure 2 gives the simulation result of the congestion window variation. We focus on the steady window period, which occurs after 3.08 second. Delay-based control provides a way to control the sending rate of a connection. Through this approach, the service rate can be optimized, and congestion window adjustment algorithm can be performed based on Round-Trip-Time of segments. Because of the aforementioned advantages, our Enhanced Vegas-Fack algorithm adopts delay-based approach as the basis of flow control.

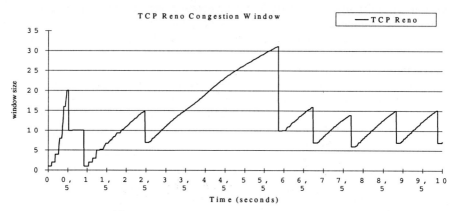

Fig. 1 Congestion window variations in TCP Reno.

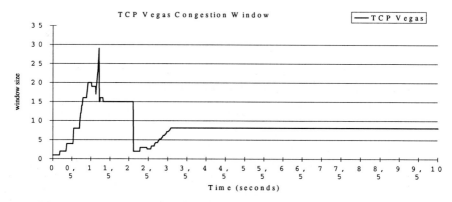

Fig. 2 Congestion window variations in TCP Vegas.

2.2 Flow Control and Congestion Control

The main purpose of our modified Delay-Based Congestion Control is to solve the throughput degradation problem under lossy network environments. Under an error-free network, the packet loss can be an indication to network congestion. However, for lossy network environments, a segment loss may be due to the error-prone nature of links. If the traditional congestion control strategy, such as Reno is adopted, the congestion window will be reduced to half whenever detecting a segment loss. In essence, if a segment was lost due to transmission error, we believe that the congestion window size should not be reduced. The concept of estimating Round-Trip-Time variation in Vegas could be a good inspiration. The real indication of congestion should be reflected by the continuous growing of the bottleneck queue, when the queue occupancy exceeds a certain threshold for a connection, it indicates that the end-to-end path is very likely congested.

One of the difference between Vegas and our implementation is regarding the calculation for numbers of outstanding segments. We uses the term (snd.nxt − snd.fack + v_retran_data) to represent the estimated outstanding segments. We demonstrate this sequence order in Figure 2. If there is no data segment loss, snd.una equals to snd.fack. Otherwise, snd.fack is updated by the SACK option to reflect the latest data received by the receiver, and the current number of outstanding segment should be (snd.nxt - snd.fack) if there is no out-of-order segment under the assumption that segments in the interval between snd.una and snd.fack were all lost.

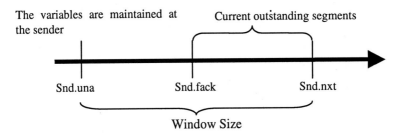

Fig. 3 TCP sender space of sequence number.

In effect, this can reduce the impact of segment loss on the calculation of current sending rate. That is, if the segment loss is due to random loss or link error, the lost segment will not be considered for calculating the current sending rate. Thus by adding the RTT as the denominator of outstanding segments, the actual current rate can be estimated precisely. We compare this value with the expected achievable rate. If the difference is larger than a certain amount of buffer occupancy, it indicates that the network is congested. We believe that the variation of bottleneck queue occupancy reflects the congestion of the network. Conceptually, since we use the BaseRTT as the minimal value of ever measured RTT, and MSS is set to one segment, the queuing time plus minimal propagation delay is considered as the BaseRTT, and MSS/BaseRTT would be the rate for one unit of queue occupancy. If the expected rate minus the actual rate is larger than the variation of one unit of queue occupancy in the bottleneck link, it does indicate the network congestion, and the congestion window should be reduced by a ratio of 1/8(in slow start phase). Otherwise, *cwnd* can be increased with the double rate in the next RTT.

In the Congestion Avoidance Phase, we follow the same decision procedure to adjust the sender's congestion window size, except that the amount of window decreasing and increasing is linear. We maintain the queue occupancy of segments in the bottleneck between one and two. This concept is same as that in Vegas. On the other hand, the idea of estimating more precise outstanding segments comes from Fack. The flow control we proposed allows the congestion window to smoothly increase under lossy networks. Since the traditional flow control immediately reduces the congestion window to half (Reno-Based version), if the random loss happens, the traditional approach will treat this situation as congestion, thus it suffers the overall throughput. Figure 3 demonstrates the state transition diagram of our proposed flow control (abbreviated as EVFack). The fast recovery phase we modified will be introduced in the next section. The following formula are defined for this figure:

<diff> = expected rate − actual rate
<actual diff> = real expected rate− (*snd.nxt−snd.fack+v_retran_data*) / RTT.
<threshold> represents the MSS/BaseRTT.

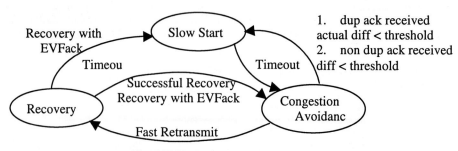

Fig. 4 EVFack transition diagram.

2.3 Fast Recovery

TCP Fack provides an accurate estimate of current outstanding packets, thus it is able to obtain the correctly received highest sequence number through the variable *snd.fack*. Based on the Sack information, Fack determines which segments have been lost, and adopts its recovery algorithm. The main approach of TCP Fack is to have a "scoreboard" to maintain the information from Sack Option. The "scoreboard" proposed in [9] is basically a data structure to maintain the required information to deal with the recovery of lost segments. The scoreboard successfully stores lost segments, and provides status of each segment, including the lost ones. However, in this Fack approach, the scoreboard has some weakness in reflecting the information of retransmitted segments. If the retransmitted segment is lost, TCP Fack will force a timeout event. The behavior thereafter is to reduce TCP congestion window to one. Under this assumption, the loss probability of retransmitted segment under the lossy environment may be high. We demonstrate through the simulation in that the retransmitted packet is lost again, and this is shown in Figure 5. At 4.57 second, packet 350 was dropped, and this packet was retransmitted later. Unfortunately, this retransmitted packet is lost again at 4.68 second because of the lossy characteristic of the link. Under such circumstance, the timeout is triggered to avoid further loss. However, the timeout only prevents congestion-generated loss, it is not suitable for this case.

From the observation of TCP Fack, it did suffer the throughput under lossy network. So we need another modification to the scoreboard to overcome the problem of throughput degradation. We define a so-called "rtxList"(retransmission list) as the current retransmission packet list. When a packet is retransmitted, an entry is added to rtxList. The structure of rtxList is shown in Figure 6:

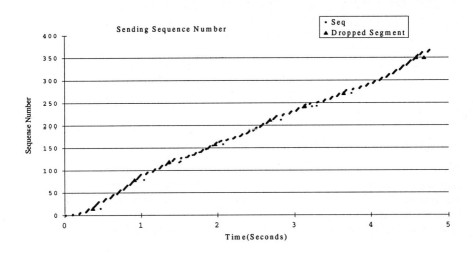

Fig. 5 TCP Fack sending sequence num with packet loss rate 0.03.

Sequence number	Sending Time	Next sequence	Success Status

Fig. 6 The structure of rtxList.

Where the "Sequence number" field stores the retransmitted sequence number, and "Sending Time" stores the time at which the retransmitted packet was sent. The "Next sequence" field maintains the expected sequence number to be sent, since in most cases the retransmitted segment will interrupt sending sequence at the source. "Success Status" indicates whether the retransmitted packet is received or not. There are two pointers, one points to the next entry, and the other points to the current reference entry, respectively, of the rtxList.

The operation of "rtxList" is simple. Initially, it has no entry for any packet. We follow the Fack's decision mechanism to detect retransmission event, then we add the retransmission entry to the rtxList. Upon receiving an Ack, the timestamp field recorded in the Ack packet will be checked if the rtxList is not empty. If the timestamp value is larger than the value in "Sending Time" field of the currently pointed entry, and the Sack option indicates that the retransmitted packet is not acknowledged, the retransmitted segment must be lost again. By now, if the *awnd* is less than *cwnd*, the lost segment, which is a retransmitted one can be retransmitted immediately. Throughout checking the "rtxList" structure, we could enhance the Fack's scoreboard for avoiding the throughput degradation, which is caused by forced timeout when a retransmitted packet is detected as lost again.

3 Simulation and Performance Evaluation

In this section, we configure different network topologies to evaluate the performance of our proposed TCP EVFack. We implement the EVFack algorithm under the LBNL simulator "ns" [11], where we added the necessary changes to the simulator.

3.1 Simulation

Simulation 1: Throughput Improvement for the lossy networks.

The first simulation uses a simple network configuration containing four nodes, as shown in Figure 7. Two of these nodes represent routers connected by a T1 link, the nominal round trip time including propagation, processing, and queuing delays between two end hosts, S1 and K1, is assumed 87ms. Thus the bandwidth*delay product is 16.3kBytes. We use an MSS (Maximum Segment Size) of 1kB in the simulation. Therefore, the required queue size in the router should be at least 17 packets. We utilize drop-tail queues to examine EVFack algorithm. The link between R2 and K1 is assumed lossy. The throughput comparisons will be made later based on environments with different loss rates.

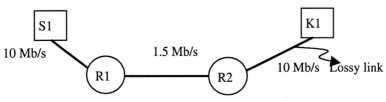

Fig. 7 Lossy network configuration for

Simulation 2: The congestion effects of EVFack Algorithm.
In the second simulation, we test our algorithm in a congestion-prone environment (Figure 8). There are eight end nodes in this topology, and four senders send data to their corresponding receivers, i.e. Sn to Kn (n=1 to 4), respectively. All senders and receivers are in LAN environments. The round-trip delay between senders and receivers is 167ms. Therefore, the total outstanding packet allowed in the end-to-end pipe is 31.31 KBytes. The main targets of observation are the buffer occupancy of routers and the window size variations.

3.2 Numerical Results and Performance Analysis

Simulation 1: The numerical results of the simulation are listed in the Table1. We compared the Reno, Fack, Vegas, and the proposed EVFack. Assume the node R2 between K1 is lossy, and the observed packet loss rate ranges from 0 to 0.3, which means the bit error rate ranging from 0 to $2.5 * 10^{-5}$. The bottleneck link is between R1 and R2, and the router buffer size is 20 packets. While the simulation time is 10 seconds.

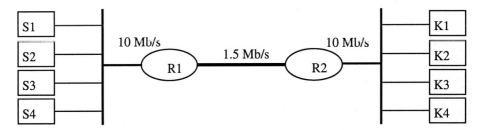

Fig. 8 Simulation 2: Congestion network topology.

Under a perfect network environment (i.e. loss rate = 0), the throughput of Reno-Based (Reno, NewReno and Fack) version is higher than the Delay-Based Versions (Vegas and EVFack). This is because the Delay-Based approach may need one more RTT to calculate the sending rate for adjusting its congestion window, thus window increment speed is slower than that in Reno-Based approaches. When the loss rate is getting higher, all the versions of TCP treat packet loss as an indication of congestion, and have improper action to the congestion window. In a very lossy environment, most implementations will "crash" to have poor performance. The "crash" means the congestion window size will be 1 forever, thus the allowed amount of output data won't be inflated. Table 1 gives the comparison of Throughputs vs. Loss Rates. The performance of EVFack is higher than the all other implementations. EVFack has tolerable performance degradation with the increasing loss rate, and the congestion window size will not crash.

Simulation 2: In the second simulation, we test the proposed EVFack by using a heavily congested network. There are two situations, one is that packet drop occurs only when the bottleneck buffer overflows, and the other is that the possible loss in the link is considered . We add different TCP agents to our senders ($S_1 \sim S_4$), and compare the throughput of each version. We run the simulations for 50 seconds with queue size being 32 packets. Table 2 lists the simulation results with error-free network, and Table 3 demonstrates the result with packet loss rate 0.03 in the target link. The throughput index is recorded by that specific observed link. The average throughput is the average of these four connections, and this value reveals the possible fairness of that connection.

Table 1 Throughputs under TCP implementations for different packet loss rates

Version \ Packet Loss Rate	0	0.001	0.002	0.005	0.007	0.01	0.02
Reno	1808	1510	1401	1295	1111	934	645
NewReno	1808	1659	1536	1328	1228	1000	704
Fack	1808	1709	1587	1411	1305	1037	751
Vegas	1749	1180	1079	783	781	779	962
EVFack	1758	1754	1751	1745	1742	1744	1732

Version \ Packet Loss Rate	0.03	0.05	0.1	0.15	0.2	0.25	0.3
Reno	462	381	190	129	97	83	48
NewReno	531	398	190	129	99	83	48
Fack	552	422	205	104	98	49	37
Vegas	856	221	91	86	67	1	1
EVFack	1715	1399	877	633	529	422	305

In the loss free case, the throughput of EVFack is similar to other versions, it is because the window adjustment of EVFack does not consider the lost segment caused by congestion drop and thus has a slower reaction to the adjustment of window size. This may cause the window size to be over-estimated, however EVFack adopts a more accurate fast recovery method to recover the lost segments. Reno and Fack repeatedly provoke self-generated segment loss and thus suffers the throughput degradation. In a lossy network environment, EVFack has the strength to fasten the recovery of the lost retransmitted packets, and sustains an expected sending rate as well.

Table 2 Congested networks with loss rate 0.

Version	Reno	Vegas	Fack	EVFack
Throughput	2356	2918	2424	2202
Average Throughput	2295	2249	2305.5	2295

Table 3 Congested networks with loss rate 0.03.

Version	Reno	Vegas	Fack	EVFack
Throughput	546	1148	548	1247
Average Throughput	2284.5	2249.25	2299	2287

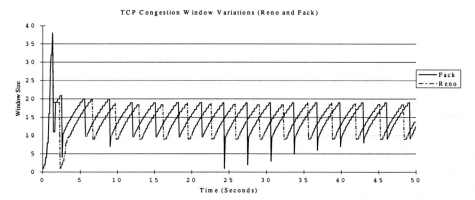

Fig. 9 Window size variations under loss rate 0 (Reno and Fack).

Figure 9 to Figure 12 show window variations and queue occupancy under loss free condition, and figure 13 to Figure 16 illustrate the conditions under loss rate 0.03. Figure 9 demonstrates the Reno and Fack window variations. Figure 10 shows these observations regarding Vegas and EVFack. In Figure 10, the window size variations of the EVFack may be over-estimated, but at the end of the sending period, it adjusts their window size to a stable value as detected. However, EVFack may inflate its window size, because the calculating formula for sending rate is different from that in Vegas. This behavior does not suffer throughput degradation because a fast and accurate recover method is invoked. Other versions of TCP are more conservative, and their window size cannot be increased, even the network has the capacity to

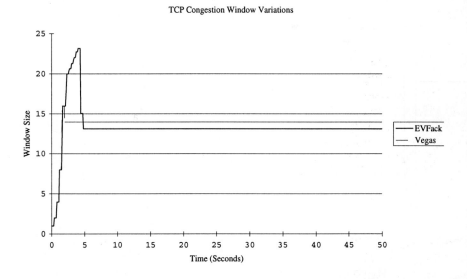

Fig. 10 Window size variations under loss rate 0 (EVFack and Vegas).

accommodate the traffic. Figure 12 shows that the EVFack has the highest bottleneck link utilization among all versions of TCP because it keeps a desirable level of link utilization. Vegas also maintains stable queue occupancy, however, its level is smaller because of the slow increasing window size for each connection, but eventually it may be as large as desired. Reno and Fack possess the natural behavior to inject too many segments to overflow the buffer size, and it leads to the loss of segments.

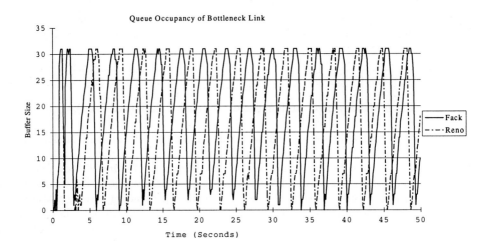

Fig. 11 Queue occupancy of bottleneck link (Loss Rate 0) : Reno and Fack.

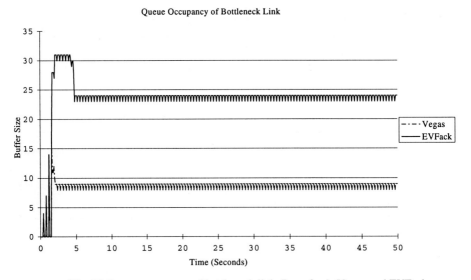

Fig. 12 Queue occupancy of bottleneck link (Loss free): Vegas and EVFack.

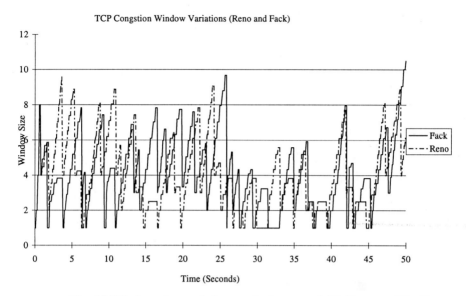

Fig. 13 Window size variations under loss rate 0.03 (Reno and Fack).

From Figure 13 to 14, the packet loss rate 0.03 is used. Both Reno and Fack feature the drastic window variations, because they may mistake random loss packet as network congestion more frequently. Besides, the queue occupancy of Reno-Based TCP switches between empty and full alternatively, thus has poor network utilization. In Figure 15 and Figure 16, EVFack has better control for the lossy networks. It sustains a higher stable queue occupancy than Vegas. Even in the lossy environments, the congestion window size in EVFack still can be maintained at an expected level, and thus achieve a full buffer utilization, as shown in Figure 16. The aforementioned two simulations demonstrate that EVFack performs much better under the lossy networks compared with other TCP versions.

4 Conclusion and Future Works

In this work, we present the EVFack algorithm which improves the TCP performance under lossy network environments. EVFack integrates the delay-based flow-control (like Vegas) and Fack's recovery method to determine the process of recovery to proceed. In our investigation, the delay-based control uses the RTT to get congestion indication, and based on that it can maintain a possible achievable window size. It takes benefits of avoiding unnecessary inflation of the congestion window. We modified the flow control which does not count the lost packets caused by random loss, and let the window size be kept as stable as possible. In the fast recovery phase, EVFack uses a "retransmission list" to record the current outstanding retransmitted packets. Thus, it successfully fast-recovers the lost retransmitted packets, and improves the performance under error-prone networks.

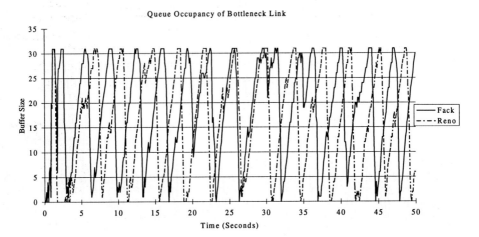

Fig. 14 Queue occupancy of bottleneck link (Loss Rate 0.03): Reno and Fack.

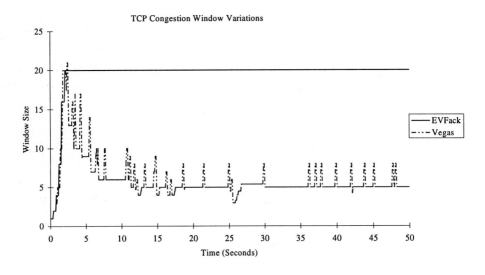

Fig. 15 Window size variations under loss rate 0.03 (Vegas and EVFack).

The simulation results demonstrate the performance improvement over the lossy links; however, it is difficult to develop a realistic Internet simulation model to observe the real lossy characteristics. We only use the simplified test environments to model our proposed implementations, so that we can reveal the phenomena of our proposed algorithm. A more appropriate simulation paradigm for lossy environment is still under investigation for the future works.

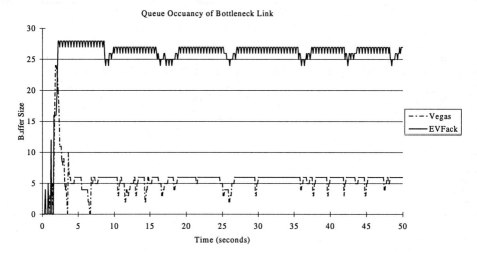

Fig. 16 Queue occupancy of bottleneck link (loss rate 0.03)

References

[1] L. Brakmo, S. O'Malley, and L. Peterson, "TCP Vegas: End-to-End Congestion Control in a Global Internet." IEEE Journal on Selected Areas in Communications (JSAC), 13(8):1465-1480, October 1995.

[2] K. Fall and S. Floyd. "Simulation-based comparisons of Tahoe, Reno, and SACK TCP." ACM Computer Communication Review, 26(3):5-21, July 1996.

[3] S. Floyd. "TCP and Successive Fast Retransmission", February 1995. ftp://ftp.ee.lbl.gov/papers/fastretrans.ps

[4] J. Hoe. "Start-up Dynamics of TCP's Congestion Control and Avoidance Schemes," Master's Thesis, MIT. June 1995

[5] J. Hoe. "Improving the Start-up Behavior of a Congestion Control Scheme for TCP,". SIGCOMM Symposium on Communications Architectures and Protocols, August 1996.

[6] V. Jacobson. Jacobson, "Congestion Avoidance and Control," in Proc. SIGCOMM'88 Symp., August 1988, pp.314-329.

[7] R. Jain, " A delay-based approach for congestion avoidance in interconnected heterogeneous computer networks," ACM Computer Communication Review, vol. 19 no. 5, pp.56-71, Oct. 1989.

[8] Phil Karn and Craig Partridge. Improving Round-Trip Time Estimates in Reliable Transport Protocols. ACM SIGCOMM, pages 2-7, August 1987.

[9] M. Mathis and J. Mahdavi. "Forward Acknowledgement (FACK): Refining TCP Congestion control." Proceedings of ACM SIGCOMM'96, pages 281-291, August 1996.

[10] M. Mathis and J. Mahdavi, S. Floyd, and A. Romanow. TCP selective acknowledgement option. Internet Draft, work in progress, May 1996.

[11] S. McCanne and S. Floyd. UCB/LBNL/VINT Network Simulator - ns(version 2) network simulator, http://www-mash.cs.berkerley.edu/ns/
[12] W. Stevens, TCP/IP Illustrated, Volume 1. Addison-Wesley, 1994.
[13] W. Stallings, High-Speed Networks, TCP/IP and ATM Design Principles. Prentice-Hall, 1998.
[14] Gary R. Wright and W. Richard Stevens. TCP/IP Illustrated, Volume II: The Implementation. Addison-Wesley, 1995.

Bandwidth Tradeoff between TCP and Link-Level FEC*

Chadi Barakat and Eitan Altman

INRIA, 2004 route des Lucioles, 06902 Sophia Antipolis, France
Email : {cbarakat,altman}@sophia.inria.fr

Abstract. FEC is widely used to improve the quality of noisy transmission media as wireless links. This improvement is of importance for a transport protocol as TCP which uses the loss of packets as an indication of network congestion. FEC shields TCP from losses not caused by congestion but it consumes some bandwidth that could be used by TCP. We study in this paper the tradeoff between the bandwidth consumed by FEC and that gained by TCP.

1 Introduction

Forward Error Correction (FEC) is widely used to improve the quality of noisy transmission media as wireless links [2, 4]. This improvement is of importance for a transport protocol as TCP [8, 16] which uses the loss of packets as an indication of network congestion. A TCP packet corrupted while crossing a noisy link is discarded before reaching the receiver which results in an unnecessary window reduction at the TCP source, and hence in a deterioration of the performance of the TCP transfer [2]. In the following, we will only focus on transmission errors on wireless links and we will call the corrupted packets *non-congestion losses* or *link-level losses* since they appear at a level below IP.

The idea behind FEC is to transmit on the wireless link, together with the original data, some redundant information so that a corrupted packet can be reconstructed at the output of the link without the need for any retransmission from TCP [2, 4]. Normally, this should improve the performance of TCP since it shields it from non-congestion losses. But, FEC consumes some bandwidth. Using much FEC may steal some of the bandwidth used by TCP which deteriorates the performance instead of improving it. Clearly, a tradeoff exists between the bandwidth consumed by FEC and that gained by TCP. We analyze this tradeoff in this paper. The question that we asked is, given a certain wireless link with certain characteristics (bandwidth, error rate, burstiness of errors), how to choose the amount of FEC so that to get the maximum gain in TCP performance. A mathematical model and a set of simulations are used for this purpose.

---

* A detailed version of this paper can be obtained upon request from the authors.

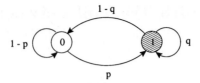

Fig. 1. The Gilbert loss model

2 Model for Non-congestion Losses

Consider a long-life TCP connection that crosses a network including a noisy wireless link of rate μ. We suppose that the quality of the noisy link is improved by a certain FEC mechanism that we will model in the next section.

Most of the works on TCP performance [10, 11, 13] make the assumption that the loss process of TCP packets is not correlated. Packets are assumed to be lost independently with the same probability P. This does not work for wireless links where transmission errors tend to appear in bursts [3–5, 9]. The model often used in the literature to analyze correlated losses on a wireless link is the one introduced by Gilbert [3, 4, 7, 9]. It is a simple ON/OFF model. The noisy link is supposed to be in one of two states: 0 for Good and 1 for Bad. A packet is lost if it leaves the link while it is in the Bad state, otherwise it is supposed to be correctly received. We use such a model in our work. A discrete-time Markov chain (Fig. 1) with two states (Good and Bad) models the dynamics of the wireless link. We focus on the loss process of link-level packets also called *transmission units*. We suppose that a TCP packet is transmitted over the wireless link using multiple small transmission units [3, 4]. A transmission unit can be a bit, a byte, an ATM cell, or any other kind of link-level blocks used for the transmission of TCP/IP packets. The state of the wireless link is observed upon the arrivals of transmission units at its output. We suppose that units cross continuously the link. If no real units exist, *fictive units* are inserted.

Let p denote the probability that the wireless link passes from Good state to Bad state when a transmission unit arrives at its output. Let q denote the probability that the link stays in the Bad state. q represents how much the loss process of transmission units is bursty. The stationary probabilities of the Markov chain associated to the wireless link are equal to: $\pi_B = p/(1-q+p)$ and $\pi_G = (1-q)/(1-q+p)$. Denote by L_B and L_G the average lengths of Bad and Good periods in terms of transmission units. A simple calculation shows that,

$$L_B = 1/(1-q), \qquad L_G = 1/p. \tag{1}$$

The average loss rate, denoted by L, is equal to

$$L = L_B/(L_B + L_G) = p/(1-q+p) = \pi_B. \tag{2}$$

3 Model for FEC

The most common code used for error correction is the *block* code [14, 15]. Suppose that data is transmitted in units as in our model for the noisy link. Block

FEC consists in grouping the units in blocks of K units each. A codec then adds to every block a group of R redundant units calculated from the K original units. The result is the transmission of blocks of total size $N = K + R$ units. At the receiver, the original K units of a block are reconstructed if at least K of the total N units it carries are correctly received. This improves the quality of the transmission since a block can now resist to R losses without being discarded.

In our work, we consider a block FEC implemented on the wireless link in the layer of transmission units. We ignore any FEC that may exist below this layer (e.g., in the physical layer). The input to our study is the loss process of transmission units which is assumed to follow the Gilbert model. In what follows, we will show how much the parameters of the FEC scheme (N, K) impact the performance of TCP transfers.

4 Approximation of TCP Throughput

Consider the throughput as the performance measure that indicates how well TCP behaves over the wireless link. Different models exist in the literature for TCP throughput [1, 10, 11, 13]. Without loss of generality, we consider the following simple expression for TCP throughput in terms of packets/s: $X = (1/RTT)\sqrt{3T/2} = (1/RTT)\sqrt{3/(2P)}$ [11]. This expression is often called the *square root formula*. $T = 1/P$ denotes the average number of TCP packets correctly received between packet losses. P denotes the probability that a TCP packet is lost in the network. In case of bursty losses, P represents the probability that a TCP packet is the first loss in a burst of packet losses [13]. This is because the new versions of TCP (e.g., SACK [6]) are designed in a way to divide their windows one time by two for a burst of packet losses. RTT is the average round-trip time seen by the connection. Note that it is also possible to use in our analysis other more sophisticated expressions for TCP throughput (e.g., [1]).

Suppose that the wireless link is the bottleneck on the path of the connection. Thus, in the absence of FEC, the throughput of TCP is upper bounded by μ. We write $X = \min\left((1/RTT)\sqrt{3T/2}, \mu\right)$.

Our objective is to express the throughput of TCP as a function of the parameters of the loss process of transmission units (p, q) and the parameters of the FEC scheme (N, K). We already have the expression of the throughput as a function of what happens at the packet level (P). What we still need to do is to relate the loss process of TCP packets to the loss process of transmission units. To simplify the analysis, we consider the best case when packets are only lost on the wireless interface whenever the wireless bandwidth μ is not fully utilize (possible since the wireless link is assumed to be the bottleneck). In the case when transmission units are lost independently of each other ($p = q$), $P = 1/T$ is simply equal to the probability that a TCP packet is lost while crossing the wireless link. This case is studied in the next section. In the case when transmission units are lost in bursts, T must be calculated as the average number of TCP packets correctly transmitted between bursts of packet losses.

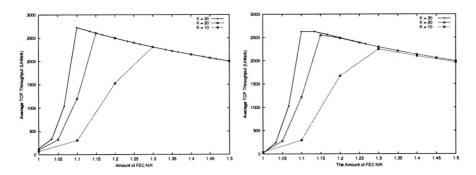

Fig. 2. Model: X vs. N/K and K **Fig. 3.** Simulation: X vs. N/K and K

This is done in Section 6 where we study the impact of the correlation of unit losses on the performance of TCP.

Now, even though it increases T, the addition of FEC consumes some bandwidth and decreases the maximum throughput the TCP connection can achieve. Instead of μ, we get $K\mu/N$ as a maximum TCP throughput. If we denote by S the size of a TCP packet in terms of transmission units, the throughput of TCP in presence of a FEC scheme (N, K) and in terms of units/s can be written as,

$$X_{N,K} = \min\left((S/RTT)\sqrt{3T_{N,K}/2}, K\mu/N\right). \quad (3)$$

5 The Case of Non-correlated Losses

Consider the case when transmission units are lost independently of each other with probability p ($p = q$). Thus, TCP packets are also lost independently of each other but with probability $P_{N,K} = 1/T_{N,K}$ which is a function of the amount of FEC (N, K). The throughput of TCP can be approximated by using (3).

5.1 The Analysis

Suppose that TCP packets are of the size of one link-level block ($S = K$ units). Given a certain block size (K) and a certain amount of FEC (N, K), the choice of the size of the TCP packet in terms of blocks is another problem that we will not address in this paper. A TCP packet is then lost when more than R of its units are lost due to transmission errors. This happens with probability $P_{N,K} = \sum_{i=0}^{K-1} \binom{N}{i}(1-p)^i p^{N-i}$.

It is clear that the addition of FEC at the link level reduces the loss probability of TCP packets. This addition improves the throughout whenever the first term of the minimum function in (3) is smaller than the second term. When these two terms are equal, the quantity of FEC added to the wireless link is sufficient to eliminate the negative effect of non-congestion losses on TCP. We say here that FEC has *cleaned* the link from TCP point of view. Any increase in FEC

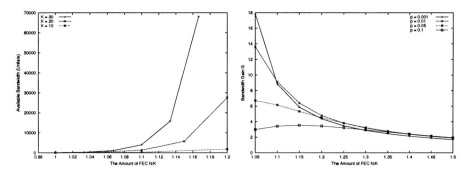

Fig. 4. Model: Optimal FEC vs. μ **Fig. 5.** Model: G vs. N/K and p

beyond this point results in a throughput deterioration. There will be more FEC than what is needed to clean the link. Given μ, K, and p, the optimal quantity of FEC from TCP point of view is the solution of the following equation,

$$(N/(KRTT))\sqrt{3/(2P_{N,K})} = \mu. \tag{4}$$

5.2 Analytical Results

We show in Figure 2 how the throughput of TCP varies as a function of the ratio N/K (FEC rate) for different values of K (10, 20, and 30 units). RTT is taken equal to 560 ms and the wireless link bandwidth μ to 3000 units/s. We can see this scenario as the case of a mobile user downloading data from the Internet through a satellite link. This value of μ is approximately equal to the maximum ATM cell rate on a T1 link (1.5 Mbps). p is set to 0.01.

It is clear that the performance improves considerably when FEC is added and this improvement continues until the optimum point given by (4) is reached. Beyond this point, any increase in FEC deteriorates the throughput. Also, we notice that for a certain quantity of FEC, an increase in K improves the performance. An increase in K results in a faster window growth. TCP window is increased in terms of packets rather than bytes [16]. The TCP source then returns faster to its rate prior to the detection of a non-congestion loss.

In Figure 4, we plot the left-hand term of (4) as a function of N/K for the same three values of K. These curves provide us with the optimal amount of FEC for given μ, p, and K. We see well how the increase in K reduces considerably the amount of FEC needed to clean the wireless link from TCP point of view. Given μ, a compromise between K and FEC rate must be done. First, we choose the largest possible K, then we choose the appropriate amount of FEC.

For $\mu = 3000$ units/s and $K = 20$, we show in Figure 6 how the throughput of TCP varies as a function of p for different values of N. It is clear that adding just one redundant unit to every FEC block results in a considerable gain in performance especially at small p. Adding more redundancy at small p deteriorates slightly the performance since the link is already clean and the additional redundancy steals some of the bandwidth used by TCP. This is not the case at high p where much redundancy needs to be used in order to get good performance. Note that even though an excess of FEC reduces the performance of TCP when losses are rare, the reduction is negligible in front of the gain in performance we obtain when losses become frequent. When the link is heavily lossy ($\log(p) > -1.7$), the three amounts of FEC plotted in the figure become insufficient and all the curves converge to the same point.

5.3 Simulation Results

Using the ns simulator [12], we simulate a simple scenario where a TCP source is connected to a router via a high speed terrestrial link and where the router is connected to the TCP receiver via a noisy wireless link. The Reno version of TCP [6] is used. The TCP source is fed by an FTP application with an infinite amount of data to send. We add our FEC model to the simulator. The transmission units on the wireless link are supposed to be ATM cells of size 53 bytes. We choose the bandwidth of the wireless link in a way to get a μ equal to 3000 cells/s. RTT is taken equal to 560 ms and the buffer size in the middle router is set to 100 packets. This guarantees that no losses occur in the middle router before the full utilization of μ.

Figures 3 and 7 show the variation of the simulated throughput as a function of the amount of FEC (N/K) and the unit loss probability p respectively. In the first figure, p is set to 0.01. We clearly notice the good match between these results and the analytical ones. The small difference is due to the fact that the expression of the throughput we used does not consider the possibility of a timeout when multiple packet losses appear in the same TCP window [6]. Also, in our analysis, we considered that RTT is always constant which does not hold when the throughput of TCP approaches the available bandwidth.

5.4 The Tradeoff between TCP Throughput and FEC Cost

We compare in this section the bandwidth gained by TCP to that consumed by FEC. Let G be the ratio of these two bandwidths,

$$G = (X_{N,K} - X_{K,K})/(X_{N,K}(N-K)/K) = (1 - X_{K,K}/X_{N,K}) \times (K/(N-K)). \tag{5}$$

This ratio indicates how much beneficial is the addition of FEC. It can be seen as a measure of the overall performance of the system TCP-FEC. A value close to one of G means that we pay for FEC as much as we gain in TCP throughput. A negative value means that the addition of FEC has reduced the performance of TCP instead of improving it.

In Figure 5 we plot G as a function of the amount of FEC for different unit loss probabilities. Again, we take $\mu = 3000$ units/s and $K = 20$. This figure shows that the gain in overall performance is important when the loss probability and the amount of FEC are small. Moreover, with small amounts of FEC, the gain decreases considerably when the loss rate ($L = p$) increases. Now, when the FEC rate increases, the curves converge approximately to the same point with a slightly better gain this time for higher loss probabilities.

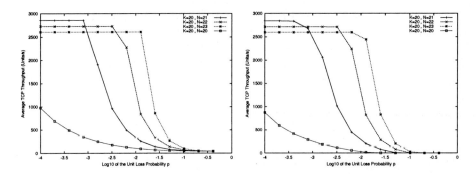

Fig. 6. Model: X vs. p and N/K **Fig. 7.** Simulation: X vs. p and N/K

5.5 Number of Connections and the Gain in Performance

We notice in Figure 5 that using a small amount of FEC gives the best gain in overall performance. Thus, in order to maintain a high gain, one can use a small amount of FEC and fully utilize the available bandwidth on the wireless link by opening multiple TCP connections. But, in practice we cannot always guarantee that there are enough TCP connections to fully utilize the available bandwidth. A TCP connection must be able to use alone all the available bandwidth. For this reason, FEC has to be added in large amounts so that to make the noisy link clean from the point of view of a single TCP connection even if the achieved gain is not very important.

6 The Case of Correlated Losses

In this section we study the influence of burstiness of transmission unit losses on the efficiency of a FEC scheme. It is clear that when unit losses tend to appear in bursts, more FEC is needed to clean the link. Packets are hurt by bursts of losses and they require a large number of redundant units per packet (R) to be corrected. But, for the same average loss rate (L), the burstiness of losses reduces the probability that the link passes to the Bad state (p decreases when q increases). This reduces the probability that a TCP packet is hurt by a burst of losses. TCP throughput may then improve and the amount of FEC could be reduced. An analysis is needed to understand these opposite effects of burstiness.

6.1 Performance Analysis

Let us calculate T, and hence the throughput of TCP using (3), as a function of the amount of FEC, K, the average loss rate, the burstiness of losses, and μ. Recall that T in this case denotes the average number of TCP packets correctly transmitted between bursts of packet losses.

Let t be the number of good TCP packets between two separate bursts. The minimum value of t is one packet and its expectation is equal to T. Let Y_n be the state of packet n. 0 is the number of the first good TCP packet between the two bursts. Y_n takes two values B (Bad) and G (Good). We have $Y_0 = G$. T can be written as $\sum_{n=0}^{\infty} P(t > n|Y_0 = G) = 1 + \sum_{n=1}^{\infty} P(t > n|Y_0 = G)$.

The computation of T is quite complicated since the TCP packets are not always transmitted back-to-back. Another complication is that $\{Y_n\}$ does not form a Markov chain. Indeed, if we know for example that a packet, say n, is of type B then the probability that packet $n + 1$ is of type G also depends on the type of packet $n - 1$. If packet $n - 1$ were G rather than B, then the *last units of packet n* are more likely to be those that caused its loss. Hence, the probability that packet $n + 1$ is B is larger in this case. This motivates us to introduce another random variable which will make the system more "Markovian" and will permit us to write recurrent equations in order to solve for T. We propose to use the state of the last *transmission unit* received, or fictively received, before the nth TCP packet. The knowledge of the state of this unit, denoted by Y_n^{-1} (which may again take the values B and G), fully determines the distribution of the state Y_n of the following TCP packet. We write T as $1 + \alpha P(Y_1^{-1} = G|Y_0 = G) + \beta P(Y_1^{-1} = B|Y_0 = G)$, where $\alpha = \sum_{n=1}^{\infty} P(t > n|Y_0 = G, Y_1^{-1} = G)$ and $\beta = \sum_{n=1}^{\infty} P(t > n|Y_0 = G, Y_1^{-1} = B)$. We shall make the following assumption,

Assumption 1: $P(Y_1^{-1} = G|Y_0 = G) \approx \pi_G$ and $P(Y_1^{-1} = B|Y_0 = G) \approx \pi_B$.

Assumption 1 holds when the time to reach steady state for the Markov chain in the Gilbert model is shorter than the time between the beginning of two consecutive TCP packets (either because the TCP packets are sufficiently large or because they are sufficiently spaced). Assumption 1 also holds when π_B and the loss probability of a whole TCP packet are small. Indeed, we can write, $\pi_G = P(Y_1^{-1} = G|Y_0 = G)P(Y_0 = G) + P(Y_1^{-1} = G|Y_0 = B)P(Y_0 = B) \approx P(Y_1^{-1} = G|Y_0 = G) \cdot 1 + P(Y_1^{-1} = G|Y_0 = B) \cdot 0$.

In view of Assumption 1, the probability that the unit preceding a TCP packet is lost can be considered as independent of the state of the previous packet. It follows that $T = 1 + \alpha \pi_G + \beta \pi_B$, with

$$\alpha = (1 - P(Y_1 = B|Y_1^{-1} = G))(1 + \alpha \pi_G + \beta \pi_B),$$
$$\beta = (1 - P(Y_1 = B|Y_1^{-1} = B))(1 + \alpha \pi_G + \beta \pi_B).$$

This yields, $1/T = \pi_G P(Y_1 = B|Y_1^{-1} = G) + \pi_B P(Y_1 = B|Y_1^{-1} = B)$. The calculation of T is then simplified to the calculation of the probability that a TCP packet is lost given the state of the unit just preceding it. Again, it is difficult to find an explicit expression for this probability. A TCP packet can be lost by a single long burst of unit losses as well as by multiple separate small bursts. To further facilitate the analysis, we assume that bursts of losses at the unit level are separated so that two bursts rarely appear within the same packet. This holds if,

Assumption 2: $(1 - q) \cdot L \cdot N \ll 1$.

Indeed, a TCP packet is supposed to be lost if it is hurt by a burst of unit losses larger than R. We don't consider the probability that multiple small and separate bursts at the unit level contribute to the loss of the packet. This is

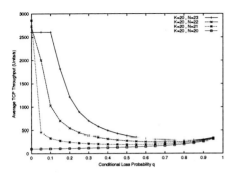 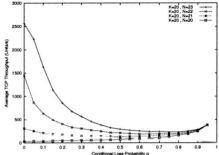

Fig. 8. Model: X vs. q and N/K **Fig. 9.** Simulation: X vs. q and N/K

possible when the sum of the average lengths of the Good state (L_G) and the Bad state (L_B) is much larger than the packet length N. Using (1) and (2), we get the condition in Assumption 2. If this condition is not satisfied, many bursts may appear within the same packet leading to a higher loss probability than the one we will find, hence to a lower throughput.

Consider first the case $Y_1^{-1} = B$. In view of Assumption 2, packet 1 is lost if its first $R+1$ units are also lost. Thus, $P(Y_1 = B | Y_1^{-1} = B) = q^{R+1}$. For the case $Y_1^{-1} = G$, packet 1 is lost if a burst of losses of length at least $R+1$ units appears in its middle. Thus,
$P(Y_1 = B | Y_1^{-1} = G) = q^R p \left(1 + (1-p) + \cdots + (1-p)^{N-R-1}\right) \simeq K q^R p$.
We used here the approximation $(1 - (1-p)^{N-R}) \simeq Kp$. Substituting p by its value as a function of q and the loss rate L (Equation (2)), we get

$$1/T = q^{N-K} L \left((1-q)K + q\right). \qquad (6)$$

6.2 Analytical Results

Using (3) and (6), we plot in Figure 8 the throughput of TCP as a function of burstiness and this is for different amounts of FEC. The burstiness is varied by varying q which is called the Conditional Loss Probability in the figure. K is set to 20 and the loss rate L to 0.01. The other parameters of the model are taken as in the previous section. We see well that a large amount of FEC gives always the best performance. The difference in performance is important for small bursts (small q). We also see that when burstiness increases, the throughput of TCP decreases drastically for the three FEC schemes we consider in the figure. This is because the length of bursts becomes larger the number of redundant units per packet (R). Here, much FEC must be added to clean the link. But, much FEC reduces the throughput of TCP when burstiness decreases given the bandwidth it consumes. A compromise must be made between much FEC to resist to bursts and a small amount of FEC to give better performance when burstiness decreases. One can think about implementing some kind of adaptive FEC that adjusts the amount of redundancy as a function of the degree of burstiness.

Now, we show in Figure 10 how the block size K can help TCP to resist to burstiness. First, we take the same amount of FEC ($N/K = 11/10$) and we

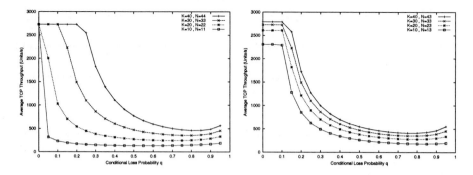

Fig. 10. Model: X vs. q at constant N/K **Fig. 11.** Model: X vs. q at constant R

vary K. Increasing K increases the number of redundant units in a TCP packet and thus helps TCP to resist to larger bursts. Better performance is obtained even though the amount of FEC is not changed. The benefit of large packets is also illustrated in Figure 11. In this figure we plot for the same R, the variation of the throughput for different packet sizes. Surprisingly, a large packet size gives better performance than a small one even though the amount of FEC is smaller. From (6), increasing K for the same R decreases T, but this decrease is small compared to the gain we get from the increase in the packet size. In other words, the throughput in terms of packets/s deteriorates when we increase K at a constant R, but it improves in terms of units/s.

6.3 Simulation Results

Our intention is to validate by simulation the analytical results we plotted in Figure 8. We consider the same simulation scenario as that in the non-correlation case. The results are plotted in Figure 9. The curves show the same behavior as those in Figure 8. But, we see some mismatch at low burstiness. This is due to our assumption that a packet can only be lost by a single burst not by multiple small and separate bursts of losses at the unit level. As one must expect, the simulation gives a lower throughput in this region given that we are overestimating T.

References

1. E. Altman, K. Avratchenkov, and C. Barakat, "A stochastic model for TCP/IP with stationary random losses", *ACM SIGCOMM*, Sep. 2000.
2. H. Balakrishnan, V. N. Padmanabhan, S. Seshan, and R. Katz, "A comparison of Mechanisms for Improving TCP Performance over Wireless Links", *ACM SIGCOMM*, Aug. 1996.
3. H. Chaskar, T. V. Lakshman, and U. Madhow, "On the design of interfaces for TCP/IP over wireless", *IEEE MILCOM*, Oct. 1996.
4. A. Chockalingam, M. Zorzi, and R.R. Rao, "Performance of TCP on Wireless Fading Links with Memory", *IEEE ICC*, Jun. 1998.
5. B. R. Elbert, "The Satellite Communication Applications Handbook", *Artech House*, Boston, London, 1997.

6. K. Fall and S. Floyd, "Simulation-based Comparisons of Tahoe, Reno, and SACK TCP", *ACM Computer Communication Review*, vol. 26, no. 3, pp. 5-21, Jul. 1996.
7. E.N. Gilbert, "Capacity of a burst-noise channel", *Bell Systems Technical Journal*, Sep. 1960.
8. V. Jacobson, "Congestion avoidance and control", *ACM SIGCOMM*, Aug. 1988.
9. A. Kumar and J. Holtzman, "Performance Analysis of Versions of TCP in a Local Network with a Mobile Radio Link", *Sadhana: Indian Academy of Sciences Proceedings in Engg. Sciences*, Feb. 1998.
10. T.V. Lakshman and U. Madhow, "The performance of TCP/IP for networks with high bandwidth-delay products and random loss", *IEEE/ACM Transactions on Networking*, vol. 5, no. 3, pp. 336-350, Jun. 1997.
11. M. Mathis, J. Semke, J. Mahdavi, and T. Ott, "The Macroscopic Behavior of the TCP Congestion Avoidance Algorithm", *ACM Computer Communication Review*, vol. 27, no. 3, pp. 67-82, Jul. 1997.
12. The LBNL Network Simulator, *ns*, http://www.isi.edu/nsnam/ns/
13. J. Padhye, V. Firoiu, D. Towsley, and J. Kurose, "Modeling TCP Throughput: a Simple Model and its Empirical Validation", *ACM SIGCOMM*, Sep. 1998.
14. L. Rizzo, "Effective erasure codes for reliable computer communication protocols", *ACM Computer Communication Review*, vol. 27, no. 2, pp. 24-36, Apr. 1997.
15. N. Shacham and P. McKenney, "Packet Recovery in High-Speed Networks Using Coding and Buffer Management", *IEEE INFOCOM*, Jun. 1990.
16. W. Stevens, "TCP Slow-Start, Congestion Avoidance, Fast Retransmit, and Fast Recovery Algorithms", *RFC 2001*, Jan. 1997.

Supporting QoS for Legacy Applications

C. Tsetsekas, S. Maniatis, and I. S. Venieris

National Technical University of Athens,
Department of Electrical and Computer Engineering,
9 Heroon Polytechniou str, 15773,
Athens, Greece
{htset, sotos}@telecom.ntua.gr,
ivenieri@cc.ece.ntua.gr

Abstract. Internet is widely known for lacking any kind of mechanism for the provisioning of Quality of Service (QoS) guarantees. The Internet community currently concentrates its efforts on mechanisms that support QoS in various layers of the OSI model. Apart from that, the Internet community is trying also to define the protocols, through which applications and users will signal their QoS requirements to the lower network layer mechanisms. The latter task, however, is not trivial, especially for legacy applications that cannot be modified and recompiled. This paper presents a framework for a middleware component that supports QoS for legacy applications. It mainly focuses on the support of a proxy-based framework for the identification of flows, the measurement of basic QoS parameters and the definition of an API that can be used by middleware components or even applications. The position of this proxy architecture in a reference network topology and the communication with other middleware entities is also discussed.

1 Introduction

The Internet has had an overwhelming effect on the way people interact and communicate. The Internet is based on the Internet Protocol (IP) that provides a simple, easily deployable and best effort in nature network service. The tremendous growth of IP has also boosted the development of various IP-based applications that include complex, quality-intensive multimedia services. As a result, the Internet community has been heavily engaged with defining the appropriate mechanisms that will provide Quality of Service (QoS) support for applications. It is foreseen that a proper combination of these mechanisms will eventually provide ubiquitous end-to-end QoS.

To be more specific, the Integrated Services (IntServ) [1] and the Differentiated Services (DiffServ) [2] are two of the mechanisms proposed by IETF. In IntServ, network resources, which are mainly defined in terms of bit rate, packet delay and maximum transfer size, are reserved in every node along the path from the sender to the receiver. In contrast to the IntServ model, which uses explicit resource reservation for every flow requesting QoS, thus raising scalability concerns, the DiffServ model is a simpler and straightforward architecture, which relies on prioritization of some flows over others. However, the simplicity of the DiffServ model and its lack of mechanisms for the systematic and automated resource allocation necessitated the introduction of the Bandwidth Broker concept [3]. The Bandwidth Broker is a logical entity, complementing the DiffServ infrastructure, which is responsible for performing policy-based admission control, managing network resources, and configuring specific network nodes, among others.

Except for the aforementioned network mechanisms, applications and users have to be able to indicate their QoS requirements to the network entities through the proper interface mechanisms. Currently, the Resource ReSerVation Protocol (RSVP) [4] provides the interface to setup and control QoS in the IntServ model. There are a lot of efforts to utilize the RSVP protocol along with other QoS technologies like DiffServ. Such protocols may be invoked with the use of APIs, like the Generic QoS API integrated in WinSock2 from Microsoft, and the QoS Application Programming Interface (API) from the Internet2 community [5]. The existence of these APIs presupposes that the applications must be modified, and compiled again in order to take advantage of them.

The main motivation behind this paper is the support of legacy applications. The individual characteristic of such applications is primarily that they cannot be modified. So they cannot directly make use of one of the aforementioned APIs. Moreover, the IP port numbers are usually not known a priori, because they are negotiated dynamically. In addition, any modification of the network QoS provisions during the lifetime of the application cannot be communicated to the application, so that it cannot react to them. In order to alleviate these inherent limitations of legacy applications, we propose to make use of a middleware component that acts as an agent between the applications and the network QoS entities. This paper presents the framework of the middleware for the support of any kind of application over a QoS-enabled IP network. The main responsibility of the middleware is to provide the mechanisms for the description and selection of QoS parameters and the forwarding of QoS requests to the appropriate network entities. Moreover, the paper describes to a great extent a proxy framework for the support of fundamental operations, like the detection of new flows and the measurement of their traffic profile, as well as additional features, like the transparent support of RSVP, and the identification of various multimedia streams (video, audio) within a Web session.

The paper is structured as follows. Section 2 gives a brief overview of the overall context within which this work is being accomplished. It presents the general Aquila [6] concept and, more specifically, the End-User Application Toolkit. Section 3 addresses the proxy framework, identifying the problems and proposing solutions. Finally, section 4 presents the conclusions.

2 The AQUILA Architecture

The Aquila project [6] aims to define, implement and evaluate an enhanced architecture for dynamic end-to-end Quality of Service support over IP networks. Existing approaches to QoS specified for the Internet, such as IntServ, DiffServ and MPLS are used as a basis, and the solutions implemented are verified and tested within trials involving end-users.

The Aquila network architecture (Figure 1) is created by the interconnection of various administrative domains controlled by different Internet Service Providers (ISPs). These domains are distinguished into two categories: core and access networks. In the core network, IP flows receive prioritized treatment over others with the adoption of the DiffServ architecture. The access network connects hosts to the core Internet, through Edge Routers that perform enhanced functionality compared to usual core routers. To be more specific, in the Aquila architecture, in the Edge Router of the ISP where an access network is connected, user packets are classified, shaped and policed. The Edge Router also marks packets and aggregates their corresponding flows into groups under the same DiffServ Codepoint.

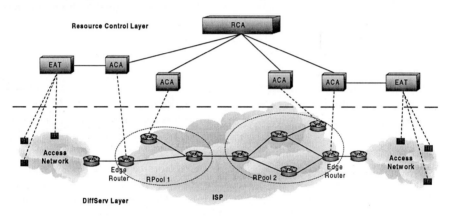

Fig. 1. Aquila Network Architecture

The main innovation of the Aquila architecture is a new layer on top of the DiffServ core ISP network, called the Resource Control Layer (RCL) [7]. The RCL is responsible for the management of network resources. It resembles a distributed Bandwidth Broker in the DiffServ architecture. However, this architecture extends the Bandwidth Broker model in two aspects: it caters for scalability by logically dividing the administrative domain in sub-areas, called Resource Pools (RPool), and it provides dynamic end-to-end support for QoS.

The Resource Pools construct a tree hierarchy of RPools. The root of this tree is the whole administrative domain, while the leaves are the edge routers. The Resource Control Agent (RCA) is the central entity of the RCL, responsible for the overall control of the administrative domain. The RCA operates on a long-term basis, by managing and distributing the resources of the domain to the Admission Control Agents (ACA), by exploiting the Resource Pool concept. An ACA is associated with each edge router of the domain, and it is always a Resource Pool Leaf. It operates on request basis, by performing policy and admission control and granting a share of resources to individual flows in response to a QoS request. In order to perform admission control, it intelligently compares the requested resources with the total amount of the available resources granted to it. The resource management is performed with the aid of an intelligent algorithm that caters for the initial distribution and re-distribution of resources during the course of the network operation [8].

The End-user Application Toolkit (EAT) [9] provides end-to-end QoS support. Applications and end users can make their own QoS requests with the use of the EAT. The major objective of the EAT is to provide a scalable and efficient approach for transparent QoS support for multimedia applications. Existing commercial applications leverage the EAT functionality without the need of any modifications. Resource reservation requests may be sent automatically, without the need of user intervention. Therefore, the end-users are shielded from the application complexity, as well as from QoS related aspects, such as traffic specification.

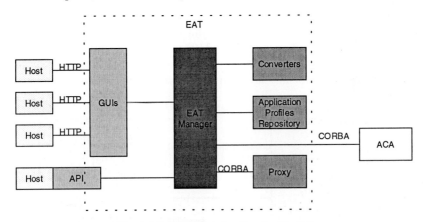

Fig. 2. The End-User Application Toolkit

The main building blocks of the EAT (Figure 2) are the EAT Manager, the Proxy module, the Converter modules, the Application Profiles Repository, the QoS Application Programming Interface (API) and the Graphical User Interfaces (GUIs). The EAT Manager coordinates all other modules, to ensure the smooth operation of the toolkit. Moreover, it is responsible for forwarding the reservation requests to the ACA. For this purpose it utilizes an interface based on CORBA [10].

The Converter modules intelligently and automatically prepare reservations for selected application sessions (e.g. a video session), with the appropriate QoS parameters. In order to be able to perform such an operation, the Converters rely on the concept of *Application Profiles*. An application profile describes the QoS requirements of a specific application based on a Data Type Definition (DTD) scheme. The concept of application profiles stems from the fact that each individual application can be thoroughly tested in order to discover its QoS requirements. After the testing of the application, the application profile is composed by the measured parameters, using the eXtended Markup Language (XML) format [11], and checked upon the defined DTD. Application Profiles are stored in a repository, which is queried by the Converters upon a QoS request.

The GUIs give the opportunity to the user to place and release reservations, as well as to monitor the current QoS conditions, through a common web browser. The API provides a library of functions that can be used by developers to provide QoS to new applications. Finally, the Proxy module addresses the inherent problems of legacy applications described in the introduction. The Proxy module is based on a well-defined framework that is briefly discussed in the rest of the paper.

The EAT modules have clear and consistent interfaces among them. This enables one module to be used outside of the context of the Aquila architecture.

3 The Proxy Framework

The Proxy module has been designed and built under the consideration of being independent of specific architectures and operating systems. It provides an Application Programming Interface (API) via CORBA [10] to allow any application or middleware to take advantage of its functionality. Therefore, it is not adequate only for the Aquila architecture, but for any architecture that needs support for legacy applications.

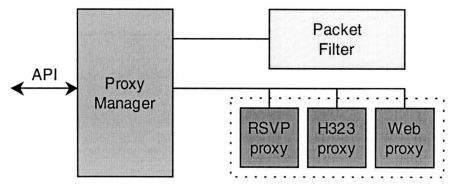

Fig. 3. The Proxy Framework

The Proxy Framework is graphically depicted in Figure 3. The Proxy Manager is the central entity that takes control of the overall operations. It co-ordinates all the components of the Proxy Framework by identifying the available Proxies and by making them available to the applications. It also keeps track of all discovered QoS flows through a *Flow Database*. The entries of this database are composed of IP flows identified from the Proxies along with measurement information.

However, the most important components of the Proxy Framework are the Application-Level Proxies and the Packet Filter. They are described in detail in the following paragraphs.

Application-Level Proxies

The main functionality of the Proxy includes the identification of flows that may need QoS treatment and the measurement of their basic QoS parameters. For this purpose, the Proxy Framework includes Application-Level Proxies that perform signaling protocol translation in an effort to extract the parameters of a flow that are important for a reservation request (addresses and port numbers). A specific Application-Level Proxy has been specified for each major protocol that is used for connection set up (SIP, H.323, RTSP etc.). The implementation of such a Proxy is based on a generic framework that specifies the interfaces it should implement and offers already established components that make development easier.

The H.323 application proxy can serve here as an example: Currently, there are many applications that make use of the H.323 protocols [12], like NetMeeting in Windows platforms. The need for such a proxy stems from the fact that the ports used for the audio and video streams are not standard, but ephemeral. The H.323 protocol makes use of the Q.931 and H.245 signaling protocols (over TCP) to set up the RTP and RTCP audio and video flows. The task of the Proxy in this case would be to intercept the exchanged H.245 control messages, in order to find out the dynamically negotiated ports used by these streams. The Session Initiation Protocol (SIP) [13] Proxy also operates in the same way. The SIP messages exchanged through the SIP Proxy who, process them and extracts information about the addresses and ports of the audio connection. The Proxy may also extract information relevant to resource reservation (see the discussion on SIP extensions [14]) and use it for the formulation of a resource request to the Resource Control Layer of Aquila.

A Web Proxy is also used for identification of Web traffic that may need QoS support. Such traffic would be links to multimedia content. The Proxy steps into the transmitted content and identifies possible audio or video streams that could need QoS treatment. Through the API, it communicates the acquired streams to the responsible entity to take care of reservations. In the Aquila context, the Proxy Manager, after being triggered by the Web proxy about specific multimedia streams, contacts the EAT Manager through the API. According to user-specified settings, the EAT Manager can either directly place a reservation with the aid of the Converter modules, or contact the end-user through the GUI to ask for confirmation.

An important component of the Proxy Framework is an RSVP proxy, responsible for identifying and interpreting RSVP protocol messages. In the Aquila architecture, this is very significant, because the core network does not support IntServ for scalability reasons. Therefore, while the PATH and RESV messages of RSVP are transparently forwarded in the Aquila core network, they are caught at the edges by the RSVP Proxy. Their content (Flowspec and Filterspec) is extracted and forwarded to the EAT for the initiation of a new reservation. In this way, legacy RSVP applications can be transparently accommodated by our architecture. It is obvious that any architecture could make use of the RSVP proxy for similar reasons, in order to decouple RSVP from core network QoS technologies.

Based on the Proxy Framework, new Proxies may easily be implemented in order to cover for the needs of other legacy applications. However, the functionality of the

Framework is not restricted to identifying flows pertaining to a specific protocol. A central component, the Packet Filter is used to detect new flows that have not already been registered by the Proxies.

Packet Filter

The Packet Filter is situated at the borders between access and core networks, usually inside a firewall. It captures all incoming and outgoing packets from the access network, checking their source and destination address. When a new flow is detected (and no Proxy has been used), the Packet Filter examines its packets, in order to extract important information: IP addresses and port numbers, as well as hints on their content, mainly through RTP headers in the case of multimedia flows. The Packet Filter can therefore provide support for flows that are not established with the use of a supported signaling protocol. Upon the detection of such a flow the EAT is contacted in order to decide whether this flow will receive QoS support or not.

However, the main functionality of the Packet Filter is to conduct periodic traffic measurements. The value of such measurements is two-fold. First, the traffic profile of QoS flow may be estimated. Upon the establishment of a new connection, measurements can provide some preliminary values for the formation of a resource reservation request. As time passes by and the flow of the IP packets reaches a more stable rate, more accurate measurements can be available and corrections to the QoS request may be made. This feature of the Packet Filter enables the *adaptation* of a QoS reservation according to the long-term fluctuations of user traffic.

The second advantage that measurements offer is that users may receive immediate feedback on their traffic usage and on whether the requested levels of QoS reservation are honored by the network. In the Aquila architecture, the flow identifiers and the measurements are passed through the Proxy Manager and the API to the EAT Manager. The EAT then displays the performance of the network in a comprehensible format to the end-user through the GUI.

The basic measurements conducted by the Packet Filter mainly include the estimation of the Tspec of a flow, as it is defined within the RSVP protocol [5]. The Tspec consists of the following parameters: Peak rate, average rate, burst size, minimum policed unit and maximum packet size. Those parameters are estimated by processing a list of packet sizes along with their corresponding timestamp. The peak rate is calculated by dividing the sum of bits sent over a small period of time by this period. The same way is used for the average rate, by using a substantially longer period of time. The burst size is estimated as the excess bytes sent over a small period of time, by subtracting the bytes that should have been sent according to the estimated average rate, from the actual bytes sent over this period. Finally, the minimum policed unit and maximum packet size is the minimum and maximum packet size that has been observed for the specific flow.

A future extension of the measurements functionality will include the estimation of the delay, jitter and packet loss values of a packet flow. Those parameters cannot certainly be measured by a standalone Packet Filter, but require the co-operation of two measurement entities. A sender and a receiver should exchange test packets with characteristics (packet marking, source and destination networks) similar to the flows under measurement. Therefore we can measure the delay and its variation. However, this solution poses new open questions, the most important being how the two Proxies will locate each other and communicate.

The Proxy API

The Proxy Manager offers an API to the EAT using CORBA. In the Aquila architecture, the EAT Manager uses this API to receive flow information and measurements. Moreover, the API provides upcalls, that enable it to signal to the EAT a significant change in the Tspec that may require the adaptation of the reservation. This approach was chosen in order not to burden the system's operation with a flood of signaling messages between the EAT Manager and the Proxy.

This API may also be used by other external entities, such as applications. Through the API, an application can query the Packet Filter about flow identifiers and measurements. Moreover, it can query, tune, or configure existing proxies, or even add new proxies to the framework to cater for individual needs.

Performance Issues

The most obvious location for the Proxy Framework is inside a firewall. In this way, it can leverage the functionality of packet filtering in the edge routers. The operation of a packet filter at the ingress of the core network is not expected to create a bottleneck. We believe that the deterioration of performance of the network will not be greater that the one introduced by a simple firewall. However, if the performance deterioration during the operation of the Proxy is significant, more than one Proxy can be introduced in the network, by serving separate sections of an administrative domain.

On the other hand, one of the advantages of the Proxy Framework concept is the provision of QoS to flows in the case of firewalls and Network Address Translation (NAT). Since the Proxy Framework is exactly in the place where address translation operations take place, it can identify the parameters of a flow (addresses and port numbers) from both sides of a firewall.

4 Conclusions

In this paper we have presented a Proxy Framework for the support of legacy applications within a QoS architecture. The Proxy Framework is an important feature of Aquila, an architecture that enables dynamic end-to-end QoS support over a DiffServ enabled core network.

The basic functionality of the Proxy includes the use of Application-Level Proxies that perform protocol translation of major connection set up protocols, in an effort to support QoS for legacy applications. Moreover, the Proxy Framework is enhanced with packet filtering capabilities that enable the performance measurements as well as estimation of the traffic profile of QoS flows.

Acknowledgement

This work was performed in the framework of IST Project AQUILA (Adaptive Resource Control of QoS Using an IP-based Layered Architecture - IST-1999-10077) funded in part by the EU. The authors wish to express their gratitude to the other members of the AQUILA Consortium for valuable discussions.

References

1. R. Braden et al., "Integrated Services in the Internet Architecture: an Overview", RFC 1633
2. S. Blake et al., "An Architecture for Differentiated Services", RFC 2475
3. K. Nichols et al., "A Two-bit Differentiated Services Architecture for the Internet", RFC 2638
4. R. Braden et al., "Resource ReSerVation Protocol (RSVP)", RFC 2205
5. B. Riddle & A. Adamson, "A Quality of Service API Proposal", http://apps.internet2.edu/qosapi.htm
6. The Aquila project: http://www-st.inf.tu-dresden.de/aquila/
7. G. Poliths, P. Sampatakos, I.S. Venieris, "Design of a multi-layer bandwidth broker architecture", Lecture Notes in Computer Science; Vol 1938, Springer Verlag, Oct 2000
8. E. Nikolouzou, P. Sampatakos, I.S. Venieris, "Evaluation of an Algorithm for Dynamic Resource Distribution in a Differentiated Services Network", Proceedings of ICN2001, June 2001
9. Ch. Tsetsekas, S. Maniatis, I.S. Venieris, "An end-to-end middleware solution for the support of QoS in the Internet", Proceedings of SoftCOM2000, Oct 2000
10. CORBA/IIOP 2.3.1 Specification, http://www.omg.org/ corba/cichpter.html
11. E.R. Harold: XML Extensible Markup Language, IDG Books Worldwide, 1998
12. ITU-T Recommendation H.323, Packet-Based Multimedia Communication Systems, 1998
13. M. Handley et al., "SIP: Session Initiation Protocol", RFC 2543
14. W. Marshall et al., "Integration of Resource Management and SIP", draft-ietf-sip-manyfolks-resource-00, November 2000

An Open Architecture for Evaluating Arbitrary Quality of Service Mechanisms in Software Routers

Klaus Wehrle

Institute of Telematics, University of Karlsruhe
Zirkel 2, 76128 Karlsruhe, Germany
Tel.: +49 721 608 6414, Fax: +49 721 388097
klaus.wehrle@acm.org

Abstract. This paper presents an open framework for building and evaluating new Quality of Service elements in a Linux based software router. Conventional QoS behavior, like bucket metering models or scheduling algorithms, have been split up to their elementary components to make the reuse and the variation of them easier. Therefore five kinds of QoS behavior types will be introduced. Based on this pool of the so called *Behavior Elements*, QoS behavior can be built by combining these basic elements. In the most cases, new QoS behaviors, i.e. new PHBs for the Differentiated Services architecture can be created from existing elementary QoS models. Furthermore, the concept of *Hook*s is introduced, which are strategic points within network protocols to extended them with QoS behavior. E.g. within the Internet Protocol five Hooks can be identified to access a certain set of packets.

1 Introduction

In the last few years, the Internet community spent a lot of research efforts in investigating several kinds of mechanisms to provide better services than the traditional best effort delivery. For both, the Integrated Services architecture [BrCS94], and as well for the Differentiated Services architecture [BBCD+98] a lot of QoS elements, like metering elements, classifiers, scheduling algorithms have been developed and a lot of them will be developed in the future. Especially the Differentiated Services architecture is designed of small building blocks (Per Hop Behaviors) composed to offer quality based services. To study the achievable quality it is extremely important to have an implementation architecture that offers the possibility to build new QoS models rapidly from elementary entities.

2 Building Quality of Service Behavior from Elementary QoS Models

The goal of the implementation architecture presented in this paper is to offer elementary QoS modules for the Internet Protocol stacks (v4 and v6), which can be combined and linked together to common QoS elements, like traffic shaper,

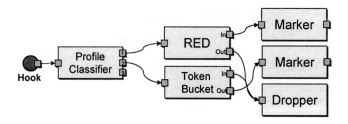

Fig. 1. Example of a QoS behavior build on basic elements

token bucket, classifier, etc. The suite also offers a variety of queues and scheduling mechanisms like priority queueing, weighted fair queueing, round robin, etc. In the following, these QoS elements are called *Behavior Elements*. In the next section a more detailed classification is given.

The main principle in building this implementation architecture was the possibility to build new QoS behavior quickly from the existing pool of elementary Behavior Elements. This can mostly be done by varying the elementary modules and connecting them in a special manner. The following example should motivate the architecture of the proposed model:

A token bucket is a widely used model to meter a certain network flow and to monitor its conformance to Service Level Agreements (SLAs). Traditionally a token bucket is metering incoming flows to their conformance. If the negotiated rate is not exceeded, the packets will be forwarded – otherwise they will be discarded. This method of metering a flow is well known, but in some scenarios (i.e. AF PHB in DiffServ networks) the packets should not be discarded. They should either be marked with a lower priority and enqueued in an alternate queue which is served with a lower priority. Or in other scenarios, two token buckets should be combined to control the peak rate and the average rate or the token bucket should be packet-oriented instead of byte-oriented. With existing implementations, new token bucket modules, a marker a priority queueing module, etc. have to be implemented.

Secondly, it is always a problem to integrate QoS elements into the right position within the protocol stack. For instance, it is important whether a token bucket is working on the IP layer – before the routing has been done – or at the output queue of a certain interface. In the first case all packets forwarded by IP will be considered in the token bucket meter, whereas in the latter case, only the packets leaving on one interface will be metered. In a third case only the packets leaving a host should be considered. It is obvious that in a protocol like IP, a lot of possible places to integrate QoS behavior can be identified (as shown in section 2.1).

As a result of this, a fast development of modules for investigating new network behavior, as i.e. new QoS behaviors, is very time-consuming with existing operating systems, since they have mostly implementations for specific QoS architectures [BSSo00,AlSK99]. Each time, new implementations have to be developed. The reuse of existing implementations is very complex.

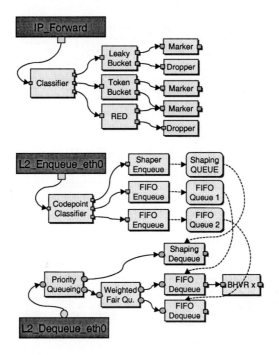

Fig. 2. Example configuration of basic QoS Behavior Elements

The presented modular architecture with its elementary QoS modules, and the individual linking of them, would solve these problems and allow to build immediately any QoS behavior for an Internet router or host – mostly without implementing new models. The existing pool of elementary QoS behaviors, and its smart integration into the Linux implementation of Internet Protocol, offer on the one hand the examination of real IP behavior and on the other hand the possibility to build and evaluate rapidly new QoS behavior in real systems.

In the next few sections the basic architecture of the KIDS (Karlsruhe Implementation of Differentiated Services) QoS architecture will be presented. First the principle of Hooks is explained, which are strategic points for including QoS elements into protocol stacks. Subsequently the five different kinds of Behavior Elements and rules to concatenate them will be introduced. It should be mentioned that the presented KIDS architecture do not need any modification within a standard Linux Kernel. Standard interfaces like traffic control and Netfilter have been chosen to create the concept of KIDS' Hooks.

Figure 2 illustrates the KIDS architecture by an example. Three service classes should be distinguished: A Premium class, offering a high priority service with low delay. The flows of the Premium class will be metered by a Leaky Bucket and shaped at the output interface. A second class should offer a better service than Best Effort with a statistical guarantee of bandwidth. This will be achieved by a weighted fair queueing scheduler. The metering will be done by

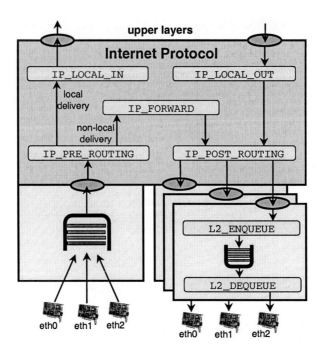

Fig. 3. Hooks for QoS behavior in the Linux OS

a token bucket. Non conforming packets will not be discarded, but degraded to the Best Effort service, which builds the third service class. The classification to the three service classes is done by a multifield classifier. This example is a possible implementation of the well-known "Two bit architecture', which is described in details in [JaNZ99]. To keep the example simple, only the Layer2-Hooks of interface eth0 and IP-Forward are shown.

2.1 Protocol Hooks

When Qualtiy of Service behavior should be introduced into an existing protocol, one basic problem is the point at which the protocol is extended with the new Behavior Elements. Regarding the Internet Protocol, five strategic points can be identified. In the following this points will be called *Hooks*. They can be distinguished in the set of packets passing the point, e.g. the IP_Post_Routing-Hook represents the set of all packets leaving the IP node on an interface - whether they have been forwarded, or created from the host:

- IP_PRE_ROUTING: All packets arriving on a network interface will pass this hook before routing is processed. Consequently all incoming packets will be processed by the Behaviors attached to this hook.
- IP_Local_In: All packets arriving for the local host will pass this hook after the routing is processed and before they leave IP for the upper protocols.

An Open Architecture for Evaluating QoS Mechanisms in Software Routers   121

- IP_Forward: All forwarded packets will pass this hook after the routing. Consequently it is the right point to perfom QoS mechanisms on routed packets.
- IP_Local_Out: This Hook is suitable for all packets leaving from upper layers, before routing is processed.
- IP_Post_Routing: The last Hook can be used to perfom any action on all packets leaving the host on a network interface card, whether they are forwarded or created from upper protocol layers.

For each network interface, two additional Hooks can be identified: L2_Enqueue_xx and L2_Dequeue_xx, where xx is the name of the network interface card in the Linux OS. They are located around the output queue(s) of each network interface card. These Hooks are the right point to add specific behaviors operating on outgoing queues, like priority queueing, traffic shaping, etc. (refer Fig. 2.1).

In this paper we only focus on the Internet Protocol Version 4 and the underlying layer. In other protocols like IPv6, TCP, UDP, etc., also Hooks are integrated yet or can be added easily at strategic places to integrate QoS behavior in the network stack.

As described above, a Hook is a place within a protocol where QoS behavior can be added. The Behavior Elements included at such a hook are elementary models offering a certain behavior. They will be described in the following section.

2.2   Behavior Elements

A Behavior Element (BE) is comparable to a black box, which offers a specific basic *behavior*. A BE consists of one in-gate, n out-gates and a certain processing behavior inside. At the in-gate a packet enters the box and receives a certain manipulation inside the module. Dependent on the calculation within the box, the packet leaves on a certain out-gate. Behaviors can be concatenated after each other. Consequently, the treatment a packet receives within a BE decides which way it will proceed and which quality it receives.

Two kinds of gates (interfaces) of Behavior Elements and Hooks can be distinguished: *packet-gates* (abbreviated as □) and *non-packet-gates* (○). The main difference between them is, that between two packet-gates IP-packets are exchanged, and between non-packet-gates only messages to request (dequeue) packets are exchanged.

The principle rule in the KIDS architecture is, that only gates from the same kind can be connected to each other. The two different types of gates and the interaction between them are described more detailed in section 2.3.

As mentiones before five kinds of Behavior Elements can be distinguished (ref. Fig. 2.2). In the following they will be introduced in detail:

- **(conventional) Behaviors** (BHVR) are elementary QoS elements which operate on IP packets. As shown in Figure 2.2, a Behavior has only one in-gate

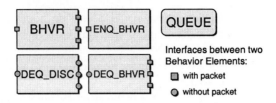

Fig. 4. Five different classes of behavior elements

and up to n out-gates, where n depends on the particular Behavior. E.g. a Token Bucket has two out-gates, one for conform packets and one for non-conform packet. Behaviors can be interconnected between one another without fulfilling other requirements. Example Behaviors are Token Bucket, Marker, Dropper, Classifier, Random Early Detection (RED), etc.

- **Queue (QUEUE):** Queues are well known packet queues. Packets can only be enqueued and dequeued with the appropriate kinds of Enqueue and Dequeue Behaviors. Several types of Queues have been implemented in KIDS, e.g. Fifo-Queue, Shaping-Queue, EDF-Queue, etc.
- **Enqueue Behavior (ENQ_BHVR)** are specialized Behavior Elements for enqueueing a packet into a queue. The queue is identified by its name and an according Enqueue Behavior should be used. One special characteristic of an Enqueue Behavior is the missing out-gate. Whether the packet is inserted into the queue, or it has to be dropped. Enqueue Behaviors can be connected to out-gates of any Behavior module. The detailed procedure of exchanging messages and packets between Behavior Elements is described in section 2.3.
- **Dequeue Behavior (DEQ_BHVR)** can be used to dequeue a packet from a certain queue. E.g. a Fifo_Dequeue module removes the first packet from the named queue and sends it to its out-gate. Dequeue Behavior modules can only be connected to an o-gate of a L2_Dequeue-Hook or a Dequeue-Discipline. After a Dequeue Behavior all kinds of Behaviors can be connected to the packet-gate.
- **Dequeue Discipline (DEQ_DISC):** A Dequeue Discipline is a strategy to choose the next Dequeue Behavior for serving a queue. Dequeue Disciplines are playing a very important role in reaching different service classes within a network. Examples for Dequeue Disciplines are Priority Queueing, Weighted Fair Queueing, Round Robin, etc.

2.3 Interactions between Behavior Elements

With the just presented five types a simple mesh of Behavior Elements can be built and connected to Hooks. But it is important to understand how a packet traverses this mesh and which interactions can occur between the Behavior Elements. The following section describes the sequence of events on two example concatenations of Behavior Elements.

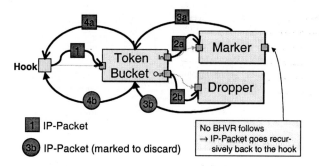

Fig. 5. Interactions on a □-junction between Behavior Elements

Interactions between packet-gates (□-junctions):

Several Behavior Elements with packet-gates can be arranged into one new model to create a new QoS Behavior. At the connections between the □-gates, IP packets are exchanged. Figure 2.3 shows an example. The Hook sends an IP packet to the Token Bucket-Behavior. When the module has completed its operations on the packet, it will be send further, when a module is connected on the dedicated out-gate. That means in the example, that a SLA-conform packet (case a) will leave on the In-gate (In-Profile) of the Token Bucket; If it is not conform the packet will leave on the Out-gate to the Dropper. If no module is connected to a Behavior on the dedicated port, or the Behavior has no port, the packet will be sent back to the previous module where the packet came from.

One can see, that a packet first traverses a chain of Behaviors and then goes recursively back to the Hook, where the normal protocol processing will be continued.

This is the normal procedure, but there are two possible exceptions. The first is when a packet has reached an Enqueue Behavior, which will enqueue it into a Queue. The second exception is a Dropper that marks the packet for discarding. In both cases, the modules will send back an IP packet to the hook, but with a return code like *Packet-Enqueued* or *Discard-Packet*. The discarding of a packet will be done in the Hook, because all Behaviors between the Hook and the Dropper have to be informed about the loss of the packet. E.g. a token bucket has to put back the tokens of the discarded packet into the bucket, because it has not consumed them.

Interactions at non-packet-gates (o-junctions):

On o-junctions (between Dequeue Disciplines or between Dequeue Disciplines and Dequeue Behaviors) no packets will be exchanged. Dequeue Disciplines will first decide which Dequeue Behavior will be asked to dequeue a packet from a queue. This mechanism will be triggered from the Dequeue-Hook sending out a Request-Packet-message to the first dequeue discipline or directly to a Dequeue Behavior, if no scheduling algorithm is used.

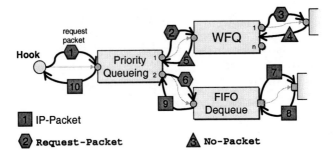

Fig. 6. Interactions on o-junctions between Dequeue Disciplines and Dequeue-Behaviors

A Dequeue Discipline decides on which of its out-gates the Packet-Request will proceed. Any combination of Dequeue Disciplines can be built, but finally a Dequeue Behavior has to be called. In Figure 2.3, the Priority Queueing module first calls on the gate with the highest priority. The Dequeue Discipline connected to that gate proceeds with its own scheduling mechanism. In Figure 2.3 the Weighted Fair Queueing module proceeds on gate 1.

Each possible chain of Dequeue Disciplines has to conclude with a Dequeue Discipline which executes the Packet-Request by dequeueing a packet from the Queue. On success and if a Behavior is connected, the packet will proceed on the □-gate of the Dequeue Behavior. This follows the same procedure as described previously about □-junctions.

When the Dequeue Behavior receives back the packet, it is sent recursively back trough the Dequeue Disciplines to the Dequeue-Hook.

If no packet can be dequeued, the Dequeue Behavior returns a No_Packet-ID to the previous Dequeue Discipline, indicating that the dequeue-operation failed. Then, the Dequeue Discipline can choose -- according to its algorithm -- another o-gate to request a packet or it returns the No_Packet-ID to its predecessor. On a successful dequeue, the hook starts the transmission of the packet on the network interface.

As mentioned above, the dequeueing is triggered by the Dequeue-Hook. Normally, Packet-Requests will be initiated by the network interface, when it has finished the transmission of the previous packet and is now able to transmit the next packet. But if the interface has been idle for a while, it would not start a new request. Therefore the Enqueue-Hook can initiate a Packet_Request, when he just inserted a packet into one of the output queues of the interface. Such an indication only starts a Packet_Request, when the NIC is in idle state.

2.4 Specification Language

To manage the creation, destroying and concatenation of Behavior Elements a special language has been developed. With this commands, arbitrary Quality of

Service Behavior can be build from existing elementary modules. The configuration can be made manually be using the specification language or by using a graphical user interface wich allows the management of the QoS Behavior Elements by simple drag 'n drop. The GUI is also able to load the current KIDS configuration from a router and able to submit the change that has been done by the administrator.

In the following the commands are shortly described:

CREATE `bhvr_class bhvr_type bhvr_name` DATA {*private data*}* END
Creates a Behavior (QUEUE, BHVR, ENQ_BHVR, DEQ_BHVR, DEQ_DISC) from a certain type (Dropper, Fifo_Queue, Marker, etc.) with the given `bhvr_name`. The first three parameters are equal to all possible Behaviors. This is followed by an optional part, where Behavior specific parameters can be set, as for example RATE = 64.000kbps.

CONNECT `bhvr_class1 bhvr_name1` TO {(`bhvr_class2 bhvr_name2 gate`) | (HOOK `hook_name`)}+ END
Appends the Behavior `bhvr_name1` after Behavior `bhvr_name2` or after the given Hook. Because the second Behavior can have more than one output gate, it has to be specified on which gate the second Behavior Element should be connected.

CHANGE `bhvr_class bhvr_name` DATA (*private data to be changed*) END
Changes the private parameters of the given Behavior. Only the listed parameters will be changed.

DISCONNECT `bhvr_class bhvr_name gate` END
Removes the connection, that leaves on the gate of the Behavior.

REMOVE `bhvr_class bhvr_name` END
Removes a Behavior from the Kernel. This is only possible, if the Behavior has no more connections.

3 Performance Evaluation

This part shows that the overhead of fine granular QoS modules is not reasonable higher than with *monolithic* QoS implementations. It is obvious that this architecture has, due to the simple and elementary QoS models, a certain overhead to switch between consecutive Behavior Elements. As measurements have shown the handover from one Behavior Element to the next takes only 30 CPU cycles at an average. The actual value varies in cause of caching and memory access. This shows that it is not very time consuming to build a QoS behavior from many basic elements instead of unsing monolithic and unflexible QoS implementations.

Another performance problem of standard PC hardware is the missing of precise clocks and timers to offer a high-resolution traffic shaping in software routers. Normally, standard PC hardware is only able to shape traffic with a precision of 100 Hz to 1000 Hz. In [RiWe00] some new techniques have been presented to realize kernel timers up to a resolution of 1.000.000 Hz. This timer, called *UKA-APIC-Timer* developed at the University of Karlsruhe has been used in the KIDS architecture to realize a high precise Earliest Deadline First queue. This queue can be combined with some other basic modules to a Leaky Bucket or a traffic shaper.

4 Conclusion

In this paper, an implementation architecture for the simple and rapid creation of Quality of Service mechanisms has been presented. The modules can easily be inserted into the TCP/IP stack of the Linux OS without the need for any kernel modification.

The creation and evaluation of Quality of Service mechanisms can easily be done by using the elementary QoS models and concatenating them in the desired way. Common models for queue scheduling (priority queueing, weighted fair queueing, round robin, etc.), metering (token and leaky bucket), classifying (multi-header-field, DS-codepoint, etc.) and forming of data flows are provided.

It is planned to publish the implementation architecture soon on our web site for the public. That should everybody offer the possibility to evaluate our modules and to increase the pool of available QoS Behaviors.

References

[AlSK99]   W. Almesberger, J. Salim and A. Kuznetsov. Differentiated Services on Linux. draft-almesberger-wajhak-diffserv-linux-00.txt, February 1999. Internet Draft.

[BBCD+98]   S. Blake, D. Black, M. Carlson, E. Davies, Z. Wang and W. Weiss. An Architecture for Differentiated Services. RFC 2475, December 1998.

[BrCS94]   R. Braden, D. Clark and S. Shenker. Integrated Services in the Internet Architecture: an Overview. RFC 1633, June 1994.

[BSSo00]   T. Braun, M. Scheidegger, G. Stattenberger and other. A Linux Implementation of a Differentiated Services Router. Proceedings of Networks and Services for Information Society (INTERWORKING'2000), October 2000.

[FeHu98]   P. Ferguson and G. Huston. *Quality of Service*. Wiley, 1998.

[JaNZ99]   V. Jacobson, K. Nichols and L. Zhang. A Two-bit Differentiated Services Architecture for the Internet. RFC 2638, July 1999.

[RiWe00]   H. Ritter and K. Wehrle. Traffic Shaping in ATM and IP networks. In *Proceedings of Intern. Conference on ATM*, Heidelberg, June 2000.

[WeBR01]   K. Wehrle, R. Bless and H. Ritter. Mechanisms for Available Rate Usage in ATM and Differentiated Services Networks. Proceedings of the 4th IEEE International Conference on ATM and High Speed Intelligent Internet (ICATM 2001), April 2001.

Measurement-Based IP Transport Resource Manager Demonstrator

Vilho Räisänen

P.O. Box 407, Nokia Research Center, FIN-00045 Nokia Group, Finland.
Vilho.Raisanen@Nokia.com

Abstract. A demonstrator of IP QoS measurement architecture is described that can be used as a part of QoS transport resource management. The use of measurements in QoS control of voice over IP (VoIP) is discussed using up-to-date concepts in standardization.

Introduction

Management of Quality of Service (QoS) in Internet Protocol (IP) networks is a prerequisite for commercial-grade IP telephony in the Internet. Voice over IP and packet video conferencing are examples of services with strict requirements for packet loss and end-to-end delays, as well as temporal correlations of these. Media stream QoS management in an IP domain can be handled with overprovisioning, but higher resource utilization can be achieved by using an IP QoS mechanism such as Differentiated Services (DiffServ) [1] or Multi-Protocol Label Switching (MPLS) [2]. Also combinations of IP QoS schemes are possible [3].

Complexity of QoS management increases greatly when the end-to-end media stream path spans multiple (logical) IP domains. The domains may be administered by different operators deploying different IP QoS technologies. Thus, call setup signalling for IP telephony requires setting up of a media bearer in a heterogeneous QoS environment. Schemes for IP transport-level signalling in heterogeneous IP QoS environment have been developed both within the academic research community and the Internet Engineering Task Force (IETF) [4]. The schemes typically involve Resource Reservation Protocol (RSVP) [5] negotiation between edge routers of IP domains. Examples of interworking issues are mapping of DiffServ classes on one domain to those of another, DiffServ/IntServ mapping and so forth.

ETSI TIPHON QoS Control Model

From commercial viewpoint, IP telephony service providers may not necessarily themselves implement actual IP transport services. This must be taken into account in designing QoS negotiation. An approach discussed within the European Telecommunication Standardization Institute (ETSI) IP telephony project TIPHON [6] is to separate service level (e.g., IP telephony service provider functionality

IPTSP) from IP transport level, and to define signalling flows firstly between two service domains, secondly between two transport domains, and thirdly between a service domain and a transport domain (Fig. 1).

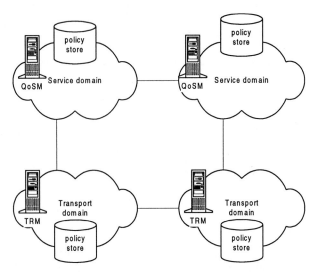

Fig. 1. The simplest layered interworking case considered in ETSI TIPHON WG5. Both service and transport domains include policy elements.

Multiple interworking scenarios between transport domains and service domains have been considered. Interested reader is referred to [7].

QoS control is handled by interworking of QoS manager (QoSM) in the service layer and a transport manager (TRM) in the IP transport layer. QoSM maps requested telephone quality class into transport-level QoS bearer characteristics such as bandwidth and limits for delay and packet loss percentage. In this mapping, service level policy aspects (e.g., what kind of QoS class the subscriber is entitled to) can be taken into account. TRM provides an authoritative answer on availability of requested bearer, and - in the case configures edge routers and actual transport layer (routers) if necessary. TRM operation, in turn, is subject to transport-level policy.

Rôle of Measurements in Transport Level Resource Management

The TIPHON framework is agnostic with respect to IP QoS mechanisms, and thus - for example – a TRM is preferably able to cooperate with any IP QoS mechanism. In what follows, measurement-based implementation of transport resource estimation availability is described. Measurements of media stream QoS at the IP level are independent of the QoS enhancements used, and thus fulfill the requirement of portability of the QoS control architecture. (Naturally the measurement system must have relevant interfaces to IP layer, such as capability of setting DSCPs in the measurement packets.) In general, the rôle of measurements in QoS management is illustrated in Fig. 2.

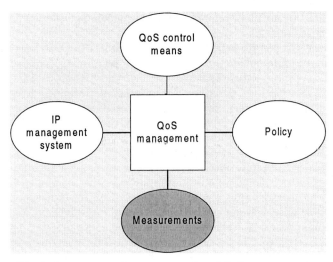

Fig. 2. Components of measurement-based QoS management.

In some cases, measurements might not be needed in present ISP implementations. However, the projected growth of data traffic is fast due to emerging markets and the advent of mobile Internet terminals (GPRS, UMTS). (These very same reasons have also driven the adoption of IPv6 with its 128-bit address space in the 3rd generation mobile networks [8].) Moreover, separation of service and transport operations and easy creation of services will make the future market highly competitive. In such a situation, high utilization levels of transport networks may be expected to be desirable, and statistical multiplexing methods based on prioritization of traffic (DiffServ), possibly used together with protocol support for traffic engineering (e.g., MPLS), may become commonplace. If interactive real-time traffic – such as VoIP – is a major source of revenue for the operator, measurement-based QoS control is a potential enabling technology for efficient use of resources.

Measurements

There are basically two ways of performing media stream QoS measurement for VoIP:
1. Passive monitoring of media streams.
2. Active measurements by transmitting test traffic.

IETF [9], ITU-T [10] and ETSI TIPHON [11] have studied both major possibilities. Also a hybrid form is possible, in which packets of user traffic are piggybacked with measurement information [12].

The pros and cons of different measurements have been compared elsewhere [13]. Quickly summarized, active measurements provide a more controlled test than passive monitoring does, but at a price of extra bandwidth. In high-capacity core networks the bandwidth of a few measurement streams for most important routes is not a hindrance, if more real-time flows can be admitted into the domain due to this.

Passive monitoring does not require extra bandwidth, but encryption and privacy issues may cause problems.

A large-scale test bed for measurements is the NLANR, which uses monitoring of network control information in addition of performing passive and active measurements [14]. Another examples include the PingER [15] and AT&T's prototype [16]. Stemm *et al.* describe an architecture performing passive application-level monitoring of network delivery quality and storing the results in local centralized repositories [17]. Elek *et al.* discuss caching for the results of active measurements between the sender and the receiver immediately prior to transmission of media stream using a short active probe [18].

These examples illustrate that it is beneficial to build a flexible measurement analysis infrastructure including a central repository with the ability to use multiple data sources. At the same time, a goal of managing QoS for IP telephony specifically makes it possible to specify in more detail the type of measurements used.

In the Nokia Research Center QoS control demonstrator, the QoS modelling functionality is isolated from actual measurement infrastructure. For demonstrator purposes, active measurements [19,20] have been used due to ease of implementation as well as portability. However, any method of obtaining measurement information can be used together with the QoS modeller functionality.

The implementation presented below makes measurements on a number of routes, the QoS situation of which is supervised. This paradigm could be implemented with passive monitoring by correlating measurement data from multiple sources. In contrast to [18], the measurements are not bound to incoming calls, but are performed periodically in the network to assess network transport QoS. This approach provides more useful management information, and also speeds up the call admission process.

In a production network implementation, active measurements have limits for scalability. Active measurements would most likely be made on major VoIP routes, and information thusly obtained be complemented with data from passive measurements as necessary.

Demonstrator Architecture

The general architecture of the demonstrator is shown in Fig. 3, and consists of QoS agent (QA), measurement control points (CPs) and actual measurement points. CPs provide an abstraction layer between QA and the measurement infrastructure, making possible free choice of measurement methods according to the need. CPs communicate per-route QoS data to QA, which performs QoS modelling as explained below. Based on the reports, QA can then provide alarms and warnings to other network elements. In the Figure, communication with a call processing server (CPS) is shown by way of an example.

The actual call admission control (CAC) algorithms are not discussed here. The measurement architecture lends itself to different usage scenarios, as discussed below.

QA controls the measurement architecture, configuring measurement parameters to CPs. The parameters include the frequency of measurements and parameters to be used in a measurement. Once a CP has been configured with the parameters, it schedules measurements automatically and reports the results back to QA.

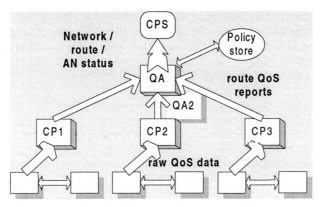

Fig. 3. Nokia Research Center demonstrator architecture. QA makes use of a policy store, and robustness is implemented with a backup QA ("QA2").

QA may also suspend measurements in a CP, activate or re-activate suspended CPs. Robustness has been implemented in the form of backup QA, into which all reports are automatically redirected if primary QA cannot be contacted. QA maintains soft state for CPs: if reports have not been received for a predefined period of time, CP is marked as temporarily unavailable.

All network elements in the demonstrator have been implemented on standard RedHat Linux / i686 platform for ease of prototyping. All software has been coded with C and standard socket interface programming language using the development tools available in normal RedHat Linux distribution. All network elements support automatic detection of the protocol version.

QoS Agent

The QoS agent implements the following functionalities:
1) Management of measurement infrastructure (add/delete measurement points, suspend/continue measurements in a CP).
2) Configure measurement parameters in CPs.
3) Receive QoS reports from CPs.
4) Perform QoS modelling based on QoS reports.
5) Communicate QoS situation to other network elements.
6) Read QA configuration from a policy store.

Management of CPs is based on socket connections, which are used for configuring and reporting measurement results.

QoS Information Format

The details of measurement parameter configuration depend on the measurement method used; a case study specific to active measurements is described below. Common to all methods is the frequency of measurement and the location of measurements: CP needs not be co-located with actual measurement entities. Finally, the form of reported QoS data may also be configured. For instance, QA may configure a number of QoS classes into CPs, with CP transmitting only QoS classification to save bandwidth. In this case, QoS class limits are defined for delay, delay jitter, packet loss percentage, and loss correlation. After a measurement, a CP returns a classification for each of the four quantities.

Alternatively, a "full" QoS report may be transmitted, an example format being shown in Fig. 4.

> Delay (in ms).
> Delay jitter (in ms).
> Packet loss percentage (in units of 1/1000).
> Average loss correlation.
> 90%, 95%, 99% quintiles of delay distribution.

Fig. 4. An example of QoS report format.

QoS Analysis

QoS modelling used in the demonstrator is based on computing per-route QoS averages. In computing the averages, discrete time exponential moving average similar to one used with the Real-time Transport Protocol (RTP) [21] is used, i.e.

(1) $\quad E_t = \alpha X_t + (1-\alpha) E_{t-1},$

where E_t is the value of the estimate at time t, X_t is the value of the observable at time t and α is a coefficient that sets the time scale of moving average. By computing estimators with multiple values of α, different timescales can be covered. Exponential averaging has been observed to correspond to temporal development of listeners' perception of QoS, termed the "recency effect" [12].

Monitoring User Interface to QoS Data

QA has multiple ways of representing the data, depending on the format of QoS reports transmitted from CPs to QA. When only a classification of QoS characteristics are reported to QA, a history view of QoS class can be provided for a monitoring user (Fig. 5):

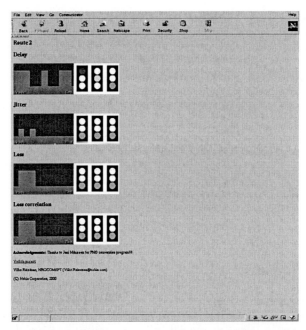

Fig. 5. A history view on QoS class on a route through an HTTP interface. Delay, delay jitter, loss percent, and loss correlation histories shown. Traffic lights indicate QoS averages for three different values of parameter α.

Another - not necessarily mutually exclusive- possible user interface, delay quintiles can be saved and plotted to study delay variations (Fig. 6).

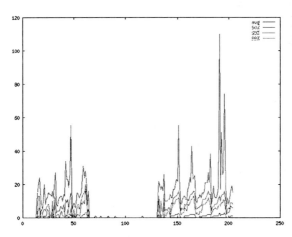

Fig. 6. An example of delay quintiles as a function of time. Red: 50%, green: 90%, blue: 95% and magenta: 99% quintile for a measured route.

Interface to Other Network Elements

In its communication with other network element (if present), QA can operate in several modes. Three examples are shown below:
1) "*Intelligent filter*": QA passes per-QoS averages to a network element with indigenous QoS modelling intelligence.
2) "*Baywatch*": provide warning and alerts to another network element based on configured profile.
3) "*TRM*": Provide authoritative answers to VoIP QoS bearer requests.

The configuration of QA is based on a configuration file, which could be viewed as a policy store in IETF parlance [22]. An external network element, such as a CPS or network management system (NMS) can configure QA by modifying the policy store contents and instructing QA to reread its configuration file. This in turn causes reconfiguration of CPs and - eventually - measurement points.

Measurement Infrastructure

The present demonstrator measurement architecture is based on active application-level measurements for VoIP [13,19]. In practical terms this means that QA configures to CPs parameters corresponding to some realistic codec such as the 3GPP standard codec AMR (Adaptive multi-rate codec), GSM, G.723.1, or G.711. Other parameters include the duration of the measurement and the temporal separation of the measurements.

Off-the-shelf i686 PCs and NICs are used as measurement hosts. A measurement host may participate to measurements on several routes. For each route, a measuring process is created in the host. In our tests, even a low-end 100 MHz Pentium PC could handle up to 30 simultaneous measurement streams. The number of streams handled by a single measurement point can be expected to be much smaller than this.

Both one-way and round-trip measurements can be supported by measurement packet format [13] (Fig. 7):

TS1	TS2	TS3	TS4	S1	S2

Fig. 7. Format of the measurement packet. Four timestamps (TS1-TS4) and two sequence number (S1, S2) are used.

The measurement program used by us performs one-way delay measurements for hosts running NTP or other suitable synchronization mechanism. IP protocol version is automatically detected, and IPv4 is used if IPv6 is not supported at both ends. The measurement program also supports setting of flow label and traffic class in IPv6. The latter is useful for verifying DiffServ performance.

Summary

A demonstrator implementation of measurement-based transport QoS monitoring system was described that can be used for managing QoS resources. The connections of the demonstrator with actual standardization topics were discussed.

The efficient utilization of information collected by the demonstrator is still a research topic, especially as to use with different IP QoS mechanisms. Thus far the working assumption has been admission control on the service level. Measurement-based robust QoS control at transport level is for further study at present.

References

[1] Cf. Differentiated Services (DiffServ) working group home page at IETF server, http://www.ietf.org.
[2] Cf. Multi-Protocol Label Switching (MPLS) working group home page at IETF server.
[3] Cf. Integrated Services over Specific Link Layers (ISSL) working group home page at IETF server.
[4] Cf. IP Telephony (IPTEL) and Session Initiation Protocol (SIP) working group home pages at IETF server.
[5] Cf. Integrated Services (IntServ) working group home page at IETF server.
[6] Cf. ETSI EP TIPHON home page at http://www.etsi.org/tiphon.
[7] ETSI standard TIPHON/TS 101329-3.
[8] Cf. 3rd generation partnership project (3GPP) home page at http://www.3gpp.org.
[9] Cf. IP performance measurements (IPPM) working group home page at IETF server.
[10] Cf. ITU-T recommendation I.380.
[11] Cf. ETSI standard TIPHON/TS 101329-5.
[12] J. Jormakka and K. Heikkinen, *QoS/GOS parameter definitions and measurement in IP/ATM networks*, in Proc. "Quality of Future Internet Services", Berlin, Germany, September 2000, Lecture Notes in Computer Science 1922, Springer, Berlin, 2000.
[13] V. Räisänen: *Measuring transport QoS for VoIP*, submitted.
[14] T. McGregor, H-W. Braun, and J. Brown, *The NLANR network analysis infrastructure*, IEEE Communications magazine **38**, p. 122 ff., May 2000.
[15] W. Matthews and L. Cottrell, *The PingER project: active Internet performance monitoring for the HENP community*, ibid., p. 130 ff.
[16] R. Cáceres et al., *Measurement and analysis of IP network usage and behavior*, ibid., p. 144 ff.
[17] M. Stemm, R. Katz, and S. Seshan, A *network measurement architecture for adaptive applications*, in Proc. Infocom'00, p.285 ff, Tel Aviv, Israel, IEEE, March 2000.
[18] V. Elek, G. Karlsson, and R. Rönngren, *Admission control based on end-to-end measurements*, in Proc. Infocom'00, p.623 ff, Tel Aviv, Israel, IEEE, March 2000.

[19] V. Räisänen and J. Rosti, *Application-level IP measurements for multimedia,* in *Proc. IWQoS'00*, p. 170 ff., Pittsburgh, U.S.A., June 2000, IEEE.
[20] V. Räisänen and G. Grotefeld, *Network performance measurement for periodic streams*, IETF IPPM WG draft draft-ietf-ippm-npmps-04.txt (work in progress).
[21] Cf. Schulzrinne *et al, the Real-Time Transport Protocol (RTP)*, IETF RFC 1889.
[22] Cf. Policy-related RFCs, e.g. RFC 2748 at IETF home page.

Packet-Size Based Queuing Algorithm for QoS Support

Myung C. Choi[1], Henry L. Owen[2], and Joachim Sokol[3]

[1] ITTL/CND, Georgia Tech Research Institute, Atlanta, Georgia 30332-0855, USA
myung.choi@gtri.gatech.edu
[2] ECE, Georgia Institute of Technology, Atlanta, Georgia 30332-0250, USA
henry.owen@eecom.gatech.edu
[3] Siemens AG, ZT IK 2, D-81730 Munich, Germany
joachim.sokol@mchp.siemens.de

Abstract. Quality of Service (QoS) support for data flows typically requires the use of a scheduling algorithm. For better support of QoS, the scheduling algorithm should be able to provide an acceptable level of performance as well as the functionality required for different service models such as Integrated Services, Differentiated Services, and ATM. This functionality includes a per-flow and an aggregated flow scheduling, shaping, priority, etc. In this paper, we described a new scheduling algorithm, Packet-size Based Queuing (PBQ) that supports the functionality in ways that are more efficient and flexible. The PBQ's scheduling discipline with a hierarchical structure provides a hierarchical class based flow management, hierarchical shaping, and priority scheme all at the same time. The hierarchical shaping functionality, unlike other traditional shaping algorithms, shapes a group of flows (or classes) in an aggregated way while providing a proper scheduling to each flow (or class) in the group locally. The priority scheme in PBQ helps to reduce the worst delay bound, which becomes larger when priority index calculation is simplified. With all this functionality, the resulting scheduling algorithm has as a primary goal implementation simplicity while still maintaining an acceptable level of performance.

1 Introduction

A packet scheduling discipline is necessary in network nodes in order to support QoS when there are insufficient resources for all connections all the time. QoS capable approaches such as Integrated Services (IntServ) and Differentiated Services (DiffServ) for IP based networks over legacy link layer technologies and ATM QoS classes require the use of packet schedulers. IntServ requires a token bucket shaping functionality with a delay bound while DiffServ needs a class based structure with different priorities. ATM requires a leaky bucket shaping functionality. There already exist a number of scheduling algorithms such as weighted fair queuing (WFQ), worst-case weighted fair queuing (WF^2Q) [6], Self-clocked fair queuing (SCFQ) [5] and Start-time fair queuing (SFQ) [4]. These existing scheduling algorithms provide "proper" fairness and delay bounds to each connection. However, WFQ and WF^2Q are known to be too complicated to implement in some environments [5]. [11] shows how WFQ implementation in FreeBSD systems (similar UNIX based system to LINUX) are unusable. Simple versions of fair queuing (FQ) such as SCFQ and SFQ reduced the complexity through simpler virtual time calculations. However, these approaches still include a monotonically increasing virtual time and a time stamping of every packet, which is not always feasible. SCFQ and SFQ require floating-point calculations, which makes them difficult to implement in some systems (for example LINUX based systems). One of the commonly implemented packet schedulers, class based queuing (CBQ) [3], uses weighted round robin (WRR) for its scheduling discipline. This does not require a floating point calculation or a time stamp. However, WRR cannot provide fairness and delay bounds to each and every connection. The lack of an easy to implement (from the standpoint of necessary computational resources) scheduling algorithm motivated us to investigate a new algorithm.

In this paper, a new scheduling discipline, Packet-size Base Queuing (PBQ) is described along with its associated fairness and delay bound analysis. PBQ is then enhanced to provide hierarchical shaping and priority functionality. Furthermore, a hierarchical class based structure, which provides an efficient way of grouping flows with similar characteristics, is established by applying the framework proposed in [1]. PBQ is feasibly implemented in environments such as LINUX based systems while still maintaining an acceptable performance level.

This paper is organized as follows. Section 2 describes the PBQ algorithm and its fairness and delay bound. Hierarchical class structure, hierarchical shaping, and priority schemes are discussed in section 3. Finally, section 4 concludes the paper.

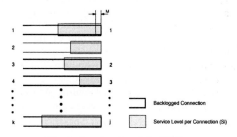

Fig. 1. Service Level in PBQ

2 Packet-Size Based Queuing: A New Scheduling Algorithm

In [2], a new scheduling discipline, PBQ was introduced. This scheduling discipline is based upon the packet size and the number of backlogged connections. PBQ uses a service level as a priority index as depicted in Fig. 1 (M in Fig. 1 is a number of bytes, which varies between the smallest packet size and a maximum packet size). The service level is updated after every packet is serviced. In order to present the details of the scheduling algorithm, a few definitions are necessary. Let S_j be the service level of connection j. Since some kernels do not support floating-point calculations, the weight w for each connection is scaled up by a factor of 100 so that integer calculations may be used. This results in a service level,

$$S_j = Jw_j MD, \quad w_j = (0,100] \tag{1}$$

where D is a precision control value. We set $D = 10^n$ where n is the number of decimal places desired to be used in the relative weights for each connection. Let C be a set of connections which have been set up; $C = \{C_1, C_2, \ldots, C_K\}$. Let B be a set of backlogged connections; $B=\{B_1, B_2, \ldots, B_J\}$ where $J \leq K$. Let E be the set of eligible connections; $E = \{E_1, E_2, \ldots, E_I\}$ where $I \leq J \leq K$. Any connection j in B becomes a member of connection set E if $j = \{j \mid S_j \geq Pksize_j*100*D\}$. As in the service level calculation, the packet size is also multiplied by 100 and the precision control factor D. The PBQ algorithm is defined as:

1. From the connection set E, choose one connection j; $j = \{j \mid \max_E(S_j)\}$.
2. Send a packet from the connection j. Then update S_j; $S_j = S_j - Pksize_j*100*D$.
3. Go to step 1 until connection set $E = \emptyset$.
4. Update S_j; $\begin{cases} S_j = S_j + JMw_j D & \forall j \in B \\ S_k = S_k + JMw_k D \text{ until } S_k \geq 100 L_{max} D & \forall k \in (C-B) \end{cases}$
5. Repeat 4 until $E \neq \emptyset$.
6. Go to step 1.

Fig. 2. PBQ Algorithm

As shown in Fig. 2, the service level when used as a priority index does not involve any complex calculations, but instead is accomplished with simple integer calculations. The precision control factor gives more granularity to users (i.e. weights < 1% are possible in an integer environment). Using a condition test on the eligible connections E, we find that $|E| \leq |C|$, and $|E|$ decreases as service continues. Thus, the complexity of sorting queues is reduced. If the value of M is too small (i.e. L_{min}), and the packet size is maximum (i.e. L_{max}), then step 4 in Fig. 2 will require more calculations in order to perform the service level update. To reduce the cost of each iteration, a larger value of M may be used. If $M=L_{max}$, then PBQ is oriented toward small size packets. Regardless of the value of M, the fairness bound will not be different. The fairness may be measured using the fairness index, F, as follow [5].

$$F = \left| \frac{W_j(t)}{r_j} - \frac{W_k(t)}{r_k} \right|, \quad j, k \in B \tag{2}$$

where $W_j(t)$ is an amount of service received by connection j before time t, and r_j is an amount of service allocated to the connection j. In [2], a fairness analysis is done for PBQ. The bound of the fairness index for PBQ is:

$$F_{PBFQ} \leq \frac{L_j^{max}}{r_j} + \frac{L_k^{max}}{r_k} \tag{3}$$

In PBQ, a backlogged connection j becomes eligible when $j = \{j \mid S_j \geq Pksize_j*100*D\}$. Let us call this condition a "Mandatory Condition" (MC). Since PBQ operates based on the eligibility condition of flows, more conditions may be added. Let us call the additional conditions "User-Specific Conditions" (USCs). Using the USCs, PBQ may be customized according to the need for users. In a sense that PBQ maintains a credit pool for established connections, one may think that PBQ is similar to the frame-based queuing [10]. However, PBQ requires a very simple unique priority index calculation based on well-known parameters such as packet size and number of backlogged connections. Virtual time, unlike [1,4,5,12], is not used in PBQ so that it is not necessary to implement such as calendar queue for the virtual time [12]. PBQ can rather easily utilize the existing queuing in a system. Therefore, this algorithm is distinct from other existing algorithms.

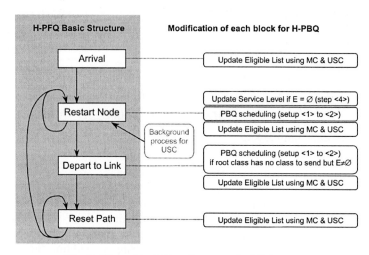

Fig. 3. Hierarchical Class Structure for PBQ

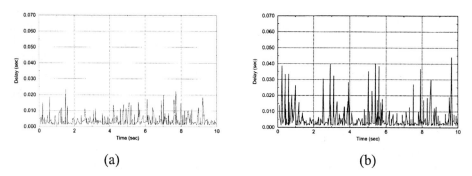

Fig. 4. H-PBQ Comparison: (a) is with constant rate sessions off, and (b) is with constant rate sessions on.

3   Hierarchical Class Structure for PBQ

A hierarchical class based structure has the ability to do flow aggregation, which is useful in QoS capable methods such as DiffServ. Flows with similar characteristics may be aggregated into a class and managed according to the class. Reference [1] provides a framework for hierarchical packet fair queuing (H-PFQ) algorithms. By applying this same framework, hierarchical class based PBQ (H-PBQ) is achieved. The basic processes from [1], ARRIVE, RESTART-NODE, and RESET-PATH are the same. However, since PBQ uses the service level as a priority index, H-PFQ is modified to accommodate the service level update and an eligibility list using Mandatory Condition (MC) and User-Specific Condition (USC). In H-PBQ, start-time, finish-time, and virtual time update are not necessary. The modified H-PFQ for PBQ is depicted in Fig. 3 In Fig. 3, "step <#>" indicates the algorithm step number in Fig. 2. USC may have some background process running. The background process may also invoke the basic process(es). However, the status of the basic process has to be checked before invoking it.

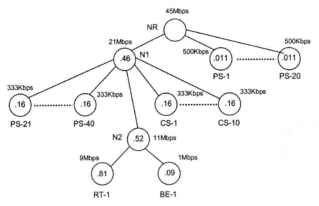

Fig. 5. Example 1 from reference [1]

We simulated the H-PBQ using the same approach as example 1 in [1] (see Fig. 5). The packet size was set to 8000 Kbyte. Poisson sources were transmitting background traffic at more than twice the speed of the guaranteed rate in order to make the Poisson sessions backlogged. As a second type of traffic we had an on/off real time traffic source transmitting at the guaranteed rate. The real-time traffic session had a 25 ms on-period and 75 ms off-period as specified in [1]. As a third traffic source type, we had a constant rate session transmitting at the guaranteed rate. Two simulations were run, one with the constant rate sessions (CS) turned on and one with it off. Fig. 4a shows the delay of the real time traffic using H-PBQ with the constant rate session off. Fig. 4b shows the results with the constant rate session on. When we compare Fig. 4 to Figures 4 and 5 in [1] respectively, we see that H-PBQ yields better results than H-SFQ [1] for both cases (CS on and off). H-SCFQ [1] performs slightly better than H-PBQ when there are CS background traffic flows. The reason for this is as follows. PBQ's service level is used as a service credit that is assigned to each connection. The service credit given to each connection is based on a connection's weight. The service credit used in H-PBQ may be thought of as a similar mechanism to using the finishing time in H-SCFQ [1]. When there is background traffic, a non-backlogged connection may just miss the service level update time and need to wait for the next turn. This may cause H-PBQ to have slightly larger delays than H-SCFQ [1]. H-WFQ and H-WF^{2}Q+[1] perform better than H-PBQ. This is an expected result since PBQ is simplified by estimating the amount of service provided in order to make implementation easy. This is a tradeoff one must make in order to be able to use a packet scheduler in systems such as LINUX based systems.

3.1 Hierarchical Shaping in H-PBQ

As described earlier, PBQ operates based on the eligibility conditions of connection flows. A connection j becomes eligible if the connection is backlogged and meets the mandatory condition (MC) and user-specific conditions (USCs). A shaping functionality may be achieved by adding a shaping condition to the USC. A typically used shaping algorithm is the token bucket shaping algorithm. For USC-shaping, the shaping process is as shown in Fig. 6.

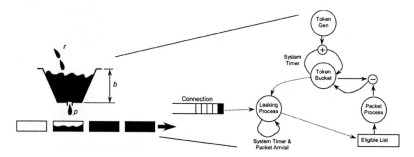

Fig. 6. USC-shaping Process

The USC-shaping functionality for a flow allows a packet to be sent out only when enough of the "leaking fluid from the bucket" is available for the packet. In Fig. 6, the fluid is leaked at a rate p (peak rate). The packet will receive service when a large enough amount of fluid has dropped for the packet. Packets are represented as leaving in the lower portion of Fig. 6. The fluid is dropped until the bucket, which has depth b, becomes empty. The bucket is filled with a rate r. The right portion of Fig. 6 shows how the USC-shaping process is implemented in PBQ. This USC-shaping process is invoked by both a system timer and packet arrival. The accuracy of this leaky bucket system depends on the system timer granularity and packet arrival. A finer granularity and higher packet arrival rate will support higher peak and average rates. The bucket implementation shown in Fig. 6 will guarantee that the peak rate will not be exceeded. The short-term average rate will be higher than rate r because of the peak rate. However, the long-term average rate will approach rate r. Note that this technique is presented only as an example shaping capability. Other shaping algorithms may also be used for the USC shape function.

With this USC-shaping and hierarchical class structure based approach, hierarchical shaping may be achieved. Fig. 7 shows the comparison between a traditional shaping architecture and hierarchical shaping architecture. The traditional shaping implementation is done on a per-flow basis. Even with aggregation capability in the hierarchical scheduling algorithm, aggregation is not effective since the shaping is done independent of the class structure. In the event that any one of the flows from the same parent class is idle, the bandwidth allocated for the idle flow is not utilized by other sibling flows because the sibling flows are shaped before they enter the scheduler. In hierarchical shaping, USC-shaping may be applied to any class in the hierarchical class structure. By enabling USC-shaping in an intermediate class as shown in Fig. 7b, the flows among the class may be shaped after aggregation. Even if one of the flows among the class is idle, the rest of sibling flows are still able to utilize their parent's entire bandwidth while being shaped. Note that this hierarchical shaping functionality is different from CBQ's bounded (non-borrowing) class. In CBQ, each bounded class limits the average bandwidth usage. This should be a distinct functionality independent from the shaping.

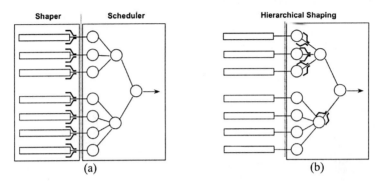

Fig. 7. Shaping Architecture Comparison – (a) Traditional vs. (b) Hierarchical

Table 1. Class Setup for Aggregation of IP data flows for hierarchical shaping

Class ID	Weight	P	r	B	Priority
11	3.3 %	5Mbps	3Mbps	84000b	1
12	1 %	1.5Mbps	900kbps	84000b	1
13	2 %	3Mbps	1.5Mbps	84000b	1
14	30 %	-	-	-	0
1	10 %	-	-	-	0
2	20 %	-	-	-	0
3	30 %	-	-	-	0
4	40 %	-	-	-	0
5	30 %	-	-	-	0
6	70 %	-	-	-	0
7	10 %	-	-	-	0
8	20 %	-	-	-	0
9	30 %	-	-	-	0
10	100 %	-	-	-	0

Hierarchical shaping may be best used at the edge of a network when different service models have to be interconnected to each other. For example, when DiffServ is used in the backbone to interconnect RSVP regions, the edge routers need to aggregate the RSVP flows into classes defined by DiffServ. Hierarchical shaping may shape the aggregated flows in terms of DiffServ's per hop behaviors (PHBs) instead of having the traditional per-flow shaping. Another example is IP flows aggregated into an ATM VC. This flow-aggregation may help in saving ATM resources when ATM is used as a backbone network for legacy networks. As in the RSVP to DiffServ transition case, hierarchical shaping may be used to shape the aggregated IP flows according to the ATM VC's traffic specification parameters.

The hierarchical shaping functionality is independent of the scheduling decision for lower level (child) classes. As long as the aggregated flows are under the aggregated traffic specification, the sibling flows may be scheduled according to the scheduling algorithm. The following simulation shows how this hierarchical shaping performs. In Fig. 8, the circles contain a class ID number. Each class was set up as shown in Table 1. Class 11, 12, and 13 were to be shaped with peak rate (p), average rate (r), and bucket depth (b) as shown in Table 1. These classes (11,12, and 13) have child classes, which result in hierarchical shaping with USC-shaping. Note that the child classes (class 1 to 9) do not have traffic specifications for shaping.

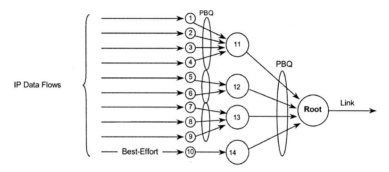

Fig. 8. Aggregation of IP data flows

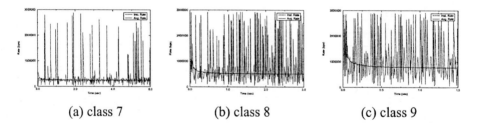

(a) class 7 (b) class 8 (c) class 9

Fig. 9. Simulation result for class 13

Fig. 9 shows the instantaneous and average rate for classes 7, 8, and 9. The jagged dotted line is the instantaneous rate, and the solid line is the average rate. These classes were aggregated under class 13. In Table 1, class 13 has traffic specification parameters of p=3 Mbps, r=1.5 Mbps, and b=8400 bits. With hierarchical shaping, the traffic through class 13 was bounded to the traffic specification parameters. As shown in Fig. 9, the instantaneous rate of class 7, 8, and 9 were all under the peak rate of 3Mbps. Since classes 7, 8, and 9 were scheduled independent of the hierarchical shaping functionality, each class was scheduled according to the weight specified in Table 1. With the weights, class 7, 8, and 9 utilized the allocated bandwidth of parent class (class 13) in the ratio of 10%:20%:30%. The allocated bandwidth for class 13 is the average rate of the traffic specification parameters, which is 1.5 Mbps. Fig. 9 shows that classes 7, 8, and 9 had processed the traffic in the average rates of 0.25 Mbps, 0.5 Mbps, and 0.75 Mbps respectively (solid lines in Fig. 9). The average rates for class 7, 8, and 9 are in the ratio of 1:2:3, and the sum of the average rates is 1.5 Mbps, which is the same as that of parent class. This simulation result shows that the hierarchical shaping was able to shape each aggregated class according to the traffic specification while fully utilizing the available bandwidth.

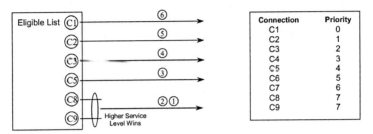

Fig. 10. Process of Priority-labeling scheme

3.2 Priority-Labeling Scheme in HPBQ

In scheduling algorithms such as SCFQ, SFQ, and PBQ, the priority index calculations are simplified in order to reduce the complexity [4,5]. This simplification causes the worst delay bound to depend on the current number of connections as shown in (4) where K is the number of connection sharing the link, σ_i and r_i are burst size and reserved rate of connection i, L_{max} is maximum packet size, and R is link speed.

$$D_i \leq \frac{\sigma_i}{r_i} + \frac{KL_{max}}{R} \qquad (4)$$

In PBQ, a priority-labeling scheme is added to shorten the worst case delay bound. The highest priority may be assigned to the real time delay-sensitive connection flow. Then, from the eligible connections, the real time flow may be serviced regardless of its priority index. Using this methodology, the worst case delay bound for the real time flow may be bounded to that of WFQ. The priority-labeling process is as shown in Fig. 10. In Fig. 10, a connection sequence with different priority levels is shown. When C1, C2, C3, C5, C8, and C9 are eligible, C8 and C9 will be considered for service regardless of the other connections' service level. Since C8 and C9 have the same priority level, the service level is used for the tiebreaker. Assuming C9 has larger service level than C8, C8 will be served after C9. Then, connections will be served in the order of priority level, which is C5, C3, C2, and C1. Once the connection finishes sending the amount of what the weight defines, the connection will be dropped from the eligibility list and will have to wait for its next turn.

This kind of priority-labeling technique cannot easily be applied to SCFQ and SFQ. Serving other flows by having a priority-label preceded the time tags influences the system virtual time for the next incoming packets. Since PBQ does not have such limitation, the PBQ algorithm works in a straightforward manner with this priority-labeling technique.

The priority-labeling based scheme applied to PBQ was simulated. The simulation was set up with eight connections. The reservation of connections 1 to 4 was mirrored in connection 5 to 8, and connections 5 to 8 were prioritized. Connections 1, 4, 5, and 8 receive their guaranteed rate of traffic while all the other connections were continuously backlogged and were being used as background traffic. Simulations were run with $M = 1500$ and the packet size was exponentially distributed with a mean of 700, minimum value of 50, and a maximum value of 1500. The simulation for priority-labeling used the set up shown in Fig. 11. Simulation results from this scenario are shown in Fig. 12. The results presented here are a cumulative density function graph of the delay that a packet experiences from the head of queue until departing to the link. With the priority-labeling based scheme, all packets from connections 5 and 8 have a smaller delay than the packets from connections 1 and 4 (refer to Fig. 12). This simulation shows that the priority-labeling scheme prioritized the traffic while providing flexibility in choosing different weights. This is possible because the weight and priority levels are decoupled in PBQ.

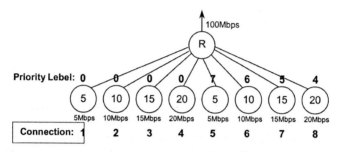

Fig. 11. Priority using Priority-Labeling

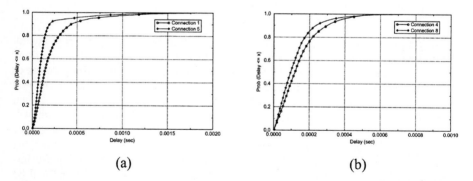

(a) (b)

Fig. 12. Priority Simulation (Delay of packets): (a) Connection 1 & 5, (b) Connection 4 & 8

4 Conclusion

In this paper, a new scheduling algorithm, PBQ, was described. With a simple service-level calculation, PBQ shows a similar performance level with other fair queuing algorithms. The enhancement of PBQ for a hierarchical class structure, hierarchical shaping, and priority was also presented along with OPNET [9] simulation results. The enhanced PBQ features provide much of the required functionality for network nodes involved in QoS, especially at the edge of heterogeneous networks. The relatively simple algorithm with its acceptable performance levels and increased functionality has been implemented in a LINUX based system, which also included an IP flow to ATM VC mapping capability. The prototyped LINUX network node has all of the H-PBQ capabilities mentioned in this paper. Measurements from this test-bed have been obtained. These results were not presented in this paper due to space limitations. These unpublished results were very encouraging in that the test bed network performed as predicted in the OPNET simulations. The desired ability of the test bed to make a reservation end to end starting from a legacy Ethernet network, crossing an ATM network, and returning to another legacy Ethernet network was enabled by the H-PBQ algorithm presented in this paper. It is this type of environment that we envisage the application of our highly implementable algorithm. Further details and results from the application of this algorithm may be obtained from reference [2].

References

1. Bernnett, J. C. R., Zhang, H., "Hierarchical Packet Fair Queuing Algorithms, " Proceedings of ACM SIGCOMM '96, August 1997.
2. Choi, M., "Traffic Flow Management for RSVP/ATM Edge Devices," Ph.D. Thesis Georgia Institute of Technology, August 1999.
3. Floyd, S., Jacobson, V., "Link-sharing and Resource Management Models for Packet Network," IEEE/ACM Transactions on Networking, Vol. 3 No. 4, pp. 365-386, August 1995.
4. Goyal, P., Vin, H. M., Chen, H., "Start-time Fair Queuing: A Scheduling Algorithm for Integrated Services Packet Switching Networks," IEEE/ACM Transactions on Networking, Vol. 5, No. 5, October 1997.
5. Golestani, S. J., "A Self-Clocked Fair Queuing Scheme for Broadband Applications," Proceedings IEEE INFOCOM '94. The Conference on Computer Communications. Networking for Global Communications, vol. 2, pp. 636-46, 1994.
6. Zhang, H., "Service Discipline for Guaranteed Performance Service in Packet-Switching Network," Proceedings of the IEEE, vol. 83, no. 10, pp. 1374-96, October 1995.
7. Georgiadis, L., Guérin, R., Peris, V., Sivarajan, K. N., "Efficient Network QoS Provisioning Based on per Node Traffic Shaping", IEEE/ACM Transactions on Networking, Vol. 4, No. 4, August 1996.
8. Georgiadis, L., Guérin, R., Peris, V. Rajan, R., "Efficient Support of Delay and Rate Guarantees in an Internet", Computer Communication Review, vol.26, no.4, pp.106-16, October 1996.
9. Mil 3 OPNET Modeler, Models/Protocols Mil 3, 1997
10. Stiliadis, D., and Varma, A. "Design and Analysis of Frame-Based Fair Queuing: A New Traffic Scheduling Algorithm for Packet Switched Networks", SIGMETRICS'96, pp. 104-115, 1996.
11. Quadros, G., Alves, A., Monteiro, E., Boavida, F., Tohme, S., Ulema, M., "How unfair can weighted fair queuing be?," ISCC 2000, p. 779-84, July 2000.
12. Stoica, I., Zhang, H., Ng, T.S.E., "A hierarchical fair service curve algorithm for link-sharing, real-time, and priority services," IEEE/ACM Transactions on Networking, vol. 8, no. 2, April 2000

Backbone Network Design with QoS Requirements

Hong-Hsu Yen and Frank Yeong-Sung Lin

Department of Information Management,
National Taiwan University, Taiwan
{d4725001, yslin}@im.ntu.edu.tw

Abstract. In this paper, we consider the backbone network design problem with a full set of QoS requirements. Unlike previous researches, we consider both the transmission line cost and the switch cost. And the QoS requirements that we considered include the average packet delay, end-to-end packet delay and node disjoint paths. We formulate the problem as a combinatorial optimization problem where the objective function is to minimize the total network deployment cost subject to the aforementioned QoS constraints. Besides the integrality constraints, the nonlinear and the nonconvex properties associated with the problem formulation make it difficult. Lagrangean relaxation in conjunction with a number of optimization-based heuristics are proposed to solve this problem. From the computational experiments, the proposed algorithms calculate creditable solutions in minutes of CPU time for moderate problem sizes.

1 Introduction

How to design a usually sophisticated backbone network with the minimum deployment and operation cost subject to various and often stringent QoS requirements is a common challenge faced by network designers and managers. Intensive research has been conducted to address this issue. However, most research tackles this backbone network design problem without considering a full set of QoS requirements.

Gavish model the network topological design problem as a nonlinear combinatorial optimization problem. The objective is to minimize the network installation cost and the queueing cost imposed on the network users. However, network installation cost and the queueing cost are two different concepts such that it is not appropriate to put them together in the objective function. In addition, end-to-end QoS constraints are not considered in [3].

A bunch of heuristics based on genetic algorithms (GA) and Tabu-Search algorithms are proposed to tackle the backbone network design problem [2, 5, 6]. However, in [2, 5, 6], the cost for network access point is not considered in the network installation cost. Furthermore, average and end-to-end delay constraints and reliability constraint are not jointly considered at the same time.

The delay QoS requirement is crucial to modern application services (e.g. VOD, tele-conferencing). And in particular, backbone network usually requires high availability, that is, redundant links and switching nodes are needed in case of failure. As a result, unlike more of previous research, the QoS requirement considered in this paper can be classified into two parts. The first part is the delay QoS (including the average packet delay and end-to-end delay for each O-D pair) and the second part is the reliability and availability of services. On the other hand, the network construction cost includes both the switch and link installation cost to reflect the cost structure of real network. This problem is a well-known difficult NP-hard problem.

This paper is organized as follows. In Section 2, mathematical formulation of the backbone network design problem is proposed. In Section 3, the dual approach for the backbone network design problem based on the Lagrangean relaxation is presented. In Section 4, the getting primal heuristics are developed to get the primal feasible solutions from the solutions to the dual problem. In Section 5, the computational results are reported. In Section 6, the concluding remarks are presented.

2 Problem Formulation

This QoS based backbone network design problem is modeled as the graph where the users and switches are depicted as nodes and the communication channels are depicted as arcs. We show the definition of the following notation.

L	The set of candidate local loop links and backbone links in the communication network.
W	The set of origin-destination (O-D) pairs in the network.
λ_w	The traffic requirement for each O-D pair $w \in W$.
$\overline{C_l}$	a capacity upper bound in the candidate capacity configurations for link $l \in L$.
P_w	a given set of simple directed paths from the origin to the destination of O-D pair w.
U_k	a set of potential incoming links to switch k.
δ_{pl}	the indicator function which is one if link l is on path p and zero otherwise.
ε_{pk}	the indicator function which is one if switch k is on path p and zero otherwise.
g_l	the aggregate flow over link $l \in L$, which is $\sum_{p \in P_w} \sum_{w \in W} x_p \lambda_w \delta_{pl}$.
D_w	the maximum allowable end-to-end delay requirement for O-D pair w.
K	the maximum allowable average cross network delay requirement.
T_w	the minimum number of node disjoint paths required for O-D pair w.
O	the set of candidate locations for switches.
H	the set of link pairs that are with the same end points but in opposite directions.
A_l	the set of candidate capacity configurations for link l.
R_k	the set of admissible switching fabric configuration for switch at location k.

E_k	the set of candidate port configuration for switch at location k.
$\varphi_l(C_l)$	the cost for installing capacity C_l on link l, including the fixed and variable cost.
ξ_l	the fixed link installation cost for link l.
$Q_k(J_k, S_k)$	the cost for installing a switch at location k with switching fabric capacity J_k and number of ports S_k.
$F_l(f_l, C_l)$	the average delay on link $l \in L$, which is a function of f_l and C_l.
$B_l(f_l, C_l)$	the average number of packets on link $l \in L$, which is a function of f_l and C_l, and by the Little's results, which is equal to $\lambda_w * F_l(f_l, C_l)$.

And the decision variables are depicted as follows.

x_p	1 when path $p \in P_w$ is used to transmit the packets for O-D pair $w \in W$ and 0 otherwise.
z_p	1 when path $p \in P_w$ is the node disjoint path for O-D pair $w \in W$ and 0 otherwise.
y_{wl}	1 when link $l \in L$ is on the path chosen for O-D pair $w \in W$ and 0 otherwise.
f_l	the estimated aggregate flow on link $l \in L$.
M_l	1 when a link is installed at location $l \in L$ and 0 otherwise.
C_l	the capacity assignment for link $l \in L$.
J_k	the switching fabric capacity assignment for switch at location k.
S_k	the number of ports for switch at location k.

We formulate the QoS based backbone network design problem as a nonlinear and nonconvex combinatorial optimization problem, as shown below.

$$\min Z_{IP} = \sum_{l \in L} \varphi_l(C_l) + \sum_{k \in O} Q_k(J_k, S_k) - \sum_{(l,\bar{l}) \in H} \xi_l M_l M_{\bar{l}} \qquad (IP)$$

subject to:

$$\frac{1}{\sum_{w \in W} \lambda_w} \sum_{l \in L} B_l(f_l, C_l) \leq K \qquad (1)$$

$$\sum_{l \in L} y_{wl} F_l(f_l, C_l) \leq D_w \qquad \forall w \in W \qquad (2)$$

$$\sum_{p \in P_w} x_p = 1 \qquad \forall w \in W \qquad (3)$$

$$x_p = 0 \text{ or } 1 \qquad \forall p \in P_w, w \in W \qquad (4)$$

$$\sum_{p \in P_w} x_p \delta_{pl} \leq y_{wl} \qquad \forall w \in W, l \in L \qquad (5)$$

$$y_{wl} = 0 \text{ or } 1 \qquad \forall w \in W, l \in L \qquad (6)$$

$$g_l = \sum_{w \in W} \sum_{p \in P_w} x_p \delta_{pl} \lambda_w \leq f_l \qquad \forall l \in L \qquad (7)$$

$$f_l \leq C_l \qquad \forall l \in L \qquad (8)$$

$$C_l \in A_l \qquad \forall l \in L \qquad (9)$$

$$M_l = 0 \text{ or } 1 \qquad \forall l \in L \qquad (10)$$

$$C_l \leq \overline{C_l} M_l \qquad \forall l \in L \qquad (11)$$

$$J_k \in R_k \qquad \forall k \in O \qquad (12)$$

$$\sum_{l \in U_k} M_l \leq S_k \qquad \forall k \in O \qquad (13)$$

$$S_k \in E_k \qquad \forall k \in O \qquad (14)$$

$$\sum_{w \in W} \sum_{p \in P_w} x_p \varepsilon_{pk} \lambda_w \leq J_k \qquad \forall k \in O \qquad (15)$$

$$\sum_{p \in P_w} z_p = T_w \qquad \forall w \in W \qquad (16)$$

$$\sum_{p \in P_w} z_p \delta_{pl} \leq M_l \qquad \forall w \in W, l \in L \qquad (17)$$

$$z_p = 0 \text{ or } 1 \qquad \forall p \in P_w, w \in W. \qquad (18)$$

The objective is to minimize network installation cost. There are three terms in the objective function. The first term is to compute the total link installation cost, including the fixed cost and the variable cost. The second term is to compute the total switch installation cost. The third term is to compute one fixed cost for each installed opposite links. The necessity of subtracting the third term is to ensure that only one rather than two fixed cost is calculated for two links with the same attached nodes but in opposite direction. Constraint (1) enforce the average cross network delay constraint. Constraint (2) enforce the end-to-end packet delay for each O-D pair. Constraints (3) and (4) require that the all the traffic for each O-D pair should be transmitted over exactly one path. The decision variable y_{wl} in Constraint (6) is an auxiliary decision variable, which is equal to $\sum_{p \in P_w} x_p \delta_{pl}$. Hence, the equality in Constraint (5) is replaced by inequality due to the ease use of the Lagrangean relaxation. Constraints (7) and (8) are the link capacity constraints. Constraint (9) determines the possible capacity configurations of all links. Constraints (10) and (11) require that the link must be installed first before link capacity assignment. Constraints (12) and (14) determine the possible switching fabric and number of ports of all switches. Constraint (13) is the switch termination constraint, which means the number of incoming links to the switch should not exceed the number of ports on that switch. Constraint (15) is the switch capacity constraint Constraints (16) and (18) are the path diversity (node disjoint) requirement for each O-D pair. Constraint (17) guarantees that link must be installed first before it could be adopted on the node disjoint path for each O-D pair.

3 Lagrangean Relaxation

We dualize Constraints (1), (2), (5), (7), (8), (11), (13), (15) and (17) of Problem (IP) to get the following Lagrangean relaxation problem (LR).

$$\min Z_D = \sum_{l \in L} \varphi_l(C_l) + \sum_{k \in O} Q_k(J_k, S_k) - \sum_{(l,\bar{l}) \in H} \xi_l M_l M_{\bar{l}} + a[\frac{1}{\sum_{w \in W} \lambda_w} \sum_{l \in L} B_l(f_l, C_l) - K]$$

$$+ \sum_{w \in W} b_w [\sum_{l \in L} y_{wl} F_l(f_l, C_l) - D_w] + \sum_{w \in W} \sum_{l \in L} c_{wl} [\sum_{p \in P_w} x_p \delta_{pl} - y_{wl}] + \sum_{l \in L} h_l[f_l - C_l] +$$

$$\sum_{l \in L} d_l [\sum_{w \in W} \sum_{p \in P_w} x_p \delta_{pl} \lambda_w - f_l] + \sum_{l \in L} e_l [C_l - \overline{C_l} M_l] + \sum_{k \in O} n_k [\sum_{w \in W} \sum_{p \in P_w} x_p \varepsilon_{pk} \lambda_w - J_k] +$$

$$+ \sum_{k \in O} m_k [\sum_{l \in U_k} M_l - S_k] + \sum_{w \in W} \sum_{l \in L} q_{wl} [\sum_{p \in P_w} z_p \delta_{pl} - M_l] \tag{LR}$$

subject to:

$$\sum_{p \in P_w} x_p = 1 \qquad \forall w \in W \tag{19}$$

$$x_p = 0 \text{ or } 1 \qquad \forall p \in P_w, w \in W \tag{20}$$

$$y_{wl} = 0 \text{ or } 1 \qquad \forall w \in W, l \in L \tag{21}$$

$$C_l \in A_l \qquad \forall l \in L \tag{22}$$

$$M_l = 0 \text{ or } 1 \qquad \forall l \in L \tag{23}$$

$$J_k \in R_k \qquad \forall k \in O \tag{24}$$

$$S_k \in E_k \qquad \forall k \in O \tag{25}$$

$$\sum_{p \in P_w} z_p = T_w \qquad \forall w \in W \tag{26}$$

$$z_p = 0 \text{ or } 1 \qquad \forall p \in P_w, w \in W. \tag{27}$$

We can decompose (LR) into five independent subproblems.
Subproblem 1: for x_p:

$$\min \sum_{w \in W} \sum_{p \in P_w} \sum_{l \in L} [\sum (c_{wl} + d_l \lambda_w) x_p \delta_{pl} + \sum_{k \in O} n_k \lambda_w x_p \varepsilon_{pk}] \tag{SUB1}$$

subject to: (19) and (20).
Subproblem 2: for C_l, y_{wl} and f_l:

$$\min \sum_{l \in L} \varphi_l(C_l) + a \frac{1}{\sum_{w \in W} \lambda_w} \sum_{l \in L} B_l(f_l, C_l) + \sum_{w \in W} b_w \sum_{l \in L} y_{wl} F_l(f_l, C_l) - \sum_{w \in W} \sum_{l \in L} c_{wl} y_{wl} -$$

$$\sum_{l \in L} d_l f_l + \sum_{l \in L} e_l C_l + \sum_{l \in L} h_l f_l - \sum_{l \in L} h_l C_l \tag{SUB2}$$

subject to: (21) and (22).
Subproblem 3: for M_l:

$$\min - \sum_{l \in L} e_l \overline{C_l} M_l + \sum_{k \in O} \sum_{l \in U_k} m_k M_l - \sum_{(l,\bar{l}) \in H} \xi_l M_l M_{\bar{l}} - \sum_{w \in W} \sum_{l \in L} q_{wl} M_l \tag{SUB3}$$

subject to: (23).
Subproblem 4: for J_k and S_k:

$$\min \sum_{k \in O}[Q_k(J_k, S_k) - n_k J_k - m_k S_k] \quad \text{(SUB4)}$$

subject to: (24) and (25).
Subproblem 5: for z_p:

$$\min \sum_{w \in W} \sum_{l \in L} q_{wl} \sum_{p \in P_w} z_p \delta_{pl} \quad \text{(SUB5)}$$

subject to: (26) and (27).

In order to deal with the nodal weight of (SUB1), the node spiltting technique [4] is used. As a result, (SUB1) could be further decomposed into $|W|$ independent shortest path problem with nonnegative arc weights. It can be easily solved by the Dijkstra's algorithm. (SUB2) could also be decomposed into $|L|$ independent subproblems. For each link $l \in L$,
Subproblem 2.1: for C_l, y_{wl} and f_l:

$$\min \ \varphi_l(C_l) + a \frac{1}{\sum_{w \in W} \lambda_w} B_l(f_l, C_l) + \sum_{w \in W} b_w \ y_{wl} F_l(f_l, C_l) - \sum_{w \in W} c_{wl} \ y_{wl} - d_l f_l + e_l C_l +$$

$$h_l f_l - h_l C_l \quad \text{(SUB2.1)}$$

subject to: $y_{wl} = 0$ or 1 $\forall w \in W$ and $C_l \in A_l$.

(SUB2.1) is a complicated problem due to the coupling of three decision variables, C_l, y_{wl} and f_l. Since the possible capacity configurations of links are finite, such as 64kbps, 128kbps, 256kbps, 512kbps, T1 and T3 for example. We can exhaustive search all different possible link configuration by finding the best y_{wl} and f_l. In [1], Lin proposed an efficient algorithm to solve y_{wl} and f_l at a given link capacity under $M/M/1$ queuing model. Therefore, the algorithm to solve (SUB2.1) under $M/M/1$ queuing model is proposed as bellow. In addition, the formulation could be extended to any non $M/M/1$ model with monotonically increasing and convexity performance metrics.

Step 1. For each possible link capacity configuration, applying the algorithm developed in [1] to solve (SUB2.1) as to find the optimal y_{wl} and f_l.

Step 2. Finding the minimum objective value of (SUB2.1) from the objective value associated with each possible link capacity configuration. Then y_{wl} and f_l can be determined from the optimal link capacity.

(SUB3) can be decomposed into $|H|$ independent subproblems. For each pair of bi-directional links $(l, \bar{l}) \in H$,
Subproblem 3.1: for M_l and $M_{\bar{l}}$:

$$\min \ -e_l \overline{C_l} M_l - e_{\bar{l}} \overline{C_{\bar{l}}} M_{\bar{l}} + G_1 M_l + G_2 M_{\bar{l}} - \xi_l M_l M_{\bar{l}} - \sum_{w \in W} q_{wl} M_l - \sum_{w \in W} q_{w\bar{l}} M_{\bar{l}}$$

$$\text{(SUB3.1)}$$

subject to: $M_l = 0$ or 1 and $M_{\bar{l}} = 0$ or 1.

In the above formulation, the G_1 and G_2 are calculated as follows.

1. If the link l is the incoming link to any potential switch, say k_1, then assign G_l to m_{k_1}, else assign G_l to zero.

2. If the link $\bar{l}$ is the incoming link to any potential switch, say k_2, then assign G_2 to m_{k_2}, else assign G_2 to zero.

In (SUB3.1), two opposite direction links are considered at the same time. As a result, the algorithm to optimally solve (SUB3.1) is proposed as follows.

Step 1. Let $N_1 = 0$, $N_2 = -e_l \overline{C_l} + G_1 - \sum_{w \in W} q_{wl}$, $N_3 = -e_{\bar{l}} \overline{C_{\bar{l}}} + G_2 - \sum_{w \in W} q_{w\bar{l}}$, $N_4 =$
$-e_l \overline{C_l} + G_1 - e_{\bar{l}} \overline{C_{\bar{l}}} + G_2 - \xi_l - \sum_{w \in W} q_{wl} - \sum_{w \in W} q_{w\bar{l}}$.

Step 2. Identify the N_i with the minimum value, where i = 1, 2, 3, 4.

Step 3. If i = 1, then assign $M_l = 0$ and $M_{\bar{l}} = 0$, else if i = 2, then assign $M_l = 1$ and $M_{\bar{l}} = 0$, else if i = 3, then assign $M_l = 0$ and $M_{\bar{l}} = 1$, else if i = 4, then assign $M_l = 1$ and $M_{\bar{l}} = 1$.

(SUB4) can be further decomposed into $|O|$ independent subproblems. For each independent subproblem, due to the number of possible switch configurations (including number of ports and switching fabric) is finite and manageable within computational time, we can exhaustively search all possible combination of switch configurations as to find the optimal J_k and S_k.

(SUB5) can be further decomposed into $|W|$ independent node disjoint shortest path problem with nonnegative arc weights. Suurballe propose an efficient algorithm to optimally solve link disjoint path problem [7]. Hence, (SUB5) could be optimally solved by the Suurballe's algorithms in conjunction with the node splitting technique.

According to the algorithms developed above to solve each subproblem, we could successfully solve the Lagrangean relaxation problem optimally. By using the weak Lagrangean duality theorem (for any given set of non-negative multipliers, the optimal objective function value of the corresponding Lagrangean relaxation problem is a lower bound on the optimal objective function value of the primal problem), Z_D is a lower bound on Z_{IP}. We could construct the dual problem to calculate the tightest lower bound and solve the dual problem by using the subgradient method.

4 Getting Primal Feasible Solutions

To obtain the primal solutions to the (IP), solutions to the (LR) are considered. We develop sophisticated getting primal heuristics to getting the primal feasible solutions. This getting primal heuristic start with the routing assignment obtained from the (SUB2.1). From the routing assignment in (SUB2.1), the aggregate flow on each link can be calculated. In order to satisfy the end-to-end delay requirement for each O-D pair, the tightest end-to-end delay for all O-D pairs is located by searching the minimum end-to-end delay requirement among all O-D pairs. From the tightest end-to-end delay, the tightest link delay can be calculated by dividing the tightest end-to-end delay to the maximum hop number in any routing path. The maximum hop number for any O-D pair is equal to the number of potential switches plus one, since the source node must home to the switch first, and then route to the other switches, and finally route to the destination node. From the tightest link delay, we can determine the minimum link capacity in order to satisfy the tightest link delay requirement. From the above statement, we could satisfy the delay requirements.

In order to satisfy the node disjoint requirement for each O-D pair, the node disjoint path assignment from (SUB5) is used. If the associated link on any node disjoint path did not install at the above procedure, the minimum nonzero capacity is installed on that link. After the link capacity is determined, the number of links incoming to each potential switch can be determined. Also from the aggregate flow on each link, the total aggregate flow incoming to each potential switch can also be determined. As a result, the minimum cost switch configuration in order to satisfy the number of ports and switch fabric constraints can be determined as well.

5 Computational Experiments

The network planning algorithms developed in Section 3 and 4 are coded in C++ and performed at PC with INTEL™ PIII-800 CPU. The input parameters include the locations for the users and the potential switches, admissible configurations and cost structures of potential switches and links, traffic requirements and survivability/connectivity requirements. And the output parameters include the switch and link configuration assignment, routing assignment, node disjoint paths assignment, average end-to-end delay and individual end-to-end delay for each O-D pair.

The maximum number of iterations for the proposed dual Lagrangean algorithm developed above are 1000, and the improvement counter is 30. The step size for the dual Lagrangean algorithm is initialized to be 2 and be halved of its value when the objective value of the dual algorithm does not improve for 30 iterations.

Two sets of computational experiments are performed. The computational time for these two sets of computational experiments are all within fifteen minutes under the network size of 30 user/switch nodes. Hence, the proposed algorithms are efficient in time complexity.

In these computational experiments, the cost of the link assignment is divided into two parts, fixed cost and variable cost. The fixed cost is calculated from the Euclidean distance between two end points that the link connected, and the variable cost is based on the link capacity configuration. There are fifteen discrete potential link capacity configurations, from 0 to 500, for the computational experiments. And the cost associated with these potential capacity configurations is a concave function to reflect the economy-of-scale effect. On the other hand, the switch installation cost is based upon the switching fabric and the number of ports on the switch. There are nineteen discrete potential switch configurations in the computational experiments. And the cost associated with these potential switch configurations is also a concave function.

In the first set of computational experiment, we want to test the solution quality when the input delay requirements are loose as compared to the output of the delay requirement. And node disjoint path requirement is not considered. In the second set of computational experiment, we want to test the solution quality when the input delay requirements are tight as compared to the output of the delay requirement. And the 2-connected node-disjoint-path requirement for each O-D pair is considered.

Table 1 depicts the computational results for the various network sizes and traffic demand without node disjoint requirement and loose delay requirements. The first column is the network size. The location, x-axis and y-axis, of user nodes and potential switch nodes are randomly distributed between the 0 and 500. The second column is the traffic demand of the user nodes, are randomly distributed between 30 to 400. The third column reports the lower bound of the primal problem. The forth column reports the upper bound of the primal problem. The fifth column reports the error gap between the lower bound and upper bound. The seventh column reports the average network delay requirement, and the sixth column reports the average network delay calculated by the proposed algorithms. As could be seen from the fifth column in Table 1, the more number of user/switch nodes the looser the error gap. And we have a tighter error gap for heavy traffic in the same network topology.

Table 1. Solution quality obtained by various network sizes and traffic demand without node disjoint requirement

# of Users / switches	Traffic Demand	Lower bound	Upper bound	Error Gap(%)	Average network delay	K
9	30~200	943.4	1639.9	73.8	0.168	10
9	60~400	1852.4	2778.0	49.9	0.068	10
12	30~200	1639.8	2664.8	62.5	0.079	10
12	60~400	3048.0	4353.1	42.8	0.069	10
15	30~200	1895.1	2918.9	54.0	0.294	10
15	60~400	3695.3	5291.5	43.2	0.131	10
30	30~200	2496.9	6042.3	141.9	0.397	10
30	60~400	4442.8	7936.1	78.6	0.117	10

Table 2. Solution quality obtained by various network sizes and traffic demand with two node disjoint requirement

# of Users / switches	Traffic Demand	Lower bound	Upper bound	Error Gap(%)	Average network delay	K
9	30~200	941.5	1980.3	110.3	0.195	1
9	60~400	1852.6	2909.0	57.0	0.054	1
12	30~200	1636.2	3220.0	90.7	0.087	1
12	60~400	3048.0	4353.1	42.8	0.069	1
15	30~200	1887.3	3369.6	78.5	0.203	1
15	60~400	3694.9	5667.0	53.4	0.130	1
30	30~200	2472.0	7098.8	187.2	0.452	1
30	60~400	4388.6	8628.5	96.6	0.130	1

Table 2 depicts the computational results for the various network sizes and traffic demand with two node disjoint requirement and loose delay requirements. As compared to Table 1, the average cross network delay and the end-to-end delay requirements are more stringent, delay = 1 instead of 10. On the other hand, there are two node disjoint requirement for each user nodes. As could be seen from the third column of Table 2, we have a looser lower bound in Table 2. However, we still have a reasonable good upper bound in Table 2.

6 Concluding Remarks

In this paper, the mathematical formulation and algorithms for the backbone network design which considers the system and user specified QoS requirements are proposed. The objective of this backbone network design problem is to minimize the total installation cost of link and switch installation cost. The system QoS requirement is the average cross network delay requirement. The user specified QoS requirements include the end-to-end delay requirement and node disjoint path requirements. Besides integrality constraints, the non-convexity of the delay performance metric makes this problem difficult. By using the Lagrangean relaxation method and the subgradient method to construct the dual problem and calculate the tightest lower bound, we provide getting primal feasible solution heuristic to obtain the primal feasible solution based on the solutions to the dual problem. Based on the solution quality and the computational time, we propose effective and efficient algorithms for this problem.

References

1. Cheng, K. T. and F. Y.-S. Lin, "Minimax End-to-End Delay Routing and Capacity Assignment for Virtual Circuit Networks," Proc. IEEE Globecom, (1995) 2134-2138.
2. Cheng, S. T., "Topological Optimization of a Reliable Communication Network", IEEE Trans. on Reliability, Vol. 47, No. 3, Sep. (1998) 225-233.
3. Gavish, B., "Topological design of computer communication networks – The overall design problem", European J. of Operational Research, 58, (1992) 149-172.
4. Lin, F. Y.-S., "Link Set Sizing for Networks Supporting SMDS", IEEE/ACM Trans. on Networking, Vol. 1, No. 6, Dec. (1993) 729-739.
5. Ombuki, B., M. Nakamura, Z. Nakao and K. Onaga, "Evolutionary Computation for Topological Optimization of 3-Connected Computer Networks", IEEE International Conference on Systems, Man, and Cybernetics, Vol. 1, (1999) 659-664.
6. Pierre, S. and A. Elgibaoui, "A Tabu-Search Approach for Designing Computer-Network Topologies with Unreliable Components", IEEE Trans. on Reliability, Vol. 46, No. 3, Sep. (1997) 350-359.
7. Suurballe, J. W. and R. E. Tarjan, "A Quick Method for Finding Shortest Pairs of Disjoint Paths", Networks, Vol. 14, (1984) 325-336.

A Multi-path QoS Routing Protocol in a Wireless Mobile ad Hoc Network

Wen-Hwa Liao[1,a], Yu-Chee Tseng[4,b], Shu-Ling Wang[1], and Jang-Ping Sheu[1c]

[1] Department of Computer Science and Information Engineering
National Central University, Chung-Li, 320, Taiwan
[a]whliao@axp1.csie.ncu.edu.tw, [c]sheujp@csie.ncu.edu.tw
[2] Department of Computer Science and Information Engineering
National Chiao-Tung University, Hsin-Chu, 300, Taiwan
[b]yctseng@csie.nctu.edu.tw

Abstract. A mobile ad hoc network (MANET) is one consisting of a set of mobile hosts capable of communicating with each other without the assistance of base stations. This paper considers the QoS (quality-of-service) routing problem in a MANET. We propose an on-demand protocol for searching for a multi-path QoS route from a source host to a destination host in a MANET, where a *multi-path* is a network with a source and a sink satisfying certain bandwidth requirement. Existing works all try to find a *uni-path* to the destination. The basic idea is to distribute a number of tickets from the source, which can be further partitioned into sub-tickets to search for a satisfactory multi-path. Through simulations, we justify that the value of our multi-path protocol is on its flexibility: (i) when the network bandwidth is very limited, it can offer a higher success rate to find a satisfactory QoS route than those protocols which try to find a uni-path, and (ii) when the network bandwidth is sufficient, it can perform almost the same as those protocols which try to find a uni-path (in both routing overhead and success rate).

1 Introduction

One research issue that has attracted a lot of attention recently is the design of *mobile ad hoc network (MANET)*. A MANET is one consisting of a set of mobile hosts capable of communicating with each other without the assistance of base stations. So multi-hop communication is sometimes inevitable, where the packets sent by the source host are relayed by several intermediate hosts before reaching the destination host.

Since MANET is characterized by its fast changing topology, extensive research efforts have been devoted to the design of routing protocols for MANETs [1, 4, 6, 9, 10]. These protocols only concerns with shortest-path routing and the availability of multitude routes in the MANET's dynamically changing environment. That is, only best-effort data traffic is provided. Connections with quality-of-service (QoS) requirements are less frequently addressed. Some works have started to focus on the QoS issue in a MANET. A ticket-based QoS routing protocol is proposed in [2] to find a route satisfying certain bandwidth and delay constrains. The basic idea is to use tickets to confine the number of route-searching packets to avoid an unwise blind flooding. In [7], a scheme to calculate the end-to-end bandwidth of a path under a TDMA-over-CDMA mechanism is proposed. The hidden-terminal problem and the bandwidth allocation problem are also addressed.

The purpose of this paper is to address QoS routing in a MANET environment. We propose an on-demand protocol for searching for a multi-path QoS route from a source host to a destination host in a MANET, where a *multi-path* is a network with a source and a sink satisfying certain bandwidth requirement. Our protocol distinguishes from the work of [2] in that they try to find a *uni-path* to the destination based on a costly *reactive* approach (namely, DSDV [9]). The basic idea is similar to the work in [2] by distributing a number of tickets from the source. However, we allow a ticket to be further partitioned into sub-tickets to search for a satisfactory multi-path. Through simulations, we justify that the value of our multi-path protocol is on its flexibility: (i) when the network bandwidth is very limited, it can offer a higher success rate to find a satisfactory QoS route than those protocols which try to find a uni-path, and (ii) when the network bandwidth is sufficient, it can perform almost the same as those protocols which try to find a uni-path (in both routing overhead and success rate).

The rest of the paper is organized as follows. Section 2 presents some background and motivation. Our protocol is developed in Section 3. Experimental results are in Section 4. Section 5 concludes the paper.

2 Background and Motivation

Existing ad hoc routing protocols may generally be categorized as *table-driven* and *on-demand*. Table-driven protocols attempt to maintain consistent up-to-date routing information from each node to every other node in the network (e.g., DSDV [9] and CGSR [3]). Contrarily, on-demand protocols create routes only when desired by the source node (e.g., DSR [1] and AODV [10]). A hybrid of these approaches is also possible (e.g., ZRP [4]). To assist routing, some protocols even adopt location information in their route discovery and maintenance procedures (e.g., LAR [5] and GRID [6]). However, all these protocols only concern the existence of a route without guaranteeing its quality. In the following, we review the QoS routing protocol by [2]. Then we will motivate our work in this paper.

2.1 QoS Routing Protocols for MANETs

In [2], a *ticket-based* protocol is proposed to support QoS routing. This protocol maintains the end-to-end state information at every node for every possible destination. This information is updated periodically by a distance-vector-like protocol (namely DSDV [9]). A source node s, on requiring a QoS route, can issue a number of probing packets each carrying a *ticket*. Each probe is in charge of searching for one path, if possible. The basic idea of using tickets is to confine the number of route-searching packets to avoid a blind flooding (flooding in a

MANET is unwise, according to [8]). Each probe, on reaching any intermediate node, should choose one outgoing path that satisfies the QoS requirements. To save the number of probing packets, several tickets may be carried by one packet and, if so, the probe can be split in the midway into multiple probes, each carrying some of the tickets and being responsible of searching a different downstream sub-path. Thus, the maximum number of probes at any time is bounded by the total number of tickets. For example, Fig. 1(a) shows a MANET, where the number associated with each link is its corresponding bandwidth. The arrows show the progress of two tickets issued from S to D. It is assumed that a path of bandwidth 3 is required, so the probe going through C fails, while that through B and E succeeds.

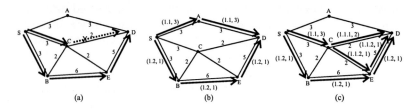

Fig. 1. Ticket-based QoS routing example: (a) searching for a route from S to D with bandwidth 3 using two tickets, and (b)-(c) two successful multi-path routing examples. In (b) and (c), the 2-tuple on each link means the ticket identity and the reserved bandwidth.

2.2 Observations and Motivations

While the ticket-based protocol in [2] solves the path-finding problem in an elegant way, the protocol may experience high failure rate when the bandwidth demand is large. For example, consider the example in Fig. 1(a) again. If we require a route from S to D with bandwidth 4, then the ticket-based protocol will fail because apparently no single route satisfies this constraint. This has motivated us to investigate the possibility of "multi-path QoS routing". Specifically, we define a *multi-path route* as one that can contain several sub-paths with the destination as the sink. Fig. 1(b) shows an example of using two paths to provide the required bandwidth of 4. Paths may even split and merge. For example, Fig. 1(c) shows a multi-path with bandwidth of 4. As will be shown later, finding a multi-path route turns out to be a quite complicated problem, which interests us from both theoretical and practical points of view.

3 Our Multi-path QoS Routing Protocol

3.1 Protocol Overview

Our protocol will follow an on-demand style to allocate bandwidth. So no global information will be collected in advance before a QoS route is required. When a source node S needs a route to a destination D of bandwidth B, it will send out some probe packets each carrying some tickets. Each ticket is responsible of searching for a multi-path from the source to the destination with an aggregated bandwidth equal to B.

On a ticket/sub-ticket arriving at a node, if the node is not the destination, some bandwidth of a qualified outgoing link will be reserved for this ticket and then the ticket will be sent out through that link. Since we allow a multi-path from S to D, if no link with a sufficient bandwidth exists, the ticket may be split into multiple sub-tickets, each being responsible of searching for a multi-path with a certain portion of bandwidth B. The destination node will, if possible, receive multiple tickets or sub-tickets. It will then pick one ticket or a set of sub-tickets forming a whole ticket and send a reply to the source node. Below we will discuss our multi-path QoS routing protocol in more details.

3.2 Ticket Format

For each bandwidth request, a number of tickets may be sent. A ticket will be denoted by $T(S, D, x, y, RID, TID, B, b)$. The meanings of the parameters in the ticket are as follows.

- S: the source host.
- D: the destination host.
- x: sender of the packet carrying the ticket.
- y: receiver of the packet carrying the ticket.
- RID: identity of a bandwidth request. This is unique for each QoS route request.
- TID: identity of a ticket. This is unique for each ticket.
- B: the required bandwidth of the multi-path from S to D.
- b: the required bandwidth of the multi-path from y to D. (So, if this is a sub-ticket, then $b < B$.)

3.3 Ticket Splitting and Inheritance Relation

As mentioned earlier, on a ticket reaching a node from which there is no outgoing link with a sufficient bandwidth, it may be split into several sub-tickets each responsible of searching for a multi-path with a partial bandwidth. The correctness of our protocol relies on a special representation of ticket identity (TID). The format of TID is a sequence of numbers separated by periods, i.e., $i_1.i_2.\cdots.i_k$. When a ticket is initiated at the source node, it will be given a unique identity i_1 (unique under the same RID). When an intermediate host receives a ticket (whole-ticket or sub-ticket) with identity TID, it may decide to split the ticket into sub-tickets. If so, each sub-ticket will be given an extension number appended after TID. Specifically, let the ticket be split into k sub-tickets. These sub-tickets will be given identities $TID.1, TID.2, \ldots, TID.k$. This is illustrated in Fig. 2. The reader may also refer to two real examples in Fig. 1(b) and (c).

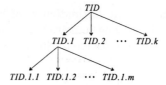

Fig. 2. Representation of ticket identities after a ticket is split twice.

It is critical in our yet-to-be-presented protocol to determine the relationship between tickets. Let $head(T)$ be the first number in a ticket T. Consider two tickets $T_1(\ldots, TID_1, \ldots)$ and $T_2(\ldots, TID_2, \ldots)$. If $head(TID_1) = head(TID_2)$, then they are two sub-tickets of the same whole-ticket. If TID_1 is a sub-string of TID_2, then T_1 is a sub-ticket split from T_2. If $head(TID_1) = head(TID_2)$ but none of them is a sub-string of the other, then they below to the same whole-ticket, but none of them is an ancestor of the other. Telling these relationships is important in our protocol. We point out some crucial points below.

- In Fig. 3(a), tickets T_1 and T_2 are two distinct tickets belonging to the same request. When they reach at the same intermediate node Y (perhaps at different time), it is not necessary to reserve separate bandwidths for them because they represent the same request. Only a bandwidth of $max(b_1, b_2)$ has to be reserved.
- In Fig. 3(b), tickets T_1 and T_2 are two sub-tickets belonging to the same whole-ticket. In this case, a total bandwidth of $b_1 + b_2$ has to be reserved at the intermediate host B.
- In Fig. 3(c), tickets T_2 and T_3 are two sub-tickets belonging to the same whole-ticket T_1, but T_2 is a sub-ticket split from T_1. In this case, a loop is detected and we should discard T_2.

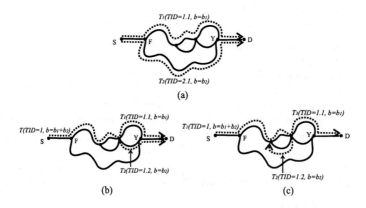

Fig. 3. Three possible relationships between tickets: (a) T_1 and T_2 are irrelevant, (b) T_1 and T_2 are irrelevant siblings, and (c) T_2 is T_1's sub-ticket. Note that all these tickets share the same RID.

3.4 Loop Avoidance

In the route discovery procedure, a probe entering a node that it has visited before typically means a failure of the search. To a protocol, this is an undesirable situation that should be avoided. In traditional approaches such as protocols [1, 5, 6, 10], whose goal is to find a path (instead of multi-path) to the destination, loop detection is a simple job by keeping records of the hosts that the probing packet has visited. In the following, we propose a method to avoid the possibility of forming loops to reduce the search failure probability.

To prevent loops from happening, we can let a mobile host collect tickets issued by its neighboring hosts, even if the tickets are not intended for itself. A host always listens to the medium and collects all tickets issued by its neighbors, no matter they are intending for itself or not. Now suppose a host receive a ticket T_1 destined to itself. The host will not forward T_1 or sub-tickets of T_1 to those neighbors who have ever sent a ticket T_2 such that T_1 is a sub-ticket of T_2 (by telling their ticket id's).

Also note that the purpose of the above loop avoidance rules is to increase the success probability of route discovery. It will not affect the correctness of our protocol. This implies that it is alright for a host to miss some tickets issued by its neighbors (perhaps due to collision or mobility).

3.5 Ticket Distribution

Our protocol follows an on-demand fashion to request a route. So when a mobile host S needs a route to a destination host D, it will issue probes carrying tickets to search for a route with a requested bandwidth B. The number of tickets to be issued may depend on many factors, such as how urgent the route is needed, how far the distance from S to D is, how much bandwidths the outgoing links from S have, and how much B is. Since there is no deterministic value for this number, we will leave it as an experimental parameter to be discovered by simulations.

This part takes care of how tickets are distributed. Consider any mobile host X. Let $Y_1, Y_2, \ldots, Y_k$ be all current neighbors of X known by X. For each $Y_i, 1 \leq i \leq k$, host X should keep the following data structures:

- b_i: the currently available bandwidth from X to Y_i.
- S_i (called the send set): the set of tickets that are sent to Y_i but are not yet confirmed.
- R_i (called the receive set): the set of tickets that are received from Y_i.
- L_i (called the listen set): the set of tickets that are issued by Y_i recently and heard by X.

The following protocol is executed by X when it hears a ticket $T(\ldots, RID, TID, B, b)$ from host Y_i (no matter it is intending for X or not).

S1. Join T into the listen set L_i.
S2. If the ticket is destined to X, join T into the receive set R_i. Otherwise, exit this procedure.

S3. If X is not the final destination, then T has to be forwarded to some of $Y_1, Y_2, \ldots, Y_k$. Let G be the set containing $Y_1, Y_2, \ldots, Y_k$ excluding those Y_i which contains a ticket T' such that T is a sub-ticket of T' (i.e., sending T to these hosts will form a loop). We sort G according to the following rules:
   a) Sort $Y_i \in G$ according to the number of tickets $T'(\ldots, RID', TID', \ldots)$ in S_i such that $RID = RID'$ and $head(TID) = head(TID')$ in an ascending order.
   b) For those $Y_i \in G$ that have a tie in the above sorting, we sort them according to the number of tickets $T'(\ldots, RID', \ldots)$ in S_i such that $RID = RID'$ in an ascending order. That is, we give priority to those links that have not been tried by the tickets under the same QoS route request.
   c) For those $Y_i \in G$ that have a tie in the above sorting, we sort them according to the remaining bandwidth from X to Y_i in a descending order, where the remaining bandwidth is defined to be

$$b_i - \sum_{\forall T'(\ldots, RID', TID', B, b') \in S_i :: RID = RID' \wedge head(TID) = head(TID')} b'.$$

That is, we give priority to those links who have higher remaining bandwidths after subtracting the bandwidths that have been reserved under the same QoS route request.

S4. Then we reserve a bandwidth of b on the link from X to the first host in G. If the remaining bandwidth from X to this host is less than b, then we reserve all its remaining bandwidth. In this case, we will go to the second host in G and reserve the deficient bandwidth. If the remaining bandwidth from X to this host is not enough, then we reserve all its remaining bandwidth. In this case, we will go to the third host in G. We repeat the above reservation, until a total bandwidth of b is reserved. For each of the above reservations, a ticket will be sent. Also, we will record this by joining the ticket into the appropriate send set S_i.

We illustrate the above steps by an example in Fig. 4(a). Suppose that there are two tickets under the same QoS request generated by the source S, who looks for a multi-path to D with bandwidth is 4. For the first ticket (with $TID = 1$), it will go to host A, which has the largest remaining bandwidth. Unfortunately, on reaching host A, the ticket will find that there is not enough bandwidth for it to proceed. So this ticket will be discarded. For the second ticket (with $TID = 2$), the link to C has the highest priority. However, its remaining bandwidth is less than 4, so we reserve all its bandwidth. Then we go to the host B, which has the second highest priority. This completes the forwarding of the second ticket. When the sub-ticket with $TID = 2.1$ reaches host C, it will be forwarded to host B according to the above rules. Finally, two sub-tickets will be received by the destination D.

3.6 Route Reply

The purpose of route reply is to confirm the bandwidth reservations that we made in the previous section. Whenever a ticket T reach the destination D, D

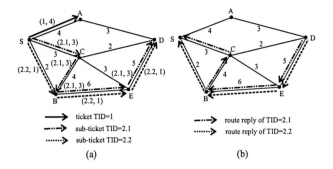

Fig. 4. (a) An example of our ticket distribution. (b) An example of route reply.

can check whether all sub-tickets under the same TID have been received or not. If a satisfactory multi-path has been found, we can send out the reply packets. Each of these reply packets should carry a sub-ticket under the same TID to the host where it was sent (this can be tracked back using the receive set R_i). Also, for each sub-ticket being sent, the corresponding entries in the receive set R_i and the listen set L_i should be deleted. These route reply packets should travel backward to confirm the reserved bandwidths, until the source host S is reached. The operations in the intermediate hosts are the same as the above. Our earlier records in the send sets and receive sets will be able to help the reply packets to track back correctly to the source S. We illustrate the route reply in Fig. 4(b) based on the example in Fig. 4(a).

4 Experimental Results

We have developed a simulator to evaluate the performance of the proposed multi-path QoS routing scheme (which is abbreviated as MP below). For comparison reason, we also simulated two other protocols: Ticket-Based (Tk) and Single-Path (SP). The Tk protocol is that proposed in [2]. Note that the Tk protocol needs the support of DSDV protocol [9] to find out the bandwidth available along each path, and thus is quite different from our on-demand design. For this reason, we also simulated the SP protocol as a referential point to our MP protocol. This protocol is similar to our MP protocol, except that a whole-ticket cannot be split while being forwarded. Thus, we can see how much benefit provided by our protocol.

A MANET in a physical area of size $1000m \times 1000m$ with 100 mobile hosts was simulated. Mobile hosts were randomly distributed in the area. Each mobile host had the same transmission range of 150 meters. Each link owned a bandwidth given by a normal distribution with mean $\mu = 1 \sim 10$ and variance $\sigma = 2$ following the probability density function $f(x) = \frac{1}{\sqrt{2\pi}\sigma} e^{-(x-\mu)^2/2\sigma^2}$. For each QoS route request, the source and destination were randomly selected, and the requested bandwidth was a fixed value of 4. For each request, the number of tickets issued could range from 1 to 10.

Fig. 5 show the success rate to find a QoS route for different protocols. Fig. 6 compares the overhead of each protocol in terms of the number of routing-related packets issued. Each such packet being transmitted for one hop is counted for 1.

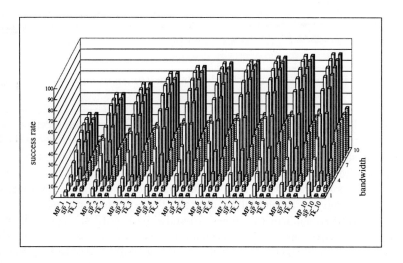

Fig. 5. Success rate under different mean link bandwidth and number of tickets issued (network size = 100 and $\sigma = 2$).

5 Conclusions

We have proposed a multi-path QoS routing protocol for finding a route with a bandwidth constraint in a MANET. As opposed to the proactive routing protocol [2], our protocol is based on an on-demand manner to search for a QoS route, so no global link state information has to be collected in advance. Our protocol flexibly adapts to the status of the network by spending route-searching overhead only when the bandwidth is limited and a satisfactory QoS route is difficult to find.

Acknowledgements

This research is supported in part by the Ministry of Education, ROC, under grant 89-H-FA07-1-4 (Learning Technology) and the National Science Council, ROC, under grants NSC89-2218-E-009-093, NSC89-2218-E-009-094, and NSC89-2218-E-009-095.

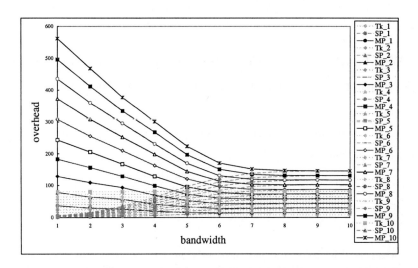

Fig. 6. Routing overhead vs. mean link bandwidth (network size = 100 and $\sigma = 2$).

References

1. J. Broch, D. B. Johnson, D. A. Maltz: The Dynamic Source Routing Protocol for Mobile Ad Hoc Networks (Internet draft) (1998)
2. S. Chen, K. Nahrstedt: Distributed Quality-of-Service Routing in Ad Hoc Networks. IEEE Journal on Selected Areas in Communications 17(8) (1999) 1488–1505
3. C.-C. Chiang: Routing in Clustered Multihop, Mobile Wireless Networks with Fading Channel. Proc. IEEE SICON '97 (1997) 197–211
4. Z. J. Haas, M. R. Pearlman: The Zone Routing Protocol (ZRP) for Ad-Hoc Networks (Internet draft) (1998)
5. Y.-B. Ko, N. H. Vaidya: Location-Aided Routing (LAR) in Mobile Ad Hoc Networks. ACM/IEEE MOBICOM '98 (1998) 66–75
6. W.-H. Liao, Y.-C. Tseng, J.-P. Sheu: GRID: A Fully Location-Aware Routing Protocol for Mobile Ad Hoc Networks. Telecommunication Systems (to appear)
7. C. R. Lin, J.-S. Liu: QoS Routing in Ad Hoc Wireless Networks. IEEE Journal on Selected Areas in Communications 17(8) (1999) 1426–1438
8. S.-Y. Ni, Y.-C. Tseng, Y.-S. Chen, J.-P. Sheu: The Broadcast Storm Problem in a Mobile Ad Hoc Network. ACM/IEEE MOBICOM '99 (1999) 151–162
9. C. Perkins, P. Bhagwat: Highly Dynamic Destination-Sequenced Distance-Vector (DSDV) Routing for Mobile Computers. ACM SIGCOMM Symposium on Communications, Architectures and Protocols (1994) 234–244
10. C. Perkins, E. M. Royer: Ad Hoc On Demand Distance Vector (AODV) Routing (Internet draft) (1998)

Study of a Unicast Query Mechanism for Dynamic Source Routing in Mobile ad Hoc Networks

Boon-Chong Seet, Bu-Sung Lee, and Chiew-Tong Lau

Network Technology Research Centre
Nanyang Technological University
Nanyang Avenue, Singapore 639798
Ebcseet@ntu.edu.sg

Abstract. This paper describes our simulation study of a new route discovery mechanism for mobile ad hoc networks. Based on the concept of unicast query and simulated with the reactive Dynamic Source Routing (DSR) protocol, the mechanism is found to perform most effectively at high node mobility, allowing routes to be acquired with significantly lower overhead. Results obtained using ns2 showed an aggregate 25% reduction of routing overhead with over 96% of data packets successfully delivered using optimal paths in most traffic sessions.

1 Introduction

Mobile ad hoc network (MANET) is a rapidly deployable, multihop wireless network where nodes communicate with each other without relying on any preplaced fixed network infrastructure. In such networks, every node must act as a router to forward packets for other peer nodes. The nodes may move about freely, causing frequent and unpredictable topology changes. In the light of such non-deterministic node behavior, a central challenge to design of any reactive MANET routing protocols is to devise an efficient algorithm to handle the frequent requests for new routes with increasing rate of network topology changes.

Dynamic Source Routing (DSR) [1,2] protocol is an extensively evaluated reactive routing protocol currently under consideration by the IETF as a proposed standard for routing in MANET. Presently, route discovery in DSR is performed using a broadcast mechanism (Fig. 1a), which results in significant network overhead and prevents DSR from scaling efficiently with larger networks.

In this paper, we investigate an alternative approach to performing route discovery. We have augmented DSR with a unicast query mechanism, and termed the modified DSR protocol 'Directed-Search DSR' (DS-DSR). The primary objective of our work is to realize a significant reduction of overhead associated with route discovery and to examine the impact of the new mechanism on the performance of DSR.

2 DS-DSR

Our solution is based on the assumption that nodes are location aware. This assumption is reasonable, given the commercial proliferation of low cost positioning devices such as GPS [3] receivers in recent years.

As an extension to the idea of Location Aided Routing (LAR) inspired by Ko [4], we further proposed that the source uses location information whenever available to make an informed selection of a route to unicast its query to a known target neighbor. That is, sending a query to a node in the vicinity of the target. If any intermediate nodes along the route have a path to the target, the node processing the query will return this path in reply to the source. If no route is found after the query reaches the neighbor, the query will be broadcast as shown in Fig. 1b.

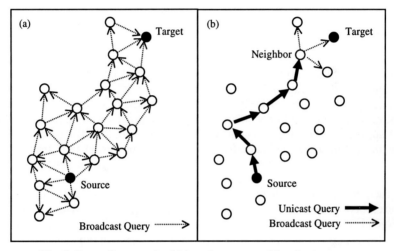

Fig. 1. Route discovery by (a) broadcast query, (b) unicast query

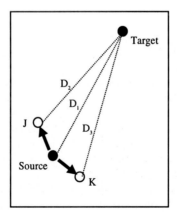

Fig. 2. Constraining the propagation of broadcast query

To further minimize the overhead associated with broadcast query, we augmented the basic operation of unicast query with a mechanism similar to [4] for constraining the propagation of broadcast query only in the direction towards the target. As illustrated in Fig. 2, D1 is the distance between the source and the target, while D2 and D3 are distances to the target from node J and K respectively. Assume that these nodes know the distance between the source and the target, then a query received from the source by node J will be propagated since D2 < D1. Similarly, a query received by node K from the source will be discarded as D3 > D1. In the event that the unicast query path is invalidated as a result of node movements, a new query will be sent either through a different unicast query path (if available) or via a broadcast.

3 Performance Evaluation

3.2 Simulation Environment

We use a detailed simulation model based on ns2 [5] with CMU wireless extensions [6] to evaluate the performance of DSR and DS-DSR. As DS-DSR is developed based on the ns implementation of DSR, we use identical traffic and mobility scenarios of previous reported results for DSR [7] to ensure fairness of comparison. A total of 50 nodes, moving over a 1500m x 300m rectangular flat space are simulated for 900s at a speed up to 20m/s. Each simulation run is modeled with 20 CBR (constant bit rate) connections, which are originated from 14 different sources. Data is sent in 64-bytes packets at 4 packets per second. The movement scenarios were generated based on random waypoint model [2], which are characterized by a pause time. 7 pause times were used in this simulation: 0s, 30s, 60s, 120s, 300s, 600s and 900s. A pause time of 0s corresponds to continuous motion, and a pause time of 900s (length of simulation) corresponds to no motion. We simulate each pause time with 10 different movement scenarios, and present the mean of each metric over these ten runs.

3.2 Simulation Results

We evaluate DSR and DS-DSR using 4 metrics:
   (i)   *Routing overhead* – The total number of routing packets sent which includes the Route Request (RREQ), Route Reply (RREP), and Route Error (RERR) packets. Each hop-wise transmission is counted as one transmission.
   (ii) *Route discovery latency* – The amount of time needed to acquire a route to the destination. This metric measures the delay between the transmission of a RREQ and obtainment of the first RREP that answers the request.
   (iii) *Path optimality* – The difference in number of hops between the route a data packet took to reach its destination and the shortest route that physically existed at the time of transmission.
   (iv) *Packet delivery ratio* – The ratio of the number of data packets successfully delivered to their destinations to those generated by the CBR sources.

3.2.1 Routing Overhead

In Fig. 3, we observed that the total number of routing packets generated by DS-DSR is consistently lower than DSR for all simulated pause times. More than 25% of the routing overhead is reduced at pause time zero when nodes were in constant motion (highest node mobility). Some 75% of the routes were discovered using unicast query, which results in fewer transmissions of RREQ and associated RREP. The margin of reduction, however, diminishes as the pause time increases. At 900s when nodes were stationary, the routing overheads incurred by both protocols were nearly comparable. This diminishment of overhead reduction with pause time is expected because when nodes were less mobile, the routes established were correspondingly more stable, and hence a lesser need for route discovery. Next, we examine how each constituent of the routing overhead (RREQ, RREP and RERR) in DS-DSR performs relative to DSR.

Route Requests

Fig. 4 shows the number of RREQ originated and forwarded for different pause times. Note that we use the term 'originated' to mean a packet that is created and transmitted by its originator, and 'forwarded' to mean a packet that is transmitted by a node other than its originator. DS-DSR achieves a 30% reduction in number of RREQ forwarded at pause time zero, above which the margin again diminishes due to a declining need for route discovery. One interesting observation is the increase in number of RREQ originated by DS-DSR. By minimizing the use of broadcast, the number of routes that a node can possibly learned is reduced. Thus, when a route is invalidated as a result of node movements, a greater tendency exists for the node to obtain a replacement route by initiating a route discovery, giving rise to the higher number of RREQ originated.

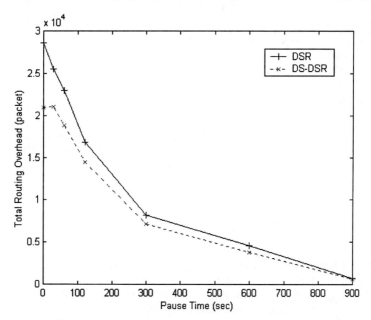

Fig. 3. Total routing overhead of DSR and DS-DSR

Route Replies

As one expects, the number of RREP originated by DS-DSR is substantially reduced as a result of fewer RREQ transmitted (Fig. 5). We also noted with interest number of RREP forwarded by DS-DSR, which seems relatively unaffected by the unicast query mechanism contrary to our intuition. This indicates an increase in average number of hops traversed by RREP, the reason being twofold:

First, the routes discovered by unicast query might not be always the most optimal. Thus, the number of hops that RREP took to return to the source could be higher. Second, the number of routes learned by each node in DS-DSR is fewer relative to DSR. Hence to acquire a route, a RREQ needs propagate further from the source and answer by RREP either from the target or from its neighboring nodes. The proximity of these nodes to the target enables them to better learn of a route to the target even when there is no direct end-to-end traffic between them. For instance, by promiscuous learning (overhearing) when both nodes and target are within wireless range of each other, and as an intermediate node that relays traffic to and from the target.

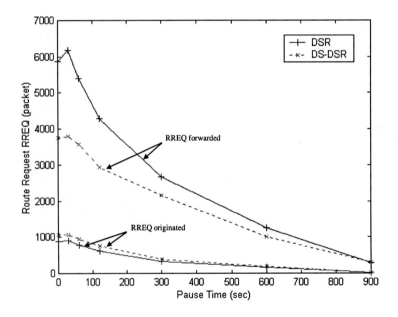

Fig. 4. Route Requests (RREQ) of DSR and DS-DSR

Route Errors

From Fig. 6, we found that DS-DSR has fewer RERR transmitted due to a fresher set of routes used for packet transmissions. The DS-DSR's characteristics of performing frequent but low-cost route discovery, and obtaining fresher routes from nodes closer to the destination (particularly when nodes were operating in promiscuous mode) are reasons that results in fewer transmission errors.

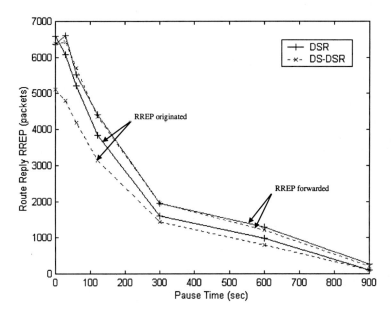

Fig. 5. Route Replies (RREP) of DSR and DS-DSR

3.2.2 Route Discovery Latency

Table 1 summarizes the latency incurred by a route discovery initiator to receive its first RREP in response to RREQ at pause time zero. We found that DS-DSR incurred a mean latency of 25.6ms, increased by nearly a factor of two over that incurred by DSR. This increased in latency is in general due to the greater number of hops over which RREQ must traverse to obtain a route.

Table 1. Latency of first RREP for DSR and DS-DSR

	Max Latency	Min Latency	Mean Latency
DS-DSR	850.4ms	1.7ms	25.6ms
DSR	956.1ms	1.3ms	14.8ms

Another contributing factor is the delay introduced when an initial attempt to acquire a route through unicast query is unsuccessful and a broadcast is performed. We found that this occurs in approximately 18% of the total unicast queries. Other factors such as network congestion plays a less dominating role in influencing latency in DS-DSR due to the fewer routing packets being transmitted. This is evidenced from the fewer occurrences of a full network interface queue that we observed in most scenarios.

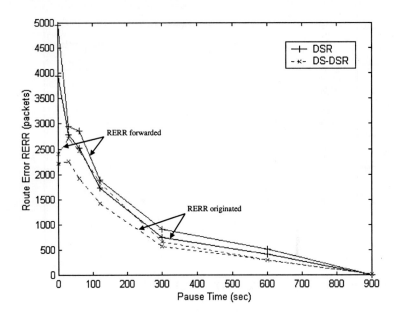

Fig. 6. Route Errors (RERR) of DSR and DS-DSR

3.2.3 Path Optimality

Fig. 7 depicts the difference between the actual number of hops taken by a data packet and the number of hops that are actually required. A difference of 0 means the packets took an optimal (shortest) route, and a difference greater than 0 reflects the additional hops taken by the packets. The statistics obtained shows that more than 80% of the packets in DS-DSR are routed through optimal paths. We also noted that 91% of the packets that did not take the most optimal paths traveled just one additional hop. If we let n be the number of additional hops incurred, J_n and K_n be the number of packets incurring n hops in DS-DSR and DSR respectively, then the average additional hops incurred per data packet can be found using Eq. (1).

$$\frac{\sum_{n=0}^{\infty}\{|J_n - K_n|*n\}}{\sum_{n=0}^{\infty} J_n} \tag{1}$$

By considering packets that fall in the range $0 \leq n \leq 3$ (we neglect those for $n > 3$ as their numbers are comparatively insignificant), we found that DS-DSR incurred on average 0.03 additional hops per data packet.

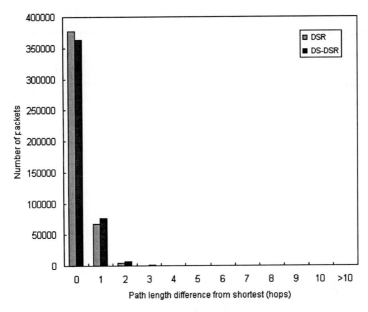

Fig. 7. Path optimality of DSR and DS-DSR

3.2.4 Packet Delivery Ratio

Fig. 8 shows that packet delivery rate for DSR and DS-DSR ranges from 96 to 100% with the latter having a slightly reduced rate of 1% at pause time zero, despite the fact it incurred fewer route transmission errors. This is due to the way packets in error are being handled in DSR. A node only salvages a packet if an alternate route exists in its route cache, otherwise the packet is dropped. Since the number of routes learned per destination per node is relatively fewer in DS-DSR, the proportion of packets in error not salvaged is envisaged to be higher.

4 Conclusion

In this paper, we investigate the performance of DS-DSR, a modified version of DSR optimized with a better, efficient route discovery mechanism based on the concept of unicast query. The major drawback that limits the current effectiveness of DSR is the high overhead it entails due to an explicit dependence on flooding for route discovery.

The new approach emphasizes on minimizing the use of broadcast as opposed to past focus on constraining the search space of broadcast. Much unique to DS-DSR is the characteristic of performing frequent but low cost route discovery, thus potentially capable of obtaining fresher routes with a lower routing overhead.

However, we also noted that broadcast is necessary for an effective distribution of routing information, which otherwise a node might take more time to obtain a route to its destination – a classic tradeoff between bandwidth and latency.

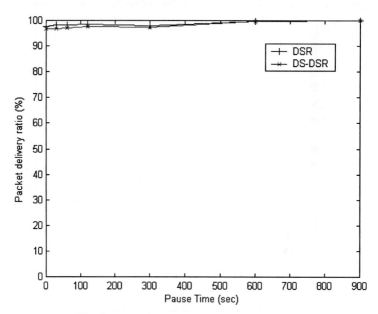

Fig. 8. Packet delivery ratio of DSR and DS-DSR

References

1. Broch, J., Johnson, D.B., Maltz, D.A.: The Dynamic Source Routing Protocol for Mobile Ad Hoc Networks. Internet draft, draft-ietf-manet-dsr-03-txt. (1999)
2. Johnson, D.B., Maltz, D.A.: Dynamic Source Routing in Ad Hoc Wireless Networks. In Mobile Computing, edited by Imielinksi, T., Korth, H., Chpt. 5, Kluwer Academic Publishers (1996) 153-191
3. Dommety, G., Jain, R.: Potential Networking Applications of Global Positioning Systems (GPS). Technical report, TR-24 (1996). Computer Science Department, Ohio State University.
4. Ko, Y., Vaidya, N.: Location Aided Routing (LAR) in Mobile Ad Hoc Networks. Proc. Fourth ACM/IEEE Int. Conference on Mobile Computing and Networking (1998) 66-75
5. Fall, K., Varadhan, K.: ns notes and documentation. The VINT Project, UC Berkeley. URL: http://www-mash.cs.berkeley.edu/ns/
6. The CMU Monarch Project's Wireless and Mobility Extensions to ns. URL: http://www.monarch.cs.cmu.edu/
7. Broch, J., Maltz, D.A., Johnson, D.B., Hu, Y., Jetcheya, J.: A Performance Comparison of Multi-Hop Wireless Ad Hoc Network Routing Protocols. Proc. Fourth ACM/IEEE Int. Conference on Mobile Computing and Networking (1998) 85-97
8. Mobile Ad-hoc Networks (MANET) working group charter, Internet Engineering Task Force (IETF). URL: http://www.ietf.org/html.charters/manet-charter.html

Ad-hoc Filesystem: A Novel Network Filesystem for Ad-hoc Wireless Networks

Kinuko Yasuda[1] and Tatsuya Hagino[2]

[1] Graduate School of Media and Governance,
Keio University, Endo 5322, Fujisawa, Kanagawa, 252–8520, Japan,
kinuko@tom.sfc.keio.ac.jp
[2] Faculty of Environmental Information, Keio University,
Endo 5322, Fujisawa, Kanagawa, 252–8520, Japan,
hagino@tom.sfc.keio.ac.jp

Abstract. In this paper, we propose a new filesystem named Ad-hoc Filesystem which targets ad-hoc wireless networks. Ad-hoc Filesystem is a serverless filesystem which automatically generates temporary shared space among multiple mobile machines when they gather in a communicable range. The generated space is kept stable as long as possible even if some machines join or leave the communicable range. We have designed an initial prototype of Ad-hoc Filesystem and have conducted preliminary evaluations on the simulator. Our design is strongly based on an assumption such that people who wish to work together would form a stable group with a single mobility behavior. Based on the assumption, our prototype distributes files among multiple machines and duplicates each on two machines, then it keeps the pairs of such machines as long as possible. We also introduced another strategy to increase file availability by exploiting each machine's client cache. The simulation results show that controlling packets used in our prototype to maintain system state hardly affects the overall system performance. Also the result indicates that our strategy that keeps two replicas for each data and exploits client caching is effective to achieve high availability.

1 Introduction

Currently, our mobile computing environment is getting split into two extremes: *private* one which is isolated on each machine and *public* one which is connected to the huge virtual world, Internet. Usually we can manage things well by choosing them appropriately. However, this also means that we cannot share information easily even with people here. We can carry our mobile machines around freely, but our machines are remain disjointed. In order to share information with people around, we have to connect to the Internet via somehow charged telephone lines and with distant providers.

Some machines have *peer-to-peer* communication functionality like IrDA, with which two machines can connect to each other. However, even with such functionality, we should also have tough time to share information with two or more machines. Things are almost same for more rich functional computers. With such computers, we may be able to set up local server to share information temporarily, but such way will significantly degrade the advantage of mobility.

Thus, our question is simple and natural: *why we cannot share information easily with people HERE?*

In order to improve such situations, we propose a new filesystem named "Ad-hoc Filesystem". Ad-hoc Filesystem is a serverless filesystem which automatically generates temporary shared space among multiple mobile machines when they gather in a communicable range. The generated space is freely available for participating machines and can be used for any temporary work there. In addition, Ad-hoc Filesystem keeps the shared space stable as long as possible, even if some of machines arbitrarily join or leave the space.

The design of Ad-hoc Filesystem is strongly motivated by the recent wireless network technologies and advances of ad-hoc network techniques [8, 12]. However, our objective intending to construct a filesystem on fully unwired networks emerges several difficulties which have not existed in existing network filesystems such as [1, 14, 11].

In this paper, we describe our prototype design of Ad-hoc Filesystem and show the simulation results which demonstrates how our design works and realizes the system.

2 Assumptions

Our design is strongly based on an assumption such that people who wish to work together would gather and form a logical group with a single mobility behavior. Such mobility model is much studied in a research area of mobility models for ad-hoc networks, and is considered far more realistic than random walk mobility assumptions [7].

In addition to the basic assumption, we assume that two (or more) machines rarely leave the group simultaneously as long as they wish to work together. Our assumption means that our fucus sits on the group virtually formed by multiple machines, rather than on each one who may leave or join. This simplifies problems we should attack and makes our design practical.

We should also take account of several issues relating wireless network characteristics. We have defined that the "communicable range" as single-hop wireless transmission range. Because our initial target is IEEE 802.11b wireless LAN devices which have a 50m nominal range, we expect that the range of single-hop would be enough to realize main concept of the virtual group in Ad-hoc Filesystem. Focusing on single-hop also greatly simplifies our design. Currently our design does not assume possibility of nonuniform propagation characteristics of wireless transmission.

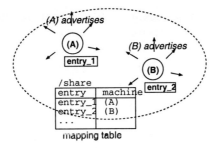

Fig. 1. Maintaining Directory Entries by Using Advertising Packets

3 Design

In the initial stage of designing Ad-hoc Filesystem, we have much focused on one point: maintaining redundancy for each file and increasing file availability. In order to realize stable shared file space among unstably networked machines, we designed an initial prototype based on a simple strategy – master and shadow.

3.1 Master and Shadow

The main keys in our initial prototype is roughly divided into following two. First, because we cannot assume a central server which should not move in or out of the communicable range, the system distributes each file among participating machines. Hence, in our system we can say that every machine is both server and client. In order to manage mappings between each top-level directory entry and its serving machine, our initial prototype uses a soft-state mechanism named *advertising*. Each machine periodically broadcasts *advertising* packet which advertises that the machine is alive and is responsible for serving the providing directory entries (figure 1).

Second, not to lose any data regardless of group reconfigurations, the system keeps redundancy by maintaining duplications for all the data. Based on our assumptions, *two machines rarely leave the group at the same time*, Ad-hoc Filesystem replicates every file on two different machines. Keeping the pair of these two machines, called *master* and *shadow*, is essential in our system.

How the shared space is generated and used is as follows. First, when two or more machines gather in a communicable range, an *empty* shared space is virtually formed and becomes available. When one of them put a file on the shared space, the machine becomes *master* (First Writer policy) and begins searching another machine to initiate *shadow* for the file. Then the machine who put the file is responsible for accepting filesystem requests and begins periodic advertising broadcast. After that, if the machine leaves the area and becomes isolated, other remaining machines can still access the file and see the shared space stably.

In order to keep master-shadow pair, each shadow is responsible for replying ack packtes to advertising packets from its master. Notice that the advertising packets are broadcast but the ack packets with them are unicast and replied by only one machine, its shadow, therefore it is not necessary to worry about packet flooding.

Every master-shadow pair obeys simple rules as following.

- If a master leaves the area,
  - the shadow becomes a new master,
  - and it begins searching a candidate for its new shadow.
- If a shadow leaves the area,
  - the master begins searching a candidate for its new shadow.

Searching candidate machines for its shadow is done by choosing random machine from the set of known machines if the master have received keep-alive advertises from enough number of other machines. Otherwise, the master sends *search shadow* request as broadcast to choose its shadow. Machines which receive this packet and have available free space for the shared space should reply *shadow propose* with it. To avoid packet collisions among some or all of these replies and naturally achieve load balancing, each machine delays slightly and performs following actions before replying with it:

1. A machine picks a delay period $d = K \times (p + r)$, where p is the number of files the machine is currently providing (including both master and shadow), r is a random number between 0 and 1, and K is a small constant.
2. The machine delays transmitting the reply for a period of d.
3. Within this delay period, the machine promiscuously receives all packets. If the machine receives a reply packet during the delay period, it cancels the reply.
4. Otherwise, the machine transmits the reply packet after the delay period.

Therefore, the master searching shadow will ideally get only one reply and can initiate shadow on the most appropriate one.

3.2 Exploiting Client Cache

Our initial prototype based on master and shadow does not assume any user activity on the filesystem nor existence of client caching. However, as observed in previous research for mobile filesystems [10, 13], client caching can effectively used to increase file availability on each client in disconnected state. Although disconnected operation on each machine is not our main goal, client cache may also be exploited to increase availability for the whole system in our prototype.

There are also several advantages if the system can exploit client cache. First, since each machine will cache files when it opens them, important files which have been opened by many machines will have got a lot of copies in the group. Second, machines which open many shared files will be supposed to be active on the group activity, hence they are also supposed to remain in the group relatively longer than others. Therefore, by exploiting client caching, important files will be able to have many cached copies on active (thus stable) machines.

Based on the above considerations, we re-designed our prototype to be more realistic. The basic strategy is hinted by an idea of disconnected operation [10] and cooperative caching [5, 4]. Each machine caches whole file data on **open**. When the master machine becomes unavailable, other machines try to locate an alternative machine which have the latest cached copy of the file.

On worst case, each machine will be able to choose the latest one by performing broadcast query and collecting answers. However, if a large number of machines perform broadcast at the same time when the master disappears, the network will be flooded easily. To avoid such broadcast storms, our prototype let machines choose one *coordinator* machine for the file. Instead of letting each machine solve the problem independently, the coordinator maintains cache status for the file and performs broadcast as a representative of every other machine if cache information is not available at that time.

To minimize broadcast overhead, each machine chooses a coordinator by non-networking operation, i.e. using consistent hash functions. Consistent hash enables multiple machines to locate one identical machine on most cases even without consistent view of the network among machines. Each machine performs as follows:

1. Each machine calculates hash values for each other machine in the group. Machines are collected by exchanging keep-alive messages such as advertising.
2. When a machine wants to open a file whose master has become unavailable, it calculates the hash value for its path name, then it chooses a coordinator whose hash value is nearest to the file's hash value.
3. The machine consults with the coordinator about the file.
4. If the coordinator has not known enough information about the file, it performs broadcast and collects information about the file's cached copies on other machines.
5. Once the coordinator has got information about the file, it can locate the new master machine which has the latest copy and can tell others about it.

While the master stays in the group, the master itself manages the multiple cached copies. The master can also inform the coordinator about the status of cache beforehand, thus the coordinator will rarely have to perform broadcast as long as the master and the coordinator leave the group simultaneously.

4 Simulation Results

To evaluate performance and validity of our design, we have implemented a simulator of Ad-hoc Filesystem using a packet-level simulation language called PARSEC [2].

The simulation parameters were chosen to model a network consisting of 10 mobile nodes moving around in a square area. The communication medium used in the simulation was IEEE 802.11b Wireless LAN with the bandwidth 2 Mbytes per second and the communication radius of 50m. Each simulation was

Table 1. Parameter Values Used in Simulations

Parameter	Value
Period between advertising broadcasts	5 sec
Max # of search shadow retransmissions	5
Max time to wait for a shadow reply	1000msec
# of allowed advertising losses before each machine assumes the master is disconnected	5
# of allowed advertising losses before shadow begins advertising	3

a run for 60 minutes in simulated clocks, and each node created and opened files at randomly selected times throughout the simulation. The average size of files created by nodes was 10 Kbytes. The parameter values used in simulations are listed in table 1.

Figure 2 shows the measured overhead of control broadcast packets. In this simulation, we used an area of 40m square and each node was initially placed at a random position within the area. During the simulation, each node moved to a random spot with a velocity between 0 and 1.0 meters per second, and paused at its current location for a short period of 60 seconds. To measure effects of broadcast to data throughput, node 0 was continuously sending data to node 1. The observed throughput between node 0 and node 1 is also shown in the figure. X-axis describes the total number of files created in each simulation.

The result shows that the broadcast packets hardly affect the observed throughput, though the traffic increases as the number of files increases. Although the measured traffic remains very small in our all simulations, it is clear that the traffic will affect the throughput if the number of files increases up to a very large number such as several hundreds. However, we expect our decision will be reasonable, because the shared directory are designed for temporary use: it is generated as empty and disappears after a while, so that it would not include such a large number of files.

Figure 3 shows effects of redundancy by shadows. Our main concern is to provide stable file space for machines within a group which is naturally formed by multiple machines. To simulate and evaluate a visible file space within such a group, each node is arranged to perform as following in the simulations:

1. enters the group (i.e. communicable range) at random time,
2. stays in the group during the random period, then
3. leaves the group at random time.

Each node created files while it stays in the group, and the total number of created files were 20 for each run. Each node leaves the group arbitrarily, but we made one restriction for the mobility: each node cannot leave the group if the group has only the one node at the time. Hence, the group remained having at least one node for the simulation period.

In figure 3, x-axis shows percentages of average time each node stays in the group and y-axis shows percentages of **open** call failures during the runs. We

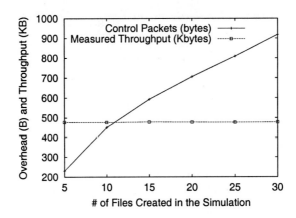

Fig. 2. Overhead of controlling packets.

have run this simulation with various random seed numbers for each case, the case without shadows, the case with shadows, and the case with shadows and client caching. In 3, each dot describes the result of each run. The result shows that our strategy using shadow and client caching can effectively increase file availability. With shadows, the open calls performed by each node hardly fails, though some failures are still observed when average time each node stays in the group is very short, e.g. less than 35%. On the other hand, if the system also exploit the client caches to increase availability, open call failures can be avoided almost at all.

5 Related Work

There are several researches in an area of serverless filesystems, but all of them have quite different assumptions for underlying machines and networks from our work. For example, xFS [1] is an early one of serverless network filesystems, but it is targeted workstation clusters tightly connected via fast switched LANs.

A technique using hash functions to locate one machine in a distributed system is much studied in various research areas. In an area of cooperative caching, PACA [4] uses a hashing technique to locate a client which should keeps a cached copy. [9] uses consistent hashing to manage distributed cooperative Web caching. Some of recent peer-to-peer mechanisms like FreeNet [6] also uses hash functions to locate each data. The philosophy of FreeNet such that important files should be naturally cached on many sites much affected our design of Ad-hoc Filesystem too.

In an area of ad-hoc networks, several multi-hop routing techniques have been studied ([8, 12]). Although currently Ad-hoc Filesystem does not assume multi-hop wireless networks, some techniques are borrowed from them, such as delaying replies with broadcast packets and promiscuously receiving other packets within the delay period.

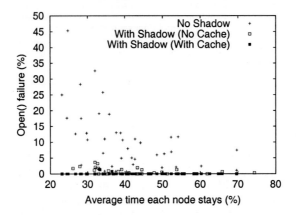

Fig. 3. Effects of shadows and effects of exploiting client cache.

6 Conclusion

In this paper, we have described design of Ad-hoc Filesystem, which is a serverless filesystem that automatically generates a temporary shared work space among mobile machines when they gather as near as they can communicate. Our system will be the first network filesystem targeting ad-hoc wireless networks and is motivated by several techniques and characteriscs of wireless networks.

Our main challenge is how to keep the shared space stable even if some of machines arbitrarily join and leave the space. We effectively simplify the problem by making one assumption such that people who wish to work together would form a stable group in an area. The system distributes all files among participating machines and intend to keep pair of *master* and *shadow* for each file. The system also exploits client caching to increase availability for the entire group, and uses a hash technology to manage multiple client caches in the unstable group.

Based on our initial design, we have implemented a packet level simulator to evaluate our prototype. The simulation results show that the broadcast packets used in our prototype to maintain system state hardly affects the overall network performance. Also the results indicate that our strategy such that keeps two replicas for each data is effective to provide enough stability and availability.

We have just finished the first step of designing Ad-hoc Filesystem. Although further investigation based on system-level implementation remains important work, we believe that the prototype has a contribution to realize the possibility of ad-hoc filesystem by synthesizing number of previous works in areas such as disconnected operation, cooperating caching, consistent hash, and ad-hoc wireless networks.

References

1. T. Anderson, M. D. Dahlin, J. M. Neefe, D. A. Patterson, D. S. Roselli, and R. Y. Wang., Serverless Network File System, In *Proceedings of 15th ACM Symposium on Operating Systems Principles*, ACM, Copper Mountain Resourt, Colorado, Dec 1995.
2. R. Bagrodia, R. Meyer, M. Takai, Y. Chen, X. Zeng, J. Margin, and H. Y. Song., PARSEC., A Parallel Simulation Environment for Complex Systems, *IEEE Computer*, 31(10).,77-85, Oct 1998.
3. A. Birrell, A. Hisggen, C. Jerian, T. Mann, and G. Swart., The Echo distributed file system. In Technical Report 111, Degital Equipment Corp., Systems Research Center, Palo Alto., CA, 1993.
4. T. Cortes, S. Girona, and J. Labarta., PACA: A Cooperative File System Cache for Parallel Machines., In *Proceedings of the 2nd International Euro-Par Conference*, I:477–486, 1996.
5. M.Dahlin, R.Wang, T.Anderson, and D. Patterson. Cooperative Caching., Using Remote Memory to Improve File System Performance. In *Proceedings of 1st OSDI*, Nov 1994.
6. The Free Network Project. http://freenet.sourceforge.net/.
7. X. Hong, M. Gerla, G. Pei, and C.-C. Chiang, "A Group Mobility for Ad Hoc Wireless Networks," In *Proceedings of ACM/IEEE MSWiM'99*, Seattle, WA, Aug. 1999, pp.53-60.
8. D. B. Johnson., Dynamic Source Routing in Ad Hoc Wireless Networks, In *Proceedings of the Workshop on Mobile Computing Systems and Applications*, pp. 158-163, IEEE Computer Society, Santa Cruz, CA, Dec 1994.
9. D. Karger, A. Sherman, and et al., Web Caching with Consistent Hasing, In *Proceedings of WWW8*, May 1999.
10. J. J. Kistler and M. Satyanarayanan., Disconnected Operation in the Coda File System., *ACM Transactions on Computer Systems*, 10(1).,3-25, Feb 1992.
11. J. Howard, M. Kazar, S. Menees, D. Nichols, M. Satyanarayanan, R. Sidebotham, and M. West., Scale and Performance in a Distributed File System. *ACM Transactions on Computer Systems*, 6(1).,51-82, Feb 1988.
12. E. Royer and C. Toh, "A Review of Current Routing Protocols for Ad-Hoc Mobile Wireless Networks", IEEE Personal Communications, Apr. 1999.
13. D. Terry, M. Theimer, K. Petersen, A. Demers, M. Spreitzer, and C. Hauser., Managing Update Conflicts in Bayou, a Weakly Connected Replicated Storage System. In *Proceedings of the Fiftheenth ACM Symposium on Operating Systems Principles*, pp.172-183, Dec 1995.
14. D. Walsh, B. Lyon, G. Sager, J. M. Chang, D. Goldberg, S. Kleiman, T. Lyon, R. Sandberg, and P. Weiss., Overview of the Sun Network File System. In *Proceedings of the 1985 USENIX Winter Conference*, Dec 1995.

A Review of Current On-demand Routing Protocols

Mehran Abolhasan[1], Tadeusz Wysocki[1], and Eryk Dutkeiwicz[2]

[1] Switched Network Research Centre, University of Wollongong, Australia, 2522
{ma11, wysocki}@uow.edu.au
[2] Motorola Australia Research Centre, 12 Lord St, Botany, NSW 2525, Australia
Eryk.Dutkiewicz@motorola.com

Abstract. Mobile ad hoc networks are data networks entirely made up of end-user communication terminals (known as nodes). Each node in the network can act as an information sink (i.e. a receiver), a source and a router. All nodes have a transmission range, which is limited by their transmission power, attenuation and interference. Mobile ad hoc networks have a number of disadvantages over wired networks. These include limited bandwidth in the wireless medium, limited power supply and mobility. The traditional routing algorithms such as DBF will not work in such networks due to lack of scalability and ability to cope with highly mobile networks. Recently, a number of routing protocols have been designed to overcome these issues. These protocols can be classified into three different categories: global, on-demand and hybrid routing. In this paper, we provide a performance comparison for on-demand routing protocols, which is based on a number of different parameters. This paper also considers which protocol characteristics will produce better performance levels in these networks, and what improvements can be made to further increase the efficiency of some of these routing protocols.

Keywords: On-demand routing, Mobile ad-hoc networks, Uni-cast routing protocols.

1 Introduction

Mobile wireless communication networks can be classified into two categories: infrastructured and infrastructureless mobile networks. The infrastructured networks have base stations that are used to provide a communication barrier (or a bridge), between the mobile nodes [1]. The infrastructureless networks (commonly known as "Ad-Hoc networks") have no centralized controller or a fixed router. Each mobile node in the network acts as a router, which means that they are capable of determining and maintaining routes to nodes in the network. Each node in such a network has a transmission range, which is limited by the transmission power, attenuation and interference. Direct communication can occur between two intermediate nodes if they are within each other's transmission range [2].

Mobile ad hoc networks have numerous types of applications, ranging from the search and rescue operations where the rescue team can combine their efforts to save victims of fire, earthquakes and other natural disasters [3], to the highly dynamic battlefield environment where different types of forces require to exchange information rapidly. The main advantage of infrastructureless networks is that they allow mobile devices to communicate in places where no centralized controller exists, such as in the middle of a jungle or a desert. However, before mobile ad hoc networks can be used successfully in the scenarios described above, an intelligent routing strategy is required to efficiently use the limited resources in these networks, namely, bandwidth, power and storage space. The aim of this paper is to provide a performance comparison for the currently proposed on-demand routing protocols. It also presents a number of ways in which the efficiency of some of these protocols can be improved. In addition, the paper discusses future research directions in designing more efficient routing protocols.

Section 2 gives an overview of the currently proposed routing protocols by classifying them into different categories and discussing the fundamental methodology used in each category. Section 3 provides a performance comparison of the currently proposed on-demand routing protocols, which is based on their theoretical routing cost and basic characteristics. Moreover, it discusses different methodologies, which can improve their performance. Section 4 concludes the paper.

2 Overview of Current Routing Protocols

Ever since the early 1970's a number of routing protocols have been proposed for mobile ad-hoc networks. Many of these protocols are direct descendents of the traditional Link State and distance vector algorithms. These routing protocols can be categorized into three different groups: Global/Proactive, On-demand/Reactive and Hybrid (reactive and proactive) routing protocols. In Global routing protocols, the routes to all destinations are determined at the start up, and they are maintained via periodic update processes [4]. In On-demand routing protocols, the routes are determined when they are required by the source. Hybrid routing protocols exhibit both the reactive and proactive properties at different times.

2.1 Global/Proactive Routing Protocols

In global routing protocols, each node stores and maintains routing information to every other node in the network. The routing information is usually kept in a number of tables. These tables are periodically updated as the network topology changes. The differences between these routing protocols exist in a way the routing information is updated, detected, and the type of information kept at each routing table (note that each protocol may also have a different number of tables) [1].

Global routing protocol can be divided into two separate categories: Flat and Hierarchical.

In networks employing flat-routing protocols, all nodes are treated equally and routing is based on peer-to-peer connection (intermediate nodes), which is restricted by the propagation condition of the network. Some examples of the flat global routing protocols include DSDV [18], WRP [17], FSR [5], GSR [11] and DREAM [6].

Hierarchical networks are made up of a number of levels (usually there are at least 2 levels). On the lowest level, the nodes with the closest geographical proximity form a cluster. In each cluster, at least one node is designed to serve as a "gateway", which usually has a higher transmission range than the other nodes. All communications between nodes in different clusters are done through the gateway node [9]. Each cluster also has a cluster-head, which controls the transmission there.

Hierarchical based routing strategies have been designed to overcome the scalability problem that exists in most of such flat global routing strategies, by introducing a structure to the network. Recently, a number of hierarchical routing strategies have been proposed. Some examples of the most recent proposals include CGSR [12], MMWN [20], HSR [10] and STAR [19].

2.2 On-demand/Reactive Routing Protocols

In on-demand routing protocols, routes are created when required by the source node, rather than storing up-to-date routing tables. The specific aim of on-demand routing is to reduce the bandwidth and storage space used in table driven protocols. In this strategy, routes are generally created and maintained by two different procedures, namely: route discovery and route maintenance. Route discovery procedure is used when a node requires a route to a certain destination. This is achieved by flooding Route Request packets through the network. When a route is found, the destination sends a Route Reply packet, which contains the hop-by-hop address from the source to destination. Once the connection is established, it is maintained by route maintenance procedures until the destination is no longer reachable or the route is not required [1].

Some examples of on-demand routing protocols include: DSR [15], AODV [13], LMR [22], TORA [14], ABR [24], SSR [23], RDMR [16], LAR [7], CBRP [25], ROAM [26]. Section 3 provides a performance comparison for these routing protocols.

2.3 Hybrid Routing Protocols

Hybrid routing protocols combine the basic properties of the two classes of protocols mentioned earlier into one. That is they are both reactive and proactive in nature. The primary objectives of these protocols are to minimize the route construction delay, which is the main problem with reactive routing protocols, and reduce bandwidth usage of the periodic update procedures in proactive protocols. This is achieved by performing some operations proactively and others reactively. That is, using the proactive and reactive operations where they are best suited with the examples Hybrid protocols being ZRP[8] and ZHLS[21].

3 Comparison of Current On-demand Routing Protocols

3.1 Characteristic Comparison

The discussion in this section is based on Tables 1 and 2. Table 1 provides a comparison of the basic characteristics and Table 2 illustrates the complexity comparison. Note that the complexity comparison represents the worst-case scenario for each protocol.

The AODV routing protocol is based on DSDV and DSR algorithm. It uses the periodic beaconing and sequence numbering procedure of DSDV and a similar route discovery procedure as in DSR. However, there are two major differences between DSR and AODV. The most significant difference is that in DSR each packet contains the full routing information, whereas in AODV the packets carry the destination address only. This means that AODV has potentially less routing overheads than DSR. The other difference is that the route replies in DSR carry the address of every node along the route, whereas in AODV the route replies only carry the destination IP address and the sequence number [1]. The advantage of AODV is that it is adaptable to highly dynamic networks. However, node may experience large delays during route construction, and link failure may initiate another route discovery, which introduces extra delays and consumes more bandwidth as the size of the network increases.

As stated earlier, the DSR protocol requires each packet to carry the full address from source to the destination. This means that the protocol will not be very effective in large networks. Since the amount of overhead carried in the packet will continue to increase as the network size increases. Therefore in highly dynamic and large networks, the overhead may consume most of the bandwidth. However, this protocol has a number of advantages over routing protocols such as AODV, LMR and TORA, and in small to medium size networks (perhaps up to a few hundred nodes), this protocol may perform better. An advantage of DSR is that nodes store multiple routes in their route cache, which means that the source node can check its route cache for a valid route before initiating a route discovery. Certainly, if a valid route is found there is no need for route discovery. This is very beneficial in network with low mobility, since the routes stored in the route-cache will be valid longer. Another advantage of DSR is that it does not require any periodic beaconing. Therefore nodes can enter sleep node to conserve their power. This also saves a considerable amount of bandwidth in the network.

The ROAM routing protocol uses internodal coordination along directed acyclic sub-graphs, derived from the routers' distance to the destination. This operation is referred to as a "diffusing computation" [26]. An advantage of this protocol is that it eliminates the search-to-infinity problem present in some of the on-demand routing protocols by stopping multiple flood searches when the required destination is no longer reachable. Another advantage is that each router maintains entries (in a route table) for destinations flowing data packets through them. This reduces significant amount of storage space and bandwidth needed to maintain an up-to-date routing table [26].

Another novelty of ROAM is that each time the distance of a router to a destination changes by more than a defined threshold, it broadcasts update messages to its neighbouring nodes, as described earlier. Although this has the benefit of increasing the network connectivity, in highly dynamic networks it may prevent nodes entering sleep mode to conserve their power.

Table 1: Basic characteristics of on-demand routing protocols.

Protocol	Routing structure	Multiple routes availability	Beaconing requirements	Route metric method	Routes maintained in	Route maintenance method
AODV	F	No	Yes, hello messages.	Freshest & SP	Route Table	Erase Route then SN
DSR	F	Yes	No	SP, or next available in route cache.	Route Cache	Erase Route then SN
ROAM	F	Yes	No	SP	Route Table	Erase Route & **
LMR	F	Yes	No	SP, or next available route.	Route Table	Link reversal & Route repair
TORA	F	Yes	No	SP, or next available route.	Route Table	Link reversal & Route repair
ABR	F	No	Yes	Strongest Associativity & SP & *	Route Table	LBQ
SSR	F	No	Yes	Strongest Associativity & stability	Route Table	Erase Route then SN
RDMR	F	No	No	Shortest relative Distance or SP	Route Table	Erase Route then SN
LAR (DSR based)	F	Yes	No	SP	Route Cache	Erase Route then SN
CBRP	H	No	No	First available route (first fit)	Route Table at cluster head	Erase Route The SN & local route repair

H = Hierarchical F = Flat SP = Shortest Path SN = Source Notification (via route error message)
LBQ = Localize Broadcast Query. * Route relaying load and cumulative forwarding delay.
** Start a diffusing search if a successor is available, else send a query with infinite metric.

The LMR protocol is another on-demand routing protocol using a flooding technique to determine its routes. The nodes in LMR maintain multiple routes to each required destination. This increases the reliability of the protocol by allowing nodes to select the next available route to a particular destination without initiating a route discovery procedure. An advantage of this protocol is that each node only maintains routing information to their neighbours. As a result, extra delays and storage overheads associated with maintaining complete routes can be avoided. However, LMR may produce temporary invalid routes, which introduces extra delays in determining a correct loop.

The TORA routing protocol is based on the LMR protocol. It uses similar link reversal and route repair procedure as in LMR, and also the creation of a DAGs, which is similar to the query/reply process used in LMR [1]. Therefore, it also has same benefits as LMR. The advantage of TORA is that it has reduced the far-reaching control messages to a set of neighbouring nodes, where the topology change has occurred. In addition TORA also supports multicasting. However, this is not incorporated into its basic operation. TORA can be used in conjunction with Lightweight Adaptive Multicast algorithm (LAM) to provide multicasting. The disadvantage is that algorithm may produce temporary invalid routes as in LMR.

ABR is another source initiated routing protocol. It also uses a query-reply technique to determine routes to the required destinations. However, route selection is primarily based on stability in ABR. To select stable route each node maintains an associativity tick with their neighbours, and the links with higher associativity tick are selected in preference to the ones with lower associativity tick. Albeit this may not lead to the shortest path to the destination, the routes tend to last longer. Therefore, fewer route reconstructions are needed, and more bandwidth will be available for data transmission. The disadvantage of ABR is that it requires periodic beaconing to determine the degree of associativity of the links. This beaconing requirement forces all nodes to stay active resulting in additional power consumption. Another disadvantage is that it does not maintain multiple routes or a route cache, which means that alternate routes will not be immediately available, and a route discovery will be required after link failure. However, ABR has to some degree compensated for not having multiple routes by initiating a localized route discovery procedure.

SSR is a descendent of ABR. SSR selects routes based on signal strength and location stability rather than using an associativity tick. As in ABR, the routes selected in SSR may not result in the shortest path to the destination. These routes, however tend to live longer resulting in less route reconstructions. One disadvantage of SSR when compared to DSR and AODV is that intermediate nodes cannot reply to route requests sent toward a destination. This may potentially create long delays before a route can be discovered, because the destination is responsible for selecting the route for data transfer. Additionally, no attempt is made to repair routes at the point were the link failure occurs. In SSR the reconstruction occurs at the source. This may introduce extra delays, since the source must be notified of the broken like before another one can be searched for. Therefore, it would be interesting to investigate the effect of introducing a localized route repair methodology in SSR. Additionally, one would have one to compare the performance between "intermediate node" replying against the "destination node only" reply strategy.

RDMR attempt to minimize the routing overheads by calculating the distance between the source and the destination, therefore limiting each route request packet to certain number of hops. This means that the route discovery procedure can be confined to localized region. RDMR also uses the same technique when link failures occur. This conserves a significant amount of bandwidth and the battery power. Moreover, RDMR does not require a location-aided technology to determine the routing patterns. However, the relative-distance micro-discovery procedure can only be applied if the source and the destinations have communicated previously. If no previous communication record is available for a particular source and destination, then the protocol will behave in the same manner as the flooding algorithms.

LAR is based on flooding algorithms (such as DSR) [7]. It attempts to reduce the routing overheads present in the traditional flooding algorithm by using location information. This protocol assumes that each node knows its location through a GPS. Two different LAR schemes were proposed in [7], in the first scheme a request zone is calculated, which defines a boundary where the route request packets can travel to reach the required destination. The second method stores the coordinates of the destination in the route request packets. These packets can only travel in the direction were the relative distance to the destination becomes smaller as they travel from one hop to another. Both methods limit the control overhead transmitted through the network and hence conserve bandwidth. They will also determine the shortest path (in most cases) to the destination, since the route request packets travel away from the source and towards the destination. One disadvantage of this protocol is that each node is required to carry a GPS. Moreover, the protocols may behave similarly to flooding protocols (e.g. DSR and AODV) in highly mobile networks.

Unlike the on-demand routing protocols described so far, in CBRP, the nodes are organized in a hierarchy. As in most hierarchical protocols, the nodes in CBRP are grouped into clusters. Each cluster has a cluster-head, which coordinates the data transmission within the cluster and to other clusters. The advantage of CBRP is that only cluster heads exchange routing information. Hence, the number of control overhead transmitted through the network is far less than in the traditional flooding methods. Of course with any other hierarchical routing protocol, there are overheads associated with cluster formation and maintenance. The protocol also suffers from temporary routing loops. This is because some nodes may carry inconsistent topology information due to long propagation delays.

3.2 Routing Cost Comparison and Scalability

Generally, most on-demand routing protocols have the same routing cost when considering the worst-case scenario. This is due to their fundamental routing nature, as they all follow similar route discovery and maintenance procedure. For example, protocols such as RDMR and LAR have the same cost as the traditional flooding algorithm in the worst-case scenario. The worst-case scenario applies to most routing protocols when there is no previous communication between the source and the destination. This is usually the case during the initial stages (i.e. when a node comes on-line). As the nodes stay longer on, they are able to update their routing tables/caches and become more aware of their surroundings. Some protocols take advantage of this more than the others. For example, in DSR when a route to a destination has expired in the route cache, the protocol initiates a network wide flooding search to find an alternate route. This is not the case for LAR or RDMR where the route history is used to control the route discovery procedure by localizing the route requests to a calculated region. Clearly, this is more advantageous in large networks, since more bandwidth is available there for data transmission.

Another method used to minimize the number of control packets is to select routes based on their stability. In ABR and SSR the destination nodes select routes based on their stability. ABR also allows shortest path route selection to be used during the route selection at the destination (but only secondary to stability), which means that shorter delays may be experienced in ABR during data transmission than in SSR. These protocols may perform better than the purely shortest path selection based routing protocols such as DSR. However, they may experience scalability problem in large network since each packet is required to carry the full destination address. This is because the probability of a node in a selected route becoming invalid will increase by $O(a.n)$, where "a" is the probability of the route failing at a node and "n" is the number of nodes in the route. Therefore, these protocols are only suitable for small to medium size networks.

Table 2: Complexity comparison of the on-demand routing protocols.

Protocol	TC[RD]	TC[RM]	CC[RD]	CC[RM]	Advantage	Disadvantage
AODV	O(2d)	O(2d)	O(2N)	O(2N)	Adaptable to highly dynamic networks.	Scalability problems, large delays, hello messages.
DSR	O(2d)	O(2d)	O(2N)	O(2N)	Route cache, promiscuous overhearing.	Scalability problems as network size grows. Control packets carry full address. Large delays
ROAM	O(d)	O(a)	O($|E|$)	O($6G_x$)	Elimination of search-to-infinity problem.	May experience large CO in highly mobile environments, as number of threshold update increase.
LMR	O(2d)	O(2d)	O(2N)	O(2a)	Multiple routes	Temporary routing loops.
TORA	O(2d)	O(2d)	O(2N)	O(2a)	Multiple routes	Temporary routing loops.
ABR	O(d + f)	O(b + f)	O(N + e)	O(a + e)	Stability of routes.	Each packet carries full address to destination.
SSR	O(d + f)	O(b + f)	O(N + e)	O(a + e)	Same as ABR	Large delays during route failure and reconstruction.
RDMR*	O(2d)	O(2d)	O(2N)	O(2N)	Localized route discovery.	Flooding used if there is no prior communication between nodes.
LAR*	O(2S)	O(2S)	O(2M)	O(2M)	Controlled route discovery.	Requires GPS, may act similar to flooding in highly mobile networks.
CBRP*	O(2d)	O(2b)	O(2x)	O(2a)	Only cluster heads exchange routing information.	Cluster maintenance, temporary loops.

TC = Time Complexity. CC = Communication Complexity. RD = Route Discovery. RM = Route Maintenance.
CO = control Overhead. d = diameter of the network. x = number of clusters (each cluster has one cluster-head)
N = number of nodes in the network. e = number of nodes forming the route reply path. a = number of affected nodes.
f = diameter of directed path of the route reply. b = diameter of the affected area. G = maximum degree of the router.
$|E|$ = Number of edges in the network. S = diameter of the nodes in the request zone. M = number of nodes in the request zone.

Reduction in control overhead can be obtained by introducing a hierarchical structure to the network. CBRP is a hierarchical on-demand routing protocol, which attempts to minimize control overheads disseminated into the network by breaking the network into clusters. During the route discovery phase, cluster-heads (rather than each intermediate node) exchange routing information. This significantly reduces the control overhead disseminated into the network when compared to the flooding algorithms. In highly mobile networks, CBRP may incur significant amount of processing overheads during cluster formation/maintenance. This protocol suffers from temporary invalid routes as the destination nodes travel from one cluster to another. Therefore, this protocol is suitable for medium size networks with slow to moderate mobility. The protocol may also best perform in scenarios with group mobility where the nodes within a cluster are more likely to stay together.

4 Conclusion

In this paper we have provided a performance comparison of current on-demand routing protocols. We have suggested a number of optimizations, which could improve the performance of some of the on-demand routing protocols, and suggested what type of network scenarios would best suite these protocols.

References

1. Royer, E.M., Toh, C-K.: A Review of Current Routing Protocols for Ad Hoc Mobile Wireless Networks, Vol. 6. No.2. IEEE Personal Communications (1999) 46-55
2. Johnson, D.B.: Routing in Ad Hoc Networks of Mobile Hosts, IEEE workshop on mobile computing systems and applications (1995) 158-163
3. Broch, J., Maltz, D.A., Johnson, D.B., Hu Y-C., Jetcheva, J.: A Performance Comparison of Multi-Hop Wireless Ad-Hoc Network Routing Protocols, MobiCom Dallas (1998) 85-97
4. Iwata, A., Chaing, C., Pei, G., Gerla, M., Chen, T.: A Scalable Routing Strategy for Multi-Hop Ad-Hoc Wireless Networks, Vol. 17. No. 8. IEEE Journal on Selected Areas in Telecommunication (1999)
5. Pei, G., Gerla, M., Chen, T-W.: Fisheye State Routing: a Routing Scheme for Ad Hoc Wireless Networks, Vol.1. IEEE International Conference on Communications (2000) 70 –74
6. Basagni, S., Chlamtac, I., Syrotivk, V.R., Woodward, B.A.: A Distance Routing Effect Algorithm for Mobility (DREAM), Proc. Of fourth annual ACM/IEEE International Conference on Mobile Computing and Networking Mobicom 98 Dallas. Tx. (1998) 76-84
7. Ko, Y-B., Vaidya, N.H.: Location-Aided Routing (LAR) in Mobile Ad-Hoc Networks, Proc. Of ACM/IEEE Mobicom 98 Dallas. Tx. (1998)
8. Hass, Z.J., Pearlman, R.: Zone Routing Protocol for Ad-Hoc Networks, Internet Draft, draft-ietf-manet-zrp-02.txt, august 99
9. Hass, Z.J.: A New Routing Protocol for the Reconfigurable Wireless Networks, ICUPC'97 San Diego CA (1997)
10. Pei, G., Gerla, M., Hong, X., Chiang, C.: A Wireless Hierarchical Routing Protocol With Group Mobility, Vol.3. Wireless Communications and Networking Conference New Orleans (1999) 1538-1542

11. Chen, T., Gerla, M.: Global State Routing: A New Routing Scheme for Ad-Hoc Wireless Networks, Vol.1. IEEE International Conference on Communication (1998) 171-175
12. Chiang, C-C.: Routing in Clustered Multihop, Mobile Wireless Networks With Fading Channel, Proc. Of IEEE SICON'97 (1997) 197-211
13. Das, S., Perkins, C., Royer, E.M.: Ad Hoc On-Demand Distance Vector (AODV) Routing, Internet Draft, draft-ietf-manet-aodv-05.txt march 2000
14. Corson, S., Park, V.: Temporally-Ordered Routing Algorithm (TORA) Version 1 Functional Specification, internet draft, draft-ietf-manet-tora-spec-03.txt November 2000
15. Johnson, D., Maltz, D., Broch, J.: The Dynamic Source Routing Protocol for Mobile Ad Hoc Networks, internet Draft, draft-ietf-manet-dsr-03.txt October 1999
16. Aggelou, G., Tafazolli, R.: Distance Micro-Discovery Ad Hoc Routing (RDMR) Protocol, Internet Draft, draft-ietf-manet-rdmr-01.txt February 2000
17. Murthy, S., Garcia-Luna-Aceves, J.J.: An Efficient Routing Protocol for Wireless Networks, Vol.1. No. 2. ACM/Baltzer Mobile Networks and Applications (1996) 183-197
18. Perkins, C.E., Bhagwat, P.: Highly Dynamic Destination Sequenced Distance Vector Routing for Mobile Computers, Vol.1. Proce. of Computer Communication Rev. (1994) 234-244
19. Garcia-Luna-Aceves, J.J., Spohn, M.: Source Tree Adaptive Routing (STAR) Protocol, Internet Draft, draft-ieft-manet-star-00.txt October 1999
20. Kasera, K.K., Ramanathan, R.: A Location Management Protocol for Hierarchically Organized Multihop Mobile Wireless Networks, Proc. of IEEE ICUPC'97 October 1997 158-162
21. Joa-Ng, M., Lu, I-T., A Peer-to-Peer Zoned-based Two-level Link State Routing for Mobile Ad Hoc Networks, Vol.17. No.8. IEEE Journal on Selected Areas in Communications (1999) 1415-1425
22. Corson, M.S., Ephremides, A.: A Distributed Routing Algorithm for Mobile Wireless Networks, Vol.1. No.1. ACM/Baltzer Wireless Networks (1995) 61-81
23. Dube, R., Rais, C.D., Wang, K-Y., Tripathi, S.k.: Signal Stability-based Adaptive Routing (SSA) for Ad Hoc Mobile Networks, Vol.41. IEEE Personal Communications (1997) 36-45
24. Toh, C-K.: Novel Distributed Routing Protocol to Support Ad Hoc Mobile Computing, Proc. of IEEE fifth Annual International Pheonix Conference (1996) 480-486
25. Jiang, M., Li, J., Tay, Y.C.: Cluster Based Routing Protocol, Internet Draft, Draft-ieft-manet-cbrp-spec-01.txt August 1999
26. Raju, J., Garcia-Luna-Aceves, J.J.: A New Approach to On-demand Loop-Free Multipath Routing, Proc. of eight International Conference on Computer Communication and Networks (1999) 522-527

Construction
of Data Dependent Chaotic Permutation Hashes
to Ensure Communications Integrity

Josef Scharinger

Johannes Kepler University, Altenbergerstraße 25,
A-4040 Linz, Austria
js@cast.uni-linz.ac.at

Abstract. Cryptographic hash functions and message digests are essential in secure communications because they aid in detecting incidental transmission errors caused by unreliable equipment and noisy environments, but also ensure message integrity in presence of intruders deliberately mounting cryptanalytic attacks.
It is the purpose of this contribution to introduce a novel approach for generating cryptographic hashes computed from input data dependent pseudo-random permutations. Essentially, input messages are processed sequentially using bytes of input data as keys to discrete chaotic Kolmogorov systems which permute an initial message digest in a cryptographically strong manner heavily depending on the input stream. As will be shown this principle can lead to very efficient and strong message digests.

1 Introduction

In communication environments integrity of the data transmitted is a major concern. Checksums and parity bits were developed at the very beginning of digital communications to cope with unreliable equipment and can offer reasonable protection to detect incidental transmission errors, but only limited protection against malicious modifications and attacks to deliberately alter the content of messages transmitted. This has led to the development of cryptographic hashes [7,120,11] (digital fingerprints, message digests) that can check communication integrity even in presence of intruders mounting cryptanalytic attacks.

Cryptographic hash functions are not limited to applications for communication integrity but constitute also an essential component in ensuring communication authenticity. Authenticity is achieved by the use of digital certificates. Once a partner certificate is trustworthy installed, any message received from that person can be checked for authenticity using his certificate. Since signature schemes available today [3,8,12] are very slow, it is impractical to sign large amounts of data. Instead a condensed representation (message digest) of the data is attached to the original data and just this message digest is actually signed, an approach that enhances performance, increases cryptographic security and allows integrity checks in a unified framework. Therefore concepts for generating strong digital fingerprints are essential components also for ensuring integrity and authenticity in secure communications applications.

It is the purpose of this contribution to introduce a novel approach for generating cryptographic hashes that provide secure digital fingerprints for arbitrary input data. These hashes are derived as 256 bit message digests computed from input data dependent pseudo-random permutations. Initially, a $16 \times 16 = 256$ bit square array is filled with pseudo-random bits. Next, input messages are processed sequentially using bytes of input data as keys to so-called discrete chaotic Kolmogorov systems which permute the square array in a cryptographically strong manner heavily depending on the input stream. Finally, the state obtained in the 16×16 array of bits is read out to provide the 256 bit cryptographic fingerprint. As will be shown in the sequel, this construction leads to very efficient and strong message digests.

2 Chaotic Kolmogorov Systems

2.1 Continuous Kolmogorov Systems

Continuous Kolmogorov systems [1,4,6,15] act as permutation operators upon the unit square. Figure 1 is intended to give a notion of the dynamics associated with a specific Kolmogorov system parameterized by the partition $\pi = (\frac{1}{3}, \frac{1}{2}, \frac{1}{6})$. As can be seen, the unit square is first partitioned into vertical strips which are then stretched in the horizontal and squeezed in the vertical direction and finally stacked atop of each other. Just after a few applications (see figure 1 from top left to bottom right) this iterated stretching, squeezing and folding achieves perfect mixing of the elements within the state space.

Fig. 1. Illustrating the chaotic and mixing dynamics associated when iterating a Kolmogorov system.

Formally this process of stretching, squeezing and folding is specified as follows. Given a partition $\pi = (p_1, p_2, \ldots, p_k)$, $0 < p_i < 1$ and $\sum_{i=1}^{k} p_i = 1$ of the unit interval $\mathbb{U}$ and stretching and squeezing factors defined by $q_i = \frac{1}{p_i}$. Furthermore, let F_i defined by $F_1 = 0$ and $F_i = F_{i-1} + p_{i-1}$ denote the left border of the vertical strip containing the point $(x, y) \in \mathbb{E}$ to transform. Then the continuous Kolmogorov system T_π will move $(x, y) \in [F_i, F_i + p_i) \times [0, 1)$ to the position

$$T_\pi(x, y) = (q_i(x - F_i), \frac{y}{q_i} + F_i). \tag{1}$$

2.2 Discrete Kolmogorov Systems

In our notation a specific discrete Kolmogorov system for permuting a data block of dimensions $n \times n$ is defined by a list $\delta = (n_1, n_2, \ldots, n_k)$, $0 < n_i < n$ and $\sum_{i=1}^{k} n_i = n$ of positive integers that adhere to the restriction that all $n_i \in \delta$ must partition the side length n. Furthermore let the quantities q_i be defined by $q_i = \frac{n}{n_i}$ and let N_i specified by $N_1 = 0$ and $N_i = N_{i-1} + n_{i-1}$ denote the left border of the vertical strip that contains the point (x, y) to transform. Then the discrete Kolmogorov system $T_{n,\delta}$ will move the point $(x, y) \in [N_i, N_i + n_i) \times [0, n)$ to the position

$$T_{n,\delta}(x, y) = (q_i(x - N_i) + (y \bmod q_i), (y \operatorname{div} q_i) + N_i). \tag{2}$$

The restriction to integral stretching- and squeezing factors is necessary to keep resultant points at integer positions within the $n \times n$ grid. Use of the div (division of positive integers a and b delivering $\lfloor \frac{a}{b} \rfloor$) and mod (remainder when dividing positive integers) operation ensures that points in $n \times n$ are mapped onto each other in a bijective and reversible manner.

It is straightforward to check that in the case $n = 16$ a total of 55 different partitions δ_i ($0 \leq i \leq 54$) can be found to define permutation operators T_{16,δ_i} on 16×16 2D arrays. Mapping byte values $b \in [0..255]$ to valid partitions δ_i (e.g. according to $i = b \bmod 55$; see [13] for more alternatives), permutations T_{16,δ_i} can be specified by bytes. Interpreting a message as a sequence of bytes, every message byte may thus be taken as a round-key to this permutation operator. This way a secure hash algorithm that generates strong 256 bit message digests as the result of message dependent pseudo-random permutations will be constructed.

2.3 Properties of Cryptographic Relevance

Kolmogorov systems tend to permute elements of the state space in a chaotic non-linear and apparently random fashion. After a sufficient number of iterations it becomes extremely hard for an observer to deduce the initial state of a Kolmogorov system from its final state. To be more specific, Kolmogorov systems offer very unique properties of cryptographic relevance that are explained in more detail in the sequel.

Ergodicity Ergodicity is important for a system that is to be applied in cryptography because it stands as a synonym for confusion. Informally speaking and expressed in terms of permutation systems, ergodicity stands for the property that almost any initial point will move to almost any other position in state space with equal probability as the system evolves in time. In other words there is no statistical way to predict the initial from the final position or vice versa.

In the following we restrict attention to the practically most relevant case of $n = p^m$ being an integral power of a prime p. The discrete Kolmogorov system T_{n,δ_r} is defined by the list $\delta_r = (n_{1r}, n_{2r}, \ldots, n_{k_r r})$ of length k_r containing the positive integers to be used as key in round r. As mentioned before there are the restrictions $1 \leq i \leq k_r$, $0 < n_{ir} < n$, $\sum_{i=1}^{k_r} n_{ir} = n$ and the constraint that all $n_{ir} \in \delta_r$ must partition the side length n.

Furthermore let the stretching and squeezing factors q_{ir} to use for vertical strip number i in round number r be defined by $q_{ir} = \frac{n}{n_{ir}}$. This results in quantities q_{ir}, $q_{ir} \geq p$ that also have to be integral powers of p because of the divisibility assumption made.

Consider an arbitrary point $(x, y) \in [N_{ir}, N_{ir} + n_{ir}) \times [0, n)$ in vertical strip number i to be transformed in round number r under the influence of the key δ_r (see equation 2 and figure 1). Coordinates x and y can then be expressed by q_{ir}-adic representations of length $t_{ir} = \lceil \log_{q_{ir}} n \rceil$ by $x = \sum_{j=1}^{t_{ir}} x_{jr}(q_{ir})^{t_{ir}-j}$ and $y = \sum_{j=1}^{t_{ir}} y_{jr}(q_{ir})^{t_{ir}-j}$. Similarly N_{ir} can be expanded according to $N_{ir} = \sum_{j=1}^{t_{ir}} Ni_{jr}(q_{ir})^{t_{ir}-j}$ and $x - N_{ir}$ may be expressed as $x - N_{ir} = \sum_{j=1}^{t_{ir}} xm_{jr}(q_{ir})^{t_{ir}-j}$. Obviously x is the sum of $x - N_{ir}$ and N_{ir}.

To clarify these relations, the following illustration should be helpful.

x					y				
$x_{t_{ir}r}$	...	x_{3r}	x_{2r}	x_{1r}	y_{1r}	y_{2r}	y_{3r}	...	$y_{t_{ir}r}$
$xm_{t_{ir}r}$	...	xm_{3r}	xm_{2r}	0	y_{1r}	y_{2r}	y_{3r}	...	$y_{t_{ir}r}$
$Ni_{t_{ir}r}$	...	Ni_{3r}	Ni_{2r}	Ni_{1r}	0	0	0	...	0

According to equation 2 application of T_{n,δ_r} will move the point (x, y) to a new position $(x', y') = T_{n,\delta_r}(x, y)$ with coordinates $x' = q_{ir}(x - N_{ir}) + (y \bmod q_{ir})$ and $y' = (y \operatorname{div} q_{ir}) + N_{ir}$, as made clear by the subsequent figure.

x'					y'				
$y_{t_{ir}r}$	...	xm_{4r}	xm_{3r}	xm_{2r}	0	y_{1r}	y_{2r}	...	$y_{(t_{ir}-1)r}$
0	...	0	0	0	Ni_{1r}	Ni_{2r}	Ni_{3r}	...	$Ni_{t_{ir}r}$

Suppose that lists δ_r are chosen independently and at random[1]. Neglecting the constraint $N_{ir} \leq x$ which follows from the fact that N_{ir} is the left border of the vertical strip containing the point (x, y) for a moment, the proof of ergodicity becomes straightforward. N_{ir} adds random q_{ir}-bits to all the q_{ir}-bits

---

[1] This is a common assumption whenever proving specific properties of iterated cryptographic schemes. Round keys are generally supposed to be random and independent.

of y' yielding a random value for the new y-coordinate in one step. Cyclically shifting the least significant position of the y-coordinate to the least significant position in the x-coordinate and shifting these random q_{ir}-bits towards more significant positions in the x-coordinate ensures that after at most an additional $\max_{i=1}^{k_r} t_{ir} \leq m$ iterations the transformed point can move to almost any other position in state space with equal probability. Thus ergodicity is achieved after at most $m+1$ iterations.

Now let us pay attention to the constraint $N_{ir} \leq x$. A moment of thought reveals that the worst non-trivial point that will need the largest number of rounds until being able to move to any position has a x-coordinate of 0 and a y-coordinate where just y_{1r} is different from zero. Then it takes at most $m+1$ iterations until the second-least significant q_{ir}-bit in the x-coordinate is set and the least significant q_{ir}-bit in N_{ir} (and also in the x-coordinate!) may assume any random value. By shifting q_{ir}-bits towards more significant positions in the x-coordinate every iteration causes one additional position in x to become random and by adding N_{ir} the same applies to the y-coordinate. This way it is guaranteed that after another at most $m-1$ iterations ergodicity is achieved after at most $2m$ steps in total.

The preceding discussion can be summarized as follows:

Theorem 1. *Let the side-length $n = p^m$ be given as integral power of a prime p. Then the discrete Kolmogorov system T_{n,δ_r} as defined in equation 2 is ergodic provided that at least $2m$ iterations are performed and lists δ_r used in every step r are chosen independently and at random.*

In the discussion above we have noted that the restriction $N_{ir} \leq x$ to observe in every step significantly increases the number of iterations necessary until an initial point can move to any other position. Particularly points with small (zero) x-coordinate need a long time until exhibiting ergodic behaviour. However, a simple trick can help a lot in reducing the number of iterations necessary to achieve ergodicity of the underlying system: after every discrete Kolmogorov permutation round just apply a cyclic shift by $\frac{n}{2}-1$ to the elements in the $n \times n$ array. This corresponds to adding $\frac{n}{2}-1$ modulo n to every x-coordinate and helps points with initially small x-coordinates to move to any other position in a reduced number of rounds. Additionally this simple trick also solves the problems associated with the fixed points $(0,0)$ and $(n-1, n-1)$ so that not just almost all points can move to almost any position but really all of the $n \times n$ points will have ergodic behaviour.

Exponential Divergence Exponential divergence is essential for a system that is to be applied in cryptography because it stands as a synonym for diffusion. Informally speaking and expressed in terms of permutation systems, exponential divergence implies that neighboring points contained in the same subspace of the state space (e.g. points of the same vertical strip corresponding to the same block of the defining partition) diverge at an exponential rate. This way even highly correlated points in input blocks will quickly loose correlations and structures present in input data will soon disappear.

Using similar arguments as applied in proving ergodicity of discrete Kolmogorv systems we have derived the the following theorem:

Theorem 2. *Let the side-length $n = p^m$ be given as integral power of a prime p. Then the discrete Kolmogorov system T_{n,δ_r} as defined in equation 2 exhibits exponential divergence of points contained in the same blocks defined by partitions δ_r ensuring that after at most $2m - 1$ iterations arbitrary non-zero deviations between initial points have propagated at least once to the most significant position in the x-coordinate.*

Mixing Property The mixing property is important for a system that is to be applied in cryptography because it stands as a synonym for confusion as well as diffusion. Informally speaking and expressed in terms of permutation systems, fulfillment of the mixing property implies that any subspace of the state space will dissipate uniformly over the whole state space. Obviously this is an even stronger requirement than ergodicity because it does not only imply that almost any point will move to almost any position in state space with equal probability but also that distances between neighboring points within certain subspaces will become random as the system evolves in time.

Combining results derived in proving ergodicity and exponential divergence of discrete Kolmogorv systems, we have shown the following theorem:

Theorem 3. *Let the side-length $n = p^m$ be given as integral power of a prime p. Then the discrete Kolmogorov system T_{n,δ_r} as defined in equation 2 is mixing provided that at least $4m$ iterations are performed and lists δ_r used in every step r are chosen independently and at random.*

2.4 Conclusion

Summarizing the preceding discussion, a simple law on the number of rounds necessary to ensure that all the essential cryptographic properties of discrete Kolmogorov systems are fulfilled can be stated as follows:

Theorem 4. *Let the side-length $n = p^m$ be given as integral power of a prime p. Then the discrete Kolmogorov system T_{n,δ_r} as defined in equation 2 fulfills the properties of ergodicity, exponential divergence and mixing provided that at least $4m$ iterations are performed and lists δ_r used in every step r are chosen independently and at random.*

Equipped with this solid knowledge it will be shown in the remainder of this contribution that discrete Kolmogorov systems can successfully be applied to develop strong and efficient hash functions to ensure communications integrity.

3 Permutation Hashes Based on Chaotic Kolmogorov Systems

To provide integrity and authenticity in secure communications applications at reasonable computational costs, efficient and strong cryptographic hash functions are needed. Our approach to compute a message digest based on discrete chaotic Kolmogorov systems runs as follows.

First a 16 × 16 square array of bits is initialized with 256 pseudo-random bits (128 zeros, 128 ones) taken from the after-comma binary expansion of some "magic" constants (π, e, golden ratio ϕ, $\sqrt{2}$, $\sqrt{5}$, etc.) as done in almost any cryptographic hash function. Taken line-by-line or column-by-column, this provides the initial 256 bit message digest MD_0.

After initialization, in every step $t = 1, 2, \ldots$ the message digest MD_{t-1} is updated by processing the message in blocks W_t of 256 bit each. Since message lengths are usually not a multiple of 256, padding the last block with arbitrary constant bits may be necessary.

Now these 256 message bits are XORed with the current 256 bit message digest to obtain $X_t = W_t \oplus MD_{t-1}$. This step ensures that any block contains approximately an equal number of zeros and ones, regardless of the message block (which could be entirely zero etc.).

To maximize input avalanche effects, the 8 32-bit words $X_t(i)$ ($0 \leq i \leq 7$) are processed according to a linear recurrence relation. First a forward dissipation step is done according to $Y_t(0) = X_t(0)$, $Y_t(i) = aY_t(i-1) + b \bmod 2^{32} \oplus X_t(i)$ with parameters a and b set accordingly (see e.g. [9] for a large variety of suitable parameter settings) to give pseudo-random sequences $Y_t(i)$. This is followed by a backward dissipation step (with index i decreasing) according to $Z_t(7) = Y_t(7)$, $Z_t(i) = aZ_t(i+1) + b \bmod 2^{32} \oplus Y_t(i)$.

After preprocessing the message block W_t to obtain the block Z_t, the actual hashing step takes place. The 256 bit of Z_t are used to provide 32 key bytes $Z_t(i,j)$ ($0 \leq i \leq 7$, $0 \leq j \leq 3$) to permute the message digest MD_{t-1} stored in the 16 × 16 array of bits using the corresponding discrete Kolmogorov system. For several reasons[2], a cyclic shift by 7 positions follows each of the 32 rounds involved in calculating the updated message digest MD_t.

Figure 2 summarizes one round when calculating data dependent chaotic permutation hashes based on chaotic Kolmogorov systems. Iterating this procedure for all blocks of the input message and finally reading the 16 × 16 2D array line-by-line or column-by-column delivers the 256 bit message digest of the message to hash in a very efficient and elegant manner as pseudo-random message-dependent permutation of the initial message digest MD_0.

4 Security Analysis

Informally speaking, a message digest scheme is called secure, if it is computationally infeasible to find a message which corresponds to a given message digest, or to find two different messages which produce the same message digest.

---

[2] It has been shown that fulfillment of essential cryptographic properties by discrete Kolmogorov systems with side length n needs iteration of at least $4*\log_2 n$ steps. This property can be achieved in less steps (see arguments given in proving ergodicity), if each iteration is followed by a cyclic shift by $\frac{n}{2} - 1$ positions. Additionally this resolves problems related to fixed points $(0, 0)$ and $(n-1, n-1)$.

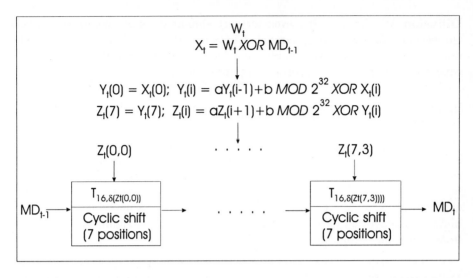

Fig. 2. One step in calculating data dependent chaotic permutation hashes based on discrete Kolmogorov systems.

Any change to a message in transit will, with very high probability, result in a different message digest, and the signature will fail to verify.

Security of cryptographic hashes calculated from discrete Kolmogorov permutations is extremely well based on the properties given in section 2.3. Additionally, we have done extensive evaluations on the cryptanalytic quality of the data dependent chaotic permutations generated by discrete chaotic Kolmogorov systems with respect to confusion, diffusion and fixed point distribution properties as well as differential and linear characteristics. These quantitative results will be summarized in the sequel.

4.1 Quantitative Security Analysis

Number of Different Message Digests Since any 256 bit message digest generated is a permutation of an initial message digest having 128 zeros and 128 ones, the number of different message digests is $\binom{256}{128} > 2^{251}$ and thus far beyond the number of particles in our galaxy [14]. Therefore it can be considered extremely unlikely that two different input messages lead to the same message digest.

Confusion We have done extensive χ^2-testing to verify the claim that data dependent chaotic permutations generated are uniformly distributed regardless of the initial distribution. Results justify the assumption that the permutations generated are statistically indistinguishable from a uniform distribution, a result expected due to the mixing property associated with any chaotic Kolmogorov system.

Diffusion We demand that even for very similar messages completely different permutations will be generated. Checking average diffusion distances the diffusion distances observed rapidly converge to the optimum diffusion distance expected for completely un-correlated permutations.

Singleton Cycles Distribution Suppose we are given a list of n elements, the resulting $n!$ permutations can be listed according to the number of singleton cycles in groups with no fixed element, one fixed element, two fixed elements, ..., all n elements fixed. Assuming that n tends to infinity the probability p_k that a random permutation has exactly k fixed elements is $p_k = \frac{1}{ek!}$. It was stunning how close the singleton cycles distributions observed matched the distribution expected for a random permutation. Therefore there was no evidence that permutations generated have a singleton cycles distribution significantly different from a random permutation.

Differential Analysis Differential cryptanalysis [2] analyzes the effect of particular differences in input pairs on the differences of the resultant output pairs. In the ideal case of a random permutation of 256 elements the most likely non-trivial differential characteristic has a probability of $\frac{1}{255}$. Deducing from our experiments we claim that the most likely differential characteristic observed rapidly converges to that optimum value.

Linear Analysis Linear cryptanalysis [5] studies linear relations between bits of inputs and corresponding outputs. When deriving linear relations one chooses a subset of the input bits and the output bits, calculates the parity (XOR) of these bits for each of the possible inputs and counts the number of inputs whose subset's parity is zero. Ideally this number does not differ from half at all. Deducing from our experiments it can be assumed that all non-trivial linear characteristics have a probability rapidly converging to the the optimum value ($\frac{1}{2} \pm 0$) that can be expected for random permutations.

5 Conclusion

Concepts for generating strong digital fingerprints are essential components for ensuring integrity and authenticity in secure communications applications. In this contribution we have introduced a novel approach for generating cryptographic hashes (digital fingerprints) that can be characterized as follows:

- Efficiency: Time-consuming operations (multiplications) are only needed in preprocessing the input block. Implementation of the chaotic permutations can be made extremely fast. Since all stretching and squeezing factors are restricted to be integral powers of 2, all the operations involved in computing a new position for points in the array can be done by just using additions, subtractions and bit shifts.

- Security: Permutation hashes generated fulfill all requirements essential for secure hashes. There is an enormous number of different fingerprints and permutations generated are indistinguishable from random permutations with respect to confusion, diffusion and fixed point distribution as well as differential and linear characteristics. Note that this observation is perfectly in line with the proven cryptanalytic properties of chaotic Kolmogorov systems (ergodic, mixing, exponential divergence).

These arguments clearly emphasize that data dependent chaotic Kolmogorov permutation hashes can offer an excellent choice when strong and efficient hash functions are needed in secure communications.

References

1. V.I. Arnold and A. Avez. *Ergodic Problems of Classical Mechanics*. W.A. Benjamin, New York, 1968.
2. Eli Biham and Adi Shamir. Differential cryptanalysis of DES-like cryptosystems. In *Advances in Cryptology – CRYPTO'90 Proceedings*, pages 2–21. Springer Verlag, 1991.
3. W. Diffie and M.E. Hellman. New directions in cryptography. *IEEE Transactions on Information Theory*, 22(6):644–654, 1976.
4. S. Goldstein, B. Misra, and M. Courbage. On intrinsic randomness of dynamical systems. *Journal of Statistical Physics*, 25(1):111–126, 1981.
5. Mitsuru Matsui. Linear cryptanalysis method for DES cipher. In *Advances in Cryptology – Eurocrypt'93 Proceedings*, pages 386–397. Springer Verlag, 1993.
6. Jürgen Moser. *Stable and Random Motions in Dynamical Systems*. Princeton University Press, Princeton, 1973.
7. US Department of Commerce/NIST. Secure hash standard. FIPS PUB 180-1, April 1995.
8. National Institute of Standards and Technology. Digital signature standard. U. S. Department of Commerce, 1994.
9. W.H. Press, B.P. Flannery, S.A. Teukolsky, and W.T. Vetterling. *Numerical Recipies in C: The Art of Scientific Computing*. Cambridge University Press, 1988.
10. RACE (Research and Development in Advanced Communication Technologies in Europe). RIPE Integrity Primitives. Final Report of RACE Integrity Primitves Evaluation, 1992.
11. R.L. Rivest. The MD5 message digest function. RFC 1321, 1992.
12. R.L. Rivest, A. Shamir, and L.M. Adleman. A method for obtaining digital signatures and public-key cryptsystems. *Communications of the ACM*, 21(2):120–128, 1978.
13. Josef Scharinger. Fast encryption of image data using chaotic Kolmogorov flows. *Journal of Electronic Imaging*, 7(2):318–325, 1998.
14. Bruce Schneier. *Applied Cryptography*. Addison-Wesley, 1996.
15. Paul Shields. *The Theory of Bernoulli Shifts*. The University of Chicago Press, Chicago, 1973.

Secure Communication:
A New Application for Active Networks

Manuel Günter, Marc Brogle, and Torsten Braun

Institute of Computer Science and Applied Mathematics (IAM),
University of Berne, Neubrückstrasse 10, CH-3012 Bern, Switzerland
http://www.iam.unibe.ch/~rvs/

Abstract. SplitPath is a new application for the easy, well-known and provably secure one-time pad encryption scheme. Two problems hinder the one-time pad scheme from being applied in the area of secure data communication: the random generation and the distribution of this random data. SplitPath exploits the flexibility of code mobility in active networks to address these problems. Especially the random generation is studied in more detail.

1 Introduction

A wide variety of encryption algorithms is in daily use to protect data communications. However, none of these algorithms is proven to be secure. A well-known algorithm exists which is very simple and perfectly secure: the one-time pad [Sch96]. It works as follows: Assume you want to encrypt a bit string P of n bits ($P, p_i \in \{0,1\}, i \in 1..n$). For that purpose you take a string of equally distributed and independent random bits $R, r_i \in \{0,1\}, i \in 1..n$ and xor (addition modulo 2) it bit-wise with P, resulting in the ciphertext $C, c_i = r_i \oplus p_i$. To decrypt C, R is xor-ed to C again. This works because $r_i \oplus (r_i \oplus p_i) = p_i$. R must be destroyed after the decryption. It is assumed that once R is generated, it is used for encryption and decryption *only once*. That's why the scheme is called one-time pad. Under these assumptions the algorithm is provably secure. It is impossible to gain any knowledge of P without knowing R, because *every possible plaintext* could have lead to a given ciphertext. Furthermore, in contrary to commercial encryption algorithms, the one-time pad needs only one very light-weight operation (xor) for encryption and decryption. However, the one-time pad is not practical for secure data communication for the following reasons:

- **Lack of random bits.** The one-time pad needs a irreproducible random bit-stream (R) of the same length as the message.
- **Ensuring single usage.** The receiver and the sender must both exclusively possess the same pad (R). How can the pad securely be established by the communication partners?

This paper presents an approach using active networking [CBZS98,TSS[+]97] to address both problems. An active network consists of active (programmable) network nodes. The data packets that are transmitted through an active network can contain code that the active nodes execute. Active networking is an instance of the mobile agents paradigm tailored to networking needs. Active network packets (also called capsules) access the networking functionalities of the nodes (e.g. forwarding and routing) and change these functionalities for packets or classes of packets.

This paper presents how (for a given application scenario) active networking enables us to use the one-time pad with its provable security and lightness for secure data communication.

In section 2 we present the basic idea how to address the distribution of the random (problem 2). Section 3 describes how to generate the necessary random (problem 1) and what the pitfalls are. An existing implementation using the well-known active networking tool ANTS [WGT98] is presented in section 4. Section 5 presents performance measurements and section 6 concludes the paper.

2 Distribution of Keys and Data

We said that with enough good random bits available, we can create an uncrackable bit stream using the one-time pad. However, the receiver must also possess the random bits. A straight-forward solution is to first deliver the random bits in a secure manner and later transmit the data. The sender could, for example, hand a magnetic storage tape to the receiver, containing the random bits. This is a secure but not very flexible application of the one-time pad. The use of a secure communication medium[1] allows the communicating parties to communicate later using an insecure communication medium. But even if the medium for sending the random is not secure, the scheme still works as long as no attacker has access to *both* random and message bits. This principle is e.g. used when encryption keys for data communication are exchanged using the postal service or the telephone system. The SplitPath idea goes one step further. The random bits (interpreted as the key) and the cipher text bits (plaintext xor-ed with the random bits) are sent along at the same time on the same media but *on different paths*. In general this scenario is not secure any more, since an attacker can eavesdrop both the random string and the encrypted message and thus easily decrypt the original message. In a network with centralised management at least the network provider will always be able to do this. However, if the network is partitioned in autonomous sub-networks (domains), as for example the Internet is, and if the two paths are entirely in different domains, an attacker will have significantly more trouble. Thus, the application of SplitPath requires the following preconditions:

- SplitPath traffic enters the untrusted networks only in split form (either random bits or xor-bits).
- The paths of corresponding random bits and xor-bits never share the same distrusted network.
- The distrusted networks do not trust each other.

---

[1] Assuming that the physical delivery of the magnetic tape is secure.

These preconditions limit the application scenario of SplitPath, but there are cases where the prerequisites are met.

- **Multi-homing.** Today's Internet service providers (ISP) compete with each other. Most of the larger ones own their own physical network infrastructure. It is not uncommon that customer networks are connected to more than one ISP.
- **Heterogeneous and redundant network technology.** A modern global data network like the Internet consists of all kinds of different network technologies (e.g. optical networks and satellite links) running different link layer protocols (ATM, Frame Relay etc.). These link layer networks are usually inter-connected on many redundant links.
- **International tension.** Unfortunately, many neighbouring countries tend to be suspicious about each other. The geographical location of-, and the relation between some nations can provide an ideal application scenario for the SplitPath scheme.

The generic situation is depicted in figure 1. At the split point the data packet is split into a random packet (the pad) and the xor result of data and pad. The resulting packets are sent along disjunct network paths. When they arrive in the trusted receiver network (at the merge point) then they are merged back into the original data packet.

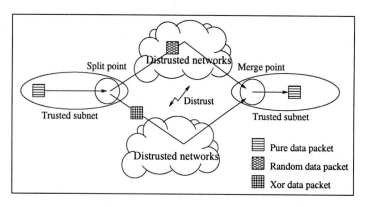

Fig. 1. The application scenario for SplitPath.

We distinguish between the sender, the receiver, a split point and a merge-point. Obviously, the split- and merge point needs to be located on a trustworthy site. Depending on the sender, receiver and the network topology the ideal location of these points varies, thus their implementation cannot be preconfigured in few network nodes. Active networking brings the flexibility we need here. With active networking the data packets can dynamically setup their private split (merge) functionality in the nodes which are appropriate for them. No router configurations are necessary. SplitPath capsules contain code that dynamically implements the split and merge capabilities and controls the routing of the packets. These mechanisms are described in section 4.

The next section shows how active networking can also help us getting the 'good' random bits which are needed in the split point.

3 Generating Random

For SplitPath, like for most crypto-systems, the availability of high quality random material is crucial. The random bits must be independent and equally distributed in order to be unpredictable. However, it is not easy to acquire such random in a computing environment. Irreproducible random values can solely be created by real world interaction. Examples are: mechanical random devices (e.g. lottery machines, dices), physical random (e.g. radioactive decay) and human behaviour (e.g. keyboard interrupt times). Multiprocessing and networking devices can also be a source of random bits [Wob98].

In SplitPath, we propose to use unpredictable traffic characteristics as seen in networking devices and generate random bits with them. Active networking allows the capsule to use its *own* varying queueing times within the network. The capsule, being autonomous, can keep information about its creation time and ask the nodes on its way about the local time. Note, that clock skew is actually good for the random generation because it introduces another factor of incertitude.

The idea is that by travelling through the net each capsule generates random bits for its own use. However, e.g. an IP packet can contain 64 KBytes of data. It needs the same amount of random bits for applying the one-time pad.

3.1 Quantity and Quality of SplitPath Random

The quantity of random bits gained by network performance measurement is limited. This is a problem but we have many options to cope with the situation:

1) Limit the payload of the capsule. Programmable capsules can fragment themselves to smaller payload sizes. This produces more packets and thus more random data per payload byte. It also adds bandwidth overhead. Note however, that congestion eases the production of random bits, since it involves a lot of non-predictable behaviour (see also section 5).

2) Generate more random bits by sending empty packets. When capsules can store data in nodes (see section 4) then we can use the idle time to send empty capsules that store their performance statistics in the split-node. Later, when payload is transported, the capsule can use the stored information to generate random.

3) Multi-hop random generation. If the capsule executes at several nodes before the split node, it can calculate its performance statistics there, too. The capsule takes the gained random bits with it and uses them at the split node. Care must be taken when distilling random bits from several nodes, because they are probably not completely independent.

These options do not affect the strength of the one-time pad, but they limit the effective throughput of data. Another approach is to use a pseudo-random function [Sch96]. This function uses the collected random data as a seed and generates a bit sequence of arbitrary length that can be used to xor the data. However, the next paragraph explains why such 'stretching' of the random bits affects the perfect security of the one-time pad.

When an attacker tries to decrypt an unknown ciphertext the most straightforward thing to do is to decrypt the message with every key and see if the result looks like a meaningful message. This is also called a *brute-force attack*. The one-time pad is immune against such attacks, because in his search through the key space the attacker will find *every* meaningful message that can be encoded in that length. However, once we start expanding the random key string by using it as a seed to a pseudo-random function, we will loose this nice property of the one-time pad. Based on Shannon's communication theory of secrecy systems, Hellman [Hel77] showed that the expected number P of different keys that will decipher a ciphertext message to some intelligible plaintext of length n (in the same language as the original plaintext) is given by the following formula: $P = 2^{H(K)-nD} - 1$. P indicates the success probability of the brute-force attack. If P is small then the brute-force attack will deliver only few possible plaintexts, ideally only one, which then can be assumed to be the original message. $H(K)$ is the entropy of the crypto-system used and D is the entropy of the encoded language. Using this formula we can show [GBB00] that for ASCII encoded English text each payload byte should be protected by at least 7 random bits. This is a very small stretching factor of 8/7.

Random Expansion in Practice. When we want to 'stretch' the random data by larger factors we can not rely any more on the perfect secrecy of the one-time pad. Instead, we have to carefully design the pseudo-random generator. First of all, the seed length must be large. We propose 128 bits. While the previous paragraph showed that a brute attack in principle will lead to the decryption of the packets (in the worst case there is only one random bit per packet), in practice the attack will not be successful, because there are too many bit combinations to try (an average of 2^{127}). If a million computers would each apply a billion decryption tries per second the average search would still last about $5*10^{15}$ years.

Second, the pseudo random generator should resist cryptanalysis. Many such generators exist and are used for so-called stream ciphers [Sch96]. Using Split-Path with expanded random is very similar to using a stream cipher in that both xor the plaintext with a secure pseudo random bit stream. However, SplitPath differs from stream ciphers in that it uses the pseudo random generator only at the sender side. Furthermore, the flexibility of an active network platform allows SplitPath to dynamically change the generator used, even during an ongoing communication. Finally, the seed of the generator is updated frequently (as soon as enough random bits have been collected by the capsules), and the seed is random. This is different from e.g. the stream cipher A5 (used for mobile telephony) which uses a preconfigured seed.

4 Implementing SplitPath in an Active Network

4.1 Implementing SplitPath with the Active Node Transfer System ANTS

We implemented the SplitPath application using the active node transfer system ANTS [WGT98]. ANTS is a Java based toolkit for setting up active networking testbeds. ANTS defines active nodes, which are Java programs possibly running on different machines. The nodes execute ANTS capsules and forward them over TCP/IP. ANTS defines a Java class `Capsule`. The class contains the method `evaluate` which is called each time the capsule arrives at a node. New capsule classes can implement new behaviour by overriding the `evaluate` method. The node offers services to the capsule such as the local time, forwarding of the capsule and a private soft-state object store (called *node cache*). Collaborating capsules can be grouped to protocols. Capsules of the same protocol can leave messages for each other using the node cache. We defined such a protocol to implement the SplitPath concept as presented in section 2 by introducing three new capsule subclasses.

1) The `Pathfinder` capsule marks nodes as splitting or merging points and sets up the split paths using the node caches. Note that several split and merge points can be set up per communication. Currently, `Pathfinders` are parametrised and sent by applications. We foresee to implement them with autonomous intelligence.

2) The `Normal` capsule is the plaintext message carrier. It checks if it is on a splitting point. If not, it normally forwards itself towards the destination. If it is on a splitting point, it applies the one-time pad computation to its payload. This results in two `Splitted` capsules, one carrying the random bits, the other the xor-ed data in the payload. The `Normal` capsule tells the node to forward the two newly produced `Splitted` capsules instead of itself, thereby using the information that was setup by the `Pathfinder`.

3) The `Splitted` capsule carries the encrypted- or the random data along a separate path. It forwards itself using the information that was setup by the `Pathfinder`. It checks if it has arrived on a merge point. If so, it checks if its split twin has already arrived. In that case it xor-s their contents (decryption) and creates a `Normal` capsule out of the result. If the twin is not there, it stores itself on the node cache to wait for it.

Applications and Interfaces. We wrote two applications that send and interact with the implemented capsules. The first application provides a graphical interface which allows the user to dynamically set up split and merge points using the `Pathfinder`. Furthermore, the user can enter and send text data using the `Normal` capsule. We also extended the nodes with a sniffing functionality. Such 'spy nodes' log capsule data to validate and visualise the SplitPath encryption. We implemented a second application which transfers a file (optionally several times) in order to test the performance by generating load.

4.2 Random Generation and the Application of the One-Time Pad

In order to be able to send large packets, we decided to use pseudo random generators to extend the collected random bits (see section 3). Each Normal capsule contains its creation time. When arriving at a split node, it uses this time and the node's current time to calculate the delay it has so far. It stores the last few bits as random seed. The number of bits is configurable and depends on the clock resolution. Unfortunately, the Java system clock resolution used by ANTS is bound to one millisecond and the delay is in the order of few milliseconds. Therefore, we used only the least significant bit. Thus, every packet stores one random bit in the split node (using the node cache). This bit can be considered as random, since it is influenced by the speed and current usage of computing and networking resources. As said before, the chosen seed length is 128 bits. So for every 128th capsule a complete seed is available. This capsule uses the seed to store a new random generator in the node cache. The next capsules use the generator for their encryption, until enough fresh random has been collected to install a generator with the new seed (key refreshing). For the bootstrapping we foresee two schemes. Either 128 empty packets are sent, or a (less secure) seed is used that can be generated by the first packet.

We have implemented two random generators. The first one is based on the secure one-way hash function MD5[2]. The seed is stored in a byte array of 16 bytes. Furthermore, there is an output buffer of 16 bytes containing the MD5 hash value of the seed buffer. The generator delivers 8 bytes of the output buffer as pseudo-random values. Then, the seed is transformed and the output buffer is updated (MD5 hash). The 'one-way' property of MD5 assures that an attacker cannot reconstruct the seed. Thanks to the avalanche property of MD5 the transformed seed produces an entirely different hash. Our seed transformation is equivalent to the increment of a long integer (8 bytes) by one. Thus, the seed only repeats after 2^{64} transformations. Long before that, SplitPath will replace the seed with freshly gathered random values. We think that for the presented reasons this pseudo random generator is reasonably secure. Nevertheless, we implemented also the pseudo random generator of RC4 as an alternative. RC4 is a stream cipher developed by RSADSI. It is used in applications of e.g. Lotus, Oracle and Netscape (for details see [Wob98]).

5 Evaluation of SplitPath

In order to evaluate SplitPath we ran the implementation on our institute network. Six ANTS nodes were set up on six different machines (sender, receiver, a split and a merge node and two sniffers; see figure 2). The split node ran on a SPARCstation 5/170. The encrypted capsules ran over two different subnets and were merged in a machine of a third subnet. The subnets are 100 Mbps Ethernets. We used the aforementioned file transfer application to generate load. Our

---

[2] ANTS includes the Message Digest 5 (MD5) [IETF RFC 1321] functionality.

interest was focussed on the quality of the encryption. We measured this by collecting all generated seeds and apply statistical tests. Furthermore, we applied statistical tests to the MD5 pseudo random generator presented in the previous section.

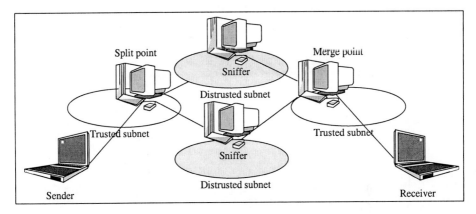

Fig. 2. The network topology for the evaluation.

In order to test the quality of the MD5 random generator, we initialised it with all seed bytes set to zero. Then we fed its output into a framework for statistical tests. Testing with samples of 40 MByte size, the produced data succeeded the byte frequency test, the run test [Knu81] and the Anderson-Darling test [IETF RFC 2330]. This is no prove that the generated pseudo-random bits are of high quality, but it shows that they are not flawed.

Seed Generation. We evaluated the generated seed bytes using statistical tests. For example we analysed 3K seed bytes (192 complete seeds, protecting 24576 packets). The seeds pass the byte frequency test (χ^2 test on the distribution of the measured byte values [Knu81]). Unfortunately, in few cases we also experienced seed generation that was not uniformly distributed. There we see some concentrations of byte values especially around the value 0. Our investigation revealed three reasons that come together to form this effect. (1) The coarse resolution of Java's clock. (2) ANTS does not send one capsule per packet, and the packet code is not send within the capsule, but dynamically loaded and cached. Thus, consecutive capsules are handled immediately after each other without delaying I/O operations. (3) The local network used has very low delay and jitter. These problems are not discouraging because normally they do not all come together and there are also countermeasures. By introducing congestion we can show that given realistic wide area delays as studied e.g. for the Internet [EM99], the seeds are equally distributed. Figures 5 and 3 show the seeds of two samples. Each sample shows the single byte values (2-complement) in the

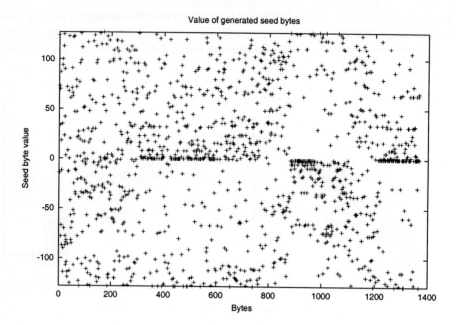

Fig. 3. Biased seed.

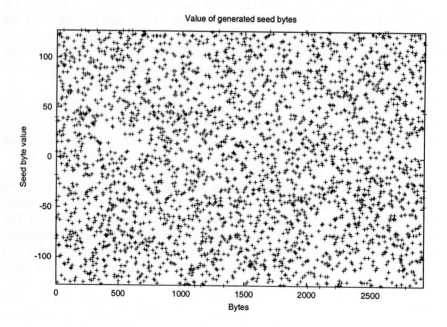

Fig. 4. Good seed.

same order as they were created. Figure 5 represents a sample suffering of the previously mentioned problems, figure 3 represents a corrected sample.

There are many more options to improve the seed quality. Additional use of other random sources for example. We foresee the exploitation of additional sources offered by active networking e.g. execution times of capsules or properties of the node cache. Finally, we can once again exploit the fact that the capsules are programmable. The procedure for collecting seed values can easily be extended to contain statistical methods to test the seed before it is used. So if for some unforeseen reason the seed is not good enough, the capsule uses the old random generator a little longer until more random bits are collected. Also, methods described in [Sch96] can be used by the capsule to distill random from biased random bits.

6 Conclusion

We presented SplitPath, a new application for the well-known and provably secure one-time pad encryption scheme. SplitPath uses the ability of active networks and the trust relations in heterogeneous networks to solve the two problems which otherwise render the one-time pad scheme useless: the random generation and the distribution of this random data. SplitPath can dynamically set up and use disjunct paths through network domains that do not collaborate, such as competitive network providers or countries. One-time pad encrypted data and random data is forwarded separately on these paths. Active networking allows SplitPath to encrypt and decrypt at any trusted node. Active networking not only allows SplitPath to dynamically set up the one-time pad splitting inside of the network, it also helps to collect good random. This can be very useful for other crypto-systems, too. SplitPath implements data capsules that use the network delays that they experience as initialisation for the random generation. With SplitPath we present an application in a (albeit specific) environment which cannot be implemented using conventional 'passive' networking, thus we promote the future study and (hopefully) deployment of active networking.

References

[CBZS98] K. Calvert, S. Bhattacharjee, E. Zegura, and J. Sterbenz. Directions in active networks. *IEEE Communications*, 36(10), October 1998.

[EM99] Tamas Elteto and Sandor Molnar. On the distribution of round-trip delays in TCP/IP Networks. In *Proceedings of the 24th Conference on Local Computer Networks LCN'99*, pages p.172–181. IEEE Computer Society, October 1999.

[GBB00] M. Günter, M. Brogle, and T. Braun. Secure communication with active networks. Technical Report IAM-00-007, IAM, 2000. www.iam.unibe.ch/~rvs/publications/.

[Hel77] M. E. Hellman. An extension to the shannon theory approach to cryptography. *IEEE Transactions on Information Theory*, IT -23(3):p. 289–294, May 1977.

[Knu81]   D. E. Knuth. *The art of computer programming, volume 2 Seminumerical Algorithms*. Addison-Wesley, 2 edition, 1981.
[Sch96]   B. Schneier. *Applied Cryptography*. John Wiley and Son, 1996.
[TSS+97] D. L. Tennenhouse, J. M. Smith, W. D. Sincoskie, D. J. Wetherall, and G. J. Minden. A survey of active network research. *IEEE Communications Magazine*, 35(1):80–86, January 1997.
[WGT98]   D. Wetherall, J. Guttag, and D. L. Tennenhouse. ANTS: A toolkit for building and dynamically deploying network protocols. In *IEEE OPENARCH '98*, April 1998. San Francisco.
[Wob98]   Reinhard Wobst. Abenteuer *Kryptologie*. Addison-Wesley, 1998.

Deployment of Public-Key Infrastructure in Wireless Data Networks

Arun K. Singh

E-Commerce Research Labs.,
Infosys Technologies Limited,
Electronic City, Bangalore, India
arunks@infy.com

Abstract. Over the last decade there has been a rapid growth in the area of wireless applications. For a wireless application to be successful, it is essential to ensure that the transactions cannot be fraudulently generated, that transactions are legally binding, and that the confidentiality of private information is adequately protected. Such assurance depends upon deployment of public-key infrastructure (PKI) technology in wireless data networks. When wireless devices are used for data communication, they have to face many environmental limitations e.g. limited computing power in wireless devices, inadequate memory space, low network bandwidth and restriction imposed by underlying communication protocols and services. This obviously creates further challenges for protocol designers, implementers and standardization body. This paper discusses the issues involved in implementing PKI in wireless data networks and possible approaches to resolve them. It also provides an overview of today's mobile system environment and current status of wireless security implementations.

1 Introduction

Wireless applications are changing the face of the Internet world. It is possible to use digital phones, PDAs, and pagers to access many commercial applications such as mobile banking, brokerage, and retailing. The emerging technology behind this is based on the Wireless Application Protocol (WAP). In order to exploit the potential of wireless Internet market, handset manufacturers such as Nokia, Ericsson, Motorola, and others have developed WAP-enabled smart phones. Using Bluetooth technology, smart phones can offer many services e.g. fax, e-mail, and phone capabilities in a single device. This is paving the way for wireless Internet applications to be accepted by ever-increasing mobile workforce. Industry estimates indicate that the number of wireless Internet clients will exceed the number of wired PC/Internet clients by 2004.

As compared to wired Internet, security concerns are much greater in the wireless network. Addressing these security concerns involve deployment of Public Key Infrastructure (PKI) technology in wireless data networks. PKI contains three common functional components: the certificate authority (CA) to issue certificates (in-house or outsourced); a repository for keys, certificates and certificate revocation lists (CRL) on an LDAP-enabled directory service; and a management function, typically implemented via a management console [1], [6]. Additionally, PKI can provide key recovery in case a user loses his/her private key due to hardware failure or other problem []. In this paper, we will discuss the issues involved in deploying PKI technology for a wireless application and list down possible approaches to handle them.

2 The Current Wireless Environment

The wireless communication model is very much similar to the Internet model with some subtle differences.

As shown in Figure 1, in Internet model the client device – PC or Laptop, uses a modem to get connected to ISP (Internet Service Provider) using PPP (Point-to-Point Protocol). The ISP authenticates the client by using any of the available authentication protocols e.g. PAP (Password Authentication Protocol), CHAP (Challenge Handshake Authentication Protocol), MS-CHAP, MD5-CHAP etc. Once the authentication is successful, it allocates an IP address to the client and provides a direct connectivity to the Internet backbone through a router or the gateway. Once the client device is connected to the Internet, it communicates with the Web Server on the Internet using Hyper Text Transfer Protocol (HTTP) [2].

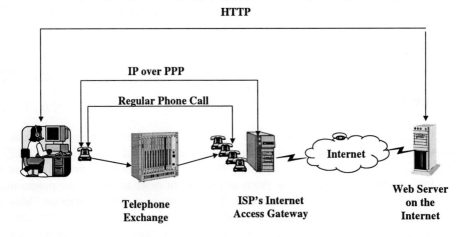

Fig.1. Internet Model

As compared to Internet model, in wireless model the client is a wireless device such as mobile phone, tablet, personal digital assistant (PDA), webTV, handheld computer and thin client terminals. These instruments have the added capability to support data in addition to voice. The connection is still made via phone call but this time using wireless communication technology such as Cellular digital Packet Data (CDPD), General Packet Radio Service (GPRS) etc.. CDPD and GPRS provide circuit switched and packet switched data services over existing wireless network using Advanced Mobile Phone Services (AMPS) and Global System for Mobile Communication (GSM) respectively [3]. The wireless client devices access the gateway using :

- Same basic protocols as Internet (HTTP/SSL/TCP/IP) e.g iMode
- Wireless Application Protocol (WAP) e.g. WAP enabled devices
- Properitery protocols e.g. PDAs and Two way pagers

iMode service is introduced by Tokyo based Mobile Communications Networks Inc. (NTT-DoCoMo) and is very popular with around 7 million user base [13]. WAP standard is defined by WapForum with some of the biggest names in the wireless devices like Motorola, Nokia, Ericsson and others behind it [12]. WAP is set to become the default standard for getting Internet data to cell phones. If the client devices are using iMode, they will connect to the iMode Gateway, which in turn provides connectivity to iMode enabled web servers over the Internet whereas WAP enabled client devices access Web servers on the Internet through WAP gateway.

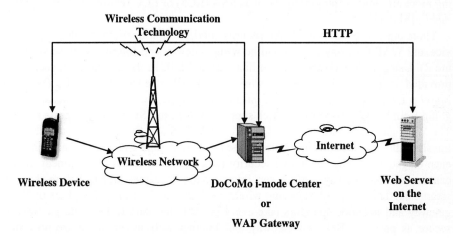

Fig. 1. The Wireless Model

In the wireless model, Figure 2, it is wireless operator rather than ISP who receives the wireless call and does the client authentication. Once the authentication is done, the wireless operator routes the data over its proprietary network to a Gateway. This gateway converts wireless protocols to http protocol and forwards the packet over Internet to the Web server that serves the wireless device complaint content.

As compared to Internet environment, the wireless environment has many limitations. On the client devices, these limitations are in terms of CPU's capabilities, available memory, battery lifetime, and user interface. From the network viewpoint, the available bandwidth is very low and due to internal structure of wireless networks there is a problem of sustained waiting time, called latency. All these constraints combined with different network configurations, business relationships, different wireless network carriers make the deployment of PKI a very challenging task [4]. Providing ordinary PKI services such as authentication of entities, digital signatures, integrity and data privacy in this constrained environment, gives rise to a number of issues.

3 Security in the Wireless Environment

On the Internet, security is provided between client and server by using TLS (Transport Layer Security) protocol, which is based on SSL3.0 (Secure Socket Layer). It is not possible to use SSL directly for the wireless networks because of low bandwidth and high latency constraints. For wireless networks, a modified protocol WTLS (Wireless Transport Layer Security), which is based on TLS1.0, is used [2]. WTLS is specifically designed to conduct secure transactions on wireless networks without requiring high processing power and memory in the mobile devices. It does this by minimizing protocol overhead, utilizing better compression techniques, and employing more efficient cryptography, such as RSA (RC5) or ECC (Elliptic Curve Cryptography) [5] & [7].

There are two parts to the WAP security model – first between wireless client device and WAP Gateway and second between Gateway and Internet Web server (Figure 2). Using WTLS protocol in the first part and SSL in the second part, security is provided in the wireless data networks. At WAP gateway, the WML (Wireless Markup Language) data must decrypt and re-encrypt itself as HTML because it is going from WTLS encryption, which supports WML, to SSL, which supports HTML. The WAP gateway works as a bridge between WTLS and SSL security protocols and translates the messages from wireless zone to Internet zone and vice versa. The presence of gateway brings a serious security issues in the wireless environment. If one can access the gateway, he/she can intercept the decrypted traffic before it is rescrambled. This is called two-zone wireless security problem [8]. Suppliers of WAP gateway and network operators take every possible measure to keep the gateway as secure as possible. Between gateway and Internet web server, there are no issues because TLS can be used exactly in the same manner as in Internet to provide security.

As compared to WAP, iMode system uses compact HTML, which allows iMode enabled wireless devices to talk directly to the Web servers using SSL protocols. Due to this two-zone security problem is not there in iMode systems. In WAP, this problem is supposed to be fixed in the next version.

4 PKI Deployment Issues

WTLS and SSL simply provide a secure tunnel between two points. Deployment of PKI in the wireless environment using these protocols gives rise to the following important issues:

    a) Public Key Algorithm Selection
    b) Key Pair Generation
    c) Storage of Private Key and Certificates
    d) Certificate Management
    e) PKI Interoperability
    f) Mobile Device Problems

4.1 Public Key Algorithm Selection

The first issue in PKI deployment is to decide about which class of crypto systems are to be used for key generation There are many cryptographic algorithms available e.g. RSA, DSA, ECC, NTRU etc.[5], [6], [7] & [14]. Selection of one of these algorithms depends on many factors such as speed, ease of implementation and public acceptance.

In the Internet environment, PKI technology is almost exclusively based on RSA cryptosystem but in wireless world a newer, alternative cryptographic technology called Elliptic Curve Cryptography (ECC) is becoming popular. ECC can perform same basic function as RSA but with fewer CPU resources and smaller Key size. This makes signatures computationally inexpensive. Deployment of PKI with ECC has the main disadvantage that it is not widely deployed and there is a little support in standard Internet protocols e.g. TLS/SSL, S/MIME etc. Due to this, it may not be possible to leverage the installed technology.

4.2 Key Pair Generation

The next important issue in PKI deployment in wireless environment is to authenticate unknown parties and establish trust. For this, a separate pair of keys needs to be generated for client as well as for server. On the server side, key-pair generation has no issue but on the client side due to limited device capability it is a big question. The main issue is to decide that who is generating the key-pair for the wireless client device? Whether it is the telecom operator or mobile device manufacturer or the subscriber himself? The security of PKI system relies on the secrecy of the private key. Whosoever generates the key-pair, he/she gets to know the private key and then non-repudiation can be guaranteed to that party only. Once the keys are generated, then there are some related issues e.g. who is installing these keys in the mobile devices? Also after the key installation we have to bother about how to update these trusted keys?

The approach to handle this key-pair generation issue depends on the trust relationship. If the subscriber trusts on telecom operator or handset manufacturer, then he/she can ask them to generate key-pair on their behalf. Generally the credit card issuers are interested in how the authentication solution they deploy is stored and used, particularly if they are financially responsible for its misuse. As a result there is an anticipated desire for issuers of their trusted agents to directly manage services such as server-based wallets, just as they closely manage the issuance of their chip cards. In server-based wallets there is no need to install private keys on the handsets otherwise, subscribers can obtain the private keys from issuers and install them on their mobile phones.

4.3 Storage of Private Key and Certificates

Private key and certificates can be stored in the software or in the hardware. From security viewpoint, storing the keys in hardware makes it tamper proof but then the key management becomes difficult. On the other hand, storing the key in software makes key management easy but replacement of phones may result in temporary loss of identities. Along with this there is an associated issue that who is responsible for installing these keys?

The current practice of key storage in mobile network is based on hardware smart cards. Telecom service provider or handset manufacturer can issue the card. In GSM world, SIM (Subscriber Identification Module) card is used whereas in WAP world, an equivalent WIM (Wireless Identification Module) card is used. These cards provide an excellent tamper resistant storage place for private/public keys and provide mobility to the personal credential. The new WIM cards are equipped with crypto processing power and they will allow users to digitally sign the message [9]. The main problem in using these new cards is posed by the wireless bandwidth. The available bandwidth is not capable of delivering strong encryption and we have to wait till 3G, the third generation wireless bandwidth. For key installation, subscribers can install their private keys either themselves or by the card issuer. The telecom operator can install trusted certificates of CAs on these cards before selling the mobile devices. Alternatively, manufacturers of handsets may install trusted certificates in the handset directly. Apart from this, at present some work is going on within IETF to make mobile credentials to be stored in software.

4.4 Certificate Management

In certificate management there are two main issues – which certificate format to use and how to check status of received certificates? For Wireless, we can't use X.509 certificates because they are somewhat more complex to decode and are larger in size. Server certificates are to be processed by the mobile client devices, which do not have the local resources or the communication bandwidth to implement revocation methods used in the wired world such as Online Certificate Status Protocol (OCSP). Due to network constraint, checking Certificate revocation list (CRL) is really a big issue.

In the wireless world, instead of X.509 there are other special purpose and more compact certificate formats are defined such as X9.68, WAP WTLS Certificates, and Bullet certificates [9], [10], [11] & [13]. WAP WTLS *mini-certificates* are functionally similar to X.509 but are smaller and simpler than X.509 to facilitate their processing in resource-constrained handsets. Bullet certificates are used with ECC. The advantage of X.509 is that it is widely deployed and complaint with Internet world but the alternative formats are neither widely deployed nor interoperable. They lock users to wireless world. For checking the status of received certificates, many vendors use *Short-lived certificates*. With this approach, a server or gateway is authenticated once in a *long-term credentials* period - typically one year - with the expectation that the one server/gateway key pair will be used throughout that period. However, instead of issuing a one-year-validity certificate, the certification authority issues a new short-lived certificate for the public key, with a lifetime of, say, 25 hours, every day throughout that year. The server or gateway picks up its short-lived certificate daily and uses that certificate for client sessions established that day. If the certification authority wishes to revoke the server or gateway (e.g., due to compromise of its private key), it simply ceases issuing further short-lived certificates. Clients will no longer be presented with a currently valid certificate, and so will cease to consider the server authenticated. The main advantage of this technique is that there is no need for revocation lists whereas use of *online checking* creates extra round trips and more network traffic.

4.5 PKI Interoperability

Internet PKI protocols such as CMP (Certificate Management Protocol), CSC (Certificate Status Check), OCSP etc. are not optimized for constrained networks and devices. Developing new protocols for this means, "split universe" because it creates need for translating all the proxies or gateways. One possible approach for this problem could be to develop and standardize a separate set of protocols. This is exactly done by WAP forum. WAP is optimized for wireless environment. One disadvantage of WAP is that it does not provide end-to-end security because plain text is available at the gateway.

For PKI interoperability, a joint work between Telecom Companies and IETF has been initiated whose focus is to amend TLS protocol and WAP for a better Internet convergence

4.6 Mobile Device Problems

The mobile handset may be lost, stolen, broken or loose power. If any of these things happen then how a subscriber authenticate and sign transactions? The possible approaches to remove this Key phobia could be use of specialized authentication/ signature token or there should be some mechanism to hold the keys without a need for recharging batteries. The other alternatives could be widespread usage in shared secret systems and small key size.

5 Conclusion and Future Trends

Security is a key issue for almost every enterprise that is looking to implement Wireless applications. At present there is a lot of uncertainty in the market as to what is available and how secure the products really are. For a specific application, there are ways of addressing the PKI deployment issues; depending on how much money is available and how important the particular issue is. But for a general application, we need a standard solution.

Future trends show that the constraints of wireless network will be reduced due to technological innovations. 3G phones and networks will offer more CPU capability, available memory and network bandwidth. Due to this, future wireless applications will be able to implement today's Internet PKI protocols but they will continue to lag wired networks and devices in terms of computational power, bandwidth etc. At present the security in Wireless application needs a specialized solution e.g. formats, protocols, storage etc. An adoption of Internet protocols (PKIX) and formats in the wireless data networks are likely to take place in the long run. Some convergence will probably occur on the wired network front too e.g. Internet protocols might be modified to accommodate wireless environments.

References

1. Singh, A.K.: E-Commerce Security. Proc. of National Seminar on E-Commerce & Web Enabled Applications at Harcourt Butler Technological Institute, Kanpur, India (2000)
2. Charles, Arehart et. al.: Professional WAP. Wrox Press Ltd. (2000)
3. Gilbert Held: NPN:The Wide World of Wireless. Network Magazine (December 1999)
4. Hasan, S., et. al.: Wireless Data Networks: Reaching the Extra Mile. IEEE Computer (December 1997)
5. Ramon, J., Hontanon: Encryption 101 – The Choices. Sys Admin (1998)
6. Anita, Karve: PKI Options for Next Generation Security. Network Magazine (March 1999)
7. Fernandez, A.D.: Elliptic-Curve Cryptography. Dr. Dobb's Journal (December 1999)
8. Epharaim, S.: Solutions on the Horizon will Relieve the Two –Zone Wireless Security Problem. InfoWorld.com (Nov. 2000)
9. http://www.ericsson.com/wapsolutions/common/p11.shtml
10. http://www.phone.com/pub/Security_WP.pdf
11. http://www.nokia.com/corporate/wap/future.html
12. http://www.wapforum.com
13. http://www.nttdocomo.com/release/index.html
14. http://www.ntru.com

A Scalable Framework for Secure Group Communication

Lih-Chyau Wuu and Hui-Chun Chen

Department of Electronic Engineering, National YunLin University
TouLiu, TAIWAN, 640 R.O.C.
wuulc@el.yuntech.edu.tw

Abstract. In this paper, we propose a scalable framework for secure group communication that not only supports the source authentication, integrity and confidentiality of the group traffic, but also enforces the join secrecy and leave secrecy. Providing security-enhanced services mentioned above for group communication is problematic without a robust group management. The framework adopts two-level group management to improve the scalability problem. To ensure secure group communication it enforces four processes: registration, auxiliary key distribution, re-keying, and secure traffic with source authentication.

1. Introduction

In the Internet, multicast [1-2] has been used successfully to provide an efficient, best-effort delivery service for group-based application. In contrary to peer-to-peer secure communication that is well developed, secure multicast remains comparatively unexplored. However, applications such as multiparty private conference, distribution of stock market information, pay per view and other subscriber services which require to embed certain security mechanisms into multicast to protect the integrity of traffic from modifications, guard for confidentiality of communication from electronic eavesdrop, and validate participant's authenticity.

In this paper, we propose a scalable framework for secure group communication that not only supports the source authentication, integrity and confidentiality of the group traffic, but also enforces the join secrecy and leave secrecy [3]. In secure group communication, *join secrecy* ensures that joining members cannot access past group data, and *leave secrecy* ensures that leaving members cannot access current or future group data.

The framework uses digital signatures [4] to provide integrity and source authentication. Providing confidentiality requires securely distributing a group key to each of a group's receivers. Having such a key allows group members to encipher/decipher the traffic within the group. As for join secrecy and leave secrecy, it must change the group key, called as re-keying, whenever member joins or leaves the group. We use auxiliary keys to encrypt the new group key to securely distributing it. Thus, each member has one shared group key and a set of auxiliary keys.

Providing security-enhanced service for group communication is problematic without a robust group management to handle membership changes and key distribution [6-7]. The framework conducts the group in two hierarchical levels: a single router domain and one or more member domains.

Each domain has its own manager and an auxiliary key distribution tree. The tree in router domain is in the form of Huffman binary tree and the tree in member domain is in the form of complete binary tree. This minimizes the number of messages exchanged during the re-keying process and storage required by each member.

The framework adopts two-level group management to improve the scalability problem. It also enforces four processes: registration, auxiliary key distribution, re-keying, and secure traffic with source authentication to ensure secure group communication. In the following, we first introduce the basic model of the framework, then explain the operation of the four protocols. Finally is the conclusion.

2. Basic Model

To initialize a group, a host called as group initiator makes an announcement. Any host, which is interested in the group, becomes a *registered member* after it executes registration process. A registered member does not have the group key. To have the group key, a registered member must become a *participant member* by executing join process. A participant member becomes a registered member when it executes leave process. For simplicity, in the remainder of this paper the term "member" is referred to participant member.

To take advantage of Internet architecture, we conduct the group in two hierarchical levels: a single router domain and one or more member domains. *Router domain* consists of multicast-capable routers, and *member domain* contains at least one member host and one multicast-capable router.

Each domain has a manager to handle local registration, membership changes and key distribution. The router that the group initiator is located is called the *group manager*, and the other routers are called the *subgroup manager*. Fig. 1. shows the general structure of the two-level group management in our framework.

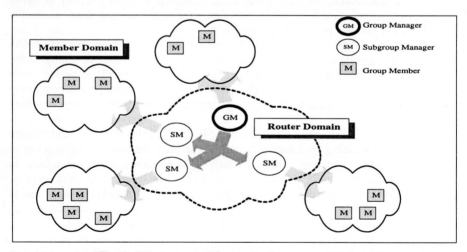

Fig. 1. The structure of the two-level group management

The symbols used in the paper are explained as follows. It is assumed that each domain has a certification authority and there exists trust-relationships among the certification authorities. Each member or router has a long-term public key pair before joining any group and generates a short-term public key pair during a group communication. Each key pair is certified.

Symbol	Meaning
G	Group name
I	Group manager
R	Subgroup manager
M	Member
GrpKey	Group key
K_A, K_A^{-1}	Long-term public key pair of user A
k_a, k_a^{-1}	Short-term public key pair of user A
$[Data] \, k_a^{-1}$	Data being encrypted by key k_a^{-1}
$\{Data\} \, k_a^{-1}$	A digital signature given by user A
$<Data> \, k_a^{-1}$	$<Data> \, k_a^{-1}$ = Data, $\{Data\} \, k_a^{-1}$
$\rightarrow$	Unicast
$\mapsto$	Multicast
$\Rightarrow$	Broadcast

3. Registration

In order to initialize a group G, the group initiator X announces the group information by the following procedure. First X sends a *Grpinit* message to its local router *I*. *I* becomes the group manager and broadcast a *Regstart* message to all routers in Internet. Each router then broadcasts a *Regstart* message to all local hosts in its networks. Note that all messages with a digital signature to ensure the message integrity and source authentication, and a field T_i is against replay attack.

$$X \rightarrow I : GrpInit(<G, Info, T_X > K_X^{-1}) \qquad (1)$$

$$I \Rightarrow all\ routers : RegStart(<G, Info, RegTime_I, T_I > K_I^{-1}) \qquad (2)$$

$$R \Rightarrow local\ hosts : RegStart(<G, Info, RegTime_R, T_R > K_R^{-1}) \qquad (3)$$

A host interested in the group must send a *Register* message to its local router R during the time $RegTime_R$. The router R becomes the subgroup manager and registers itself to the group manager I after the time $RegTime_R$ is expired. The *Register* message, sent by the router R, has a field *RekeyingWeight* to denote the number of registered number in this Member Domain. Note that at registration process, each M and R must produce its short-term public key pair to be used during the group communication and put the public key k_x on the *Resister* message.

$$M \rightarrow R : Register(<G, k_m, T_M > K_M^{-1}) \qquad (4)$$

$$R \rightarrow I : Register(<G, k_r, RekeyingWeight, T_R > K_R^{-1}) \qquad (5)$$

Rekeying Weight = Number of Registered Member

4. Auxiliary Key Distribution

After the time $RegTime_t$ is expired, the group manager establishes a Huffman binary tree and each subgroup manager establishes a complete binary tree. Each tree denotes a set of auxiliary keys distribution. Although the logical structure of a group is a two-level, the group key is shared by all of the participant members and the auxiliary keys are used to securely distribute a new group key in re-keying process.

4.1 Complete Binary Tree in Member Domain

In the Member Domain, we employ the approach in [8] to distribute the auxiliary keys in a complete binary tree. As shown in Fig. 2(a), the square leaf nodes in the tree denote the registered members or null in the domain. The round internal nodes denote the auxiliary keys in the Member Domain. Each participant member holds the auxiliary keys on the branch from the leaf representing itself to the root of the tree. For example, member m6 holds the auxiliary keys K_{Top}, K_2, K_1 and $\overline{K}_0$.

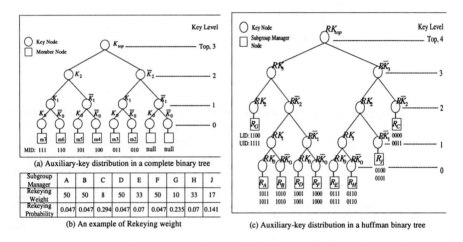

Fig. 2. Auxiliary key distribution in Member Domain and Router Domain

Each registered member is assigned a unique binary string as its member ID (MID), denoted by $b_{n-1}b_{n-2}...b_0$, where n is the depth of the binary tree and b_i can be either 0 or 1. The MID corresponds to auxiliary-keys held by a participant member. A participant member (MID=$b_{n-1}b_{n-2}...b_0$) will hold K_{Top} and K_i if $b_i=1$ or $\overline{K}_i$ if $b_i=0$, where $0 \leq i \leq n-1$.

4.2 Huffman Binary Tree in Router Domain

In the Router Domain, we distribute the auxiliary keys held by the subgroup managers in a Huffman binary tree. Fig. 2(b) gives an example with 10 Member Domains. The group manager first computes the rekeying probability of each Member Domain X as follows:

$$rekeying\ probablity_x = \frac{1/MemberDomainSize_x}{\Sigma 1/MemberDomainSize_j}$$

Then the group manager establishes a binary tree by executing Huffman algorithm.

As shown in Fig. 2(c), the square leaf nodes in the tree denote the subgroup managers. The round internal nodes denote the auxiliary keys in the Router Domain. Each subgroup manager holds the auxiliary keys on the branch from the leaf representing itself to the root of the tree.

Every subgroup manager is assigned two unique binary string: upper ID (UID) and lower ID (LID), denoted by $u_{n-1}u_{n-2}...u_0$ and $l_{n-1}l_{n-2}...l_0$ respectively, where n is the depth of the Huffman tree and u_i (l_i) can be either 0 or 1. The UID and LID of a subgroup manager are assigned as follows:

On the tee level i ($0 \leq i \leq n-1$)

If the subgroup manager holds RK_i then $u_i = l_i = 1$.

If the subgroup manager holds $R\overline{K}_i$ then $u_i = l_i = 0$.

Otherwise $u_i = 1$, $l_i = 0$.

For example, in Fig. 2.(c) the UID and LID of the subgroup manager R_j are 0101 and 0100 respectively since R_j holds $R\overline{K}_3$ on level 3, RK_2 on level 2, $R\overline{K}_1$ on level 1 and nothing on level 0.

During the group communication, a participant subgroup manager (UID= $u_{n-1}u_{n-2}...u_0$, LID=$l_{n-1}l_{n-2}...l_0$) will hold RK_{Top} and RK_i if $u_i = l_i = 1$ or $R\overline{K}_i$ if $u_i = l_i = 0$, where $0 \leq i \leq n-1$. For example, in Fig 2(c) the UID and LID of R_c are 0011 and 0000, R_c will hold RK_{top}, $R\overline{K}_3$ and $R\overline{K}_2$.

5. Re-keying Mechanism

To have confidential group communication, the group key must be changed periodically, or whenever some registered member joins or participant member leaves the group. It is the responsibility of the group manager to generate a new group key and distribute it securely to the group. It is noted that after a join or periodic key changing, the new group key can be sent by using the previous group key to encrypt it and no auxiliary key needs to be changed. However, after a leave, the previous group key can not be used to encrypt the new group key. Thus, the group manager must use the auxiliary key in Huffman tree to encrypt the new group key then multicast to the subgroup managers first. Then each subgroup manager uses the auxiliary key in its complete binary tree to encrypt the new group key then multicast to all its local members. Subsequently, some member or manager must update its new auxiliary key by itself if necessary. The updating auxiliary key is as follows:

$newAuxiliaryKey = f(AuxiliaryKey, newGrpKey)$, where $f()$ is an one-way hash function.

In the following, we explain how the group/subgroup manager distributes the new group key securely whenever membership changes.

5.1 Single Registered Member Joins

A registered member joins a group by sending a *JoinRequest* message to its subgroup manager. The remaining procedure depends on if the subgroup manager has joined the group or not.

Case I: the subgroup manager has not joined the group yet.
If the subgroup manager receives the *JoinRequest* message for the first time, it means that the subgroup manager has not joined the group yet that it must join the group by sending a *JoinRequest* message to the group manager. The group manager first generates a new group key (denoted by *newGrpKey*), then encrypts it by the old group key (denoted by *GrpKey*) and multicast it to the group. After that, the group manager sends a *JoinConfirm* message to the subgroup manager with the new group key and the corresponding auxiliary keys needed by the subgroup manager. Finally, the subgroup manager sends a *JoinConfirm* message to the joining member with the new group key and the corresponding auxiliary keys needed by the member.

$$M \rightarrow R : JoinRequest(< G, T_M > k_m^{-1}) \tag{6}$$

$$R \rightarrow I : JoinRequest(< G, T_R > k_r^{-1}) \tag{7}$$

$$I \mapsto G : Rekey([< newGrpKey, T_I > k_i^{-1}]GrpKey) \tag{8}$$

$$I \rightarrow R : JoinConfirm([< k_i, newGrpKey, auxiliarykeys_R, T_I > K_I^{-1}]k_r) \tag{9}$$

$$R \rightarrow M : JoinConfirm([< k_i, k_r, newGrpKey, auxiliarykeys_M, T_R > K_R^{-1}]k_m) \tag{10}$$

Case II: the subgroup manager has joined the group.
When the subgroup manager has joined the group, it sends a *RekeyRequest* message to the group manager after receiving a *JoinRequest* message. The group manager first generates a new group key, then encrypts it by the old group key and multicast it to the group. After receiving the *Rekey* message, the subgroup manager sends a *JoinConfirm* message to the joining member with the new group key and the corresponding auxiliary keys needed by the member.

$$M \rightarrow R : JoinRequest(< G, T_M > k_m^{-1}) \tag{6}$$

$$R \rightarrow I : RekeyRequest(< G, "join", T_R > k_r^{-1}) \tag{7a}$$

$$I \mapsto G : Rekey([< newGrpKey, T_I > k_i^{-1}]GrpKey) \tag{8}$$

$$R \rightarrow M : JoinConfirm([< k_i, k_r, newGrpKey, auxiliarykeys_M, T_R > K_R^{-1}]k_m) \tag{10}$$

5.2 Single Participant Member Leaves

A participant member leaves the group by sending a *LeavingRequest* message to its subgroup manager. The remaining procedure depends on if the subgroup manager needs to leave the group or not.

Case I: the subgroup manager does not need to leave the group.
The subgroup manager can not leave the group when there are other participant members in the domain. Thus, the subgroup manager sends a *RekeyRequest* message to the group manager. The group manager first generates a new group key, then encrypts it by the auxiliary key in the root of the Huffman tree (denoted by RK_{Top}) and multicast it to all the subgroup managers. For the subgroup manager whose member leaves, it encrypts the new group key by the auxiliary key in its binary tree and not being held by the leaving member, and the non-leaving members in this domain must change their auxiliary keys after receiving the new group key. As for the other subgroup manager, they use the auxiliary key in the root of their binary tree to encrypt the new group key. The members in those domains do not need to change their auxiliary keys.

$$M \rightarrow R : LeavingRequest(< G, T_M > k_m^{-1}) \quad (11)$$

$$R \rightarrow I : RekeyRequest(< G, "leaving", T_R > k_r^{-1}) \quad (12)$$

$$I \mapsto \forall R_x : Rekey([< newGrpKey, T_I > k_i^{-1}]RK_{top}) \quad (13)$$

For R with member leaving $(MID = b_{n-1}b_{n-2}...b_0)$ $\quad (14)$

$for(j = 0 \text{ to } n-1)$

$\quad if(b_j = 0)$ then $R \mapsto \forall M_R : Rekey([< newGrpKey, T_R > k_r^{-1}]K_j^R)$

$\quad if(b_j = 1)$ then $R \mapsto \forall M_R : Rekey([< newGrpKey, T_R > k_r^{-1}]\overline{K}_j^R)$

end for

All members in the member domain R change their auxiliary keys :

$\quad newAuxiliaryKey = f(AuxiliaryKey, newGrpKey)$

For R_x without member leaving $\quad (15)$

$$R_X \mapsto \forall M_{R_X} : Rekey([< newGrpKey, T_{R_x} > k_{r_x}^{-1}]K_{top}^{R_x})$$

Case II: the subgroup manager needs to leave the group.
When the last participant in this domain is leaving the group, the subgroup manager must leave too. It sends a *LeavingRequest* message to the group manager. The group manager first generates a new group key and encrypts it by the auxiliary key in the Huffman tree and not being held by the leaving subgroup manager, then multicasts it to all the other subgroup managers. After receiving the new group key, the non-leaving subgroup managers encrypt it by the auxiliary key in the root of their binary tree, then multicast it to their local members. After that, the non-leaving subgroup managers change their auxiliary keys.

$$M \rightarrow R : LeavingRequest(< G, T_M > k_m^{-1}) \quad (11)$$

$R \to I : LeavingRequest(< G, T_R > k_r^{-1})$ (12a)

For the group manager I   (13a)

(assume the leaving sugroup manager $UID = u_{n-1}u_{n-2}...u_0$, $LID = l_{n-1}l_{n-:}$

for ($j = 0$ to $n-1$)

    if $(u_j = l_j = 0)$ then $I \mapsto \forall R_X : Rekey([< newGrpKey, T_I > k_i^{-1}]RK_j)$

    else if $(u_j = l_j = 1)$ then $I \mapsto \forall R_X : Rekey([< newGrpKey, T_I > k_i^{-1}]R\bar{K})$

       else break the forloop

end for

For each nonleaving subgroup manager R_X (15a)

$R_X \mapsto \forall M_{R_X} : Rekey([< newGrpKey, T_{R_X} > k_{r_x}^{-1}]K_{top}^{R_x})$

R_X changes its auxiliary keys in Huffman tree

    $newAuxiliaryKey = f(AuxiliaryKey, newGrpKey)$

5.3 Multiple Registered Members Join

It's more efficient to aggregate several join requests to reduce the number of re-keying. Thus, no matter whether the group manager or any subgroup manager who can batch the join/re-keying requests to reduce transmission and encryption cost. The join procedure (6)-(10) stated above still can be applied in multiple registered members joining.

5.4 Multiple Participant Members Leave

To aggregate several leaving requests efficiently, it is desired to minimize the number of messages sent to the group when executing re-keying step (13a) or (14). To do that, it first needs to decide some auxiliary keys, called as *communal auxiliary keys*, which are common to subsets of the non-leaving members/subgroup managers but not being held by the leaving members/subgroup managers. Then those keys are used to encrypt the new group key. No matter in Member Domain or Router Domain, to determine the communal auxiliary keys is based on the Boolean algebra minimization method. The approach is explained as follows.

Case I. Multiple members leave from a Member Domain

When multiple members leave from the same Member Domain, the subgroup manager can batch the leaving requests then send one "*RekeyRequest*" message (12) instead of several ones. After receiving the new group key, the subgroup manager performs a minimization of the Boolean membership function [8] to decide the communal auxiliary keys.

The input of the Boolean membership function is the MID of each leaf node in the complete binary tree, and the corresponding output is as follows.
- The field containing a 0 corresponds to the member that leaves the domain now.
- The field containing a 1 corresponds to the remaining participant member.
- The field containing an X corresponds to the registered member or the leaf node has not been assigned to any member.

For example, assume that the leaf node m6 in Fig. 2(a) is a registered member, and m0 and m1 have not been assigned to any member. Now the participant member m3 and m7 leave the domain at the same time. The Boolean membership function is shown as Fig. 3(a). More details about minimizing the Boolean function can be found in [5]. Fig. 3(b) shows a Karnaugh map representation of figure 3(a), and Fig. 3(c) identifies the largest possible rectangles containing only 1 and X.

The minimization output is $\bar{b}_1 + \bar{b}_0$. Thus, the subgroup manager encrypts the new group key by $\bar{K}_1$ and $\bar{K}_0$ respectively, and multicast the encrypted messages to its Member Domain. The message encrypted by $\bar{K}_1$ can be decrypted by m5 and m4 only, and the message encrypted by $\bar{K}_0$ can be only decrypted by m2, and the leaving member m3 and m7 can not decrypt the new group key anyway.

Note that the result of minimizing a Boolean function is a sum of product expression. Each product term represents a key used to encrypt the new group key. However, a product term may consist of more than one literal. In such a case, the key, called as *conjoint key*, must consist of more than one auxiliary key. For example, if some minimization output is $\bar{b}_1 + \bar{b}_2 b_0$, then the subgroup manager encrypts the new group key by the auxiliary key $\bar{K}_1$ and the conjoint key $K_{\overline{2}0}$ respectively, where $K_{\overline{2}0} = f(\bar{K}_2, K_0)$.

Input(MID) $b_2b_1b_0$	Output
000	X
001	X
010	1
011	0
100	1
101	1
110	X
111	0

(a) Boolean membership function

(b) Karnaugh Map

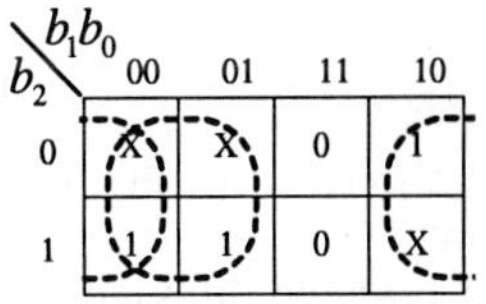

(c) Prime term selection

Fig. 3. Boolean algebra minimization in Member Domain

Case II. Multiple subgroup managers leave from the Router Domain

When multiple subgroup managers leave the Router Domain, the group manager can batch the leaving requests and execute one Re-keying instead of several ones. After generating a new group key, the group manager performs a minimization of the Boolean membership function to decide the communal auxiliary keys.

Since the minimization is based on the MID of a complete binary tree, we first expand the Huffman binary tree to a complete binary one. After that, some subgroup managers may have more than one MID. For example, Fig. 4 shows the expanded complete binary tree from the Huffman tree in Fig. 2(c). The subgroup manager R_G has MID 1111~1100, and R_J has MID 0101 and 0100.

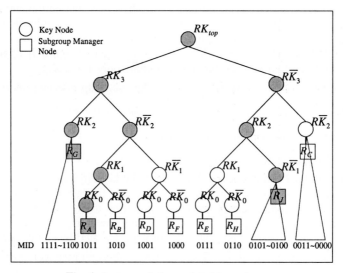

Fig. 4. An expanded complete binary tree

The input of the Boolean membership function is the MID of each leaf node in the expanded complete binary tree, and the corresponding output is as follows.
- The field containing a 0 corresponds to the subgroup manager that leaves the group now.
- The field containing a 1 corresponds to the remaining participant subgroup manager.
- The field containing an X corresponds to the subgroup manager has not join the group yet.

For example, assume that the subgroup manager R_A, R_G, R_J will leave the group. The corresponding Boolean membership function and minimization process is shown as Fig 5. The minimization output is $\bar{b}_3 b_1 + \bar{b}_2 \bar{b}_1 + \bar{b}_2 \bar{b}_0$. That means the group manger can use the three conjoint keys $RK_{\overline{31}} = f(R\bar{K}_3, RK_1)$, $RK_{\overline{21}} = f(R\bar{K}_2, R\bar{K}_1)$ and $RK_{\overline{20}} = f(R\bar{K}_2, R\bar{K}_0)$ to encrypt the new group key, and guarantee that the leaving subgroup manager R_A, R_G, R_J can not decrypt the new group key. Although the Boolean minimization method works well in a complete binary tree, it is incomplete in the Huffman tree. For example, the remaining subgroup manager R_C can not get the new group key since it has only the auxiliary keys $R\bar{K}_3$ and $R\bar{K}_2$. It's impossible for R_C

to generate any one of the conjoint keys $RK_{\overline{31}}$, $RK_{\overline{21}}$, $RK_{\overline{20}}$. Thus, after performing the Boolean minimization, the group manager must execute a complementary procedure described in Fig. 6 to make sure the remaining subgroup managers can get the new group key.

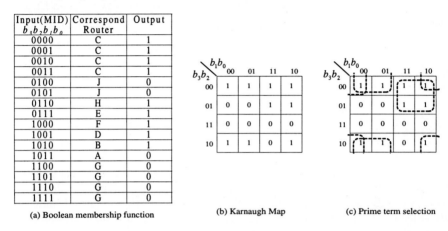

Input(MID) $b_3b_2b_1b_0$	Correspond Router	Output
0000	C	1
0001	C	1
0010	C	1
0011	C	1
0100	J	0
0101	J	0
0110	H	1
0111	E	1
1000	F	1
1001	D	1
1010	B	1
1011	A	0
1100	G	0
1101	G	0
1110	G	0
1111	G	0

(a) Boolean membership function   (b) Karnaugh Map   (c) Prime term selection

Fig. 5. Boolean algebra minimization in Member Domain

Input:
  COMK = $\{C_1,C_2,...,C_n\}$, C_i is the corresponding communal auxiliary keys
    set of the product term i of Boolean minimization output
  REMSG = $\{SG_1,SG_2,...SG_m\}$, SG_j is the remaining subgroup manager j
  AUXK = $\{AK_1,AK_2,...,AK_m\}$, AK_j is the auxiliary keys set of SG_j
Output:
  COMPK : the complement keys set
Procedure:
  Begin
    COMPK = ϕ
    for j = 1 to m
      if UID and LID of SG_j are different then
        if all C_i such that $C_i \not\subset AK_j$ then
          add AK_j to COMPK
    End for
  End

Fig. 6. Complementary Procedure

Continuing the above example, COMK={{ $R\overline{K}_3, RK_1$ },{ $R\overline{K}_2, R\overline{K}_1$ }, { $R\overline{K}_2, R\overline{K}_0$ }} REMSG={$R_B, R_D, R_F, R_E, R_H, R_C$}, AUXK={$AK_B$, AK_D, AK_F, AK_E, AK_H, AK_C}, where AK_B={ $RK_3, R\overline{K}_2, RK_1, R\overline{K}_0$ }, ..., AK_C={ $R\overline{K}_3, R\overline{K}_2$ }. Since only the R_C whose UID is not equal to LID and there does not exist a C_i in COMP which is a subset of AK_C. Thus COMPK={{ $R\overline{K}_3, R\overline{K}_2$ }}, the group manger must multicast four re-keying messages being encrypted by $RK_{\overline{31}}, RK_{\overline{21}}, RK_{\overline{20}}$ and $RK_{\overline{32}}$ respectively.

6. Secure Traffic with Source Authentication

In fact, any message transmitted in the group is required with a sender's digital signature to provide the source authentication and message integrity. In addition, a sender can use the group key to encrypt the group traffic to afford message confidentiality. Thus, any member must use the following format to transmit the group traffic.

$$M \mapsto G : GroupData([<Data> k_m^{-1}]GrpKey \tag{16}$$

7. Comparison with Previous Works

In this section, the proposed framework is compared with the works of [9], [10], [11], [12], [8]. The structure of key management will affect the number of messages needed in a re-keying process and the storage needed by each member to keep the auxiliary keys. In [9] and [10], the group is divided into multiple subgroups with a hierarchy of group security agents. A star structure for auxiliary key distribution is in each subgroup, that is, each member has an auxiliary key. In [9] there does not exist a key shared by all the group members. Thus, the group traffic must be decrypted/encrypted whenever through an agent. That increases the data transmission time a lot and the overhead of each agent.

The scheme proposed in [11] uses a hierarchy of keys to solve the scalability problem. The binary tree of auxiliary keys distribution is proposed in [12] and [8] which reduces the number of auxiliary keys needed from $O(N)$ in [11] to $O(\log N)$, where N is the number of group members. They use Boolean function minimization to reduce the number of messages needed in a re-keying process.

Table 1 and 2 show the number of keys needed in a group and the number of messages needed in a re-keying process respectively. Assume that the number of group members is N and there exists one router domain and $m+1$ member domains (or subgroups) and M_i denotes the size of a member domain i. Table 1 shows that the number of keys needed by ours is less than [9], [10] and [11], and is approximately the same as [12] and [8]. Table 2 shows that our scheme needs less number of messages than the others in a re-keying process.

Table 1. The number of keys needed in a group

	[9]	[10]	[11] with degree d	[12]	[8]	Ours
Group Key	0	1	1	1	1	1
Subroup Key	m+1	M+1	0	0	0	m+2
Auxiliary Key	N+m	N+m	$\frac{d}{d-1}(N-1)$	$2\log_2 N$	$2\log_2 N$	$\sum_{i=0}^{m} 2\log_2 M_i + 2(m-1) \sim 2(\log_2 m)$

Table 2. The number of messages needed in a re-keying process

		[9]	[10]	[11] with degree d	[12], [8]	Ours
Requested Subgroup Manager	Join	2	2			$\log_2 M_i + 2$
	Leave	$M_i - 1$	$M_i - 1$			$\log_2 M_i$
Non-requested Subgroup Manager	Join	0	0			0
	Leave	0	1			1
Group Manager	Join		1	$2(\log_d N - 1)$	$\log_2 N + 2$	1 or $((\log_2 m + 2) \sim m)$
	Leave		m	$d(\log_d N - 1)$	$\log_2 N$	1 or $(\log_2 m \sim (m-1))$

8. Conclusion

In this paper we propose an efficient two-level management for secure group communication. The group is divided into one router domain and several member domains. Such a division can solve the scalability problem and is in accordance with Internet architecture. A group key is shared by all of the members. This makes the group can use any multicast protocol to transmit the group traffic and only the sender/receivers need to do encryption/decryption.

The framework establishes a complete binary tree at each member domain, but a Huffman tree at the router domain. It lets the subgroup manger with more registered members to have more auxiliary keys, and the subgroup manager with less registered members to have less auxiliary keys. This is because the probability that the subgroup manager with less members leaves the group is higher than the one with more members. The group manager uses auxiliary keys, which can not be held by the leaving subgroup manager, to encrypt the new group key. Thus, it is preferred that a subgroup manager with high probability of leaving the group has less auxiliary keys to make the decision of communal auxiliary keys easy.

References

1. N.F.Maxemchuk, "Video Distribution on Multicast Networks", IEEE Journal on Selected Areas in Communications, Vol. 15, No. 3, April 1997, pp.357-372.
2. B.M.Waxman, "Routing of multiple connections", IEEE Journal on Selected Areas in Coom, 1988, pp.1617-1622.
3. M.J.Moyer, J.R.Rao, P.Rohatgi, "A survey of security issues in multicast communications", IEEE Network, 1999, pp.12-23.
4. B.Schneier, Applied Cryptography , second edition, John Wiley & Sons, Inc., 1996.
5. E.J.McCluskey, "Minimization of Boolean Functions", Bell System Tech. Journal, Vol. 35, No. 6, Nov., 1956, pp.1417-1444.
6. D.Wallner, et al., "Key Management for Multicast: Issues and Architectures", RFC 2627.
7. P.S.Kruus, J.P.Macker, "Techniques and issues in multicast security", Military Communications Conference, 1998. MILCOM 98. Proceedings, IEEE, 1998, pp.1028-1032.
8. I.Chang, et al., "Key management for secure Internet multicast using Boolean function minimization techniques", INFOCOM '99, pp.689 -698.
9. S. Mittra, "Iolus: A Framework for Scalable Secure Multicasting", Proceedings of ACM SIGCOMM'97, Cannes, France, 1997, pp.277-288.
10. T.Hardjono, B.Cain, "Secure and Scalable Inter-Domain Group Key Management for N-to-N Multicast", 1998 International Conference on Parallel and Distributed Systems, 1998, pp.478-485.
11. C. K. Wong, et al., "Secure Group Communications Using Key Groups", Proceedings of ACM SIGCOMM'98.
12. G.Garonni, et al., "Efficient security for large and dynamic multicast groups", Seventh IEEE International Workshops on Enabling Technologies: Infrastructure for Collaborative, 1998, pp.376-383.

Authenticating Multicast Streams in Lossy Channels Using Threshold Techniques

Mohamed Al-Ibrahim[1] and Josef Pieprzyk[2]

[1]School of IT & CS
University of Wollongong
Wollongong, NSW 2522
Australia
ibrahim@network.kuniv.edu.kw

[2]Department of Computing
Macquarie University
Sydney, NSW 2109
Australia
josef@ics.mq.edu.au

Abstract. We first classify the state-of-the-art stream authentication problem in the multicast environment and group them into Signing and MAC approaches. A new approach for authenticating digital streams using Threshold Techniques is introduced. The new approach main advantages are in tolerating packet loss, up to a threshold number, and having a minimum space overhead. It is most suitable for multicast applications running over lossy, unreliable communication channels while, in same time, are pertain the security requirements. We use linear equations based on Lagrange polynomial interpolation and Combinatorial Design methods.

1 Introduction

Communication in computer network may be established as a unicast or multicast connection. In a unicast connection, messages are flowing from a single sender to a single recipient. In a multicast connection, however, a single sender conveys messages to many receivers. Distribution of Pay TV channels is typically done using multicast communication. Protection of multicast messages includes normally both their confidentiality and authenticity. However, in majority of network services, confidentiality seems not to be of the main concern. Authentication of messages is normally the main security goal. This is true in the Internet environment where most web pages are publicly available while their authentication is the major security worry – the lack of proper authentication is always an attractive target for hackers with inclinations for practical jokes.

Source authentication is the main theme of this work. This problem has been already addressed for unicast connections. For instance, the IPSec protocol suit or IP version 6 on the network layer supports source authentication. Higher layers on the OSI reference model, make use of authentication services provided by Secure Socket Layer (SSL). Authentication is normally based on message authentication codes

(MACs) generated using private-key cryptography where the sender and receiver share the same secret key. This approach cannot be easily extended to cover authentication for multicast connections.

Public-key cryptography offers an alternative solution for authentication which is ideal for multicast communication – the transmitted message is digitally signed by the sender (holder of the secret key) and every body can verify the validity of the signed message by using the matching public key. Unfortunately, both generation and verification of digital signatures are very expensive as they consume a lot of computing resources. Additionally, digital signatures are excessively long normally more than 778-bits consuming extra channel bandwidth.

Multicast is commonly used for many real-time applications such as multimedia communication for distant teaching, dissemination of digital video, etc. Some of distinct features of multicasting include:
- possible packet loss at some (or perhaps all) destinations, were there is no standard reliable multicast IP protocol yet,
- the application tolerates some packet loss but does not tolerate an excessive delay.

Authentication of multicast messages (streams) has to:
- provide strong security – no outsider is able to insert packets with forged contents. All packets accepted at destinations have to be genuine,
- tolerate loss of packets – any outsider or/and noisy channel may erase a strictly defined number of packets. The number is typically determined by the application. The application refuses to work correctly if the number of accessible packets drops below the limit,
- be fast and efficient.

The work is structured as follows. Section 2 reviews related work. Section 3 describes the model and states the assumptions for a multicast environment in which the proposed solutions are be based on. In section 4 we present two schemes using linear equations and in section 5 we present two schemes based on Combinatorial Designs.

2 Related Work

There are many interesting papers published recently that are dealing with multicast stream authentication (see [1,2,3,4,5,6]). The solutions proposed there can be divided into two broad classes. The first class includes solutions based on MACs (private-key cryptography). The second one employs digital signatures (public-key cryptography).

Genaro and Rohatgi [1] used digital signatures to authenticate multicast streams. They considered two cases: on and off-line. In the off-line version, they used a chaining of packets. The digest of packet P_i depends on its predecessor P_{i+1}. The first packet is signed using a one-time signature. This technique is efficient and introduces a minimum packet redundancy but does not tolerate packet loss. In the on-line version, each packet is signed using one-time signature and the packet contains the public key that is used to verify the signature of the preceding packet.

Wong and Lam [3] applied Merkle signature trees to sign a stream of packets. In this scheme a block of packets are buffered to form the leave of a tree, in which the message digests of the packets are computed and the root of the tree is signed. The root is the hash of all message digests of the block. Each packet attaches the hashes of other packets in the block as well as the root of the tree to form ancillary information. The number of attached hashes depends on the degree of the tree. The scheme is designed to run at sufficient speed for real-time applications, add a bearable space overhead from authentication data, and tolerate packet loss gracefully. It was argued in [4] that this scheme increases per-packet computation with packet loss rate especially in the case of mobile receivers that have less computational power and thus have higher packet loss. In their work they tried to overcome this shortcoming by proposing an enhanced scheme.

Rohatchi [5] extended idea of [3] by computing hash values ahead of time to reduce the time delay for signing a packet. They extended the one-time signature scheme to k-time signature scheme and reduced the communication and space overhead of one-time signature scheme. The scheme uses 270 bytes for signature and the server has to compute 350 off-line hash values and the client needs 184 hash values to verify the signature.

MACs were used in [2] and [4]. As we mentioned earlier that using MACs for multicasting needs special care. The scheme of Canettie at el. [2] uses asymmetric MAC. The idea behind this technique is that the sender holds n keys and shares half of these keys with each receiver of a group such that no two receivers hold the same set of keys that are held by another receiver. The keys are distributed in such a way that it guaranties that no receivers up to W could collide to reveal the n keys hold by the sender, otherwise it would violates the security condition of the scheme. The sender sends a packet attached with n number of MAC's per packet, while the receiver verifies only the MAC's in the packet that it holds its corresponding keys and ignores the rest. This solution does not suffer from packet loss problem, and each packet is authenticated individually, but it suffers from the space overhead incurred from extra digest messages attached to each packet.

Another proposal based on MACs is described in [4]. It uses MACs with delayed key disclosure sent by the sender using chaining technique on periodic basis. The idea is to let the sender to attach to each packet a MAC, which is computed using a key k known only to it. The receiver buffers the received packet without being able to authenticate it. On period time d, which is set in advance, the sender discloses the key k that is used to authenticate the buffered packet on the receiver side. Packets that received after the period time are discarded. Actually, the key k is used to authenticate all the packets in the period interval d. The sender and receiver have to synchronize their timing to ensure correct key disclosures.

3 Model

Messages we would like to authenticate are coming from a single source and are distributed to multiple destinations (multicasting). Authentication by signing the whole message is not an option when the message is very long or alternatively, if the transmission of the message may take unspecified period of time (the length of message is not known before hand). Real-time live TV communication is of this type.

On the other hand, a vendor who offers video channel normally knows the length of the transmitted message but the verification of message authenticity could be done at the very end of the movie and it fails with a very high probability due to some lost or/and corrupted packets (multicast connections are not reliable).

Given a long message M (it is also called a stream). The message is divided into blocks of well defined size and each block consists of n datagrams (packets). Assume that stream is authenticated by signing blocks. There are two possibilities:

(1) blocks in the stream are signed independently – attackers may change order of blocks without receivers noticing it. To prevent this, blocks may contain the sequence number or other timestamping information. This possibility is suitable for applications that tolerate loss of blocks. For instance for live TV, loss of a block is typically perceived as a momentary freeze of the frame,

(2) blocks are signed in a chaining mode – the signature of the current block depends on its contents and on all its predecessors. Attackers are no longer able to manipulate the order of blocks. Loss of a block, however, makes authentication of all blocks following the lost, impossible. This option suits applications that do not tolerate loss of blocks.

Consider a block B which consists of n datagrams so $B = (m_1, \cdots, m_n)$. Our attention concentrates on authentication of a single block so that:

1. verification of authenticity of blocks is always successful if receivers obtained at least t datagrams (out of n sent) – authentication tolerates loss of up to $d = n - t$ packets,
2. expansion of packet contents is minimal – redundancy introduced to the block is minimal under the assumption that authentciation tolerates loss of d datagrams,
3. generation of signatures and their verification is efficient, i.e. computing resources consumed by them is reasonable (or in other words, the delay caused by them is tolerable by the applications run by receivers).

The last point needs further ellaboration. The sender (source) is normally much more powerful than recepients. If this is true then authentication can be performed as follows:

- a powerful sender generates (expensive) digital signatures independently for each block,
- a receiver applies batch verification of signatures, i.e. merges signatures of blocks and the result can be verified as a single signature.

This solution is quite attractive especially for applications which tolerate longer delays, in which the validity of blocks can be asserted only after receiving the last signature in the batch. Also this solution must be applied with an extreme caution as attackers may inject invalid signatures and false messages which will cancelled each other in the batch (verification process gives OK when, in fact, false blocks have been injected to the stream).

More formally, stream authentication scheme is a collection of two algorithms:

- $A=GEN(B, secret\_key)$ – generates authenticator A for a block B with the aid of the sender secrret key,

- *VER(B,A,secret_key)* – verifies verify authenticator *A* attached to the block *B* using sender's public key. The result is binary and can be either OK or FAIL.

The scheme must :
- tolerate loss of packets within the block – any *t* out of *n* packets allows to run the algorithm *VER* which can generate either OK or FAIL,
- be secure – any malicious/accidental change of either contents or order of packets within the block must be detected by receivers,
- introduce minimal redundancy to the block *B*, or in other words, the bandwidth expansion should be minimal,
- be efficient – execution of both algorithms *GEN* and *VER* should consume small amount of computing resources (if the sender is powerful than efficiency of *GEN* is not crucial).

The above requirements are typically referred as the design goals.

4 Stream Authentication Based on Linear Equations

We start from the following authentication scheme. It authenticates a single block *B* using a digital signature generated for the hash value of the block *B*. The verification algorithm can be run if all hash values of packets are available. In our scheme, *t* hash values are obtained directly from packets (we assume that *t* packets from the block have arrived safely to the destination). The missing *d* hash values are computed from the information attached to an extra packet m_{n+1}.

Stream Authentication (Scheme 1)

GEN(B, secret_key):

1. Divide the message block *B* into *n* datagrams $m_1 \ldots m_n$.

2. Create $h_i = H(m_i)$ for $i = 1, \ldots, n$ where *H* is a collision resistant hash function such as MD5 or SHA1.

3. Form a polynomial: $\quad F(x) = \sum_{i=1}^{n} h_i x^{i-1}$

   and compute *d* control values $F(j)$ for $j = 1, \ldots, d$.

4. Compute the message digest $D = H(h_1, \ldots, h_n)$

5. Sign the computed value *D* using a strong cryptography signature scheme *S*.

6. Return datagram: $\quad m_{n+1} = F(1) \| \ldots \| F(d) \| S(D, secret\_key)$, where $S(D, secret\_key)$ is the digital signature generated in step 5.

$VER(\tilde{B}, m_{n+1}, public\_key)$
:

1. The block $\tilde{B}$ must contain at least t datagrams. If this does not hold or m_{n+1} does not exist, the algorithm returns FAIL.
2. If the number of packets in $\tilde{B}$ is at least t, then order the datagrams correctly so the sequence is identical with the one in the source with at most d missing datagrams.
3. Find all the hash values h_i for the packets from $\tilde{B}$. Note that at least t hash values are generated.
4. Compute the missing hash values for lost datagrams from the d control values $F(j)$; where $j = 1, \ldots, d$. In other words, solve the system of linear equations generated from the control values.
5. Verify the signature using a native (to signature) verification algorithm and the public key of the sender. If the signature is correct return OK, otherwise exit FAIL.

Consider the design goals. The first goal – tolerance for packet loss – is not achieved. The scheme works only if the $(n+1)$-th packet containing control values arrives to the destination. Consider the case when t packets arrived safely to the destination together with the last packet m_{n+1}. Without the loss of generality, assume that these packets are $(m_1, \cdots, m_t)$. The receiver can run the verification algorithm VER if they are able to compute the digest D of the whole block. To do this, they must be able to recover hash values of all packets. First, they compute values $h_1, \cdots, h_t$ directly from the available packets. Next, they compute the missing ones from the control values accessible in the $(n+1)$-th packet. In other words, they solve the following system of linear equations:

$$\begin{aligned} F(1) - a_1 &= h_{t+1} + \cdots + h_n \\ F(2) - a_2 &= h_{t+1} \times 2^t + \cdots + h_n \times 2^{n-1} \\ &\vdots \\ F(d) - a_d &= h_{t+1} \times d^t + \cdots + h_n d^{n-1} \end{aligned}$$

where

$$a_r = h_1 + h_2 r + \cdots + h_t r^{t-1}$$

for $r = 1, \cdots, d$. The above system of linear equation has always solutions as it is characterized by a Vandermonde matrix that is always nonsingular (whose determinant is nonzero). Having all hash values, the receiver computes the digest D of the block B and verifies the signature (using the sender public key).

We claim that the security of the scheme is equivalent to the security of the underlaying digital signature (assuming that the hash function is collision resistant). To see this, it is enough to note that the only difference between the underlaying signature and the signature used in the stream authentication scheme is the way the verification is being performed. For stream authentication, the verifier computes at least t hash values from packets that safely arrived to the destination while d (or less) missing ones are computed from the control values. For the original signature scheme, the verifier computes all hash values directly from messages.

Data expansion is indeed minimal as the receivers are getting the minimum necessary information to successfully run the verification algorithm. This statement is true under the assumption that hash values are much smaller than the payload of datagrams. In practice, the length of a hash value is 128 bits (for MD5) or 180 bits (for SHA1) while the maximum length of packets in X25 is 8192 bits.

As noted, the scheme considered above is not tolerant for the loss of packets. More specifically, the absence of the $(n+1)$-th packet precludes the receiver from getting the system of linear equation, which are necessary to generate the digest D of the block B.

Now we present the second scheme that does tolerate a loss of d packets. The main idea is to distribute the content of the $(n+1)$-th packet among other packets using a system of linear equations (in a very similar fashion as it has been done for hash values).

Stream Authentication (Scheme 2)

GEN(B, secret_key):

1. Divide the message block B into n pieces $m_1 \ldots m_n$.

2. Compute $h_i = H(m_i)$ for $i = 1, \ldots, n$ and the block digest $D = H(h_1, \ldots, h_n)$, where H is a collision resistant hash function.

3. Create a polynomial $F(x) = \sum_{i=1}^{n} h_i x^{i-1}$.

4. Form a redundant message $R = (F(1) || F(2) || \ldots || F(d) || S(D, \text{secret\_key}))$ and split it into t pieces of the same size, i.e.
$$R = (b_0, \cdots, b_{t-1})$$
where S is a strong cryptography signature scheme.

5. Create a polynomial $R(x) = b_0 + b_1 x + \cdots + b_{t-1} x^{t-1}$
and define n control values $R(1), \cdots, R(n)$ that are assigned to respective datagrams.

6. Return the sequence of datagrams each of the form $(m_i, R(i))$ for $i = 1, \ldots, n$.

VER($\tilde{B}$, public_key):

1. The block $\tilde{B}$ must contain at least t datagrams. If this does not hold, the algorithm returns FAIL.

2. If the number of packets in $\tilde{B}$ is at least t, then order the datagrams correctly so the sequence is identical with the one in the source with at most d missing datagrams.

3. From each datagram extract the contents m_i and the corresponding control value $R(i)$.

4. Find all the hash values h_i for the packets from $\tilde{B}$. Note that at least t hash values are generated directly from m_i.

5. Assemble a system of linear equations from the available control values $R(i)$, solve it and determine the redundant message $R = (b_0, \cdots, b_{t-1})$.

6. Extract the control values $F(i_j)$ for the missing j and reconstruct the hash values of missing datagrams.

7. Verify the signature using a native (to signature) verification algorithm and the public key of the sender. If the signature is correct return OK, otherwise exit FAIL.

The only aspect of the second scheme, which needs some clarification, is how the receiver reconstructs the redundant message R or equivalently finds the polynomial $R(x)$. Note that the receiver knows at least t points laying on the polynomial $R(x)$. As the polynomial is of the degree $(t-1)$, any t different points uniquely determine a polynomial, which contains that points. This can be done very efficiently using the Lagrange interpolation.

5 Stream Authentication Based on Combinatorial Designs

In this scheme, combinatorial design methods will be used to distribute message digests in each of the packets in such a way that *all* message digests are retrieved when at most d packets are lost. Two methods are represented here:

5.1 Balanced Incomplete Block Design

The arrangement of the message digests into the packets uses the combinatorial technique of Balanced Incomplete Block Design (BIBD) [12]. A BIBD (v, b, k, r, λ) is an arrangement of v distinct objects into b blocks such that each block contains exactly k distinct objects, and each object occurs in exactly r different blocks. In BIBD every (unordered) pair of distinct objects occurs together in exactly • blocks. In our case, the objects are the message digests and the blocks are the packets, and the problem is to distribute the v message digests into the b packets in such a way that each message digest appears in r packets and each packet holds k different message digests. By the very definition of BIBD, if up to $d = r-1$ of such packets are lost, we can still retrieve all the k message digests from the remaining $t = v - d$ received packets.

As an example, suppose we have $v=7$ message digests that need to be distributed over $b=7$ packets. Then, a BIBD $(7, 7, 3, 3, 1)$ will allow us to retrieve all the 7 message digests from any $t=5$ of the received packets, as the following listing of the packets shows (the message digests are coded to integers 0...6):

P_0: 0, 1, 3 P_1: 1, 2, 4 P_2: 2, 3, 5 P_3: 3, 4, 6
P_4: 4, 5, 0 P_5: 5, 6, 1 P_6: 6, 0, 2

This special case when $v = b$ is called the Symmetric Balanced Incomplete Block Design (SBIBD). The necessary condition for this sort of designs is $(v-1) = k(k-1)$. Since in our applications, the number of message digests are always equal to number of packets, as each packet has a corresponding message digest, we will always exclusively deal with SBIBD in these applications. Note that, it is possible to further reduce the number of message digests in the above example in each packet to 2 messages, since it is possible to compute the message digest of the holding packet directly from the packet itself. The enhancement is as follows:

P_0: 1, 3 P_1: 2, 4 P_2: 3, 5 P_3: 4, 6
P_4: 5, 0 P_5: 6, 1 P_6: 0, 2

There are many ways for constructing SBIBD designs that is found in the literature and for different choices of parameters. Our choice was focused on the design that yields highest ratio of k to v, i.e. the minimum number of message digests that need to be carried by each packet, relatively to the total number of packets need to be present to authenticate a block. In other words we are looking for high ratio of packet loss while still are able to authenticate the block. Colbourn and Dinitz [13] list over 12 families of SBIBD($v, k, \bullet$) designs for various values of parameters. One of these families refer to Hadamard designs, and they correspond to designs with parameters $v=2^n-1$, $k=2^{n-1}-1$ and $\bullet= n-1$. A Hadmard SBIBD design exists only and only if $H(4n)$ exits, where H is a *Hadamard* matrix of order $n \times n$. The ratio of k to v is high as approximately the half, which means it is possible to authenticate the block in the presence of half of the packets of the block.

To construct a Hadamard matrix for a group of 2^n packets, the matrix is normalized by suitable row and column negations. Removing the first row and column leaves a $2^{n-1} \times 2^{n-1}$ matrix, the core of Hadamard matrix. Replace all -1 entries with 0 entries, the result is an SBIBD($2^n-1, 2^n-1, n-1$). The 1 entries actually represents the arrangements of the v entries (message digests) over b (packets). These matrices designs could be prepared according to the proper sizes and selected during the implementation. To construct such method a number of packets of size 2^n are grouped to form a block. Then, (choose) Hadamard matrix of order $2^n \times 2^n$ is constructed. Finally, message digests are assigned to the corresponding packet index in the block according to Hadamard matrix [14].

One advantage of the SBIBD approach is that there is no additional computations overhead cost involved in computing the threshold parameters and retrieving them by sender and receiver respectively. The additional space overhead is minimum compared for to other technique as in [2,3].

5.2 Rotational Designs

Another dynamic, simple and easy way in which it is possible to control the degree of the threshold is by using Rotational Designs. The following example is the distribution of 6 message digests over a block of 7 packets:

Table 1.

Packet number

0	1	2	3	4	5	6
1	2	3	4	5	6	0
2	3	4	5	6	0	1
3	4	5	6	0	1	2
4	5	6	0	1	2	3
5	6	0	1	2	3	4
6	0	1	2	3	4	5

In this distribution, $v = b = 7$ and $k = r = 6$. The column represents the packet number and message digests of packets it holds, and the rows represent the message digests distribution over packets. Now, if we want to apply the threshold scheme and require only one space overhead per packet but with high availability then (i.e. one packet loss maximum), then the first row only of Table 1 will be distributed. If however, two message digests are attached to each packet, then maximum two packets are tolerated to loss, therefore the first and the second row distribution will be applied only and so on. The distribution in Table 1 is the worst case and typically the one used in Wong and Lam scheme [3]. Formally speaking, each packet holds the message digests of rest of packets in the block and consequently each block is authenticated individually. Obviously, their solution suffers from space overhead of the message digests, worst when counting the block signature space.

6 Conclusion and Future Work

Multicasting is an emerging communication mode, in which a single sender conveys messages to many receivers. Authenticating of multicast source is a major concern. The problem has been already addressed in a number of proposals. There is no single scenario that fits divers scenarios. In this paper, we first classified these proposals, then we investigated the problem in an unreliable communication environment. Different schemes based on threshold methods have been introduced. The schemes try to rid off the drawbacks of other proposals. Mainly, they have less overhead space and can tolerate packet loss up to a threshold number. The continuation of this work is to develop a fast batch verification algorithm for a digital signature algorithm that considers the order of received packets.

Acknowledgment

We would like to thank Prof. Jennifer Seberry and Dr. Abdul Hamid Al-Ibrahim for their consultations in the area of Combinatorial Designs.

References

1. R. Gennaro and P. Rohatchi, " *How to Sign Digital Str*eams", *CRYPTO'97*, Lecture Notes in Computer Science 1249, Springer-Verlag, 1997, pp 180-197
2. R. Canetti, J. Garay, G. Itkins, D. Micciancio, M. Naor and B. Pinkas, " *Multicast Security: A Taxonomy and some efficient Constructions*", *IEEE INFOCOM'99*.
3. C. Wong, S. Lam, "*Digital Signatures for Flows and Multi*casts", Proceedings IEEE ICNP '98, Austin TX, Oct 1998.
4. A. Perrig, R. Canetti, J.D. Tygar, D. Song, " *Efficient Authentication and Signing of Multicast Streams over Lossy Channels*", IEEE 2000 online.
5. P. Rohatchi, " A Compact and fast hybrid signature scheme for multicast packet Authentication", In the 6th Conference on Computer and Communications Security, November 1999.

6. P. Golle, " *Authenticating Streamed Data in the presence of random packet loss*", extended abstract, internet draft 2000.
7. S. Even, O. Goldreich, S. Micali , " *On-Line/Off-Line Digital Signatures*", Journal of Cryptology, 9(1): 35-67, 1996.
8. J. Pieprzyk, " *A Book in Computer Security* ", internet edition, 2000
9. SMUG web site, http://www.securemulticast.org
10. R. Canetti, B. Pinkas, "*A taxonomy of Multicast Security Issues*" , <draft-irtf-smug-taxonomy-01.txt> August 2000
11. A. Shamir, " *How to share a Secret*" , Communications of ACM; 22: 612-613,Nov97
12. G. R. Blakley, "*Safeguarding Cryptographic Keys*", in the proc. AFIP 1979 National Computer Conference, p. 813- 817. AFIPS 1979.
13. 12 Marshall Hall, " *Combinatorial Theory*", A Wiley-Inter-science Publication John Wiley and sons, 1986
14. C. J. Colbourn , J. H. Dinitz, " *The CRC Handbook of Combinatorial Designs*", CRC press 1996
15. Wu-chi Feng, " *Buffering Techniques for Delivery of Compressed Video in video on Demand Systems*", Kluwer Academic Publisher, 1997
16. L. McCarthy, " *RTP for source Authentication and Non-Repudiation of Audio and Video conferences*", draft-mccarthy-smug-rtp-profile-src-auth-00.txt, May 1999

Tuning of QoS Aware Load Balancing Algorithm (QoS-LB) for Highly Loaded Server Clusters

Kimmo Kaario[1], Timo Hämäläinen, and Jian Zhang[2]

[1] Honeywell Industrial Automation & Control
Control System Development
Ohjelmakaari 1
FIN-40500 Jyväskylä
Finland
kimmo.kaario@honeywell.com
[2] University of Jyväskylä
Faculty of Information Technology
Department of Mathematical Information Technology
Telecommunication
P.O.Box 35, FIN-40351 Jyväskylä
Finland
timoh@cc.jyu.fi and zhang@st.jyu.fi

Abstract. This paper introduces a novel algorithm for content based switching. A content based scheduling algorithm (QoS Aware Load Balancing Algorithm, QoS-LB) which can be used at the front-end of the server cluster is presented. The front-end switch uses the content information of the requests and the load on the back servers to choose the server to handle each request. At the same time, different Quality of Service (QoS) classes of the customers can be considered as one parameter in the load balancing algorithm. This novel feature becomes more important when service providers begin to offer the same services for customers with different priorities.

1 Introduction

One well-accepted design principle for the next-generation Internet architecture is to push packet processing complexity to the network edge so that backbone routers can be designed to be simple and operate very fast. However, this architectural decision also places more demanding requirements on edge router designs in that they have to not only forward generic IP packets at a rate scalable with the link speed, but also support a wide array of packet processing functions for value-added services, examples of which range from low-level router implementation enhancements to improve end-to-end performance such as active queue management [16] and congestion state sharing [14], to adding application-specific payload processing such as content-aware forwarding [3] and media transcoding [2]. Consequently, edge routers become I/O-intensive as well as compute-intensive systems.

There have been a lot of research that addresses content based switching. Several products are available or have been announced for use as front-end nodes in cluster servers [5, 9]. The request scheduling strategies used in the cluster front-ends are mostly variations of weighted round-robin, and do not take into account the request's target content. One of the exceptions is the Dispatch product by Resonate, Inc., which supports content-based request distribution [15]. The product does not appear to use any dynamic distribution policies based on content and no attempt is made to achieve cache aggregation via content-based request distribution. A TCP option designed to enable content-based load distribution in a cluster server is proposed in [8]. The design has not been implemented and the performance potential of content-based distribution has not been evaluated as part of that work. Also, no policies for content-based load distribution were proposed. [7] reports on the cluster server technology used in the Inktomi search engine. The work focuses on the reliability and scalability aspects of the system and is complementary to our work. The request distribution policy used in their systems is based on weighted round-robin.

Loosely-coupled distributed servers are widely deployed on the Internet. Such servers use various techniques for load balancing including DNS round-robin [4], HTTP client redirection [1], smart clients [18], source-based forwarding [6] and hardware translation of network addresses [5]. Some of these schemes have problems related to the quality of the load balance achieved and the increased request latency. A detailed discussion of these issues can be found in [8], [9] and [6]. None of these schemes support content-based request distribution. IBM's Lava project [11] uses the concept of a hit server. The hit server is a specially configured server node responsible for serving cached content. Its specialized OS and client-server protocols give it superior performance for handling HTTP requests of cached documents, but limits it to private Intranets. Requests for uncached documents and dynamic content are delegated to a separate, conventional HTTP server node.

Here, we present a content based scheduling algorithm (QoS Aware Load Balancing Algorithm, QoS-LB), which can be used at the front-end of the server cluster. The front-end switch uses the content information of the requests and the load on the back servers to choose the server to handle each request. At the same time, different Quality of Service (QoS) classes of the customers can be considered as one parameter in the load balancing algorithm. The rest of this paper is organized as follows. Chapter two describes our QoS-LB algorithm. In Chapter 3 we present simulation environment and Chapter 4 analyzes obtained results. Finally, Chapter 5 summarizes our main results and gives some ideas to the future work.

2 The Algorithm

In QoS-LB, the algorithm has a strategy that consists of three phases. First, the algorithm tries to find a server that is included in the set of preferred servers for the requested service and its QoS needs. If this kind of server is found and its load is under the limit $T_{high}(class, service)$, it is chosen as the server for the requested connection.

We have function $f(m)$ to assign each server m a maximal QoS class that it is preferred to serve. By using this function, the set of preferred QoS classes in server m becomes

$$q_{pref}(m) = \{1, ..., f(m)\}$$

Choosing of relevant function $f(m)$ is one topic of discussion in this paper.

In the second phase of the algorithm, i.e. in the case when all the preferred servers are too busy to achieve the requested QoS level, the algorithm tries to find a server that has a load less than T_{low}. The limit T_{low} must satisfy the rule

$$T_{low} < \min\left(T_{high}(c,s)|c \in C, s \in S\right) \quad (1)$$

where C is the set of all supported QoS classes in the server cluster, and S is the set of all the supported services in the cluster.

If the first two phases of the algorithm don't give us the server, the last phase of the algorithm just chooses the least loaded node of the cluster as the server.

In the algorithm, the maximal number of concurrent connections to the back-end nodes is limited to

$$T_S = (m_{max} - 1) \cdot \max(T_{high}(c,s)|c \in C, s \in S) + T_{low} - 1 \quad (2)$$

This limit is very important parameter in QoS-LB. Tuning of T_S is another main topic in this paper. In fact, tuning of T_S equals to tuning the parameters T_{low} and $T_{high}(c,s)$ (where $c \in C$ and $s \in S$).

3 Simulations

Let $M \in \mathbf{N}$ be the set of servers in the server cluster, $S \in \mathbf{N}$ the set of supported services (note that each service must be identified uniquely by a number in this set), and $Q \in \mathbf{N}$ the set of supported QoS classes in the server cluster (again, QoS classes must be identified by these numbers). If m_{max} is the number of servers, s_{max} the number of services, and q_{max} the number of QoS classes, then $M = \{1, ..., m_{max}\}$, $S = \{1, ..., s_{max}\}$, and $Q = \{1, ..., q_{max}\}$.

Now, we can define $f(m)$ e.g. by using a linear approach

$$f(m) = \begin{cases} \max\{1, \left\lfloor \frac{q_{max}}{m_{max}} \right\rfloor + \left\lceil \frac{-q_{max}}{m_{max}} m + q_{max} \right\rceil\}, & \text{when } m_{max} \leq q_{max} \\ \max\{1, \left\lceil \frac{-q_{max}}{m_{max}} m + q_{max} \right\rceil\}, & \text{when } m_{max} > q_{max} \end{cases} \quad (3)$$

where $m = 1, ..., m_{max}$. Here, $\lceil x \rceil$ means the ceil of x, and $\lfloor x \rfloor$ is the floor of x. Linear approach means that the number of preferred QoS classes decreases quite linearly when the server number increases. In some cases, however, there might be need for different weighting between the QoS classes. In this paper, we have also tried a bit different $f(m)$, namely

$$f(m) = \min(q_{max}, \max(1, \lfloor -\frac{m^2}{2} + q_{max} \rfloor + \lceil \frac{q_{max}}{m_{max}} \rceil)) \quad (4)$$

In most of the simulations, inter-arrival times were created by using simple Poisson process. The service times were distributed as Pareto process defined by

$$F_X(x) = P(X \le x) = 1 - \left(\frac{\beta}{x+\beta}\right)^\alpha, x > 0$$

where $0 < \alpha < 2.5$ and $\beta > 0$. Some of the most important properties for Pareto distribution are that for $0 < \alpha \le 1$, $E(X) = \infty$, for $1 < \alpha \le 2$, $E(X) < \infty$ and $Var(X) = \infty$ and for $\alpha > 2$, both $E(X)$ and $Var(X)$ are finite. With Pareto distribution, we can create self-similar service times.

In some of the simulations, we also created inter-arrival times to be self-similar. The existence of self-similarity in network traffic can be studied in more detail e.g. from [10, 13] and [17].

4 Results

In the following, we show a set of simulations that reveal the importance of T_S and $f(m)$ in the performance QoS-LB. In Figure 1, we have presented the performance of simple Round-Robin with our simulated, self-similar traffic load. It is used as an reference for an algorithm that is not taking QoS needs or server loads as input parameters.

In Figure 2, the performance of LARD [12] is depicted. It is a good reference to a well-known and accepted locality-aware load balancing algorithm. We would like to underline that the aim of this paper is not to directly compare the performance of LARD and QoS-LB, but to show the potential of having QoS information as one input parameter in the load balancing algorithm. This is done by comparing LARD and QoS-LB.

Figures 3 and 4 show the cases when parameter T_{high} is unsuccessfully dimensioned. In Figure 3 we have used the linear function of Equation 3, and in Figure 4 the unlinear version of Equation 4. The Figures 5 and 6 show the successful cases, respectively.

In all the figures, the load on the vertical axis is normalized to the original (unsuccessful) limit of T_{high}. When using Equation 3 in Figures 3 and 5, we had

$$q_{pref}(1) = \{1, ..., 10\},$$
$$q_{pref}(2) = \{1, ..., 7\},$$
$$q_{pref}(3) = \{1, ..., 5\}, and$$
$$q_{pref}(4) = \{1, 2\}.$$

In the nonlinear case of Figures 4 and 6, we had

$$q_{pref}(1) = \{1, ..., 10\},$$
$$q_{pref}(2) = \{1, ..., 10\},$$
$$q_{pref}(3) = \{1, ..., 8\}, and$$
$$q_{pref}(4) = \{1, ..., 5\}.$$

As we can see, there is only a small difference on the server loads with different strategy of $f(m)$. Most clearly the impact of $f(m)$ is seen in Figures 5 and 6, where the nonlinear choice of $f(m)$ in Figure 6 loads the Server 4 less than the linear $f(m)$ in Figure 5. The importance of $f(m)$ increases, when we are using the option to set the preferred servers not only by the QoS class of the request but also by the service type of it. This gives us an interesting application of declustering methods - the aim is to assign load between servers unequally while maintaining high cache hit rates. In this case, the complexity of tuning the algorithm increases.

The impact of T_{high} on the loads is the key factor on the performance of QoS-LB. If we choose too small value for T_{high} (compared to the current load on the servers), all the servers are equally loaded as in Figures 3 and 4. If we choose too big value for T_{high}, all the load will be on the first server when having quite low total load on the cluster. Both of the scenarios do not give us good performance. That is why we are going to implement the algorithm with adaptive tuning of these parameters.

You should note that we have implemented the algorithm for these simulations by limiting the maximal number of concurrent sessions by Eq. (2). In real implementations, there is a need for parameter $T_{max}(m)$, $m \in M$, to define the maximal request processing capacity in each server. In other words, $T_{high}(c, s)$ for all $c \in C, s \in S$ equals $T_{max}(m)$ for all $m \in M$ in this paper. This should be kept in mind when analyzing the figures.

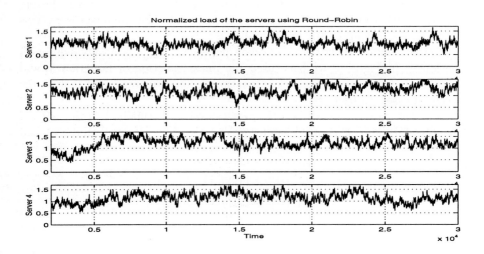

Fig. 1. Round-Robin.

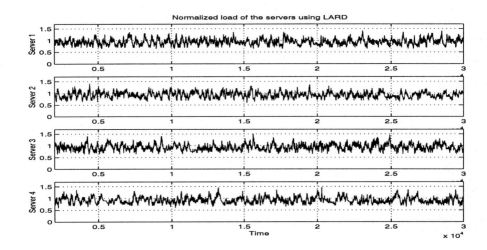

Fig. 2. LARD.

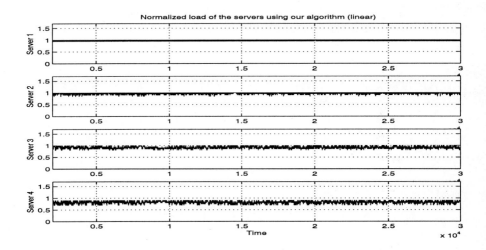

Fig. 3. Unsuccessful dimensioning of QoS-LB (linear $f(m)$).

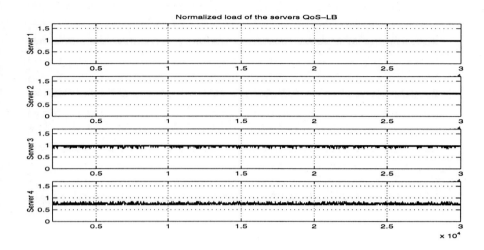

Fig. 4. Unsuccessful dimensioning of QoS-LB (nonlinear $f(m)$).

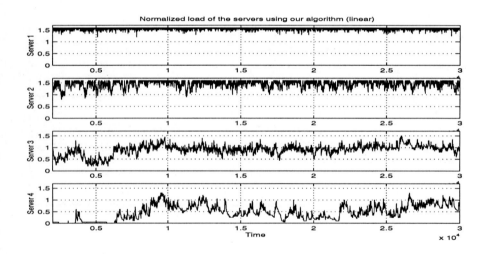

Fig. 5. Successful dimensioning of QoS-LB (linear $f(m)$).

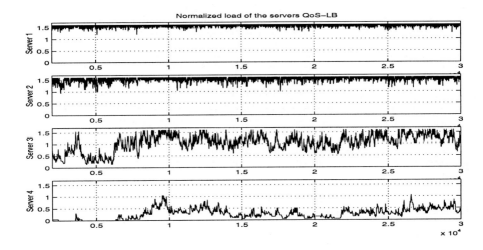

Fig. 6. Successful dimensioning of QoS-LB (nonlinear $f(m)$).

5 Conclusions

In this paper we have developed a new algorithm for distributed computer systems that allows not only load balancing, but also QoS control at each host. As illustrated by the simulations in Chapter 4, our algorithm will lead to significant improvement in system performance. There are several interesting variations of the combined load-balancing and QoS problem that merit further investigation. For example, it may be appropriate to analyze the cache hit rates and find rules for assigning services to preferred servers (in this paper, all the servers were preferred servers for any service). Other important future work will be to study the need for adaptive tuning of T_{high}. The methodology introduced in this paper may also give some insight to solving the end-to-end QoS problem in integrated service networks at the server side.

References

1. D. Andresen et al. SWEB: Towards a Scalable WWW Server on MultiComputers. In Proceedings of the 10th International Parallel Processing Symposium, Apr. 1996
2. E. Amir, S. McCanne, R. H. Katz: An Active Service Framework and Its Application to Real Time Multimedia Transcoding. Proceedings ACM SIGCOMM, 1998.
3. G. Apostopoulos, D. Aubespin, V.Peris, P. Pradhan, D. Saha: Design, Implementation and Performance of a Content-Based Switch. Proceedings of IEEE INFOCOM 2000.
4. T. Brisco. DNS Support for Load Balancing. RFC 1794, Apr. 1995.
5. Cisco Systems Inc. LocalDirector. http://www.cisco.com.

6. O. P. Damani, P. Y. E. Chung, Y. Huang, C. Kintala, and Y. M. Wang: ONE-IP: Techniques for hosting a service on a cluster of machines. Computer Networks and ISDN System
7. A. Fox, S. D. Gribble, Y. Chawathe, E. A. Brewer, and P. Gauthier: Cluster-based scalable network services. In Proceedings of the Sixteenth ACM Symposium on Operating System Principles, San Malo, France, Oct. 1997. s, 29:1019–1027, 1997.
8. G. Hunt, E. Nahum, and J. Tracey: Enabling content-based load distribution for scalable services. Technical report, IBM T.J. Watson Research Center, May 1997.
9. IBM Corporation. IBM interactive network dispatcher. http://www.ics.raleigh.ibm.com/ics/isslearn.htm.
10. W. Leland, M. Taqqu, W. Willinger and D. Wilson: On the self-similar nature of Ethernet traffic (extended version). IEEE/ACM Tran. Networking, Vol. 2, 1994, pp. 1-15.
11. J. Liedtke, V. Panteleenko, T. Jaeger, and N. Islam: High-performance caching with the Lava hit-server. In Proceedings of the USENIX 1998 Annual Technical Conference, New Orleans, LA, June 1998.
12. V. S. Pai, M Aron, G. Banga, M. Svendsen, P. Druschel, W. Zwaenepoel, E. Nahum: Locality-Aware Request Distribution in Cluster-Based Network Servers. In Architectural Support for Programming Languages and Operating Systems, 1998.
13. V. Paxson and S. Floyd: Wide Area Traffic: The Failure of Poisson Modeling. IEEE/ACM Transactions on Networking, Vol. 3, No. 3, June 1995, pp. 226-244.
14. P. Pradhan, T. Chiueh, A. Neogi: Aggregate TCP Congestion Control Using Multiple Network Probing. Proceedings of ICDCS 2000.
15. Resonate Inc. Resonate dispatch. http://www.resonateinc.com.
16. B. Suter, T.V. Lakshman, D. Stiliadis, A.K Choudhury: Buffer Management Schemes for Supporting TCP in Gigabit Routers with Per-Flow Queueing. IEEE Journal in Selected Areas in Communications, August 1999.
17. B. Tsybakov, N. D. Georganas: On Self-Similar Traffic in ATM Queues: Definitions, Overflow Probability Bound, and Cell Delay Distribution. IEEE/ACM Transactions on Networking, Vol. 5, No. 3, June 1997, pp. 397-409.
18. B. Yoshikawa et al.: Using Smart Clients to Build Scalable Services. In Proceedings of the 1997 Usenix Technical Conference, Jan. 1997.

The Incremental Deployability of Core-Stateless Fair Queuing

Yannick Blanpain, Hung-Yun Hsieh, and Raghupathy Sivakumar

School of Electrical and Computer Engineering
Georgia Institute of Technology, Atlanta, GA 30332, USA
gte463w@prism.gatech.edu, {hyhsieh,siva}@ece.gatech.edu
http://www.ece.gatech.edu/research/GNAN

Abstract. In this paper, we study the *incremental deployability* of the Core-Stateless Fair Queuing (CSFQ) approach to provide fair rate allocations in backbone networks. We define incremental deployability as the ability of the approach to gracefully provide increasingly better quality of service with each additional QoS-aware router deployed in the network. We use the *ns2* network simulator for the simulations. We conclude that CSFQ does not exhibit good incremental deployability.

1 Introduction

The growing diversity of Internet applications has motivated the need for supporting service differentiation inside the network. A real time multimedia application streaming live video, and a file transfer application, require very different network-level quality of service. While today's best-effort Internet model will not differentiate between the two applications, the Internet is slowly moving towards a pay-per-use model wherein applications can "buy" the specific network-level service they require. Parameters of such a service can include bandwidth, delay, jitter, loss, etc.

In recent years, the Internet Engineering Task Force (IETF) has developed two different quality of service (QoS) models for the Internet. While the Integrated Services (*intserv*) model provides fine-grained per-flow quality of service, it is not considered as a solution for backbone networks due to its inability to scale to a large number of flows [4]. The Differentiated Services (*diffserv*) model, on the other hand, is scalable at the expense of providing only a coarse-level quality of service that does not support any assurances to individual flows.

Of late, newer QoS models that attempt to bridge the scalability of the diffserv model, and the service richness of the intserv model have been proposed [2-5]. We refer to these new models as *Core-Stateless QoS models* in keeping with the terminology introduced by the authors of [2]. While the core-stateless approaches offer great promise in terms of both scalability and the service models they can support, not much work has been done in terms of evaluating the feasibility of their practical deployment. Specifically, given the enormous size of the present Internet, any solution requiring replacement of routers in the Internet has to be *incrementally deployable*. In the context of quality of service architectures, we define incrementally deployability as *the ability to provide increasingly better quality of service with increasing number of QoS-aware routers*.

In this paper, we study the incremental deployability of core stateless fair queuing (CSFQ) [2], a QoS model that attempts to provide rate fairness without maintaining per-flow state at core routers. The contribution of our work is twofold: *(i) We study the performance of CSFQ through simulations, and conclude that it has poor incremental deployability. (ii) Based on the insights gained through our study of CSFQ, we present some "guidelines" for the deployment of an incrementally deployable core-stateless QoS model.*

The rest of the paper is organized as follows: In Section 2, we provide a brief overview of the CSFQ mechanism. In Section 3, we describe the simulation model, and in Section 4 we discuss the simulation results. In Section 5 we conclude the paper.

2 Background

2.1 Core Stateless Fair Queuing

CSFQ attempts to emulate the behavior of fair queuing [1] at core routers without maintaining any per-flow state. Combined with an end-to-end adaptation scheme (like that of TCP), it approximately achieves max-min fairness [4]. We provide a quick overview of the CSFQ mechanism in the rest of the section. CSFQ estimates the *fair share* of each link without maintaining any per-flow state in the core router. The fair share α at a core router represents the share of the output link capacity that is allotted to each flow that traverses the router. In CSFQ, each packet has the rate r - of the flow to which the packet belongs - stamped in its header by the *ingress* edge router. When the packet arrives at a core router, the router drops the packet with a probability of $max\{0, 1 - \alpha/r\}$. If the packet is not dropped, it is accepted for transmission.

If A represents the aggregate arrival rate, F represents the aggregate accepted rate (where the two variables are updated after the arrival of every packet), and C represents the link capacity, the fair share α is updated as follows:

if $(A > C)$ $\alpha_{new} \leftarrow \alpha_{old} * C/F$
else $\alpha_{new} \leftarrow$ largest rate of any active flow

The combination of fair share estimation and probabilistic dropping of packets for those flows whose rate exceeds the fair share enables CSFQ to enforce fair sharing of a link without maintaining any per-flow state in the router.

2.2 Incremental Deployability

In Section 4 we evaluate the incremental deployability of the CSFQ mechanism by studying the fairness properties of a network in which only a fraction of the nodes are CSFQ routers. Specifically, we investigate incremental deployability from two perspectives: the core of the network and the edges. In other words, we study both the incremental deployment of CSFQ routers in the core of a backbone network (assuming that all the edge routers are CSFQ-aware) and the incremental deployment of CSFQ-aware routers at the edges of the network. For both cases, we use First In First Out (FIFO) routers for the non-QoS aware routers. Note that incremental deployment of QoS-aware routers at the edges can also be seen as increasingly more number of flows in the network being QoS-aware flows. In the rest of the paper, we refer to QoS-aware routers as *fair routers* and the non-QoS-aware routers as *legacy routers*. Likewise, we refer to QoS-aware flows as *fair flows*, and the default flows as *legacy flows*.

3 Simulation Model

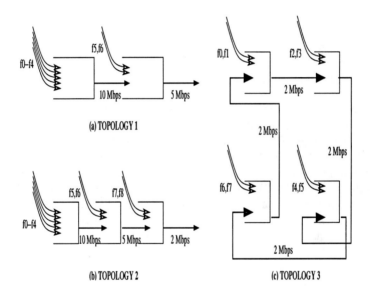

Fig. 1. Small Topologies Used For Simulations

We use the *ns2* network simulator [7] for our simulations. The *ns2* extensions for CSFQ were downloaded from http://www.cs.cmu.edu/~hzhang/csfq/. Although several topologies were used for the study, in this paper we illustrate our arguments using results for the 3 simple topologies shown in Figure 1 and the large topology shown in Figure 6. The four topologies are described in the next section. For each of the topologies and for the two scenarios of core and edge deployment, we start from a scenario where all routers use the FIFO scheme. The subsequent scenarios are obtained by incrementally changing one router at a time to use the CSFQ mechanism or be CSFQ-aware, till all the routers in the scenario are changed. For each of the scenarios, we measure the fairness offered by the network as a whole. *We use Jain's fairness-index to demonstrate the fairness achieved among the flows in the network in terms of their end-to-end throughput.* We plot the fairness index against the configuration of the network defined by the number of CSFQ (or CSFQ-aware) routers.

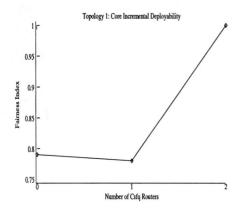

Fig. 2. Topology 1: Core Incremental Deployability

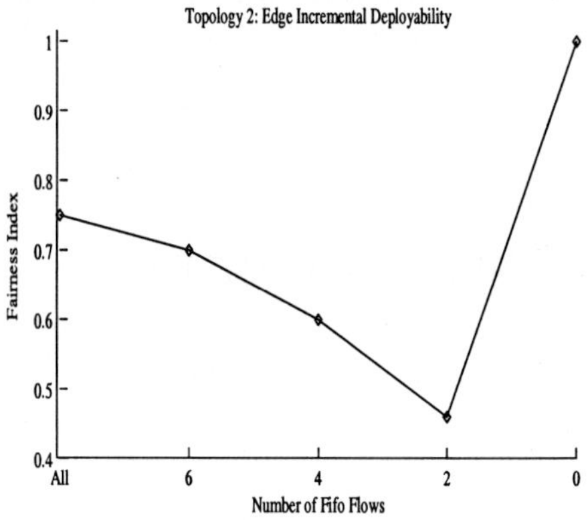

Fig. 3. Topology 1: Edge Incremental Deployability

The fairness-index plotted is an average over multiple simulations using the same number of QoS-aware routers, but with different placements for the QoS-aware routers. The flows use UDP and the traffic is generated using Poisson distribution. The labels on the x-axis of the graphs represent the number of QoS-aware routers used. The fairness-index is plotted on the y-axis.

4 Simulation Results

4.1 Simple Topologies

Figures 2, 4, and 7 show the fairness demonstrated by CSFQ when CSFQ routers are incrementally deployed in the core for Topologies 1, 2, and 3 respectively. Figures 3, 5, and 8 show the fairness when CSFQ-aware routers are incrementally deployed at the edges. As can be seen from Figures 2, 4, and 7, CSFQ does not exhibit good incremental deployment when being deployed in the core of the network. Note that each datapoint in the graphs shown was averaged over multiple simulations for all possible configurations with that many number of CSFQ routers. For example, in Figure 2, the second datapoint represents the average of the fairness indices for the two possible configurations (CSFQ-FIFO and FIFO-CSFQ) of Topology 1 with one CSFQ router. Also, for each possible configuration, simulations were run with varying rates for the flows, and an average taken over all the simulation runs.

Similarly, from Figures 3, 5, and 8, it can be observed that CSFQ does poorly in terms of edge router incremental deployment. As mentioned earlier,

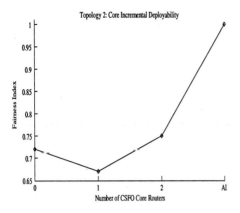
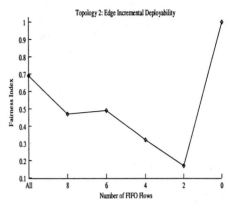

Fig. 4. Topology 2: Core Incremental Deployability

Fig. 5. Topology 2: Edge Incremental Deployability

with fewer number of CSFQ-aware routers at the edges, the number of CSFQ-aware flows in the network decreases. Hence, the unfairness stems from CSFQ flows sharing links with best-effort flows (legacy flows) in the core of the network. An interesting aspect of the results shown in the figures is the "dip" in the curve for all the scenarios. Specifically, the fairness index goes down with increasing number of fair-routers at the edges and rises only when all the edge routers are fair-routers. This anomaly can be explained as follows: When legacy-flows share a congested CSFQ link with fair-flows, CSFQ will be significantly unfair toward the fair-flows: When the fair share estimate is initially reduced, CSFQ will not observe any reduction either in the arrival rate (because legacy-flows cannot be assumed to adapt to fair share estimates). This will result in a further reduction in the fair share estimate. However, since the rate of the legacy-flows by itself is more than the link capacity, the presence of sustained congestion will ensure that the fair share estimate is cut down to zero, at which stage none of the packets belonging to the fair-flows will be accepted. However, if all the flows on a CSFQ link happen to be legacy-flows, "better" fairness can be expected as CSFQ will not interfere on behalf of any particular flow, leaving it for the indiscriminate (but not unfair) allocation of the drop-tail mechanism to decide the per-flow shares. Hence, it can be observed that the fairness achieved when there are no fair-flows in the network is better than the intermediate cases.

4.2 Large Topology

For the results shown in Figures 9 and 10, the topology shown in Figure 6 was used. All link capacities were set to 10Mbps. Each core-router in the topology was connected to exactly one edge-router. Simulations were performed with 1100 flows. Each edge-router serves as an ingress for 100 flows. The 100 flows at an ingress-router is divided into 10 sets of equal number of flows, each set having a unique egress-router among the other 10 edge routers. The rate of each flow was

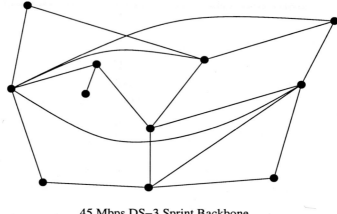

45 Mbps DS–3 Sprint Backbone

Fig. 6. Topology 4 - Sprint Backbone

set to 200 kbps. Although, simulations were performed using other flow rates, we do not present them here for lack of space. The results demonstrate that even for more realistic topologies, CSFQ performs poorly in terms of incremental deployability, both at the core and at the edges.

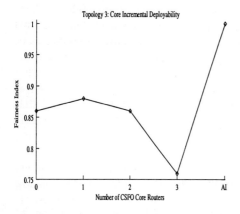

Fig. 7. Topology 3: Core Incremental Deployability

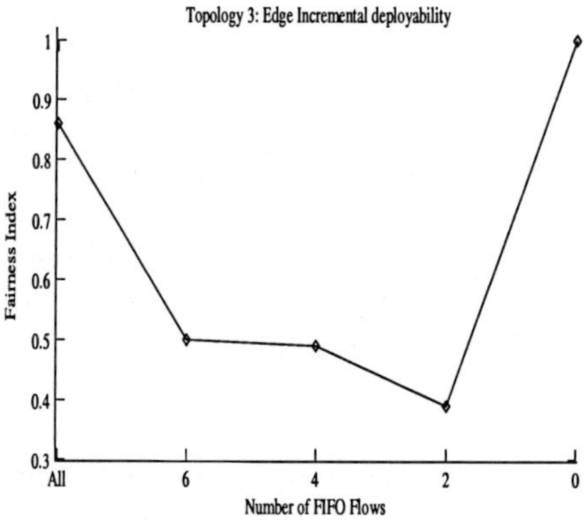

Fig. 8. Topology 3: Edge Incremental Deployability

4.3 Discussion and Insights

Our simulation results demonstrate that CSFQ is not an incrementally deployable approach to achieve fair rate allocation. However, the following insights can be gathered from conducted study:

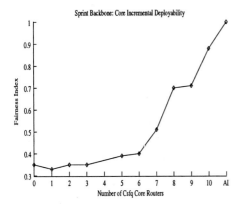 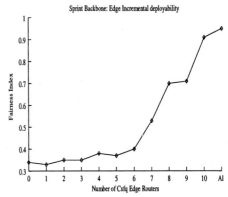

Fig. 9. Topology 4: Core Incremental Deployability

Fig. 10. Topology 4: Edge Incremental Deployability

Impact of Legacy Core Routers The impact of the presence of legacy-routers on the performance of fair-routers depends upon the specific allocation scheme employed at the fair-routers. In core-stateless approaches, fair allocation of link bandwidth at core routers is achieved by making use of *dynamic state information* about the flows. Edge routers pass the dynamic state to core routers through varying schemes including fields in the packet headers (dynamic packet state) [2], or specialized packets in the flow (dynamic flow state) [4]. The core routers rely solely on the dynamic state carried by the flow, and hence do not perform any per-flow processing. However, in the presence of legacy-routers, such approaches encounter the following problems that adversely affect the fair service they offer:

- Legacy-routers drop packets indiscriminately[1]. Hence, flows traversing such routers will inherently receive unfair service. While this by itself cannot be completely overcome, it leads to unfair service even at fair routers. We elaborate upon this phenomenon next.
- While legacy-routers drop packets indiscriminately, being unaware of the fair allocation scheme, such routers will, in addition, fail to update the dynamic state of flows when dropping packets. Hence, the dynamic state at downstream fair routers can be inconsistent with the actual flow state. When fair routers use such inconsistent dynamic state to perform rate allocation, the allocation will be unfair.
- While the unfair allocation because of inconsistent state can be plausibly perceived as a transient phenomenon (if edge routers are assumed to adapt to the fair share feedback they receive, it can be shown that flows will eventually receive their fair allocations at the fair routers), this is true only if the unfair allocations at the drop-tail routers remain stable. In other words, given an arbitrarily indiscriminate rate allocation at legacy-routers, that fluctuates

---

[1] Recall that we have assumed the use of the drop-tail mechanism at legacy-routers.

with time, the fair share computation at fair routers will fail to converge, causing unfair allocation at fair routers to become a persistent phenomenon.

Impact of Legacy Edge Routers While core routers are responsible for fair allocation, edge routers in core-stateless approaches are responsible for conveying to the core routers, the dynamic state used in their rate allocation schemes. In the event that an edge router is a legacy router, it will fail to convey any such dynamic state information to the core. Hence, the presence of such edge routers will result in *legacy-flows* co-existing with *fair-flows* in the core network. While this does not have any impact on legacy-core-routers (where no fair allocation schemes exist anyway), it obviously has a severe impact on the fairness achieved at fair-routers. Specifically, given that the legacy-flows carry no dynamic state, how should the fair routers treat the legacy-flows? Aggregating all legacy-flows into one logical flow might result in unfair allocations to legacy flows. A traffic engineering solution (wherein, the capacity of the network is partitioned between legacy- and fair-flows) might be possible, but would not be desirable in a pay-per-use service model [8].

The challenge then is to determine dynamically how the capacity at a fair router should be divided between legacy and fair-flows. Once the split is determined, it is sufficient to then provide fairness only among the fair-flows (within their allocation), as the legacy-flows do not expect any fair allocation in the first place. However, it is critical for legacy flows not to be penalized in any way due to the upgrade of a part of the network[2].

5 Summary

We study the incremental deployability of the core-stateless fair queuing (CSFQ) mechanism. Based on our simulations, we conclude that CSFQ is not incrementally deployable. However, to be fair to its authors, CSFQ was not designed to be incrementally deployable [2]. Our motivation for the study was to gain insights that can help in the design of an incrementally deployable core-stateless QoS model. We present some of the insights in Section 4.

Acknowledgments

We thank the Yamacraw organization (http://www.yamacraw.org) for their generous support and funding part of this work.

References

1. H. Zhang, "Service Disciplines For Guaranteed Performance Service in Packet-Switching Networks", Proceedings of the IEEE, 83(10), Oct 1995..

---

[2] Note that we perceive "unfair allocation" and "indiscriminate allocation" differently, with the former being clearly more undesirable than the latter.

2. I. Stoica, S. Shenker, H. Zhang, "Core-Stateless Fair Queuing: A Scalable Architecture to Approximate Fair Bandwidth Allocations in High Speed Networks", Proceedings of ACM SIGCOMM, 1998.
3. I. Stoica, H. Zhang, "Providing Guaranteed Services Without Per Flow Management" Proceedings of ACM SIGCOMM, 1999.
4. R. Sivakumar, T. Kim, N. Venkitaraman, and V. Bharghavan, "Achieving Per-flow Rate Fairness in Core-Stateless Networks", Proceedings of International Conference on Distributed Computing Systems, 2000.
5. T. Kim, R. Sivakumar, K-W. Lee, and V. Bharghavan, "Multicast Service Differentiation in Core-Stateless Networks", Proceedings of International Workshop on Network Group Communication, 1999.
6. Rossi, H. Peter, J. D. Wright, and A. B. Anderson, "Handbook on Survey Research", 1983.
7. K. Fall and K. Vardhan, "*ns* notes and documentation," available from http://www-mash.cs.berkeley.edu/ns/, 1999.
8. Lucent Technologies, "Network Quality of Service: Survey Report," .

A New IP Multicast QoS Model on IP Based Networks

Hussein Sherif Eissa and Tarek Kamel

Electronics Research Institute, Computers & Systems Dept.,
El-Tahrir st., Dokki, 12622, Cairo, Egypt
hussein@eri.sci.eg, tkamel@idsc.gov.eg

Abstract. In this paper, a new proposed QoS (Quality of Service) model has been presented for the MS NetMeeting videoconferencing application on IP based networks. It saves more than 44% of the consumed bandwidth, and 72% of the consumed buffers inside the core routers than the Qbone model. Section 2.2, presents the packet capturing experiment for the MS NetMeeting to design the inputs' generators. Sections 3.1 and 3.2 highlight on the architecture and the implementation of the new proposed model. Section 4, discusses different simulation scenarios applied to this model and the results obtained. Section 4.1 describes the No Loss real-time meter design. Sections 4.2 and 4.3 present the aggregation, and IP Multicast results from node delay and bandwidth/buffers savings points of view. Finally, sections 5 and 6 summarize the model performance analysis and the future work that should be done to modify this model.

1 Introduction

Real-time traffic creates the environment of sensation and integration where text, images, audio and video integrate and act together to give a high degree of visualization. To enhance the quality of received audio/video streams the delay, the packet loss, and the MTU (Maximum Transfer Unit) should be minimized, but this needs a huge amount of bandwidth and buffers. So, there is a trade off between the real-time applications requirements and the network resources requirements and should be compromised. The Diffserv. (Differentiated services) Model [1] is a scalable model, using simple blocks, but it does not make any assumptions on the Packet Loss values, Queuing Delay, and Jitter Delay. So, there is a trend to design a simple, scaleable, and delay-sensitive model that provides acceptable quantified values for the Packet Loss, Queuing Delay, and Jitter Delay in order to maintain an expected quality level of the audio/video streams across the IP based Networks but with an acceptable consumed amount of the bandwidth and buffers (networks resources saver) specially at the core routers in which the congestion usually occurs.

2 MS NetMeeting Videoconferencing Application Characteristics

2.1 Overview

It has been discovered that audio and video streams could not be represented by an exponential PDF curves as the most data streams across the Internet. So, in this paper, the MS (Microsoft) NetMeeting v.2.11 application [2] has been used as an example of the audio/video streams source across the Internet to have the required packet size, and packet inter-arrival PDF curves.

2.2 Packet Capturing Experiment

The Packetboy Analyzer Software [3] has been used to capture the audio/video packets, which were sent between two fully equipped PCs inside the same Ethernet (UPENN. Multimedia Lab.). The Packetboy analyzer/decoder Analyzer Package is a complete real-time visual Internet/LAN network monitor suite. Multiple captures can be loaded concurrently and packet traces can be loaded and saved to disk. The inputs of the NetMeeting Software were Speech, Audio, and still/moving person video. Then the capture process has been done on assigned filter at the Analyzer Software to capture the generated packets from the MS NetMeeting (link between source address and destination address). In this experiment, the packets have been captured between the destination address (IP: 130.91.61.2, Ether: 00:80:3e:79:6b:e4) and the Source address (IP: 130.91.62.2, Ether: 00:50:04:78:59:fe) by using protocols of Ethernet/IP/UDP/data. The experiment has been done for 20 times each for 10 minutes with different inputs (speech, played audio file, video of still/moving person). From the previous, it is found that both audio and video packets were sent on the top of the UDP, and two different source/destination ports. Also, it was found that, the audio packet size is constant, and the audio inter-arrival time variance between the packets is having a small value. But, for the video, the packet size and the inter-arrival time are varying with greater values than the audio packets. So, the audio/video packet size/packet inter-arrival time PDF curves could be presented.

3 A New IP Multicast QoS Proposed Model

3.1 The Model Architecture

The new proposed model guarantees the required quality by using some of the Diffserv Simple blocks at edge devices only (to decrease the core routers' overheads). In this model, only three types of traffics are proposed; the real-time traffic, the non-real-time traffic, and the best-effort traffic (instead of AF and EF [4],[5],[6]). This model relies on metering the input traffics, then police the out-of-profile one by dropping it in case of real-time streams, and shape it in the case of the non-real-time streams. The model architecture is represented as shown in figure 1. It also, uses the IP Multicast mechanism [7] to save more networks resources. So, each sender in this model must support the IP Multicast mechanism.

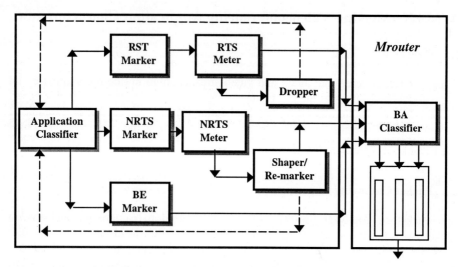

Fig. 1. A New Proposed Model Architecture.

- *Application Classifier:*
It classifies the streams based on the used applications. For example, if the used application is a Videoconferencing application (such as MS NetMeeting), then it passes the packets to the RTS Marker to mark it with real-time applications codepoint, and if the used application is a Web Browser, then it passes the packets to the BE (Best-Effort) Marker to mark it with the Best-Effort codepoint.
- *Marker:*
Based on the used application, it marks the packets with three different values in the precedence field (3 bits) located inside the TOS (Type Of Service) field in the IPv4 [7]. Then the Markers pass the packets to the BA Classifier.
-*Meter:*
The meter here (RTS, or NRTS) does not delay the incoming packets, but it checks if it is conformed to some defined traffic profile or not. It relies on the No-queue single token bucket measurement mechanism [8], which controls the rate and the burstness of data transfers into the network. In this token bucket (but without the packet queue), the tokens are inserted into the bucket with token generation rate (r) in kb/s. The depth of the bucket is the burst size (b) in bits. When the traffic arrives at the bucket, if sufficient tokens are available then the traffic is said to *conform/In-profile* and the corresponding number of tokens are removed from the bucket. If insufficient tokens are available then the traffic is said to *exceed/Out-of-profile*.
-*Dropper:*
The dropper here is acting as a policer to *RTS exceed/Out-of-profile* packets. It drops them immediately and sends an *Error_Drop message* to the source Application to re-adjust the traffics settings to the defined traffic profile.
-*Shaper/Re-marker:*
The shaper/dropper block gets the *NRTS exceed/Out-of-profile* packets. It delays the incoming packets to change the status of these packets from *exceed/Out-of-profile* to *conform/In-profile* by using a Single token bucket mechanism [8]. Then the packets will be re-marked with the Best-Effort precedence value (0) and send back an *Error_Drop message* to the source Application to re-adjust the traffics settings.

-BA Classifier (inside the Mrouter):
The Classification is based on the value of the precedence field only. So, it classifies the outgoing packets into the same three categories: RTS, NRTS, and BE. Then the Classifier, as shown in the figure 1, passes the RTS packets to *highest priority* queue, the NRTS packets to the *medium priority* queue, and the BE streams to the *lowest priority* queue at the Mrouter. Note that the priority queuing has been used, because of its simplicity, and scalability.

3.2 The Model Implementation.

In this paper, for simplicity, only the RTS, and BE blocks has been considered, because the scope here only at the Real-time streams, and NRTS queue approximately will not affect RTS which have the highest priority queue at the Mrouters.

To implement the new proposed model, the *OPNET MODELER Network Simulator version 6.0* [9] has been used. After having the PDF curves (as shown at section 2.2), the *OPNET PDF Editor* has been used to draw these curves. The *Packet Generator* in the *Node Editor* has been used to create the audio/video packet generators. The generator's generation rate attribute, generation distribution attribute, average packet size attribute, and packet size distribution attribute must be set in the generator module. The *inter-arrival PDF attribute* has been used to set the inter-arrival times of generated packets to be audio/video Inter-arrival distribution. The *pk size pdf attribute* has been used to set the size of generated packets to be audio/video Packet Size distribution. For the Mrouter implementation, the *process model attribute* will be set to *Priority Queuing* and the *service_rate attribute* will be set to the value of reserved bandwidth to the RTS streams. At last, the *subqueue attributes* will be used to set one new queue (for the BE streams).

4 Simulations Scenarios Results

4.1 RTS Meter Design

In this section, it is desired to find out the r and b values for the audio/video generators that will give an acceptable packet loss and consumed resources.
- *Audio Meter Design*
The following are MS Net-meeting audio rates:
Peak rate = 23.11 kb/s, Average rate = 19.72 kb/s, and Minimum rate = 17.83 kb/s.
From the previous, the Qbone model reserves bandwidth equals to 23.11kb/s (peak rate) [10]. So, the following simulation scenarios are trying to prove that it should take less than the above values to save considerable amount of reserved resources. The simulation scenarios are depending on generating the audio streams from the Audio Generator and pass them to the Meter block. As shown in table 1, the result for the first scenario has Packet Loss equals to 0% with average values of both r and b.

Table 1. Simulation Scenarios results for the audio Meter.

r (bits/sec)	b (bits)	Packet Loss
19,500	624	0
19,400	624	5.132E-3 (not acceptable)
19,00	624	2.565E-2 (not acceptable)

- *Video Meter Design*

The following are MS Net-meeting video rates:
Peak rate = 282.61 kb/s, Average rate = 167.73 kb/s, and Minimum rate = 94.6 kb/s.
Two groups of simulation scenarios have been implemented. The first group has set the values of token generation rate equals to the average rate (fixed), and has increased the bucket size from the average to the maximum by 10% for each step. As shown in figure 2, it is clear that, the packet loss is improving, but by the end of the simulations (at the maximum acceptable bucket size) the value of the packet loss equals to 4%, which is not acceptable (should be equal or less than 1E-3).

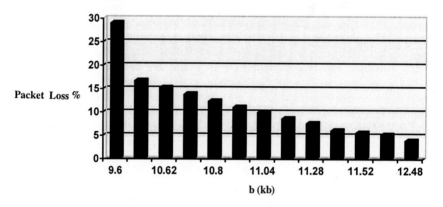

Fig. 2. Group_1: Video Meter Simulation Scenarios.

The Second group has set the values of the bucket size equals to the average rate and has increased token generation rate from the average to the maximum acceptable value by 10% for each step. As shown in figure 3, the desired value of the token generation rate equals to 216,667 bits/sec that gives 0% packet loss.

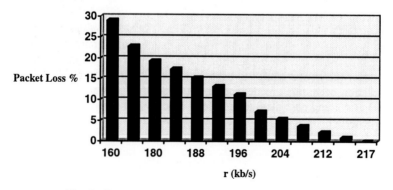

Fig. 3. Group_2: Video Meter Simulation Scenarios.

This solution (others could be found) saves around 23.4% of the reserved bandwidth and saves also 57.4% of reserved buffers than the values of the Qbone model [10].

4.2 Different Aggregations Scenarios

In this group of scenarios (50 scenarios), audio/video streams have been generated from 10 similar nodes. This paper presents Saved Bandwidth percentage, Node delay, and Node Buffer consumption statistics. The input traffics have been aggregated into every node, as shown in figure 4. Note that for example, scenario_30 means for example; 30 audio streams, 30 video stream, and 30 BE stream have been used.

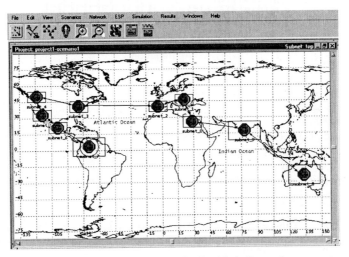

Fig. 4. OPNET diagram for the 10 similar nodes.

-Bandwidth Savings
As shown in figure 5, it is found that the bandwidth saving percentages are increasing from scenario_2 to scenario_5 then it reach a saturation level at which it ends at the saving percentage equals to 21%. The scenario_1 saves 0% of consumed bandwidth.

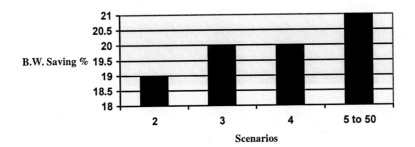

Fig. 5. Bandwidth savings

-Node Delay
Figure 6 represents the minimum/maximum of the node delays values during all the simulations with number of different aggregations. The value of node delay is decreasing by increasing the aggregations. Also, it clear that the node delay value is aggregation position dependent.

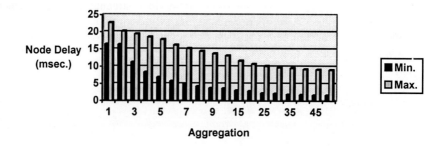

Fig. 6. Node Delay ranges.

-Buffers Consumption
Table 2 represents the of required buffers (buffer size= 1500 bytes) with number of different aggregations. The value of buffer consumption is increasing by increasing the number of aggregations. Also, it is clear that the buffer consumption value is aggregation position dependent. Finally, the new proposed model is saving more than 72% of consumed buffers than the Qbone model.

Table 2. Required Buffers inside the core routers.

Aggregations (Audio+ Video)	Min. required buffers	Max. required buffers	Av. required buffers	Min. buffers Saving %
1-10	1	3	2	72%
11-20	1	5	3	83%
21-30	2	7	5	86%
31-40	2	9	6	87%
41-50	2	11	7	88%

4.3 IP Multicast Scenarios

In this section, 12 simulation scenarios (6 for Unicast, and 6 for Multicast) have been described to study the advantages and the disadvantages of the IP Multicast mechanism usage on the proposed model. In this group of simulations, each real-time session (videoconference event) consists of 10 subscribers (In this example, Unicast Streams consume bandwidth 10 times more than the Multicast streams).
-Node Delay:
From table 3, it is clear that the Unicast mechanism is better at the Node delay values than the Multicast mechanism. This is due to the reserved bandwidth amount for the Unicast is 10 times the bandwidth reserved for the Multicast. But, it is found also that, by increasing the number of sessions, the Multicast mechanism improves the values

of the Node delay. This means that the new proposed model will act better and better by having more and more Multicast sessions.

Table 3. Node delay for both Multicast and the Unicast Mechanisms

	Multicast (msec.)	Unicast (msec.)	Unicast Saving (msec.)
1 Session	22.6	3.8	18.8
2 Sessions	16.4	1.9	14.5
3 Sessions	11.2	1.3	9.9
4 Sessions	8.4	1	7.4
5 Sessions	6.9	0.8	6.1
6 Sessions	5.7	0.6	5.1

N.B.: The savings in the Node delay does not matter, because it is lower than the limit of losing the packet (too late to be played back). So, the saving here will not be an effective factor in the real-time streams performance.

-*Buffer consumption:*
From table 4, it is clear that the Unicast aggregations are facing serious problems from the buffer consumption point of view. The Multicast mechanism is saving much more buffers than the Unicast mechanism. This means that, the Unicast mechanism will consume all the available buffers by having a small number of real-time sessions (specially if it has larger number of contributors).

Table 4. Buffer Requirements for both Multicast and the Unicast Mechanisms.

	Multicast (No. of buffers)	Unicast (No. of buffers)	Multicast buffers Saving %
1 Session	1	3	66.7%
2 Sessions	1	5	80%
3 Sessions	1	6	83.4%
4 Sessions	2	8	75%
5 Sessions	2	10	80%
6 Sessions	2	11	81.8%

5 Summary and Conclusions

In general, by adding more aggregations, the values of the Node delay are decreasing and the values of the Buffers consumption from node-to-node point of view are decreasing, but from scenario-to-scenario are increasing by slight normal values. It needs on the average, only 1 extra buffer for each extra 10-audio/video sessions (i.e. it could support up to 380-audio/video NetMeeting sessions at the same time by consuming only 40 buffers at any core router). Also, this model has zero packet loss, and it saves 44.4% of consumed bandwidth (23.4% from the Meter design and 21% from the aggregations) and more than 72% of consumed buffers more than the Qbone model.

So, as a conclusion, this model is network resources (Bandwidth, and Buffer consumption) saver, and has an acceptable Node delay, and zero packet loss for the audio/video NetMeeting traffics. But, it is more complex than the Qbone model, because of its input traffic analyzing stage that should be done before using the new proposed model. Also, in this model, the Best-Effort traffics may suffer at high QoS traffics load case. But, the Buffers consumption savings and the load balance ratios between QoS and Best-Effort traffics (40% for QoS, and 60% for the BE) should decrease the probability of losing Best-Effort packets at the core routers.

6 Future Work

Basically, there are some different directions to modify the proposed model in the future. The first is to have a routing protocol, which it should concern with the IP Multicasting mechanism and the QoS parameters. This protocol should route the packets dynamically instead of having static routing technique. The second is to implement a cost function, which will help to have an optimal design for the Meter (single token bucket) parameters (r & b). The third that this model needs more input traffics generators (real-time and non real-time applications) to be studied and analyzed. After having such generators, a more general formula could be concluded from bandwidth/buffers savings points of view. Finally, the load balance ratios between the QoS traffics and Best-Effort traffics should be re-calculated to have an optimal ratio that should not suffer the Best-Effort traffics.

References

1. S. Blake, D. Black, M. Carlson, E. Davies, Z. Wang, and W. Weiss: An Architecture for Differentiated Services. RFC 2475, December 1998.
2. MS NetMeeting Application Web site: http://www.micosoft.com/windows/netmeeting
3. Packetboy Analyzer Software Web site: http://www.ndgsoftware.com/
4. K. Nichols, S. Blake, F. Baker, D. Black: Definition of the Differentiated Services Field (DS Field) in the IPv4 and IPv6 Headers. RFC 2474, December 1998.
5. J. Heinanen, F. Baker, W. Weiss, J. Wroclawski: Assured Forwarding PHB. RFC 2597, June 1999.
6. V. Jacobson, K. Nichols, K. Poduri: An Expedited Forwarding PHB. RFC 2598, June 1999.
7. Douglas E. Comer: Internetworking With TCP/IP: Principles, Protocols, and Architecture. 3rd edn. Prentice-Hall, 1995.
8. Andrew S. Tanenbaum: Computer Networks. 3rd edn. Prentice-Hall, 1996.
9. OPNET Software Web site: http://www.mil3.com/
10. Ben Teitelbaum: Qbone Architecture. Internet 2, QoS Working Group, May 1999.

Acknowledgment

Special thanks to Prof. M. El-Zarki, and Prof. R. Guerin for the support and help during the implementations of the experiments and the simulations at Multimedia & Networking Lab., EE Dept., University of Pennsylvania, USA.

Integrated Management
of QoS-Enabled Networks Using QAME

Lisandro Zambenedetti Granville, Liane Margarida Rockenbach Tarouco,
Márcio Bartz Ceccon, Maria Janilce Bosquiroli Almeida

Federal University of Rio Grande do Sul – UFRGS – Institutes of Informatics
Av. Bento Gonçalves, 9500 – Block IV – Porto Alegre, RS – Brazil
{granville, liane, ceccon, janilce}@inf.ufrgs.br

Abstract. Providing QoS-guaranteed services in current installed networks is an important issue, but only the deploying QoS services is not enough to guarantee their success: QoS management must also be provided. Nowadays, police-based management addresses this need, but such management is not enough either. Network managers often deal with QoS tasks that cannot be performed using only policy-based management. This paper describes six important QoS management-related tasks (QoS implementation, operation maintenance, discovery, monitoring, analysis and visualization) and shows solutions that can help managers proceed with these tasks. Unfortunately, these solutions are independent from each other, leading to a scenario where integration is difficult. To solve this lack of integration, QAME (QoS-Aware Management Environment) has been developed. QAME provides support to allow the execution of the defined QoS tasks in an integrated fashion.

Keywords: QoS management, policy-based management, management environment

1 Introduction

The great majority of today's networks operate based on a best-effort approach. There is no warranty concerning traffic delay, jitter and throughput. Several applications operate properly on this environment, but several others can only be delivered if network QoS (Quality of Service) warranties are present.

Managers should be aware of QoS features implemented on the network. QoS architectures can only be effective and provide guaranteed services if QoS elements are adequately configured and monitored. Thus, in addition to the management of traditional elements (router, switches, services, etc.), managers must also manage QoS aspects. In this scenario, it's not a surprise to realize that the overall management will be more complex.

Effective QoS management can only be achieved if QoS management tasks are properly identified. Based on these tasks, management mechanisms can be defined to help managers to deal with QoS elements. Besides, such mechanisms must allow the replacement of the current device-oriented management approach by a network-oriented approach. This approach replacement is necessary because the amount of management information is even greater when QoS-related data are present. Device-specific management takes too long and does not provide a global view, which is specially needed when monitoring QoS parameters in critical flows.

QoS management must also take place within the same environment used to manage standard network elements. Management platforms should be aware of QoS to help managers proceed with QoS related tasks. Unfortunately, today's management platforms are not QoS-aware and managers have to investigate each QoS-enabled device to check QoS parameters. Again, a network-oriented approach is needed, with features that explicitly present QoS information to managers in an intuitive fashion.

This work presents our ongoing QoS management project named QAME (QoS-Aware Management Environment). The paper shows current analysis on QoS tasks and related software developed. The final goal is to provide an environment where managers are able to deal with explicit QoS information in a higher abstraction level, and with a global network view, implementing the network-oriented approach mentioned before.

The paper is divided as follows. Session 2 presents QoS-related tasks that should be performed to allow the deployment and maintenance of QoS architectures. Session 3 introduces our proposed environment. Finally, session 4 concludes the paper and also shows future works to be developed.

2 QoS Management-Related Tasks

QoS-enabled networks can only present reliable QoS services if managers are aware of QoS. Each network can use one or more QoS solutions and protocols. For instance, one could use MPLS [1] within an administrative domain, and have agreements with neighboring domains using DiffServ [2]. Another could use RSVP [3] to dynamically reserve network resources for a scheduled videoconference. Every QoS-related element (routers, brokers, hosts, classifiers, markers, etc.) must be managed to guarantee proper QoS service operation.

In order to manage QoS elements, managers must perform several tasks, besides the tasks applied to traditional management. QoS-related tasks must be performed by using facilities provided by the network management environment, but today's platforms do not provide any explicit QoS facility. To start providing such facilities, firstly we need to identify QoS tasks and then check how facilities should be created to help QoS management. Thus, we have classified QoS management tasks as described in the following sub-sessions.

2.1 QoS Installation

We call "QoS installation" the task of choosing QoS solutions and installing such solutions into a network. The result of QoS installation is a QoS-unique architecture that provides QoS services to a particular network. The resulted QoS architecture is a collection of QoS solutions and protocols that, together, offer services to network users.

Nowadays, the main QoS solutions offered are those developed under IETF: MPLS, DiffServ and IntServ [4]. Actually, each solution can be deployed by using different configurations. For instance, one could use bandwidth brokers [5] to allow dynamic DiffServ resource reservation. Another could use DiffServ with no brokers installed on the network.

There is a need for software that advises managers about possible QoS solutions, configurations and problems. One solution can be appropriate for a particular network, but totally inappropriate for another network. The final architecture must result from the analysis of the following elements:

- **Network topology**. Each network has a particular topology and for each topology a particular architecture is more suitable;
- **Network traffic**. Even with identical topologies, each network faces different traffic. One could have more downstream than upstream traffic, or more internal than external traffic;
- **Network application priorities**. Network traffic is mainly generated by applications (network itself doesn't generate too much traffic). The level of desired application priorities vary from network to network;
- **Network available solutions**. The final architecture can only be applied if selected solutions are supported by network devices. Otherwise, the final architecture has to be computed again.

Deriving the final architecture could be helped by QoS simulation solutions. Managers would describe its network topology, traffic and application priorities, and start checking each available solution. Different solutions could exist in the same environment at the same time, to collaborate with each other. For instance, a manager could use RSVP reservation internally and DiffServ on the network boundaries.

After deciding on a final architecture to be used, managers should start configuring devices, changing inappropriate ones, updating software on others, etc. Such procedures can be done in two different ways: locally or remotely. Local procedures are time consuming. Thus, we must provide tools able to keep local procedures minimal. Configuring and updating software can be remotely done on almost all cases. Except for IP number configuration and a few other operations, configuration is remotely driven, mainly through Telnet command line interface, HTTP in modern devices, and by using SNMP [6] agent/manager interaction. Changing equipment is an intrinsically local operation, and cannot be done remotely.

2.2 QoS Operation Maintenance

After QoS architecture is defined and installed, QoS services are ready to be offered to network users. At the same time QoS architecture is serving user needs, the manager must define appropriate operational parameters. We qualify any procedure taken to define QoS service behavior "on the fly" as "QoS operation maintenance".

Procedures often taken in QoS operation maintenance are those related to traffic classification, marking and prioritizing. Bandwidth static reservation, SLAs management [7] and policing are also examples of operation maintenance.

Today, the promising solution in QoS operation maintenance is policy-based QoS management. By using policies, managers can determine, at a higher abstraction level, how the QoS architecture should proceed to meet a desired behavior. Policy-based management actually constitutes a whole architecture itself, with dedicated elements (PEP and PDP [8], for instance). One important procedure in policy-based management is, for example, the location of policy enforcement points. Tools should also help managers in defining such points.

2.3 QoS Discovery

Some network equipment is overfilled with features. It is not rare to see complex routers with several services being used only for packet forwarding. Although several features could be used to help QoS provisioning, they are not, since managers cannot handle so many device features in large, complex networks.

We define "QoS discovery" as the task of searching the network for features that can help or improve QoS provisioning. Normally, QoS discovery is done through SNMP messages sent to devices, or by using any other proprietary method. QoS discovery can also be performed checking equipment documentation, but that is not an operation that could be automated.

QoS discovery is helpful in two important moments: in QoS installation and in QoS operation maintenance. In QoS installation the new discovered features can be used to decide among QoS architectures. In QoS operation maintenance, it can report new added equipment with QoS features while the QoS architecture is running.

QoS discovery can be effective if discovering mechanisms are installed. Such mechanisms can be simple (polling equipment trying to find MIBs related to QoS architectures) or complex (distributed network monitoring to check traces of critical protocols, such as IGMP [9] and RSVP [3]).

2.4 QoS Monitoring

Managers have to be up-to-date about the difference between the desired QoS and the faced QoS. This difference can never be greater than a defined critical value. If the faced QoS is too different from the desired QoS, user applications would degrade indicating that the contracted services are not running properly.

To check the current QoS, managers should collect data in the network through QoS monitoring. This task has to be able to determine two related pieces of information:
- **End-to-end achieved QoS**. The QoS parameters of particular segments are important, but end-to-end QoS is crucial. If end-to-end achieved QoS is not correct, managers must be warned to fix the problem. End-to-end QoS

degradation is the sum of the degradation of each segment on the end-to-end path. If just one segment is introducing degradation, that will be noticed in the end-to-end QoS.
- **Points of degradation**. If end-to-end QoS degradation is noticed, QoS monitoring should be able to determine where in a flow path the degradation points are.

Today, most QoS monitoring solutions are only able to satisfy the first item. It is simple to check if there is end-to-end degradation, by using RTP/RTCP [10] protocols, for example. However, identifying degradation points is a more complex procedure, and requires more complex processing on the network [11].

2.5 QoS Analysis

In a proactive management, managers should anticipate future problems and attack them as soon as possible. To achieve proactive QoS management, QoS analysis tasks must be performed.

Cataloged historical behavior can show, for example, the number of refused RSVP sessions due to lack of resources. If the number of refused sessions increases too much, it indicates that the manager should update network resources. Analysis of a QoS-dependent monitored client/server operation could show the frequency of QoS degradation, and the frequent degradation points. In this case, the manager should check the critical point to see link problems.

One crucial part of QoS analysis is QoS visualization. We consider QoS visualization such an important procedure that it is defined separately from QoS analysis, although it is part of such analysis.

2.6 QoS Visualization

Today's network management platforms are topology-oriented, i.e., they show information from a network topological perspective. Managers browse network topology and check each desired device. Some facilities can be found, allowing managers, for example, to ask the platform for a map listing every printer.

For QoS management tasks, current visualization is poor. Managers often search for each important device in maps, and check to see if such device is QoS-enabled. This is a time-consuming task, and should be replaced with an automated search procedure.

QoS visualization is not a task itself. Rather, it is a feature that helps managers to proceed with QoS tasks. Tools should provide visualization based on QoS characteristics. We list here some helpful QoS visualizations.

- **Colored link utilization**. Each link shows, instead of a single black connecting line, a colored set of lines describing link utilization by each flow/aggregate (figure 1, left);

- **Selected end-to-end sessions**. A topology map should have selected end-to-end sessions highlighted through colored lines. Managers could ask to visualize only sessions that match some pre-determined features;
- **QoS enabled devices**. A topology map should highlight QoS-enabled devices. Different colors show devices that implement different solutions. For example, green boxes indicate DiffServ-enabled routers, whereas red boxes indicate RSVP-enabled ones;
- **Segments with QoS degradation**. Segments with QoS degradation could be shown in red or be highlighted (figure 1, right), to indicate degradation. Orange segments could indicate probable degradation coming.

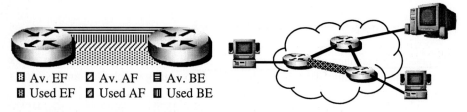

Fig. 1. QoS visualization examples

Several other visualization facilities could be created to help managers identify QoS-related information. Today, QoS visualization can only be found on separate software that has no integration. The next session shows some current solutions used in QoS visualization and other tasks.

3 QAME

Previous sessions presented QoS-related management tasks and some solutions that help managers perform such tasks. Each solution provides functionalities to attack a particular problem, but that is done independently from other solutions. Thus, there is not any QoS task integration.

In addition to the lack of integration, current network management platforms are device-oriented and not QoS-aware, i.e., even if they have QoS support they do not show QoS information properly. A more appropriate environment should allow a network-oriented view to the management, also allowing explicit QoS information.

This session presents the current state in the development of a QoS-integrated management environment, called QAME. The environment is QoS-aware in the sense that it takes QoS information explicitly and shows that information more properly. Besides, QAME provides support for managers to proceed with the six QoS tasks previously defined (QoS installation, operation maintenance, discovery, monitoring, analysis and visualization).

3.1 QAME Architecture

QAME architecture extends the policy-based solution defined in [12] by introducing some new elements. The architecture is initially divided into active elements and databases. Active elements perform management and QoS provisioning tasks. They also store, retrieve and update information in databases. Active elements are subdivided in an upper element (the User Environment), intermediate elements (Policy Consumer, QoS Monitor and Target Finder) and lower elements (Targets). Figure 2 shows QAME elements and databases, which, on their turn, are presented in the following sub-sessions.

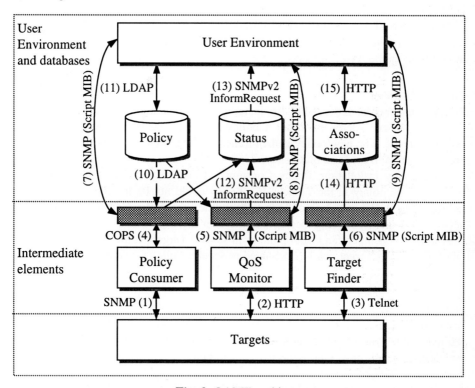

Fig. 2. QAME architecture

Targets

Targets are active elements that influence the final QoS observed in the network. Each device can have several Targets that can influence the network parameters. For example, in a router each interface is a Target. Targets are the final elements that effectively implement a QoS architecture.

Network manager has indirect access to Targets throughout Policy Consumer, QoS Monitor and Target Finder elements. The interface between these elements and Targets is device-specific, and different protocols should be used to access different Targets. A router from vendor A, for instance, can be accessed via a Telnet client, while another router from vendor B can be accessed via HTTP.

Target Finder

In the network, searching each device to identify its Targets is a time-consuming task. Also, new devices just attached to the network must have their Targets cataloged for use. Finally, if QoS discovery is an important task, automatic Target finding will be necessary.

Target Finders are special agents that search the network for current and new Targets. Each Target Finder recognizes at least one specific Target, using a specific Target finding algorithm. For example, a DiffServ Target Finder is the one that looks within routers and checks the existence of packet prioritization based on the IP DS field. To do that, the DiffServ Target Finder can open a Telnet session or check for DiffServ MIB implementation.

Since each Target can have different capabilities, Target Finders are also responsible for classifying new discovered Targets. Target Finders store any Target information on the Associations database.

Policy Consumer

Policy Consumer is responsible for installing a policy into Targets. Each Policy Consumer, when ordered to install a policy, retrieves the policy by accessing the Policy database. The new policy is then translated into device-specific instructions to program Target to conform that policy.

After policy installation, Policy Consumer is also responsible for checking the success of the policy in the Targets. If a policy could not be installed either due to failure in the Target or to lack of Target capabilities, the Policy Consumer notifies the network manager by sending messages to the User Environment.

A Policy Consumer can install policies in several Targets at the same time. On the other hand, each Target can be associated to several Policy Consumers, even though only one can be the current active consumer.

QoS Monitor

Installed policies might not behave as stated in the policy definition. The QoS resulted from a policy installation can be different from its specification. Critical policies must then have their expected QoS monitored. The element responsible for doing that is the QoS Monitor.

The network manager defines which policies must be checked and QoS Monitors are then associated to the Targets that implement those policies. QoS Monitors access policy definitions also in the Policy database and compare the effective behavior on the network with the one defined in the policy. If degradation is observed, QoS Monitor notifies the network manager by sending special messages to the User Environment, too.

The greater the number of QoS Monitors used, the more accurate the monitoring process will be. Also, the greater the number of QoS Monitors used, the greater the information analysis overhead will be.

User Environment

QAME graphic user interface is implemented in the User Environment, which uses Web technology to show management information. User Environment is responsible for running analysis processes that complement the functionality presented in the Policy Consumer, Target Finder and QoS Monitor. For example, User Environment receives special messages from Policy Consumer telling a policy could not be installed, and messages from QoS Monitor when the observed QoS is different from the expected QoS.

User Environment also interacts with the three databases in order to define their contents. Users define policies that are stored in the Policy database by using the environment interface. Policies can also be modified or removed from the database. Network topology is shown by accessing the Associations database information. Users check network topology on their Web browsers and order actions to be installed in Targets.

Databases

The three databases shown in figure 2 are defined to store different classes of information. Every Target, Policy Consumer, QoS Monitor and Target Finder is registered in the Associations database. With appropriate searching of the base, the User Environment can derive the network topology, existing QoS solutions, and resource capabilities. Furthermore, the Associations database stores the associations between Targets and the other elements. For example, a Target named A uses Policy Consumer B for policy translation and installation, QoS Monitor B for policy performance checking and Target Finder C for discovering possible new capabilities in Target A.

The policies are stored in the Policy database. Since policies themselves do not define on which Targets they should be installed, policies can be stored separately from Targets. We do that because the database needed for Targets and the database needed for policies have different requirements. Policies once defined have little or no change. On the other hand, Targets can have their capabilities extended, associations updated, and have more changeable data overall.

Even more changeable are data used to represent the status of a deployed and/or monitored policy. QoS Monitors and Policy Consumers change data in the Status database every time a deployed policy has its status altered. Thus, Status database binds information stored in the Policy database (the policies) with that stored in the Associations database (the Targets), and introduces new information about this relationship: the status of a policy.

3.2 Elements Location

The previous sub-session described each element of the QAME architecture. This present sub-session explains where in the network infrastructures these elements are located and how many elements of the same type can be used.

The more obvious element location is the Target location. Targets are located within network devices that play any active role in QoS provisioning. Target examples are routers and switch interfaces in their queue disciplines. Marking, policing and shaping processes are also examples of Targets. Targets can be located in hosts, too. RSVP-enabled applications, or DiffServ marking processes in end systems [13] are Targets, since they influence the end-to-end QoS.

We leave intermediate elements location for the next paragraphs, since this is a more complicated issue. On the other hand, User Environment location is almost as obvious as Target location was. We use a central point that runs QoS analysis processes and generates HTML pages showing the results. A Web server is also used to allow remote access to the environment. The central point could generate too much traffic on larger networks. Distributed User Environment can be used since databases are detached from the User Environment. Several environments would access the same databases.

Databases can be located on the same device that implements User Environment, or on separate devices. Since there are three databases, some can be found together with User Environment, and others separately. Although figure 2 shows only one copy of each database, for security reasons we could have more copies of the same base and use database replication for consistency. Also, more copies of the same database would facilitate the distribution of network traffic generated by QoS Monitors, Policy Consumers and Target Finders when they need to update databases information.

A Trickier aspect is the location of Target Finders, QoS Monitors and Policy Consumers. First of all, since they are independent elements they can be located in different places. QoS Monitors are very tightly related to their Targets. Thus, it is expected that QoS Monitors are located within the same devices that contain the monitored Targets. However, depending on the installation of the QoS Monitors, they can also be located close to devices, but not inside. For example, a monitor created to check the bandwidth traffic of a router interface could access the MIB-II interface group and realize that an interface is facing overflow, even though the monitor is not located within the router.

Policy Consumers are often located outside devices, but modern equipment is expected to have built-in policy consumer implementations. Even more, Policy Consumers can be located together with the User Environment, thus improving User Environment and Policy Consumer communication.

Finally, Target Finders are often located together with User Environment, acting as special plug-ins that search the network for QoS-enabled devices. Target Finders can also be located in network segments other than the User Environment segment. They would act as segment monitors, looking for QoS traffic generated by QoS-enabled devices. The less suitable location for a Target Finder is within devices, since devices and their Targets are the objects of the finding process. One exception is when devices are able to announce themselves and their capabilities to the network, registering on the databases. Up to now, authors are unaware about devices with this feature.

Table 1. QAME elements location. Rows list QAME elements and colums list possible locations. Cells marked with an "x" denote that the QAME element in the row can be present in the equipment of the colum. "Devices" are network equipment (routers, switches, bridges, etc.). "Proxies" are network equipment used to host some active elements that act on different equipment (e.g., a QoS Monitor located within a host used to monitor a router). "Hosts" are listed to explicitly define elements located and acting in a host. Finally, "management stations" are used to denote the hosts where QAME User Environment and databases are placed.

	Devices	Proxies	Hosts	Management stations
Targets	x	--	x	Only if target plays active role in QoS provisioning
QoS Monitors	x	x	x	x
Policy Consumers	x	x	x	x
Target Finders	--	x	--	x
User Environment	--	--	--	x
Databases	--	--	--	x

One important issue about location is the management interface between the User Environment and Target Finder, QoS Monitor and Policy Consumer. This interface is depicted in figure 2 by the gray rectangles connecting User Environment and intermediate elements. These interfaces can be found together with the elements if the elements implement such interfaces. Otherwise, the interfaces are located in the User Environment and translate requests into element-specific commands. This separation between element implementation and interface is important because modern devices could implement elements with interfaces different from those used by QAME. In this case an interface translation is needed to allow the use of built-in device elements. Table 1 summarizes the possible location of QAME elements.

3.3 Protocols

This sub-session describes the protocols used to provide communication between QAME elements. In the protocol definition phase of the project, we decided to use standard and open protocols to implement such communication. Even though some standard protocols might not be the best choice for some critical tasks, we believe that this choice makes our architecture open, making future implementation of new modules easier.

Targets Protocols
The protocols used to communicate with Targets are actually defined by the devices that contain the Targets. These protocols are then dependent on the device's provider, which could choose to use standard protocols or implement its own proprietary protocol.

The most common Targets protocols available are Telnet and SNMP, but modern devices also use HTTP for configuration management (figure 2, labels 1, 2 and 3). Fortunately, proprietary protocols are more rare.

Despite the diversity of possibilities, Targets protocols are not a critical issue since intermediate elements are responsible for protocol translations. Thus, when two different routers, using different protocols, should be programmed to prioritize a defined flow, that programming task "sees" the different routers as equal because of the protocol translation executed in the Policy Consumer. This translation also occurs in the QoS Monitor and Target Finder elements (figure 2, labels 1 and 4, 2 and 5, 3 and 6).

QoS Monitors, Policy Consumers and Target Finders Protocols
Protocols used to communicate with intermediate elements can be divided into two groups: those implemented by the elements, and those used as interface with elements (gray rectangles in figure 2). Protocols implemented by the elements are element-dependent and defined by the element developer. For example, a vendor implementing a Policy Consumer within its new router could use COPS [14] to transfer policies. Another vendor could choose Script MIB [15] definitions to have access to the policies (figure 2, labels 4, 5 and 6).

On the other hand, protocols used in the access interface are always the same. This is a requirement to allow access to the same interface in the User Environment. Thus, interface actually implements only protocol translation, from User Environment interface access into the intermediate elements-specific interface (figure 2, labels 4 and 7, 5 and 8, 6 and 9). The current QAME implementation uses Script MIB to communicate with intermediate elements interface.

Database Protocols
The protocols used to access database information are different because the nature of each base is different. Policy database is reached using an LDAP [16] protocol (figure 2, labels 10 and 11). Since policies are information that has few updates but can be accessed several times, a write-once read-several times protocol like LDAP is more suitable.

Status and association information are more dynamic, and LDAP protocol should be avoided. The current implementation uses SNMPv2 InformRequest message to update status information in the Status database (figure 2, label 12). InformRequest messages can be faced as an SNMPv1 trap message with confirmation. Thus, QoS Monitor perceiving QoS degradation can update the status of a monitored flow or aggregate trapping the Status database, and still have confirmation of the update operation due to the reply of the InformRequest message. Policy Consumers can also notify the Status database when a policy deployment fails. The InformRequest message can be forwarded to the User Environment when critical monitoring tasks are performed (figure 2, label 13). Finally, User Environment and Target Finders reach Associations database through HTTP and queries to a PHP4 engine [17] (figure 2, labels 14 and 15).

4. Conclusions and Future Works

Providing QoS services in networks is currently very important because time-dependent application cannot be deployed using best-effort based services. QoS architectures must be installed, but should also be properly managed to be effective. Traditional network management cannot be applied to QoS management because the amount of available information is much larger and complex.

Current efforts try to find a way to allow complexity abstraction by using policy-based management. But policies are not sufficient to allow total QoS management. Other QoS management-related tasks are performed by network managers, and support should be available for those tasks.

This paper has defined six main QoS management related tasks: QoS installation, operation maintenance, discovery, monitoring, analysis and visualization. Policy-based management only addresses QoS operation maintenance. Other tasks have no advantage of policy-based management. For these tasks we have presented current solutions that help managers install them. Unfortunately, there is no integration between solutions, and managers often have to deal with too many tools, increasing the management complexity.

To make management easier we have built a QoS management environment called QAME where the six important QoS tasks can be performed in an integrated fashion. We have described the QAME architecture, its elements, and where these elements can be located in the network (e.g., within routers, switches, hosts, or integrated in the management station). QAME elements exchange information to allow integration. This information exchange is done by communication protocols. We have shown that the QAME environment uses LDAP, SNMPv2, Script MIB and proprietary protocols to implement communications.

Protocol translation is a required feature to allow better interaction between user environment (where management information is presented) and lower-level elements (where management information is gathered and processed). The protocol translation is possible because QAME architecture detaches lower-level elements implementation from their interface. The element implementation can be located, for example, in a router, while the element interface can be located in the management station.

Future works may address security, database replication and distributed management issues. Since SNMP messages are not encrypted, the use of SNMPv3 is a natural choice. Database replication deserves a more accurate research because single database instances could be inadequate to manage very large networks. Because management traffic will be greater, database access at a single point could prevent better management performance. Also, in larger networks, a central point of management should be avoided, and distributed management with more than one User Environment would be preferable.

References

1. Rosen, E., Viswanathan, A., Callon, R.: Multiprotocol Label Switching Architecture. Internet draft <draft-ietf-mpls-arch-07.txt>. Work in progress (2000)
2. Blake, S., Black, D., Carlson, M., Davies, E., Wang, Z., Weiss, W.: An Architecture for Differentiated Services. Request for Comments 2475 (1998)
3. Braden, R., Zhang, L., Berson, S., Herzog, S., Jamin, S.: Resource ReSerVation Protocol (RSVP) - Version 1 Functional Specification. Request for Comments 2205 (1997)
4. Shenker, S., Wroclawski, J.: General Characterization Parameters for Integrated Service Network Elements. Request for Comments 2215 (1997)
5. Nichols, K., Jacobson, V., Zhang, L.: A Two-bit Differentiated Services Architecture for the Internet. Request for Comments 2638 (1999)
6. Case, J., Fedor, M., Schoffstall, M., Davin, J.: A Simple Network Management Protocol (SNMP). STD 15, Request for Comments 1157 (1992)
7. McBride. D.: The SLA Cookbook: A Recipe for Understanding System and Network Resource Demands. Hewlett-Packard Company. Available via URL http://www.hp.com/openview/rpm (1996)
8. Yavatkar, R., Pendarakis, D., Guerin, R.: A Framework for Policy-Based Admission Control. Request for Comments 2753 (2000)
9. Fenner, W.: Internet Group Management Protocol, Version 2. Request for Comments 2236 (1997)
10. Schulzrinne, H., Casner, S., Frederick, R., Jacobson, V.: RTP: A Transport Protocol for Real-Time Applications. Request for Comment 1889 (1996)
11. Jiang, Y., Tham, C.K., Ko, C.C.: Providing Quality of Service Monitoring: Challenges and Approaches. NOMS 2000 - IEEE/IFIP Network Operations and Management Seminar (2000)
12. Mahon, H., Bernet, Y., Herzog, S.: Requirements for a Policy Management System. Internet draft <draft-ietf-policy-req-02.txt>. Work in progress (2000)
13. Granville, L., Uzun, R., Tarouco, L., Almeida, J.: Managing Differentiated Services QoS in End Systems using SNMP. IPOM 2000 - IEEE Workshop on IP-oriented Operations & Management (2000)
14. Chan, K., Durham, D., Gai, S., Herzog, S., McCloghrie, K., Reichmeyer, F., Seligson, J., Smith, A., Yavatkar, R.: COPS Usage for Policy Provisioning. Internet Draft <draft-ietf-rap-cops-pr-02.txt>. Work in progress (2000)
15. Quittek, J., Kappler, C.: Remote Service Deployment on Programmable Switches with the IETF SNMP Script MIB. DSOM 2000 – IFIP/IEEE International Workshop on Distributed Systems: Operations and Management. Springer Werlag (1999)
16. Arlein, R., Freire, J., Gehani, N., Lieuwen, D., Ordille, J.: Making LDAP active with the LTAP gateway: Case study in providing telecom integration and enhanced services. In Proc. Workshop on Databases in Telecommunication (1999)
17. Hypertext Preprocessor – PHP4. Available via WWW at URL: http://www.php.net (2001)

On Web Quality of Service: Approaches to Measurement of End-to-End Response Time

Mike Tsykin

Systems Engineering Research Centre, Fujitsu Australia Limited
1230 Nepean Hwy. Cheltenham Vic. 3192 Australia
mike.tsykin@fujitsu.com.au

Abstract. Tools for direct measurement of Quality of Service (QoS) as represented by End-To-End Response Time (ETE RT) are gaining acceptance. However, classifications of such tools were developed before the wide commercial adoption of the Web and are, thus, out of date. In this paper, author presents an update to Tsykin-Langshaw '98 classification, with the specific emphasis on the measurement of the Web-based activities. Review of the existing tools is included and specific difficulties of operation in the Web-based environment are discussed. Future trends are indicated.

1. Introduction

Measurement of Service Levels and particularly End-To-End Response Time (ETE RT) is one of the most necessary facilities for Capacity Management practitioners. The subject is much debated, but practical approaches and tools for Open Systems were lacking for a long time. Recently, the situation changed. Therefore, the objectives of this paper are:
- To review the history of measurement of ETE RT
- To review and classify the measurement methods and tools
- To discuss the direct measurement of Service Levels, including problems encountered in the Web environment and their solutions

2. Historical Perspective

Measuring computer service has been achieved in many different ways over the past 40 or more years. Initially, with single tasking, batch systems, the measure of service was simple - how long did it take to execute a particular program. This metric naturally progressed to the concept of throughput - how many programs or jobs could be executed in a given amount of time. With this progression, and the birth of performance tuning, came the idea of measuring the performance or efficiency of various critical parts of the system such as the CPU, input/output devices, and memory usage. Over a period these performance/efficiency metrics came to be regarded as indicators of how well the system was doing its job, and operating systems were instrumented as a matter of course to provide various metrics, ranging from simple measurements to the results of complex analyses.

As operating systems improved, and multi-tasking and online systems became the norm, these system metrics became the common indicators of how well the system was servicing the organization. They were there, and it was easier to use what was available than to make alterations to applications to record service level information directly.

Slowly, however, sophisticated ways of determining service levels began appearing. Hardware and software suppliers started to "instrument" their products to provide measures of operation of their equipment. Using those, DP / IS departments started to provide guarantees of service to the organizations concomitant with their spending. And so the Service Level Management was born. Towards the end of the so-called 'mainframe era', it even became possible to directly measure terminal response time. Thus, measurement of transaction-based ETE RT became possible.

In mainframes, a transaction is assumed to be a single protocol block (SDLC protocol is the dominant one in such an environment) or a message as seen by a Transaction Monitor (e.g. CICS). Either one of these closely relates to a meaningful user interaction, therefore the approach works. Attempts to combine such transactions into more meaningful units (Natural Business Units or NBUs) started, but the environment changed radically. Open Systems arrived in force.

The roots of these are in scientific and personal computing. This means that various derivations of TTY protocol became dominant. As opposed to SDLC, this is a byte-mode protocol. Therefore, the correlation between a user action and protocol block disappeared. Furthermore, complex Transaction Monitors went out of fashion. The result of this on the measurement of ETE RT was devastating – it simply became too hard.

However, the need remained. Indeed, it even increased due to the advent of Outsourcing (both external and internal). In addition, end-user computing was evolving, with its' emphasis on service. And so, development continued.

First tools started appearing on the market in 1997-98. Rapidly, the numbers of available tools increased, but the acceptance remained cautious. This is still the case, although the recent focus on Internet Quality of Service (QoS) may change the trend in the future.

3. ETE RT – Approaches to Measurement and Tools

3.1 General

It is useful to formally define ETE RT. In the words of Maccabee [2], ETE RT is: 'The time between the start of users request (indicated by depressing of a key or a button) and the time when the user can use the data supplied in response to the request'. As such, it adequately describes the level of interaction between a user and a computer system he / she uses. Traditionally, the concept is applied to a single interaction between a user and a system - that is, a transaction. It is assumed that a transaction is a self-contained unit; therefore ETE RT thus derived represents the service level. For the purposes of this chapter, let's accept this concept of a transaction with one qualification: a transaction may or may not be directly measurable. If it is not, there is no option but to emulate or simulate it.

ETE RT may be broken into discrete components. The most common breakdown is into client, network and server components. This may become complicated if a network of servers is involved in servicing one transaction (multi-tier client-server implementations). Further, depending upon a protocol, each of these components may be further broken down (connection time, Therefore, there are several ways in which transaction-based ETE RT may be measured. McBride [3] proposed one such. Snell [5] identified three types of ETE RT measurement tools. In 1998, Tsykin and Langshaw [6] expanded and modified Snell's classification. At the same time, independently, Scott and Conway published a similar one. This paper modifies and expands Tsykin / Langshaw classification to reflect the current developments. It is presented below.

The four fundamental approaches to measurement of ETE RT are presented below. Known tools are summarized in Figure 1.

3.2 Approaches to Measurement

3.2.1 Benchmarking and Emulation
- Description:
  - Benchmarking.
    This approach involves three essential steps:
    1. Intercept user transaction (for instance, access to a Web page) and modify as required. Alternatively, an artificial transaction may be scripted
    2. Re-issue this transaction / transactions using 'robots' from one or more locations. Robots would measure ETE RT and it's components
    3. Collate and analyze results
  - Emulation
    This is a variation of 'benchmarking, except that 'packets' of data (e.g. issued by *ping* command) are used instead of intercepted or scripted transactions
- Advantages
  - These approaches are intuitive, conceptually simple and acceptable to most people
  - They are 'active', in that they work regardless of whether there are actual users on a system being measured – or not. Therefore, they are suitable (and widely used) for measurement of availability, either throughout a period or at a certain time (e.g. before the start of a morning shift)
  - They support quite a sophisticated approach to troubleshooting, including identification of problem segments via triangulation
  - They are available as either services or products
- Disadvantages
  - They are 'intrusive'. This means that they impose explicit additional load on the system. It need not be large, but may not be philosophically acceptable to many managers

- Penetration: Lately, these emerged as the dominant form of ETE RT measurement. Emulation is, understandably the preferred approach in a scientific community. Benchmarking is the mainstay of the WWW QoS measurement

3.2.2 Instrumented Applications
- Description
  In this case, transaction points are inserted into user applications, which are then monitored. The most widely known version is ARM (toolkit available), but other versions exist also particularly in the Web area
- Advantages
  This is the most suitable approach for application tuning. It enables tracing of application (depending upon positioning of transaction points within code).
- Disadvantages
  - It requires changes to source code of applications and, therefore commitment by suppliers and customers
  - Roll-out implications impede acceptance by users
  - Commercial benefits to suppliers are (at best) unclear, which impedes development
- Penetration : at present, minimal number of applications is instrumented. This is not expected to change rapidly.

3.2.3 'Wire Sniffers'
- Description
  This approach involves interception of network traffic, analysis of the packets and attribution of those to certain transactions. The rest is conventional analysis. Two possibilities exist for traffic interception:
  - 'Raw' network traffic is monitored, decoded and analyzed.
  - Communications traffic on a server or a client is used. This approach, strictly speaking, should be categorized as an Instrumented Client (see below). However, it is included here due to the obvious similarities with the other 'Wire Sniffer' approach
- Advantages
  This is a technically clever approach, which provides a comprehensive analysis of network performance and some idea of application response time.
- Disadvantages
  - Hardware-based probe (either dedicated or a board in 'promiscuous' mode) or instrumentation of Operating System is required. In this case, either a 'shim' or LSP (Layered Service Provider) maybe used.
  - Architecture is limited in not supporting data on server performance activity
  - Results are particularly dependent upon placement of the probes within the network
- Penetration
  Small as yet.

3.2.4 Instrumented Clients

- Description
   This approach involves interception, decoding and analysis of Operating System events on a client workstation. Given the penetration and messaging architecture, the logical choice of a client platform is Windows. However, the approach is as useful for other 'intelligent' (instrumentable) clients. Of particular interest are the Web-based applications
- Advantages
  - An architecturally advanced approach, because it allows tracing of all users of OS services, be they applications or systems products
  - Does not require special hardware, nor instrumentation of applications
  - Within the framework of this approach, ETE RT is easily available as one of the monitored metrics
  - The most flexible approach of all
  - Particularly suitable for product development and debugging
- - Disadvantages
  - Requires instrumentation of Operating Systems
  - Presents logistical problems in multi-platform, multi-OS environment
  - Presents problems with measurement of Web activity, due to the coding of popular browsers (neither IE nor Navigator use Windows events much). To overcome that, a combination of 'wire sniffing' and instrumented client has proved useful
- Penetration
   Small, but growing

3.3 Summary

All the four approaches described above are in use now. The first one – Benchmarking – is the most popular one at this stage. This is likely to continue. The second (Instrumented Applications) and the third (Wire Sniffing) became 'niche offerings' and are likely to remain this way. The fourth (Instrumented Client) enjoyed the initial burst of popularity, and then faltered somewhat due to the technical limitations. These are being overcome and the popularity should return in the future.

4. Direct Measurement of Service Levels

4.1 Objectives

Objective of Service Level Management (SLM) is to manage service provided to user. Therefore, the first and necessary step is to measure such service.

As McBride points out in relation to SLM: 'What cannot be measured cannot be managed' [3]. However, he then proceeds to discount client-based measurement in favor of application-based one, and suggests that ARM (Instrumented Applications) should be used for measurement. The author disagrees. It stands to reason that measurement of levels of service to a User should be done from the user's perspective

- that is, on a user's workstation and for user's units of work - that is, a business transaction, NBUs or equivalent.

Naturally, it is desirable to trace the transaction across the network, as is 'skips' from server to server. However, from the point of view of MEASUREMENT of service levels though, it is not, strictly speaking, mandatory.

Product name	Supplier	Bench -mark	Instr. App-n	Wire Sniffer	Instr. Client
SLM	Fujitsu			*	*
EnView	Fujitsu	*			
Quoetient	Fujitsu	*			
Tivoli	IBM	*	*		*
?	Hewlett Packard				*
MeasureWare	Hewlett Packard				
FSE	Concord			*	*
ARO	Computer Associates				*
TNG Response Option	Computer Associates			*	
ETEWatch	Candle				*
EcoScope	Compuware			*	
Sniffer	Network Associates			*	
Chariot	Ganymede			*	
Hypertrack Performance Monitor	Trio Systems			*	
Netscout Application Flow Management	Netscout			*	
VitalSuite	Lucent				
BEST1	BMC		*		
Patrol	BMC		*		
Info Vista	Info Vista		*		
Trinity	Info Vista	*			
S3	Nextpoint		*		
Keynote	Keynote	*			
Proctor	Dirig	*			
PingER	SLAC	*			
	AlertSite.com	*			
@watch	Quicksand	*			
?	Envive	*			
NetScore	Anacara	*			
Netcool	Micromuse	*			

Fig. 1. Known Products.

4.2 Metrics

Metrics are best grouped by their intended usage. In the case of Service Level Measurement, there are two main applications:
- Monitoring and reporting
- Troubleshooting

4.2.1 Monitoring and Reporting. Thus far, two metrics have proved the most popular ones in measurement of service levels: ete rt and availability. These are looked at below.
- ETE RT - traditional end-to-end response time measure defined above. It may be successfully measured by Benchmarking, Instrumented Applications or Instrumented Clients methods
- Availability – the critical metric, defined as time during which an application is available to be used. Therefore, only active methods (i.e. Benchmarking) may be used to measure it. Other methods are passive, that is they rely on actual user activity. This means that if there's no activity – there's no measurement.

4.2.2 Troubleshooting. In troubleshooting, composite metrics are not sufficient. In order to be able to find a problem, one must know the details of ete rt. It was mentioned before that ete rt might be split into one or more (depending upon the level of client-server nesting) groups of three functional components:
- Server part
- Network part
- Client part

Server and network parts were explored in detail previously and are well enough understood. However, a client component presents challenges. These stem from the fact that it too is a complex, powerful, multi-tasking platform. Therefore, for troubleshooting purposes, we must be able to measure the interrelations between applications on a client platform. Some metrics suitable for this purpose are listed below:
- Application Busy-Switch Time (ABST). The time period between an application becoming busy and user switching away from it. This metric describes a ' level of patience' by a user. Consistently low ABST (say, 1 - 2 seconds) may indicate that a user expects a delay and does not wish to wait.
- Application Cycle Time (ACT) The time period between successive switches of focus to an application. Low ACT would indicate a consistently and frequently (but not continuously) used application.
- Inter-Application Time (IAT) ... the time period from when focus switches away from a specific application until that application receives focus again.
- Application Active Time (AAT). The time period from when an application receives focus until focus is switched away from that application. (Note: IAT + AAT = ACT)

- Consecutive Wait Count (CWC). A count of the number of time the wait cursor was active when focus switched from an application, and still active the next time focus switched to the application.
- Application Switch Table (AST). A table with applications as row and column headings containing counts of the number of time focus switched from one application (row) to another application (column). It is useful in identifying user work profiles. An example is presented below in Figure 2.

	Application A	Application B
Application A	x	23
Application B	9	x

Fig. 2. Example of Application Switch Table

Focus switched from application A to application B 23 times. Focus switched from application B to application A 9 times.

Author believes that only the Instrumented Client approach is suitable for this task. Please refer to [6] for more details.

4.3 Methods of Data Capture by Instrumented Client

Discussion below will refer to Windows systems. In principle, there is no reason why similar things may not be done on Unix. However, it was not explored so far.

4.3.1 Hooks. Windows gives us a relatively easy ability to intercept various Operating System events and messages. This is done via the so-called *hooks* – documented and stable Windows interfaces. Detailed discussion on hooks is beyond the scope of this paper. Please refer to [4] for detailed instructions and examples. To illustrate the point, Figure 3 below provides an example of a GUI used to implement the interception of hooks. These are sufficient to calculate the metrics discussed above for Windows GUI applications.

4.3.2 Layered Service Provider. Non-GUI Windows programs (e.g. services) and both Netscape and Internet Explorer Web Browsers either do not use hooks at all, or use them 'sparingly'. Therefore, if Service Level measurement is required for these programs, network traffic interception and interpretation must be relied upon. LSP (Layered Service Provider) is available for the purpose. This facility allows one to incorporate user code into the protocol stack with minimal danger of damaging the system. Please refer to [4] for detailed instructions and coding examples and to Figure 3 for an example of a GUI.

4.4 Trouble with the Web

4.4.1 Problems. The approach described in Section 4 worked well enough for Windows GUI applications. However, Web-based ones proved initially to be troublesome. There are three reasons:
- Web Browsers do not use hooks for internal processing. Thus, it proves impossible to identify URLs using hook information;
- Web Browsers use caching. Thus, a requested URL may not appear in network traffic
- In secure environments, network traffic may be encrypted. Therefore, even if intercepted by an LSP, it may not be of any use

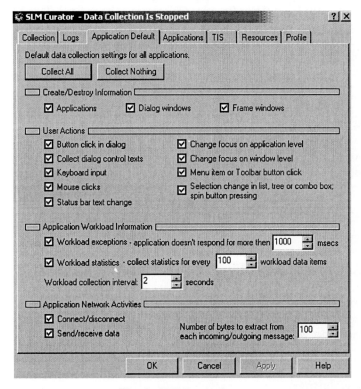

Fig. 3. GUI Example

4.4.2 Current Resolution.
Use of Hooks and Caching. These two problems are combined because they have the same solution – interpretation of hook data (user actions) and network traffic together. User actions supply timing, invaluable in the analysis of network traffic. For instance, it allows the cache problem to be overcome by identifying the first REFER clause of a GET statement after a certain user action. The method is not foolproof, but yields definite results.
Security. There is no obvious solution, which is how it should be – proper security should block the monitoring of the kind involved in direct measurement of Service

Levels. Therefore, if the entire traffic is encrypted, instrumented clients and Wire Sniffers simply do not work. In such a case, the only viable method is benchmarking. However, the overheads of Secure Sockets and their effect on response time generally ensure that only the most sensitive parts of each session are encrypted. If that's the case, all is not lost, because the representative sample may still be acquired for each session.

4.5 Futures

Web Browsers MAY be instrumented – but the author is not aware of an approach that would work for both Internet Explorer (IE) and Netscape.

In case of IE, HBO (Help Browser Objects) may be utilized. This MAY provide a way to acquire information directly, without recourse to network traffic analysis.

HBOs do not exist for Netscape. However, Mozilla Open Source does! This, in principle, allows any instrumentation to be developed.

Research on both subjects is in its infancy at the date of writing.

5. Conclusion

This paper reviewed the history of the development and existing approaches to measurement of Service Levels, as well as the currently available tools. Objectives and the concepts of direct measurement of Service Levels, appropriate metrics and market penetration were also looked at. It is hoped that all this will be of use to the readers.

6. Acknowledgement

Author is indebted to Chris Langshaw and Colin Makewell (both of the Systems Engineering Research Centre, Fujitsu Australia Limited). Without their unstinting help, this paper would not have been possible.

7. Disclaimer

All brand names are Registered Trademarks and Requested Trademarks of their respective owners.

References

1. D. Scott, B. Conwey Tools for Measuring Application Availability and Performance Gartner Group; COM-03-8396; 28/4/98
2. M. Maccabee Cline/Server End-To-End Response Time: Real Life Experience Proceedings of CMG96
3. D. McBride Toward Successful Deployment of IT Service Management in the Distributed Enterprise Proceedings of CMG95
4. Microsoft Developer Network (MSDN) http://msdn.microsoft.com/default.asp
5. M. Snell Tools Solve Mysteries of Application Response Time LAN Times, Vol. 14, Issue 5 http://www.wcmh.com/lantimes/97/97mar/703a034b.html
6. M. Tsykin, C. D. Langshaw End-To-End Response Time and beyond: direct measurement of Service Levels Proceedings of CMG98

Path Computation for Traffic Engineering in MPLS Networks

Gargi Banerjee and Deepinder Sidhu

Maryland Center for Telecommunication Research
Department of Computer Science and Electrical Engineering
University of Maryland, Baltimore County
1000 Hilltop Circle, Baltimore, MD 21250
Fax:410-455-1220
Tele:410-455-3063/2860
gargi@mctr.umbc.edu, sidhu@umbc.edu

Abstract. We consider the problem of computing traffic engineered paths for bandwidth requests, when these requests arrive in the network independent of one another. Reservation of bandwidth along pre-selected paths have become important in networks providing service differentiation and bandwidth guarantees to applications. Service providers are looking at traffic engineering to automate path selection procedures and to maintain network loading at an optimal level. Sophisticated path selection algorithms are being developed which deviate from the "shortest path" philosophy in traditional IP networks. While these algorithms perform well under moderate network loads, their behavior under high load conditions often leads to risks of network instability. In addition, these sophisticated algorithms are often computationally intensive. In this paper we provide an $O(nlogn)$ algorithm that improves network utilization under moderate load and also maintains stability under high load conditions. We show that the algorithm reduces the complexity of a competitive algorithm and achieves better performance.

Keywords: traffic engineering, path computation, K-shortest path, MPLS

1 Introduction

The exponential growth of the Internet has led to the increasing importance of network management and control functions. It is evident today that adding more bandwidth to networks is not the solution to all congestion problems. At the same time, more and more providers are showing interest in making revenues from offering differentiation of services in their networks. This requirement has increased the importance of gaining control over networks via automated Traffic engineering (TE). TE reduces congestion, improves network utilization, satisfies diversified requirements and thus leads to an increase of revenue.

TE is identified as the most important application of MPLS networks [1]. MPLS traffic engineering (MPLS-TE) enables a MPLS backbone to expand upon traffic engineering capabilities by routing flows across the network based on the resources the flow requires and those currently available. Since the essence of traffic engineering is mapping traffic flows onto a physical topology, it implies that at the heart of MPLS-TE resides the problem of path computation.

Traditional best-effort IP networks have always favored "shortest-path" algorithms because of their efficiency and stability under all load conditions. However, as networks become more Quality of Service (QoS) aware it becomes necessary to find paths between a source destination pair (*s-d*) that are different from the shortest path(s) returned by a minimum distance algorithm (Min-Dist[1]). This is because the shortest path may be highly congested while alternate paths between *s-d* are running at relatively lower utilization. In addition, since flows now have different QoS requirements, a single path may not be able to satisfy requirements of all flows between *s-d*.

Previous path computation algorithms were mainly hop-by-hop distributed algorithms optimizing static metrics (e.g. link cost/length etc.). Among the new generation of TE algorithms, [6] considers dynamic weight settings for links but assumes knowledge about future requests. MIRA [4] discusses a source-routing scheme (entire path is computed at a centralized point e.g. source), which considers another dynamic constraint (available maximum flow in the network) to identify *critical links*. Paths returned by the algorithm tries to avoid these critical links.

In this paper, we propose a source-routing algorithm that uses static and dynamic link constraints. The algorithm improves upon MIRA by first bounding path lengths so that they do not exceed a threshold T, and then considers max-flow computations to find a minimal blocking path. We also note that the max-flow computations in MIRA are computationally intensive to run for each request in a large network. As an alternative, we propose using feedback from the network to dynamically identify critical links.

The outline of the paper is as follows. In section 2 we describe our TE based path computation approach. Section 3 describes some initial results. In section 4, we present our conclusions.

2 Traffic Engineering Path Computation

In this paper we consider QoS requests of the form (A, B, Bw) where A: source node, B: destination node and Bw: minimum bandwidth the application requires. Using online TE path computation algorithm we then compute a path P_{A-B} that satisfies Bw. We assume Bw units of bandwidth are reserved along P_{A-B} by some TE signaling mechanism like CR-LDP or RSVP [2,3] and are available for the application. No assumption is made about future requests.

---

[1] Min-Dist: Dijkstra/Bellman-Ford type shortest-path algorithm

2.1 Maximizing Open Flow among a Set of Bounded Length Paths

The MIRA heuristic was proposed to find a path between a pair of nodes that blocks least amount of available flow between all other source-destination (src-dest) nodes. The interested reader is referenced to [4] for details. For convenience, some of the highlights of the algorithm are given below.

$N (n, m)$: Network of n nodes and m links
L : Link set
S : Given set of all src-dest pairs
θ_{ab} : Max-flow between pair $(a, b) \in S$
$l(i, j)$: Link from node i to j
$C_{a,b}$: Critical links[2] between pair (a, b) (determined by max-flow algorithm)
$r(i, j)$: Unreserved Bandwidth on $l(i, j)$

The following function assigns dynamic link weight $w(i, j)$:
- Assign_Dynamic_Link_Weights():

$$w(i, j): \sum_{(s, d): l(i,j) \in C_{s,d}} \alpha_{sd} \quad \forall \, l(i, j) \in L, \text{ where, } \alpha_{sd} = 1/\theta_{sd} \qquad (1)$$

Algorithm:
For each QoS request (p, q, Bw):
   Compute the set of critical links $C_{a,b}$, $\forall \, (a, b) \in S$
   Assign_Dynamic_Link_Weights()
   Prune off links that have $r(i, j) < Bw$
   Select the shortest path based on $w(i,j)$
Output: Path that blocks the least max-flow between all other src-dest pairs
Complexity: $O(n^2 \sqrt{m})$ using Goldberg Tarjan preflow push algorithm.

Fig. 1. MIRA

None of the static constraints (e.g. path length, path cost) is considered in MIRA. While at low to moderate network load this might be acceptable, longer paths selected at high load conditions, stresses an already overloaded network by occupying more resources. This may cause network instability. In addition, with this class of algorithms, the path lengths can become long enough to make the path practically unusable.

The first algorithm we propose is a variation to MIRA, which we call *Bounded-MIRA*. Let $u(i, j)$ be a static link metric. This new path computation scheme takes into account the dynamically assigned $w(i, j)$ as well as static $u(i, j)$. We use link length as

---

[2] Critical links defined as union of all minimum cut links [4]

our static metric in all our experiments. The highlight of Bounded-MIRA is that it first finds a candidate set of paths by the *K-shortest* paths[3][5] algorithm ($O(m + nlogn + K)$) on the static link metric $u(i, j)$. This returns a set of K paths on which MIRA type computations finds the path that blocks the least amount of max-flow among other source-destination pairs. Thus, the path returned is never longer than the *K-th shortest* path.

Algorithm :
For each QoS request (p, q, Bw):
    $A = K$-*shortest* path set computed based on static metric $u(i, j)$
    Compute the set of critical links $C_{a,b}$, $\forall (a, b) \in S$
    Assign_Dynamic_Link_Weights()
    Prune off links that have $r(i, j) < $ Bw
    From among K paths in A, select shortest path from based on dynamic wt. $w(i, j)$
Output : Path in A that blocks least max-flow among all src-dest pairs.
Complexity: $O(m + nlogn + K) + O(n^2 \sqrt{m}) = O(n^2 \sqrt{m})$

Fig. 2. Bounded-MIRA

2.2 Using TE Information to Reduce Algorithm Complexity

Max-flow computations open up the possibility of using dynamic constraints in path computation and by using Bounded-MIRA type algorithms we can find length constrained minimum blocking paths. However, the commonly known max-flow algorithms have a complexity of $O(n^2 \sqrt{m})$ whereas the simplest shortest path algorithms are of $O(nlogn)$ complexity. Max-flow computations are thus expensive especially when they have to be run for millions of requests. We propose using *link_load*, another dynamic constraint that does not need such expensive computations. Instead, we can rely on existing traffic engineering infrastructure to provide this information. The basic assumption is that most networks run link state Interior Gateway Protocols (IGP) for routing purposes. [7] suggests simple extensions to IGPs by which dynamic TE information like reserved link bandwidth can be periodically fed-back to the source nodes. This information can be used in place of max-flow computations for identification of critical links. We define link loading as:

$$link\_load_{(i,j)} = (Reserved\ Bw\ on\ l(i, j)\ /\ Total\ Reservable\ Bw\ on\ l(i, j)) * 100 \quad (2)$$

---

[3] Network optimization problem where K-shortest paths are ranked between an initial and a terminal node.

A critical link is now defined as a link which has its load running above a threshold percentage U. We use the following function and parameter definitions in our algorithm.

- Identify_C_N():
  C_N : Set of critical links of a network N.
  $l(i, j) \in C_N$ if $link\_load_{(i,j)} > U$. (3)

- Assign_Dynamic_Link_Weights():
  Each link $l(i, j)$ is assigned a dynamic critical weight $w(i, j)$, where:
  $w(i, j) = 0 \qquad if \; l(i, j) \notin C_N$ (4)

  $w(i, j) = \alpha(i, j) \quad if \; l(i, j) \in C_N$, where $\alpha(i, j) \propto link\_load_{(i,j)}$. (5)

- Criticality of a path P is defined as:
  $W(P) = \sum w(i, j) \quad \forall \; l(i, j) \in P$ (6)

We propose the Least-Critical-K-Shortest (LCKS) path algorithm in which the computation takes into account dynamically assigned link weight $w(i, j)$ as well as a static link metric $u(i, j)$. LCKS and Bounded-MIRA differ in their definition of criticality and hence in their computational complexity.

Algorithm:
For each QoS request (p, q, Bw):
    Prune off links that have $r(i, j) < Bw$
    A = K-shortest path set computed based on static metric $u(i,j)$
    Identify_C_N()
    Assign_Dynamic_Link_Weights()
        From among K paths in A, select least critical path based on W(P)
Output: Least critical path that is no longer than *K-th shortest* path.
Complexity: $O(m + nlogn + K) + O(n) + O(K) = O(nlogn)$

Fig. 3. Least-Critical-K-Shortest path (LCKS)

3 Performance Studies

We present some initial results of comparisons of the path computation algorithms. All experiments have been run on random networks consisting of 10, 15 and 20 nodes with an average node degrees varying in the range of 4-6. Experiments with higher number of nodes and node degrees are in progress. All links are symmetric and link weights assigned randomly. Link capacities vary from moderate (12000 units of reservable Bw) to high (48000 units of reservable Bw). The ratio of moderate to high capacity links in all networks was fixed at 3:1. For each network, there was a fixed set S of all possible source-destination pairs. Requests were generated by an offline utility and were of the form (src, dest, Bw). The source and destination nodes were picked randomly from S The bandwidth requirement varied uniformly from 1 to 4 units.

The results reported are for a 20-node network with maximum node degree of 10. Fig. 4 shows the bandwidth blocking (Bw request refused) obtained from MIRA, Bounded-MIRA and Min-Dist routing. We note that blocking (high load condition) begins only after about arrival of 20000 requests. Blocking in Min-Dist is relatively more. We note that as load increases, both MIRA and Bounded-MIRA perform equally well.

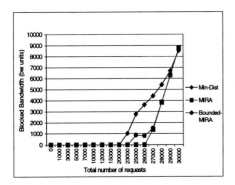

Fig. 4. Bandwidth blocking

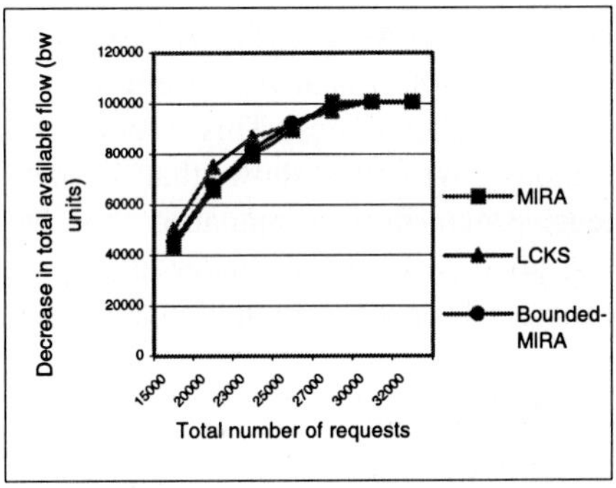

Fig. 5. Decrease in total available flow at high load

Fig. 5-7 presents comparisons of MIRA, Bounded-Mira and LCKS. In Fig. 5, we note that as load conditions increase, all three algorithms have comparable performance in terms of minimizing reduction in network flow. Thus, our scheme based on TE feedback information is as efficient as one that uses expensive max-flow computations.

The load of a path is measured by the maximum load of its component links. Fig. 6 shows that under high load condition the average loading of LCKS paths is the lowest. Bounded-MIRA returns lower-loaded paths than MIRA. For Fig. 7, we define stretch factor of a path $P(x, y)$ as $\Delta(x, y) = T(x, y)/ C(x, y)$, where $T(x, y)$: length of $P(x, y)$, $C(x, y)$: length of shortest-path returned by Min-Dist. As shown, LCKS paths have the least average Δ even at highest network load conditions. MIRA paths have the highest Δ. The least loaded shorter length LCKS paths returned under high load conditions demonstrates the stability of the algorithm.

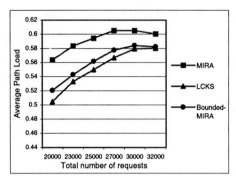 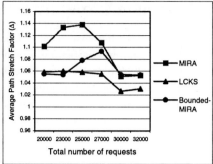

Fig. 6. Increase in avg. path load at high load **Fig. 7.** Avg. path stretch factor (Δ) at high load

4 Conclusion

W conclude that while dynamic link metrics can help us maintain network loading at an optimal level, it is always useful to use static constraints like link length to bound the path characteristics. This avoids using up critical resources during high load conditions. We also show that utilizing TE feedback information rather than expensive max-flow computations can help to reduce algorithmic complexity while giving good performance under high load. From this perspective, LCKS qualifies as an efficient TE path computation algorithm.

References

[1] D. Awudche, J. Malcolm, J.Agogbua, M. O'Dell, J. McManus, "Requirements for Traffic Engineering Over MPLS", *Request for Comments:2072*, September 1999.
[2] Bilel Jamoussi, "Constraint-Based LSP Setup using LDP", *Work in Progress*, July 2000.
[3] D. Awudche, L. Berger, D. Gan, T. Li, V. Srinivasan, G. Swallow, "RSVP-TE: Extensions to RSVP for LSP Tunnels", *Work in Progress*, August 2000.
[4] M. Kodialam and T. Lakshman, "Minimum Interference Routing with Applications to MPLS Traffic Engineering," *Proceedings of IEEE INFOCOM'2000*, March 2000.
[5] D. Eppstein, "Finding the k shortest paths," *SIAM Journal on Computing*, 28: 652-673, 1998.
[6] B. Fortz, M. Thorup, "Internet Traffic Engineering by Optimizing OSPF Weights", *Proceedings of IEEE INFOCOM'2000*, March 2000.
[7] D. Katz, D. Yeung , "Traffic Engineering Extensions to OSPF", *Work in Progress*, Oct. 2000.

Minimum Regret Approach to Network Management under Uncertainty with Application to Connection Admission Control and Routing*

Vladimir Marbukh

National Institute of Standards and Technology
100 Bureau Drive, Stop 8920
Gauthersburg, MD 20899-8920, USA
marbukh@nist.gov

Abstract. This paper proposes a framework for network management intended to balance network performance under normal steady operational conditions with robustness under non-steady, and/or adverse conditions. Working in conjunction with anomaly and/or intrusion detection, the proposed framework allows the network to develop a set of measured responses to possible anomalies and external threats by minimizing the average maximum network losses, i.e., regrets or risks, due to uncertainty. Loss maximization guards against uncertainty within each scenario. Averaging of the maximum losses reflects an available information on the likelihood of different scenarios. The proposed framework includes Bayesian and minimax approaches as particular cases. The paper demonstrates how the proposed framework can alleviate high sensitivity of a cost-based admission and routing scheme to uncertain resource costs and possible presence of excessive and/or adversarial traffic. Specific examples include competitive and minimum interference admission and routing schemes.

1 Introduction

Different approaches to providing Quality of Service (QoS) guarantees by a system of shared resources include using Traffic Controlling Devices (TCD) for traffic shaping and policing, Fair traffic Scheduling Mechanisms (TSM), and Measurement Based Control (MBC). TCD and TSM allow the network to provide QoS guarantees to a source complying with its service agreement by making admission control decisions based on the worst case scenario. However, due to a wide range of traffic patterns generated by current and especially future applications, this approach runs a risk of significant over provisioning of the network resources under normal operating conditions. This risk can be mitigated by MBC, which extracts control information from on-line measurements [1]. This real-time control information can be used by the resources to control usage and/or by users to modify their behavior according to

---

* This work was supported in part by DARPA under NMS program.

the smart market concept [2]-[7]. Reliable extraction of the control information from the measurements and optimal behavior of smart market are achievable in equilibrium. However, even in equilibrium MBC faces a problem of making decisions under uncertainty due to the statistical nature of measurements and state aggregation needed to reduce amount of signaling information. The uncertainty becomes a critical problem for a network operating under non-steady and/or adversarial conditions. These conditions may be a result of natural events, i.e., fiber cuts, flash crowds, and/or adversary attack, i.e., denial of service attacks [8]. Since anomaly and intrusion detection procedures are based on statistical inferences, such as hypotheses testing, clustering, discriminant analysis, etc., the result of the detection is typically a set of possible scenarios and their likelihood. Uncertainty within a scenario may be a result of interval rather than point estimates for the state of environment specified by the scenario. For example, for flash crowds and distributed denial of service attacks a scenario represents a subset of flows - an aggregate, responsible for congestion [8]. Uncertainty results from difficulty to identify the aggregates and inability to distinguish between legitimate and excessive traffic within each aggregate.

Currently commercial networks including the Internet may carry mission-critical applications with wide range of bandwidth and QoS requirements. Possibility of anomalies caused by sudden increase in offered traffic, fiber cut and/or an adversary attack necessitates developing network management schemes that balance cost efficiency under normal operating conditions with robustness under non-steady and/or adversarial conditions. The right balance between cost efficiency and performance can be achieved by adaptive sharing of the network management responsibilities between MBM, TCD and TSM. The critical role in this collaboration belongs to MBM which determines the optimal "weight" of each scheme by adjusting parameters of the TCD and/or TSM algorithms.

Section 2 of this paper proposes a framework for network management under uncertainty. Working in conjunction with anomaly and/or intrusion detection, the proposed framework allows the network to develop a set of measured responses to possible anomalies and external threats by minimizing the average maximum network losses, i.e., regrets or risks, due to uncertainty. Loss maximization guards against uncertainty within each scenario. Averaging of the maximum losses reflects an available information on the likelihood of different scenarios. The proposed framework includes Bayesian and minimax approaches as particular cases. Section 3 demonstrates how the proposed framework can alleviate high sensitivity of cost-based admission and routing strategies to the uncertain resource costs and possible presence of excessive and/or adversarial traffic. Section 4 considers specific examples of competitive [9] and minimum interference admission and routing schemes [10].

2 Risk Assessment and Management

Throughout the paper we assume that the network utility $W(u|\theta)$, typically the revenue generated by the network, is a known function of the network control action

u and environment θ. Vector u may describe pricing, resource allocation, admission control, routing, scheduling etc., and vector θ may characterize topology, resource capacities, traffic sources, etc. We consider a problem of selecting the network control action u under incomplete information on the state of environment θ. This section discusses approaches to assessment and minimization of the risks due to uncertainty.

2.1 Risk Assessment

If the state of environment θ is known, the optimal control action $u = u^*(\theta)$ maximizes the network utility $W(u|\theta)$:

$$u^*(\theta) = \arg\max_{u \in U} W(u|\theta). \tag{1}$$

Formula (1) determines the best network response to given θ. In presence of other players besides the network, e.g., users, each player determines his best response to the other player strategies and the state of environment θ by maximizing his utility function. Harsanyi transformation of sequential games with incomplete information [11] assumes that the state of environment θ is selected by "Nature" at the beginning of the game according to some probability distribution $p_0(\theta)$ known to all players. Players may have private information regarding selected vector θ and they update their information on the selected vector θ according to Bayesian rules as game progresses. According to this approach, the best network response is

$$u^* = \arg\max_{u \in U} \sum_{\theta} W(u,\theta) p(\theta) \tag{2}$$

where $p(\theta)$ is the updated distribution $p_0(\theta)$. According to the theory of non-cooperative games, once the best responses by players are determined, reasonable players should select their equilibrium strategies. Note that a number of equilibrium concepts exist, including the most widely used Nash equilibrium.

In this paper we are interested in a case when the "Nature", or as we prefer to call it "environment" may have malicious intend directed towards some players, i.e., the network and some users. Due to space constraints, in this paper we only consider a case when this malicious intend may be directed towards the network. Following Savage [12] we characterize the network loss in performance resulted from non-optimal selection of the control action u due to uncertain environment θ by the following regret or loss function:

$$L(u|\theta) = \max_{u' \in U} W(u'|\theta) - W(u|\theta). \tag{3}$$

Local maximums $\theta_i^* = \theta_i^*(u)$ of the loss function (3) over θ represent different risk factors for the network due to uncertain environment (for example see [13]). We propose to model uncertain environment as a player with utility function (3) and certain restrictions on the set of strategies $\theta \in \Theta$. This assumption leads to a framework for network management under uncertainty, which, in effect, is a systematic approach to balancing different risk factors. Bayesian and minimax approaches are particular cases of this framework.

Note that there is a certain degree of freedom in selecting loss function $L(u|\theta)$. This selection reflects the desired balance between different risk factors. Using loss function (3) in networking context has been proposed in [13]. Competitive approach to admission control and routing [9] guards the network against the worst case scenario with respect to the state of environment θ by containing the losses

$$L(u|\theta) = \frac{1}{W(u|\theta)} \max_{u' \in U} W(u'|\theta) - 1 \qquad (4)$$

where vector of control action u represents admission control and routing, and vector θ represents completely unknown sequence of future request arrivals. The competitive approach assumes that the holding times are known upon request arrival, at least in probabilistic sense. Another basic assumption is that exact information on the instantaneous link utilization is available.

2.2 Risk Management

Network operating under non-steady and/or adversarial conditions cannot rely on extensive historical data to estimate the state of environment θ. However, some aggregated information on θ is often available. We assume that this information can be quantified in terms of probabilities $p_i = \Pr ob(\theta \in \Theta_i)$, $\sum_i p_i = 1$ for some partition of the region of all possible vectors $\Theta = \{\theta\}$ into set of mutually exclusive regions Θ_i: $\bigcup_i \Theta_i = \Theta$, $\Theta_i \bigcap \Theta_j = \varnothing$. Given scenario $\{p_i, \Theta_i\}$, the optimal network response is

$$u^* = \arg\min_{u \in U} \hat{L}(u) \qquad (5)$$

where the performance loss (regret or risk) due to uncertainty is

$$\hat{L}(u) = \sum_i p_i \max_{\theta \in \Theta_i} L(u|\theta) . \qquad (6)$$

Procedure (5)-(6) can be interpreted as a game between the network and environment with constraints on the set of environment strategies. In an extreme case when each subset Θ_i is a singleton, i.e., consists of a single point θ, procedure (5)-(6) reduces to the Bayesian procedure (2). In another extreme case when Θ is partitioned into itself and the empty set $\emptyset$, procedure (5)-(6) can be interpreted as a zero sum game between the network and environment. Since anomaly and/or intrusion detection typically results in a set of possible scenarios $s = \{1,..,S\}$ rather than one scenario, the network faces a task of balancing risks associated with different scenarios. This task can be formulated a multi criteria optimization problem

$$\min_{u \in U} (R_1(u),..,R_S(u)) \qquad (7)$$

where risk associated with scenario s is $R_s(u) = \hat{L}_s(u) - \min_{u' \in U} \hat{L}_s(u')$, and the loss in performance for scenario s is $\hat{L}_s(u)$. Pareto frontier yields a reasonable set of the network control actions $u \in U$ in this situation. Any additional information can be used to reduce the set of Pareto solutions. For example, if anomaly and/or intrusion detection can identify (subjective) probabilities π_s of different scenarios s, $\sum_s \pi_s = 1$, then the network may minimize the average (Bayesian) risk

$$\min_{u \in U} \sum_s \pi_s R_s(u) \ . \qquad (8)$$

In a case of unknown π_s and/or adversarial environment the network may prefer to minimize the maximum risk

$$\min_{u \in U} \max_{s=1,...,S} R_s(u) \ . \qquad (9)$$

Working in conjunction with anomaly and/or intrusion detection the network should perform two tasks. The long time scale task is maintaining and updating a library of possible scenarios s and subroutines for calculating the corresponding risks $R_s(u)$. The short time scale task is risk minimization (7).

3 Connection Admission Control and Routing under Uncertainty

Admission of a request, on the one hand, brings revenue to the network, but, on the other hand, ties up the network resources until the service is completed, and thus may cause future revenue losses due to insufficient resources for servicing future requests. The implied cost of a resource represents this potential revenue loss, and the surplus value is the difference between the revenue brought by the admitted request and the implied cost of the occupied resources. An incoming request should be accepted if the surplus value is positive, and should be rejected otherwise. This section demonstrates how proposed approach to risk assessment and management can be used to balance the performance and robustness of a cost based connection admission control and routing scheme under uncertainty.

3.1 Risk Associated with Connection Admission and Routing

Consider an arriving request for bandwidth b on a route $r \in \{r_1,..,r_k\}$, where the set of feasible routes $\{r_1,..,r_k\}$ is determined by the origin-destination of the request, availability of the bandwidth, maximum allowed number of hops, QoS requirements, etc. Let the implied cost of tying up bandwidth b along route r be c_r where we dropped b from notations. Given the set of feasible routes and route costs, the optimal route r_* and admission condition for an arriving request willing to pay rate w are as follows:

$$c_* \equiv c_{r_*} = \min_{r \in \{r_1,..,r_k\}} c_r \leq w . \qquad (10)$$

The implied costs c_r are determined by future events such as future request arrivals, holding times, availability of resources, topology, etc. Inability of the network to predict these future events, especially in non-steady and/or adversarial environment, is a source of uncertainty in the implied costs c_r. Uncertainty in w may be caused by presence of excessive and/or malicious traffic such as in flash crowds or denial of service attack [8]. Utility of the admission and routing decisions can be characterized by the surplus value

$$W(r|w,c) = \begin{cases} w - c_r & \text{if } r \neq \varnothing \\ 0 & \text{if } r = \varnothing \end{cases} \qquad (11)$$

where the request is accepted on route r if $r \neq \varnothing$, and is rejected if $r = \varnothing$. For utility function (11) the loss function (3) takes the following form:

$$L(r|w,c) = \begin{cases} \max\{0, w - c_*\} + c_r - w & \text{if } r \neq \varnothing \\ \max\{0, w - c_*\} & \text{if } r = \varnothing \end{cases} . \qquad (12)$$

3.2 Approximation of Separable Route Costs

Computational feasibility to minimize losses (12) over r critically depends on the range Θ of possible vectors of implied costs $(c_r : r = r_1,..,r_k)$. In this subsection we consider a case of separable route costs:

$$\Theta = \bigcap_{r=r_1,...,r_k} [\check{c}_r, \hat{c}_r]. \tag{13}$$

Note that (13) is a "first order" approximation to more realistic scenarios since implied costs c_r of different routes strongly correlate with each other due to the global nature of implied costs and route overlapping.

In a symmetric case: $c_r \in [\check{c}, \hat{c}]$, $\forall r = r_1,..,r_k$, the network has two pure strategies: reject the request, and accept the request on a randomly selected route $r = r_1,..,r_k$. The adversarial environment has two pure strategies $c_r = \check{c}$ and $c_r \in \hat{c}$ for $r = r_1,..,r_k$. The corresponding payoff matrix (12) is:

$$\begin{array}{lll} & reject: & accept: \\ \check{c}: & \max\{0, w-\check{c}\} & \max\{0, w-\check{c}\} + \check{c} - w. \\ \hat{c}: & \max\{0, w-\hat{c}\} & \max\{0, w-\hat{c}\} + \hat{c} - w \end{array} \tag{14}$$

Game (14) has different solutions in the following three cases. In cases $w \le \check{c}$ and $w \ge \hat{c}$ game (14) has the saddle point and the network has pure optimal strategy to reject and, respectively, accept the request. In a case $\check{c} < w < \hat{c}$ game (14) does not have saddle point and optimal strategy for the network is mixed: reject the request with probability $1-\alpha$, and accept with probability α, where $\alpha = (w-\check{c})/(\hat{c}-\check{c})$.

In a general case of separable route costs (13) if $w \le \check{c}_* \equiv \min_r \check{c}_r$ or $w \ge \hat{c}_* \equiv \min_r \hat{c}_r$, then the corresponding game has the saddle point and the network has pure optimal strategy to reject and, respectively, accept the request on a route $r^{opt}: \hat{c}_{r^{opt}} \equiv \min_r \hat{c}_r$. In a case $\check{c}_* < w < \hat{c}_*$ the corresponding game does not have the saddle point and the optimal strategy is closely approximated by the following mixed strategy: reject the request with probability $1-\alpha$, and accept with probability α on route r^{opt}, where α and r^{opt} are determined by the following optimization problem:

$$\alpha = \max_{r=r_1,...,r_k} \left\{ \frac{w - \check{c}_r}{\hat{c}_r - \check{c}_r} \right\}. \tag{15}$$

4 Examples

This section presents examples, including an aggregated version of the competitive scheme under uncertain holding times, and an aggregated version of the Minimum Interference Routing Algorithm (MIRA).

4.1 Aggregated Competitive Scheme under Uncertain Holding Times

A competitive admission and routing scheme [9] strives to contain loss (4) under the worst case scenario sequence of future request arrivals θ. It was shown [9] that these maximum losses can be bounded by $O(\log v\tau)$ assuming that the holding time τ becomes known upon arrival of the request, and where v is some constant. This bound can be achieved with the cost-based admission control and routing strategy with additive route costs, i.e., the cost c_r of a route r is a sum of the costs c_j of the links j comprising the route r: $c_r = \sum_{j \in r} c_j$. The cost of a link j is a function of the instantaneous load carried by the link x and holding time τ: $c_j = c_j(\tau, x)$. In a case of throughput maximization [9] the link j cost is

$$c_j(\tau, x) = [(2Fh\tau + 1)^{x/B_j} - 1]B_j \qquad (16)$$

where F is some constant, the maximum number of hops allowed in a route is h, and link j bandwidth is B_j.

The proposed in this paper framework can be used to mitigate assumption of known, at least probabilistic sense [9], holding times, and address the need to aggregate the real-time information on instantaneous link utilization x. As an illustration consider a case of uncertain holding time $\tau \in [\check{\tau}, \hat{\tau}]$, and two aggregates: link j is said to be in the aggregate state $y_j = 0$ if the link instantaneous utilization $x_j \in [0, B_{1j}]$ and $y_j = 1$ if $x_j \in [B_{1j}, B_j]$ where $B_{1j} \in [0, B_j]$ is some threshold. The uncertainties in τ and x_j cause the following uncertainty in the route r cost c_r:

$$\sum_{j \in r} \check{c}_j(y_j) \equiv \check{c}_r \le c_r \le \hat{c}_r \equiv \sum_{j \in r} \hat{c}_j(y_j) \qquad (17)$$

where $\check{c}_j(0) = 0$, $\check{c}_j(1) = c_j(\check{\tau}, B_{1j})$, $\hat{c}_j(0) = c_j(\hat{\tau}, B_{1j})$, $\hat{c}_j(1) = 2Fh\hat{\tau}B_j$. Under approximation of separable route costs (13), subsection 3.2 results define the aggregated version of the competitive scheme under uncertain holding times.

4.2 Aggregated Minimum Interference Routing Algorithm

Minimum Interference Routing Algorithm (MIRA) [10] does not assume any knowledge of future request arrivals or holding times, but takes advantage of the known network topology by selecting routes that do not "interfere too much" with a route that may be critical to satisfy future demands. The problem was motivated by the need of service providers to set up bandwidth guaranteed path in their backbone or networks. An important context in which these problems arise is that of dynamic Label Switching Path (LPS) set up in Multi-Protocol Label Switched (MPLS) networks. In MPLS packets are encapsulated at ingress points, with labels that are then used to forward the packets along LPSs. Service providers can use virtual circuit switched, bandwidth guaranteed LPSs as component of an IP Virtual Private Network (VPN) service with the bandwidth guarantees used to satisfy customer service-level agreements (SLAs).

MIRA is a heuristic, on-line, state-dependent, cost-based routing algorithm with additive route costs: $c_r = \sum_{j \in r} c_j$. The link j cost c_j is a function of the vector of residual link capacities $z = (z_i)$ in all links i:

$$c_j(z) = \sum_{(n,m)} w_{nm} \delta_{nm}^{(j)}(z) \tag{18}$$

where (n, m) are all possible ingress-egress pairs, w_{nm} is the revenue generated by providing an unit of bandwidth to ingress-egress pair (n, m), and $\delta_{nm}^{(j)}(z) = 1$ if link j is critical for ingress-egress pair (n, m), and $\delta_{nm}^{(j)}(z) = 0$ otherwise. Link j is critical for ingress-egress pair (n, m) if the maximum flow for pair (n, m) decreases whenever the residual capacity of link j decreases, and is not critical otherwise. Paper [10] discusses effective algorithms for calculating sets of critical links, i.e., indicators $\delta_{nm}^{(j)}(z)$, and reports simulation results.

The proposed in this paper framework can be used to mitigate the following problems of MIRA. The first problem is inability to utilize an available incomplete information on the expected loads. The second problem is high sensitivity to the current set of critical links, which is likely to change in the future due to fluctuations in the vector z. The third problem, related to the second one, is need to aggregate the residual capacities. Due to space constraints we briefly address the second and third problems. Assuming that available information on residual link i capacity z_i is characterized by aggregates $z_i \in [\breve{z}_i, \hat{z}_i]$, the range for the link j cost c_j is

$$\sum_{(n,m)} w_{nm} \delta_{nm}^{(j)}(., \hat{z}_{j-1}, \breve{z}_j, \hat{z}_{j+1},..) \leq c_j \leq \sum_{(n,m)} w_{nm} \delta_{nm}^{(j)}(., \breve{z}_{j-1}, \hat{z}_j, \breve{z}_{j+1},..) . \tag{19}$$

Under approximation of separable route costs (13), subsection 3.2 results define the aggregated version of MIRA.

References

1. Courcoubetis, C., Kelly, F.P., and Weber, R., Measurement-based usage charges in communication networks. http://www.statslab.cam.ac.uk/Reports/1997/
2. Bonomi, F., Mitra, D, and Seery, J., Adaptive algorithms for feedback-based flow control in high-speed wide-area networks. *IEEE JSAC*, 13 (1995) 1267-1283
3. Low, S.H., and Varaiya, P.P., A new approach to service provisioning in ATM networks. *IEEE Trans. On Networking*, 1 (1993) 547-553
4. MacKie-Mason, J.K. and Varian, H.R., Pricing congestible network resources. *IEEE JSAC*, 13 (1995) 1141-1149
5. Shenker, S., Fundamental design issues for the future Internet. *IEEE JSAC*, 13 (1995) 1176-1188
6. Gibbens, R. and Kelly, F.P., Resource pricing and evolution of congestion control. http://www.statslab.cam.ac.uk/~frank/PAPERS/evol.html
7. Kelly, F.P., Mauloo, A., and Tan, D., Rate control in communication networks: shadow prices, proportional fairness and stability. J. of the Oper. Res. Soc., 49 (1998) 237-252
8. Mahajan, R., Bellovin, S.M., Floyd, S., Ioannidis, J., Paxson, V., and Shenker, S., Controlling high bandwidth aggregates in the network. http://www.research.att.com/~smb/papers/
9. Plotkin, S., Competitive routing of virtual circuits in ATM networks. *IEEE JSAC*, 13 (1995) 1128-1136
10. Kar, K., Kodialam, M., and Lakshman, T.V., Minimum interference routing of bandwidth guaranteed tunnels with MPLS traffic engineering applications. *IEEE JSAC*, 18 (2000) 2566-2579
11. Harsanyi, J.C., Papers in Game Theory. Reidel Publishing Company (1982)
12. Blackwell, D. and Girschick, M., Theory of Games and Statistical Decisions. Wiley, New York (1954)
13. Marbukh, V., Network management under incomplete information on the operational environment. Intern. Symp. on Inform. Theory and its Appl. (ISITA2000) 637-640

MPLS Restoration Scheme Using Least-Cost Based Dynamic Backup Path

Gaeil Ahn[1] and Woojik Chun[2]

[1]Network Security Department, Electronics and Telecommunications Research Institute
(ETRI), 161 Kajong-dong, Yusong-gu,Taejon, 305-350, Korea
agi63053@etri.re.kr
[2]Department of Computer Engineering, Chungnam National University
220, Goong-Dong, Yusong-Gu,Taejon, 305-764, Korea
chun@ce.cnu.ac.kr
http://flower.ce.cnu.ac.kr/~chun

Abstract. The path restoration in MPLS is the technique to reroute traffic around a failure in a LSP. The existing path restoration schemes have some difficulty in solving problem such as resource utilization and protection of backup path. This paper proposes a dynamic path restoration scheme using the least-cost based backup path, which may increase resource utilization and protect backup path without requiring longer restoration times. In the proposed scheme, each node on working path has candidate nodes that may become an egress node of backup path. The node that detects a failure finds a node out of the candidate nodes that is used as splicing node of backup path at the lease cost. And then it establishes new backup path up to the node in order to reroute traffic on working path to the backup path. Through simulation, the performance of the proposed scheme is measured and compared with the existing scheme in terms of packet loss, re-ordering of packets, concurrent faults, and resource utilization.

1 Introduction

The explosive growth and the advent of sophisticated services of Internet have reduced the actual carried traffic and service quality. So IETF(Internet Engineering Task Force) is developing Multi-Protocol Label Switching(MPLS)[5][12], which combines flexibility of IP routing and efficiency of link-level switching.

MPLS router, called Label Switching Router(LSR), enables to forward IP packet directly to a next hop by using label inserted in the packet as index into table which specifies the next hop. Label Distribution Protocol(LDP)[10] is defined to distribute labels and to establish Label Switched Path(LSP). MPLS also provide capability that can establish a Constraint-based Routed LSP(CR-LSP), which is Explicit-Routed Label Switched Path(ER-LSP) based on QoS. For it, CR-LDP(Constraint-based Routing LDP)[1] and RSVP(ReSerVation Protocol)-TE[3] was proposed.

The ability to protect traffic around a failure or congestion in a LSP can be important in mission critical MPLS networks. The path restoration[13] is to reroute traffic around a link/node failure or congestion in a LSP; that is, packets will be followed along the backup path in case of working path failure.

Such a backup path can be established after a working path failure is detected or, alternatively, it can be established beforehand in order to reduce the path switchover time. The former is called a dynamic protection and the latter is called pre-negotiated protection. According to how the repairs are affected upon the occurrence of a failure on the working path, two possibilities exist: global repair and local repair. In global repair, protection is always activated on end-to-end basis, irrespective of where a failure occurs. But, in local repair, protection is activated by each LSR that detects a failure.

Even if dynamic protection can increase resource utilization, most of the existing schemes have used pre-negotiated protection[2][4][8]. This is because dynamic protection may require longer restoration times and even fail in setup of backup path. However, pre-negotiated configuration also has some difficulty in solving problem such as resource utilization and protection of backup path.

This paper proposes a dynamic path restoration scheme using the least-cost based backup path, which may increase resource utilization and protect backup path without requiring longer restoration times. In our scheme, there are two main ideas in establishing backup path. One is candidate nodes that may become an egress node of backup path. The other is Least-Cost Backup-Path algorithm used to find a node out of the candidate nodes that is used as splicing node of backup path at the lease cost.

The rest of this paper is organized as follows. Section 2 overviews and analyzes the existing path restoration schemes. Section 3 explains our scheme. In section 4, the performance of the proposed scheme is measured and compared with the existing scheme in terms of packet loss, re-ordering of packets, concurrent faults, and resource utilization through simulation. Finally conclusion is given in Section 5.

2 Related Research

In this paper, two prevailing schemes are introduced; those are Haskin's scheme[4] and Makam's[2]. Haskin's scheme uses pre-negotiated protection and local repair. Makam's scheme uses pre-negotiated protection and global repair.

Fig. 1 shows Haskin's scheme and Makam's. In Fig. 1, the straight line between LSR1 and LSR9 is the working path.

In Haskin scheme, a backup path is established as follows:
1. The initial segment of the backup path is established between PML(Protection Merging LSR) and PIL(Protection Ingress LSR) in the reverse direction of the working path. In Fig. 1, the dashed line between LSR9 and LSR1 illustrates such a segment of the backup path.
2. The second and final segment of the backup path is established between PIL and PML along a transmission path that does not utilize any working path.
3. The initial and final segments of the backup path are linked to form an entire backup path.
4. When a failure occurs, the node that detects the failure reroutes incoming traffic by linking the upstream portion of the working path to the downstream portion of the backup path. In Fig. 1, when the node LSR7 fails, the working traffic is rerouted along the backup path, LSR5-3-1-2-4-6-8-9

The merit of Haskin's scheme is that there is almost no packet loss during link/node failure. However, Haskin's scheme introduces re-ordering of packets in case that traffic is switched back from the backup path to the working path after a link/node goes up.

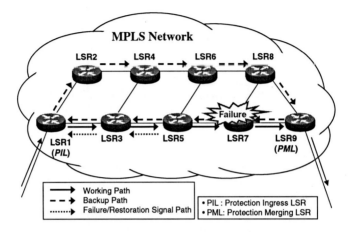

Fig. 1. Path restoration: Haskin's scheme and Makam's

In Makam's scheme, a backup path is established as follows:
1. A backup path is established between PIL and PML along a transmission path that does not utilize any working path.
2. When a failure occurs, the node that detects the failure sends a failure notification message toward its PIL. On receiving the message, PIL reroutes the incoming traffic through the backup path. In Fig. 1, when the node LSR7 fails, the working traffic is rerouted along the backup path, LSR1-2-4-6-8-9.

The merit of Makam's scheme is that there is almost no problem in reordering of packets during link/node failure. However, Makam's scheme introduces packet loss because PIL doesn't execute the protection switching until it receives the failure notification message from a node that detect a link/node failure.

3 Dynamic Path Restoration Scheme

The proposed scheme uses dynamic protection and local repair. In the proposed scheme, each node on working path has candidate nodes that may become an egress node of backup path. A backup path is established as follows:
1. The node that detected a failure calculates an explicit route between itself and a node out of the candidate nodes that is used as splicing node of backup path at the lease cost.
2. In setup of new backup path, the calculated explicit route is used as the ER(Explicit Route) of MPLS signaling message(e.g. CR-LDP, RSVP). For the purpose of splicing the existing working path and the backup path to be

established, LSPID of working path is also used as a ER hop. The holding priority of working path may be used as setup priority of backup path.
3. As soon as the backup path is established, traffic on working path is switched to backup path.
4. If setup of backup path fails, go to 1.

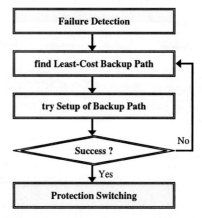

Fig. 2. Procedure of path restoration

```
procedure Least-Cost-Backup-Path (Src_node, Candidate_nodes, TParam)
    // Src_node is a node that detected a failure //
    // Candidate _nodes is a set of nodes that exist from the downstream node of
       failed node to the destination node in the direction of the working path //
    // TParam is the traffic parameter value of working LSP //

    least_cost  ← ∞
    best_backup_path ← nil
    for ( each splicing_node of Candidate_nodes ) do
         backup_path ← Get-CR-Path (Src_node, splicing_node, TParam)
         cost ← Get-Cost (backup_path)
         if ( cost ≤ least_cost ) then    best_backup_path ← backup_path
                                          least_cost ← cost
         end
    end
    return least_backup_path
end Least-Cost-Backup-Path
```

Fig. 3. Least-Cost-Backup-Path algorithm

Fig. 2 illustrates the procedure of restoration function briefly. Fig. 3 shows Least-Cost Backup-Path algorithm. In Fig. 3, the parameter *Candidate_nodes* can be easily found in ER of CR-LDP(or ERO of RSVP) or Path Vector of LDP(or RRO of RSVP). The function *Get-CR-Path* can be easily implemented by using the existing algorithm such as Widest-Shortest Path Algorithm[6], Shortest-Widest Path Algorithm[14], and

Shortest-Distance Path Algorithm[11]. The function *Get-CR-Path* is a function that calculates links delay or hop count of backup path.

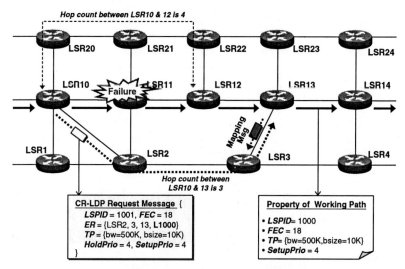

Fig. 4. Example for the proposed scheme

Fig. 4 is an example for the proposed scheme. When LSR11 fails, LSR10 that detected the failure calculates an explicit route for new backup path. The candidate nodes for the working LSP(LSPID 1000) in LSR10 are LSR12, 13, and 14. Assuming that hop count is used in calculating cost of backup path, cost from LSR10 to LSR12 is 4, cost to LSR13 is 3, and cost to LSR14 is 4. Because cost from LSR10 to LSR13 is the least of them, the explicit route for new backup path become LSR10-2-3-13.

The calculated explicit route is used as ER of CR-LDP Request Message. The LSPID of working path, 1000 is used as its ER hop in order to splice the existing working LSP and new backup LSP. As soon as LSR10 receives a CR-LDP Mapping Message for the backup LSP, it reroutes traffic on working path to new backup path. If the link between LSR10 and LSR11 fails, the working traffic is rerouted along the backup path, LSR10-20-21-11.

The proposed scheme has the following advantages:

- High Resource Utilization -- because resource is not reserved beforehand.
- Fast Path Restoration -- because a backup path with least cost can be selected by using the proposed algorithm.
- Protection of Backup Path -- because backup path is handled in the same way as working path.

4 Simulation and Performance Evaluation

In order to simulate the proposed scheme, Haskin's and Makam's, we have extended MPLS Network Simulator[7]. MPLS Network Simulator supports the setup of CR-LSP based on QoS as well as basic MPLS core functions such as label switching, LDP, CR-LDP, and various options of label distribution.

4.1 Simulation Environment

Simulation environment is shown in Fig. 5. Node0 is IP node. The rest of the nodes are MPLS nodes. Each node is connected with a duplex link with the bandwidth 2Mbps, a delay of 10ms, and a CBQ queue.

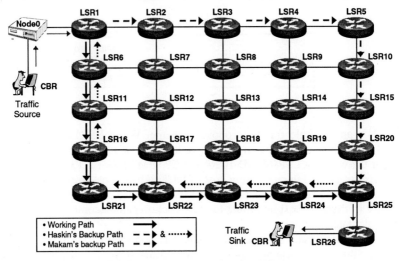

Fig. 5. MPLS networks for simulation

There is one pair of UDP traffic, called CBR. Traffic is injected into Node0 and escape from LSR26. CBR generates 256-byte-long packets at 1Mbps and at constant bit rate. Each node is monitored every 10ms to protect the working path.

In this simulation, we considered node failure instead of link failure as failure type for the purpose of simulating our scheme in worse environment because its performance is worse in node failure than in link failure

4.2 Simulation Results and Evaluation

Several metrics are proposed for performance evaluation of path restoration schemes by [9][13]. Each scheme is measured and evaluated in terms of packet loss, reordering of packets, concurrent faults, and resource utilization.

In our simulation, two kinds of traffic, best-effort traffic and QoS traffic are used as traffic to be protected. A working LSP and a backup LSP for best-effort traffic are setup by using an ER-LSP, which does not consider resource reservation but can utilize the rest bandwidth of the link. A working LSP and a backup LSP for QoS traffic are setup by using a CR-LSP, which is guaranteed the bandwidth required but can not utilize the rest bandwidth of the link.

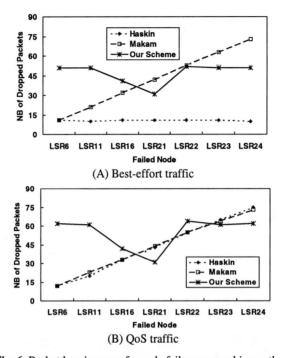

Fig. 6. Packet loss in case of a node failure on working path

Fig. 6 shows a packet loss comparison of each scheme by the location of a failed node. Haskin's scheme and Makam's have a problem that the number of the dropped packets increases more in proportion to the distance between PIL(i.e. LSR1) and a LSR that detected a node failure, in QoS traffic and in best-effort traffic and QoS traffic, respectively.

The proposed scheme also has a problem in packet loss. However, it has almost no connection with failure location in performance as shown in Fig. 6.

Fig. 7 shows a packet re-ordering comparison of each scheme by the location of a failed node. Haskin's scheme has a problem that the number of the re-ordered packets increases more in proportion to the distance between PIL and a LSR that detected a node failure, in best-effort traffic and QoS traffic. Makam's has almost no problem in reordering of packets. The proposed scheme also has almost no problem in packet re-ordering as shown in Fig. 7.

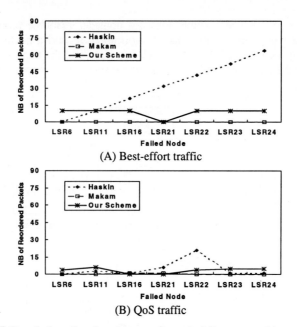

Fig. 7. Reordering of packets in case of a node failure on working path

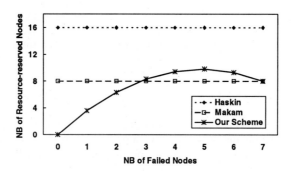

Fig. 8. Resource utilization: Resource-reserved nodes for backup path

Fig. 8 shows resource utilization comparison of each scheme by the number of failed nodes. In this simulation, resource means label, bandwidth, and buffer reserved for backup path. The proposed scheme shows the best performance than any other schemes in resource utilization.

Fig. 9 shows a comparison of each scheme by the number of sequential and concurrent faults. In this simulation, node failures occur from PML toward PIL concurrently and sequentially. As shown in Fig. 9, both Haskin's scheme and Makam's have no connection with concurrent faults. In both schemes, the more the number of failed node increases the more the number of dropped packet decreases. This is because the more the number of failed node increases the closer to PIL the location of failed node is. The proposed scheme shows poor performance in concurrent faults. The reason is that the more the number of failed node increases the more the try count of LSP setup for backup path may increase.

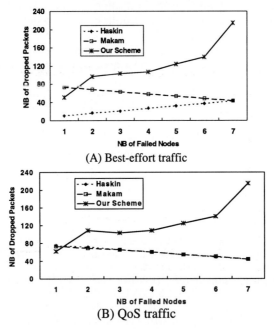

(A) Best-effort traffic

(B) QoS traffic

Fig. 9. Concurrent faults on working path

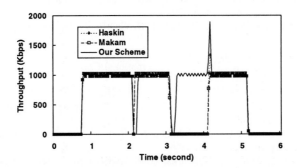

Fig. 10. Node failure on both working path and backup path

Fig. 10 shows throughput of each scheme in case of both working path and backup path failure. In this simulation, a node failure on working path occurs at 2 seconds and a node failure on backup path occurs at 3 seconds. Those failures are recovered at 4 seconds. As shown in Fig. 10, only our scheme protects traffic of backup path as well as that of working path.

5 Conclusion

This paper proposed a dynamic path protection scheme using the least-cost based backup path, which may increase resource utilization and protect backup path without requiring longer restoration times.

In this paper, two ideas in establishing backup path were proposed. One is candidate nodes that may become an egress node of backup path. The other is Least-Cost Backup-Path algorithm used to find a node out of the candidate nodes that is used as splicing node of backup path at the lease cost. Simulation results show the performance of our scheme is better than or similar to that of the existing schemes in case of single failure.

We are planing to simulate and evaluate the proposed scheme by using more comparison criteria such as 1:n protection in large MPLS networks.

References

1. Bilel Jamoussi, "Constraint-Based LSP Setup using LDP," Internet Draft, Oct. 1999.
2. Changcheng Huang, Vishal Sharma, Srinivas Makam, Ken Owens, "A Path Protection/Restoration Mechanism for MPLS Networks," Internet Draft, July 2000.
3. Daniel O. Awduche, Lou Berger, Der-Hwa Gan, Tony Li, Vijay Srinivasan, George Swallow, "RSVP-TE: Extensions to RSVP for LSP Tunnels", draft-ietf-mpls-rsvp-lsp-tunnel-08.txt, Feb. 2001
4. Dimitry Haskin, Ram Krishnan, "A Method for Setting an Alternative Label Switched Paths to Handle Fast Reroute," Internet Draft, May 2000.
5. Eric C. Rosen, Arun Viswanathan, Ross Callon, "Multiprotocol Label Switching Architecture," Internet Draft, April 1999.
6. . Apostolopoulos, D.Williams, S. Kamat, R. Guerin, A. Orda, and T. Przygienda, "QoS Routing Mechanisms and OSPF Extensions," RFC2676, August 1999.
7. Gaeil Ahn and Woojik Chun, "Design and Implementation of MPLS Network Simulator Supporting LDP and CR-LDP," IEEE International Conference on Networks(ICON2000), Singapore, Sep. 2000, pp. 441-446
8. Hae-Joon Shin, Youn-Ky Chung, Young-Tak Kim, "A Restoration Scheme for MPLS Transit Network of Next Generation Internet," in Proceeding of the 14th ICOIN, Taiwan, Jan. 2000, pp. 3D1.1-3D1.5.
9. Loa Andersson, Brad Cain, Bilel Jamoussi, "Requirement Framework for Fast Re-route with MPLS," Internet Draft, June 1999.
10. Nortel Networks, Ennovate Networks, IBM Corp, Cisco Systems, "LDP Specification," Internet Draft, June 1999.
11. Q. Ma and P. Steenkiste, " On Path Selection for Traffic with Bandwidth Guarantees," In Proc. of IEEE International Conference on Network Protocols(ICNP), Oct. 1997.
12. R. Callon at al., "A Framework for Multiprotocol Label Switching," Internet Draft, Sep. 1999.
13. Tellabs, Nortel Networks, AT&T Labs, "Framework for MPLS Based Recovery," Internet Draft, March 2000.
14. Z. Wang and J. Crowcroft, "Quality of Service Routing for Supporting Multimedia Applications," IEEE JSAC, 14(7):1228-1234, Sep. 1996.

Connection Management in MPLS Networks Using Mobile Agents

Sakir Yucel[1] and Tuncay Saydam[2]

[1]Marconi Communications, 1000 Fore Dr. Warrendale, PA 15086 USA
sakir.yucel@marconi.com
[2]University of Delaware, Department of Computer and Information Sciences,
Newark, DE 19716 USA
saydam@cis.udel.edu

Abstract. In this paper, the use of mobile agent technology for creating QoS-driven MPLS tunnels and LSP's is investigated and an architectural framework to integrate related service and network management activities is proposed. Using a MIB definition, details of an example implementation for configuring MPLS network elements over CORBA and JAVA facilities are discussed. Critical evaluation of the proposed approach together with its advantages, issues and comparison to traditional connection management approaches is presented.

1. Introduction

MPLS (Multiprotocol Label Switching) emerged as a new technology that sits between the network and the data link layers in the Internet architecture, enabling traffic-engineering capabilities to provide both qualitative and quantitative QoS. MPLS creates a connection below the IP layer, called the LSP (Label Switched Path), through which the traffic flows. MPLS will also offer better reliability and availability through the protection and restoration of LSPs. With these key capabilities and state of the art management architectures, as we propose in coming sections, service providers will be able to offer value-added services such as the IP based Virtual Private Networks (VPN's) with better than best-effort, in some cases, with guaranteed quality [1] [2].

Service management architectures are needed to effectively create, provision and manage services over different networking domains. Our methodology to manage complex services over MPLS networks will be based on distributed component oriented software engineering and on mobile agent technology. Our architecture is influenced by the CORBA, WEB, TMN and TINA-C frameworks. CORBA and WEB are the middleware environments for providing and accessing the value-added services easily with a comparable low-cost. TMN (Telecommunication Management Network) approach separates the service management from network management. TINA-C (Telecommunication Information Networking Architecture – Consortium) framework promotes development of reusable components for management of services over heterogeneous networks. Mobile code/agent technology enables to bring the management functionalities flexibly at the network element abstraction level. These functionalities include configuration of network elements with software plug-ins and mobile agents on the fly, creation of protocol layers and adapters, and even composition of protocol stacks actively with respect to the QoS requirements of the services running on top of the network elements. The Java Programming Language has become the basic choice for the mobile agent paradigm as a means to migrate the functionality into a network element and from one platform into another, due to its portability and dynamic loading features. However, use of mobile agent technology for creating MPLS tunnels and LSPs, to our knowledge, has not been investigated by researchers yet. In this research, we will propose an architecture that employ mobile agents to manage distributed services with different QoS constraints in MPLS networks and investigate the feasibility of using mobile agent technology in creating MPLS tunnels and LSPs based on service level requirements (SLAs).

2. General Architecture

The architecture we are proposing is a collaborative framework bringing together the object-oriented technologies, mobile agent paradigm, telecommunication management standards such as TMN and TINA-C, and the networking technologies around the middleware solutions, as shown in Figure 1. The middleware solutions are CORBA, WEB and Java. Attached around the middleware are the end user systems, management systems, basic middleware services, agent repository servers and their managers, event servers and their managers, security servers and their managers, and the network elements. Each of these elements is accessible to one another through the use of either CORBA or WEB or both. End user systems belong to the customers who access and use the value-added services. The management systems belong to the value-added service and/or network service providers and help them offer services on top of the physical networks and network elements. The agent repository servers keep the directory of software units that can be dispatched by management systems and sent to network elements to perform dynamic management operations. The event collection and correlation server listens to the events emitted by the network elements, correlates them and notifies the management systems. The security servers provide fine-grained security mechanisms and security services over the middleware. The managers for the mobile component repository server, the event server and the security server perform the configuration of the respective servers. These managers, although not shown in Figure 1 due to space limitations, are also attached around the middleware bus. In other words, our architecture is a collaborative environment having everything as an object around the middleware bus, whereby each element runs the middleware software and is exported into the naming service. To support the seamless communication over middleware, each element contains Java Runtime Environment and runs a CORBA ORB application (usually via the Java Run-time environment). In addition to the CORBA process, management systems and the servers run Java servlets via the interface components to offer their services over the web using the HTTP protocol [7].

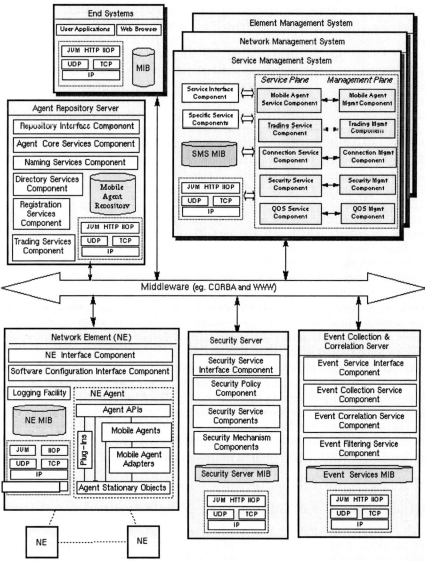

Fig. 1. A Detailed Architecture for QoS-Driven Connection Management Using Mobile Agents

The network elements (NEs) are physical devices such as an IP router, over which the network and value-added services are realized. As shown in Figure 1, an NE contains the device hardware, device drivers, the NE operating system, communication protocol stack, a management agent software (i.e. SNMP) and a MIB (Management Information Base). The latter contains the NE related attributes that could be monitored and modified. In this paper, we will assume that all the NEs are MPLS capable. To handle the diversity of today's NE devices, our framework places a unified abstraction on top of the basic NE so that we can handle different kinds of them. For this purpose, the assumption we make about the network elements is the availability of Java Virtual Machine (JVM). By downloading some packages and classes dynamically into the JVM, we can make an NE a part of the middleware pool and implement variety of management tasks on it in a uniform manner, regardless of its hardware and software specifics. This way we can tackle the scalability and heterogeneity issues of the networking devices. On top of the NE abstractions, the management systems create network-level and service-level abstractions as well, that improves the scalability and the robustness of the resulting system. Most today's networks elements, however, are not accessible through CORBA and are not capable of running JVM. Access to such network elements will be achieved via the SNMP, LDAP (Light-weight Directory Access Protocol), signaling protocols or command line interfaces, instead of CORBA. These communication mechanisms form a "network management middleware" for communicating with the network element in addition to the common middleware like CORBA and WEB. For NEs incapable of running JVM, element management systems will provide "agencies" to execute dynamic Java code, which will communicate with NEs using network management middleware communication mechanisms.

Agent repository servers store many different kinds of objects. Three particularly important to be used in this paper are stationary agents, mobile agents and agent systems. A *stationary agent (plug-in)* is an object that can be uploaded to a management system or to an MLPS network element and stay at the platform where they were installed. A *mobile agent (mobilet)* is an autonomous active agent that travels over some managed platforms acting on behalf of a management system to perform some management tasks. Mobilets are autonomous and can be intelligent to make decisions on their own, i.e. what to do at the managed platform, where and when to move to another platform. Along the way, mobilets may report to the manager about the problems in the local managed platforms that they encounter. An agent system is an application or a thread with a set of Java language classes that can receive, authenticate, run and relocate agents. Both plug-ins and mobilets can carry serialized Java bytecode or they are described in XML. When plug-ins and mobiles are described in XML, agent systems instantiate them out of the XML description.

3. Connection Establishment Using Mobile Agents

We are developing a proof-of-concept prototype system to investigate the research ideas with respect to the advantages, disadvantages, limitations and problems of mobilets in the field of distributed service management in general. Specifically for the purpose of this paper, the prototype should provide an environment for testing the suitability of mobile agent technology in establishing MPLS connections. The prototype system is specified and designed by using UML (Unified Modeling Language) and implemented with Java2 platform APIs, RMI and IDL-RMI packages, Java Reflection API, JavaBeans component model and the InfoBus. The repositories at the management systems and at the repository servers are implemented with JNDI (Java Naming and Directory Interface) over LDAP compatibly with DEN (Directory Enabled Network) and CIM (Common Information Model) design initiatives. The architecture makes extensive use of dynamic object and agent creation on demand employing document driven programming (DDP) techniques for constructing objects out of XML specifications [8]. Stationary and mobile agents are specified using XML. JAXP API is used for processing XML and implementing the DDP techniques.

We are developing a number of connection management related plug-ins and mobile agents. These plug-ins and mobilets provide the functionality for connection establishment, connection reestablishment, connection information collection and connection release (teardown). In our previous work, we sketched the detailed algorithm for creating an ATM PVC using a Connection Establishment Mobilet (CE-Mob) [7] [9]. Similar to our ATM work, a CE-Mob can be programmed to create an LPS, a tunnel or a hierarchy of tunnels in an MPLS network. For the connection establishment in the following example, end-to-end LSP over a number of MPLS domains is to be created, with one CE-Mob for each MPLS domain creating an LSP within that domain. Currently, there is no standard MIB, similar to ATOMMIB, for MPLS router. There are, however, Internet drafts [3] [4] [5], as "work in progress", that define managed objects for MPLS LSRs, for MPLS traffic engineering and for MPLS LDP (Label Distribution Protocol) respectively. We decided to base the implementation of a CE-Mob on the first two Internet drafts. New CE-Mobs can easily be programmed to work on different MIB definitions, or, an adapter object is installed to map the [4] MIB definitions into the proprietary MIB definition. First, we outline creation of LSP using in an MPLS domain using the [4] MIB definition.

When the connection establishment mobilet arrives in an LSR, it goes through the following steps: (1) it specifies traffic parameters for both input and output segments, (2) it reserves and configures in and out segments, (3) cross connects the in and out segments, and (4) specifies label stack actions, if any.

To perform the first action, mobilet should make sure there is enough bandwidth available in the router by comparing the required bandwidth with the mplsInterfaceAvailableBandwidth object of the mplsInterfaceConfTable table. If this comparison fails, the mobile agent may report this to its manager via the *communicateMgmtInfo()* method, depending on the error checking and reporting policy. For example, if the itinerary (the path) of the mobile agent was assigned by the connection management component, the mobile agent reports the situation and waits for the new itinerary from the manager. On the other hand, if the mobile agent is permissible for changing the connection path, it backtracks to the previous LSR to try another LSR. If there is enough bandwidth, the mobilet selects a new mplsTrafficParamEntry with {mplsTrafficParamIndex, mplsTrafficParamMaxRate, mplsTrafficParamMeanRate, mplsTrafficParamMaxBurstSize, mplsTrafficParamRowStatus, mplsTrafficParamStorageType} attributes, assigns a unique id to the mplsTrafficParamIndex, assigns the traffic parameters to the next three attributes and assigns createAndGo(4) to the 4th attribute. mplsTrafficParamStorageType is left alone. Mobilet checks if this operation is successful or not. If not successful, it handles it likewise. If successful, it continues with the set up of mplsTrafficParamEntry for the output direction.

After the traffic parameters have been specified, the mobilet creates in and out segment entries. For the in segment, mobilet creates an mplsInSegmentEntry with {mplsInSegmentIfIndex, mplsInSegmentLabel, mplsInSegmentNPop, mplsInSegmentAddrFamily, mplsInSegmentXCIndex, mplsInSegmentOwner, mplsInSegmentTrafficParamPtr, mplsInSegmentRowStatus, mplsInSegmentStorageType} attributes. It assigns the incoming interface number and the incoming label value to the first two attributes, respectively. mplsInSegmentNPop is assigned the number of labels to pop from the incoming packets, mplsInSegmentOwner is assigned with part of the certificate of the mobilet to indicate that the owner of this LSP is the connection management component that deployed this mobilet. mplsInSegmentTrafficParamPtr is assigned with the index of the mplsTrafficParamEntry created for the in segment. Finally, mobilet assigns createAndGo(4) to the mplsInSegmentRowStatus attribute. Other attributes are left alone to their default values. The mplsInSegmentXCIndex attribute will be assigned the index of the cross connect entry later on. Mobilet handles the failure of this operation, if any, otherwise continues with the creation of out segment entry in a similar way.

The next operation is the creation of cross connect entry for switching in and out segments. Mobilet creates an mplsXCEntry with { mplsXCIndex, mplsXCLspId, mplsXCLabelStackIndex, mplsXCIsPersistent, mplsXCOwner, mplsXCRowStatus, mplsXCStorageType, mplsXCAdminStatus, mplsXCOperStatus} attributes. mplsXCIsPersistent is assigned true if the LSP is desired to be automatically restored after a failure. Mobilet assigns values to appropriate attributes for associating in and out segments that it created. When done with it, mplsInSegmentXCIndex in the in segment and mplsOutSegmentXCIndex in out segment will be assigned to point to the newly created cross connect entry. Mobilet handles the failure of this operation similarly to above cases. If the operation is successful, it continues with specifying label actions, if any, otherwise it is done at this LSR and migrates to the next one in its path.

If any label actions are to be specified, mobilet assigns the index of the mplsLabelStackEntry to the mplsXCLabelStackIndex attribute of the cross connect entry. An mplsLabelStackEntry is created with mplsLabelStackIndex and assigning the label value to mplsLabelStackLabel attribute. mplsLabelStackRowStatus is assigned createAndWait(4) similarly to create the entry. Upon success of this operation, the mobilet is done with this LSR and is ready to migrate to the next one with the help of the agent system.

Operations in an LER is very similar to LSR except that the connection management component handles the last step for assigning label operations in LERs, as opposed to mobilet performing it in the core of the network in LSRs.

Another CE-Mob is programmed to create an MPLS tunnel operating on the managed objects defined in [5]. Basically, this CE-Mob configures the attributes the objects such as mplsTunnelTable, mplsTunnelResourceTable, mplsTunnelHopTable in similar way to the example above.

4. Critical Evaluation and Conclusions

Looking at the implementation example, we see that it is very difficult to justify the use of mobile agent technology based on dynamic code for creation of LSPs and tunnels as of today, as opposed to our conclusions for the ATM technology. Mobile agent approach is justifiable if mobile agent brings intelligence into the NE. As we see, example implementation on a proposed MIB definition doesn't exhibit much intelligence other than making some failure handling. Obviously, we cannot see many of the benefits of applying mobile agent technology for PVCs here in creating LSPs. The reason is basically MPLS is in a very immature stage compared to ATM traffic management. MPLS doesn't support delay constraints like CTD, CDV, CVDT as in

ATM QoS specification and doesn't impose queuing, buffering, scheduling, policing and shaping mechanisms (RSVP supports guaranteed service and controlled-load service via admission control, classifier and packet scheduler modules for applications requiring bounded delay or enhanced best effort services. However, these are still incomparable to ATM traffic management). In ATM, these are well-understood and well presented as configurable MIB objects, and mobile agent technology make use of these to create a PVC with optimum utilization of local switch resources. On the other hand, it seems the best a mobile agent can do in MPLS using an available MIB definition is not much different from what a signaling protocol does, which is essentially creating an MPLS with just the bandwidth reservations.

Only few advantages over signaling and classical management provisioning are notable with this implementation. One is that mobile agent approach is flexible with respect to error checking and decision making within the network elements. For example, the mobilet can make adaptive decisions by itself, as to which route to select based on the requested QoS parameters and the availability of local resources. This minimizes the number of primitives passed between the management system and the network element, and therefore reduces the overall bandwidth requirement needed for connection provisioning. Mobile agent can make intelligent decision based on the characteristics, policies, constraints of the provisioned service and the available resources in the local element in selecting its route. Another is the improved scalability, by delegating the connection provisioning to an autonomous mobilet. Mobile agent approach for service management is usually remarkable, as in this case, in reducing the load of the management systems when they would be involved in many provisioned connections at the same time.

Management system's decision on whether to use signaling, classical provisioning or a mobilet is also based on the type of the LSP to be created. Although mobilets can be used in place of manual provisioning as well as signaling, their use should be justified since they introduce overhead. For P-LSP (permanent LSP spanning over long terms like days, weeks, more) and for SP-LSP (signaled permanent LSP), it would choose to deploy a mobilet that would map the customer SLA onto NE as well as the management policies if implementing some policies is desired. For S-LSP (Signaled LSP spanning over short term like minutes, hours) type of connections that don't require QoS, it would use signaling protocols. If DiffServ is running over MPLS, management system (NMS) would create two types of LSPs, E-LSPs and L-LSPs, using mobilets if bandwidth allocation is required during establishment. The first one, E-LSP, supports up to eight behavior aggregates for a given FEC with any number of ordered aggregates, where an ordered aggregate is a set of behavior aggregates that share an ordering constraint. NMS configures the EXP bits of the MPLS header to point to correct PHB for E-LSP type connections. With the second one, L-LSPs, a separate LSP is established for each of the forwarding equivalence class to ordered aggregate pair. Specifically implemented mobilets make these operations transparent to the management system.

One of the many advantages of stationary and mobile agents is the ability to add new functionality on demand into a network element by downloading dynamic code from a directory server, i.e. repository server. In spite of many efforts, downloading mobile agent in the form of a dynamically loadable code as in Java at the network element level is still unfeasible at this time. Basically, network elements are not designed and engineered to support such agent execution environments (i.e. agent system) at all. Main control processors implement the control and management plane functionalities in the most efficient way running specialized software on proprietary (i.e. CISCO IOS) or real-time operating systems (i.e. VxWorks, real-time Linux). For scalability purposes, destination lookup (based on IP, MPLS label, ATM VC/VP ids), prefix matching and packet classification as well as policing, marking, shaping and RM aggregation operations are implemented in the ingress hardware by special ASICs and/or network processors. The code for network processors is usually developed in microcode in its assembly language and optimized for efficiency reasons. Similarly, queuing, scheduling and other traffic management functions at the output side are performed usually by special ASICs. A dynamically loaded agent in an NE cannot do much, other than changing the configuration parameters of these ASICs. Besides, the idea of running a dynamically loadable code of Java or spawning a new object out of XML definition in the main switch/router processor is nonsense in the views of equipment vendors because their main concern is to be able to run their own code in the hardware rather than someone else's dynamically loadable code. On the other hand, command line scripting is usually supported by the equipment vendors and is widely utilized by service providers for configuration and diagnostics of network elements.

We believe, an agent approach based on command line scripting rather than dynamically loadable code for configuration of LSRs and LERs to create LSPs and tunnels is a more feasible approach than having dynamic Java code. This approach can be realized in the form of uploading scripts on designated LEs, or migrating a script around the NEs to be configured. Scripting eliminates the operator effort to manually create tunnels and LPSs over a set of LSRs. Scripts are written for equipments of different vendors to configure their proprietary features and configuration parameters as opposed to programming mobile agent code in Java to work with those proprietary characteristics. Basically, it doesn't come with the overhead of providing an agent execution environment and the security issues related to dynamic code loading. Our architecture efficiently supports storing, querying and deploying such mobile agents as well. If we go into this direction, we can still specify the plug-ins and mobilets in XML but it is required to translate the action part of agent's XML description into the command line scripts of the destination platforms.

Scripting, however, doesn't introduce new functionality but utilizes the available ones in the NE. Placing agencies that can provide agent execution environment near network elements is another way of introducing dynamic functionality into the network. Element Management stations and hosts in equipment rooms are perfect candidates for these agencies. This way, agent execution environments are created in a more flexible operating system (i.e. Linux and UNIX stations found in the equipment rooms). The mobile agent in the form of dynamically loadable Java bytecode travels over the agencies in a system and communicates with the network elements via some communication mechanism (i.e. SNMP, LDAP, signaling, scripting). The agencies make it possible to introduce new functionality by installing some stationary agents, such as an agent implementation of bandwidth broker and exporting them to CORBA naming service. Employing policy agents into the network is another immediate application of this approach. Our framework also perfectly supports this approach.

Mobile agent based connection establishment has a direct impact on performance. Some of the main issues regarding performance are:
- For time-critical and short-term connection management, our architecture's added overhead may not be suitable. In such as situation, a signaling protocol based connection establishment would be preferred.
- Management systems interact with the servers during the service creation. However, some servers will not be used for every time a connection requested. The specific deployment of the architecture should be done in such a fashion that the communication overhead could be reduced. For example, anticipated mobile agents should be downloaded and cached in advance to reduce the need to access the repository server.
- Mobile agents distribute functionalities and bring them to nodes where they are needed. They reduce network traffic around the management systems compared to classical management-agent paradigm. On the other hand, mobile agent paradigm adds overhead and complexity in terms of the management of the mobile agents. The more autonomous the agents are, the heavier they become. Mobile agents should be leanly constructed, and their data and protocol needs should provided by the agent systems.
- It is feasible to create agencies at the EMSs (in Figure 1) providing agent execution environments for both stationary and mobile agents, thus, introducing dynamic functionality into the network. Also, it is feasible to deploy mobile agents in the form of command-line scripts to configure the LERs and LSRs in an MPLS enabled network for traffic engineering and management purposes.

However, our architecture has several important advantages over the classical connection management approaches. Some observations are:
- In a service involving multiple networks, parallel and independent execution of mobile agent, in separate domains, improves the performance significantly. There is also the flexibility in providing different QoS requirements in different networks by use of mobile agents.
- Our architecture is not optimal for a small simple network. However, nowadays, end-to-end applications traverse multiple complex and heterogeneous networks, thereby requiring services that can only be supplied through hierarchical architectures.
- The agent-based connection approach makes the implementation of complex QoS-dependent service contracts much simpler.
- QoS driven connections add distinct value to services. However, they need extra service interactions among the user, the network and the service providers.
- Mobile agents provide local flexibility in selecting routes, evaluating node resources and aggregating QoS. The mobile agents could be implemented to map the given service level QoS parameters onto the MIB objects of the network elements in the most optimum way. The required knowledge for the node's specific needs shift from the classical manager onto a mobile agent. Therefore the specific details of network nodes become more transparent to management systems.

References

1. MPLS Working Group, http://www.ietf.org/html.charters/mpls-charter.html
2. MPLS Forum, http://www.mplsforum.org/
3. Cucchiara, J., Sjostrand, H., Luciani, J., "Definitions of Managed Objects for the Multiprotocol Label Switching, Label Distribution Protocol (LDP)", Internet draft, draft-ietf-mpls-ldp-mib-07.txt
4. Srinivasan, C., Viswanathan, A., Nadeau, T. "MPLS Label Switch Router Management Information Base Using SMIv2", Internet draft, draft-ietf-mpls-lsr-mib-07.txt.
5. Srinivasan, C., Viswanathan, A., Nadeau, T. "MPLS Traffic Engineering Management Information Base Using SMIv2", Internet draft, draft-ietf-mpls-te-mib-06.txt.
6. Java Products and API, http://java.sun.com/products/
7. Yucel,S. Saydam,T., Mayorga,A. "A QoS-driven Connection Management Architecture Using Mobile Agents", under review.
8. Yucel,S. Saydam, T. "SLA Specification Using XML for Services Over IP Technologies", under review.
9. Yucel,S. Saydam.T., "QOS Management of Multimedia Services Using ATM MIB Objects", *Journal of Networking and Information Systems, vol2, n2,* July 1999.

General Connection Blocking Bounds and an Implication of Billing for Provisioned Label-Switched Routes in an MPLS Internet Cloud

George Kesidis[1] and Leandros Tassiulas[2]

[1] EE and CS&E Depts, Pennsylvania State University
University Park, PA, 16802, USA
kesidis@engr.psu.edu

[2] E&CE Dept, University of Maryland
College Park, MD, 20742, USA
leandros@eng.umd.edu

Abstract. Two related problems of traffic shaping are considered in this paper. Both are applicable to the context of provisioned MPLS label-switched routes. The first involves a novel application of traffic shaping to connection-level arrival processes. A type of Erlang-B bound is derived. The second involves the transmission of secondary best-effort data traffic into a provisioned variable bit-rate (VBR) "channel" handling a primary (real-time) flow.

1 Introduction

In this paper, we consider an MPLS label-switched route (LSR) with explicitly provisioned bandwidth. In this first part of this paper, we suppose that a number of separate connections are using the LSR. A traffic shaper is applied to the *connection* arrival process to control connection blocking. A bound similar in function to the Erlang-B is derived for an M/GI/∞ queue and its utility is explored.

In the second part of this paper, we suppose that the provisioned LSR is used by a single (perhaps aggregated) connection with two leaky bucket constraints on the *packet*-level flow, as would be the case in ATM's VBR or diffserv's Assured Forwarding (AF) traffic classes. A simple method for moving best-effort (BE) packets *within* (i.e., "inband") the LSR is described. A condition on billing which makes this practice profitable is also given.

2 Connection-Level Shaping to Obtain Connection Blocking Bounds

We consider the problem of reducing the number of blocked connections in an Erlang-type loss network. One extreme is not to inhibit the arriving traffic (connection set-up requests) at all; this leads to blocking probabilities given by Erlang's formula for the case of a network of $M/M/C/C$ queues [6, 14]. The other extreme is to put a queue at each source (network access point) and have the connections wait until a circuit is available. In this case, a connection is never blocked. However, this requires complex network monitoring of available circuits and constant communication with the network access points. Also, connection set-up delay may be excessive.

The approach proposed in this paper is somewhat of a compromise. We suggest the use of traffic shapers at the network access points that make "smooth" the arrivals so that the servers receive connection requests at more regular intervals, thus decreasing the chances of call rejection. This method does not need any network information. However, connection requests may experience an additional delay in the shaper queue. We study the tradeoff between queuing delay and blocking probability via simulations.

In Figure 1, a "single node" situation is depicted. Let $M(s,t]$ be the number of connection set-up requests (simply called "connections") that arrive to the shaper during the continuous time interval $(s,t]$. The shaper can queue up to K connections. If a connection arrives to the shaper to find it full (i.e., K connections queued), the connection is blocked. Let $N(s,t]$ be the number of connections departing the shaper over $(s,t]$.

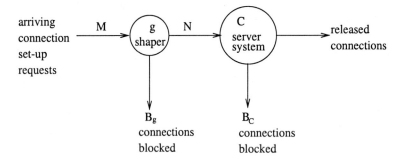

Fig. 1. A traffic shaper followed by a C-server loss system

The shaper operates so that

$$N(s,t] \leq g(t-s) \quad \text{a.s.}$$

for all $t \geq s$ where g is a strictly increasing function on $\mathbf{R}^+$ such that $g(0) = 1$ and $g(0-) \equiv 0$. Using tandem leaky buckets we can, for example, implement $g(x) = \max\{1 + \pi x, \sigma + \rho x\}$ for $x \geq 0$ where $\sigma > 1$ and $0 < \rho < \pi$.

The scenario of Figure 1 is also relevant to ATM virtual path (VP) switching and to label-switched routes (LSRs) in a MPLS Internet domain. In both cases, connections may be set up by "servers" in the context of limited available capacity. For a specific example, consider a LSR which has been set-up through an

MPLS domain using an RSVP-type protocol so that the LSR is assigned a finite amount of bandwidth B. The LSR is used by a "service" provider for connections each requiring a bandwidth $b \ll B$. The number of "circuits" or "servers" of this LSR is simply $C = B/b$.

The use of the traffic shaper also allows us to study the problem of connection blocking for a very wide variety of connection arrival processes. In modern communication network settings, the incident traffic M is often unknown and difficult to predict. Note that the shaper does not take into account the potential holding times of the connections. When a connection is blocked by the shaper, the corresponding user would quickly receive a "fast busy signal". A less desirable situation occurs if the user's connection set-up request passes through the shaper (at the expense of some delay) only to be blocked by the "C-server system".

We now focus on the C-server system. This system has no associated queue. For the time being, we consider a $G/G/C/C$ system with arbitrary interarrival and service distributions. The ith connection arrives to this system at time T_i and has duration (holding time) S_i, i.e., if this connection is not blocked and obtains a server, it will occupy the server for S_i seconds. Further suppose that all the connections have a maximum holding time (duration) $S^{\max} \leq \infty$, i.e., for all i, $S_i \leq S^{\max}$ a.s. Finally, assume that all holding times are identically distributed. If $g(S^{\max}) \leq C$, then the C-server (G/G/C/C) system will never block connections [3, 10].

Now consider a G/GI/∞ system with connection arrival process N (identical to that of our C-server system in Figure 1) and i.i.d. connection holding times. Let Q be distributed as the steady-state occupancy of this infinite server system. We approximate the connection blocking probability of the C-server system by $P(Q > C)$. In [10] we showed that $EQ \leq Eg(S)$ and, using the Chernoff inequality, we further showed that

$$P(Q > C) \leq \exp\left(-\sup_{\theta > 0}\{\theta C - \int_0^\infty \log(\Phi(\gamma)e^\theta + 1 - \Phi(\gamma))d\gamma\}\right) \quad (1)$$

where $\Phi(x) \equiv P(g(S_i) > x)$.

It was demonstrated in [2] that an improper choice of shaper parameters could lead to *greater* blocking and/or excessive set-up delay; indeed, in an extreme situation, one can arrange for the shaper queue to be unstable. To avoid this, sample measurements of the arriving traffic (M) could be used to estimate parameters such as the mean arrival rate λ. These estimated parameters could then be used to tune the shaper parameters into an appropriate regime which is stable in particular. We also tested a modified version of the leaky bucket shaper with finite connection buffer in which connections were presented to the C-server system without tokens whenever the shaper itself was about to block a connection. With this modified shaper, overall blocking probabilities were obtained that were close to the "overflow" probabilities of the infinite server case (1) and the delay performance was similar to the "unmodified" shaper with a finite connection buffer.

3 An Implication of Pricing on a Single Provisioned Connection

We now shift our focus to a single connection (circuit). We assume that the packets of this connection are "delay sensitive" and that they are transmitted at a variable bit-rate. We further assume that bandwidth and buffer memory has been explicitly provisioned for this connection and that the traffic is "shaped" as it enters the network. Unlike the previous section, the traffic is taken to be a stream of fixed-length *packets* instead of a stream of connection set-up requests.

More specifically, suppose a user has purchased a rate-regulated channel (perhaps as an MPLS LSR) specified by the traffic descriptors (σ, ρ, π) and quality-of-service (QoS) parameters (ε, δ) which are interpreted as follows. For *any* interval of time (s, t), the total number of packets transmitted by the user in that interval, $A(s, t)$, must satisfy

$$A(s,t) \leq \min\{\sigma + \rho(t-s), 1 + \pi(t-s)\} \quad (2)$$

where $\sigma > 1$ has dimensions packets and $0 < \rho < \pi$ have dimensions packets/s, i.e., A is both (σ, ρ) and π-peak-rate "constrained".[1] In addition, the end-to-end delay D experienced by a typical packet of the connection must satisfy $P(D \geq \delta + \Pi) \leq \varepsilon$ where $\Pi + \delta > 0$ is a critical delay threshold and Π represents the deterministic end-to-end propagation and processing delays.

Let $d = D - \Pi$. Assume that the cumulative *queueing* delay experienced by the packets in the network backbone is negligible [12], i.e., the only significant queueing delay is at the network access point. Alternatively, assume that the cumulative queueing delay in the network backbone can be bounded by a small constant (that would be a component of Π) as is the case when "guaranteed-rate" scheduling is used in the backbone switches, see [8, 9] and the references therein. Let c be the *allocated* amount of bandwidth for the connection. Therefore, we can represent the QoS requirement of the connection as

$$P(d \geq \delta) \leq \varepsilon \quad (3)$$

where d is interpreted as the delay through a queue with constant service rate c packets/s.

Invoking a recent "worst-case" bound for a corresponding "fluid" queue (see [11] and the references therein), we get the following inequality:

$$P(d \geq \delta) \leq \frac{\sigma - \zeta^{-1}\delta c}{\frac{c}{\rho}\sigma - \delta c} \quad (4)$$

where $\zeta = (\pi - c)/(\pi - \rho) < 1$ (the queueing delay d is deterministically bounded above by $\zeta\sigma/c$). We can therefore equate

$$\frac{\sigma - \zeta^{-1}\delta c}{\frac{c}{\rho}\sigma - \delta c} = \varepsilon \quad (5)$$

---

[1] In IETF guidelines, σ and ρ are known as the token buffer size and token rate, respectively. In ATM standards, σ and ρ are known as the sustainable packet rate (SCR) and (roughly) the maximum burst size (MBS), respectively.

and solve for the *worst-case* bandwidth allotment c of the VBR channel specified by (2) and (3).

3.1 General Pricing Framework for Connections

Now consider an arbitrary connection in a communication network. The user's cost in dollars is [7, 16]

$$\alpha + \beta T + \gamma N = \alpha + (\beta + \gamma m)T$$

where $\alpha \geq 0$ is the fixed connection set-up cost in dollars, T is the total time duration of the call in seconds, N is the total number of packets, and $m = N/T$ is the mean rate of transmission in packets/s. The quantities β and γ will also vary according to the time of day, day of the week, etc. In addition, the quantity β [$/s] may depend on the declared traffic descriptors. Indeed, let r represent the "worst-case" *a priori* required allocation of bandwidth resources, e.g., obtain r from the solution c to (5). In this paper, we will use the more general pricing formula $\alpha + f(r, m)T$ where the function f has the simple property that it is jointly nondecreasing in both arguments.

3.2 Transmission of Data Inband

The user transmits a *primary* delay-sensitive traffic stream A_1 that satisfies (2). Assuming no usage-based costs, the user has an incentive to add a secondary best-effort traffic stream A_2. For example, the primary stream might consist of a sequence of video frames and the secondary stream might consist of data (e.g., email). In the following, we will explain how the user can add the secondary stream so that: the aggregate stream $A_1 + A_2$ also satisfies (2), the QoS of A_1 is not affected, and the duration of the connection T does not change.

A simple condition on f will now be found that will determine whether the best-effort process A_2 is cheaper to transmit as a secondary process in a rate-regulated channel or as a primary process in a separate best-effort connection. Suppose that the average primary transmission rate is m_1 cells/s and that the total primary transmission is N_1 cells, i.e., $A_1(0,T) = N_1$ and $m_1 T = N_1$. Assuming that the set-up cost of a best-effort connection is negligible, the total cost quantities that are to be compared are $\alpha + f(r_1, m_1 + m_2)T$ and

$$\alpha + f(r_1, m_1)T + f(0, m_2)N_2 = \alpha + (f(r_1, m_1) + f(0, m_2))T.$$

Thus, if and only if the following condition is holds, it is cheaper to transmit best-effort cells as a secondary stream in a shared rate-regulated channel:

$$f(0, m_2) > f(r_1, m_1 + m_2) - f(r_1, m_1). \tag{6}$$

For the case where $f(0, m) \equiv \gamma m$, this condition can be expressed as:

$$\gamma > \frac{f(r_1, m_1 + m_2) - f(r_1, m_1)}{m_2} \approx \frac{\partial f}{\partial m}(r_1, m_1) \tag{7}$$

where the last approximation holds when m_2 is relatively small. Note that if the cost per unit time does not depend on the usage (m), the right-hand-side of (7) is zero and, consequently, (7) holds. There is significant debate as to whether usage-based pricing for the Internet is feasible or "scalable".

Under (7) we will now explain how the user can add the secondary stream to the primary one without affecting the QoS of the primary cells or incurring additional cost. Take time to be slotted where the time-duration of each slot is π^{-1} seconds. By ensuring that no more than one cell per slot is transmitted, we ensure that the peak-rate constraint is not violated.

Now consider an *emulated* queue operating in discrete time with the unit of discrete time being π^{-1} seconds. The arrival process is $A = A_1 + A_2$ and the constant service rate is ρ/π cells per unit time (ρ cells/s). Let X be the occupancy of this queue at slot n. Let d_i the departure time (in slots) of the ith cell to arrive to the queue and let a_i be the arrival time of this cell. We interpret the constant service rate of the queue X in the following way:

$$d_i = \lceil \mathcal{F}_i \rceil \text{ where } \mathcal{F}_i = \max\{a_i, \mathcal{F}_{i-1}\} + \frac{\rho}{\pi}$$

and $\mathcal{F}_0 = 0$. Note that, if the ith cell arrives to an empty queue, $d_i = a_i + \lceil \rho/\pi \rceil$. We also have that [8]

$$X(n) = \left\lceil \max_{0 \leq m \leq n} A[m,n] - \rho(n-m+1) \right\rceil.$$

Similarly, let X_1 be the occupancy of the queue with arrivals A_1 and constant service rate ρ.

The quantity $\sigma - X_1(n)$ is the *slackness* in the (σ, ρ) constraint of A_1 at time n. We want to add the secondary cells so as not to cause a cell of the primary process A_1 to violate the (σ, ρ) constraint sometime in the future, i.e., we do not want the addition of a secondary to cause a future primary cell to arrive when $X = \sigma$ (when the emulated queue is full).

Consider the busy periods and idle periods of X_1. Let I represent the total length of all the idle periods of X_1. A simple approach would be to add secondary cells during the *idle periods* of X_1 so that the busy periods of X_1 are not "affected". Note that this requires knowledge of when the *next* busy period of X_1 is to begin and, consequently, is a "slightly" noncausal rule for inserting secondary cells. Because the *allocated* bandwidth in the network $c > \rho$, the busy periods of the primary cells will be contained in those of the corresponding queue serviced at rate c; therefore, the primary cells will otherwise not be affected by the insertion of secondary cells in this manner. Let A_2^0 represent the maximal secondary arrival process constructed this way.

Claim: For *any* secondary arrival process A_2 that does not cause the duration T of the VBR connection to increase,

$$N_2 \equiv A_2(0,T) \leq A_2^0(0,T) = \rho I.$$

Proof: Suppose a secondary cell is inserted during a busy period at (empty) slot n. The busy period of X will consequently be extended by $\lceil \pi/\rho \rceil$ slots

($1/\rho$ seconds) and X will increase by one from n to just before the end of the (extended) busy period, see Figure 2. Since we require that the total duration of the VBR connection not be extended, we can insert at most ρI secondary cells. □

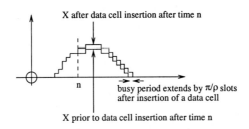

Fig. 2. Effect of inserting a best-effort cell

So, the simple method of inserting A_2^0 is optimal, requires minimal knowledge of future arrivals of primary cells, and does not change the maximum of X, i.e., does not cause the traffic constraint (2) to be violated (assuming the primary transmission satisfies (2)). Therefore, transmission of A_2^0 is recommended. To summarize, our two required conditions for inserting a secondary cell at time n are:

1. no primary cell is transmitted in the nth slot.
2. $X_1(m) = 0$ for all $m \in [n, d]$ where d is the departure time of the cell inserted at time n, $d = n + \lceil \rho/\pi \rceil$.

Note that by transmitting A_2^0, the queue X (with arrival process $A = A_1 + A_2^0$) has a single busy period encompassing all of $[0, T]$. So, even if the connection was allowed to be slightly extended in duration by the insertion of best-effort cells, only an additional σ cells could be inserted in $[0, T]$ before the maximum value of X exceeds σ, i.e., the (σ, ρ) constraint is violated. We also note that the emulated queue mechanism (X_1) required for the above two conditions may already be present in the user's traffic shaper.

Finally, best-effort data can also be transmitted in-band in a CBR channel. The rule is simpler because only step 1 above is required. The received QoS of the primary traffic using the CBR channel is not affected because the network basically allocates peak rate, i.e., $c = \pi$.

To evaluate the throughput of best-effort traffic using in-band insertion, consider again the queue X_1 defined above. A classical result of queueing theory states that [1] $P(X_1 = 0) = 1 - m_1/\rho \approx I/T$ in steady-state, where the approximate equality is basically an "ergodicity" assumption on X_1. So, over this duration, roughly $\rho I = \rho T(1 - m_1/\rho) = T(\rho - m_1)$ best-effort cells are inserted in-band by the method described above. Thus, the throughput of best-effort cells is $\rho - m_1$ cells/s. Clearly, by (2), $m_1 \equiv A_1(0, T)/T \leq \rho + \sigma/T$. The quantity σ/T

is expected to be negligibly small since σ represents the a maximum number of primary cells in *short term* "burst". Thus, $m_1 \leq \rho$.

For the "extremal" process A_1 [11] that achieves the bound (4) subject to (2) and (3), $m_1 = \rho$ and $\mathsf{P}(X_1 = 0) = 0$. So, as expected, an extremal primary transmission would leave no room for the insertion of best-effort cells since there is no unused bandwidth capacity. Some authors have considered the problem of choosing suitable standard traffic descriptors for a given trace of VBR (typically video) traffic, see, e.g., [17]; but, for pre-recorded video, piecewise-CBR service has been argued [13, 15]. Of course, for real-time flows, the user may not have a priori knowledge of the mean transmission rate m_1. Moreover, the high degree of statistical complexity of real-time VBR traffic [5, 4] is often not capturable by the standard traffic descriptors. Thus, the user may choose ρ conservatively larger than m_1 and, thereby, allow a substantial amount of best-effort cells to be transmitted in-band.

Finally, we note that existing model-based techniques for connection admission control of VBR flows may be compromised by inband data transmission. With inband data, VBR flows are "smoother" and, therefore, less statistical multiplexing gains may be achievable by the network.

4 Summary

Two related problems of traffic shaping were considered. Both are applicable in the context of MPLS label-switched routes. The first involved a novel application of traffic shaping to connection-level arrival processes. A type of Erlang-B bound was derived. The second involved the transmission of secondary best-effort data traffic into a provisioned variable bit-rate (VBR) "channel" handling a primary (real-time) flow.

Acknowledgements

We wish to thank K. Chakraborty for the simulation results of Section 2 and M. Beshai of Nortel Networks for his sponsorship.

References

1. F. Baccelli and P. Bremaud. *Elements of Queueing Theory.* Springer-Verlag, Application of Mathematics: Stochastic Modelling and Applied Probability, No. 26, New York, NY, 1991.
2. K. Chakraborty, L. Tassiulas, and G. Kesidis. Reducing connection blocking likelihoods by traffic shaping. Technical report, University of Maryland I.S.R. Technical Report, 2000.
3. R.L. Cruz. Quality of service guarantees in virtual circuit switched networks. *IEEE JSAC*, Vol. 13, No. 6:pages 1048–1056, Aug. 1995.
4. M.W. Garret and W. Willinger. Analysis, modeling and generation of self-similar VBR video traffic. In *Proc. ACM SIGCOMM*, pages 269–280, 1994.

5. D. Heyman, A. Tabatabai, and T.V. Lakshman. Statistical analysis and simulation study of video teleconferencing traffic in ATM networks. *IEEE Trans. Circuits and Systems for Video Tech.*, Vol. 2, No. 1:pages 49–59, March 1992.
6. F.P. Kelly. Effective bandwidths of multi-class queues. *Queueing Systems*, Vol. 9, No. 1:pp. 5–16, 1991.
7. F.P. Kelly. Charging and accounting for bursty connections. In *Internet Economics (Editors L.W. McKnight and J.P. Bailey)*, pages 253–278. MIT Press, 1997.
8. G. Kesidis. *ATM Network Performance.* Kluwer Academic Publishers, Boston, MA, Second Edition, 1999.
9. G. Kesidis. Scalable resources management for MPLS over diffserv. In Proc. SPIE ITCom 2001, Denver, CO, Aug. 2001.
10. G. Kesidis, K. Chakraborty, and L. Tassiulas. Traffic shaping for a loss system. *IEEE Communication Letters*, Vol. 4, No. 12:pp. 417–419, Dec. 2000.
11. G. Kesidis and T. Konstantopoulos. Extremal traffic and worst-case performance for a queue with shaped arrivals. In *Analysis of Communication Networks: Call Centres, Traffic and Performance*, edited by D.R. McDonald and S.R.E. Turner, Fields Institute Communications/AMS, ISBN 0-8218-1991-7, 2000 (proceedings of the Nov. 1998 conference at the Fields Institute, Toronto).
12. E. Livermore, R.P. Skillen, M. Beshai, and M. Wernik. Architecture and control of an adaptive high-capacity flat network. *IEEE Communications Magazine*, pages 106–112, May 1998.
13. J.M. McManus and K.W. Ross. A dynamic programming methodology for managing prerecorded VBR sources in packet-switched networks. *Telecommunications Systems*, 9, 1998.
14. K.W. Ross. *Multiservice Loss Models for Broadband Telecommunication Networks.* Springer-Verlag, London, 1995.
15. J. Salehi, Z.-L. Zhang, J. Kurose, and D. Towsley. Supporting stored video: Reducing variability and end-to-end resource requirements via optimal smoothing. In *ACM SIGMETRICS*, May 1996.
16. D. Songhurst and F.P. Kelly. Charging schemes for multiservice networks. In *ITC-15, Washington, DC, V. Ramaswami and P.E. Wirth (Editors)*, pages 781–790. Elsevier Science, B.V., 1997.
17. P.P. Tang and T.-Y. C. Tai. Network traffic characterization using token bucket model. In *Proc. IEEE INFOCOM'99, New York*, April 1999.

FPCF Input-Queued Packet Switch for Variable-Size Packets

Peter Homan and Janez Bester

University of Ljubljana, Faculty of Electrical Engineering,
Trzaska 25, SI-1000 Ljubljana, Slovenia
{Peter.Homan, Janez.Bester}@fe.uni-lj.si

Abstract. The existing forward planning conflict-free (FPCF) packet switches are designed for fixed-size packets and synchronous operation mode. In this paper an asynchronous FPCF packet switch for variable-size packets is proposed. The key problem of efficient packet insertion procedure was solved by introduction of a so-called quasi-synchronous record array arrangement. A performance evaluation based on simulations is provided in terms of throughput, delay and loss probability. The limited maximum switch throughput is a deficiency of the presented asynchronous FPCF packet switch. In most cases, utilization of over 90% can be achieved by using multiple traffic flows, while preserving proper packet order. Maximum packet delays are comparable to those produced by output-queued packet switches. By comparing FPCF and virtual-output-queued packet switches a conclusion is drawn that FPCF switch produces both, lower maximum packet delays and lower packet delay variations.

1 Introduction

Output-queued (OQ) packet switches are most suitable for quality-of-service (QoS) deliverance [1]. There is however a high price to pay for fast switching fabric [2]. Input-queued (IQ) packet switches can be used to avoid switching-fabric speedup [1], [3]. In most existing IQ packet switches first-in first-out (FIFO) queues are used. In order to prevent head-of-line blocking events, a so-called virtual-output-queuing (VOQ) queue configuration is usually implemented – rather than maintain a single FIFO queue for all packets, each input maintains a separate FIFO queue for each output [2].

Another way to achieve high throughput in IQ packet switches is by using forward planning conflict-free (FPCF) queuing policy [4]. Fig. 1 depicts a generic FPCF packet switch architecture. It was designed for fixed-size packets and synchronous operation mode. The essential idea is that by using an FPCF algorithm, the output-contention problem at the input of the switching fabric can be entirely eliminated. Any possible conflicts are resolved before the newly received packets are inserted into appropriate FPCF queues. Newly received packets can be inserted at any position inside FPCF queues. All packets inside FPCF queues are periodically shifted toward the switching fabric at the end of each time-slot. FPCF switches have one distinctive advantage in comparison to FIFO IQ switches – the total packet delay is known immediately after the packet is inserted into the FPCF queue and the worst case packet delay is limited by the FPCF queue size B and the switching fabric transfer speed.

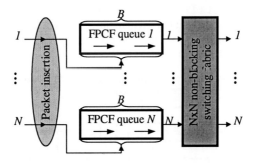

Fig. 1. Generic FPCF packet switch architecture

In this paper we propose a modified version of FPCF queuing policy that is capable of switching variable-size packets. A complete performance evaluation based on simulations is provided in terms of throughput, delay and loss probability. The reminder of this paper is organized as follows. Sect. 2 introduces the modified version of FPCF queuing policy. Sect. 3 presents packet switch simulation. First the methodology in use during the switch performance analysis is revealed, followed by simulation results. Sect. 4 concludes this paper.

2 Asynchronous FPCF Queuing Policy for Variable-Size Packets

Basic characteristics of synchronous FPCF algorithm for fixed-size packets, as described in [4], apply here as well. However, there are two important modifications: the asynchronous operation mode and variable packet sizes. Asynchronous operation mode implies that each newly received packet is inserted into the appropriate FPCF queue immediately after being received. In synchronous operation mode all packets form all inputs were inserted at one instance of time on the rotation priority basis. In asynchronous operation mode packets are inserted on a first-come first-serve (FCFS) basis. Asynchronous operation mode was selected due to variable-size packets. Different packets require different amount of FPCF buffer size and different amount of time for transfer across the switching fabric, which makes synchronous operation mode difficult to implement.

Packets can be inserted into any position inside FPCF queues. Moreover, packets can occupy any size between specified minimum and maximum values. Time and size granularities are virtually eliminated by using arbitrary real-time numbers. Fig. 2 depicts an example of asynchronous FPCF packet switch with 3 input and 3 output ports. Numbered rectangles indicate already inserted packets, while shaded areas indicate free space. Numbers inside rectangles determine the packets' output ports. With variable-size packets the packet insertion procedure complicates. FPCF algorithm keeps records of all occupied places inside each FPCF queue. In order to expedite the packet insertion procedure, additional records of already booked transmission periods are kept at each output port [4]. To find an appropriate place for newly received packet, cross sections of both, FPCF record array at appropriate input port and output record array at appropriate output port must be derived. The search is started at the beginning of FPCF queue i.e. at the switching fabric input ports, and is executed in the direction of packet switch input ports i.e. toward the end of FPCF queue. New packet is inserted into the first sufficiently wide cross-section.

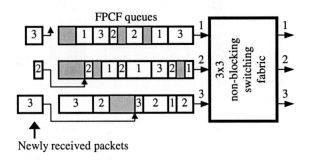

Fig. 2. Asynchronous FPCF packet switch for variable-size packets, $N = 3$

There are two new problems, which did not appear in synchronous FPCF packet switches for fixed-size packets. The first one is decrement of maximum switch throughput. It is a direct consequence of unexploited areas inside FPCF queues. Due to both, variable-size packets and asynchronous operation mode, many cross-sections are too small to be appropriate for packet insertion. Thus, maximum switch throughput of 100% cannot be expected. The second new problem is packet overtaking. As can be readily seen, smaller packets can swiftly find a suitable cross-section. Larger packets are usually compelled to search further toward the end of FPCF queue to find an appropriate cross-section. This is especially true when switch is heavily loaded. We could say that FPCF switch favors smaller packets – a traffic flow with smaller average packet size will likely to achieve higher throughput, lower delays, and lower packet loss probability. Packet overtaking can be prevented by keeping position records of last inserted packets. Unfortunately, by doing so the maximum switch throughput is additionally decreased. As a compromise, traffic can be logically separated into several traffic flows, while maintaining a single FPCF queue per input port. Overtaking is then allowed only between different traffic flows. Within each traffic flow, proper packet order is preserved. This way both, high maximum switch throughput and proper packet order can be achieved.

2.1 Fast Packet Insertion Algorithm

Finding an appropriate cross-section for variable-size packets is not a trivial task. At each input port an array of records is preserved. One record for each queued packet is maintained. Each record holds two notes. The first one indicates the time when packet's transfer across the switching fabric will commence and the second one indicates the time when this transfer will be concluded. Similarly, at each output port an array of records is preserved. One record for each packet to be transferred to this output port is maintained. Just as with input ports, each record holds two notes with the above mentioned data. A straightforward search algorithm can be used to find a suitable cross-section. As is turns out, this algorithm is too slow, even for non-real-time simulation purposes.

We developed a new algorithm to expedite the packet insertion procedure. The key invention is a so-called quasi-synchronous record array arrangement, as presented in Fig. 3. Like before, an array of records is preserved at each input and output port. However, the number of records in each array is fixed set by the FPCF queue size B and thus independent of the number of currently queued packets. Packet sizes, though variable, are usually limited with minimum and maximum values. Let us assume that the minimum packet size is set to 1 size unit. Let us also assume that the switching-fabric transfer speed is set to 1 size unit per time unit, i.e. a packet with size 1 would require 1 time unit to be completely transferred across the switching fabric. Let us imagine to parcel out the FPCF queue into a number slots with size equal to the minimum-sized packets. For example, if FPCF queue can concurrently accommodate at most 100 minimum-sized packets, it would be parceled out into 100 slots. Naturally, the real FPCF queues are left intact. However, this slotted arrangement is applied in record arrays. To look back to the previous example, each array would contain 100 records. Each record contains *status* and *time* information. Status can be either *free*, *bottom*, *top* or *occupied*. *Free* indicates that the slot is completely empty. *Bottom* indicates that the slot is partially empty, from the beginning of the slot until the specified time. *Top* also indicates a partially empty slot, from the specified time to the end of the slot. *Occupied* indicates either that a slot is completely full or that the slot is partially filled from both upward and downward directions and thus unfit for additional use. The *time* information is used in combination with *top* and *bottom* states, to demarcate the boundary between free and occupied space.

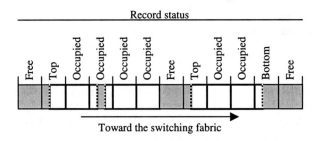

Fig. 3. Quasi-synchronous record array arrangement

Packet insertion algorithm maintains its own clock that is adjusted to the minimum packet size and switching-fabric transfer speed, e.g. in our example the clock is set to 1 time unit. At the end of each time slot, the records are shifted by one step toward the switching fabric. Records in front are discarded and new, empty records are added at the end of arrays. Records in front are then marked as transmitting and thus unavailable (*occupied*).

The packet insertion procedure is invoked upon the arrival of each new packet. In the first step the required number of slots to accommodate the newly received packet is determined, e.g. for a packet with size 4.14, at least 5 slots are required. Then a simple linear search algorithm is invoked to find cross-section of appropriate size. First slot can be either *free* or *top*. Intermediate slots must all be *free*. The last slot can be either *free* or *bottom*. Then the actual cross-section size is determined by considering *time* information in the first and last record. If the actual size is sufficient, packet is inserted and records' statuses are updated. Otherwise, the procedure is continued toward the last record of the array.

3 Packet Switch Simulation

3.1 Methodology

The primary goal of simulation is to study the behavior of our proposed FPCF queuing policy for variable-size packets. For comparison reasons, VOQ and OQ packet switches are simulated as well. Oldest packet first [2] scheduling mechanism is selected for scheduling purposes in VOQ packet switch. A simple queue configuration with a single FIFO queue per output port is used in QO packet switch. A switch configuration with N input and N output ports is selected, and a non-blocking switching fabric is assumed in all switch architectures. All investigated packet switches operate in asynchronous mode – when a packet is completely transferred across the switching fabric a new path through the switching fabric is established and the transfer of a new packet is started immediately.

New packets arrive at inlets following N identical and independent source processes. A uniform reference pattern of arrivals is assumed, i.e. the destinations of packets are uniformly distributed over the set of N outlets. A uniform packet size distribution is used. Packet inter-arrival times are exponentially distributed around mean value, which is calculated with regard to the specified offered switch load and average packet size. It should be noted however that packet arrival rate is limited by the switch input-line rate. Therefore, if input line is already occupied, a newly generated packet is delayed inside a traffic source until the input line becomes available. Limited input-line rate can be considered as a simple traffic shaper.

Throughout our simulation the normalized units are used. Normalized transmission speeds of input lines, input ports, switching fabric, output ports, and output lines are all set to 1. A single exception is the switching-fabric transfer speed in OQ packet switch, which is set to infinity to completely eliminate the output-contention problem. The minimum and the maximum packet sizes are set to 1 and 10, respectively. Therefore, a newly generated packet can occupy any size between 1 and 10. Queue sizes are also expressed in normalized units. For example, a queue with a normalized size 2000 can concurrently deposit at most 200 packets with normalized size 10. Normalized time unit can be readily defined by using normalized transmission speed and normalized packet size. For example, a packet with normalized size 1 requires 1 normalized time unit to be fully transmitted across the output line. The following performance measures are used:

- (normalized) throughput, defined as the sum of normalized sizes of all received packets per outlet, divided by the normalized duration of the simulation,

- packet loss probability, defined as the probability that a new packet arrives at an inlet when the queue for that inlet is full,
- mean packet delay, defined as the average of the time intervals form when a new packet arrives at an inlet to when it is delivered to its outlet,
- maximum packet delay, defined as the maximum of the time intervals form when a new packet arrives at an inlet to when it is delivered to its outlet.

All simulation results presented in this paper were obtained using Modsim III, an object-oriented simulation language from CACI. Simulation results were obtained by simulating the steady-state behavior of these systems and by applying the replication method, as explained in [5]. Simulations were repeated until switch throughput and mean packet delay measures achieved or exceeded the required level of relative precision (1%) at the 95% confidence level. During each simulation execution at least 10^6 packets were generated, setting the minimum detectable packet loss probability to 10^{-6} or smaller.

3.2 Maximum Throughput Results

The purpose of the first set of experiments was to study the influence of various parameters on the maximum throughput of the proposed asynchronous FPCF packet switch. In the first experiment the influence of FPCF queue size parameter B was investigated. Switch size parameter N was fixed set to 10. Packet overtaking was allowed. The results are presented in Table 1. As expected, while increasing the FPCF queue size, the maximum switch throughput is increased as well. However, for large FPCF queues the throughput increment is small. For example, when FPCF queue size is increased from 100 to 1000, the maximum switch throughput is increased for 13%, and when FPCF queue is increased from 1000 to 3000, the maximum switch throughput is increased for only 2.1%.

Table 1. Maximum switch throughput versus FPCF queue size B, $N = 10$, overtaking allowed

FPCF queue size:	100	300	1000	3000
Max. throughput (%):	81.5	87.9	92.1	94.0

Next, the influence of switch size parameter N was investigated. FPCF queue size parameter B was fixed set to 1000. Packet overtaking was allowed. The results are presented in Table 2. The highest maximum switch throughput is obtained for small switch sizes. While increasing the switch size, the maximum switch throughput is decreased. However, for large switch sizes the throughput decrement is small. For example, when switch size is increased from 3 to 10, the maximum switch throughput is decreased for 4.4%, and when switch size is increased from 30 to 100, the maximum switch throughput is decreased for only 0.8%.

Table 2. Maximum switch throughput versus switch size N, $B = 1000$, overtaking allowed

Switch size:	3	10	30	100
Max. throughput (%):	96.3	92.1	90.4	89.7

Finally, the influence of packet overtaking prohibition was investigated. It can be readily seen that the chances for successful packet insertion are higher when packet overtaking is allowed, since packet can be inserted into any appropriately wide cross-section. With packet overtaking prohibited, a newly received packet must be inserted after the previously inserted packet of the same traffic flow. Overtaking can still occur between packets destined to different destinations. In our simulation a traffic flow is defined as a sequence of packets originating at the same input port and destined to the same output port. This traffic flow can be further divided into several independent traffic flows. In this case, although packets have the same origin and destination, a packet overtaking can occur between different traffic flows. Within each traffic flow a proper packet order is still preserved, however. In the extreme case, when the original traffic flow is divided into infinite number of independent flows, each packet becomes independent, which is exactly the situation if packet overtaking is allowed.

Table 3. Maximum switch throughput versus number of traffic flows, $N = 10$, $B = 1000$, overtaking prohibited

Number of flows:	1	3	10	30	100	∞
Max. throughput (%):	82.6	87.0	90.0	91.4	92.0	92.1

Simulation results are presented in Table 3. The highest maximum switch throughput is obtained when infinite number of traffic flows are used, i.e. when packet overtaking is allowed. While increasing the number of flows, the maximum switch throughput is increased. However, for large number of flows the throughput increment is small. For example, when the number of flows is increased from 1 to 10, the maximum switch throughput is increased for 9.0%, and when the number of flows is increased from 30 to 100, the maximum switch throughput is increased for only 0.7%.

3.3 Switch Comparison Results

The purpose of the next set of experiments was to compare the characteristics of different switch types: OQ, VOQ, and FPCF. Three different scenarios were selected to study the behavior of FPCF packet switch:

- FPCF with single traffic flow and overtaking allowed (FPCF, 1, ON),
- FPCF with five traffic flows and overtaking prohibited (FPCF, 5, OFF),
- FPCF with single traffic flow and overtaking prohibited (FPCF, 1, OFF).

A switch configuration with 10 input and 10 output ports was selected. The total queue size was set to 20.000 units in all simulated models. For VOQ this meant 200 units per queue, and for FPCF and OQ it meant 2000 units per queue. A performance evaluation was provided in terms of throughput, delay and loss probability versus offered switch load.

Both, OQ and VOQ packet switches achieve near 100% switch throughput. Following maximum switch throughput results were obtained for FPCF packet switches: 93.8% when using single traffic flow and packet overtaking allowed, 88.8%

when using five traffic flows and packet overtaking prohibited, and 82.7% when using single traffic flow and packet overtaking prohibited.

Packet loss probability was investigated in the next experiment. It was already mentioned that during each simulation run, any packet loss probability of 10^{-6} or higher was detected. Any smaller packet loss probability would produce zero loss results. Indeed, no packet loss was detected right until the offered switch load approached the maximum switch throughput values to a few percents. In cases where offered switch load was higher than maximum switch throughput, the measured packet loss probability was in accordance to the surplus of offered switch load.

Mean packet delay results are presented in Table 4. At light and medium loads, below 70%, FPCF packet switch delivered mean packet delays comparable to those generated by OQ or VOQ packet switches. At higher loads but before the maximum switch throughput was reached, FPCF mean packet delays were about 3 to 5 times higher in comparison to those produced by OQ or VOQ packet switches. A distinctive increase of packet delays appeared when the offered switch load exceeded the maximum switch throughput. Nevertheless, packet delays inside FPCF packet switches were explicitly limited by the FPCF queue size and the switching fabric transfer speed. Therefore no packet could be delayed for more than 2000 time units inside our simulated FPCF packet switch. This axiom is even more evident in Fig. 4, where simulation results for maximum packet delays are presented. Indeed, to assure QoS for delay-sensitive traffic, maximum, i.e. worst-case packet delays must be kept under control.

Table 4. Mean packet delay versus offered switch load

Switch type	Offered switch load (%)			
	30	50	70	80
OQ	7.5	9.4	13.7	19.2
VOQ	8.0	11.7	20.6	32.2
FPCF, 1, ON	9.1	16.4	45.0	104
FPCF, 5, OFF	9.1	16.5	45.7	111
FPCF, 1, OFF	9.1	16.5	49.0	176

Table 5. Maximum packet delay versus offered switch load

Switch type	Offered switch load (%)			
	30	50	70	80
OQ	54	70	119	214
VOQ	80	126	350	507
FPCF, 1, ON	58	84	159	330
FPCF, 5, OFF	58	84	170	357
FPCF, 1, OFF	58	84	171	532

Maximum packet delay results are presented in Fig. 4 and Table 5. At medium loads, FPCF packet switches delivered maximum packet delays comparable to those generated by OQ packet switches. FPCF maximum packet delays were thus only half the maximum packet delays produced by VOQ packet switches. At higher loads but before the maximum switch throughput was reached FPCF packet switches still delivered lower maximum packet delays in comparison to those generated by VOQ packet switches. While FPCF maximum packet delays were explicitly limited, VOQ maximum packet delays could rose to very high values, as shown in Fig. 4.

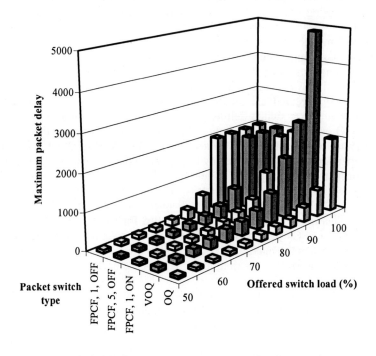

Fig. 4. Maximum packet delay versus offered switch load

4 Conclusions

We started this paper by describing the motive for application of input-queued packet switches. We further presented the properties of synchronous FPCF packet switches for fixed-size packets. These switches have an inherent capability to support delay-sensitive traffic, which is a feature rarely seen by input-queued packet switches. The main purpose of this paper was the introduction of asynchronous FPCF packet switches for variable-size packets and the investigation of their characteristics. By using a discrete-event simulation engine maximum switch throughput values were delivered. By using the same method, packet loss probability, mean packet delay and maximum packet delay values were delivered as well.

One of the major problems that had to be resolved was to devise an efficient packet insertion algorithm for asynchronous FPCF packet switches for variable-size packets. The problem was solved by introduction of a so-called quasi-synchronous record array arrangement. The devised packet insertion algorithm is sufficiently fast for simulation purposes. Its feasibility for implementation in real FPCF packet switches is yet to be investigated.

Simulation results confirmed our predictions that FPCF packet switches could be modified to operate in asynchronous mode and to switch variable-size packets. Truly, mean packet delays are higher in comparison to those produced by OQ and VOQ packet switches. However, maximum packet delays of FPCF packet switches are comparable to those produced by OQ packet switches and thus lower in comparison to delays generated by VOQ packet switches. This is especially beneficial for delay-sensitive traffic. By comparing FPCF and VOQ packet switches we conclude that FPCF switches produce both, lower maximum packet delays and lower packet delay variations. It must not be forgot that the maximum packet delays inside FPCF switches are always explicitly limited. This is especially important when switch congestion occurs, since the possibility of excessive packet delays is virtually eliminated.

The limited maximum switch throughput and thus the limited switch utilization is one of the drawbacks of the presented asynchronous FPCF packet switches. Actual switch utilization depends on various parameters: switch size N, FPCF queue size B, and number of traffic flows. Packet overtaking is another new problem that must be eliminated, unfortunately at the cost of additional switch utilization decrement. However, a switch utilization of over 90% can usually be achieved by using multiple traffic flows, while preserving proper packet order. We presume that using more sophisticated packet insertion algorithms could increase the switch utilization. This subject is for further study.

We conclude this paper with the statement that FPCF packet switches are well suited for transfer of delay-sensitive information, regardless of whether fixed-size packets or variable-size packets are employed.

References

1. Nong, G., Hamdi, M.: On the Provision of Quality-of-Service Guarantees for Input Queued Switches. IEEE Commun. Mag. **12** (2000) 62-69
2. Krishna, P., Patel, N.S., Charny, A., Simcoe, R.J.: On the Speedup Required for Work-Conserving Crossbar Switches. IEEE J. Select. Areas Commun. **6** (1999) 1057-1065
3. McKeown, N., Mekkittikul, A., Anantharam, V., Walrand, J.: Achieving 100% Throughput in an Input-Queued Switch. IEEE Trans. Commun. **8** (1999) 1260-1267
4. Yau, V., Pawlikowski, K.: An Algorithm That Uses Forward Planning to Expedite Conflict-Free Traffic Assignment in Time-Multiplex Switching systems. IEEE Trans. Commun. **11** (1999) 1757-1765
5. Akimaru, H., Kawashima, K.: Teletraffic: Theory and Applications. 2nd edn. Springer-Verlag, Berlin Heidelberg New York (1999)

A Cost-Effective Hardware Link Scheduling Algorithm for the Multimedia Router (MMR)*

M.B. Caminero[1], C. Carrión[1], F.J. Quiles[1], J. Duato[2], and S. Yalamanchili[3]

[1] Department of Computer Science.
Universidad de Castilla-La Mancha.
02071 - Albacete, Spain.
{blanca, carmen, paco}@info-ab.uclm.es
[2] Department of Information Systems and Computer Architecture.
Universidad Politécnica de Valencia.
46071 - Valencia, Spain.
jduato@gap.upv.es
[3] School of Electrical and Computer Engineering.
Georgia Institute of Technology.
Atlanta, Georgia 30332-0250.
sudha@ece.gatech.edu

Abstract. The primary objective of the Multimedia Router (MMR) project is the design and implementation of a compact router optimized for multimedia applications. The router is targeted for use in cluster and LAN interconnection networks, which offer different constraints and therefore differing router solutions than WANs. One of the key elements in order to achieve these goals is the scheduling algorithm. In a previous paper, the authors have proposed a link/switch scheduling algorithm capable of providing different QoS guarantees to flows as needed. This work focuses on the reduction of the hardware complexity necessary to implement such algorithm. A novel priority algorithm is presented, and its hardware complexity is compared to that of the original proposal.
Keywords: Communications switching, multimedia communications, Quality of Service, performance evaluation

1 Introduction and Motivation

The Multimedia Router (MMR) project is aimed at the design and implementation of a compact router optimized for multimedia applications. The goal is to provide architectural support to enable a range of quality of service (QoS) guarantees at latencies comparable to state-of-the-art multiprocessor cut-through routers.

At the core of the MMR, there is one key element to provide QoS guarantees to the multimedia flows: the link and switch scheduling algorithm. The authors have recently proposed a link and switch scheduling algorithm, based on the

---

* This work was partially supported by the Spanish CICYT under Grants TIC97-0897-C04 and TIC2000-1151-C07-02

concept of biased priorities [1, 4, 10] that is well suited to parallelization and pipelining [11].

The key point in this scheme is that priorities are biased according to the ratio between the QoS a flit is receiving and the one it should receive. This approach combines the effect of the scheduler (measured as the delay or the jitter experienced by a flit in the queue) with the QoS requirements (measured as the bandwidth requested by the connection). One of the proposed biasing functions was *Inter-Arrival Biased Priority (IABP)*. With this algorithm, the priority of a flit is computed as the ratio between the queuing delay, and the inter-arrival time for the flits in the connection. The effect is that the priority grows as queuing delay grows. Moreover, priority grows faster for those flits belonging to high-bandwidth consuming connections, that is, there are more chances that they will be forwarded sooner through the switch.

IABP has been shown to be able to guarantee QoS to flows with a wide range of bandwidths requirements, and timing constraints. But its main drawback is the complexity needed to implement this algorithm. The reason is that a division must be performed to compute priority. The purpose of this work is to introduce a simpler version of the IABP algorithm. This new algorithm, known as *SIABP (Simple-IABP)* retains the idea behind IABP, but makes its implementation much simpler and compact. As a priority computing engine is needed for each virtual channel in the MMR, this reduction will significantly contribute to the overall simplicity and speed of the router.

The rest of the paper is organized as follows. Next section describes the main architectural issues of the Multimedia Router. Then, the resource scheduling algorithms are outlined. Next, a description of the new proposed algorithm to compute priority is presented, followed by a comparison of the hardware requirements for the original and new algorithms. The paper concludes with some performance results obtained with CBR and VBR traffic, and some conclusions.

2   Multimedia Router Architecture

The general organization of the Multimedia Router is shown in Figure 1-a. The key issues regarding its architecture are summarized in the following paragraphs. The interested reader is referred to [3] for a more detailed description of the MMR, as well as the design trade-offs.

Switching technique. The MMR uses a hybrid approach, where the most suitable switching technique is used for each kind of traffic: a connection-oriented scheme (Pipelined Circuit Switching [5, 6]) for the multimedia flows, and Virtual Cut-Through for best-effort messages.

Input buffers. For each connection, a virtual channel is provided. This means that a large amount of buffers will be needed. In order to optimize their implementation, the buffers are organized as modules of RAM memory, interleaved with a simple scheme (see Figure 1-b). HOL-blocking [8] is also avoided in this way.

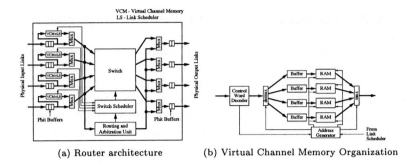

Fig. 1. MMR Architecture

Flow control. The MMR uses *credit-based flow control* [9], to avoid the need for large buffers. InfiniBand [7] has also selected this flow-control method.

Multiplexed crossbar. Due to the large number of virtual channels, a multiplexed crossbar is the most feasible implementation. Arbitration is needed at the input side (*link scheduling*), to select one virtual channel from each physical channel, and within the switch (*switch scheduling*), because several input channels might request the same output link. Arbitration is made concurrently with flit transmission, and considers the QoS requirements of the connections. Flits are synchronously forwarded through the crossbar.

Large flits. Large flits are needed to pay off for the overhead introduced by flow control, and will also help to hide arbitration delay and amortize crossbar reconfiguration delay.

3 Resource Scheduling Algorithm

3.1 Bandwidth Allocation

Link bandwidth and switch port bandwidth are split into flit cycles. Flit cycles are grouped into *rounds* also referred to as *frames*. The number of flit cycles in a round is an integer multiple K ($K > 1$) of the number of virtual channels per link. Bandwidth for a connection is allocated as an integer number of flit cycles. The allocated flit cycles will be assigned to the requesting connection every round.

3.2 Scheduling Algorithm

Link and switch scheduling are partitioned into a set of basic decisions: *candidate selection, port ordering,* and *arbitration.*

    Candidate selection is performed during link scheduling. The candidates are the C virtual channels from a physical link ($1 \leq C \leq N$, N = number of ports) whose head flits have the highest priorities. These candidates are forwarded to the *switch scheduler*, which computes a conflict-free matching among input and output ports, in a way motivated by chances for parallelization and pipelining.

All the candidates selected by every port are arranged into a *selection matrix*, according to their priority value. A *conflict vector* is used to help to select the output port with less non-matched higher priority requests in the *port ordering* phase. In the *arbitration* step, if there are several potential matches for an input port, the candidate with the highest priority is selected.

See [11] for more details.

4   The Simple-IABP (SIABP) Link Scheduling Algorithm

In our previous switch scheduling proposal [11], link scheduling was performed by using the concept of *biased priorities* [1, 4, 10]. The priority value for each flit was related to the QoS requested by the connection, and was biased depending on the QoS received by it. One of the proposed biasing functions was IABP (Inter-Arrival Based Priority). With IABP, the priority value was computed as the ratio between the queuing delay experienced by the flit, and the Inter-Arrival Time (IAT) specified for the connection. Results showed that this approach provides good performance for CBR traffic, differencing the QoS received by each type of connection. The main problem with the IABP algorithm is the way priorities are computed, since a division is needed for every virtual channel. Hardware implementations of dividers are slow and expensive, and hardly fit into our fast, compact router. So, an alternate algorithm has to be devised. The idea is to apply the same rationale introduced by IABP, that is, relate the bandwidth required by the connection with the experienced queuing delay, but replacing the division with some other operation faster and cheaper to implement.

In the new algorithm, called *SIABP (Simple-IABP)*, the priority value for each head flit is computed in a different way. The queuing delay for the flit is stored as a counter that is incremented in each router cycle, that is, the queuing delay is expressed as a number of router cycles.

The priority biasing function is performed as follows. The initial value for the priority is the bandwidth required by the connection. Instead of representing this by the IAT of the connection, the number of slots per frame reserved to service the average bandwidth of the connection is used. The advantage of this approach is that this magnitude is an integer value.

It must be noted that computing the ratio between the queuing delay and the IAT (as IABP does) is equivalent to compute the product between the queuing delay and the bandwidth requirements. But if the product could be replaced with some shift operation, the implementation would be much simpler. So, the priority value is shifted to the left (that is, it is multiplied by two) each time the queuing delay value becomes greater than 1, 2, 4... 2^n, that is, each time a bit in the queuing delay counter is set for the first time since it was last reset. In this way the QoS needed (represented by the initial priority value) is also related to the QoS received by the flit (the queuing delay).

4.1 Hardware Requirements

To assess the simpler implementation of the SIABP algorithm, a VHDL [17] description has been done for both IABP and SIABP algorithms.

Figure 2 shows the hardware blocks necessary for both priority computing engines. Every virtual channel in the router needs one of these blocks to compute the priority of their head flits. Both algorithms use a counter to store the queuing delay, which is incremented in every router cycle, and reset when a new flit enters the input buffer. MMR routers with a single flit buffer per virtual channel have been considered [2]. Extension to consider multiflit buffers is straightforward, since the only element to be replicated is the counter that stores the queuing delay (one per flit is needed). A register to store the bandwidth requirements of the connection is also needed. This will be represented by the IAT (Inter-Arrival Time) of the connection for the IABP algorithm, and by the number of flit cycles reserved within a frame for that connection, in the case of the SIABP algorithm.

In order to compute priorities, the IABP algorithm uses a standard divider module, while the SIABP module uses a special module called *SIABP module*. The pin-in and pin-out of this module and an outline of its internal logic are shown in Figure 3 for 4-bit priority values. Its purpose is to take into account the bandwidth requirements and perform the shift operation when needed. Results have been computed for 16 bit-wide priority values, which suffice to represent the range of priorities needed.

These modules have been synthesized with the Synopsys design tool [16], in order to get implementation details. The synthesis process has used the libraries provided by Europractice, MIETEC 0.35 μm. The implementation of the IABP algorithm was based on a 32-bit integer divider, provided by the design kit DesignWare [15], in order to get a lower bound estimation of the complexity of the IABP algorithm. The IABP algorithm is formulated to work with floating point magnitudes. Estimated delay for the IABP logics is 155.87 nsecs, while the SIABP implementation shows a delay of 4.07 nanosecs, that is, the SIABP algorithm shows a delay around 38 times lower than that achieved by the IABP algorithm. In terms of silicon area, the reductions are also dramatical: 272.56 μm for the SIABP algorithm versus the 6101.33 μm needed by the IABP implementation. This means a reduction of around 22 times in silicon area.

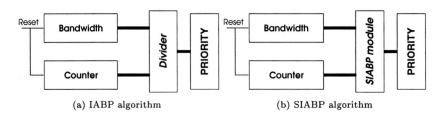

Fig. 2. Block diagrams for the studied algorithms

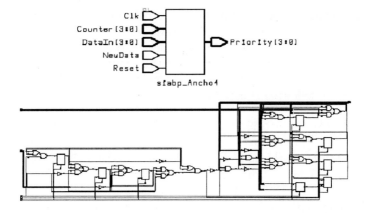

Fig. 3. 4-bit SIABP module pin-in and pin-out and internal logic

Now, it is necessary to assess that QoS guarantees are not threatened by this reduction in hardware complexity. Performance analysis is presented in next section.

5 Performance Evaluation

The configuration used in simulations is composed of a single MMR router, with a Network Interface Card (NIC) attached to each input link. The traffic sources inject their flits into buffers located in the Network Interface Card (NIC) (see Figure 4). These buffers are considered to be infinite because the host main memory can also be used if the NIC buffers become full. Credit-based flow control governs the transmission of flits from the NIC buffers to the corresponding MMR virtual channel buffer.

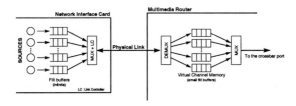

Fig. 4. Conceptual representation of the NIC and MMR buffers

Simulations have been run for a 4 × 4 router, with 128 virtual channels per physical link. 1.24 Gbps 16 bit-wide links have been considered. This imposes a router cycle of 12.9 nanosecs. The switch scheduler is based on a selection matrix, with 4 levels of candidates [11]. Buffer size is set to 1 flit per virtual channel inside the MMR [2]. NIC buffers, on the other hand, are regarded as infinite, since the host memory could be used in case of necessity.

Tests using CBR and VBR traffic have been carried out. All the connections are considered to be active throughout all the simulation time. Their destination is chosen randomly, among the output ports of the router. The results are presented in the following subsections.

5.1 Evaluation with CBR Traffic

For the CBR tests, workload was composed of a random mix of connections belonging to three different types, depending on their bandwidth requirements:{64 Kbps, 1.54 Mbps, 55 Mbps}. That is, connections with low, medium, and high bandwidth requirements have been considered. Simulations were run for 400 scheduling cycles.

Results related to the QoS received by the connections are shown in Figures 5, 6, and 7. The first ones plot respectively the average flit delay and jitter for the most and least bandwidth consuming connections, while the latter one shows the distribution of flit delays, in order to know how many flits suffer from delays shorter than a certain deadline. The considered deadlines are multiples and submultiples of the IAT (Inter-Arrival Time) of every flow.

Both algorithms are able to provide QoS guarantees with workload level up to 81% of link bandwidth. What is more, when the router enters saturation, performance degrades more gracefully with SIABP than with IABP.

It can be seen that the overall performance of the SIABP algorithm is similar to the one achieved by the IABP algorithm for the 55 Mbps connections, whereas medium and low bandwidth connections get better performance. This is due to the fact that with SIABP, lower bandwidth connections get their priority value increased faster than with IABP, specially while queuing delay is not too big. So, the average delay experienced by their flits is lower, and they are likely to be considered sooner for scheduling. Anyway, this does not affect significantly the performance achieved by the other flows. Recall that every time a priority value is updated, it is multiplied by two, and that priority values are doubled less frecuently as queuing delays increase. With IABP, priority is increased more continuously but more slowly when queuing delays are not too big.

These results are rather encouraging, because not only the complexity of the priority computing engine is reduced, but also performance is enhanced. This must be assessed by testing the algorithm with VBR traffic.

5.2 Evaluation with VBR Traffic

Performance evaluation of the proposed algorithm with VBR traffic is presented in this subsection. The VBR traffic models used in our simulations are based on MPEG-2 video traces. This is a typical type of multimedia flow.

MPEG-2 video coding standard [14] encodes the video streams as a sequence of different frame types, I, P, and B, ordered with a prefixed and repetitive pattern, called GOP (Group Of Pictures). Every 33 milliseconds, a frame must

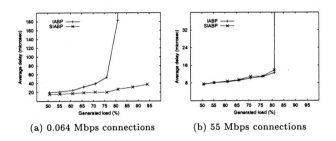

(a) 0.064 Mbps connections   (b) 55 Mbps connections

Fig. 5. Average flit delay since generation

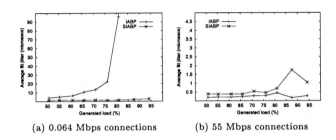

(a) 0.064 Mbps connections   (b) 55 Mbps connections

Fig. 6. Average flit jitter

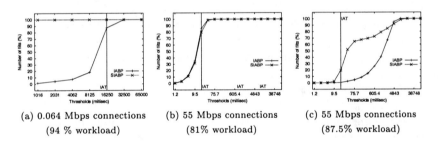

(a) 0.064 Mbps connections (94 % workload)   (b) 55 Mbps connections (81% workload)   (c) 55 Mbps connections (87.5% workload)

Fig. 7. Distribution of flit delay since generation for 0.064 and 55 Mbps connections

be injected. A frame is composed of a number of flits. The number of flits that compose every frame has been extracted from real MPEG-2 video traces. Some data about their frame sizes are shown in Table 1.

Two ways of injecting these flits into the NIC buffers have been considered:

Back-to-Back (BB) model. In this model, all the flits are transmitted at a peak bandwidth, common to all the connections. The *peak bandwidth* is such that it allows the injection of the largest frame among all the connections within 33 millisecs. Transmission of each frame starts at a frame time boundary. All the flits that compose the frame are injected at the selected peak rate, and then, the source becomes idle until the next frame time boundary. The peak bandwidth in the experiments is 50 Mbps for all the connections.

Table 1. MPEG-2 video sequence statistics: Image sizes in bits

Video Sequences	Ayersroc	Hook	Martin	Flower Garden	Mobile Calendar	Table Tennis	Football
Max.	535030	454560	444588	900139	970205	933043	590532
Min.	148755	159622	116094	308411	412845	260002	340246
Average	232976	272738	199880	497126	600742	440547	441459

Figure 8-a depicts this model. IAT_p stands for the Inter-Arrival Time (IAT) related to the peak bandwidth.

Smooth-Rate (SR) model. In this model, the flits that compose a frame are transmitted with a different IAT for every frame, in such a way that the flits of a frame are evenly distributed within the frame time. The IAT has been computed as the ratio between 33 msecs and the number of flits that compose the frame. A graphical representation is shown in Figure 8-b.

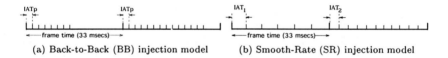

(a) Back-to-Back (BB) injection model   (b) Smooth-Rate (SR) injection model

Fig. 8. VBR injection models

Simulations have been run until four complete GOPs (Group Of Pictures) from every connection have been forwarded through the router. The connections sharing the same physical link have been randomly aligned, that is, they start at a random time within a GOP time.

In order to assess performance, and as the considered application data unit is the video frame, the presented results are related to the delay and jitter experienced by the frames. Figure 9 shows the average frame delay since generation. It can be noticed that before saturation, both approaches behave quite similarly, and when workload is near saturation (around 80 %), SIABP performs slightly better than IABP. This holds for both flit injection models.

Regarding frame jitter, SIABP achieves clearly better results than IABP for the SR injection model, specially with workload levels greater than 70%, while with BB model SIABP performs slightly worse. Anyway, average jitters are always under 4 microsecs, which is a quite bearable value. Note that even jitter values of several millisecs can be absorbed at the destination by using jitter absorption techniques [12].

Lastly, the distribution of frame delays since generation has been computed. The deadlines considered are multiples and submultiples of a frame time (FT, 33 milliseconds), ranging from $FT/16$ (2.06 millisecs) to $8192 \times FT$ (270 seconds).

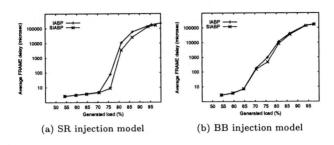

(a) SR injection model   (b) BB injection model

Fig. 9. Average frame delay since generation

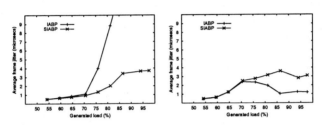

Fig. 10. Average frame jitter

Results are plotted in Figure 11 for both injection models. Only results for the workload levels right before and right after saturation are shown.

With the SR injection model, all the frames experience delays lower than 66 millisecs for both algorithms, with a workload of 81%. Tighter thresholds can be met for more flits when using the SIABP algorithm. Even when workload has entered saturation (workload = 87%), both algorithms make frame delays fit into deadlines of around 500 milliseconds. Again, if SIABP is used, more frames are able to fit into lower delay bounds. On the other hand, with the BB injection model, results are quite similar for both algorithms. Deadlines of 132 millisecs right before saturation, and around 500 millisecs right after saturation can be met both with IABP and SIABP.

These are quite reasonable values, since a typical deadline for MPEG-2 video transmission is one second between endpoints. This is the value for the CTD (Cell Transfer Delay) recommended by the ATM Forum for video distribution services, using MPEG-2 [13].

The reason for this behavior is that the flits injected by SR sources usually have to spend less time in buffers, because they are injected at a lower rate than the ones from BB sources. Within that period of time, their priority increases quicklier. Note that as queue delay continues increasing, priority is doubled less frequently. On the other hand, BB sources inject their flits as bursts with a fixed peak rate. So, their flits have usually to wait longer, and their priority values get to increase slower and slower, in a more similar way to what IABP does. This effect is most noticeable for workloads over 70%.

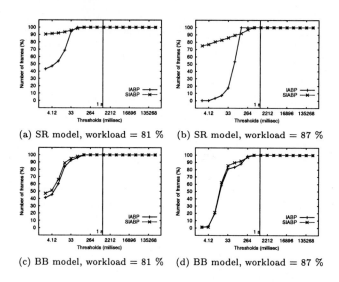

Fig. 11. Distribution of frame delay since generation for the SR injection model

6 Conclusions

The Multimedia Router is aimed at providing QoS guarantees to different classes of traffic in a LAN environment, while remaining compact and fast. One of the key elements to achieve these goals is the link/switch scheduling algorithm. In this algorithm, priorities have to be computed for the head flit in every virtual channel. Thus, as many priority computing engines as virtual channels are needed in the MMR router. In this paper, a novel priority biasing algorithm has been presented. This algorithm, called SIABP (Simple-IABP), considerably simplifies the hardware implementation of the mechanism used to compute priorities with respect to the previous proposal. Results obtained with the Synopsys design tool show that silicon area is reduced 22 times, and the estimated delay is 38 times lower. Simulations where carried out with both CBR and VBR traffic, and show that performance does not suffer with this hardware simplification. What is more, overall performance is enhanced with the new algorithm.

References

1. A. Chien, J.H. Kim, "Approaches to Quality of Service in High Performance Networks," *Proceedings of the Workshop on Parallel Computer Routing and Communication*, Lecture Notes in Computer Science, Springer-Verlag, pp.1-19, June 1997.
2. B. Caminero, C. Carrión, F. J. Quiles, J.Duato, and S. Yalamanchili, "Tuning buffer size in the Multimedia Router (MMR)," To appear in *Proceedings of the Workshop on Communication Architecture for Clusters*, held in conjunction with the *2001 International Parallel and Distributed Processing Symposium (IPDPS'2000)*, April 2001.

3. J. Duato, S. Yalamanchili, M.B. Caminero, D. Love, and F.J. Quiles, "MMR: A high-performance multimedia router. Architecture and design trade-offs," *Proceedings of the 5th Symposium on High Performance Computer Architecture (HPCA-5)*, pp. 300-309, January 1999.
4. D. Garcia, D. Watson, "ServerNet II," *Proceedings of the Workshop on Parallel Computer Routing and Communication*, pp. 119-136, June 1996.
5. P. T. Gaughan and S. Yalamanchili, "Adaptive routing protocols for hypercube interconnection networks," *IEEE Computer*, vol. 26, no. 5, pp. 12–23, May 1993.
6. P. T. Gaughan and S. Yalamanchili, "A family of fault-tolerant routing protocols for direct multiprocessor networks," *IEEE Transactions on Parallel and Distributed Systems*, vol. 6, no. 5, pp. 482–497, May 1995.
7. D. Pendery, J. Eunice, "InfiniBand Architecture: Bridge Over Troubled Waters," *Research Note*, available from the web page: http://www.infinibandta.org
8. M. J. Karol, M. G. Hluchyj and S. P. Morgan, " Input versus output queuing on a space division packet switch," *IEEE Transactions on Communications*, December, 1987.
9. M. G. H. Katevenis, et al., "ATLAS I: A single-chip ATM switch for NOWs," *Proceedings of the Workshop on Communications and Architectural Support for Network-based Parallel Computing*, February 1997.
10. J.H. Kim, "Bandwidth and latency guarantees in low-cost, high-performance networks," Ph. D. Thesis, Department of Computer Sciences, University of Illinois at Urbana-Champaign, 1997.
11. D. Love, S. Yalamanchili, J. Duato, M.B. Caminero, and F.J. Quiles, "Switch Scheduling in the Multimedia Router (MMR)," *Proceedings of the 2000 International Parallel and Distributed Processing Symposium (IPDPS'2000)*, May 2000.
12. M. Perkins and P. Skelly, "A hardware MPEG clock recovery experiment for variable bit rate video transmission," ATM Forum, ATM94-0434, May 1994.
13. M Schwartz and D. Beaumont, "Quality of service requirements for audio-visual multimedia services," ATM Forum, ATM94-0640, July, 1994.
14. "Generic coding of moving pictures and associated audio," Recommendation H.262, Draft International Standard ISO/IEC 13818-2, March, 1994.
15. "Synopsys DesignWare Guide," version 1998.02.
16. "Synopsys Reference Manual," version 1998.02.
17. "IEEE Standard VHDL; Language Reference Manual," IEEE STD1076, 1993.

The Folded Hypercube ATM Switches

Jahng S. Park[1] and Nathaniel J. Davis IV[2]

The Bradley Department of Electrical and Computer Engineering
Virginia Polytechnic Institute and State University
Blacksburg, VA 24061
[1] jahng@vt.edu, (540) 231-2295, fax: (540) 231-3362
[2] ndavis@vt.edu, (540) 231-4821, fax: (540) 231-3362

Abstract. With the increasing demand and growth of high-speed networks, it is imperative to develop high performance network switches that also have low complexity. Over the years, many high performance ATM switches have been proposed. The majority of these switches has high performances but also have very high hardware complexities. Therefore, there is a need for switch designs with lower complexity and high performance. The authors are proposing the use of the folded hypercube (FHC) as a new infrastructure for ATM switches. This paper presents two novel ATM switches that are implemented from the FHC networks. It presents the design of the two switches, discusses how the simulation models are developed, and gives the analysis of the simulation results. Finally, the performances of the two switch models are compared to other switches in the literature.

Introduction

For broadband access to the Integrated Services Digital Network (ISDN), CCITT has chosen Asynchronous Transfer Mode (ATM) as the switching and multiplexing technique. ATM is being widely accepted by common carriers as the mode of operation for future communication systems, transporting and switching various types of traffic (voice, audio, video, and data). The ATM switches for Broadband ISDN (BISDN) differs from the existing ATM LAN switches in terms of network capacity, throughput, and delay performance. Many high performance ATM switches have been proposed in the literature, and comprehensive surveys on ATM switches can be found in [1, 2]. The authors have designed new ATM switches based on the folded hypercube (FHC). This paper presents how the two switches are designed using the FCH as their fabrics, discusses their performances.

Hypercube and Folded Hypercube Networks

For the high-speed cell switching requirement of ATM networks, 150Mbps or higher, interconnection networks would represent the best solution to implement the I/O interconnection device of an ATM switching fabric. This can be attributed to their fast, self-routing packet property as shown by [3]. Furthermore, the parallel processing capability of these networks that exploits their self-routing property makes supporting the load values envisioned for an ATM switch feasible. Most of the interconnection network-based ATM switches in the literature are based on different types of multistage interconnection networks [1, 2].

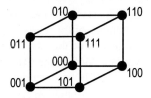

Fig. 1. A 3-dimension Hypercube Network.

However, a few researchers have studied ATM switches based on the hypercube [4, 5]. The hypercube (Fig. 1), also known as the Cube [6], is a single-stage network that offers a rich, low-complexity interconnection structure with large bandwidth, logarithmic diameter, high degree of fault tolerance, and inherent self-routing capability. An n-dimension hypercube has $N=2^n$ nodes. The two hypercube-based ATM switches proposed by [4] and [5] were shown to have performances comparable with other ATM switches in the literature, but lower hardware complexity. They also provided a different direction in the design of the interconnection network based ATM switches.

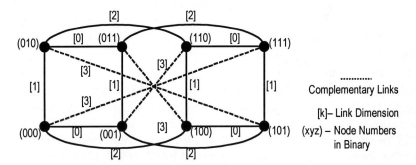

Fig. 2. The Structure of an FHC(3) [ElL91].

In order to further improve the characteristics and performance of the hypercube networks, some variations of the hypercube structure have been reported in the literature. These variations further reduce the diameter and traffic congestion of the hypercube with little or no hardware overhead. One of these variations is the folded hypercube (FHC) proposed by El-Amawy and Latifi [7, 8]. A folded hypercube of dimension n, FHC(n), can be constructed from a standard n-dimension hypercube by connecting each node to the unique node that is farthest from it (Fig. 2). This extra links are the *complementary links*. The hardware overhead is almost $1/n$, which is negligible for large n. The FHC is shown to have a self-routing capability, and this is an important feature that avoids unnecessary communication delays and message congestion. [7, 8] gave a routing algorithm that performs the routing between any pair of nodes.

The analytical study by [7, 8] showed that the FHC has better performance measures in addition to the smaller diameter when compared to the hypercube. The average distance[1] in the FHC is at least 15% less than that of the same size hypercube. The cost factor[2] for the FHC is also appreciably less than that of the n-cube and tends to half for large n. Finally, the FHC saturates at a higher loading value compared to the hypercube. A detailed presentation of the analytical result can be found in [6, 7]. To exploit the performance advantages of the FHC, we are using the FHC as a new infrastructure for two ATM switches. Before the FHC is modified into the new ATM switches, a simulation model of the FHC is developed and studied. The validity of the model is verified by comparing the simulation results with the analytical results of El-Amawy and Latifi [9]. Based on this validated model of the FHC, two new ATM switches are developed.

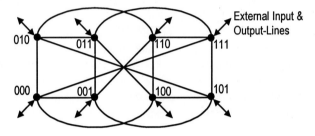

Fig. 3. An (8×8) FAS-1.

New FHC ATM Switches

The two new ATM switches are developed using the folded hypercubes (FHC) [ElL91, LaE89] as their switching fabrics. A brief description of the architectures of the switches is given here. The first proposed architecture is called FAS-1 (FHC ATM Switch-1), and it uses a FHC(n) to implement an $N \times N$ switch where $N = 2^n$ (Fig. 3). Our second proposed architecture, called FAS-2 (FHC ATM Switch-2), uses a FHC(n+1) to implement an $N \times N$ switch where $N = 2^n$ (Fig. 4). The nodes with addresses from 0 to $N-1$ are called *input-nodes*, and the nodes with addresses from N to $2N-1$ are called *output-nodes*. The set of input-nodes (output-nodes) is called *input-cube* (*output-cube*). Fig. 5 shows the internal structure of a FAS-1 node. All the nodes of FAS-1 and the input-nodes of FAS-2 have store-and-forward (SAF) buffers that temporarily store the cells from the links and the external line while they wait to be routed. Finally, all the nodes of FAS-1 and the output-nodes of FAS-2

---

[1] The average distance is the summation of distances of all nodes from a given source node over the total number of nodes.

[2] The cost factor can be defined as the product of the diameter and degree of the node (i.e., the number of links per node) for a symmetric network.

have the output buffers. The output buffer is needed since more than one cell can arrive at the same time. Also, re-sequencing of the cells is performed while the cells are queued in the output buffers.

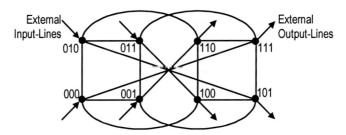

Fig. 4. A (4×4) FAS-2.

To minimize the buffers needed in the network, the two new switches use *deflection routing*. [10, 11] have shown that the deflection routing performs well in hypercube networks of different sizes for the whole range of light to heavy loading. In deflection routing, congestion causes packets in the network to be temporarily misrouted rather than buffered or dropped. This allows good performance with smaller number of buffers compared to the non-deflection routing. It has performed well when [4, 5] have used it in their switches.

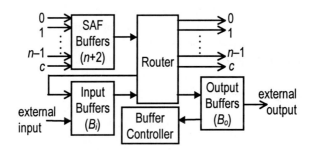

Fig. 5. Structure of a node in FAS-1.

The two new switches use a three-level priority scheme in their routing algorithm. The highest priority level is the *class-priority*. An ATM network has two general classes of traffic, time-sensitive and time-insensitive. A time-sensitive cell will have a higher priority than a time-insensitive cell during the link selection process. This is to guarantee the Quality of Service (QoS) of the time sensitive ATM traffic. The next priority level is the *age-priority*. The oldest cell among all the cells requesting a particular link wins during the link selection process. Giving a higher priority to the older cell preserves the cell sequence as much as possible since keeping the cell sequence is as important as providing high throughput, low loss switching in ATM environment. Furthermore, the destination node will also use this cell numbers to re-order the cells. The lowest priority used is called *dist-priority*. The cell closer to its destination will have a higher priority because a cell closest to its destination has the fewest possible candidate links and is most likely to encounter link contention. With deflection routing, no cells are dropped while they are routed. A cell is dropped when the external input-line buffer overflows or when a cell reaches its destination and it is older than the last cell transmitted to the external output-line. These two methods, deflection routing and priority scheme, keep the cell loss rate to a minimum. When a cell enters the switch, information, such as the class field, cell number (CN), the distance-to-destination (DD), and the destination address, is tagged in the header of the cell. This information is used by each node to route and resequence the cells.

The routing algorithm of [7, 8] has been extensively modified for the two proposed switches. The new algorithm, *Distributed Dynamic Routing Algorithm*, is now dynamic and can be performed at each node. When a node receives cells at the beginning of a cell cycle, the node first sorts the cells based on the priority scheme, and routes them by using the routing algorithm. If the cells have reached the destination, the cells with the higher class-priority are allowed to enter the output buffer first. During the routing process, link cells are considered before the line cells for the following reasons. It is impossible to guarantee an output link for every cell in a node of FAS-1 and input nodes of FAS-2 since up to $m+1$ cells may arrive at a time when there are only m links to choose from in these nodes. However, since the number of input and output inter-node links is equal, we can guarantee at least one output link for each link cell if the link cells are routed before the line cells. A new line cell is dropped if the input buffer is full, so the switches should have enough input buffers for input cells that experienced link blocking for all the preferred and detour links. Different input buffer sizes were studied to find a buffer size that gives the best cost and cell loss trade off. Under a normal traffic and unsaturated load level, however, the probability that at least one cell has reached its destination is high. Therefore, a link can be guaranteed for a line cell as well.

Each node of FAS-1 and each output node of FAS-2 will have an *output buffer controller* (OBC) that manages the cells queued in the output buffer. The OBC maintains N virtual queues for cells from each of N external input ports. Each cell is placed in the corresponding virtual queue by checking the external input port address of the cell's header. We chose a shared buffering strategy over a non-shared strategy since the shared buffering strategy was shown to require fewer total number of buffers than physically separated (partitioned) buffers [12, 13, 14]. One of the two functions of the OBC is monitoring whether the buffers are shared in a fair manner. This function is needed without a good, fair buffer controller, the shared buffers may perform worse than the non-shared buffers [14]. The second task of the OBC is re-sequencing the cells that originated from the same external input port before they are transmitted through the external output line. The OBC does this by using the cell numbers mentioned above. The two switches are modeled and simulated with OPNET Modeler[3] to study their performances.

---

[3] An object-oriented communications network modeler and simulator developed by OPNET Technologies.

Simulation Results

In simulating the two switches' message traffic, the uniform distribution is used to generate destination addresses to study the network performance under different conditions. If the distribution of destination addresses is not uniform, some parts of the network will be more congested than others, contributing to a decrease in throughput. Also, the packets are generated by N independent random processes, one for each of N nodes. These assumptions lead to an analysis of the maximum performance that a network can support [15].

In addition to the uniform distribution, the normal distribution is used to generate the destination addresses. Even though the uniform destination address distribution gives the maximum throughput of a network, it is common in real world applications where one or few destinations may be more favored by the sources. To analyze the network performance in a more practical situation, the normal distribution is also used to generate destination addresses or hot-spots. Hot-spots cause a disproportionate number of packets to be routed to a subset of the destinations [16, 17]. This is done by generating packets over a range of destination addresses with a preferential concentration around a mean destination address. For the simulations with hot-spots, packet destinations are chosen using a normal distribution with a specified mean and standard deviation. The mean is chosen randomly, but the value chosen for the standard deviation is one-quarter of the network size, based on the study by [18].

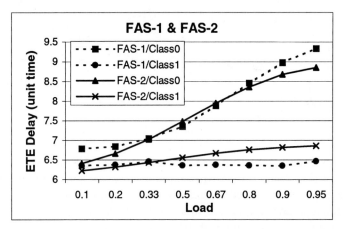

Fig. 6. Average end-to-end Delay for FAS-1 & FAS-2 (32×32, uniform dist.)

The simulations were run until about 10 million cells are submitted into the switch. Among these cells, 30% of them were randomly selected to be Class 1 (time-sensitive) cells and the rest were Class 0 (time-insensitive) cells. The simulation study was done under different traffic patterns (uniform and normal distributions), switch sizes (8×8 to 1024×1024), and loads (10% to 100%)[4]. The simulation results

---

[4] Cells are generated at a maximum rate of one cell per simulation time unit.

show that both FAS-1 and FAS-2 have low cell loss rate and low average end-to-end delay while having low hardware complexity. Two sample cases of the simulation results are given here.

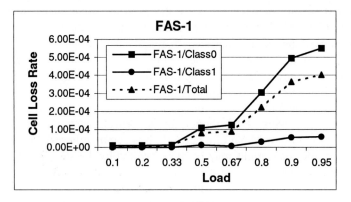

Fig. 7. Cell loss rate for FAS-1 (32×32, uniform dist.)

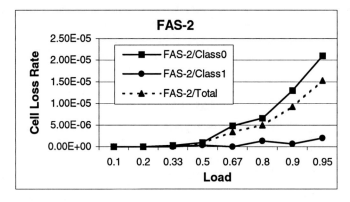

Fig. 8. Cell loss rate for FAS-2 (32×32, uniform dist.)

Fig. 6 shows the average end-to-end cell delay for 32×32 FAS-1 and FAS-2. The Class 1 cells of FAS-1 and FAS-2 experienced much lower end-to-end delay through the switches than the Class 0 cells because of the time-priority implementation. Also, because of the time-priority, the cell loss rate for the Class 1 cells was much lower than that of the Class 0 cells for both FAS-1 and FAS-2 (Fig. 7 and 8). The time priority scheme of FAS-1 and FAS-2 was able to route the Class 1 cells faster even when the switches were under high loads. Similar results were obtained for different switch sizes and for normal distribution. As can be seen in Fig. 9 and 10, FAS-1 and FAS-2 maintained low cell loss rates for the high priority Class 1 cells at high loads by sacrificing the cell loss rates for the Class 0 cells. At 95% load with normal distribution, 32×32 FAS-1 achieved a cell loss rate of 5.5×10^{-4} for the Class 1 cells, and 32×32 FAS-2 achieved a cell loss rate of 2.1×10^{-5}.

The switch designed by [4] uses an n-dimensional hypercube to implement an $N \times N$ switch ($N = 2^n$). [4] gave simulation results for a 128×128 switch (7-dimensional hypercube, a total of 128 nodes) with five different routing algorithms. The performance comparison between FAS-1/FAS-2 and [4] is summarized in Table 1. With the same number of nodes and a slight increase in the number of links ($1/n$), FAS-1 gave slightly higher total cell loss rate. However, FAS-1 gave higher priority to time-sensitive cells, and the Class 1 cells experienced a much lower cell loss rate than the cells of [4]. FAS-2 used twice the number of nodes, but its cell loss rate was much lower than the switch of [4] and even FAS-1.

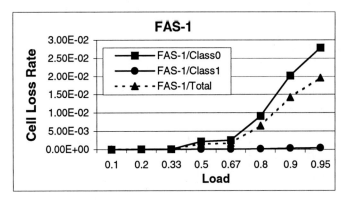

Fig. 9. Cell loss rate for FAS-1 (32×32, normal dist.)

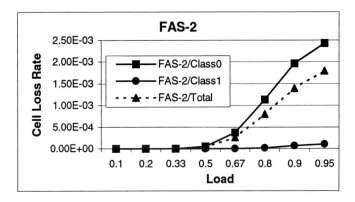

Fig. 10. Cell loss rate for FAS-2 (32×32, normal dist.)

The switch designed by [5] uses an $n+1$-dimensional hypercube to implement an $N \times N$ switch ($N = 2^n$). [5] gave simulation results for a 32×32 switch (5-dimensional hypercube, a total of 64 nodes). The performance comparison between FAS-1/FAS-2 and [5] is summarized in Table 2. FAS-1 may have a slightly higher cell loss rate than the switch by [5], but needs smaller number of total nodes (32 nodes) and smaller sized output buffers. Also, FAS-1 is able to keep the cell loss rate and the average cell delay of the time sensitive cells (Class 1) very low. FAS-2 also had a total of 64 nodes, but gave much lower cell loss rate with smaller sized output buffers. The cells of the switch by [5] experienced delay of about 25 time units as they traversed through the switch from input to output under uniform traffic pattern and 0.95 loading. However, the cells of the FAS-1 experienced average delay of about 8.5 time units--the Class 1 cells experienced only about 6.5 time units. The cells of FAS-2 experienced lower end-to-end delay. The total cell loss rate of FAS-2 was lower than that of [5], and the cell loss rates of the time-sensitive Class 1 cells was lower by an factor of 100.

Table 1. Performance Comparison of FAS-1 and FAS2 with [4] (128×128 switch under uniform traffic and 0.9 loading)

	[4]	FAS-1			FAS-2		
		Total	Class 0	Class 1	Total	Class 0	Class 1
Cell Loss Rate	1.5×10^{-5}	4.1×10^{-5}	5.5×10^{-5}	8.3×10^{-6}	4.7×10^{-6}	4.5×10^{-6}	5.2×10^{-7}

Table 2. Performance Comparison of FAS-1 and FAS2 with [5] (32×32 switch under uniform traffic and 0.95 loading)

	[5]	FAS-1			FAS-2		
		Total	Class 0	Class 1	Total	Class 0	Class 1
Cell Loss Rate	2.0×10^{-4}	4.0×10^{-4}	5.5×10^{-4}	5.9×10^{-5}	1.5×10^{-5}	2.1×10^{-5}	2.0×10^{-6}
Ave. Cell Delay	25	8.47	9.33	6.47	7.72	8.85	6.86
Input Buffer Size	8	2			1		
Output Buffer Size	40	38			35		

Conclusion

Based on the folded hypercube network (a variation of the hypercube network), two new ATM switches FAS-1 and FAS-2 were proposed in this paper. Simulation study of the two switches showed them to have very good performances in terms of end-to-end cell delay and cell loss rate. The performances of the two switches were compared to two hypercube switches in the literature. FAS-1 and FAS-2 outperformed the switches by [4, 5] with only slight increase in hardware cost. While the switches by [4, 5] did not implement any time-priority, FAS-1 and FAS-2 were implemented with two-level time-priority giving higher priority to the time-sensitive cells. The benefit of this priority scheme was shown in the simulation results. The cells loss rate and the average cell delay of the time-sensitive cells were much lower than those of the time-insensitive cells. Even when the loading neared 100%, both FAS-1 and FAS-2 still managed to achieve very good cell loss rate for the time-sensitive cells. Also, the time-sensitive cells experienced low end-to-end delay even when the switches were heavily loaded. These results indicated that FAS-1 and FAS-2 are good candidates as new ATM switches for the fast growing high-speed networks.

References

1. H. Ahmadi and W. E. Denzel, "A Survey of modern high-performance switching techniques," *IEEE Journal on Selected Areas in Communications*, vol. 7, no. 7, pp. 1091-1103, Sep. 1989.
2. R. Y. Awdeh and H. T. Mouftah, "Survey of ATM switch architectures," *Computer Networks and ISDN Systems*, vol. 27, no. 12, pp. 1567-1613, Nov. 1995.
3. J. H. Patel, "Performance of processor-memory interconnection network," *IEEE Transactions on Computers*, vol. 30, no. 10, pp. 771-780, Oct. 1981.
4. T. Matsunaga, "Sorting-based routing algorithms of a photonic ATM cell switch: HiPower," *IEEE Transactions on Communications*, vol. 41, no. 9, pp. 1356-1363, Sep. 1993.
5. D. C. W. Pao and W. N. Chau, "Design of ATM switch using hypercube with distributed shared input buffers and dedicated output buffer," *International Conference on Network Protocols*, pp. 92-99, 1995.
6. H. J. Siegel, "Analysis Techniques for SIMD Machine Interconnection Networks and the Effects of Processor Address Masks, " *IEEE Transactions on Computers*, vol. C-26, pp. 153-161, Feb. 1977.
7. A. El-Amawy and S. Latifi, "Properties of folded hypercubes," *IEEE Transactions on Parallel and Distributed Systems*, vol. 2, no. 1, pp. 31-42, Jan. 1991.
8. S. Latifi and A. El-Amawy, "On folded hypercubes," *International Conference on Parellel Processing*, vol. 1, pp. 180-187, 1989.
9. J. S. Park and N. J. Davis IV, "Modeling the Folded Hypercube Network with OPNET," *The Proceedings of OPNETWORK '99*, Washington, DC, August 1999.
10. A. G. Greenberg and B. Hajek, "Deflection routing in hypercube networks," *IEEE Transactions on Communications*, vol. 40, no. 6, pp. 1070-1081, Jun. 1982.
11. E. A. Varvarigos and D. P. Bertsekas, "Performance of hypercube routing schemes with or without buffering," *IEEE/ACM Transactions on Networking*, vol. 2, no. 3, pp. 299-311, Jun. 94.
12. F. B. Chedid and R. B. Chedid, "A new variation on hypercubes with smaller diameter," *Information Processing Letters*, vol. 46, no. 6, pp. 275-280, Jul. 26 1993.
13. B. R. Collier and H. S. Kim, "Efficient analysis of shared buffer management strategies in ATM networks under non-uniform bursty traffic," *IEEE INFOCOM'96*, pp. 671-678, 1996.
14. S. C. Liew, "Performance of various input-buffered and output-buffered ATM switch design principles under bursty traffic: simulation study," *IEEE Transactions on Communications*, vol. 42, no. 2/3/4, pp. 1371-1379, 1994.
15. D. M. Dias and J. R. Jump, "Packet switching interconnection networks for modular systems," *IEEE Computer*, vol. 14, pp. 43-53, Dec. 1981.
16. A. L. DeCegama, *The Technology of Parallel Processing, Volume 1*, Prentice Hall, Englewood Cliffs, NJ, 1989.
17. G. F. Pfister and V. A. Norton, "Hot spot contention and combining in multistage interconnection networks," *IEEE International Conference on Parallel Processing*, pp. 790-797, 1985.
18. J. S. Park, Performance analysis of partitioned multistage cube network and adaptive routed single-stage cube network, Master's Thesis, Virginia Polytechnic Institute and State University, 1994.

Open Software Architecture for Multiservice Switching System

Ho-Jin Park, Young-Il Choi, Byung-Sun Lee, Kyung-Pyo Jun

Network Technology Lab., Electronics & Telecommunication Research Institute(ETRI)
161 Kajong-Dong, Yusong-Gu, Taejon, 305-600, Korea
{hjpark, yichoi, bslee, kpjun}@etri.re.kr

Abstract. To meet the rigorous demands of today's fast-moving telecommunication environment, every carriers want to deploy new services smoothly with minimal impact on the underlying infrastructure. Also it is necessary to create an open environment that enables the deployment of multi-vendor switching systems with optimum functionality and performance. The MSF(Multiservice Switching Forum) is an industrial forum to develop the network infrastructure that meets the above requirements. The MSF has been focusing on reaching and promoting consensus on the implementation of MSS(Multiservice Switching System) that realizes the vision of multiservice switching and defines a set of open intra-switching system interfaces. In this paper, we review the system architecture and characteristics of the MSS based on MSF architecture. We propose the open software architecture for the implementation of the MSS, the control scenarios for ATM SVC(Switched virtual Connection) service and management functions in the software architecture. The environment for software development is also considered.

1 Introduction

Because of rapid changes in customer's service requirement, telecommunication market environment and technology, every carriers are concerned about the network infrastructure that reduces risk by enabling the smooth deployment of new services with minimal impact on the underlying infrastructure. Since it can be risky to lock into a single vendor, proprietary solution for network infrastructure, it is necessary to create an open environment, where carriers and network operators can select the best of breed for each type of telecom equipment. This network infrastructure is totally different from the traditional one and requires flexible multiservice switching technology with open interfaces.

At present, there are some organizations making the consensus on multiservice switching technology with open interfaces: such as MSF(Multiservice Switching Forum)[1][2], IEEE PIN(Programmable Interface for Network)[3], ISC(International Softswitch Consortium)[4], etc. The MSF has been founded in December 1998 for reaching and promoting consensus on implementation of MSS(Multiservice Switching System)[5][6] that realizes the vision of multiservice switching and defines a set of open intra-switching system interfaces. The goal of MSF is to develop an open switching system platform with both IP and ATM based services

In this paper, we review the system architecture and characteristics of MSS based on MSF architecture. The open software architecture for the implementation of MSS is proposed. We discuss the control scenarios for ATM SVC(Switched virtual Connection) service and management functions in the software architecture. The environment for software development is considered. Finally, we conclude our work.

2 MSS(Multiservice Switching System)

MSS is a distributed switching (based on frame, cell or packet) system designed to support voice, video, and data such as ATM, Frame Relay, and Internet Protocol services.

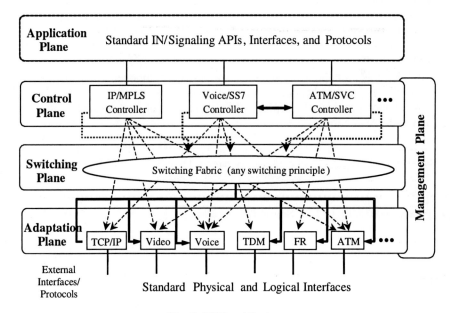

Fig. 1. MSS architecture

2.1 Multi-plane System Model

A multi-plane system model has been selected as the basis of MSS architecture. An essential characteristic of the architecture is its flexibility in terms of the technology chosen to implement functionality of each plane. New technologies are allowed to evolve independently from each other in all three planes.

As shown in the Fig. 1, MSS architecture is composed of 5 planes. The adaptation plane includes the physical interface to a user or another network element. The switching plane supports the actual switching fabric by which physical interfaces are connected. The control plane provides the generic capability to manage network service events and provides control over both the adaptation and switching planes.

Standard protocols are used in communicating between the control plane and the switching/adaptation planes. The application plane provides services that use the capabilities of the control plane. It also provides enhanced services which control the services within the control plane. The management plane performs FCAPS(Fault, Configuration, Accounting, Performance, Security) functions against each plane.

2.2 Characteristics of MSS Architecture

The main characteristics of MSS architecture are as follows:
- MSS architecture realizes the vision of Multiservice switching by allowing logical partitioning of common resources between an ever-changing set of services.
- It separates the control processing from the switching and adaptation aspects of a MSS to establish a framework which can be easily extended to support new Adaptation, Switching and Control Plane functions. Standard protocol is used at the interface between controller and switch, such as GSMP(General Switch Management Protocol)[7] or MEGACO(Media Gateway Control Protocol)[8], COPS-PR(Common Open Policy Service – Policy Provisioning)[9].
- It divides the monolithic switch into specialized components and defines open standardized interface between components, as shown in Fig. 2. These interfaces allow service providers to deploy MSSs composed of best-of-breed components from multiple vendors.

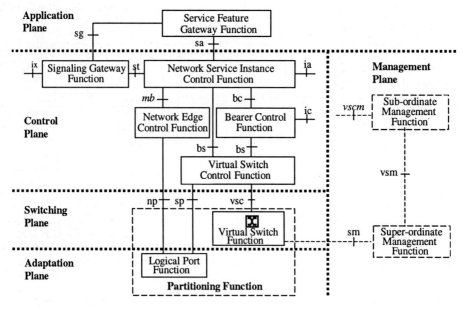

Fig. 2. Reference architecture of MSS

3. Open Software Architecture of MSS

In this section, it is proposed the open software architecture for the implementation of MSS for ATM services.

3.1 Separation of Complexes

For physical implementation, MSS is decomposed into three components: switching complex, processor complex and management complex, as shown in Fig. 3.

The switching complex contains ports implementing adaptation plane functions connected via a switching fabric. The switching complex connects to a cluster of one or more controller processes in a processor complex. The processor complex includes computers running the controller software that implement functions in the control plane. The management complex covers functions for monitoring and control of the switching, adaptation and control planes.

GSMPv3 protocol is used between the processor complex and the switching complex, SNMP(Simple Network Management Protocol) protocol between the management complex and the switching complex, the proprietary interface based on the middleware, CORBA(Common Object Request Broker Architecture) between the management complex and the processor complex.

In the following subsections, the building blocks in each complex are defined and their roles and interactions are discussed.

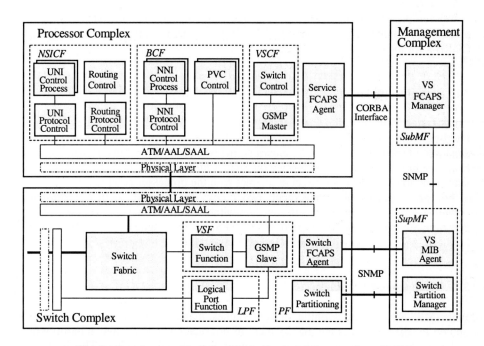

Fig. 3. Complexes and building blocks for the implementation of MSS

3.2 Processor Complex

For the control of network service instances, two building blocks are defined: *UNI Control* and *Routing Control*. *NNI Control* and *PVC Control* are defined for the control of bearer connection between end points, *Switch Control* and *GSMP Master* for the control of virtual switch and logical ports, *Service FCAPS Agent* for the management of services in the processor complex.

UNI(User Network Interface) Control contains the logic and information to establish, maintain, modify, and release a network service instance. It provides the information for billing and performance and statistics analysis. ATM Forum UNI protocol 3.1 and 4.0 are used at UNI interface. *Routing Control* determines the end point address and routing options for reaching the address. It identifies the control signaling, bearer signaling, address requirements of the associated network service instance. ATM Forum PNNI(Private Network Node Interface) routing protocol is used to manage and maintain routing information with peer entities.

NNI(Network Node Interface) Control establishes, manages and maintains the state of the required switched bearer path(s) for a specific network service instance. It also manages and maintains the state of links under its control. PNNI signaling protocol is used to perform signaling to peer entities. *PVC(Permanent Virtual Connection) Control* establishes, manages and maintains the state of the required permanent virtual bearer path(s) for a specific network service instance.

Switch Control controls and monitors the virtual switch and logical ports within a partition. It manages the cross-connection between the logical ports in the virtual switch. *GSMP Master* performs the master functions of GSMP protocol to control the switching complex.

Service FCAPS Agent performs FCAPS functions against the services in the processor complex. The interaction with the management complex is done through CORBA for location and access transparency.

3.3 Switch Complex

Cross-connect functionality between logical ports is realized by the following three building blocks. *Logical Port Function* provides media mapping and service specific adaptation functions related to the incoming media stream. Switching of media stream from one port to another is performed by *Switch Function*. *GSMP Slave* performs the slave functions of GSMP protocol to interact with the processor complex.

For partitioning and management, two blocks are defined. *Switch Partitioning* creates a virtual switch by specifying the switch resources that are to make up the partition. *Switch FCAPS Agent* performs FCAPS functions against the physical switch hardware and interacts with the management complex.

3.4 Management Complex

In the MSS architecture defining virtual switches, management functions are required on two levels: the first is the management of physical switch and switch partitioning, and the second is the management of individual virtual switches. The former is referred to as the Super-Ordinate Management Function(SupMF), whilst the latter as the Sub-Ordinate Management Function(SubMF).

SupMF is composed of two building blocks: *Switch Partition Manager, VS MIB Agent(Virtual Switch Management Information Base Agent)*. *Switch Partition Manager* controls the partition of virtual switch by creating, modifying, and deleting virtual switches. Four kinds of switch resources are partitioned into each virtual switch. The first is the bandwidth, buffer space, and queue schedulers per logical port assigned to a virtual switch. The second is the label space resources per logical ports (e.g., VPI/VCI for an ATM port, DLCI for a frame relay port, etc.). The third is the routing table space of the switch per virtual switch. The last is the bandwidth of communication link between the processor complex and the switch complex, and the processor cycles of controller in the processor complex. *VS MIB Agent* partitions and virtualizes the management information and functions presented to the SubMF corresponding to each virtual switch. It allows the manager to perform FCAPS functions against the switch complex by managing the physical hardware.

SubMF is mapped into 5 building blocks as follows. *VS Configuration Manager* manages the configuration of logical ports in the virtual switch. The configuration information of logical ports is received from the switch complex at the initialization phase of controller in the processor complex with all port configuration message procedures of GSMP. *VS Fault Manager* performs supervision and test functions on logical ports. Changes in port state are reported by the switch complex with event message of GSMP. It performs management functions against logical ports with the help of Switch FACPS Manager in SupMF. *VS Account Manager* is responsible for billing to the call/connection services in the virtual switch. It gathers the billing information from UNI Control in the processor complex. *VS Performance Manager* measures and analyzes, monitors call traffics and the performance of connections and ports. Call traffic data is gathered from the building blocks in the processor complex, and performance data from the switch complex via Switch FACPS Manager in SupMF. *VS Security* Manager identifies and authenticates users. It controls resource access to make sure that users are authorized to perform the functions they request.

4 Control Scenarios for Service and Management

In this section, the simplified control scenarios for ATM SVC service and management functions in the software architecture are considered.

4.1 ATM SVC Service

In this scenario (see Fig. 4), it is assumed that a customer using ATM protocol accesses a network carrier's ATM transport service to establish a SVC connection.
1. *UNI Control* creates an originating process to control the call when it receives SetUp message from the signalling channel of the ATM subscriber that is permanently connected to a controller in the processor complex via the switch complex at system initialization.
2. *UNI Control* process requests an available VPI/VCI label and bandwidth on the input port to *Switch Control* and receives the reply.

3. If the reply is OK, *UNI Control* process requests number translation and routing for the call to *Routing Control* and receives the reply with the output port id.
4. *UNI Control* process sends SetUp_Internal message to *NNI Control*, which creates an outgoing process to control the call.
5. *NNI Control* process requests an available label and bandwidth on the output port to *Switch Control* and receives the reply.
6. *NNI Control* process notifies the self-process id. to *UNI Control* process, and send SetUp message to the peer switching system through the signalling channel.
7. When *NNI Control* process receives Connect message from the peer switching system, it relays the message to *UNI Control* process.
8. *UNI Control* process requests through-connection between the input port and the output port to *Switch Control*.
9. *Switch Control* requests through-connection via *GSMP Master* to the switch complex with AddBranch message of GSMP.
10. After the successful connection in the switch complex, *UNI Control* process sends Connect message to the originating ATM customer and terminates the call setup phase.

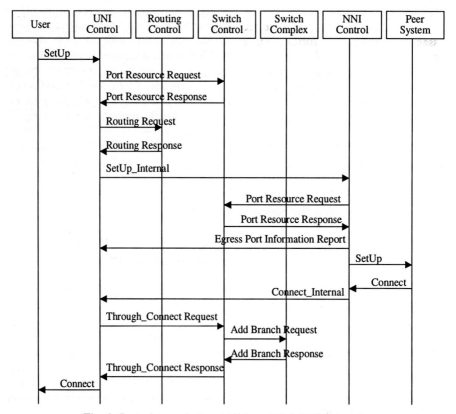

Fig. 4. Control scenario to establish an ATM SVC connection

4.2 Switch Partition

At the initialization of system, switch partition procedure to create virtual switches is done as follows (see Fig. 5):

1. *Switch Partition Manager* gets the switch resource information from the switch complex by doing the SNMP "get" operation. The switch resource information includes the number of ports, the label type and range, the equivalent bandwidth, the performance parameters (e.g., packet loss ratio, maximum packet delay) for each port.
2. *Switch Partition Manager* creates virtual switches by partitioning the switch resources. It also configures the communication interface between the switch complex and controllers in the processor complex, such as controller id., control protocol (e.g., GSMP, MEGACO, COPS-PR), control interface (e.g., VPI/VCI for ATM interface, IP address and port number for Ethernet interface) for each controller.
3. *Switch Partition Manager* commits switch partitioning to the switch complex by doing the SNMP "set" operation on the variables in switch partition MIB.
4. *Switch Partitioning* in the switch complex setups switch partitioning and sends the virtual switch configuration information to each controller with GSMP.
5. With the virtual switch information, each controller controls calls and connections within the virtual switch. The virtual switch information is shared with SubMF to manage the virtual switch.

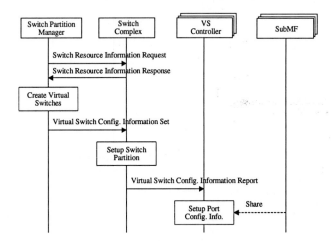

Fig. 5. Control scenario for switch partitioning

4.3 FCAPS Management

The scenario to perform FCAPS management functions against switch hardware is done as follows (see Fig. 6):
1. The building block in SubMF requests management function for the virtual resources(e.g., logical port) to *VS MIB Agent* by doing the proper SNMP operation.
2. *VS MIB Agent* checks whether the SubMF are authorized to perform the functions against for the virtual resources. If not authorized, the refusal for the request is replied. Else, *VS MIB Agent* maps the virtual resource id. to the physical by referring the switch partition information and requests management function for the physical resources to the switch complex by doing the proper SNMP operation.
3. *Switch FCAPS Agent* in the switch complex performs management function and replies the result to *VS MIB Agent*. *VS MIB Agent* relays the response to the building block in SubMF.

When VS *MIB Agent* receives the event on physical switch resources from the switch complex, it identifies the related controllers by referring the switch partition information and relays the event on virtual switch resources to the controllers.

The procedures to perform management functions against controllers in the processor complex are done between the building blocks in SubMF and *Service FCAPS Agent* in the processor complex through CORBA.

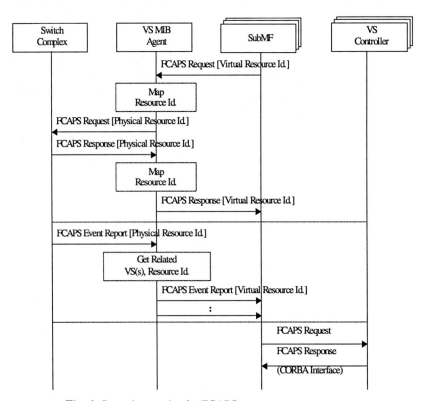

Fig. 6. Control scenarios for FCAPS management

5 Software Developing Environment

The software structure is specified and designed by SDL(Specification and Description Language) that is a graphical language recommended by ITU-T. C codes are automatically generated from SDL code by CASE tool. We estimate that with SDL method the development cycle time is reduced by 50% comparing with the hand-written coding. The performance of the execution file is degraded by about 10%, which is considered not critical.

In implementing the prototype of MSS, commercial workstation with Solaris OS is used for a control platform which accommodates the processor complex and the management complex. For the switch complex, it is used HANBit ACE 256 which is an ATM switching system developed in Korea.

6 Conclusion

We review the system architecture and characteristics of MSS based on MSF architecture. The open software architecture for the implementation of the MSS is proposed and the control scenarios for ATM service and management functions in the software architecture are discussed. The environment for software design and the prototype implementation is also considered.

In 2000, which is the first year of our project to develop network components for open multiservice networks, ETRI with 6 telecommunication equipment companies in Korea had developed the prototype of MSS for ATM services. In the following years, we are going to add MPLS(MultiProtocol Label Switching) functions to the prototype for Internet Protocol services. Media Gateways for VoATM(Voice over ATM) and VoIP(Voice over Internet Protocol) services with an open architecture will be also developed to interwork with the existing circuit switched networks.

References

1. N. Bjorkman et al., "The Movement from Monoliths to Component-Based Network Elements," IEEE Communications Magazine, January 2001.
2. MSF(Multiservice Switching Forum): http://www.msforum.org
3. IEEE PIN(Programmable Interface): http://www.ieee-pin.org
4. ISC(International Softswitch Consortium): http://www.softswitch.org
5. MSF, "Multiservice Switching Forum System Architecture Implementation Agreement 1.0," MSF-ARCH-001.00-FINAL IA, May 2000.
6. MSF, "White Paper: Physical Realization of the MSF Functional Architecture," MSF 2000-105, Jun. 2000.
7. IETF GSMP: http://www.ietf.org/html.charters/gsmp-charter.html
8. F. Cuervo et al., "Megaco Protocol Version 1.0," IETF RFC 3015, November 2000.
9. K.H. Chan et al., "COPS Usage for Policy Provisioning (COPS-PR)," IFTF Internet Draft, October 2000.

A Multicast ATM Switch Based on PIPN[1]

Sema F. Oktug

Department of Computer Engineering
Electrical-Electronics Faculty
Istanbul Technical University
Maslak Istanbul 80626 Turkey
oktug@itu.edu.tr

Abstract. This paper presents a multicast ATM switch based on the Plane Interconnected Parallel Network (PIPN) which was proposed earlier. In order to convert the PIPN structure into a multicast ATM switch, input port buffers, internal input buffers, and resequencers are deployed. Moreover, to prevent packet loss in the internal stages, the backpressure mechanism is used. Copies of multicast packets are generated by using the recycling technique which employs multicast tree generation. The performance of the proposed multicast ATM switch is studied by simulations and the results are given in terms of throughput.

Introduction

As B-ISDN and ATM find their places in our lives, it has been seen that multicast switches are inevitable parts of this technology. With the introduction of various applications, such as teleconferencing, distributed data processing, and video on demand, the percentage of multicast traffic in aggregate network load is increasing day by day. It is also clear that multicast switches are causing efficient use of network resources. As a result, new multicast ATM switches and multicast versions of the previously proposed ATM switches have been introduced [1].

In multicast ATM switches, one of the important issues is to generate multiple copies of a multicast cell. For this purpose, three techniques are used mainly: 1) Employing a dedicated copy network, 2) Generating copies while multicast cells are kept in input buffers, and 3) By using recycling based on the multicast tree approach. First two techniques generate all multicast cells at once or by using a cell splitting mechanism. In the multicast tree approach, multicast cells are generated according to a tree hierarchy arranged at call set up. In this manner, instant fluctuations of the internal switch traffic can be prevented. However, it may take longer for all the multicast cells to leave the switch.

This paper introduces a new multicasting ATM switch based on the *Plane Interconnected Parallel Network* (*PIPN*) architecture. The PIPN which exploits the

---

[1] This work is financially supported by Turkish Scientific and Technical Research Council under grant number EEEAG 197E027 and the Istanbul Technical University Research Fund under grand number 1076.

properties of banyan networks by alleviating the drawbacks of them was introduced in [2,3]. Its structure was mainly based on the idea of parallel interconnected routing planes. In the design of the PIPN, the main aim was to distribute the incoming heterogeneous traffic evenly throughout the interconnection structure in order to increase the performance under heterogeneous traffic. In this work, the PIPN structure is turned into a multicast ATM switch with the help of input port buffers, internal input buffers, and the backpressure mechanism. The recycling technique is used to generate multicast packets by using a multicast tree for each connection. It is observed that the number of recirculations is dropped significantly due to the outlet grouping property of the PIPN. However, due to the internal buffering and multiple paths between each input-output port pair, deployment of resequencers at the outlets of the switch have been inevitable in order to prevent out-of-order departures.

This paper is organized as follows: Section 2 introduces the structure of the enhanced PIPN. Section 3 discusses the copy generation in multicasting and explains the recycling technique. The function of the input ports and the deciders in multicasting is explained here. The performance comparison of the multicast PIPN and a Banyan network based multistage load distributing multicast switch is given in Section 4. Finally, Section 5 concludes the paper.

Structure of the Multicast PIPN

The PIPN is composed of three main units, namely the *distributor*, the *router*, and the *output-port-dispatcher*, as explained in [2]. The performance enhancement obtained after adding internal input buffers to the elements in the distributor and the router is presented in [3]. In this work, the structure of the PIPN is enhanced some more in order to support multicast traffic. There are three techniques used in the performance enhancement, namely input port buffering, internal input buffering, and the backpressure mechanism. Moreover, in order to support multiple paths between inlets and outlets, resequencers are deployed. Figure 1 shows the general structure of the multicast PIPN switch. In the figure, the packet recirculation structure is shown only for input port N-1 not to make it very complicated.

Enhanced Distributor

The distributor is composed of 2^{n-1} identical *distributing elements* (*DE's*). Each DE is connected to two consecutive input ports and has two outlets. There is an internal input buffer of one packet size at either inlet of a DE. Packets entering the distributor are separated into two groups by assigning them either to the upper or to the lower DE outlets with equal probability. The upper/lower outlets of DEs are connected to the *back plane/front plane* of the router. DEs complement the destination address fields of the packets going into the front plane and set their complement fields to logical one. The complement fields of the packets going into the back plane are set to logical zero and their destination address fields remain the same. Backpressure lines are not shown in the figures not to make them very complicated.

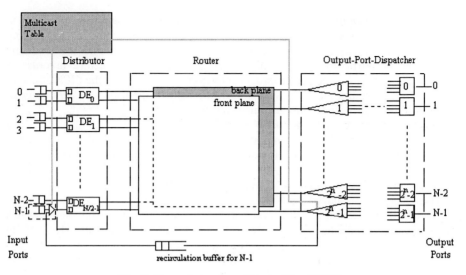

Fig. 1. General structure of the multicast PIPN

Enhanced Router

The router is made of two $2^{n-1} \times 2^{n-1}$ planes which are called the *front plane* and the *back plane*. The upper outlets of the DEs are connected to the back plane and the lower ones are connected to the front plane. Since sending a packet into the back plane or the front plane is equally probable, assignment of the distributor outlets to the planes has no significant importance. Each plane has $n-1$ stages and each stage has 2^{n-2} SEs. There are internal input buffers of size one at both inlets of SEs. The deployment and operation of SEs are the same as those in a baseline network. However, the interconnection structure of SEs in the PIPN is different than that of a baseline network. The internal structure of a 16x16 router is shown in Figure 2. The details related to the interconnection structure of the router planes can be found in [2].

Packets entering the router are transferred through the interconnected planes by using the information kept in their destination address fields just like in a baseline network. However, here, the destination address fields are inspected starting with the second most significant bit since the router planes have $n-1$ stages. When a packet reaches a paired SE, it moves to the other plane if it is not the loser side of a contention. The loser packet stays in the internal input buffer to be tried in the next time slot. With the use of input port buffering, internal input buffering, and the backpressure mechanism, packet loss at the internal stages of the switch is prevented. While moving packets from one plane to the other, the destination address and the complement fields are not modified. Packets continue their ways in the transferred plane without modifying their destination address fields.

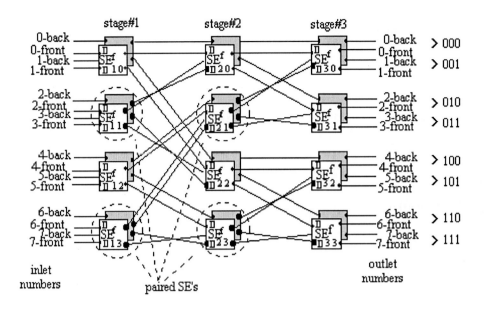

Fig. 2. Internal structure of a *16x16* router

Enhanced Output-Port-Dispatcher

The packets reaching the outlets of the router are sent to the output-port-dispatcher in order to be routed to the requested output ports. The output-port-dispatcher is composed of the units of two types called *deciders* and *collectors*. If it is necessary (i.e. it is a multicast packet), a packet can be recycled back to an input port. Each inlet of the output-port-dispatcher can carry packets destined to four different output ports as shown in Figure 3. This is called the *outlet grouping property* of the PIPN. For unicast packets, inspecting their complement fields and the most significant bit of the destination address fields is enough to determine their actual destination addresses. However, for a multicast packet, the decider first checks if it needs to be recirculated by using its *recirculation_tag* which will be explained in the next section. If it is necessary, a copy of the packet is generated and send to the corresponding recirculation buffer. Then, the packet is dispatched to the collectors of the destined output ports which are in the same outlet group. It is possible to have a multicast packet destined to more than one output port which are in the same outlet group. In such a case, additional copies are generated in the deciders.

There is a collector per output port. A collector can process four packets arriving within the same time slot, however, sends only one of them to the corresponding output port. Due to the multiple paths between an inlet-outlet pair and internal input buffering, out-of-order arrivals may occur at the collectors. To prevent out of order departures, a resequencer is deployed at each collector. As the resequencer structure the one proposed in [4] is selected. *Fixed-waiting-time* is used to set the minimum amount of time slots spent by the packets in the switch. Those packets that arrive outlets earlier wait in the resequencers.

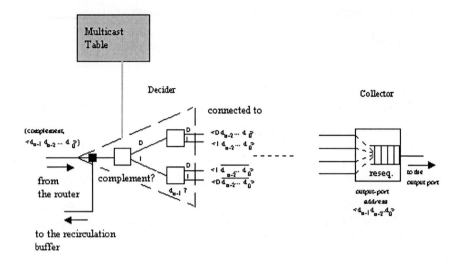

Fig. 3. Internal structure of the enhanced output-port-dispatcher

Operations Related to Multicasting

The structure and the routing logic of the PIPN is very appropriate for multicast traffic. A packet which reaches a decider can be sent to four different output ports in the same group. So, if there is a packet destined to all four or some of these output ports in the same group, necessary copies can be generated from only a single packet arriving at the output-port-dispatcher. In this way, the load in the distributor and the router is reduced significantly. This property of PIPN is very meaningful when multicast traffic is applied. Multicast packets can be delivered to the requested outlets without making the environment overcrowded.

Multicast Table and Multicast Packet Structure

Whenever a new multicast connection is established, a new entry in the multicast table is created. This entry contains two important fields: *multicast-outlets, and multicast-groups*. *Multicast-outlets* is a binary array of size N showing the multicast outlets for a connection. Zeros and ones are used to represent if outlets take part in a multicast connection or not. *Multicast-groups* is a binary array of size $N/4$. It is obtained from multicast-outlets considering the outlet grouping property of the PIPN. Outlet groups are numbered from zero to $N/4-1$. If, at least, one outlet of an outlet group involves in the multicast connection, corresponding bit in multicast-groups is made one, zero otherwise.

Packets in the multicast PIPN have three additional tags: *packet-type, multicast-id,* and *recirculation-tag*. *Packet-type* is a one bit field used to show if a packet carries a multicast or a unicast cell. *Multicast-id* is necessary to find the corresponding entry of a multicast packet in the multicast table. *Recirculation-tag* is used in the multicast tree generation. It specifies the packets to be recirculated in the deciders.

Multiple Cell Generation

The employed multiple packet generation technique is derived from the one used in [5]. In [5], Turner introduced the recycling technique in order to generate multiple copies of multicast packets.

This technique is adapted to the multicast PIPN architecture after some modifications. In the modified version, a binary tree is constructed with the source switch inlet at its root and *outlet group numbers* at its leaves. Internal nodes also represent *outlet group numbers* and act as relay points by recycling packets. In [5], one of the packets generated regarding the tree structure is sent to an output port and the other is recycled. However, in the proposed switch, both copies reach the output ports to leave. If necessary, one of them (the one with a set *recirculation-tag*) can be used to generate additional copies as shown in Figure 3.

Whenever a multicast connection is established, corresponding input port takes action to create a multicast table entry. Then first level branches of the tree are decided randomly by using *multicast-groups*. With the help of recycling packets, the multicast tree can be formed dynamically.

An example multicast tree for a 16x16 multicast PIPN is given in Figure 4. Initially, the output ports are grouped. At each iteration, at most two copies of the multicast cell are generated. Outlet groups are selected randomly. The destination address field of such a packet keeps the address of an output port which is in that outlet group and involves in the multicast connection. If necessary, that packet can be marked to be recirculated, such as G2 in Figure 4.

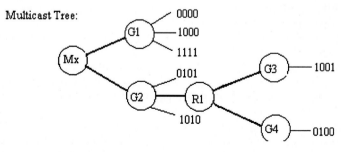

Fig. 4. An example multicast tree for the proposed switch of size 16x16

Results

Performance of the enhanced PIPN is studied by simulation. The simulation environment is created on a SUN Ultra 10 by using the C programming language. The results obtained under unicast and multicast traffic are compared with those of other ATM switches. In order to use in comparisons, a banyan network based load distributing multicast ATM switch, similar to [5], is also simulated. The general structure of this switch is given in Figure 5.

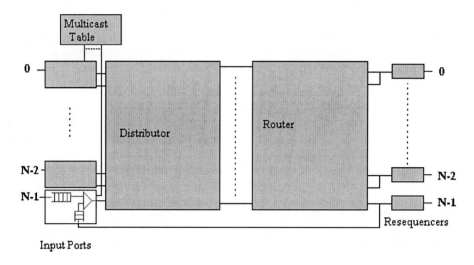

Fig. 5. General structure of the banyan network based load distributing multicast switch

Enhancing PIPN causes increase in performance not only under multicast traffic but under unicast traffic also. In order show the performance gain obtained, the performance of the enhanced PIPN and the enhanced banyan network of size 128x128 is compared. It is also crosschecked that the simulation results obtained under uniform traffic confirm those presented in [6]. As the heterogeneous traffic pattern, *communities-of-interest* traffic, which can range from uniform traffic to extremely heterogeneous traffic by changing the community size, is used [7]. It is observed that enhanced PIPN performs better than the banyan network based switch which is enhanced using similar techniques.

The arrivals in the multicast traffic are simulated by using independent and identically distributed Bernoulli processes where input load parameter may change between 0.0 and 1.0. The multicast fanout is calculated by using a model similar to the one used in [8]. There, the multicast fanout is defined as a random variable F with distribution density function $P(F=f \text{ destinations}) = 1/(f*f)$ where f is between 2 and the switch size, N. This function is normalized in order to cover switch size.

The performance of the multicast PIPN is compared with that of the enhanced banyan network based load distributing multicast switch which also uses multicast tree approach to generate multicast packets and employs resequencers [5]. Currently, simulations are performed for the networks of size 16x16, and 32x32. It is calculated that the average fanout values for 16x16, and 32x32 switches are 4.08, and 4.98, respectively when the above mentioned multicast traffic is applied. Some results obtained for the networks of size 32x32 are shown in Figure 6.

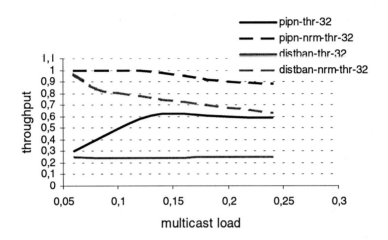

Fig.6. Throughput/normalized throughput obtained by the enhanced PIPN and the enhanced banyan network under multicast traffic for size 32x32

In this test *fixed-waiting-time*s for the enhanced PIPN, and the enhanced banyan network are taken as 28 time slots, and 832 time slots, respectively. *Fixed-waiting-time*s are adjusted according to the performance of these switches under uniform traffic and the employed multicast tree approach. It is shown that as a result of the outlet grouping property of the PIPN, internal switch traffic does not increase tremendously. A single packet going through the router network is enough to generate multiple copies for the outlets in the same outlet group. Due to this property, multicast packets circulate a few times in the enhanced PIPN and fixed-waiting-time is kept smaller.

Moreover, the delay experienced by networks of various sizes under various traffic types are also studied.

Conclusions

In this paper a new multicast ATM switch which is based on the PIPN structure introduced earlier is proposed. The PIPN architecture is enhanced by employing input buffers, internal input buffers, the backpressure mechanism, and the recycling technique. Out of order departure of packets is prevented by deploying resequencers. It is shown that according to the multicast structure of the proposed architecture, multicast cells can be generated without overloading the interconnection network.

Currently, the performance of the multicast PIPN is being compared with the banyan network based multicast ATM switches under various traffic patterns. It is shown that as the multicast traffic becomes heterogeneous, the performance of the proposed multicast ATM switch becomes better than many other multicast ATM switches due to its internal structure.

References

1. Guo M.H., Chang R.S, "Multicast ATM Switches:Survey and Performance Evaluation," ACM Comp.Comm.Rev., Vol.28, No.2, Apr.1998
2. Oktug S.F., Caglayan M.U., "Design and Performance Evaluation of a Banyan Network Based Interconnection Structure for ATM Switches" IEEE JSAC, Vol.15, No.5, Jun 1997.
3. Oktug S.F., Caglayan M.U., "Evaluation of the Performance Enhancements for The Plane Interconnected Parallel Network," Proc. IEEE Globecom'97, Nov.1997.
4. Turner J.S., Data Packet Resequencer for a High Speed Data switch, United States Patent, No.5,260,935, Nov.9, 1993.
5. Turner J.S., "An Optimal Nonblocking Multicast Virtual Circuit Switch," Proc.IEEE Infocom'94, Jun.1994.
6. Theimer T.H., Rathgeb E.P, Huber M.N, "Performance Analysis of Buffered Banyan Networks," IEEE Trans. on Comm., Vol.39, No.2, Feb.1991.
7. Chiussi F., Tobagi F.A., "A Hybrid Shared Memory/Space Division Architecture for Large Fast Packet Switches," Proc.IEEE ICC'92, 1992.
8. Uzun N., Blok A., "Ten Terabit Packet Switch with SRRM Scheduling," Proc. BSS'99, Jun. 1999.

Concurrent Access to Remote Instrumentation in CORBA-Based Distributed Environment

Armin Stranjak[1], Damir Kovačić[2], Igor Čavrak[2], Mario Žagar[2]

[1] Lucent Technologies, Cherrywood Science & Technology Park,
Cabinteely, Co. Dublin, Ireland
astranjak@lucent.com

[2] Faculty of Electrical Engineering and Computing, University of Zagreb,
Unska 3, 10000, Zagreb, Croatia
{damir.kovacic, igor.cavrak, mario.zagar}@fer.hr

Abstract. This paper describes a system architecture for a distributed measurement process. Base systems provide only low-level bus oriented communication protocol (realized with General Purpose Interface Bus hardware and software). Distribution and remote control [1][2] requirements led to the design of CORBA [3][4] (Common Object Request Broker Architecture) middleware objects [5][6] that abstract underlying hardware/software installation issues. This design addresses problems of authentication, authorization, provision of exclusive instrument access from one client by enabling instrument locking mechanism, and session oriented communication due to CORBA's inability to provide satisfactory solutions to these matters. The central point of the design is the introduction of a special GPIB arbiter object. An arbiter object represents implementation of the factory pattern providing access to instrument objects for each client connecting to the arbiter. In order to secure uniqueness of connection session and allow authorization and authentication of remote clients, session oriented connection has been introduced. It was found as necessity to describe proposed solution in the form of a software design pattern[7][9].

1 Introduction

The concept of remote instrument access has followed a rapid development of computer networks and modern technology. Measurement systems represent only a small part of distributed applications or are used to offer acquisition services to interested parties or remote users. The remote instrumentation has grown from the need to overcome the obvious disadvantages of limited control and monitoring possibilities when the measurement process does not involve the separation of the physical location of measurement and instrumentation equipment and the location where data analysis is performed. The acquired data should often be distributed to multiple targets such as databases, monitoring and control applications etc. In order to satisfy the new requirements, one possible solution leads to the instruments equipped with a network interface and appropriate supporting software infrastructure necessary for that kind of connection. The concept introduces not only unwanted overhead on instrument manufacturers but it also excludes the existing measurement equipment as non-compatible without the additional support of a hardware module providing direct network access. A solution involving an independent component equipped with a computer network interface and GPIB interface is more appropriate for the task since it does not require changes or the replacement of the existing measurement equipment. The first interface allows network connection with remote clients interested in acquisition, controlling or just collecting the measured data. The second interface is a GPIB based connection to the attached instruments. Such a solution should be based on embedded computer system [6] with limited memory space.

2   Distributed Measurement System

Distributed measurement systems (DMS) offer some widely recognized advantages such as increased performance due to exploiting concurrency and better resource utilization using shared devices. Distribution of tasks to multiple machines also improves scalability and reliability. The architectural part of DMS is usually based on the client/server architecture, which is a widespread concept in distributed environments and basically offers a simple approach in the establishment of a remote connection. Shifting from the two-tier architecture to the three-tier architecture reflects the need for a separate abstract layer between remote instrumentation and clients or data/information consumers due to improved and increased management of the measurement process as well as allowing easier monitoring of users rights in establishing connections with remote servers. In fact, the middle tier is populated with remote measurement servers representing the needed measurement processes. Fig 1 shows the main concept of the three-tier architecture.

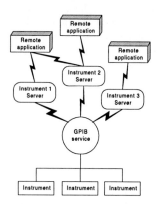

Fig. 1. Distributed three-tiered measurement architecture

The problem of defining a remote instrument interface to a certain level common to all instrument types represents a serious problem due to the diversity of equipment that can be part of such a measurement system. The problem complexity level can be reduced by using remote measurement servers primarily due to a simpler interface representation of a measurement process and measured data. Another major achievement using remote measurement servers is in handling large number of concurrent client connections. The problem of authentication, authorization or securing of exclusive instrument access can be easily solved by using the above-mentioned middle-tier component.

Instrument servers are directly responsible for GPIB-based communication with the attached measurement equipment. Instrument servers are handled by measurement servers. The abstraction levels of the instrument server interface are lower than the measurement server interface since they are abstracted GPIB instruments only and do not support high-level actions related to the GPIB command set.

The selection of the conceptual and implementation platform was a straightforward decision. CORBA[3][10] covers most of our requirements and satisfies all the tasks needed by complex distributed measurement system. Nevertheless, the mechanism of granting rights to clients or authorization and authentication should be solved outside of CORBA since there is no standardized or unified way to recognize remote clients. It is more appropriate to develop a separate module for the measurement server, which handles those activities. That concept will be described later.

3   Session-Oriented Remote Connection Using Arbiter Module

In general, an instrument connected to the public network can be viewed as a device shared between multiple remote clients. Such a device cannot satisfy client requirements concurrently, since each measurement usually requires some kind of setup procedure, which transforms the state of the device. If completely concurrent access to the same device would be allowed, that would lead to the possible corruption of measurement process and would yield unpredictable results. This problem is not unique to the measurement process. Similar problems arise in different areas, from traffic control to database systems.

A simple approach is to grant exclusive access to the device for a limited period of time (to lease a device). The client who was granted exclusive access is responsible to configure and set up the device for the measurement process it desires. In order to prevent the corruption of the process, other clients are forbidden from accessing the device during that time. In other words, the measurement system establishes a session between the client and the instrument. Such a concept is well known in digital hardware design, where multiple devices are connected to the same system bus. In order to coordinate access to shared resource, the design includes the concept of bus arbiter (or bus controller), which grants and revokes the right to use the system bus from the devices that request it.

The first step in establishing a session between the client and the target instrument is client authentication and authorization, which represents only a part of the overall security system. Although the Object Management Group (OMG) proposed an answer to this problem by defining the Security Service as a standardized CORBA service, this concept to date has not been implemented in an interoperable way. Thus, truly distributed CORBA based system has to implement access control through some other means. In the system described here, client authorization can be easily achieved since access to instruments is granted by a single Arbiter object. However, to ensure integral system security, it is required to perform this kind of communication through an encrypted channel. The underlying CORBA protocol, IIOP, is not cryptographically protected, but it is possible to tunnel the IIOP communication through a secure channel, for example by using SSL (Secure Socket Layer) or SSH

(Secure Shell) tunneling. Since these issues are common to all CORBA based distributed systems, they are not further elaborated here.

The authorized client still cannot access the instrument if it is already in use. Therefore, the measurement system has to have a way of distinguishing between different clients. Unfortunately, CORBA views object clients as anonymous entities. This prevents object implementation to query any information about the client so it basically disables the server side to distinguish clients that invoke operations on the server object. Such problem is frequently solved by applying Factory pattern, where the central component generates a unique object for each client requesting access to a particular object type.

We have found that concept inappropriate for the distributed measurement system, because the instruments are typically located on different locations and attached to different computers. Each instrument in the system has been abstracted by a separate CORBA object that exports high level interface to facilities provided by the instrument. Such an instrument server process has to be instantiated on the computer to which the instrument has been attached, because it will use a low level API to perform the actual communication with the device. The implemented instrument servers are persistent, and would run independently of any arbitration logic. Such an approach was taken in order to improve the reusability of developed objects by decoupling client management (Arbiter object) from client-instrument communication.

To provide session-oriented communication between the client and the instrument, it was necessary to introduce a special kind of information that would ideally be known only to the authorized client and the instrument. Such information should be provided by some external authorization service (in our case, from the Arbiter object) only to the two parties participating in the communication. This piece of information is called a session key.

A similar idea is frequently used in dynamic WWW based applications where it is called a *cookie*. HTTP protocol itself is stateless, which presented serious problems for server based CGI applications. Due to the stateless nature of the underlying protocol, the server side application cannot distinguish between the large number of clients invoking operations that generate dynamic content displayed in client browser. Therefore, HTTP protocol was enhanced so that the server would be able to store small piece of information in the client memory space, and later retrieve it in order to recognize the client. Such an approach proved to be very successful, since it did not require major changes to the existing protocols and applications.

Unfortunately, the introduction of session key bears changes in interface specifications to instrument servers. It is obvious that the instrument server has to retrieve the session key from the client for every operation invocation on the server. There are three possible ways to carry this information from client to server:

- The session key can be transferred by using a *context* clause in IDL interface specification. The usage of this IDL feature has been deprecated[4] and not all CORBA tools support it.
- The client can be a fully qualified CORBA object with a predefined interface, so that the server can perform callback operation and get the session key. This solution brings additional problems, because the server implementation has to take care about misbehaved clients (for example, when the client crashes during processing a server callback). It is especially

hard to avoid such problems if the server has to be realized as single threaded process.
- The server interface specification can include the session key as parameter in every operation parameter list. This solution is straightforward. The only drawback is a more complex server interface specification.

The instrument server implementation is responsible for verifying the client supplied session keys. The only question remained is which object is responsible for key generation. Obviously, the arbiter object has to have some kind of "private" interface to instrument servers, since clients initially communicate with the arbiter. This private interface is used to change session keys after each event. Events are generated for provided granting and revoking access right from client to instrument.

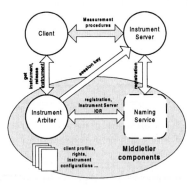

Fig. 2. Component responsibilities in distributed measurement system

4 'Arbiter Proxy' Pattern for Concurrent Remote Access

We found it necessary to describe the mentioned concept in terms of software design patterns[7]. The usage of the described concept could be expanded to a various set of problems encountered in systems which are based on concurrent access to common and shared devices.

It is often inappropriate to access a component directly and not recommended to hard-coded physical location into clients due to possible modifications or decreased capability to reuse and distribute components across several distributed platforms, so additional control mechanisms are needed. A solution should address the following forces:
- Access exclusivity and scalability - only one client may use an instrument at the same time and access to the component should be transparent and simple.
- Sharing capabilities - different clients are allowed to access different instruments at the same time on the same GPIB bus.
- Centralized management - there should be only one additional CORBA object which takes care about the clients and the instrument management.

The client will be establishing a connection with the remote instruments via the proxy component. This representative offers the interface of the component but performs additional pre- and post-processing such as client authorization and authentication, allowing exclusive instrument access etc. The descriptions and CRC cards of all CORBA interfaces involved in this design are shown in the Table 1.

Table 1. CRC cards of all interfaces

Class: Arbiter
Responsibilities
▪ Provides the interface to instrument servers
▪ Takes care of authorization and authentication of clients
▪ Takes care of exclusive access to the instrument using the session key mechanism
Collaborators
▪ InstrumentLock (through ORB)

Class: AbstractInstrument
Responsibilities
▪ Provides an interface for all instruments
▪ Ensures the usage of minimum actions needed for every instrument (such as identity etc.)
Collaborators

Class: Instrument
Responsibilities
▪ Provides an interface for any specific instrument
▪ Implementation of GPIB commands
Collaborators
▪ Client (through ORB)

Class: Lock
Responsibilities
▪ Provides an interface for setting session key
Collaborators

Class: InstrumentLock
Responsibilities
▪ It is inherited from the Instrument and the Lock interface to allow the Arbiter to set the session key
Collaborators
▪ Arbiter (through ORB)

Class: Client
Responsibilities
▪ Uses interface to request IOR and session key from the Arbiter
▪ Invocation of instrument server methods
Collaborators
▪ Instrument (through ORB)
▪ Arbiter (through ORB)

The *Arbiter* interface describes the management component of all instrument servers and presents the proxy object for clients. It takes care of the authorization and authentication of clients using the session key for every single session established between the client and instrument server. On the client's request for an object reference (IOR), the Arbiter will supply the client with an IOR of the requested instrument server if:

- instrument is not in used by any other client
- client is allowed to send requests to the remote instrument represented with the instrument server

The *AbstractInstrument* interface provides the interface implemented by the Instrument object (interfaced by *Instrument*) and is only an abstract representation of the remote instruments.

The *Instrument* interface represents the interface for implementing and wrapping GPIB commands for every single instrument involved in the measurement process.

The *Lock* interface allows setting of session key by Arbiter.

The *InstrumentLock* interface is inherited interface from the *Instrument* and the *Lock* interface. Through this interface, communication between the Arbiter and the Instrument is established due to setting the session key for any specific session. The client will establish a connection using the *Instrument* interface so it will be unable to set the session key on its own and skip the Arbiter component.

Fig. 3 shows the relationships between the interfaces, while the sequence diagram will bring a better view to described concept as shown in Fig 4.

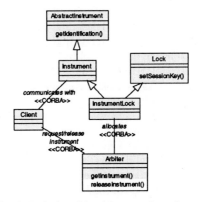

Fig. 3. Relationships between interfaces

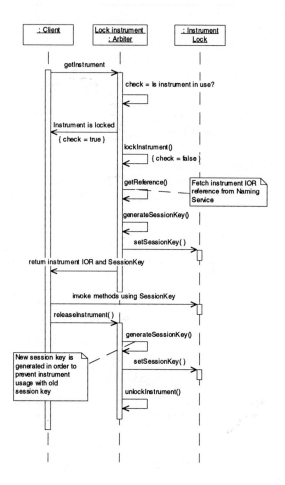

Fig. 4. Typical dynamic scenario of Arbiter pattern structure

The client should send a request for the IOR reference (actually, for pointer to the remote CORBA object on the server side) of the Instrument Server to the Arbiter. First, the Arbiter will check if it is allowed to arbitrate the requested instruments in the Arbiter config file. If it does, Arbiter will check if any other client is using the Instrument. If the Instrument is not in used by any clients, the Arbiter will lock the instrument (set internal variable), retrieve the IOR reference from the Naming Service (The Naming Service is not shown on this sequence diagram), generate a unique session key which will be used by the Client and Instrument Server for preventing any other client which already knows IOR reference to establish communication with the Instrument Server because Instrument Server will not perform any method invocation without the correct session key. All mentioned actions are hidden in *getInstrument* method. After that, the Arbiter will send the generated session key to the Instrument Server and will send IOR reference as well as session key to the Client. Now, the Client is capable to invoke methods of the Instrument Server using this session key.

After Instrument Server is no longer needed, the client should unregister itself through the Arbiter to allow other clients to get access. The Arbiter will generate a new session key and send it to the Instrument Server. In this way previous clients are prevented from accessing the instrument with an expired session key.

There is no risk if one client has got the IOR of an instrument but has released the connection, it still cannot reconnect to the instrument because it does not know the session key which is placed on the Instrument Server. But, what if the client tries to set its own session key? To prevent that, it was necessary to create derived IDL interface (*InstrumentLock* interface) from the IDL interface of instrument (*Instrument* interface) and that derived interface is used only by the Arbiter side. The derived interface will have one additional method just for setting the session key from the *Lock* interface. Clients are using the Instrument interface only and will not be able to "see" that additional method. The Fig. 5 represents that concept.

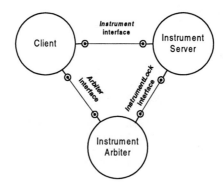

Fig. 5. Principal overview of interconnection between main components

5 Conclusion

This article describes the problems encountered during the design and development of a distributed measurement systems in heterogeneous environment. Therefore this paper is not focused on primary tasks performed by the measurement system since they are usually solved in a straightforward manner. We have found that building such systems frequently brings similar secondary requirements that have significant impact on the architecture of the measurement system. The issues mentioned are not specific for this area, since they appear in many different areas of computing. The described concept could be expanded to a various set of problems encountered in systems based on concurrent access to common and shared devices. Therefore we found it appropriate to describe this concept in the form of a software design pattern, in order to reuse knowledge in other specific areas of computer science.

We believe that introducing CORBA-based mobile agents could bring even more benefits to the existing model leading to a completely new concept of distributed measurement systems. The current model relies heavily on network availability and performance, while disabling possibility of offline operation. This promising technology will be the focus of our future work.

References

1. Žagar M., Oršulić J., Čavrak I.: Remote Instrument Access Using Remote Method Calls. 4th IEEE Mediterranean Symposium on Control & Automation, Chania, Crete, June 10-14, 1996.
2. Čavrak I., Stranjak A., Žagar M.: Remote Access to GPIB-Based Equipment Using Public Computer Networks. 19th International Conference on Technology Interfaces, Pula, June 17-20, 1997.
3. -: CORBA 2.0/IIOP Specification. Object Management Group, 492 Old Connecticut Path, Framingham, 1996.
4. Seán Baker: CORBA Distributed Objects – Using Orbix. Addison Wesley, 1997.
5. Čavrak I., Stranjak A., Žagar M.: CORBA-based Remote Instrumentation. IEEE MELECON '98, 9th Mediterranean Electrotechnical Conference, Tel Aviv, Israel, May 18-20, 1998.
6. Žagar M., Kovačić D., Stranjak A., Kurešević A.: Embedded CORBA-based System For Remote Instrument Access. 2nd International Workshop on Design, Test and Applications, Dubrovnik, Croatia, June 14-16, 1999.
7. Buschmann F., Meunier R., Rohnert H., Sommerlad P., Stal M.: A System of Patterns. John Wiley & Sons, 1996.
8. Larman C.: Applying UML and Patterns. Prentice Hall PTR, 1998.
9. Gamma E. et all, Design Patterns, Addison-Wesley, 1994.
10. Henning M., Vinoski S., Advanced CORBA Programming in C++, Adison Wesley, 1999.
11. Jacobson I., Booch G., Rumbaugh J., The Unified Software Development Process, Addison Wesley, 1999.

Design and Implementation of CORBA-Based Integrated Network Management System

Joon-Heup Kwon[1] and Jong-Tae Park[2]

[1]HyComm Incorporated, 2674 N. 1st St. Suite 210, San Jose, CA 95134, U.S.A.
andrew@Hycomm.net
[2]School of Electronic and Electrical Engineering, Kyungpook National University 1370
SanKyug-Dong, Buk-Gu, Taegu, 702-701, Korea
park@ee.knu.ac.kr

Abstract. There have been several research works for the CORBA-based network management system (NMS). After the JIDM 's work, ITU-T's SG4, Committee T1's T1M1.5, and a number of working groups are working on the development of CORBA-based TMN framework. In this paper, we design and implement a integrated NMS platform in which manage both pure CORBA-based NMS system, and legacy SNMP/CMIP-based NMS system revising JIDM CORBA/CMIP, CORBA/SNMP gateways. For the efficient implementations, we propose a new CORBA naming service, Smart Naming (SN) service. We also design the Multiple-Object Operation (MOO) service according to T1M1.5's specification. Finally, the performance of the system has been evaluated by measuring the response time and the system resource usage to demonstrate the efficiency of the proposed system.

1. Introduction

Recently, the Common Object Request Broker Architecture (CORBA) is increasingly noticed as the base technology for the realization of higher layer functions of network management architecture.

There have been several research works for the realization of a CORBA-based NMS system. The most significant work of the approach was that of the Joint Inter-Domain Management (JIDM) task force: [1] [2] [3] [4] [5]. JIDM Specification Translation (ST) describes the translation algorithm for mapping of the CMIP/SNMP management information to the CORBA IDL [4]. According to JIDM ST, GDMO/ASN.1 to IDL and SNMP SMI to CORBA IDL translators have been implemented [6], [7]. JIDM Interaction Translation (IT) describes the dynamic converting mechanisms for management information between CORBA and other legacy network management protocols [5]. However, since JIDM most focused on the interoperability between CORBA-based system and legacy CMIP or SNMP system, the capability of CORBA was not fully utilized.

Working Group T1M1.5 of Committee T1 and Study Group 4 (SG4) of ITU-T are presently defining the specification on pure CORBA-based TMN framework. T1M1.5 is defining CORBA-based network management framework with two draft standards: [8], [9]. [8] deals with framework requirements, CORBA COS usage, non-standard COS for the framework, and information modeling guidelines, and [9] specifies information models, defining a set of generic interfaces and constants based on

M.3100. SG4 is currently revising the recommendations M.3000 and M.3010 for adopting CORBA as one of standard communication architecture of TMN [10], [11]. However, T1M1.5's work has some limitations to realize integrated NMS system in the sense that it focuses only on the management of telecommunication network.

In this article, we propose platform architecture for integrated network management. To do so, we revise JIDM's CORBA/CMIP and CORBA/SNMP gateway according to T1M1.5's framework, propose a new CORBA naming service, the Smart Naming (SN) service for efficient gateway system resource usage, and design and implement T1M1.5's Multiple-Object Operation (MOO) service for our platform [8].

In section 2, we present the basic architecture of the platform introducing Virtual Network Element (VNE) concept, and design naming tree architectures for each gateway. We design the SN service and the MOO service in Section 3. In Section 4, the implementation details and the performance evaluation are described, and, finally, we present our conclusion in Section 5

2. Platform for Integrated Network Management

In this section, we introduce the platform architecture and design the naming trees in each CORBA/CMIP and CORBA/SNMP gateway.

Fig. 1 shows the network element (NE) architecture within the CORBA-based TMN framework defined by T1M1.5 and SG4. According to the framework, there are CORBA-based MOs within an NE. The CORBA manager and CORBA MOs communicate each other directly using CORBA Common Object Service (COS) and framework support services defined by T1M1.5. The containment relationships between MOs are represented by the naming tree, which is formed by using CORBA naming service.

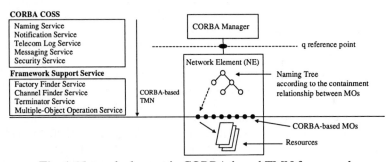

Fig. 1. Network element in CORBA-based TMN framework

2.1 The Virtual Network Element (VNE)

To represent each legacy CMIP/SNMP- based NEs, we form Virtual NEs (VNEs) that consist of a set of proxy objects. Fig. 2 shows the basic architecture of the VNE.

A VNE in the gateway consists of a set of Proxy Managed Objects (PMOs) for CORBA/CMIP gateway, or Proxy Leaf Nodes (PLNs) for CORBA/SNMP gateway. A PMO represents a MO in legacy CMIP based NE, and a PLN represents a Leaf Node of the MIB tree of SNMP agent. A CORBA manager interacts with PMOs or PLNs, and the actual interactions with legacy CMIP or SNMP agent are performed through Proxy Agent (PA) object. A PA represents a CMIP or SNMP agent. PA actually performs the protocol conversion task.

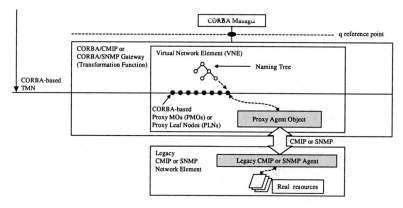

Fig. 2. The basic concept for VNE

The IDL definitions for PMOs are made according to [8], and the IDL definitions for PLNs are made by SNMP SMI to CORBA IDL translator [6].

2.2 Design of Naming Tree for CORBA/CMIP Gateway

Using CORBA naming service, a CORBA manager object can locate and access a PNE or PLN object instance with its meaningful name. In this section, we design the naming tree for the CORBA/CMIP gateway according to [8].

Within a CORBA/CMIP gateway, there are several VNEs that represent each target CMIP-based NE. and a VNE consists of a number of PMOs. These PMOs are organized by a naming tree according to the containment tree of the target CMIP-based NE. Fig. 3 shows the naming tree formed in CORBA/CMIP gateway.

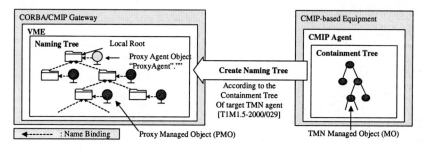

Fig. 3. Naming Tree in CORBA/CMIP Gateway

Besides those PMOs, a Proxy Agent (PA) object is also instantiated in a VNE. We use the IDL definition of PA object defined in JIDM IT [9]. Multiple object operation like CMIP scoping/filtering operations are performed by Multiple-Object Operation (MOO) service defined by T1M1.5 [8]. The PA instance is bounded to local root naming context so that MOs in the VNE easily locate and perform operations from CORBA manager.

2.3 Design of Naming Tree for CORBA/SNMP Gateway

In this section, we design the naming tree for CORBA/SNMP gateway in our platform. Since T1M1.5 does not consider the management of the Internet, we form a naming tree for CORBA/SNMP gateway according to the MIB tree information of a target agent. Since there is no IDL interface definition for SNMP, we use the SNMP SMI to CORBA IDL translator that performs the translation task [6]. For the actual interaction with SNMP agent, we also place PA in VNE. We use the IDL definition of PA defined in [5].

In CORBA/SNMP gateway, there are PLNs that represent leaf nodes of the MIB tree of SNMP agent. In this section we define the naming convention for these PLNs. Every leaf nodes in the SNMP MIB tree has their own OID.

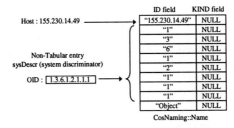

Fig. 4. Example OID conversion to CORBA name

In Fig. 4, the name of PLN, which represents the leaf node, `sysDescr` (system discriminator), is mapped into a CORBA name.

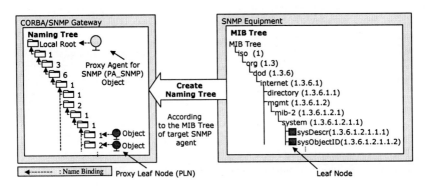

Fig. 5. Naming Tree in CORBA/SNMP Gateway

To address the SNMP agent, the IP address of the agent is inserted in the first name component of the CORBA name. And every component of the OID value of the system discriminator (sysDescr, 1.3.6.1.2.1.1.1) is inserted into each name component. The last name component contains "Object"."" indicating the PLN object instance itself [8]. Fig. 5 shows the naming tree architecture for CORBA/SNMP gateway. Naming context objects form a naming tree according to the MIB tree of VNE agent, so that CORBA manager can access a PLN according to the OID value of a target leaf node.

3. Design of the Smart Naming (SN) Service and the Multiple-Object Operation (MOO) Service

In this section, we design the new naming service, SN service, in order to utilize the gateway system resources more efficiently, and the MOO service, which is defined by T1M1.5 but is not a standard OMG CORBA service, for the platform [8].

3.1 Design of SN Service

Before designing SN service, we first present problems occurred when the conventional OMG Naming Service is utilized.

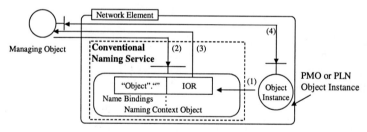

Fig. 6. Resolving an managed object with the conventional naming service

With the conventional naming service, PMO or PLN object should be instantiated [12]. The reason is that the conventional naming service uses the Interoperable Object Reference (IOR), whish is standard object reference of CORBA object 'instance' to build a name binding [3], [12]. Fig.6 shows the procedure of getting an IOR to a CORBA-based managed object instance. The procedure in the figure is as follows: (1) the object should be instantiated providing its IOR to build the name binding. (2) When a CORBA manager asks the naming context to return the IOR of the instance, (3) The naming context locate the name binding, and then, return the IOR to the manager. (4) With the IOR, manager access to the instance and perform operations.

In gateway system, there are millions of proxy objects. With conventional naming service, all proxy objects should be instantiated at the same time to make a naming tree. However, the object instances do not participating in network management at the same time. It means that, part of the system resource is assigned to instances that are not working at a moment. So, there must be other way for the efficient use of the system resources.

Fig. 7 shows the IDL definition of SN service. SN service adds one data structure, `Inst_Info`, one interface, `IS`, and one method in naming context, `bind_sns`, to the IDL definition of OMG naming service.

The name binding of the conventional naming service consists of two components: the name and IOR of an object. But, the name binding of SN service has one more component for information for instantiating an object. The instantiation information is defined as a structure named `Inst_Info` (in Box A). `MO_Class_Name` contains the class name of the object, and `Real_Resource_ID` indicates the actual MO or leaf node.

```
// SNS.idl
#include "CosNaming.idl"
module SNS
{
  typedef CosNaming::Istring Istring;
  typedef CosNaming::Name Name;
  typedef CosNaming::BindingList BindingList;
  interface BindingIterator : CosNaming::BindingIterator;

  struct Inst_Info{
      string         MO_Class_Name;                                    A
      unsigned long  Real_Resource_ID;
  };
  interface NamingContext : CosNaming::NamingContext{
      void bind_sns(in Name n, in Inst_Info i)                          B
         raises( NotFound, CannotProceed, InvalidName, AlreadyBound);
  };
  interface IS{
      Object inst(in Inst_Info order)                                   C
         raises( NotFound, CannotProceed, InvalidName, AlreadyBound);
  };
};
```

Fig. 7. IDL definition of SN service

The definition of the naming context of SN service is shown in box B. It inherits from that of conventional OMG naming service. Therefore, a CORBA manager can resolve a managed object instance in the same way of conventional OMG naming service [13]. The method, `bind_sns`, is for building a naming context with the instantiation information. With this method, the name binding of a PMO or PLN is built without instantiating it. The definition of the Implementation Storage (IS) is in box C. IS is an object factory. The method, `inst`, is for instantiating the object specified in the parameter, `order`.

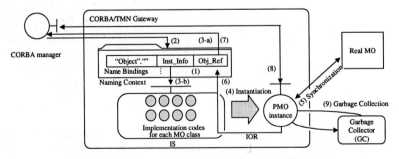

Fig. 8. Resolving an managed object with the SN service

Fig. 8 shows the procedure of the `resolve` operation in CORBA/CMIP gateway. The function block, Garbage Collector (GC) is to release the system resources assigned to instances that is not used for a long time. It periodically checks the PMO instances list. When it finds an instance that has not participated in any network management operation, it releases the system resources assigned to the instance, and sets the value in the `Obj_Ref` field of the name binding of garbage to NULL. Since GC is an internal function block of SN service, so there is no IDL definition of GC in the `SN service.idl`.

The procedure is as follows: (1) The name binding of the PMO is built with the instantiation information. (2) When a CORBA manager request naming context, the naming context checks the existence of the instance checking the `Obj_Ref`. (3-a) If exist, the naming context should just return the IOR of the instance. (3-b) If not, the naming context should pass the instantiation information in the `Inst_Info` field IS to instantiate the object. (4) With the `MO_Class_Name`, the factory object instantiates the class. (5) the PMO is synchronized with the real managed object specified by the `Real_MO_ID`. (6) The naming context inserts IOR of the new instance in the `Obj_Ref`. (7) The naming context object returns IOR. (8) With the IOR returned, the CORBA manager performs operation. (9) GC releases the system resource assigned to instances that are not used for a while.

3.2 The Multiple-Object Operation (MOO) Service

Since CORBA does not have the multiple object access capability, operations such as CMIS scoping and filtering operation (scoped get, scoped update, scoped delete) are supported by the Multiple-Object Operation (MOO) Service defined by T1M1.5 [10]. MOO service is not standard OMG CORBA COS. MOO is the one of them.

T1M1.5 defines two MOO service interfaces [8]: `BasicMooService` - scoped get operation, `AdvancedMooService` - scoped update and delete operations.

CORBA manager invoke operations MOO to perform multiple-object operation. MOO service receives scoped operations from CORBA manager, selects appropriate MOs according to the naming tree, and invokes operations of those instances. Fig. 9 shows the procedure of the scoped get operation defined on the `BasicMooService` interface.

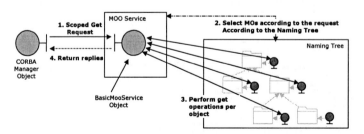

Fig. 9. The Procedure of The MOO service [scoped get]

The procedure is like this; (1) CORBA manager object send scoped get operation to the `BasicMooService`. (2) The object selects a set of object according to the parameters form CORBA manager object and the naming tree, and (3) performs get operation on each MO. (4) `BasicMooService` object make the response that contains every response returned from MOs.

We design MOO service applicable for both CMIP and SNMP-based system. MOO service is placed in the same computing platform with CORBA/CMIP and CORBA/SNMP gateways for performance reason [12]. For handling replies from CMIP or SNMP agents, we used Multiple Reply Handler (MRH), which was introduced in JIDM [9]. Fig. 10 shows the procedure of the CMIP scoped get operation of the `BasicMooService` object in CORBA/CMIP gateway. Other scoped operations (scoped update and scoped delete) have similar procedure.

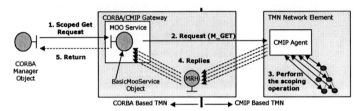

Fig. 10. The MOO service in CORBA/CMIP gateway [scoped get]

The procedure is like follows; (1) A CORBA manager object send `scopedGet` operation to the `BasicMooService` specifying the name of base object, the scope, names of attributes, the filtering expression, etc. (2) The object converts the request to the CMIS scoped get request, and send it to the target TMN agent. (3) The CMIP agent performs the scoped operations and sends the results to MRH. (4) MRH object receive these replies from TMN agent, and send it to the `BasicMooService` Object. (5) `BasicMooService` object make the response that contains every response returned from MOs.

The GET-BULK operation of SNMPv2 is the only multiple operation in SNMP domain. The operation is supported by `scopedGet` operation of `BasicMooService` object. Note that the first component of the name of VNE leaf node is IP address of a computer. With it, the `BasicMooService` object tells that the request is the SNMP bulk operation. The `BasicMooService` object directly send request to SNMP agent instead of invoking operation of each PLN. Fig. 11 shows the procedure of the scoped get operation of the `BasicMooService` interface.

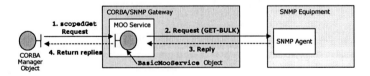

Fig. 11. The MOO service in CORBA/SNMP gateway [SNMPv2 get-bulk]

The procedure is like follows: (1) A manager object sends `scopedGet` operation to the `BasicMooService` specifying the name of base object, which corresponds to the OID of a node. (2) The object converts the request to the SNMPv2 GET-BULK operation, and sends it to the target SNMP agent. (3) The agent performs the operation, and a reply is sent to the `BasicMooService` object. (4) The `BasicMooService` object converts the response to CORBA response format, and sends it to the manager.

4. Implementation and Performance Evaluation

SN service: Name binding is internal data structure of naming service. So there is no IDL level definition for name binding. The structure of the name binding of SN service contains a name, instantiation information, and IOR of an object instance. Fig. 12 shows the structure of name binding in SN service.

```
// SNS_Impl.h
#include "SNS.hh"
// . . .
struct NameBinding{
  SNS::Name_var           name;
  SNS::Inst_Info_var      inst_info;
  CORBA::Object_var       Obj_Ref;
};
// . . .
```

Fig. 12. The structure definition of SN service Name Binding

In the structure, `Obj_Ref`, is to indicate that the object already exists or not. If the object is not instantiated, the value in the field is NULL. Otherwise, the field contains the IOR of the object. As described in Section 3.1, GC is an internal function block of SN service. GC has its own table which contains the list of the instances that was created by IS, periodically checks the time stamps for each instance inspecting the difference between the time-stamp and the current time exceeds the predefined time-out value. Once GC finds an object instance timed-out, GC releases the resource that has been assigned to the object instance, and set the value of `Obj_Ref` field to NULL.

Interaction through the Gateway: Fig. 13 shows the interaction procedure through CORBA/CMIP gateway. The Multiple Reply Handler (MRH) object, defined by JIDM for CORBA/CMIP gateway, receives one or multiple replies from TMN agent, converts the replies to CORBA form, and passes them to the appropriate PMO. The operation procedure is as follows: (1) the CORBA manager first locates the appropriate PMO through the naming tree, and (2) sends a request. (3) PMO invokes the appropriate operation of PA. (4) PA converts the request into CMIP/CMIS format, and (5) sends the request to the target agent. (6) The agent performs the operation, and (7) sends the response to MRH. (8) MRH convert the response to CORBA format and pass it to PMO. (9) The PMO sends the result to the manager.

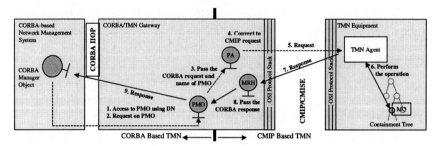

Fig. 13. Interaction through CORBA/CMIP Gateway

Fig. 14 shows the interaction procedure through CORBA/SNMP gateway. A CORBA manager access to a PLN through the naming context using the OID value of and perform network management operation. The operation procedure is like follows: (1) a manager first locates the appropriate PLN, and (2) invokes an operation supported by the PLN. (3) the PLN invokes the appropriate operation of PA. (4) PA converts the request into SNMP format, and (5) send it to SNMP agent. (6) The agent performs the operation, and (7) send the result to PA. (8) The PA converts the reply to CORBA format and passes it to PLN, and (9) The PLN object send the response to the manager.

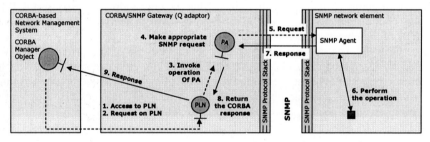

Fig. 14. Interaction through CORBA/SNMP Gateway

Experimental Results: Fig. 15, Fig. 16 and Fig. 17 show the result of our performance evaluation. The performance of SN service and conventional naming service are comparatively evaluated. We use Orbix 2.3 [23] and implemented in SUN SPARC 1000 system.

As the first phase, we assumed that there are 1000 MOs in a managed system. With the conventional naming context, 1000 MOs and 1001 naming context objects are instantiated, and, otherwise, with SN service, only 1001 naming context objects are instantiated initially. The manager object accesses 100 objects at a time, and the target objects are selected by uniform random sampling. With 2 seconds time interval, the manager object perform the operation 50 times. The time out value for GC is set to 10 seconds.

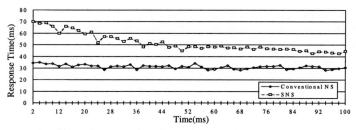

Fig. 15. response time of conventional naming service and SN service: the average response time required to get the IORs of 100 MO instances comparing that of the conventional naming service and that of SN service. As time goes on, the average response time merge.

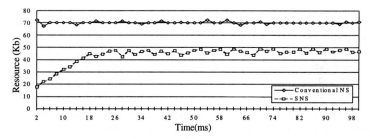

Fig. 16. The volume of memory of conventional naming service and SN service: Though the memory required for SN service is increased, it remains shorter than conventional NS.

Though the response time of SN service is longer than that of the conventional naming service at first, as time goes on, the response time get reduced, and finally it becomes slightly longer than conventional naming service. The reason is that once an object is instantiated, it lasts before the time stamp in the instance expires. However, the volume of the system resources assigned to network management operations with SN service was quite smaller than that of the conventional OMG naming service.

As the second phase, we checked the response time while increasing when there is relatively large number of proxy objects. We increased the number of proxy objects and checked the average response time of the proxy objects. The average response time is the time when SN service is in the steady state. We increased the number of objects from 100 to 100,000. Fig. 19 shows the result of the experiment.

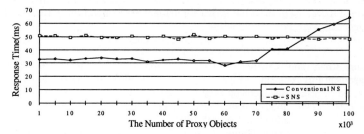

Fig. 17. Response time of conventional naming service and SN service

When the number of proxy objects exceeds 70,000, in other words, the volume of proxy objects exceeds the system memory space, the system starts swapping. When the number of proxy objects is around 85,000, the response time of each naming service eventually same. As the number of proxy object is increased, the response time of SN service is shorter than that of the conventional naming service. As long as the system memory of gateway system is not boundless, the swapping occurs whenever the volume of proxy objects exceeds the volume of system memory.

5. Conclusion

In this paper, we have developed a new CORBA-based network management platform in which pure CORBA-based management application and traditional SNMP or CMIP agent systems can be interoperable very effectively. It is noted that the CORBA/CMIP or CORBA/SNMP gateway specification proposed by JIDM does not deal with the construction of the naming tree, so that they cannot be inter-operable with the pure CORBA-based approach proposed by ANSI T1M1.5.

We have designed the system, which can operate with robustness even in the case where there are thousands of managed objects to be handled. With the conventional OMG CORBA naming service, all of those managed objects may need to be instantiated all at the same time. These may shut down the system, or may deteriorate the performance of the system so. Smart Naming (SN) service not only provides tools to avoid these situations, but also it saves the memory space required for the platform. SN service provides an efficient way to build a naming tree within CORBA/CMIP and CORBA/SNMP. In addition, we have also designed the Multiple-Object Operation (MOO) service defined by T1M1.5. While directly invoking operations supported by PA, the MOO service reduces the number of interactions between MOO service objects and PMO or PLN, and efficiently performs the CMIS scoping/filtering operations and SNMPv2 GET_BULK operation within CORBA/CMIP and CORBA/SNMP gateway. We have designed the operational procedures of network management operations in CORBA/CMIP or CORBA/SNMP gateway. It is found that the platform system is operating satisfactorily.

References

1. N. Soukouti and U. Hollberg, "Joint Inter Domain Management: CORBA, CMIP and SNMP," Proceeding of the 5th IFIP/IEEE International Symposium on Integrated Network Management, May 1997, pp.153-164.
2. PROSPECT Consortium, D33A: Concepts for CORBA/TMN interworking, Deliverable Number: AC052/GDM/WP3/DS/S/003//b1, May 1997.
3. Graham Chen and Qinzheng Kong, "Integrated TMN Service Provisioning and Management Environment," Proceeding of the 5th IFIP/IEEE International Symposium on Integrated Network Management, May 1997, pp.99-112.
4. X/Open and NMF, Inter-domain Management: Specification Translation, Open Group Preliminary Specification P509, March 1997.
5. The Object Management Group (OMG), "Interworking Between CORBA and TMN Systems Specification," version 1.0, formal/2000-10-01, October 2000.

6. Jong-Tae Park, Moon-Sang Jeong and Seong-Boem Kim, "A platform architecture for the Integration of CORBA Technology within TMN Framework," IEICE Transactions on Communications, Vol.E82-B, No.11, pp.1770~1779, November 1999.
7. ORBYCOM, GDIDL: a GDMO/ASN.1 to IDL translator, 1998, See also http://www.orbycom.fr
8. Committee T1 – Telecommunications Working Group T1M1.5, "Working Document for Draft Standard ANSI T1.2xx-2000, Framework for CORBA-Based Telecommunications Management Network Interfaces," August 2000.
9. Committee T1 – Telecommunications Working Group T1M1.5, "Working Document for Draft Standard ANSI T1.2xx-2000, CORBA Generic Network and NE Level Information Model," June 2000.
10. ITU-T – Study Group 4, " Overview of TMN Recommendations," Oct. 1999.
11. ITU-T – Study Group 4, "Draft Revised Recommendation M.3010 – Principles for a Telecommunications Management Network," Oct. 1999.
12. The Object Management Group (OMG), "Event Service Specification," OMG Document formal/2000-06-15, version 1.0, June 2000.
13. The Object Management Group (OMG), "Naming Service Specification," OMG Document formal/2000-06-19, Version 1.0, April, 2000.

Framework for Real-Time CORBA Development

Z. Mammeri[1], J. Rodriguez[1], and P. Lorenz[2]

[1]IRIT, Toulouse, France
{mammeri, rodrigue}@irit.fr
[2]IUT de Colmar, Colmar, France
lorenz@colmar.uha.fr

Abstract. Object-oriented real-time applications require enforcement of end-to-end timing constraints on service requests from clients to servers in a distributed computing environment. Thus, timing aspects should be integrated and handled particularly by CORBA which is considered as a standard to support distributed object-oriented applications. In spite of the release of a (real-time) RT-CORBA specification, a few vendors have ported their ORBs (Object Request Brokers) to real-time operating systems. In this paper we analyze existing RT-CORBA implementations, and we propose a framework that integrates the main components (scheduling, real-time communication, end-to-end quality of service providing, etc.) necessary for implementation of RT-CORBA to support static and dynamic real-time applications.

1. Introduction

With the advent of increasingly complex applications that depend on timely execution, many distributed computer systems must support components capable of real-time distributed processing [3, 22, 36, 39]. That is, these applications require enforcement of end-to-end timing constraints on service requests from clients in a distributed computing environment.

In the last years, the object-oriented (OO) technology became increasingly popular. This technology has prevailed over non-real-time computing as it reduces the development complexity and maintenance costs of complex applications and facilitates reuse of components. To deal with the complexity of design, analysis, maintenance and validation of real-time applications, the real-time systems engineering community is more and more interested in using OO technology at different levels, mainly: design level, programming level and middleware level. Thus, timing aspects should be integrated and handled at different levels: at specification and design levels, at programming language level and at middleware level. To take into account these needs, various works are particularly undertaken within the OMG to extend UML, JAVA and CORBA to make them suitable for real-time applications. The main extensions focus on scheduling, memory management, concurrency, and communication management to guarantee end-to-end quality of service [36]. This paper deals with Real-Time CORBA (RT-CORBA).

The ability to enforce end-to-end timing constraints, through techniques such as global priority-based scheduling, must be addressed across the CORBA standard. The approach is to assume a networked architecture that consists of an underlying system with its network, an ORB, and Object Services, all with real-time requirements. The real-time requirements on the underlying systems include the use of real-time operating systems on the nodes in the distributed systems (supporting applications) and the use of adequate protocols for real-time communication between nodes in these distributed systems. Thus, several aspects have to be apprehended, mastered and developed. However an important step towards distributed real-time systems supported by CORBA is the introduction of concepts related to the time constraints in CORBA, without modifying basically the original CORBA.

In spite of the release of an RT-CORBA specification [27,29], a few vendors have ported their ORBs (Object Request Brokers) to real-time operating systems. Nevertheless, some experimental implementations of RT-CORBA are being (or have been) developed. In this paper we analyze existing RT-CORBA implementation, and we propose a framework that integrates the main components (scheduling, real-time communication, end-to-end quality of service providing, etc.) necessary for implementation of RT-CORBA to support static and dynamic real-time applications. The development of RT-CORBA necessitates many mechanisms resulting from real-time systems, networks, distributed systems, and the object-orientation. We emphasize the difficulties and complexity of this development.

In section 2, an overview of the real-time CORBA architecture is given. In section 3, the main work in RT-CORBA is summarized and discussed. The framework we propose for the development of RT-CORBA is presented is section 4. Some conclusions are given in section 5.

2. Real-Time CORBA Architecture Overview

Real-time CORBA deals with the expression and enforcement of real-time constraints on end-to-end execution in a CORBA system [23]. This section presents the key features most relevant for researchers and developers of distributed real-time systems [35]; it focuses on RT-CORBA with static priority scheduling; the dynamic scheduling is not integrated yet to RT-CORBA.

In 1995, a Special Interest Group (SIG) at OMG was formed to initiate the RT-CORBA and assess the requirements and interest in providing real-time extensions to CORBA model by defining the common terminology and concepts for use in developing real-time extension, introducing the needs of RT-CORBA market showing what technology extensions are necessary, and indicating what real-time extensions to CORBA are feasible [23]. The need for a common specification for RT-CORBA arises for the same goals that have driven standard CORBA specification: transparent distribution, heterogeneity, portability and interoperability.

In January 1998, the OMG released the first request for proposal (RFP) for extending CORBA to make it suitable for real-time systems [24]. The responses to this RFP were analyzed and an RT-CORBA specification (version 1.0) have been issued and integrated in CORBA 2.4 [29]. RT-CORBA 1.0 focuses only on static priority scheduling; dynamic scheduling is under work at OMG. A dynamic CORBA environment is one in which clients and servers may be added and removed, and where constraints may change [28].

Developing CORBA systems that meet functional requirements within real-world timing constraints requires that the ORB exhibits a level of predictability (bounded latency of operation invocations, bounded message transfer delays), with respect to the timings for its various behaviors. Given that an ORB has to perform more than one activity at time, the allocation of the resources (processor, memory, network bandwidth, etc.) needed for those activities also has to be controlled in order to build predictable applications [24]. RT-CORBA 1.0 specification defines features that support end-to-end quality of service (QoS) for operations in fixed-priority CORBA applications [27]. RT-CORBA is defined as extension to CORBA [26] and the messaging specification [25].

RT-CORBA based systems will include the following four major components, each of which must be designed and implemented taking into account the need for end-to-end predictability [27]:

- scheduling mechanisms in the operating system (OS),
- real-time ORB,
- communication transport handling timing constraints,
- applications specifying timing constraints.

Not to over constraint no-real-time application development, RT-CORBA is positioned as a separate extension to CORBA (see Fig. 1). The set of capabilities provided by RT-CORBA is optional. An ORB implementation compliant to RT-CORBA 1.0 must implement all RT-CORBA except the scheduling service which is optional.

Thread pools. Many distributed applications have complex object implementations that run for variable duration. To avoid unbounded priority inversion and deadlock, real-time applications often require some form of preemptive multithreading. RT-CORBA addresses these concurrency issues by defining a standard *thread pool* model. This model enables server developers preallocate pools and set some thread attributes (default priority, and so on). Server developers can optimally configure thread pools to buffer or not buffer requests, thus providing further control over memory usage.

Priority mechanisms. RT-CORBA defines platform-independent mechanisms to control the priority of operation invocations. Two types of priority are defined: CORBA priorities (handled at CORBA level) and native priorities (priorities of the target OS). RT-CORBA supports two models for the priority at which a server handles requests from clients: *Server declared priority model* (the server dictates the priority at which object invocations are executed), and *Client propagated model* (the server honors the priority of the invocation set by the client).

Scheduling. Scheduling is of prime importance to guarantee timing constraints of requests issued by clients. RT-CORBA 1.0 specification targets fixed-priority scheduling. RT-CORBA *scheduling service* defines a high level (i.e., different of the scheduling service of the operating systems) scheduling service so that applications can specify their scheduling requirements (worst case execution time, period, and so on) in a clear way independent of the target operating system.

Managing inter-ORB communication. Contrary to CORBA standard, which supports location transparency, RT-CORBA lets applications control the underlying communication protocols and endsystems to guarantee a predictable QoS. This can be achieved by two mechanisms: selecting and configuring protocol properties, and explicit binding to server objects. An RT-CORBA endsystem must integrate mechanisms that support protocols to guarantee timeliness of communications (i.e., bounded transfer delays and jitter). According to used network (ATM, TCP/IP, FDDI, and so on), the mechanisms are very different.

Interfaces to specify timing constraints. A real-time ORB must provide standard interface so that applications can specify their resources requirements and timing constraints. The policy framework defined by CORBA messaging specification lets applications configure ORB end-system resources to control ORB behavior.

Real-time ORB services. RT-CORBA ORBs must preserve efficient, scalable and predictable behavior for high level services and application components.

Use of timeouts. RT-CORBA application may set a timeout on an operation invocation in order to bound the time that the client application is blocked waiting for a reply. This can be used to improve the predictability of the system.

3. Related Work

One attended many work on CORBA and real-time the five last years. Thuraisingham et al. [42] were among the first to discuss the main extensions to the object model, the ORB, object adapters and IDL to take into account timing constraints. The work in real-time CORBA has taken two major directions:
1) Real-time above CORBA without CORBA extensions: proposals were developed trying to provide mechanisms to support real-time applications above CORBA without any modification or extension to standard CORBA [6, 7, 9, 10, 21, 22, 30, 31, 43]. Some of the previously mentioned work proposed to deal with real-time as follows: install CORBA on real-time OS (Operating System), and realize faster version of ORBs by removing features like dynamic invocation, use protocols with fixed connections between clients and servers. This approach is necessary in an RT-CORBA, but it is not sufficient for predictable behavior guarantee. The work in this direction targeted soft real-time applications (sometimes with best effort), and considered that CORBA should remain a general-purpose middleware and it should not be extended to be suitable for real-time applications. Since CORBA 2.4 version, new mechanisms suitable for real-time have been integrated in CORBA, thus such a direction of research is considered nowadays inappropriate for real-time applications that require some guarantee of timing constraints.

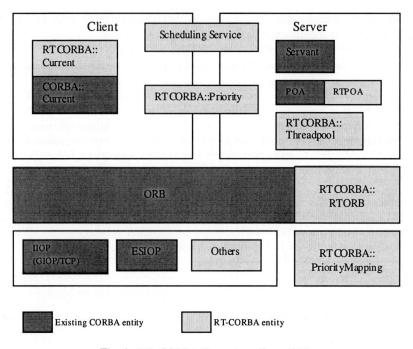

Fig. 1. RT- CORBA Extensions (from [27])

2) Real-time with CORBA extensions: the objective of the work in this direction is to investigate how existing CORBA can be extended to provide support for real-time computing. The extensions considered IDL (to specify end-to-end QoS execution of applications), ORB (resource management algorithms), and client and server [5, 20, 40, 44, 45, 46, 47]. This work has contributed to the establishment of the RT-CORBA specification as it is integrated in CORBA 2.4.

Some RT-CORBA implementations are being developed:
- *TAO* at the university of Washington (it is the most advanced implementation of RT-CORBA, and it is also the most used) [34],
- *MITRE Real-time CORBA* at MITRE corporation [41],
- *ARTDOM* project at MITRE corporation [44],
- *RapidSched,* a trademark of Tri-Pacific [46],
- *Epiq* project extending TAO at University of Illinois at Urbana-Champaign [2],
- *Realize* at University of California, Santa Barbara [6, 21],
- *UCI TMO* project at university of California (Irvine) [8],
- *Dynamic Real-time CORBA* at US Navy Research and Development laboratories and the University of Rhode, Island [47].

Some of the proposed implementations contain many of the real-time ORB extensions defined in RT-CORBA specification [27, 35]. The main differences between the previously cited work lie in aspects like targeted applications (hard or soft real-time applications), timing constraints and QoS, IDL extensions, scheduling techniques, resource protocol management, and communication protocols. All proposed implementations started whereas the RT-CORBA specification was not known yet and consequently, they are not compliant with CORBA 2.4. Their great merit is that they contributed to the development of RT-CORBA specification within the OMG.

4. A Framework for RT-CORBA Development

Real-time applications cover a very broad spectrum of domains (ranging from very critical systems, like nuclear installations, to multimedia applications for children) and require methods, tools and platforms very different from a domain to another. In particular, to support real-time applications by CORBA several types of RT-CORBA products are necessary. We use the term "RT-CORBA profile" to indicate an RT-CORBA product which meets the needs of a particular application class.

In addition, the development of complex systems and applications can be made efficiently only if one can use (re-use) "components" that have been developed and tested. Nowadays, there is an increasingly marked tendency towards the use of commercial-off-the-shelf components. In this section, we show the main components for RT-CORBA profile development.

Existing work on RT-CORBA (see section 3) brings solutions on particular points, for particular contexts. This work (often) lacks total and generic sight to approach RT-CORBA, thus making it very hard (even impossible) to build RT-CORBA with components which fulfil the needs of various specific applications. To fill this gap, we propose a general framework for the development of RT-CORBA allowing to better see the elements (components) which are parts in RT-CORBA development. Such a framework is adaptable and can switch scheduling algorithms, communication protocols, and other policies and system components depending on the needs and the environment.

Our framework is based on two concepts: RT-CORBA profiles (which are instantiation of the generic framework) and components (which enable reuse and abstraction during RT-CORBA development). Developing RT-CORBA profiles requires to consider several aspects discussed hereafter, and shown in Fig. 2 and 3.

4.1 Classification of Real-Time Applications

Real-time applications to run above RT-CORBA are very numerous and span a large spectrum of domains. Application requirements lead to differentiate hard and soft real-time constraints [15,38]. Applications imply hard real-time constraints when a single failure to meet timing constraints may result in an economical, human or ecological disaster. Soft real-time constraints are the case when timing faults cause damages which cost is considered tolerable under some conditions on fault frequency. Within these two classes, there may be periodic, aperiodic and sporadic operations (i.e., object requests in the case of an OO application). For soft real-time applications, there may be firm or non-firm deadlines. Some real-time applications require deterministic QoS. Others require statistical QoS. And others require both deterministic and statistical QoS. The degree of validation/verification/test depends heavily on the criticalness of real-time applications. Thus the means to deploy (their costs, their complexity) are dependent on the application nature (criticalness).

The classification of real-time applications enables to elaborate the models of the constraints and QoS to be handled by RT-CORBA (i.e., what should be handled? and what is the required QoS?). An RT-CORBA profile should be associated with each class of real-time applications.

4.2 RT-IDL

Standard (i.e., non-real-time) IDL has been extended to handle only *priority* of client requests [29]. Nevertheless, such an extension is not sufficient, for almost real-time applications. New extensions are necessary to specify different constraints and QoS according to the class of considered applications. The new RT-IDL should specify: application type (hard or soft), type of requests (periodic, aperiodic, sporadic), deadlines, type of deadlines (firm or non-firm), worst execution times, relationships between requests, distribution functions for aperiodic requests, minimum interarrival times for sporadic requests, quality of service (response time bounds, jitter bounds, and so on), ...

4.3 Mapping of Timing Constraints and QoS onto RT-CORBA

Once the timing constraints and QoS parameters are specified using RT-IDL, their QoS have to be mapped onto RT-CORBA mechanisms. The mapping leads to the definition of attributes used by underlying mechanisms, particularly threads and messages scheduling.

For some hard real-time applications, static binding of objects is necessary. For others, the load is not known, and binding is dynamic.

For some applications, a call admission control procedure has to be defined to enable a server to test if the load associated to a client requests may be guaranteed or not. For some applications, a step of QoS negotiation between client and server are necessary before any object operation invocation. There may be renegotiations according to changes occurring at client and/or server level.

One challenge is to specify the various mappings of timing constraints and QoS parameters onto RT-CORBA mechanisms according to each class of RT-CORBA profile. The work is hard!

4.4 Scheduling, Concurrency Control and Schedulability Analysis

Scheduling of threads that treat client requests is of prime importance to guarantee (eventually with some probability) the timing constraints associated with requests. In the present version (i.e., 1.0) of RT-CORBA, only static scheduling is supported. Nevertheless, in most complex real-time systems (RTSs), static scheduling is not sufficient, since many non-predictable resource conflicts influence execution times. Therefore, dynamic scheduling is required for such RTSs.

In the context of hard real-time applications, Rate Monotonic algorithm [13] and Deadline Monotonic algorithm [11] are currently used for static scheduling. Earliest Deadline First [13], and Least Laxity First [1] are used for dynamic scheduling. Nevertheless, these four algorithms are somewhat ineffective for soft real-time applications with numerous constraints, it is why other techniques (such as techniques based on generic algorithms or neural networks) are used in practice.

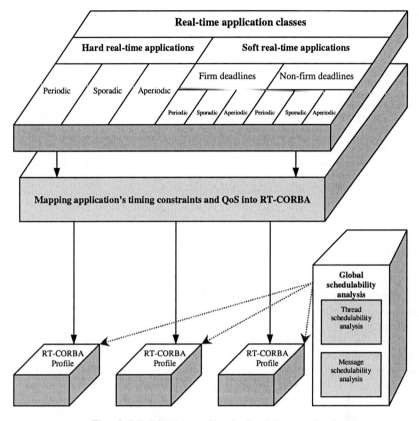

Fig. 2. RT-CORBA profiles for Real-time applications.

To develop powerful RT-CORBA profiles, it is necessary:
- to classify existing scheduling algorithms and define their performances,
- to select scheduling algorithms according to the constraints to be handled,
- to classify existing resource management protocols and define their performances,
- to select a management protocol according to the constraints to be handled,
- to achieve a schedulability analysis.

In a distributed system running above an RT-CORBA, client sites and server sites use scheduling algorithms (note that a client site may use a scheduling algorithm different from the one used by the server site). Many work in real-time scheduling exist [15], and has to be adapted to develop RT-CORBA. The ideal would be to have a tool that automatically selects a scheduling algorithm given the constraints of the client requests. To achieve such an ideal, existing algorithms should be classified and their performances specified; the algorithm selection should be based on algorithm performances, and not on their internal functioning. Thus, scheduling algorithms will be considered as components to develop RT-CORBA profiles. Much work remains to be made to clearly specify scheduling components.

Real-time object-oriented computing poses the need to control concurrent access to objects under timing requirements (prevent deadlock, minimize priority inversion). As proposed in [37], it is necessary to combine semantic locking and resource management protocols (such as Priority Inheritance Protocol, Priority Ceiling Protocol, Stack Resource Policy) for concurrency control in real-time object-oriented systems. As for scheduling, resource management protocols should specified as reusable components.

Once the scheduling algorithms and resource management protocols selected, it is important for major real-time applications to analyze the schedulability of the threads (tasks) composing an application. Commercial-off-the shelf tools, like PERTS [14] and TimeWiz -a trademark of TimeSys corporation-, may be used to know the requests for which the timing constraints will be met, and those for which the constraints will not be met. According to the results of the analysis (number of guaranteed requests, number of timing faults, maximum timing faults for each periodic request, and so), some adjustment or modification may be necessary.

4.5 Real-Time Communication

CORBA was created targeting applications in general, which desire transparency in the distribution and in the way that resource are managed. Besides, the CORBA interoperability (IIOP – Internet Inter-Orb Protocol) is designed for general-purpose systems, using TCP connections. Standard CORBA GIOP/IIOP interoperability protocols are not well suited for real-time applications with stringent latency and jitter constraints because TCP lacks predictability.

The issues of scheduling and load balancing can never be fully resolved without thorough knowledge of communication protocols, which allow participating parties to exchange information. Real-time communication is particularly tricky due to uncertainties in the network, such as loss of messages or node failures, so that guaranteeing message delivery is difficult unless special measures are taken.

Much work has been carried out to analyze the performances of networks and the provided QoS. RT-CORBA profiles should benefit from existing and approved results. In particular, much papers have been published concerning how to set up, and parameterize existing networks and protocols to deal with real-time communications in the Internet [3], in ATM [18], or in LANs [16].

Based on the existing work in real-time communication, we propose to handle timing constraints and QoS in RT-CORBA profiles as follows: 1) build a real-time messaging component, 2) Specify and negotiate traffic parameters and QoS, and 3) analyze schedulability of messages.

Real-time messaging component

As shown on Fig. 3, an RT-CORBA profile uses a *real-time message component*. Such a component may be obtained in different ways combining different components:

- At the first level, the RSVP -Resource ReSerVation Protocol- [32] or another reservation protocol may be used for the bandwidth reservations to provide some QoS. No protocol should be used, if the reservation problem is dealt with by the upper level.
- At the second level, a transport protocol, a network protocol or directly a data link protocol has to be chosen, and a mapping with previous choice has to be done.
- At the third level, a physical network (Ethernet, FDDI, token bus, ...) should be chosen, and a mapping with previous choice has to be done.

Traffic parameters and QoS specification and negotiation
Several real-time message components may be defined; each one has its own internal components. It is necessary to have an interface that gives an abstraction of the real-time messaging component for the ORB. Such an interface achieves the following functions:
- It enables the ORB to specify the traffic parameters and QoS independently of the chosen real-time messaging component.
- It maps the traffic parameters and QoS onto the chosen real-time messaging component. Connections (of TCP/IP or ATM, for instance) parameters are defined at this level. Some connections are statically established (for applications with hard real-time applications). Some others are dynamically negotiated.

Message schedulability analysis
The (absolute or statistical) predictability of message transfer delays must be established by using automatic tools of message scheduling analysis. Such tools depend on the hierarchy of used protocols (ATM, IP, ...). For each real-time messaging component an analysis tool may be defined or a commercial-off-the-shelf tool (like PERTS, TimeWiz, Opnet, etc.) may be acquired.

Note that the translation of traffic parameters and QoS of client requests into parameters of communication protocols requires a number of decisions to be made, it is a complex problem which is a challenge to take up in the future by real-time and network research communities. Saksena [33] gave an example of automatic synthesis of QoS in a restricted context (a uniprocessor architecture with hard real-time constraints) and showed the impact of choosing one or several threads on QoS.

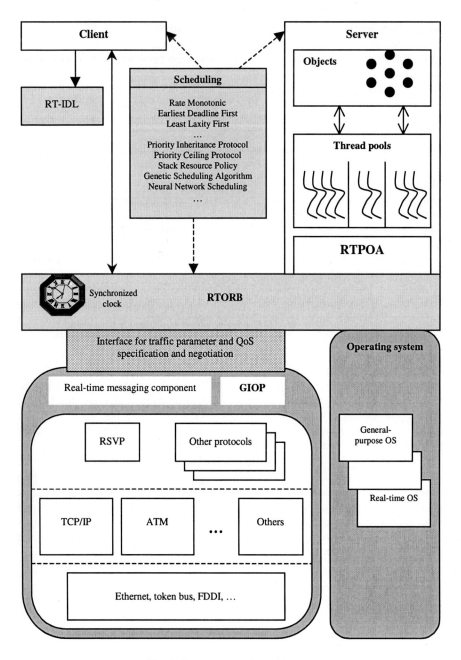

Fig. 3. Components of RT-CORBA profile.

4.6 Operating System

One important component for RT-CORBA profiles is the operating system supporting these profiles. An OS may be a general-purpose OS (like Unix), an RTOS (like VxWorks), or another type of OS. The OS performances (number of priorities, context switch overhead, dispatching, resource locking mechanisms, etc.) considerably influence the performances of RT-CORBA profiles. The choice of an OS to build an RT-CORBA profile is crucial.

Levine et al. [12] proposed the following recommendations to decrease non-determinism and limit priority inversion in OSs that support real-time ORB middleware:

- Real-time OS should provide low, and deterministic context switching.
- To reduce latency, OS should execute system calls in the calling thread context, rather than in a dedicated I/O worker thread in the kernel.
- Real-time OS should integrate the I/O subsystem with ORB middleware (i.e., integrate the scheduling performed by the I/O subsystem with other OS resource management strategies to process packets according to the priorities of threads destined to receive them).
- Real-time OS should support QoS specification and enforcement (they should integrate admission control and so on).
- Real-time OS should provide tools to determine sources of overhead experienced by threads.
- Real-time OS should support priority inheritance protocols to handle priority inversion.

4.7 Clock Synchronization

In a distributed real-time application running over an RT-CORBA, the timings constraints (like deadlines specified by the clients and guaranteed by the servers) refer to a global time. Thus each RT-CORBA profile should integrate a mechanism (an algorithm) to synchronize clocks. The most commonly used clock synchronization algorithm is NTP [19]. Nevertheless, NTP may be not suitable for some real-time applications requiring a high clock synchronization accuracy, and it is necessary to use other algorithms. To facilitate RT-CORBA profiles, clock synchronization should by specified as components with their performances [4, 17]. Clock synchronization algorithm selection is driven by the properties (synchronization accuracy, fault-tolerance and overhead of the algorithms) of global time required by applications.

4.8 Global Schedulability Analysis

The schedulability analysis of threads and messages taken individually is necessary but it is insufficient. One must have tools to achieve a global schedulability analysis, i.e., an analysis taking into account all the components of a real-time system (a set of clients, a set of servers, a set of communication protocols, a set of scheduling algorithms, etc.).

5. Conclusion

The demand for real-time applications that encompass everything from video-on-demand and real-time telephone system management to traditional embedded systems is increasing significantly. We believe a new generation of real-time applications is now possible due to two parallel streams of research. First, the development of widespread standardized object-oriented middleware that supports the development of applications built from components using high-level abstractions of the underlying computational and network infrastructure. This is realized in the CORBA ideal, as it allows for an application to use components that are identified dynamically and without regard to location. Second, support for quality-of-service constrained computations by the underlying infrastructure has also risen. This is realizable with the introduction of real-time POSIX, ATM, RSVP, and IPV6. Until very recently, these two developments have been progressing along non-intersecting trajectories, however convergence could result from real-time extensions to CORBA.

Some extensions necessary for static real-time applications have been accepted by the OMG and are part of CORBA 2.4. Other extensions are under study to fulfill the needs of dynamic applications. The RT-CORBA defines which are the basic components to develop a system that actually meets real-time requirements. Delivering end-to-end QoS for distributed real-time OO applications remains a significant challenge for the next decade(s).

As far as we know, TAO is the most advanced tool that implements RT-CORBA. Software developing companies are still mistrustful of RT-CORBA, to invest time and money in order to put on the market RT-CORBA software. The mains reasons to this mistrust are: 1) RT-CORBA is not yet completely defined within the OMG, and 2) the OO approach is not commonly accepted, in practice, by engineers and developers of the real-time domain.

The difficulties of developing RT-CORBA are that much concepts have to be integrated and combined to achieve an RT-CORBA. In this paper, we have presented a generic framework to study the different mechanisms (components) used to build an RT-CORBA. The spectrum of real-time applications is so large that it is necessary to define and implement a large number of RT-CORBAs. We introduced the concept of *RT-CORBA profile* to take into account the diversity of the constraints to handle.

The field of OO distributed real-time systems is young and it is growing quickly because it provides such a wide spectrum of applicability in many domains. RT-CORBA is one of the success key, others such as RT-UML, and RT-JAVA, have to be considered.

One could think that having resolved the issues with real-time scheduling, load balancing, and real-time communication, would let the researchers and practitioners sleep quietly, but that is not so. Knowing what algorithms and protocols to use is just a tip of an iceberg; one needs to apply these concepts in real circumstances. In other words, intensive development and verification procedures are needed to prove that at least some of these concepts work.

References

1. Dertouzos M.L., and Mok A.K.L.: Multiprocessor on-line scheduling of hard real time tasks. IEEE Trans. on Soft. Eng., Vol. 15, 12. (1989) 1497-1506
2. Feng W., Syyid U., and Liu J.W.-S.: Providing for an open, real-time CORBA. In Proceedings of the IEEE Workshop on Middleware for Real-Time Systems and services, San Francisco, December (1997)
3. Ferguson P, and Huston G.: Quality of service: delivering QoS on the Internet and in corporate networks. John Wiley & Sons Inc. (2000)
4. Fonseca P., and Mammeri Z.: A framework for the analysis of non-deterministic clock synchronisation algorithms. In Lecture Notes in Computer Science n° 1151 (1996) 159-17
5. Hong S., Min D., and Han S.: Real-Time Inter-ORB protocol on distributed environment. In Proceedings of the 1st IEEE International Symposium on Object-Oriented Real-Time Distributed Computing (ISORC'98). Kyoto (20 - 22 April, 1998). 449-456
6. Kalogeraki V., Melliar-Smith P., and Moser L.: Soft real-time resource management in CORBA distributed systems. In Proceedings of the IEEE Workshop on Middleware for Real-time Systems and Services. San Francisco (Dec. 1997)
7. Kalogeraki V., Melliar-Smith P., and Moser L.: Dynamic scheduling for soft real-time distributed object systems. In Proceedings of the 3rd IEEE International Symposium on Object-Oriented Real-Time Distributed Computing, ISORC'00. Newport Beach, California (15 - 17 March, 2000) 114-121.
8. Kim K.H.K.: Object structures for real-time systems and simulators. IEEE Computer, Vol. 30, No. 8 (August 1997) 62-70
9. Kim K., Shokri E., and Crane P.: An implementation model for time-triggered message-triggered object support mechanisms in CORBA-compliant COSTS platforms. In Proceedings of the 1st IEEE International Symposium on Object-Oriented Real-Time Distributed Computing (ISORC'98). Kyoto, Japan. (20 - 22 April, 1998) 12-21
10. Kim K., Beck D., Liu J., Miyazaki H., and Shokri E.: A CORBA service enabling programmer-friendly object-oriented real-time distributed computing. In Proceedings of the Fifth International Workshop on Object-Oriented Real-Time Dependable Systems. Monterey, California (18 -19 November, 1999) 101-107
11. Leung J., and Whitehead J.W., On the complexity of fixed priority scheduling of periodic real-time tasks.: Performance Evaluation 2(4) (December 1982) 237-250
12. Levine D., Flores-Gaitan S., and Schmidt D.C.: An empirical evaluation of OS support for real-time CORBA Object Request Brokers. In Proceedings of the MultiMedia Computing and Networking Conference (MMCN'2000), San Jose (January 25-27, 2000)
13. Liu C., and Layland J.W.: Scheduling algorithms for multiprogramming in a hard real-time environment. Journal of ACM, Vol 20, n°1. (1973) 46-61
14. Liu J.W.S., Rendondo J.L., Deng Z., Tia T.S., Bettati R., Silberman A., Storch M., Ha R., and Shih W.K.: PERTS: A prototyping environment for real-time systems, Technical report UIUCDCS-R-93-1802, Department of computer science, University of Illinois, (May 1993)
15. Liu J.W.S.: Real-time systems. Prentice Hall (2000)
16. Malcolm N., and Zhao W.: Hard real-time communication in multiple-access networks. Journal of Real-Time Systems (8) (1995) 35-77

17. Mammeri Z., and He J.: Modeling and timing performance analysis of deterministic clock synchronization algorithm. 9th International Conference on Parallel and Distributed Systems (PDCS'96). Dijon, France (September 25-27, 1996) 219-224

18. Mammeri Z., Delay jitter guarantee for real-time communications with ATM network. Second IEEE International Conference on ATM. Colmar, France (June 21-23 1999) 146-155

19. Mills D.L.: Network Time Protocol (NTP), RFC 1305 (March 1992)

20. Montez C., Fraga J., Oliveira R., and Farines J.-M.: An adaptive scheduling approach in real-time CORBA. In Proceedings of the 2nd IEEE International Symposium on Object-oriented Real-time Distributed Computing (ISORC'99) Saint-Malo, France (May 2 - 5, 1999) 301-309

21. Moser L.E., Narasimhan P., and Melliar-Smith P.M.: Object-oriented programming of complex fault-tolerant real-time systems. In Proceedings of the 2nd IEEE Intern. Workshop on Object-Oriented Real-time Dependable Systems (WORDS'96) Laguna Beach, California (February 1996) 116-119

22. Nett E., Gergeleit M., and Mock M.: An adaptive approach to the object-oriented real-time computing. In Proceedings of the 1st IEEE Intern. Symposium on Object-oriented Real-time Distributed Computing (ISORC'98) Kyoto, Japan 20-22 April 1998.

23 Object Management Group.: Real-Time CORBA. A white paper - Issue 1.0. OMG December 5, 1996.

24 Object Management Group.: Real-Time CORBA 1.0 – Request for Proposal. Document/Orbos 97-09-31. January 19, 1998.

25. Object Management Group.: CORBA Messaging Specification. OMG Document ORBOS/98-05-05 (May 1998)

26. Object Management Group.: The Common Object Request Broker: Architecture and specification. 2.3 (June 1999)

27. Object Management Group.: Real-Time CORBA Joint Revised Submission. Document orbos/99-02-12 (March 1999)

28. Object Management Group.: Dynamic scheduling, – Request for Proposal. Document/Orbos/99-03-32 (October 25, 1999)

29. Object Management Group.: The Common Object Request Broker: Architecture and Specification v2.4. OMG October 2000

30. Polze A., and Sha L.: Composite objects: real-time programming with CORBA. In Proceedings of the 24th Euromicro conference, Vaesteras, Sweden (August 25-27, 1998) 997-1004

31. Polze A., Wallnau K., Plakosh D., and Malek M.: Real-time computing with off-the-shelf components – The case for CORBA. Parallel and Distributed Computing Practices. Vol. 2, N° 1 (1999)

32. R. Braden et al.: Resource ReSerVation Protocol (RSVP) 31 September 1997

33. Saksena M.: Towards automatic synthesis of QoS preserving implementations from object-oriented design models. In Proceedings of the 5th International Workshop on Object-Oriented Real-Time Dependable Systems (WORDS'99). Monterey, California (18 - 19 November, 1999) 93-99

34. Schmidt D.C., Levine D.L., and Mungee S.: The Design of the TAO real-time object request broker. Computer Communications 21 (1998) 294-324

35. Schmidt D.C., and Kubns F.: An overview of the real-time CORBA specification. Computer (June 2000) 56-63

36. Shokri E., and Sheu Ph.: Real-time distributed object computing: an emerging field. Computer (June 2000) 45-46.

37. Squadrito M., Esibov L., DiPippo L.C., Wolfe V.F., Cooper G., Thuraisingham B., Krupp P., Milligan M., and. Johnston R: Concurrency control in real-time object-oriented systems: the affected set priority ceiling protocols. In Proceedings of the 1st IEEE International Symposium on Object-Oriented Real-Time Distributed Computing (ISORC'98) Kyoto, Japan (20 - 22 April, 1998) 96-105

38. Stankovic J.A.: Misconceptions about real-time computing. Computer (1988) 21 10-19

39. Stankovic J.A.: Distributed real-time computing: the next generation. Technical report TR92-01. Dept. of computer science, University of Massachusetts (1992)

40. Sydir J.J., Chatterjee S., and Sabata B.: Providing end-to-end QoS assurances in CORBA-based system. In Proceedings of 1st IEEE International Symposium on Object-Oriented Real-Time Distributed Computing (ISORC'98) Kyoto, Japan (20 - 22 April, 1998) 53-61

41. Thuraisingham B., Krupp P., Schafer A., and Wolfe V.F.: On real-time extensions to the Common Object Request Broker Architecture. In Proceedings of the Object Oriented Programming, Systems, Languages, and Applications (OOPSLA'94) Conference (Oct. 1994)

42. Thuraisingham B., Krupp P., and Wolfe V.F.: On real-time extensions to Object Request Brokers: position paper. In Proceedings of the 2nd IEEE Workshop on Object-Oriented Real-Time Dependable Systems (WORDS '96) (1996) 182-185

43. Usländer T., Lebas F.X.: OPERA: A CORBA-based Architecture Enabling Distributed Real-Time Simulations. In Proceedings of 2nd d IEEE International Symposium on Object-oriented Real-time Distributed Computing (ISORC'99). Saint-Malo France (May 2 - 5, 1999) 241-244

44. Wohlever S., Wolfe V.F., Thuraisingham B., Freedman R., and Maurer J.: CORBA-based real-time trader service for adaptable command and control systems. In Proceedings of the 2nd IEEE International Symposium on Object-oriented Real-time Distributed Computing (ISORC'99). Saint-Malo France (May 2 - 5, 1999) 64-71

45. Wolfe V.F. and al.: Real-time method invocations in distributed environments, Technical Report. University of Rhode Island (1995) 95-244

46. Wolfe V.F., DiPippo L., Bethmagalkar R., Cooper G., Johnston R., Kortmann P., Watson B., and Wohlever S.: RapidSched: static scheduling and analysis for real-time CORBA. In Proceedings of the 4th IEEE International Workshop on Object-Oriented Real-Time Dependable Systems (WORDS'99). Santa Barbara, California (27 - 29 January, 1999) 34-39

47. Wolfe V.F., DiPippo L.C., Ginis R., Squadrito M., Wohlever S., Zykh I., and Johnston R.: Expressing and enforcing timing constraints in a dynamic real-time CORBA system. International Journal of Time-Critical Computing Systems, 16 (1999) 253-280

Development of Accounting Management Based Service Environment in Tina, Java and Corba Architectures

Abderrahim Sekkaki [1], Luis Marco Cáceres Alvarez [2], Wagner Tatsuya Watanabe [2], and Carlos Becker Westphall [2]

[1] Department of Mathematics and Computer science, University Hassan II, Faculty of sciences, Ain chok, P.O Box 5366 Maarif, Casablanca, Morocco
sekkaki@facsc-achok.ac.ma

[2] Network and Management Laboratory, Federal University of Santa Catarina, Caixa Postal 476, 88040-970 Florianópolis –SC – Brazil
{caceres, wagner, westphal}@lrg.ufsc.br

Abstract. TINA (Telecommunications Information Networking Architecture) concepts and principles are introduced with the objective of correcting problems of centralized service control and service data model existing in IN (Intelligent Network). Now, TINA has been designed with the goal to offer an universal vision of telecommunications services, it answers to the increasing needs of fast developments of news services e.g., multimedia, multi-party conferencing, etc., will need to be rapidly and efficiently introduced, deployed, operated and managed. In this context, TINA developed a comprehensive architecture for multi-service networks that support multimedia service. On the other hand, the provisioning of all the service management context functionality for TINA services (i.e. FCAPS) is still an open research question.
In this paper, we discuss accounting features and requirements, security services for TINA management context. We propose and implement a model of accounting and security management middleware architecture. A prototype uses the environment CORBA object and JAVA to validate the concepts.

1 Introduction

TINA might be considered an architecture that is relevant to the future and existing telecommunications services infrastructure (i.e. the IN). This is mainly due to one of the TINA assumptions, which is based on the presence of the DPE (Distributed Processing Environment), and CORBA (Common Object Request Broker Architecture) as a signaling infrastructure. Services in TINA are carried out as distributed applications and provided over a distributed computing environment. While the TINA Consortium has coordinated works to refine and extend the specifications to ensure internetworking with legacy systems, and explain how TINA applies to IN and answers the IN evolution challenges, the specifications concerning the management domain remain incomplete.

TINA has defined the Network Resource Architecture (NRA) [1] that covers the management areas of connection, configuration, fault and accounting management. Some of these areas has been defined in great detail (e.g. connection management[2]) while some others areas have been addressed much less (e.g. fault and accounting management architecture[3]), and some others are not yet defined at all (e.g. performance management). Security management is also not yet fully addressed in TINA. Hence, to begin with, we have been working on the framework of the TINA accounting management architecture, as one of service management areas that is an indispensable part of all the TINA services. We also discuss the relevance of the security of accounting management and how to protect the accounting information.

This paper is organized in the following way: Section 2 presents a TINA overview: service architecture, service management and TINA accounting architecture. Section 3 presents the accounting management system, the main components of TINA accounting management and their features. Section 4 propose a model of security management, discusses security services, mechanisms and presents security considerations in accounting. Section 5 describes the specifications and visualization of the prototype. Finally, the conclusion is given in Section 6.

2 TINA Overview

2.1 TINA Service Architecture

TINA defines a flexible architecture which consists of several modeling concepts and specifications. As well as the service architecture[4], the main specification involves the business model [5], the computational model, and the service component specification [6].

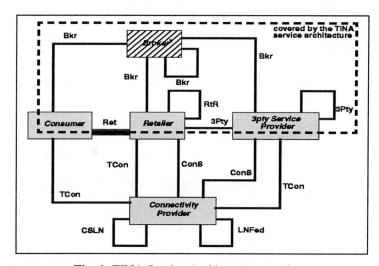

Fig. 1. TINA Service Architecture overview.

In the business model, the participants (or stakeholders) in a service are categorized into five groups according to the roles they play in the service. The roles defined within TINA are Consumer, Retailer, 3rd Party Service Provider, Connectivity Provider and Broker (Fig. 1). The Broker and 3rd Party Service Provider roles have not been described in detail in TINA. The Consumer, Retailer, and Connectivity Provider roles, however, have been specified in TINA with sufficient detail, including the interfaces between them. The interface among the participants are defined as References Points. The most important reference point for service management is the Ret-Reference Point [7] defined between the Consumer and the Retailer.

The service architecture also defines two types of sessions, an access and a service session. The access session is responsible for the identification and authentication of the user. It is also used to search for available services and executes the services by creating a service session. The service session is responsible for managing the current state of the service. It also creates and controls stream bindings, which are an abstract representation of end-to-end connection among applications. The functionality of the service session is defined by a number of functional groups called Feature Sets. The Feature Sets carry out various functions, such as: joining a service, inviting parties in the service session, establishing stream connectivity requirements, issuing voting, etc.

Subscription subsystem and subscription information model defined in TINA allow an user to discover the new services and eventually subscribes to them. This is independent of the access technology used because the end-users require more and more flexibility, and demand the ability to activate and deactivate service subscription themselves, and alter the related profiles[8].

2.2 TINA Service Management

Service management is a very broad discipline and it is also an important part of TINA service architecture. It involves management functions in TINA service architecture. It controls the end-to-end transactions for TINA services to provide FCAPS properties dictated by binding management contexts. There are four conceptual axes associated with service management:

i. Partitioning axis: TINA is partitioned into three layers (service, resource and DPE).
ii. Functional axis: is represented most notably by FCAPS functions. To support the FCAPS integrity of a service session, construct such as management context and service transaction are provided.
iii. Computational axis: represents computational support for management needs.
iv. Life cycle axis: represents the life cycle issues, including service life cycle management and user (consumer) life cycle management.

To develop the management functions on a DPE (Distributed Processing Environment) platform, TINA uses a common object interface description language (IDL or ODL). It uses also the Network Resource Information Model (NRIM), which enables us to describe systematically the network management information using unified abstract models [9].

2.3 TINA Accounting Architecture

In addition to the existing standards such as X'742 [10], TINA dedicated one document to this topic[3]. Considering TINA, a higher level concept is necessary, which maintains service management context across multi-domains (stakeholders), and separates service management from service as clearly as possible[11], and on-line charging is still a challenging problem.

TINA accounting management consists of four cycles namely: Metering, Classification, Tariffing and Billing. It introduces a concept of Accounting Management Context (AcctMgmtCtxt) associated with Service Transaction (ST). The purpose of AcctMgmtCtxt is to guarantee that accountability be preserved through a set of activities of distributed objects, which constitute the service. We want to emphasize that accountability is not a property or an attribute of a single object. It is rather a set of quantities measured or calculated over a set of activities of distributed objects throughout the service. When a ST is activated by a Service Session Manager (SSM), its AcctMgmtCtxt is interpreted in accordance with its Session Component description (unit of service to be accounted) and with the tariff structure within the SSM component. Then, the SSM passes control and necessary parameters to resource or computational level mechanism such as Communication Session Manager (CSM), notification server, metering managers and service management.

It can be seen that the Service Transaction (ST) concept consists of three phases:
  i. Set-up phase : it is a phase of negotiation between the user and the provider (e.g. Tariff structure). The agreed schema is submitted as AcctMgmtCtxt of the ST.
  ii. Execution phase :The service is being offered to the user, as it was specified in the first phase. The description of information of the service session is specified as a part of AcctMgmtCtxt.
  iii. Wrap-up phase : The service is concluded, account record or charging information may be sent to the user at the conclusion of the transaction, if it is specified in AcctMgmtCtxt.

3 TINA Accounting Management Architecture and Components

In [12], an extension to the OSIMIS management platform through implementing the Usage Metering Function defined in the OSI functional model has been developed and validated. Our present work is the continuity of activities undertaken in our laboratory.

3.1 Architectural Model

Figure 2 shows the architectural model of accounting management context with the interaction between the main components of TINA service architecture. Components are interpreted here as the units of reuse. TINA AcctMgmtCtxt consists of separate components, covering different aspects of service, and network control and management.

Accountable Event Collecting : The Accountable Events Collecting component receives, collects and collates the accounting events associated with the Session Components status or Session Component state changes, and events associated with charging information generated by the SSM. It is divided into two modules: *EventManagememt* and *SessionComponent*.

Tariffing: Tariffing component in TINA AcctMgmtCtxt can be represented by two different modules:
  i. *Recovery* Module, which gives charging-rate as a function of the current Session State. It converts the collected accountable events to a charging record and stores it in the charging record database. In addition, it allows the restoration of the information collected when failure in the service.
  ii. *Tariff* Module, which gives accumulated current charges as a function of a sequence event. It calculates the tariff, using the charging formula according to the contract of subscription and stores it into the billing record.

Billing : The billing component may be automatically issued at the end of a billing period as negotiated and defined in the customer's contract. It generates the bill based on charging information from billing records. There are four billing configuration :On-line charging, Shared billing, Third-party billing, and Credit-debit billing. The billing implementation in our prototype is configured in the on-line charging.

Usage Metering: During the life time of the connection, the usage metering component collects and controls data acquisition concerned with the use of network resources generated by the LNC (Layer Network Coordinator). It also registers the collected data for future processing (fault and performance management).

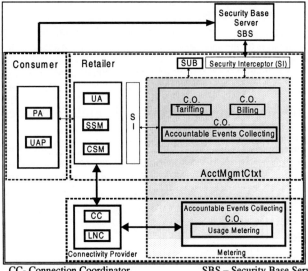

CC- Connection Coordinator
CSM - Communication Session Manager
LNC - Layer Network Coordinator
PA - Provider Agent
SI - Security Interceptor
SBS – Security Base Server
SSM – Service Session Manager
SUB – Subscription Management Component
UA - User Agent
UAP - User Application

Fig. 2. Architectural model.

3.2 Definition the AcctMgmtCtxt Components

The model implementation uses the CORBA objects on Visibroker 3.04 [13]. The objects are defined such as interfaces in IDL, so they can access and be accessed by any other CORBA object independently of the programming language and underlying computing platforms used [14]. The definitions in IDL of the AccMgmtCtxt components are defined in Figure 3.

The building of an IDL for implementation the prototype allows the creation of the interfaces that communicate between them. The data-bases defined in IDL are divided: "*SubscripInfo.mdb*" which defines the subscription user, the "*TariffInfo.mdb*" which registers the events of Tariffing component while the service execution is carried out, and "*BillingInfo.mdb*" which defines the charging-billing of the user.

The module AcctMgmtCtxt is the most important component in this class, it is generated in the set-up phase of the ST and it is destroyed when the Service Transaction is concluded. The interface *SessionComponent* is used for controlling and managing the SC (Session Component), as also start and stop metering actions in the accounting object. The SC refers one service to the specified user. *EventManagement* defines the events management parts (for example: delivery).

In the prototype implementation, one session is used to test the service session: the "digital video-audio conference" between the users and a provider. The components forming parts of the session are extended to generate accountable events towards the AcctMgmtCtxt components and allow to each consumer to retrieve on-line billing information about an ongoing shared service session. In the beginning of the execution, a graphical interface *"Client"* is visualized (see Fig. 5 in Sect. 5.1), subdivided into the following modules:

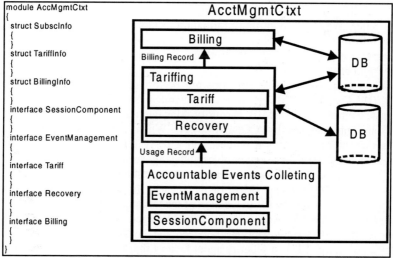

Fig. 3. Definition the AcctMgmtCtxt components.

- **Subscription**: login as a known user: authentication and verification of the user features; and display data user;
- **Up-date Subscribed**: up-date of user:(Insert, Modify, Delete, etc..);
- **Monitoring service**: display the video conference service for each user; and
- **Error messages**: show the errors.

4 Security of TINA Accounting Management

In distributed systems, security is an essential and a natural part of it. Of course accounting is not an exception. Security of accounting implies: guaranteed accountability, trustful accounting, integrity of accounting information, etc.

i. Guaranteed accountability refers to the service transaction that should ideally offer a mechanism to guarantee integrity of the service. This means that " you do not pay if you do not get the agreed service".

ii. Trustful accounting means that accounting information should be assured. The requirement come from the openness of TINA. An user and a server provider which are totally unknown to each other can connect by using an on-line Yellow Page or by using a trader (situation where security mechanism is urgently needed). How can user trust the service provider and conversely how can a service provider trust the user? The first question is related more to accounting management whereas the second is related more to security management. It is similar to the case where an unknown phone company sends an expensive bill based on questionable service usage. Openness is not necessarily a good thing unless both the user and the service provider are properly protected. Accounting information should be trustworthy and should be able to be recorded on both sides in a non-modifiable and non-refutable manner.

iii. Integrity of accounting information refers that integrity of accounting information should be preserved over network failures and over disruption of services, considering the service through over different management domains.

In our system, the security functions are performed by the SBS (Security Base Server), (see Fig. 4) [15]. It provides an authentication service, a ticket distribution, an access control and a SMIB (Security Management Information Base). These services are implemented by CORBA objects. When a principal invokes a service on a server object, there is a request and a reply interacted between them. This provides security services, secure objects having security functionality using the Security Interceptor(SI) module to intercept the request and the reply. The mechanism is transparent to the principal. Most of the security functions are performed by the Security Base Server during the invocation of the target object.

In our model, the SBS acts as a Security Manager. It controls and manages the security context of the interactions between the consumer and the retailer. The Security Interceptor (SI) is a security agent factory that helps in the access control of the service in the retailer domain, including the AcctMgmtCtxt.

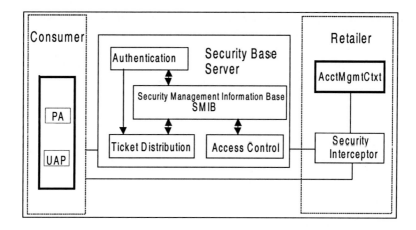

Fig. 4. Architecture of the Security Base Server.

Figure 2 illustrates a relation between the model (SBS) with AcctMgmntCtxt, the components are seen by the SBS as services objects. Then they need secure transactions. The SI present in the Retailer domain will intercept invocation to AcctMgmtCtxt and provide security.

5 Description and Visualization of the Results

5.1 Description Mechanism

The prototype is divided into two phases: subscription phase and monitoring phase.
a) Subscription phase
   The prototype begins a service when it is solicited from user (module Service, click button "Start" in Fig. 5) that registers his log, entering his Id and password (module Subscription in Fig. 5).
   Immediately, a verification (access to database "SubscripInfo.mdb") and an authentication are carried out through the security management (SBS in Fig. 4). The authorized user presents himself under two types:
   1. the user client which has a contract with provider and priority of full service uses: audio (speaking-listening) , video (vision-image), i.e. A+V;
   2. the user, who uses the service for the first time, and does not have a contract with provider, gets only a partial service and he is seen as an Audio participant. The new user will be registered in database (module UpDate: "Add New User" in Fig. 5) to ensure that his amount of available credit is enough to make use of the service or in another way, giving a logout, not permitting the use of the service.
   Notice that the security mechanism is present during a session processing. All requests destined to the provider are transparently intercepted by the Security Interceptor (SI) that redirects it to SBS (see Fig. 4), making possible the principal

authorization by checking his identity and rights in SMIB. Data exchanged will be encrypted and it will guarantee security in DPE like TINA [15].

b) Monitoring phase
- For each user authorized, the proper AcctMgmtCtxt is created, starting service (module Service, click button "Enter" in Fig. 5).
- Automatically, the AcctMgmtCtxt creates the objects (interface *SessionComponent* and interface *EventManagement*) that identify the management events during the execution.
- At the same time, a Session State Vector is generated (SSVec) where each position registers the information relating the events of service:
- (1): initial time (it); (2): time time (ft); (3): accumulated time "full service use": A+V (fs) or "partial service use": (ps); and (4): accumulated time "shared service use" of the users (ts).
- Click on button "Enter", the "it" is initialized while activating "fs" or "ps". The vector is seen as a dynamic and volatile structure, any failure that occurs during the service is registered as information in database. The back-up is necessary to store the information (events) in the database (*"TariffInfo.mdb"*) because its interaction with module Tariffing (interface *Tariff*) is periodic.
- During the period of the service, the provider must be verify that the time expended for the user does not exceed 80% of the available limit. If it occurs, the provider must notifies the user and consults him if he desires to modify his available quota (module UpDate: "Change Max Amount" in Fig. 5), or logout of service (active the interface *Billing* for its charging).
- The user that makes use of the service can leave the session (module Service, click button "Leave" in Fig. 5). In this moment, the user is in the state auditing (A). He has also the option to retake to the service (module Service, click button "Enter" in Fig. 5) or end of session (module Service, click button "Stop" in Fig. 5).
- The option also exists for several users to share the service at the same time. This happens when the button "Enter" is activated and then it actives "ts" and stop "fs". To leave the shared state, click button "Leave" to active "fs" or "Stop" to quit.
- If the user is in the state "Stop", the timers are deactivated, and "ft" is registered. Finally, the on-line charging (interface *Billing*) is activated for the supplied service of the user and the information is stored in database (*"BillingInfo.mdb"*).

5.2 Scenario Execution

We consider in Figure 5 three users who share the same session of "multi-party conferencing service" provided by the retailer. Each user in the execution of the service has its proper graphical interface "*Client*", that defines the Accounting Management Context, which allows the management of the available resources in the session to start.

Accounting Management Based Service Environment in Tina, Java and Corba   447

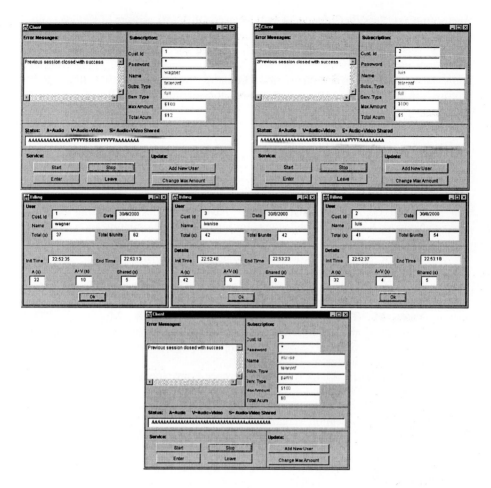

Fig. 5. Example of scenario

When the session is finished, the graphical interface "*Billing*" is displayed, that shows the on-line charging of the service supplied for the Retailer. The users have three options of access to the service according to their contracts: V (Video+Audio), A (Audio) and S (Video+Audio Shared). The use of the "V option" costs 3$/unit, the "A option" costs 1$/unit and the "S option" costs 2$/unit.

6 Conclusion

CORBA facilities and services offer a set of interface that supports TINA service management. This paper provides first a brief overview of the TINA concept an the focuses on accounting management. We described accounting features, requirements, and other accounting issues. Then, we proposed an accounting management architecture specifying an architectural model which contains components such as

AcctMgmtCtxt, Session Component, etc., covering different aspects of service control and management. We also integrated security in TINA accounting management through, security services and mechanisms, based on secure objects. In order to realize our model, we described the accounting components in IDL, their roles and their interactions to accomplish the accounting functionality. We illustrated a prototype in TINA-based service environment considering on-line billing.

References

1. Steegmans, F. (ed.): TINA Network Resource Architecture. TINA-C, http://www.tinac.com, 1997.
2. Pavlou, G., Mota, T., Steegmans, F., Pavon, J.: Issues in Realizing the TINA Network Resource Architecture. In: Interoperable Communication Networks Journal, Vol. 2, No. 1, pp. 133-146, Baltzer Science Publishers. March 1999.
3. Hamada, T.: Accounting Management Architecture. TINA-C, http://www.tinac.com, 1996.
4. Kristiansen, L. (ed): TINA Service Architecture 5.0. TINA-C, http://www.tinac.com, 1997.
5. Yates, M., Takita, W., Demoudem, L., Jansson, R., Mulder H.: TINA Business Model and Reference Points. TINA-C, http://www.tinac.com, 1997.
6. Abarca, C. (ed) : TINA Service Component Specification. v. 1.0b, TINA-C, http://www.tinac.com, 1998.
7. Farley, P. (ed) : TINA Retailer Reference Point Specification. v. 1.0, TINA-C, http://www.tinac.com, 1998.
8. Pampaey, M., Couturier, A.: Using TINA Concepts for IN Evolution. In: IEEE Communication Magazine – June 2000.
9. Steegmans, F. (ed) : TINA Network Resource Information Model. TINA-C, http://www.tinac.com, 1997.
10. ITU-T Recommendation X.742, Information Technology – Open Systems Interconnection – Systems management: Accounting meter function- July 1994.
11. Hellemans, P., Redmond, C., Daenen, K., Lewis, D: Accounting Management in a TINA-Based Service and Network Environment: the Sixth International Conference on Intelligence in Services and Networks , Barcelona, April 1999.
12. Kormann, L.F., Westphall, C.B., Coser, A.: OSI Usage Metering Function for OSIMIS Management Platform. In Journal of Network and Systems Management, Vol. 4, No. 3, 1996.
13. Visibroker 3.04 http://www.inprise.com
14. Object Management Group. The Common Object Request Broker Architecture 2.0/IIOP – specification Revision 2.0 – OMG Document – August 1996.
15. Sekkaki, A., Nguessan, D., Müller, M.D., Westphall, C.B.: Security within TINA Accounting Architecture Management. IEEE International Conference on Communication , June 11-15, 2001 Helsinki, Finland.

A QoS System for CaTV Networks

Juan Leal and José M. Fornés

Área de Ingeniería Telemática, Universidad de Sevilla
Camino de los Descubrimientos s/n, E-41092 Sevilla
{jleal, fornes}@trajano.us.es

Abstract. This paper proposes an architecture for the integration of QoS enabled systems in the context of a CaTV network. The proposed architecture makes possible to use several signaling protocols, to add QoS signaling capabilities to systems that would otherwise only support static QoS provisioning and to integrate the QoS mechanisms of the different technologies used in CaTV networks.

This paper also describes an implementation of the proposed architecture using Java Beans, where CORBA based signaling is used to provide on demand QoS to the Com21 cablemodem system.

1 Introduction

Recently, boosted by the growth of the Internet, the need for IP to provide Quality of Service (QoS) mechanisms has become patent in order to support new services demanded by users. In this direction, the IETF is carrying on important efforts that lead to a number of technologies, mainly Integrated Services [1], Differentiated Services [2] and Policy Based Admission Control [3].

At the same time, work is being done in the integration of IP with other technologies that support QoS, like ATM [4]. MPLS [5] is an example of this effort.

In parallel, conscious of the importance of QoS, many network equipment manufacturers incorporate in their devices their own static priorization or reservation mechanisms.

Putting it all together, there are a number of technologies that address the problem of QoS, and their integration is necessary to get the best out of each one of them.

This work consists in the design of an architecture that integrates different QoS enabled systems in the context of a Common Antenna TV (CaTV) network [6]. CaTV networks have a number of particularities that make them interesting for this work. In the first place, they use several different technologies (cablemodem, IP, ATM...) that must be integrated. Secondly, cablemodem access systems support QoS at a MAC level but at the moment only provide statically configurable QoS [7–10]. Finally, they are managed by one single organization, which makes possible to use solutions that are not applicable to the whole Internet.

Therefore, this paper is only addressed to CaTV networks and the solutions proposed are not applicable to the Internet.

The key benefits of this work are: providing a transient solution that allows the provision of QoS until a definitive standard is adopted; integrating in the system existing network elements which are widely deployed and whose QoS features are not being efficiently used at the moment (e.g. cablemodem systems) and integrating different existing partial solutions in order to extend QoS guarantees to the highest number of systems possible, in the CaTV environment.

A testbed of the proposed architecture has been implemented and tested on a real CaTV network. at the moment, the system operates only on the Com21 [8] cablemodem system and receives subscribers' signaling through a CORBA [11] interface.

The remainder of this article is organized as follows. Section 2 introduces CaTV from the data transmission point of view and cablemodem systems. Section 3 explains the proposed architecture for the integration of QoS technologies in CaTV networks. Later, Section 4 describes the implemented testbed and finally conclussion remarks are given in Section 5.

2 Introduction to CaTV Networks

Bi-directional CaTV networks based on the Hybrid Fiber Coaxial (HFC) technology [12] consist of a number of so called Primary Nodes linked together and to the network's Headend through an IP backbone supported on a WAN network such as SDH or ATM. The Primary Nodes are also connected to HFC networks that reach the subscribers' homes. The network's Headend contains other important network elements such as the web and mail servers, the O&M segment and the Internet connection.

The cablemodem system consists of the subscriber access devices, called cablemodems, and a number of CableModem Termination Systems (CMTS), located at the Primary Nodes, which control the cablemodems and forward their traffic through de HFC network to the cable operator's network and vice versa. An O&M workstation installed in the Headend manages the cablemodem system.

There are several cablemodem systems. Motorola's and Com21's are among the most deployed proprietary systems [13]. The Cable Television Laboratories (CableLabs) has defined a standard and interoperable cablemodem system called Data Over Cable Service Interface Specifications (DOCSIS) [9, 10]. Current version is 1.0 and version 1.1 is in process of specification (interim status).

3 Proposed Architecture

The IP QoS technologies in use nowadays [1, 2] use a decentralized paradigm, where each network element receives in-band signaling (that is, signaling packets follow the same route as data packets) and takes the appropriate actions to fulfill the QoS requests. This is obviously the best solution to provide end to end QoS in a context of different networks managed by different administrators, and is consistent with Internet's philosophy.

However, in the context of CaTV we are interested in adding QoS signaling to elements, such as cablemodem systems, which currently have only static QoS mechanisms. In order to do this, because we can not modify existing network elements, we need a centralized element that receives out-of-band signaling and modifies the otherwise statically assigned QoS in these elements.

It is evident that a centralized QoS architecture is not convenient for the Internet and that it has important drawbacks, mainly related to scalability. However, in the context of CaTV networks, which are managed by an unique administration, this is not a problem. All the contrary, because of the diversity of technologies used (cablemodems, ATM, routers, SDH...), a centralized structure can be the most efficient manner to achieve the integration of the different QoS mechanisms used in these technologies.

Therefore, this paper proposes a centralized QoS architecture for achieving end to end QoS in CaTV networks by means of efficiently integrating the different QoS technologies used in this networks and adding QoS signaling to elements that do not support it.

The primary element of this architecture, shown in figure 1, is a QoS server that would likely be located at the network Headend, connected to both the backbone and the O&M network.

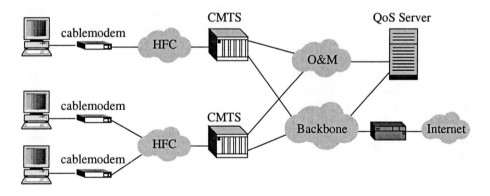

Fig. 1. Centralized architecture

3.1 Server Architecture

The architecture for Integrated Services in the Internet [1] is a decentralized one where every router receives in-band QoS signaling, performs admission control and modifies the forwarding of the packets accordingly.

The reference model for Integrated Services routers is shown simplified in figure 2. The reservation setup agent receives user's QoS signaling, checks admission control to decide whether the request should be granted and modifies the forwarding behavior accordingly.

The proposed centralized architecture is based on a server with an internal architecture which is a generalization of this reference model. Changes have

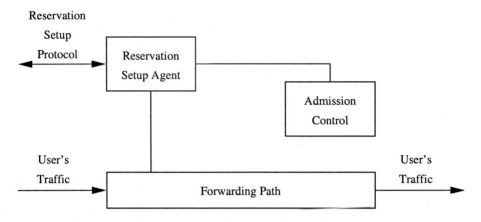

Fig. 2. Reference model for routers defined in RFC 1633

been made in order to manage the different elements of the CaTV network and support several QoS signaling protocols.

As it has been explained, IP QoS signaling protocols used at the moment, remarkably RSVP [14], are in-band protocols, which are not suitable for a centralized architecture. Therefore, a different protocol has to be used. In order to be able to use different signaling protocols until there is a standard one, it is desirable that the architecture allows support for several signaling protocols simultaneously. For this purpose, instead of a single reservation setup agent, the proposed architecture can contain several modules (which we call "agents"), each of which implements a QoS signaling protocol.

The equivalent in the proposed architecture of the forwarding path in the router reference model are a number of components called "managers". Because the server is centralized and does not forward user's traffic, a forwarding path makes no sense. Instead, the server must take the appropriate actions on the managed devices in order to provide the requested QoS. Therefore managers act as proxies of network elements performing, with respect to agents, the same functions as the forwarding path.

Finally, admission control functions are divided in two different blocks. Admission control based on resource availability is dependent of the network element and must be therefore performed by managers. However, policy based admission control (that is, administrative authorization) is better done in a centralized way and is therefore implemented in an "authorization component" in the server.

The rest of functions performed by the server, like initialization, logging and accounting are implemented by other components of the server which we refer to as "Auxiliary Elements".

This architecture, depicted in figure 3, allows the integration of the different technologies in a CaTV network from a centralized server, with an structure similar to the router reference model but generalized and adapted to a centralized

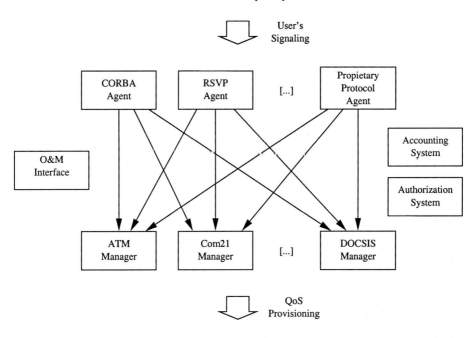

Fig. 3. Server architecture

system. Using this architecture, the server can receive user's QoS signaling using several protocols and can reserve the requested QoS in all of the network elements of a CaTV network.

4 Testbed

The proposed architecture has been implemented and the resulting system has been tested on the HFC network of Supercable Andalucía, the cable operator of southern Spain. A testbed consisting of one dedicated Com21 CMTS, one management workstation, four cablemodems, one PC acting as QoS server and two multimedia client PCs was employed. The testbed also used the operator's SDH backbone and a private portion of the real HFC network.

4.1 Server Implementation

The principal design issue for the centralized QoS server is upgradability, in order to support new network elements and signaling protocols. Consequently, an architecture based on Java and the Java Beans [15] conventions is proposed. Thanks to using out-of-band signaling, the server does not suffer the load of forwarding users' traffic. Therefore, the main shortcoming of Java, performance, is not a central issue for the server. In fact, as will be proven later, server's processing time is a minimal part of the overall response time of the system.

The server components described in the proposed architecture are implemented as Java Beans and the communication among them is performed through Java events. The purpose of this architecture is that new Agents, Managers and Auxiliary Elements can be implemented when required and incorporated to the server simply by linking its events to the rest of the components of the server, which can be done in a graphical Integrated Development Environment.

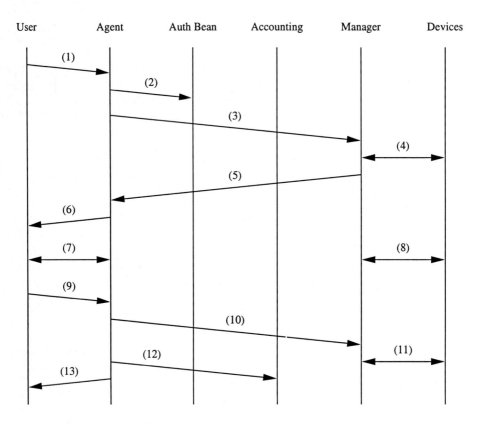

Fig. 4. Processing of a QoS request

A typical QoS request from a user is processed as shown in figure 4. First of all, the request is received by the agent that implements the signaling protocol used (1). This agent decodes the request and translates it to a common format understood by the rest of components in the server. Then the agent sends an authorization request event to the registered authorization beans (2). If no exception is raised, meaning authorization is granted, the agent issues reservation request events to all of the managers and waits for their responses (3).

Each manager then tries to fulfill the request performing the necessary operations on the network elements that it manages (4). Three possible results can be returned by each manager to the originating agent, OK in case the request is successfully completed, NOK if an error occurred (for example due to lack of resources) and IGNORE if the managed network elements are not related to the request.

Upon receiving all of the responses (5), the agent calculates the aggregate result (typically failure if at least one NOK is returned and success otherwise) and returns it to the requesting user (6). The agent is also responsible for maintaining any necessary state information on the request, possibly exchanging keepalive messages with the user (7), and the managers must ensure the reservation remains active on the managed devices (8).

When the reservation is torn down by the user (9), the agent sends cancellation events to the appropriate managers (10), which are responsible for terminating the reservation (11), so no confirmation is necessary.

Finally, the agent sends a Call Detail Record (CDR) event to the accounting beans (12) and reports the result to the user (13).

4.2 Implemented Components

At the moment, only one agent that uses CORBA [11] for signaling and one manager for the Com21 cablemodem system have been developed. Also, in order to test the system, a videoconference application has been used.

The agent implements a CORBA interface to communicate with users. This interface consists of one object containing methods for users to make, release and maintain QoS reservations. The choice of CORBA for signaling is based on simplicity: instead of designing a completely new protocol, it was considered easier to use CORBA, which relieved us from the communication and platform interoperability issues. Although CORBA would definitely not suit for an inband protocol, being the server centralized it appeared as a good alternative. Even performance could be argued against CORBA, the tests of the implemented QoS Server, described later in this paper, showed CORBA's overhead was not a concern.

The implemented manager performs the necessary SNMP [16] operations on the Com21 cablemodem system's CMTSs in order to modify the otherwise statically assigned QoS of the cablemodems. When this manager receives a request from an agent, it searches the affected cablemodems through all the CMTSs owned by the cable operator and temporarily modifies its QoS level, changing it back when the reservation ends.

The auxiliary elements are very simple in this early stage of development. The O&M bean starts and shuts down the server. The authorization bean always grants the requests. The accounting bean writes a CDR to disk.

VoIP and videoconference are some of the main applications that benefit QoS. They have therefore been chosen for demonstration and testing of the system. For this purpose, an application that performs QoS signaling on behalf of Microsoft NetMeeting has been developed.

4.3 Performance Results

The results of the tests were satisfactory in that the QoS level of the pertinent cablemodems was actually modified and the clients were able to perform a QoS enhanced videoconference.

The performance measurements showed that the maximum number of requests per minute that the system could serve without congestion was around 40, and was mainly limited by the response time of the CMTS to the SNMP operations. In fact, the response time of the server excluding the SNMP operations was about 90 ms. which allows to estimate the maximum load in over 600 requests per minute. This also proves the hypothesis that Java and CORBA performance issues were not a concern.

5 Future Work and Conclussions

Strictly speaking, to provide QoS end to end, it has to be supported in all of the elements in the path between the end hosts. Therefore, the final objective of this work is to interface with all the necessary systems in order to offer end to end QoS guarantees. For this purpose, it will be necessary to develop new managers for other devices. New agents that implement other signaling protocols will be needed to interface with the clients. Also, as described above, for the system to be useful in the practice, authorization and accounting beans that interface to policy and billing systems are necessary.

At this moment, this work allows to provide on-demand QoS over Com21 cablemodem systems by means of a custom CORBA interface. Assuming the operator's backbone is sufficiently overprovisioned, the system allows to maintain high quality VoIP calls and videoconferences between cable subscribers.

Presently, we are studying how to interface the system with the new DOCSIS 1.1 cablemodem platform.

References

1. IETF RFC 1633. Integrated Services in the Internet Architecture: an Overview (6-1994)
2. IETF RFC 2475. An Architecture for Differentiated Services (12-1998)
3. IETF RFC 2753. A Framework for Policy-based Admission Control (1-2000)
4. Eichler, G., et al.: Implementing Integrated and Differentiated Services for the Internet with ATM Networks: A Practical Approach. IEEE Communications Magazine. Vol 38-1 (2000)
5. IETF RFC 3031. Multiprotocol Label Switching Architecture (1-2001)
6. Ciciora, W.: Cable Television in the United States. An Overview. Cablelabs (1995)
7. Rabbat, R., Siu, K.: QoS Support for Integrated Services over CaTV. IEEE Communications Magazine. Vol. 37-1 (1999)
8. Com21 Inc.: ComUNITY Access System - 2.3 Release - Technical Reference Manual (1998)
9. Cablelabs: Data-Over-Cable Service Interface Specifications. Radio Frecuency Interface Specification. Version 1.1, revision I05 (7-2000)

10. Fellows, D., Jones, D. DOCSIS Cablemodem Technology. IEEE Communications Magazine. Vol. 39-3 (2001)
11. Object Management Group (OMG): The Common Object Request Broker: Architecture and Specification, Revision 2.2 (2-1998)
12. Bisdikian, C., Maruyama, K. Seidman, D., Serpanos, D.: Cable Access Beyond the Hype: On Residential Broadband Data Services over HFC Networks. IEEE Communications Magazine. Vol. 34-11 (1996)
13. www.cable-modems.org/articles/market
14. IETF RFC 2205. Resource ReSerVation Protocol (RSVP) (9-1997)
15. Sun Microsystems: The JavaBeans API Specification (7 1997)
16. IETF RFC 1157. Simple Network Management Protocol (SNMP) (5-1990)

Towards Manageable Mobile Agent Infrastructures

Paulo Simões, Paulo Marques, Luis Silva, João Silva, and Fernando Boavida

CISUC, University of Coimbra
Dep. Eng. Informática, Pólo II
P-3030 Coimbra, Portugal
{psimoes}@dei.uc.pt

Abstract. This paper addresses the problem of managing distributed mobile agent infrastructures. First, the weaknesses of current mobile agent implementations will be discussed and identified from the manageability viewpoint. The solutions devised and experimented in order to alleviate these weaknesses in our own agent platform will then be presented. These solutions are generic and could easily be applied to the majority of existing mobile agent implementations. The paper will finish with the discussion of a new approach we are following in the M&M Project, based on a rather different architecture that significantly reduces the manageability requirements.

1 Introduction

A mobile agent (MA) is a software program that is able to migrate to some remote machine, where it is able to execute some function or collect some relevant data and then migrate to other machines in order to accomplish another task. The basic idea of this paradigm is to distribute the processing throughout the network: that is, send the code to the data instead of bringing the data to the code. MA systems differ from other mobile code and agent-based technologies because increased code and state mobility allow for even more flexible and dynamic solutions. Telecommunication applications and network management are part of the broad range of application fields for MA systems [1].

Mobile agent implementations are based on some kind of distributed supporting infrastructure, that provides the execution environment for the agents in each location, controls the agent migration and lifecycle, and usually delivers additional services such as security and multi-agent coordination. This infrastructure typically consists of mobile agent platforms designed as extensions of the host's operating system. Each network node hosts one platform where agents from different applications and different users coexist and (hopefully) cooperate with each other.

One of the problems with current mobile agent systems is the considerable overhead required to properly install, configure and manage this large, distributed and remote infrastructure. This overhead often counterbalances the advantages of applying mobile agents in the first place, and seriously affects the usability of MA technology. For this reason, effective infrastructure management solutions are crucial for the success of mobile agent systems.

2 Agent Management and Infrastructure Administration

Manageability of mobile agent systems includes two distinct domains: the mobile agents themselves, and the supporting infrastructure.

Agent management is by now fairly well covered. Almost every known MA implementation allows adequate control of the agent lifecycle, the agent location, inter-agent coordination, migration and security. Although current implementations rely on proprietary and internal mechanisms, ongoing standardization initiatives - like MASIF [2] and, to some extension, FIPA [3] - are expected to provide the desired means of interoperability, as more implementations start to comply with these standards.

The management of the supporting distributed infrastructure, on the other way, is usually considered as a side-problem not directly related with MA technology and, therefore, not deserving the same level of attention. Although this might be so for small-scale prototypes, when one tries to deploy real world distributed MA platforms, the costs of installation and administration easily rise to unacceptable levels.

In the last couple of years we have developed the JAMES platform for mobile agents [4,5], which was later used by our industrial partners to produce and deploy several MA-based applications for telecommunications and network management. This scenario provided us with a good perspective on infrastructure manageability. Three key problems were identified in current systems: (i) they are too much focused on mobile agents; (ii) they do not decouple the management functionality from the infrastructure itself; and (iii) they lack effective support for remote maintenance of geographically distributed platforms.

The excessive focus on mobile agents results in the assumption that applications are 100% based on mobile agents. However, most applications should be composed by a mix of "static" modules (user interfaces, databases, system specific modules, etc.) and, just where appropriate, specialized mobile agents. This excessive focus on the MA technology, instead of the overall MA-based application, leads to a general lack of clear and powerful interfaces between the mobile agent infrastructure, the agents and the "static" components of agent-based applications. This results either in complicated and inefficient ad-hoc interfaces between the agents and the applications and in unbalanced application design, with more and more functionality pushed into mobile agents just because there is no easy way to interface with external modules.

The lack of decoupling between the management functionality and the platform itself is the second major drawback with current systems. Although they already entail some kind of global infrastructure management, with basic monitoring and control, this functionality is hidden in the platform internals and, at most, only available through closely attached user interfaces. General-purpose management applications are thus unable to integrate the management of MA infrastructures into its global system management framework.

The remote management factor is also an important but overlooked issue. One of the key advantages of MA systems is the ability to dynamically upgrade remote services installed on mobile agents. However, this advantage is dependent on the robustness of the underlying agent platform, which demands support for remote operations like installation, upgrading, monitoring, rejuvenation and tuning. With few exceptions, current platforms do not provide such services and require either local installation and maintenance (with increased running costs) or the usage of general remote desktop applications like Microsoft's Systems Management Server or Intel's Landesk, with increased complexity and poor integration.

3 Platform Manageability in the JAMES Project

In order to tackle with these key issues, three simple but effective solutions were introduced in the JAMES platform: a low-level service for remote upgrade and control of MA platforms; a high-level API for external applications interfacing with the platform or the platform's mobile agents; and an SNMP service that allows legacy management applications to monitor and administer the MA infrastructure. Fig. 1 shows the integration of these three solutions into the JAMES framework. Their implementation uses very simple and pragmatic approaches, and other MA systems should also be able to support this management model without requiring major system redesign.

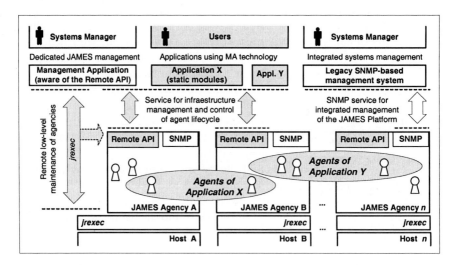

Fig. 1. JAMES Threefold Support for Platform Management

3.1 Low-Level Management of Remote Agent Platforms

The lower level of management for remote agent platforms (i.e. agencies) is provided by *jrexec*, a small and very stable service that runs at the host operating system level and controls the execution of the MA platform providing several services from remote locations: agency start and stop, agency rejuvenation, agency monitoring and upgrade of the agency software. This service represents a hook that avoids expensive local interventions, even in the case of unexpected MA platform crashes.

The installation and operation of *jrexec* is dependent of the host's operating system. It might be installed, for instance, as an Windows NT service or a Unix daemon. However, the interface to the management services provided by the several versions of *jrexec* is homogeneous, resulting in uniform MA infrastructure management across heterogenous networks. Since it is even possible to run the agencies without *jrexec* (relying in general-purpose remote management tools or using local interventions whenever necessary) the portability of the JAMES platform is not affected.

3.2 Remote Interface for External Applications

JAMES includes a single unified interface for communication between the mobile agent system and external applications. The implications of this interface, that we name Remote API, are manifold:

- the core of the platform became much simpler. Several user interfaces for basic operations and maintenance tasks (system monitoring, agent lifecycle control, system maintenance), previously attached to the platform's core, were pushed to the outside and redesigned as standalone applications that interact with the platform using the Remote API. In fact the current version of the platform has no graphical user interface at all;
- interaction between mobile agents and its associated "static" modules (applications) was enhanced. Each application directly launches, controls and communicates with its agents in a straightforward fashion. This simplifies the application design and allows more efficient usage of agent technology;
- and MA technology can be totally hidden bellow MA-based applications that, through the Remote API, completely control and make use of JAMES without requiring direct contact between the user (i.e. the customer) and MA related technology. This opens the way for deployment of MA-based applications into mainstream markets.

The current implementation of the Remote API is based on Java RMI, and consists on the interface itself, on the side of platform, and a set of Java classes to be included on the external applications (Fig. 2). These classes provide to the application developer with a high-level interface to the agent infrastructure, either to control/access their own agents or to manage the whole infrastructure.

The administrative model of the Remote API considers several types of entities: Applications, Application Licenses, Users, Mobile Agents, Instances of

4 One Step Further: Maintenance of Application-Centric Mobile Agents

As already mentioned, the solutions devised for the JAMES platform tackle with some of the identified problems with the maintenance of MA systems simply by adding ad-hoc management support to an already established architecture, common to most MA implementations. In the context of the M&M project [8] we are now working on a new architecture for mobile agent support. The most distinctive characteristic is that in this new approach there are no agent platforms. Instead, agents arrive and leave from the applications they are part of. The application is central and MAs are just a part of the system playing specific roles (see Fig. 5). The applications are able of sending, receiving and interacting with mobile agents by using well-defined binary software components (Javabeans or ActiveX components). Management applications are developed using the current industry best-practice software methods and can become agent-enabled by integrating mobility components. There is one small component that provides the basic support for mobile agents (migration, lifecycle control) and, using a flexible extension mechanism, more sophisticated services (agent coordination and communication, agent tracking, security, persistence, infrastructure management, etc.) are added as components, if and where needed.

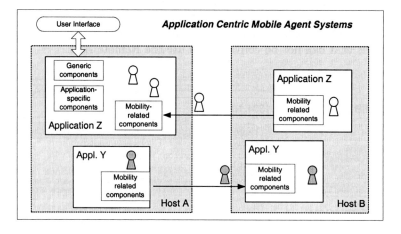

Fig. 5. The M&M Architecture for Application Centric Mobile Agents

A more detailed description of this novel perspective on mobile agents, that brings potential advantages in several fields, is presented in [9, 10]. From a strict manageability viewpoint, the key differences are:
- there is no agent platform to install and maintain. Although there are still distributed applications to install and manage, this is much simpler than managing a separate infrastructure shared by a large number of distributed applications with different policies and requirements;

- agents interact directly with the applications from the inside. This eliminates the need to set up interface agents and configure and manage its security policies. This also partially eliminates the need for mechanisms like the Remote API;
- for each application only the required components (including even the management services) are installed, resulting in a simpler and lighter framework to manage.

This framework results, therefore, in a significant reduction on the specific infrastructure management requirements and in a shift towards the use of more generic application management tools. The manageability problem becomes less complex and more generic. Nevertheless, some agent and infrastructure management services are available in the form of extension service components.

These services provide an additional service layer for managing the agent support components, available both from within the agent-enabled application and from external applications (see Fig. 6). Available functionality includes agent management (instantiation, monitoring, shutdown, etc.) and mobility component control (start/shutdown, monitoring, configuration, resource management, etc.). Additional services, such as agent tracking, are also available through similar service components. This architecture results in a simple but very flexible

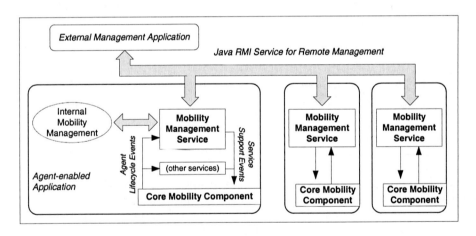

Fig. 6. M&M Management Architecture

framework for maintenance of the mobility-related components of the application. New interfaces (such as SNMP, Corba, WBEM or HTML) can easily be added building new components, and the management functionality can even be extended to entail the administration of the whole agent-enabled application. The implementation of the M&M framework is partially available for download at [8], including a prototype external management application for demonstration purposes.

One important point is that the component infrastructure must be manageable itself. For instance, some applications require runtime instantiation or installation of new services. The need to change certain component parameters at runtime (e.g. the maximum number of running agents, a listening port, or the logging level) is also quite common. In our case this is accomplished by having a management component service that performs several functions:

- it listens for external management requests;
- when a service instantiation is requested it analyses the Service Deployment Descriptor that was issued and then instantiates and configures the service. This involves close interaction with the core Mobility Component for registering the service with the appropriate configuration and security permissions;
- when it receives a request to modify a component parameter it propagates the request to that component. A unique identifier distinguishes each component, and each component understands certain administration tasks. For instance, the Mobility Component has a property named *listen port*, which represents the port for the incoming agents. A request can be made to change the *mm.mob.Mobility.listenPort* property to a new value. The request first arrives at the management component, which identifies the target component and then propagates the request to that component through a callback interface;
- finally, the management component also allows to broadcast a property change request to all the components. For instance, most of the components use a property called *mm.mob.logLevel* that determines which calls and actions must be sent to the log file. It is quite easy to request all the running components to change their log level by issuing a broadcast that will notify each of the components of the change.

Despite its simplicity, this management component provides a high degree of flexibility. Its strength resides in its generality, that allows any changes to be propagated to the running components, even when those components were not known at the time of development or deployment. The management interface of each of the components is not hard-coded in a class but propagated to the components themselves. Each component includes and understands its specific set of management properties and actions.

Another interesting point of the framework is that it allows the runtime deployment of new services not only using the management interface but also through administration agents. This means that an agent with the appropriate permissions has the ability to examine which components are available at a host, to modify their properties, and to change that configuration by shutting down services or instantiating new ones. This feature is especially interesting in applications where monitor agents roam through the network examining the state of the machines, quickly reacting to malfunctions (e.g. crashes, failures and environment changes) by autonomously recovering the applications.

5 Conclusions

The high costs associated with the installation and administration of distributed mobile agent infrastructures are an important but often overlooked obstacle to widespread deployment of mobile agents. In this paper we present two approaches to reduce those costs. In the first approach, validated in the JAMES project, several ad-hoc solutions are added to a classic platform-based architecture, resulting in enhanced communication with external applications, integration with legacy SNMP-based management applications and remote control of the distributed infrastructure. The second approach is based on the M&M framework, where a significant change in the agent-system architecture results in much simpler management requirements. Using the M&M integration framework, a small management service component is available both for internal and external management of the agent-support components. In the future new interfaces can be added - for instance for SNMP support - and new management functionality can be provided, such as application management.

Acknowledgements

The M&M project is partially funded by CISUC (R&D Unit 326/97) and by FCT (Project Reference POSI/33596/CHS1999). JAMES was an Eureka Project (Reference Σ!1921) partially funded by ADI (*Agência de Inovacão*).

References

1. Pham, V., Karmouch, A.: Mobile Software Agents: An Overview. IEEE Communications Magazine, pp. 26-37, July (1998)
2. Mobile Agent System Interoperability Facilities Specification. OMG TC Document orbos/97-10-05 (1998)
3. Foundation for Intelligent Physical Agents, http://www.fipa.org/
4. Silva, L., Simões, P., Soares, G., Martins, P., Batista, V., Renato, C., Almeida, L., Stohr, N.: JAMES: A Platform of Mobile Agents for the Management of Telecommunication Networks. Proceedings of IATA'99, Springer-Verlag LNCS 1699 (1999)
5. University of Coimbra, JAMES Project Homepage, http://james.dei.uc.pt/
6. Simões, P., Silva, L., Boavida, F.: Integrating SNMP into a Mobile Agents Infrastructure. Proceedings of DSOM'99, Springer-Verlag LNCS 1700 (1999)
7. Simões, P., Lourenco, E., Pereira, P., Silva, L., Boavida, F.: J.AgentX: a Tool for Dynamic Deployment of Open Management Services. Proceedings of 2000 International Conference on Software, Telecommunications and Computer Networks (SoftCOM'2000), Split (2000)
8. University of Coimbra, M&M Project Homepage, http://mm.dei.uc.pt/
9. Marques, P., Silva, L., Silva, J.: Going Beyond Mobile Agent Platforms: Component-Based Development of Mobile Agent Systems. Proceedings of the 4th International Conference on Software Engineering and Applications (SEA'2000), Las Vegas (2000)
10. Marques, P., Simões, P., Silva, L., Boavida, F., Silva, J.: Providing Applications with Mobile Agent Technology. Proceedings of the 4th IEEE International Conference on Open Architectures and Network Programming (OpenArch'01), Anchorage (2001)

Dynamic Agent Domains
in Mobile Agent Based Network Management

Robert Sugar and Sandor Imre

Department of Telecommunications
Technical University of Budapest
Pazmany Peter set. 1/D, Budapest, Hungary, H-1117
E-mail: {sugar, imre}@hit.bme.hu

Abstract. Today's network management systems suffer from scalability problems and involve the transmission of large amounts of raw data towards the centralized network management station. Therefore mobile agent (MA) based solutions were presented to enhance the efficiency of network management tasks. The paper investigates the existing MA delegation schemes and migration policies, and proposes a solution for effective agent deployment using dynamic agent domains. The size of the domains is altered during the trading process, where agents exchange nodes in order to equalize their workload using a lightweight communication model. Cloning and merging operations can be initiated to modify the number of agents, providing adaptivity to changing network conditions. The presented method includes the population and load control for network management agents along with robustness and fault tolerance mechanisms to achieve a guaranteed visiting frequency for the managed hosts.

Keywords: Mobile agents, Dynamic agent domains, Network management, Adaptivity

1 Introduction

Network management systems are currently based on a centralized model. These systems are hard to re-configure, and experience serious flexibility and scalability problems. The major drawback of the centralized approach is the transfer of large amounts of data, such as SNMP tables, which can create a bottleneck at the central management station and unnecessarily consume bandwidth resulting in the ineffective utilization of network resources. Besides, the increasing heterogeneity of the networks greatly hinders the unified handling of all hosts. Therefore the centralized NMS usually becomes a large and complicated entity difficult to alter or maintain. Another issue is that a human network operator has increased duties as the number and the heterogeneity of the network elements grow and needs more and more knowledge of the new technologies.

These problems strive for a more automated, distributed and unified network management system. A possible solution for the aforementioned weaknesses is using mobile agents [1] for network management tasks [3,4]. A mobile agent is a software agent with the ability to transport itself on the network and continue its execution on a new location [5]. The agent technology is not naturally dominant over the traditional client/server based solutions, since almost anything that can be solved with mobile agents can be implemented with static objects, but these solutions are usually more difficult to implement, less efficient and often clumsy. MA technology seems to be a promising paradigm in large decentralized systems and also has advantages over the traditional solutions in network management. The network heterogeneity can be hid with a Java virtual machine and a unified management API. As the agents are autonomous their deployment can lower the number of necessary human interventions. Furthermore, by their distributed functioning the load of the centralized management station can also decrease. Thus, using mobile agents for network management tasks provide [6] flexibility, autonomy, robustness and bandwidth saving. A performance analysis and the trade-off between the traditional and the agent based management can be read in [7].

This paper is organized as follows: Section 2 describes the existing agent delegation schemes currently used or proposed. Section 3 presents a new mobile agent delegation model using dynamic agent domains. Section 4 contains the evaluation of the presented method and simulation results concerning the stability and the dynamic behavior of the algorithm as well as a comparison to other methods.

2 Mobile Agent Delegation Schemes

The agent deployment scheme and the agents' itinerary have a large effect on the generated network traffic and therefore on the efficiency of the whole network management system. Several approaches exist in assigning mobile agents to hosts. In [9] two delegation schemes are presented: the Get 'n' Go and the Go 'n' Stay method. At the first one the agent travels through several managed hosts whereas at the latter one an agent is exclusively delegated to the targeted host and only returns when its task is fulfilled. Similar distinctions have been made in [2] between deglets (delegation agents), which are sent out for a special task for a limited time, and netlets (network agents), which are continuously present on the network. In the remaining part of the paper we only consider the Get 'n' Go scheme for further investigations.

Applications where the netlet approach can be successfully applied include network monitoring [8] and executing of recurrent tasks [9,10] in fault management or in configuration management. The semi-permanent presence of MAs is also desirable by the concept of "repair agents". Repair agents are tiny programs designed for a specific task [2]. For example, network congestion can be detected and routers could be reconfigured to handle the situation. By using agent colonies, the repair can be very fast and does not require the involvement of a human operator resulting in a higher degree of autonomy. Repair agents can also be used to prevent more severe problems to evolve by detecting and handling the problems in an early stage.

The size and the location of the MAs along with the number of active agents on the network dominantly effect the response time of the system (e.g. the frequency of visiting a node) and the consumed resources. In [11] domains are predetermined for the agent at launch time, and the injection of mobile programs is exclusively performed by the mobile agent manager.

In [10], swarm intelligence is used for controlling agent migration. Swarm intelligence is a biologically inspired completely distributed way of problem solving. The principle is based on individually simple and unintelligent entities showing intelligent behavior as a whole. Communication is indirect, accomplished by a set of small pieces of data called "pheromones" laid down at certain places.

3 MA Delegation with Dynamic Agent Domains

In this section a new agent delegation and migration control mechanism is described using dynamically changing agent domains. The goal was to design a flexible and robust agent delegation scheme that guarantees a minimal visiting frequency of the hosts for time-critical applications. The proposed method also provides population control by using cloning and merging functions to further increase adaptivity.

The agents are considered to be homogeneous and replaceable. Two types of communication are used: a direct communication between neighboring agents and an indirect communication through data structures in the managed objects - similar to the swarm in the [10] case. The latter model is used by exploration and agent fault detection described in Section 3.2. Decisions are taken in a stochastic fashion according to the current workload of the agents.

Through the paper the term of agent workload is defined as follows: The amount of time required to return to a host using a sequential visiting order. Besides, the workload of a host is meaning the amount of management tasks remaining there also measured by the time it needs to complete. I used time as a measure of workload because this is in accordance with the desired visiting frequency and it also provides a unified view of every task the agent might do. If I chose a MIPS-like unit as workload indicator the CPU power and the current load of the hosts due to other than management processes could not been taken into account.

For the effective operation of the agents the workload must be properly estimated. This can be achieved by measuring and remembering the previous execution and migration times and placing indications about the former visits in every host the agent visits. In addition to this distributed data structure the agent also must be aware of the workload in the whole domain since this estimation is needed to make reasonable decisions.

The workload estimation is updated when arriving to or leaving from hosts based on the migration and execution time measured by the agent and the previous data stored in host. By a "forgetfulness" parameter the proportion of the newly measured and the previously stored data can be set. This parameter affects the dynamic behavior as well as the stability of the algorithm. The problem is that a sudden and permanent change in network conditions is hard to be immediately distinguished from a simple jitter. Thus, to avoid the overreacting a discrete delay and still have fast reaction time to an overload situation the forgetfulness parameter should have a fair and reasonable value.

3.1 The Functioning of an Agent

The lifecycle of an agent consists of three stages: Normal operational phase, when agents roam their domains performing the regular tasks; trading phase, where domain connections are initiated and cloning/merging phase, where heavily loaded agents can multiply, or two underloaded agents can merge.

3.1.1 Normal Operational Phase

In the normal operational phase agents perform their management tasks by sequentially visiting the hosts of their domain. These management tasks can be either recurrent or single tasks depending on the type of the agent. Round Robin visiting order is used for intra-domain migration. The method was chosen, besides its simplicity, because in that case the agent migration is predictable and starvation of nodes can be avoided, which is essential to provide a minimum visiting frequency for the hosts. The routines of performing the actual tasks on the nodes are application-specific and out of scope of the paper.

3.1.2 Trading

In our perspective trading is a transaction between two agents where one of the parties shifts and the other one obtains the responsibility over a specified node. The trading is only initiated with a limited scope – between neighboring domains – to reduce communication costs. The trading process starts with the domain workload measurement where the agents evaluate their domain size according to predefined parameters – optimal, maximum and minimum load. When the MA decides whether to receive or dispose nodes, it contacts the neighbors searching for possible trading partners. The trading strategy can vary from "greedy", where only mutually advantageous transactions are completed to "altruistic", where asymmetric deals can occur between a greatly overloaded and a slightly overloaded agent. Altruism can help to avoid getting stuck in local optima and achieve a globally acceptable performance, but can lead to slow degradation when overused.

3.1.3 Cloning/Merging

If the neighbors reject trading offers and the agent load is still unsatisfactory, the MA can choose to "clone" itself or "merge" with another agent with a certain probability depending on how critical its state is. Cloning means that the previous domain is split into two parts for the two newly born agents, which are considered to be equal. These agents start the trading process right after birth to harmonize the domains with the surrounding MAs. Cloning is essential when dynamic changes occur in the network such as a newly opened subnetwork that needs to be populated, or overload on the nodes or on the links that slow down the operation of MAs.

During the merge process two agents unite their domains resulting in a single MA that immediately starts the trading process to deal with its enlarged domain size. The process is required to reduce the number of agents when too many of them are present of the network, especially after a recovery from a fault or a critical network situation.

3.2 Robustness

The system implements several procedures to provide error-tolerant functioning. During the normal operational phase the MA regularly inspects the neighboring hosts of the domain, searching for faults such as expired time stamps or a missing host responsible. If an error is detected the agent may try to contact with the host owner, explore the nearby area or initiate the node seizure process, where the MA can unilaterally take the responsibility over a managed object. In case of an agent crash, congestion in the network or trashing, the domain of the MA is soon to be populated by the neighbors providing fault tolerant and robust functioning.

3.3 Starting an Agent Colony

The distribution of the initial agent colony can be accomplished several ways. When a new agent type is created at the NMS the goal is to have them spread out on the whole network. This can be left to the standard agent mechanisms or different techniques could be used to boost up the initial agent spawning process.

3.3.1 Starting with One Agent
In that case the mobile agent generator spawns only one single agent with a domain of one single host. The agent realizes that the neighboring nodes have no owner and starts to explore the network. As its domain grows it quickly multiplies spawning other agents that will soon continue the exploration and will eventually populate all of the network. To speed up the creation of the initial population spawning behavior can be set to aggressive in the beginning of the populating process.

3.3.2 Starting with Multiple Agents
It is also possible to initiate the colony with several mobile agents. They can either start from one single host or from multiple hosts. The first case is similar to the previously described process, it only gives a startup boost to agent spawning. If we use more than one host during initiation we must synchronize the startup and make sure that all the agents have the same code. Using multi-host initiation smoothes up the network traffic due to excessive agent multiplication and can be more effective on a network with rare connectivity especially when the subnetworks are connected by only one node or not connected at all.

3.4 Physical and Logical Topology

First we have to define the idea of logical topology. Logical network topology is a virtual set of connections stored in the hosts. If the physical topology of the managed network is rare in connections the use of logical topology can facilitate the correct functioning of the algorithm. The logical topology should follow the physical topology, because setting up an arbitrary set of connections will greatly increase the migration time.

Let us consider an FDDI ring for instance. If we used the physical topology only with coherent domains the domain of the agents would be sectors with only two neighbors. In that case the trading was only possible for the two side nodes of every sector. By defining a logical topology that introduces extra virtual connections between the ring members can allow effective trading and merging.

3.5 Overpopulation Control

A very undesirable situation is that when due to incorrect workload estimation, agent malfunction or fraud, the agents start to multiply in an uncontrollable fashion and flood the network. A solution for that is to maximize the number of agents present on the network. It should be designed not to interfere with the normal functioning of the agent but to set up an upper bound for the number of agents instead. It can be based on credits as in [14].

It means that the original initial agent has the credit of the maximum number of agents, which is conservatively divided through every spawn and integrated through every merge. If the credit of the agent is one it cannot spawn anymore. In order to provide some flexibility, it can obtain also credits from neighbors. The credit sharing is executed during trading as well to prevent the bulk of the credits concentrating in a relatively peaceful segment of the network when there is a great need of spawning on the other side.

4 Evaluation and Simulation Results

A simulator was designed to evaluate the algorithm. The system was tested in several simulated network conditions and numerous parameters were introduced to control the behavior of the agents. The most important ones were: the pre-defined optimal load according to the desired visiting frequency; the probability functions for trading, spawning and merging; and the selfishness of the agents in accepting trading and merging offers. During simulation the normal behavior of the agents was analyzed on a stable network investigating whether the MA domains and the workload of the agents tend to equalize. Another question was that how the adjustment of parameters affects stability. The analysis also targeted the dynamic performance of the system in a sense that how fast it can respond to sudden changes on the network. In the next figures we present a set of simulation results concerning the functioning of the algorithm.

Fig. 1 shows the number of agents on the network in a dynamic network situation. The optimal agent population is calculated by dividing the workload on the whole network with the optimal workload of the agent. As it shows after a stable network condition the workload starts to rise and then peak on a certain level followed by a gradual fall of work amount. On a stable network (such as in timeframe 10-20 and 70-90) the number of agents is very close to the optimal value. This stands for most of the test cases investigated. In rapidly changing conditions (timeframe 20-50 and 100-120) the actual population of the agents follows the change in workload by increasing the spawning or merging activities and thus modifying the population. The delay of the reaction, and in parallel with it the overhead caused by the trading process, can be adjusted by the agent behavior parameters.

Agent population

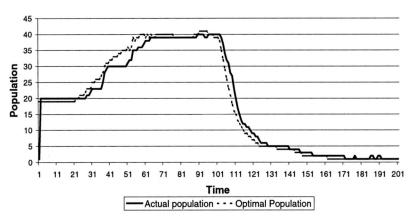

Fig. 2. The current and optimal agent population in dynamically changing network conditions

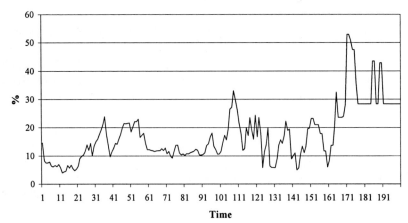

Fig. 1. Agent workload concerning the average relative standard deviance from the predefined optimal value

Fig. 2 shows that on a stable network the deviation of the agent workload can be about ten percent of the pre-defined optimal value. This deviance greatly depends on the deviation of work on the hosts, which cannot be influenced by the agent. This means that most of the deviation comes from the imprecise estimation of the workload, not from the incorrect functioning of the load-sharing algorithm.

During simulation we experienced that on a stable network, with correct configuration of the parameters, the agent workloads tended to equalize over time and showed stability in the long term in almost every simulation case. The parameter that had the largest effect on the stable agent distribution was of course the pre-defined optimal workload value, which determined the average size of agent domains and thus the average visiting frequency for a host.

We also investigated the dynamic functioning of the algorithm. Comparing to the previous case the parameter configuration has a larger effect on the behavior of the system. The most vital parameter was the frequency of the trading process and the pre-defined critical workload values.

4.1 Comparison to Other Delegation Methods

We also made a comparison to the previously existing agent delegation schemes. We found that static agent domains can be effective for a relatively stable network, but when dynamic changes occur – which are very likely to happen by network faults – the performance of the management can significantly decrease. Constant domains work best on highly hierarchical networks where the physical topology can be used to determine agent domains exploiting that intra-subnetwork migration is much cheaper than inter-subnetwork agent transport. Our method looks dominant when robustness and flexibility are taken into account but also adds some additional costs introduced by the agent cooperation.

Comparing to the swarm intelligence based solution the greatest advantage is the more predictable agent behavior that allows the assured visiting frequency. Considering that agents in the Swarm tend to migrate towards areas with higher pheromone density, hot spots can form on the network, while other areas remain virtually unpopulated. Nodes do not have a clear responsible, thus double-checking of hosts can easily occur when certain nodes can be ignored. We found that the swarm can be effective with non time-sensitive applications, but the minimum visiting frequency is hard to be assured without further enhancing the communication scheme.

5 Conclusions and Future Work

In the paper a new MA delegation and migration control scheme was presented for semi-permanent network management agents. Dynamically changing agent domains were used to provide flexible, adaptive and robust operation. The trading process enables the agents to exchange nodes allowing the dynamic transformation of agent domains towards the equal work sharing between the MAs. Cloning and merging is used for controlling the agent population to approach a globally optimal agent distribution.

The advantages of the method are the decentralized operation and limited scope of communication over centralized management models, guaranteed visiting frequency and a more calculable behavior over the swarm based solutions and flexibility and fault-tolerance over constant agent domains. During simulation stability and convergence were analyzed and results show that with an appropriate selection of agent behavior controlling parameters the workload of agents tend to equalize over time. The system can rapidly react to dynamic changes on the network such as a network fault, an unpredicted congestion or overload, and also provides robustness in the case of an agent malfunction or termination.

In the future the focus will be placed on enhancing the algorithm with a hierarchical control, and investigating other application areas in agent based simulation and in biologically inspired multi-agent systems. Another way of advancement is to deploy agent colonies on originally flat networks such as ad-hoc mobile networks for different network management tasks or even facilitating smooth packet transmission or routing.

References

1. Pham V., Karmouch A., "Mobile Software Agents: An Overview", IEEE Communications, Vol. 36, No 7, pp. 26-37, 1998.
2. Andrzej Bieszczad, Bernard Pagurek, and Tony White. Mobile agents for network management. IEEE Communications Surveys, Vol. 1. No.1. September 1998.
3. Gavalas D., Greenwood D., Ghanbari M., O'Mahony M., "An Infrastructure for Distributed and Dynamic Network Management based on Mobile Agent Technology", Proceedings of the IEEE International Conference on Communications (ICC'99), pp. 1362-1366, 1999.
4. Goldszmidt G., Yemini Y., "Delegated Agents for Network Management", IEEE Communications Magazine, Special Edition on Network Management Paradigms, March 1998.
5. Lange D., "Mobile Objects and Mobile Agents: The Future of Distributed Computing?", Proceedings of The European Conference on Object-Oriented Programming (ECOOP'98), 1998.
6. A.Puliafito, O. Tomarchio, "Advanced Network Management Functionalities through the use of Mobile Software Agents", Proceedings of the 3rd International Workshop on Intelligent Agents for Telecommunication Applications (IATA'99), LNCS vol. 1699, pp. 33-45, August 1999.
7. M. Rubinstein, O. C. Duarte, "Evaluating the Performance of Mobile Agents in Network Management", Proceedings of the IEEE Global Communications Conference (Globecom'99), December 1999.
8. Liotta A., Knight G., Pavlou G., "On the Performance and Scalability of Decentralised Monitoring Using Mobile Agents", Proceedings of the 10th IFIP/IEEE International Workshop on Distributed Systems: Operations & Management (DSOM'99), October 1999.

9. Gavalas D., Greenwood D., Ghanbari M., O'Mahony M., "Using Mobile Agents for Distributed Network Performance Management", accepted to the 3rd International Workshop on Intelligent Agents for Telecommunication Applications (IATA'99), 1999.
10. T. White, A. Bieszczad and B. Pagurek, Distributed Fault Location in Networks Using Mobile Agents. In Proceedings of the Second International Workshop on Agents in Telecommunications Applications (IATA '98), pp. 130-141, July 4th-7th, 1998.
11. D. Gavalas, D. Greenwood D, M. Ghanbari, M. O'Mahony, "Advanced Network Monitoring Applications Based on Mobile/Intelligent Agent Technology", Computer Communications Journal, special issue on "Mobile Agents for Telecommunication Applications", publication in January 2000.
12. Armin R. Mikler: Agent-Based Wave Computation: Towards Controlling the Resource Demand. Department of Computer Science University of North Texas, Denton, Texas 76203, USA. mikler@cs.unt.edu

The terminal renders the user commands in the supported user interface format, i.e., HTML, textual or graphical user interface. Method executeCommand handles the command parsing and execution.

```
public TextMessage executeCommand(String command)
  if(command.startsWith("remove")) {
  try {
    index = Integer.parseInt(
        command.substring("remove ".length()
           ,command.length()));
  } catch(Exception e) {
    buffer.append("No such message:"+command+
                                    " "+e+"\n");
  }
  if(index>-1&&titles.size()>0&&titles.size()>index){
     messageBox.removeMessage(
         (String)titles.elementAt(index));
     buffer.append("Removed message:"+
                              titles.elementAt(index));
  } else buffer.append("Could not remove message:"+
                                              index);
    return new TextMessage(buffer.toString());
}
else if(command.startsWith( //process other commands
```

When the command is executed, both the user agent and the service task could be remotely accessed using the proxies. The executeCommand method returns a textual message indicating the results of the command, which is shown to user. In order to allow user to access the service via terminal gate, it must

- display the service list, and open the selected service into the user interface,
- provide user login and logout functionality.
- execute selected commands and display the results into the user interface,
- keep track of ServiceData objects and update the user interface when new data arrives and
- select one of the user interface formats, modify and display the user interface with a selection of service based commands.

Networking required by different user devices only has an effect at the service and terminal gate layers of the agent system. It is also important that service developers can customise user interfaces by specifying user interface layouts and service-specific commands to provide more functionality to the user interface. Task and terminal gate-based development in DAN requires the learning of only a few interfaces (Agent, Gate, Task, ServiceData, DataFormat, ServiceProvider) to develop services for different devices.

In the DAN service development, the most of the work is required by the service user interface, because we must create three initial layouts for different types of user interfaces (graphical, HTML and text-based) to meet even simple look-and-feel requirements. Despite that fact, a simple service with single user interface format (textual) and user commands is usable via SMS, HTML or graphical user interfaces. We considered developing a script system or using XML to define the service user interfaces, but we later decided to use these separate formats, because they matched to available user devices.

5. Performance

Performance of remote method calls in the DAN system has been measured using 100 user agents. In the tests, each agent made hundred remote method calls to the task of the service agent resulting in 10 000 requests.

In the first test, the average method call time was 325 ms., while the service agent was at the same computer (NT 4.0, Pentium II 400 MHz, 196Mb, JDK 1.2 with JIT) with the user agents. Minimum and maximum method call times were 50 ms and 1600 ms. In the second case, the service agent was moved to another computer (NT 4.0, Pentium Pro 220 MHz, 128 Mb, 100 Mbit LAN) and the test was repeated. Remote method calls were executed at 1667 ms on average, while the maximum was 4060 ms and the minimum 120 ms.

Testing with 10, 20 and 50 agents and with different message sizes indicated that remote method calls are approximately four times slower over LAN. Typical message size was three kilobytes without compression. It should be noted that both computers were swapping heavily, indicating a lack of memory during the tests and the encryption was turned off.

It is well known that Java interpretation, garbage collection and compilation techniques affects performance [10]. In our case, Java JIT compiler improved the remote method call execution time by 30 %. We used a larger heap size in the Java execution environment (128 Mbytes) and compiled the code using optimisations (the /o parameter).

6. Related Work

There are many different Java-based agent platforms available, Pham and Karmouch, for example, provides an overview and comparison of a number of systems [11] and Green et al. reviews some mobile agent systems [12]. A lot of software agent work is related to telecommunications [13]. Proxies are also used in other agent systems to enable remote access [14,15]. In DAN and other agent systems, the services are made virtually available for the user agents, which makes the system inherently insecure. However, these systems rely on external security such as SSL (Secure Sockets Layer) to encrypt communication between agent systems. In DAN, the proxy-based communication between agents is also encrypted using RSA [16] implementations called Cryptix [17] and Forge [18].

One agent platform for telecommunication systems is Grasshopper [19]. With Grasshopper as well as many other agent systems, service development requires that new agents are created into the system. We believe that this requires a deeper understanding of the agent system's internal functionality than is necessary in service construction for different devices. By using DAN, the service provider can construct new services on-top-of agents without knowing the details of the underlying agent system. New devices are not directly supported by other agent systems either even graphical user interfaces are usually made in a similar way to DAN graphical terminal gates.

Developing agent systems has become popular with the Java programming language [20]. Java Remote Method Invocation (RMI) uses similar proxy objects to DAN, but RMI lacks mobility support. When the RMI object is moved, its name (network address) is changed and all users of that object must rediscover the access to that object. In our approach, the original object can move with its agent to different location without requiring any actions from the remote users. The agent system transfers method calls using global names and location information from the agent system repository.

7. Conclusions

Implementing network support for different devices is a difficult task due to new emerging handheld terminals with different service access methods. DAN is a platform for service development in such environment. DAN eases the development of services by abstracting underlying technical details such as network addresses, user terminal features, data persistence and encrypted communication from the service developers. New services are added to the network by allocating task objects to agents. Developers can focus on developing service interfaces, and implementing the service on-top-of agents by creating agent tasks and remote user interfaces.

Proxy-based communication makes new services immediately available to all supported terminal types via their respective terminal gates. DAN also provides a feature, which allows service providers contact the customer using the customer-preferred device via software agents. Basically sending information from operator to user requires two remote method calls. Agent system routes information from the operator user interface to the user's agent who updates the terminal gate. Currently, system supports public key cryptography, which means that there are four encryption and decryption operations in each method call. Unfortunately, this slows down the system and encrypted communication is not sufficient for services with real-time requirements. Typical services could be personal and office services using any device, e.g., e-mail access and notifications, travel or other request-response type of service. Security is also deteriorated by external servers such as WAP gateways, e-mail and www servers, which transport data from the terminal gate to the user device.

Acknowledgements

The work presented in this paper has been accomplished in a national research project called CTI (Computer Telephone Integration). The CTI project is funded by the Technology Development Centre of Finland (TEKES), Nokia Internet Communications, CCC Data Ltd., Nokia Networks Ltd., Koillismaan Yrityspalvelukeskus, the University of Oulu and VTT Electronics. The support of the project consortium is greatly acknowledged.

References

1. Bradshaw, D., Sheina, M., Glassman, S., O'Loughlin, M.A.: Computer Telephone Integration: From Call Centre to Desktop. Ovum Reports, Ovum Ltd. (1997)
2. Shoham, Y.: An Overview of Agent-oriented Programming. In Bradshaw, J.M. (ed.): Software Agents. AAAI Press, Menlo Park CA (1997) 271-290
3. Anonymous: Mobile Agent System Interoperability Facilities Specification. Joint Submission. OMG TC Document, orbos/97-10-05 (1997)
4. Palola, M., Heikkinen, M., Kaksonen, R.: Improving Telecom Service Access Convergence using Software Agents. The 1st Asian-Pacific Conference on Intelligent Agent Technology, Hong Kong, China (1999)
5. Brenner, W., Zarnekow, R., Wittig, H.: Intelligent Software Agents: Foundations and Applications. Springer-Verlag, Berlin Heidelberg (1998)
6. Baumann, J., Hohl, F., Rothermel, K, Schwehm, M., Straβer, M.: Mole 3.0: A Middleware for Java-Based Mobile Software Agents. In Davies, N., Raymond, K., Seitz, J. (eds.): Middleware'98. Springer-Verlag, London (1998) 355 - 370
7. Tripathi, A., Karnik, N., Vora, M., Ahmed, T., Singh, R.: Mobile Agent Programming in Ajanta. The 19th International Conference on Distributed Computing Systems. (1999)
8. Palola, M., Heikkinen, M.: Constructing Mobile Web Services on a Software Agent Platform. In Graham, P., Maheswaran, M. (eds.): The International Conference on Internet Computing. Las Vegas, Nevada, USA (2000)
9. Shapiro, M.: Structure and Encapsulation in Distributed Systems: the Proxy Principle. The Sixth International Conference on Distributed Computer Systems (ICDCS). (1986) 198-204
10. Dikaiakos, M., Samaras, G.: Quantitative Performance Analysis of Mobile Agent Systems - A Hierarchical Approach. Technical Report TR-00-2. Department of Computer Science, University of Cyprus (2000)
11. Pham, V.A., Karmouch, A.: Mobile Software Agents: An Overview. IEEE Communications Magazine, 31(7). (1998) 26-37
12. Green, S., Hurst, L., Nangle, B., Cunningham, P., Somers, F., Evans, R.: Software Agents: A Review. Technical report http://www.cs.tcd.ie/publications. University of Dublin (1997)
13. Hayzelden, Alex, L.G., Bigham, J.: Agent Technology in Communications Systems: An Overview. Knowledge Engineering Review, Vol.14, No. 3 (1999) 1-35
14. Biskup, J., Freitag, J., Karabulut, Y., Sprick, B.: A Mediator for Multimedia Systems. Third International Workshop on Multimedia Information Systems. Como, Italia (1997) 145-153
15. Chess, D., Grosof, B., Harrison, C., Levine, D., Parris, C., Tsudik, G.: Itinerant Agents for Mobile Computing. IEEE Personal Communications, vol. 2, no. 5. (1995) 34-49
16. Rivest, R.L., Shamir, A., Adleman, L.: A Method for Obtaining Digital Signatures and Public-Key Cryptosystems. Communications of the ACM, Vol.21, Nr.2 (1978) 120-126
17. Anonymous: Cryptix JCE 3.1.0. http://www.se.cryptix.org/products/jce/index.html (1999)
18. Anonymous: Forge JCE 1.32. http://www.forge.com.au/Products/crypto/index.html (2000)
19. Breugst, M., Choy, S., Hagen, L., Höft, M., Magedanz, T.: Grasshopper–An Agent Platform for Mobile Agent-based Services in Fixed and Mobile Telecommunication Environments.

Realizing Distributed Intelligent Networks Based on Distributed Object and Mobile Agent Technologies

Menelaos K. Perdikeas[1], Odysseas I. Pyrovolakis[1], Andreas E. Papadakis[1], and Iakovos S. Venieris[1]

[1] National Technical University of Athens – Dept. of Electrical & Computer Engineering
9 Heroon Polytechniou, 157 73 Athens, Greece
{perdikea,ody,apap}@telecom.ntua.gr ivenieri@cc.ece.ntua.gr

Abstract. An architecture enhancing the Intelligent Network to a Distributed IN is described. To this end, Distributed Processing Environment and Mobile Agent technologies have been employed. The added value, in the context of flexibility and manageability of the exertion of the aforementioned technologies, as well as the extent to which they can be utilized are discussed. The configuration architecture is presented and an implementation – oriented, detailed description of the Service Execution Node and its constituents is provided.

1 Introduction

The Intelligent Network (IN) approach for providing advanced services to end users aims primarily at minimizing changes in network nodes by locating all "service-related" (and thus likely to change) functionality in dedicated "IN-servers", named as "Service Control Points" (SCPs) [1], [2]. These servers are in a sense "external" to the core network which in this way needs to comprise only a more or less rudimentary switching functionality and the ability to recognize IN call requests and route them to the specialized SCPs. Since the time of its standardization by ITU-T [3], IN has become the default way for telecom operators to enhance the ability of their core network to provide value added services to their customers in a cost effective manner.

In this paper we are considering the use of distributed processing technologies such as CORBA (Common Object Request Broker Architecture), DCOM (Distributed Component Object Model) and RMI (Remote Method Invocation) along with mobile code ones in order to enhance IN's potential. We are mainly focusing on issues concerning the design and implementation of an IN architecture that would be more flexible, distributed and facilitating object oriented service creation and efficient services life cycle management.

2 Applying New Software Technologies in Intelligent Network

Before going on into describing the actual architecture, a more abstract and general discussion containing some of our experiences and issues emerging when applying novel software technologies in the context of IN precede. Specifically, the impact of the introduction of well known distributed object practices (such as CORBA [4]) and more radical solutions (such as Mobile Code [5]) are discussed.

2.1 Enhancing IN with CORBA

Commercial IN implementation rely on traditional protocol stacks for dispatching messages between elements of the architecture. These protocols rest on the asynchronous exchange of messages and are thus termed "message-based" protocols. Problems with such protocols and the relative merits of the more advanced and generic remote method invocation paradigm have been recognized by many researchers and are well documented.

CORBA is a standardized solution to provide distributed processing capabilities to an environment and the closest to being a "de jure" one. Microsoft's DCOM is a more proprietary technology having as main advantage its tighter binding with the MS-Windows operating systems. Given the diversification that characterizes telecommunication systems, CORBA is more or less the only distributed object technology that can seriously been contemplated in this context.

Distributed object technologies extend the notion of a memory or an object pointer to include pointers to remote address spaces. Using CORBA's terminology these are termed "Interoperable Object References" (IORs). IORs are not only opaque handles to the objects they represent but can also serve as a generic network addressing scheme. An object doesn't have to be aware of the IP address and the port number on which its peer object "listens" – it is simply responsible for obtaining its "handle" and passing it to the CORBA runtime layer (known as ORB – Object Request Broker) whenever it wants to communicate with it. This allows the architecture to be transport protocol independent. Irrespectively of whether the underlying transport mechanism is TCP/IP or ATM or ISDN, as long as a CORBA layer can operate on top of it, the communicating objects only have to perceive the semantics and the notational conventions used by CORBA. The architecture can be reconfigured with no changes to the hosted objects at all. Communicating objects can be thought of as "plugging" on a generic method-invocation relaying layer (the ORB) and from that point on being able to invoke arbitrary methods on other objects "plugged" onto it as well. In contrast, reconfiguration of an architecture based on traditional "message-based" protocols is much more difficult.

2.2 Mobile Code in Intelligent Networks

In the traditional IN architecture Service Logic Programs (SLPs) run in centralized SCPs. This approach naturally renders the network vulnerable to "single point of failure" types of problems as well as makes it susceptible to performance penalties associated with the intrinsic bottlenecks that any centralized architecture has. When utilizing mobile code, SLP distribution among the core network nodes, Service Switching Points, (SSPs), can come very naturally if SLPs are implemented as mobile code objects [6]. Once a mobile code platform is installed on each of these nodes, SLPs can be downloaded on demand or in an off-line fashion for load-balancing purposes, to respond to a failure in a node they were previously residing or to preserve network resources. The fact that mobile code platforms are relatively generic infrastructures hosting a large number of SLPs and being oblivious to their functionality and operations, facilitates the introduction of new and withdrawal of old services. Because of being generic, these components may never (or in any case very infrequently) need to be modified. The ability of the SLPs to migrate to SSPs and continue their operations from their new location, augments each SSP with SCF functionality. Service provisioning can be offered directly by the switches' physical entities removing entirely the need for a centralized SCP. This should not be confused as reverting to switch-based services implementation as now, services are not statically bound to the switches but retain the flexibility of mobile code objects. As a result, all advantages offered by IN with respect to speedier introduction of new services are maintained. We refer to the resultant IN architecture as Distributed IN (D-IN).

It is important to note that the use of mobile code in IN, is not separated from the use of distributed technologies. In fact, distributed technologies and mobile code are used in concert and are of equal importance to realizing the D-IN concept. First of all, they are both IT-based paradigms and have predominant Object-Oriented characteristics. Both technologies' offerings have to a significant extent to do with non-functional characteristics[1], which are moreover complementary. Distributed technologies are concerned with location transparency and increase programmer's productivity whereas mobile code aims at offering performance benefits by moving client (or server) code closer to where it is actually executed, instead of engaging in network spanning and bandwidth consuming remote dialogues. [7] CORBA's approach to "objectizing" even procedural pieces of code through wrapping them in Interface Definition Language interfaces and plugging them to the ORB using an object abstraction is well suited for the inherent autonomy and object-oriented characteristics of mobile code. Location transparency and the use of generic network addressing schemes (as are the opaque IOR handles) facilitates object mobility since the actual location of the objects is not important. More importantly yet, and for a variety of technical reasons, mobile code's way to introduction in telecommunication environments needs to be paved by the previous introduction of distributed object technologies. Simply stated, CORBA's, DCOM's or RMI's adoption will have the potential of transforming the telecommunication world into a more "permissive environment", one in which mobile code will be more easily and naturally employed.

3 Realizing Distributed IN Systems

In our design of a Distributed IN system we considered as a starting point the well defined IN conceptual model. We have aimed for an architecture that would allow a coherent and smooth transition from the classical, SS7 (Common Channel Signaling System 7) based, IN architectures towards open, CORBA based distributed ones. Towards this direction, new correspondences between Functional Entities (FEs) and Physical Entities (PEs) were defined, allowing the possibility to duplicate some FEs in heterogeneous systems. In particular, the Service Control Function (SCF) is used in both service control nodes and switching systems. In addition, the Specialized Resource Function (SRF) is integrated and combined with the SCF. As a result,

---

[1] Implementation transparency however is a functional enhancement.

advanced switching systems are introduced where SCP functions are added to the ones of a SSP, as well as Service Nodes, derived from the integration of SCPs with Intelligent Peripherals. An architecture implying a more loose mapping of functional entities onto physical ones, affords us the ability to locate SLPs indifferently in the switches systems or in the service nodes in a distributed manner.

Mobile code's role is critical as it gives objects the opportunity to roam through similar but remote FEs (as an example, a SCF located in a service control point and a similar one located in a switching system). Following this approach, the binding of objects to FEs contained in PEs is not static. Instead dynamic allocation can take place according to an observed pattern of service requests or in response to congestion or network failures.

3.1 Reference Architecture

Based on the considerations of the previous section, the physical entities of the proposed architecture extend the traditional capabilities of the Service Control Points and the Service Switching Points. They are both required to adopt the following extensions:

- Introduction of a CORBA ORB in the software platform
- Capability to host mobile code downloaded from other network elements.

Within our reference configuration, these systems are named Service Execution Nodes (SENs) and Broadband Service Switching and Control Points (B-SS&CPs) respectively. Figure 1 presents the reference configuration [8], [9].

In every element of the core network an Agent Execution Environment (AEE) is installed. This environment is composed of an ORB and the necessary components to provide the runtime capabilities of hosting, monitoring, interacting with and executing agents. The AEE is the key element that allows the distributed execution of the IN based services. In the classical IN the allocation of service logic and data is static. Within our approach, the adoption of distributed and mobile code technologies allows a responsive network reconfiguration on the basis of real time needs. Such needs can include those arising from situations of congestion or overloading in terms of processing or signaling load in some areas or during certain periods of time or from the characteristics of the services themselves.

Services with a low rate of invocations can be stored and executed in the SEN. When a given threshold is overcome, the service logic and data can be downloaded in the B-SS&CPs reducing processing load on the SEN or signaling load along the B-SS&CP – SEN connection. This behavior is completely transparent to the end users and to the call control software modules.

Mobile code technologies are also introduced at the service design and deployment level. Using this approach, services are designed as Java-based Mobile Agents in appropriate Service Creation Environments (SCEs) and then transferred to the Service Execution Nodes utilizing the migration capabilities provided by mobile code platforms. The basic IN Application Protocol (INAP) semantics governing information flows between IN FEs are not affected however. Service providers can thus benefit from a flexible service provisioning environment, which enables them to adopt OO techniques for software design (minimizing the service development time) and to make use of mobile code facilities in order to apply immediate and sophisticated policies for service release distribution, update and maintenance.

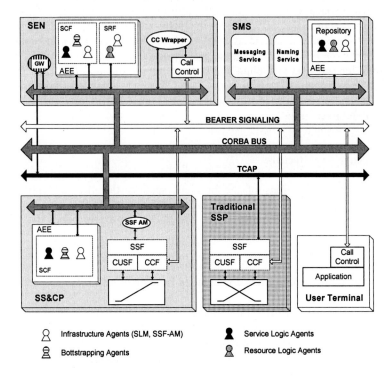

Fig. 1. D-IN Reference Architecture

3.2 The Service Execution Node

The Service Execution Node is derived from the integration of the B-SCP and the B-IP [10]. Additionally CORBA and a Mobile Agent platform are introduced which together play a key role in the service provision.

The SEN initially enables the switch to access the IN agent-based services and provides an environment to locally execute the stationary or mobile agents that compose the services in case this can result in performance gains.

It comprises the Service Control and Data (SCF/SDF), Mobility Service Control and Data (SCF/SDF), Specialized Resources (SRF), and Call Control Agent (CCAF) functions. The potential introduced by the introduction of distributed object and mobile code technologies is best exemplified when considering migration operations.

The service logic programs are deployed as stationary or mobile agents and the latter can be distributed and executed among different physical entities. The architecture is open enough to allow for the use of SLPs implemented in a native language if stringent performance requirements are to be met. In the latter case the SLPs are stationary objects. An important property that is maintained with the imposition of various access points within the architecture is that the SCF cannot tell if it is interacted with a natively compiled SLP or a Java-based mobile code one. In fact the SCF is oblivious to the actual location of the SLP with which it communicates since the logic responsible for locating, accessing or downloading the SLPs is orthogonal with respect to the core SCF functionality. Several B-SS&CPs, some of which may generate a high number of service requests and consequently a high processing and communication load, can concurrently access a service in a SEN. A possible migration of the appropriate agents can improve network performance.

Agent migration is also employed to support mobility management procedures, for example location registration/deregistration, user registration/deregistration, local authentication processing and so on. In these cases the mobile agents are more passive objects representing mobility management data and simply offer appropriate methods for accessing or modifying these data.

Objects and Agents within SEN. The object model of the SEN considers three major parts. The fist parts contain the objects for the interaction of SEN with the B-SS&CP. The second part contains the core of the Service Logic and the third part contains the objects responsible for the SEN to user interaction. Figure 2 reports a high level view of the SEN architecture.

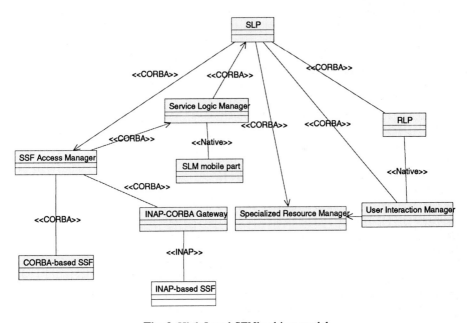

Fig. 2. High Level SEN's object model

The *SSF Access Manager* is the SEN's gateway object. All SSF-SCF interaction is passing through this object. When the system is bootstrapped SSF needs only obtain a CORBA reference at the SSF Access Manager. Technically it consists of two CORBA servers one for the SCF side (supporting SCF-initiated information flows) and one for the SSF side (for the SSF-initiated ones). The INAP information flows have been mapped into equivalent method invocations so for instance the SCF side interface contains such methods as "DropParty", "RequestReportSSMChange" and so on.

The *Service Logic Manager* (SLM) is the core SCF entity. It is responsible for locating the SLPs for each particular service that is requested and for initiating migration procedures either triggered by the management system or by the migration criteria that are applied. The SLM is however not aware of any service-level details of the SLPs it manages. All its volatile data are embodied in its mobile part, which is created at the network management center and migrates on the SCF in order to attach to the stationary part. Since the interaction between the stationary and the mobile part is very intense, for performance reasons the interface between them is not based on CORBA but on Java. So these two objects hold direct pointers to each other allowing native method invocations between them. A similar approach has been adopted for all other mobile-stationary part pairs.

The SEN, as far as UNI signaling is concerned, is viewed as a terminal. The Call Control Agent Function (CCAF), being the user side of the UNI signaling, provides access to the network resources. It interfaces with the CCF, which resides in the switch, and can either establish a new connection or accept an incoming one. The user invokes an IN service and afterwards the Service Logic connects the user terminal to the SEN, in order to elicit user input. Further service execution is significantly based on this info.

The *CCAF Wrapper*, being a CORBA server, provides a CORBA interface to the signaling stack wrapping the call control functionality. This functionality is implemented in native code and can be viewed as a legacy component. This interface can provide a flexible way to establish new connections, to modify existing ones, to retrieve the total number or specific characteristics of each of the active ones. The CCAF Wrapper basically resolves each active session, based on a unique sessionID, to the connection. For this purpose the SessionId parameter is used to retrieve the VPI/VCI values (in the case of an underlying ATM broadband network) associated to a User-Plane connection. The SessionId is included in the Broadband High Layer Information (BHLI) information element of the Q.2931 SETUP message sent from the B-SS&CP (CCF) to the SEN (CCAF) [11].

The *Specialised Resource Manager* is a "hub" for containing volatile service and user related info. It can be used to contain user subscription and authentication information, information about available services, about service providers (e.g. video servers for the Information Retrieval service) or about FTP servers that host content raw data. The entire body of information contained in the Specialised Resource Manager is populated in the form of mobile objects that are dispatched there by the management center. An agent encapsulating content descriptive attributes (such as the location of the content server or the category that this new content should appear to user terminals) will be created at the management center. Then it will be instructed to migrate to the Specialized Resource Manager premises where using the Region Registry's facilities he will attach natively to it (perhaps replacing in the process obsolete information). After this attachment has taken place, new users connected to the SRF and browsing through its catalogues for movies will find that movie available for selection.

The *User Interaction Manager* (UIM) intermediates between the SLPs and the basic functionality of the IP. Exactly like the Service Logic Manager and the SSF Access Manager are the SLPs' "connectivity" mediators to the SSF (and vice versa), the UIM is the SLPs' mediator to the SRF. The interface between the SLPs and the

User Interaction Manager is also based on CORBA. In this way all SLP references to their environment as well as the environment's references to the SLPs are distributed allowing location transparent communication to take place. Belonging to the SCF, the UIM models the way the user interacts with the system for service selection and content preview/selection. It offers service-specific user interaction functionality. This functionality is implemented by the *Resource Logic Programs* (RLP). The RLPs contain the logic and processing capability to receive and send user information and convert information received from users.

Each RLP is activated by the UIM after it has been invoked by the service logic and when the connection between the terminal equipment and the IP has been established. The RLP resumes the control of the service session for handling an user-plane connection. While the RLP can actually use the established ATM connection, only if it knows the corresponding VPI/VCI pair, the service logic is aware only of the pertinent SessionId. This SessionID is sent to the CC signaling stack in the SETUP message.

To provide service specific functionality, there should be one to one correspondence between the RLPs and the available services. Taking as an example the Interactive Multimedia Retrieval (IMR) service, the pertinent RLP is responsible for verifying which sub-service interests the user (e.g. video on demand, music on demand, news on demand) and enabling him to preview content from a variety of service providers. When the user interaction ends, the SLP retrieves the result (i.e. kind of sub-service, chosen service provider) and, after processing it (e.g. retrieving the physical address of the SP), instructs the establishment of one or possibly more physical connections between the user and the appropriate SP.

At last but not least, the user interaction is not restricted to service specific context. The SRF can offer service-independent functionality, as well. This can include the authentication and the authorization of a user, the registration and de-registration for services and, in general, the interrogation and modification of the user profile. An alternative way to perform these operations is via the Call Unrelated Signaling Function (CUSF) implemented in the TE and interfacing with the CCF. However this way presupposes that the TE supports additional signaling stacks (Q.2932 in the broadband case) and that the corresponding information is encoded in "facility" Information Elements (IE) of the signaling messages. These IE are afterwards decoded by the service logic. Furthermore the updating/enhancement of the available user interaction scenarios when they are implemented via the SRF seems to be more flexible and user-friendly. In such case the Specialized Resources Manager is populated not only by agents bearing SP specific information but also user-specific in order to render more efficient the execution of the above mentioned operations. This way checking and updating profile, registering, authentication, authorizing can be provided locally in the IP.

The implementation of RLPs is static. Another approach could be the implementation of the RLPs as a combination of static and mobile parts. The mobile service-dependent parts could migrate to the IP from a management center. This way the functionality of generic IPs, offering a service-neutral API, employed for playing announcements and collecting user info, could be enhanced and be service specific. Furthermore these mobile parts could migrate to the TE, however, the inevitable necessary enhancements of the TE software platform may overpower the potential added flexibility.

4 Conclusions

The IN architecture represents an effective solution to the problem of providing advanced services to end-users. New software technologies which are becoming available allow to increase the efficiency of this architecture by enabling the distribution of services in the network and reducing bottleneck problems deriving from the use of a centralized architecture.

In adopting distributed object technologies, the proposed architecture emphasizes the notion of Functional Entities engaged in a location transparent exchange of "Information Flows" between them (the concept on which the IN is based) and allows increased programmer's productivity in the introduction of new concepts and services. Interoperability with traditional IN is also preserved by means of IN/CORBA bridges.

Mobile code technology affords a greater flexibility in dynamically composing the network or parts of it at runtime and in keeping the infrastructure as generic (and thus as unlikely to need changes) as possible. Mobile code is also used to increase the robustness of the system and even to seek out and exploit performance gain opportunities in cases of intense remote interactions that could be made local. Overall, the combined distributed object / mobile code layer provides significant advantages to service design and management, allowing for high reutilization of components and for simplified schemes for release deployment and handling.

References

1. Magedanz, T., Popescu-Zeletin, R.: Intelligent Networks - Basic Technology, Standards and Evolution, Thomson Computer Press, ISBN:1-85032-293-7, London (1996)
2. Venieris, I.S., Hussmann, H. (eds): Intelligent Broadband Networks, John Wiley, ISBN: 0-471-98094-3, Chichester (1998)
3. ITU-T Recommendations – Intelligent Network, Series Q.12xx, Geneva
4. OMG, CORBA/IIOP 2.3 Specification (1998), www.omg.org/docs/ptc/98-12-04.pdf.
5. Perdikeas, M., Chatzipapadopoulos, F., Venieris, I., Marino, G.: Mobile agent standards and available platforms, Computer Networks and ISDN systems
6. Breugst, M., Magedanz, T.: On the Usage of Mobile Agent Platforms in Telecommunication Environments, IS&N98, Antwerp (1998)
7. Perdikeas, M., Chatzipapadopoulos, F., Venieris, I.: An Evaluation Study of Mobile Agent Technology: Standardization, Implementation and Evolution, ICMCS '99 Florence – Italy
8. Venieris, I., Zizza, F., Magadanz M. (eds): Object Oriented Software Technologies: From Theory to Practice, J. Wiley & Sons, ISBN: 0-471-62379-2, Chichester (2000)
9. Chatzipapadopoulos, F., Perdikeas, M., Venieris, I.: Mobile Agent And CORBA Technologies In The Broadband Intelligent Networks, IEEE Communications Magazine, pp. 116-124, Vol.38, No. 6 (2000)
10. Prezerakos, G., Pyrovolakis, O., Venieris, I.: Service Node Architectures Incorporating Mobile Agent Technology and CORBA in Broadband IN, IEEE MMNS'98, France (1998)
11. ITU-T Recommendation Q.2931, Digital subscriber signaling system No. 2 (DSS2) – User Network interface (UNI) layer 3 specification for basic call/connection control (1995)

A New Cut-Through Forwarding Mechanism for ATM Multipoint-to-Point Connections

Athanasios Papadopoulos[1], Theodore Antonakopoulos[2], and Vassilios Makios[2]

[1]Computers Technology Institute, Riga Feraiou 61, 26100 Patras, Greece
[2]Department of Electrical Engineering and Computers Technology,
University of Patras, 26500 Rio Patras, Greece
Tel: +30 (61) 997 346, Fax: +30 (61) 997 342,
antonako@ee.upatras.gr

Abstract. This paper presents a new Cut-through forwarding mechanism for multipoint-to-point communications over ATM networks. Initially, we present the cell interleaving problem and a concise description of current ATM multicast proposals. Then, we present an improvement to existing Cut-through forwarding schemes for providing a good level of fairness among the connections competing for the available bandwidth of the output link, while maintaining good throughput performance. This scheme is referred as Conditional Cut-through forwarding (CCT) and as simulation results show, it supports multipoint-to-point connections efficiently.

1 Introduction

ATM networks use various operational modes, like unicast, broadcast and multicast in order to support different application requirements. Multicast serves as a communication abstraction, allowing message delivery to multiple destinations in a single step and thus reduces overall bandwidth consumption, helps control network traffic and decreases the amount of processing at every host. Efficient implementation of multicast is useful in many applications, such as distributed computing, parallel discrete event simulation (PDES) and multimedia applications.

ATM multipoint communications have been studied at various international organizations, like ATM Forum and ITU. The Internet Engineering Task Force (IETF) also studies the mapping of IP multicasting to ATM networks [1], but direct ATM multicast service is still in its early phase of definition. The ATM User-to-Network Interface (UNI 3.1 or 4.0) signalling supports source-based tree approach with point-to-multipoint virtual channels (VCs) [2]. In current ATM implementations, AAL5 does not have any provision within its cell format for the interleaving of cells from different packets to a single connection. Therefore, point-to-multipoint connections are unidirectional, permit the root (a single source) to send data to the leaves (multiple destinations) but do not permit the leaves to transmit to the root on the same connection. In order to reduce the number of virtual connections maintained for each group, ATM switches must support multipoint-to-multipoint connections.

The point-to-multipoint connections supported by UNI achieve multipoint-to-multipoint communication using either a centralized multicast server, or many point-to-multipoint VC connections in order to completely connect hosts in a mesh topology. Both of these approaches have efficiency limitations, since either they have to perform data retransmission (in the multicast server case), or they have to use many network resources in order to establish the required connections (in the case of overlaid VCs). In order to use multipoint-to-point connections, where many input links are mapped to an output link, we must solve the cell interleaving problem: when traffic is merged on a multipoint-to-point connection, cells that belong to packets from different sources use the same VPI/VCI and may interleave at the receiver site and the AAL5 entity cannot reassemble the data.

2 ATM Multicast Proposals

AAL5 is the mostly employed ATM protocol. Several approaches have been proposed for solving the cell-interleaving problem over AAL5. These mechanisms include VC merging, VP switching, AAL5 modifications, use of resource management cells and the use of sub-channels within a VC.

VC merging uses buffering of cells at the network switches and individual forwarding per packet. The MPLS (MultiProtocol Label Switching) and SEAM (Simple and Efficient ATM Multicast) proposals follow this technique. MPLS [3] implements a Store-and-Forward technique, while SEAM [4] aims to improve the performance of MPLS with a Cut-through forwarding algorithm. VP switching uses the VCI field to identify the sender and the VPI field to forward the cells. In schemes like DIDA (Dynamic IDentifier Assignment) [5], a value identifies the cells of a specific frame, while in the improved VP Switching proposal [6], the VCI identifies the sender, not the frame.

Modifications of AAL5 add new fields in the cell for multiplexing or change the current fields of the cell header. Simple Protocol for ATM Multicast (SPAM) [7] inserts a 15 bits MID (Message Identifier) field in the cell payload to distinguish the cells coming from different senders. In the AAL5+ method [8], the MID field is 16 bits long and its value is assigned per packet by using a uniform probabilistic function.

Other proposals use Resource Management (RM) cells. In SMART [9], a host must hold a token (control message) before it starts transmitting data to a tree. In CRAM [10], each group of cells belonging to the same sender is preceded by a Resource Management (RM) cell, which contains a number of Sender Identifiers. In the Sub-channel Multiplexing technique [11], 4 bits from the Generic Flow Control field in the ATM cell header are used to carry the multiplexing identifier (sub-channel ID) to distinguish between multiple sub-channels in a VC.

All approaches require some modifications either to the ATM switches or to the edge devices. Some of them suffer from excessive overhead (SPAM, AAL5+) and high complexity (SMART, CRAM). VPs should not be used by edge-devices (DIDA, Improved VP), because they are used for the accumulation of traffic in the backbone. Also, the sub-channel multiplexing technique is not very scalable, since only fifteen simultaneous senders can use the VC. Proposals based on Store-and-Forward and Cut-through algorithms (MPLS, SEAM) add small delay to data traffic [12], are scalable and simple to implement in order to support multipoint communications. As an additional benefit, these approaches can enhance intelligent discarding schemes (Early Packet Discard or Partial Packet Discard – EPD/PPD) by reducing the number of partial transmitted packets. This capability reduces further the wasted bandwidth since the EPD/PPD packet dropping policies reduce congestion and transmission of traffic that would inevitably be retransmitted [13].

2.1 Cut-Trough Forwarding

The Cut-through mechanism avoids the cell-interleaving problem in multipoint-to-point VCs by forwarding the first cell of a packet before receiving the End-of-Packet (EOP) cell, whenever the outgoing VC is idle and continues to transmit cells until the last cell of the packet. The ATM switch, which keeps separate buffers for each sender, has to buffer other incoming packets for the same outgoing VC, until the current packet has been forwarded completely. However, if a long packet from a slow source is forwarded or if the last cell has been lost, the delay increases, since cells belonging to other packets must wait in the input queues until the EOP cell of the first packet arrives, as shown in Figure 1.

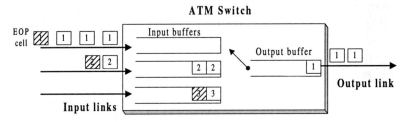

Fig. 1. Large packets from slow sources increase the delay in Cut-through forwarding

SEAM proposes a timer to overcome the loss of an EOP cell and the Store-and-Forward mechanism for slow links. But the value of timer is really critical and could significantly impact buffer lengths into switches. If the duration of the timer is too short, it would lead to an undesirable discard of good packets, and if it is too long, the delay of buffered packets would increase seriously. Additionally, slow links do not exploit the advantage of the cut-through mechanism, which is the immediate transmission of a new cell, and slow sources from faster links can still block the procedure. The modifications of the Cut-through discipline that are proposed by Stolyar in [14] do not solve these issues.

3 The Conditional Cut-Through (CCT) Forwarding Scheme

The design goal of the Conditional Cut-through (CCT) forwarding scheme is to minimize the delay of buffered packets that is due to large packets from slow incoming links and sources or due to the loss of an EOP cell. In CCT, we define the variable "time in buffer" for each incoming packet. The variable "time in buffer" measures the number of arrived cells and is increased by 1, whenever a new cell arrives into the input buffer. When an EOP cell arrives into the buffer, the packet is marked as "Ready". Then, the "time in buffer" has its maximum value and is equal to the total number of cells for that specific packet. It is obvious that each packet has its own "time in buffer".

When packets come from a source that generates cells at a rate less than the maximum supported rate (defined by the link speed), contiguous cells within a packet will be spaced by idle slots. We assume that the switch measures and stores the cell input rate for each partially forwarded packet and for every new packet that arrives into any empty buffer. Additionally, the switch measures and stores the mean and max values of the "time in buffer" for each input buffer separately.

When the switch receives the first cell of packet X and if there are no cells from other sources queued or being forwarded, it directly forwards all cells on the output buffer without queuing. The switch copies each cell of the partially forwarded packet into a backup buffer. During the transmission of packet X, all the cells of a new packet Y arrive also into an empty buffer. Then, the switch will stop the packet forwarding and will start transmitting the new packet Y, if all the following conditions are satisfied:

a. the "Ready" packet Y has less "time in buffer" than the current "time in buffer" of the partially forwarded packet X,

$$Y_R < X_C \quad (1)$$

b. the "Ready" packet Y has arrived with higher cell input rate than X,

$$CIR_Y > CIR_X \quad (2)$$

c. the "time in buffer" of the current forwarded packet is less than the mean value of the corresponding input buffer,

$$X_C < X_{mean} \quad (3)$$

The switch transmits a "null cell" to indicate to the receiver that the previous cells have to be rejected. The stopped packet X is stored into the backup buffer until the EOP cell of packet arrives. Then, the switch transmits the packet X, irrespective of the status of other packets.

If the transmission of a packet has been completed successfully, the next packet is selected by the "time in buffer" of the awaiting "Ready" packets. The switch compares all the first-in-buffer "Ready" packets among them. The packets with smaller "time in buffer" (smaller packets) have higher priority. The switch must transmit all the "Ready" packets from the results of the current comparison, before it moves on to a new comparison with new incoming "Ready" packets.

In case the previous conditions are not satisfied, the loss of an EOP cell could block the procedure. Then, if there is an input buffer Y with two or more "Ready" packets and their total "time in buffer" satisfies the condition:

$$Y_R + Y_{R+1} + Y_{R+2} + \ldots > \max\{X_{max}, Y_{max}\} \quad (4)$$

then the first packet of this buffer will stop the forwarding action. According to Condition (4), where Y_{R+1} is the "time in buffer" of the second incoming "Ready" packet in buffer Y, CCT gives to packet X enough time to be forwarded successfully, taking into account the case when the size of packets between two buffers differs significantly. When an EOP cell loss event occurs, the packet would be eventually discarded by switch. Consequently, the other input links are not blocked waiting for the output link to be released.

A system using the CCT forwarding scheme is shown in Figure 2. We assume that packet 1 is currently being served and buffers 2 and 3 have no complete packets. Packet 1 is forwarded even though it has not arrived completely and it is copied into the backup buffer. During the transmission, the EOP cell of packet 3 arrives. According to the CCT algorithm, if conditions (1) to (3) are satisfied, the switch generates the "null cell" to indicate the rejected cells and it starts transmitting packet 3.

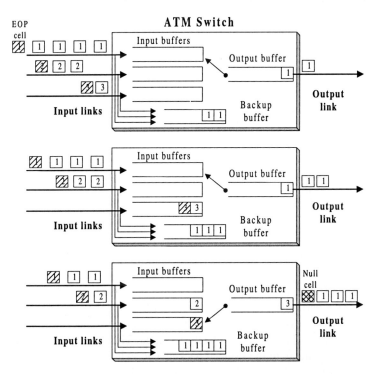

Fig. 2. The Conditional Cut-through (CCT) forwarding scheme

A generic flow chart for CCT forwarding scheme is shown in Figure 3. CCT gives higher priority to smaller packets from faster links or sources. The algorithm does not

network. Moreover, it provides a policing mechanism for misbehaving users and facilitates the pricing process.

This paper is organised as follows: section 2 provides a brief review of related work in the literature. Section 3 describes the round robin service discipline for packet switching networks. Section 4 deals with round robin in ATM environment. The end-to end delay bounds and resources needed in network environments are described in section 5. Details of the Cell-by-Cell Round Robin (CRR) for deterministic, statistical and best-effort services are given in section 6. In section 7, the connection admission control tests are described. Evaluation and benefits of using CRR in ATM environments are give in section 8.

2. Related Work

Several service disciplines have been proposed in the literature. In general, we can divide them into three main types: FCFS, static priority and dynamic priority scheduling disciplines.

The FCFS scheduler, which is used in most conventional networks, is the simplest. It may be capable of providing some performance guarantees (it can support only one delay bound for all streams), but only under a number of constraints and with a very inefficient use of network's resources [3].

A static priority (SP) [3] scheduler supports a fixed number of priority levels for connections, supporting a limited number of delay bounds. In addition, its exact admission test is computationally very complex.

Dynamic priority schemes that aim to provide different qualities of service to different connections have been proposed. Examples of these schemes are Delay Earliest-Due-Date [4] and Stop-and-Go [5]. These solutions are based on either a time-framing strategy, or a sorted priority queue mechanism. In time framing schemes, it is difficult to support connections with both low delays and low throughput requirements efficiently since framing introduces dependencies between scheduling priority and bandwidth allocation granularity. A scheme based on sorted priority queue is difficult to implement in high speed network and requires a complicated schedulability test at connection establishment time [3].

3. Round Robin Service Discipline

In 1987, Nagle [2] proposed a simple service discipline for packet-switching networks. In this scheme each connection has its own separate FIFO queue. The different queues are serviced in a round robin manner. This service discipline has several merits. It provides fairness among the different connections and isolation between them. It prevents the misbehaving users from degrading the service of the other clients. However, a drawback of this algorithm is that it does not take into account variable packet sizes. By combining smaller packets into one large packet and sending large packets, a user can gain an unfairly large fraction of the bandwidth compared with other users. Moreover, it doesn't provide any performance guarantees.

Now, as Asynchronous Transfer Mode (ATM) networks, which have fixed packet size, are widely used, we would like to re-visit the idea of using the round robin service discipline in ATM based networks. A successful implementation would need to answer the following questions:
1. How can the round robin discipline guarantee the different performance requirements of the different connections?
2. What are the admission tests according to this service discipline? And what are the traffic descriptors required for these tests?
3. How can this discipline be implemented in ATM environment?
4. What are the benefits of using this discipline in ATM networks?

In the next few sections, we will try to address these questions.

4. Round Robin in ATM Environment

We suggest a new service discipline based upon the round robin idea for ATM networks where all packets or cells have the same size. We call it cell by cell round robin (CRR). In this scheme, each connection has its own separate FCFS queue. The different queues are serviced according to a pre-determined transmission order so that each connection has some weight according to its performance requirements.

An important issue is how to choose this transmission order in such a way that guarantee the different performance requirements for all connections. On the one hand, if the weight of a connection is assigned according to its performance requirements (e.g. according to the delay bound), then the problem of the coupling between delay and bandwidth allocation granularity will occur so that connections with both low delays and low throughput requirements cannot be supported efficiently.

On the other hand, if the weights are assigned according to the traffic characteristics (e.g. average and peak rates) the performance requirements can not be guaranteed.

To avoid these problems, the order in which the different queues are serviced must depend on both the traffic characteristics and the performance requirements. One solution is to describe the traffic using the effective bandwidth and buffer-size needed to guarantee the performance requirements on a single-node basis, and to assign the weights according to these effective bandwidths. Now, if each connection is guaranteed its required effective bandwidth at a single switch, its performance requirements will be guaranteed.

However, a single node analysis is needed to convert the performance requirements (delay bound, cell loss ratio) into required effective bandwidth and buffer-size notations. This is needed to avoid the dependencies between the bandwidth allocation and the required delay.

Several stochastic per-connection effective bandwidth models are suggested in the literature [6, 7]. Few deterministic models based on a per-connection effective bandwidth notion exist [8]. However, it is possible to evaluate the required effective bandwidth and buffer-size using the different suggested deterministic traffic models. Moreover, a new deterministic model, which describes the traffic streams in terms of the minimum required bandwidth and buffer-size, is suggested in a sequel paper.

Throughout this paper, we assume that traffic streams are described using a per-connection required effective bandwidth and buffer-size; and these required resources guarantee the performance requirements on a single node basis.

5. Delay Bounds and Resources Required on a Multiple Node Basis

The main idea behind using round robin in ATM networks is that each connection can guarantee a minimum required service rate. This provides fairness and isolation between traffic streams. It also simplifies the pricing process as the link capacity is plainly divided between connections. The major advantage is that if the technique is applied on a multiple-nodes network environment the worst-case end-to-end queuing delay is reduced to a single node worst-case delay. Moreover, the resources at all nodes are the same as those needed in the single node case. This is proved by the following result.

Proposition 1:

Provided that the following conditions are satisfied:
- A variable bit-rate stream is characterised using a worst-case deterministic traffic model that is based upon calculating the required effective bandwidth and buffer-size.
- The effective bandwidth and buffer-size are calculated on a single node basis; i.e., the performance requirements are guaranteed if the stream is serviced at a rate greater than or equal to the effective bandwidth, and is allocated the required buffer-size at a single switch.
- The stream is transmitted through a network (multiple node environment) and the same effective bandwidth and buffer-size are guaranteed to the stream; i.e., at all switches the stream is serviced at a minimum rate equal to the effective bandwidth and allocated its exact required buffer-size (irrespective of other streams multiplexed with it).

Then, even if the scheduling discipline is work-conserving , the following holds:
1. The end-to-end queuing delay in the networking environment is equal to the worst-case queuing delay introduced by a single switch.
2. The buffer-size required in the following switches is the same as that required in the single switch; i.e., no more resources needed to compensate for the network load fluctuations at the different switches.

Proof:

A worst-case representation of any traffic may be described as follows. If the actual traffic of a connection is given by a function A such that $A[\tau,\tau+t]$ denotes the traffic arrivals in the time interval $[\tau,\tau+t]$, an upper bound on A can be given by a function A^* if for all times $\tau \geq 0$ and all interval lengths $t \geq 0$, the following holds:

$$A[\tau,\tau+t] \leq A^*(t) \quad (1)$$

We refer to function $A^*(t)$ as a time invariant traffic constraint function. An example $A^*(t)$ is given in fig.1.

If the single-node bound delay = dreq, then the effective bandwidth (ρ) can be calculated using the following equation:

$$dreq = 1/\mu * \max(A^*(t) \quad \rho t) \quad (2)$$

$$\text{worst-case queue length} = \text{effective buffer-size} = \max(A^*(t) - \rho t) \quad (3)$$

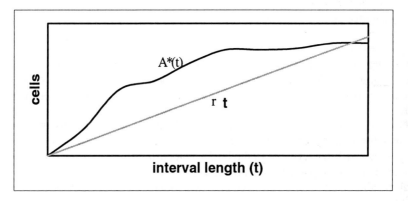

Fig. 1. The constraint function $A^*(t)$

If the stream is serviced at a rate greater than or equal to ρ and allocated the required buffer-size at all switches, we can evaluate the end-to-end queuing delay and the buffer size to be allocated in the following switches to ensure zero loss-rate.

i. The end-to-end queuing delay

Let 'D' denote end-to-end delay and 'di' denote delay at switch i, then:

$$D = \Sigma\, di \quad (i=1,2,..m) \quad (4)$$

Where m is the number of nodes in the path between source and destination.
Now we will analyse the traffic pattern after the first switch:

As the worst-case bound delay introduced by the 1st switch = dreq, then any cell arriving at the 1st switch at time t_o can be delayed by no more than dreq; i.e., we can assume that each cell in the output stream of the 1st switch has two limiting times: initial time and final time (t_o, t_o+dreq).

time traffic streams. It achieves flexibility in supporting different QoS requirements, efficiency in using network resources and simplicity of admission control tests. It provides isolation between the different connections and fairness among them. Most importantly, it reduces the end-to-end queuing delay to the worst-case delay introduced by a single switch. The CRR service discipline was evaluated and it was shown that it achieves the most important aims of QoS service disciplines. We have developed a traffic model to be used in conjunction with CRR scheduling, which will be presented in a sequel paper.

References

1. J. Nagle, "*On Packet Switches with Infinite Storage,*" RFC 896, 1985.
2. J. Nagle, "*On Packet Switches with Infinite Storage,*" IEEE Trans. on Communications, April 1987, Vol. **35**(4): p. 435-438.
3. H. Zhang and D. Ferrari, "*Rate-Controlled Static-Priority Queueing,*" Proceedings of INFOCOM'93, San Francisco, CA, April 1993.
4. D. Ferrari and D. C. Verma, "*A Scheme for Real-Time Channel Establishment in Wide-area networks,*" IEEE Journal on Selected Areas in Communications, Apr. 1990, Vol. **8**(3): p. 368-379.
5. S. J. Golestani, "*A Stop-and-Go Queueing Framework for Congestion Management,*" In Proceedings of ACM SIGCOMM'90, Philadelphia, Pennsylvania, Sep. 1990 : p. 8-18.
6. A. I. Elwalid and D. Mitra, "*Effective Bandwidth of General Markovian Traffic Sources and Admission Control of High Speed Networks,*" IEEE/ACM Trans. Networking, June 1993, Vol. **1**(3): p. 329-343.
7. C. Chang and J. A. Thomas, "*Effective Bandwidth in High-Speed Digital Networks,*" IEEE Journal on Selected Areas in Communications, Aug. 1995, Vol. **13**(6): p. 1091-1100.
8. H. Mokhtar, R. Pereira, and M. Merabti, "*Resource Allocation Scheme for Deterministic Services of Video Streams,*" accepted for publication in ISCC'01, July 2001.

Delay and Departure Analysis of CBR Traffic in AAL MUX with Bursty Background Traffic

Chul Geun Park[1] and Dong Hwan Han[2]

[1] Department of Electronics, Information and Communications Eng.
Sunmoon University, Asan-si, Chungnam, 336-708, Korea
cgpark@sunmoon.ac.kr
[2] Department of Mathematics
Sunmoon University, Asan-si, Chungnam, 336-708, Korea
dhhan@sunmoon.ac.kr

Abstract. This paper models and evaluates the AAL multiplexer to analyze delay and departure distributions of CBR traffic in ATM networks. We consider an AAL multiplexer in which a single periodically deterministic CBR traffic stream and several variable size bursty background traffic streams are multiplexed and one ATM cell stream goes out. We model the AAL multiplexer as a $B^X + D/D/1$ queue and analyze departure process of CBR cells by using queueing model. We represent the cell delay distribution and the cell interdeparture time distribution.

1 Introduction

The use of Asynchronous Transfer Mode (ATM) as the key transport technique for B-ISDN networks has motivated engineers to study ATM systems by using the queueing model. Many papers[1,2] have studied the queueing model of ATM MUXs (multiplexers) to analyze the ATM layer protocol. But it has been uncommon to make a study on the analysis of the ATM Adaptation Layer (AAL) protocol[3]. It has been also uncommon to make a study of queueing model of the AAL MUXs in order to analyze the Cell Delay Variation (CDV) of a cell stream.

In this paper, we model and evaluate the AAL MUX to analyze the AAL protocol and CDV of a delay sensitive traffic stream in ATM networks. We concentrate on the performance measures of both packet-level (AAL) and cell-level (ATM). There are two packet streams entering the AAL MUX. The first one is a fixed size periodic and delay sensitive Constant Bit Rate (CBR) traffic stream while the other is bursty background traffic stream. This system is modelled as a single First In First Out (FIFO) MUX. This system is applied to the Video on Demand (VoD) service using a CBR transport stream with the background traffic such as Internet Protocol (IP) and signaling data packets from other connections[4].

Periodically arriving CBR transport packets are multiplexed with data packet from other connections. The passage of a packet across the AAL is modelled as a bulk arrival of cells due to the segmentation process in the AAL. The number

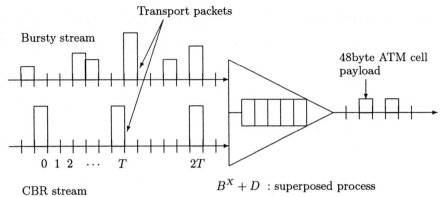

Figure 1. Input traffic and queueing model of AAL MUX

of cells in the bulk is determined by the packet size in cells. The outgoing cells of AAL segmentation process are passed to the ATM layer. The segmentation processing time devoted to each cell corresponds to the segmentation algorithm and requires several ATM cell time[5], which is represented by the service time S in the queueing model of the AAL MUX.

We consider an AAL MUX in which a single CBR stream (D : deterministic process) and several variable size bursty traffic streams (B^X : batch Bernoulli process) are multiplexed and one ATM stream goes out. Thus we can derive a $B^X + D/D/1/\infty$ queueing model for illustrating the cell departure process of AAL MUX in order to analyze the CDV of CBR traffic.

The overall organization of this paper is as follows. In the section 2, we describe the input traffic and system model of the AAL MUX. In the section 3, we analyze the system by using an MMBBP/D/1 queue, where the Markov Modulated Batch Bernoulli Process (MMBBP) is a superposition of bursty background traffic and deterministic CBR traffic[6]. In the section 4, we obtain the waiting time distribution of a CBR cell. In the section 5, we represent cell delay time and cell interdeparture time distribution. In the section 6, we take some numerical examples for the performance measures of our queueing model. Finally, we have a conclusion in the section 7.

2 Input Traffic Model and System Model

Periodically arriving CBR traffic streams are multiplexed with signaling, ABR, or UBR traffic stream from other connections. On many ATM interface cards, the multiplexing effects in AAL MUX should be considered. We consider an AAL level MUX in which a single CBR (D) stream and several variable size bursty traffic (B^X) streams are multiplexed and one ATM cell payload stream goes out. We derive a $B^X + D/D/1$ queueing model for the evaluation of delay and delay variation experienced by the CBR stream (see Fig. 1).

We consider a system model that operates on a discrete time basis. Time is divided into intervals (called slots) which are equal in length (a cell transmission time). Integer times $n \in \{0, 1, 2, \cdots\}$ are assigned to individual slot boundaries. We assume that packets of a bursty traffic streams arrive in various batch sizes of cells just prior to the end of slot boundaries and packets of a single CBR stream arrive in a constant batch size d of cells in each period T such that the interarrival time is always equal to T. The two cell streams are assumed to be independent.

Let B_n denote the number of cell payloads of the bursty stream arriving just prior to the end of the n-th slot, that is, arrival batch size at time n. Then the bursty stream can be modelled by a batch Bernoulli process, that is, in each time slot n, arrivals occur in independent and identically distributed batches of size B_n with a common distribution $b_k = P\{B_n = k\}$. We define $B(z)$ by the probability generating function (PGF) of the distribution b_k.

Now, to describe the superposition of a bursty stream and a CBR stream, we assume that the superposed input process is a doubly stochastic point process in which the number of arriving cells depends on the underlying arrival phase process. The arriving phase for each slot varies on the states $\{1, 2, \cdots, T\}$. Let Y_n denote the arrival phase of the n-th slot. It then follows that Y_n forms a discrete time Markov chain having state space $\{1, 2, \cdots, T\}$. The arrival process of the superposed input process and the underlying Markov process are assumed to be independent of each other. Let X_n denote the number of cells arriving from the superposed process at time n. Then this input process $\{(Y_n, X_n)\}$, which is called an MMBBP, is characterized by the transition matrix P and the arrival matrix $\Lambda(z)$ of the superposed process

$$P = \begin{pmatrix} 0 & 1 & 0 & \cdots & 0 \\ 0 & 0 & 1 & \cdots & 0 \\ \vdots & \vdots & \vdots & \ddots & \vdots \\ 0 & 0 & 0 & \cdots & 1 \\ 1 & 0 & 0 & \cdots & 0 \end{pmatrix}, \quad \Lambda(z) = \begin{pmatrix} B(z) & 0 & \cdots & 0 & 0 \\ 0 & B(z) & \cdots & 0 & 0 \\ \vdots & \vdots & \ddots & \vdots & \vdots \\ 0 & 0 & \cdots & B(z) & 0 \\ 0 & 0 & \cdots & 0 & z^d B(z) \end{pmatrix}. \quad (1)$$

3 Queueing Analysis of AAL MUX with Infinite Buffer

In this section, we return our attention to the queueing situation where a single server system operates on a discrete time basis. We consider a discrete time single server queue with MMBBP input, infinite buffer and constant service time S. The buffer receives cells according to a process modelled as an MMBBP, giving rise to an MMBBP/D/1 queue. Let $J(t)$ be the phase of the MMBBP and $N(t)$ be the number of arriving cells during the time interval $(0, t]$ respectively. We denote the transition probabilities of the process $\{(N(t), J(t)), t \geq 0\}$ by

$$P_{jj_1}(k, t) = P\{N(t) = k, \, J(t) = j_1 \,|\, N(0) = 0, \, J(0) = j\},$$

and we denote $P(k, t)$ as the $T \times T$ matrix of the probabilities $P_{jj_1}(k, t)$. Then $P(k, t)$ has the PGF as follows

$$\tilde{P}(z, t) = \sum_{k=0}^{\infty} P(k, t) z^k = [P \cdot \Lambda(z)]^t, \quad |z| \leq 1,$$

where P and $\Lambda(z)$ are given in (1).

Let $Q(t)$ be the number of cells in the data buffer at time t and let t_n be the n-th embedded (just after departure) point. Let $Q_n = Q(t_n+)$ and $J_n = J(t_n+)$ be the state (level) of the queueing system and the phase of the MMBBP just after t_n, respectively. Then $\{(Q_n, J_n),\ n \geq 0\}$ forms a Markov chain with state space $\{0, 1, 2, \cdots\} \times \{1, 2, \cdots, T\}$. The one step transition probability matrix $\tilde{Q}$ is given by

$$\tilde{Q} = \begin{pmatrix} C_0^* & C_1^* & C_2^* & C_3^* & C_4^* & \cdots \\ C_0 & C_1 & C_2 & C_3 & C_4 & \cdots \\ O & C_0 & C_1 & C_2 & C_3 & \cdots \\ O & O & C_0 & C_1 & C_2 & \cdots \\ \vdots & \vdots & \vdots & \vdots & \vdots & \ddots \end{pmatrix},$$

where $C_k = P(k, S),\ k = 0, 1, 2, \cdots$, and O is a $T \times T$ matrix with all zeros.

Now we must find the probability matrix $C_k = P(k, S),\ k = 0, 1, 2, \cdots$. To do this, we define the matrices R_1 and R_2 as follows

$$R_1 = \begin{pmatrix} O & I_1 \\ O_1 & O \end{pmatrix}, \quad R_2 = \begin{pmatrix} O & O_2 \\ I_2 & O \end{pmatrix},$$

where the block I_1 is a $(T-S) \times (T-S)$ identity matrix, O_1 is an $S \times S$ zero matrix, the block I_2 is an $S \times S$ identity matrix, O_2 is a $(T-S) \times (T-S)$ zero matrix, and the blocks O's are zero matrices. Let $I_{i,j}$ be the $T \times T$ matrix with all entries zeros except (i,j)-entry one. Then, we can find $C_k,\ k = 0, 1, 2, \cdots$, entries as follows

$$C_k = \begin{cases} (R_1 - I_{T-S,T} + I_{T,S})\, b_k^{(S)}, & k \leq d-1, \\ (R_1 - I_{T-S,T} + I_{T,S})\, b_k^{(S)} + (R_2 - I_{T,S} + I_{T-S,T})\, b_{k-d}^{(S)}, & k \geq d, \end{cases}$$

where the superscript means the S-fold convolutions of the batch size distribution. To find $C_k^*,\ k = 0, 1, 2, \cdots$, let $A(z)$ be a PGF of $\{A_k\}$ as follows

$$A(z) = [I + P\Lambda(0) + (P\Lambda(0))^2 + \cdots + (P\Lambda(0))^{T-1}]\, P\, [\Lambda(z) - \Lambda(0)], \quad (2)$$

where the term $(P\Lambda(0))^k P[\Lambda(z) - \Lambda(0)],\ k = 1, 2, \cdots, T-1$, means that the first arrival to empty system occurs at time $k+1$, given that the idle period started time 0. Then the matrix $C_k^*,\ k = 0, 1, 2, \cdots$, can be obtained from

$$C_k^* = \sum_{l=1}^{k+1} A_l\, C_{k-l+1},\quad k = 0, 1, 2, \cdots.$$

To find the steady-state probability vector π of the matrix $\tilde{Q}$, let $\pi_{i,j}$ be the limiting probability of $(Q_n, J_n),\ n \geq 0$, that is, for $i = 0, 1, 2, \cdots,\ 1 \leq j \leq T$,

$$\pi_{i,j} = \lim_{n \to \infty} P\{(Q_n, J_n) = (i, j)\},$$

and let $\pi = (\pi_0, \pi_1, \pi_2, \cdots)$ with $\pi_i = (\pi_{i,1}, \pi_{i,2}, \cdots, \pi_{i,T})$. Then the steady-state probability vector π at embedded points is obtained by the equations

$$\pi \tilde{Q} = \pi, \quad \pi e = 1,$$

where e is a column vector with all ones.

Let G be a matrix satisfying the following equation

$$G = \sum_{k=0}^{\infty} C_k G^k.$$

Then we have the unique solution of $g = gG$, $ge = 1$, which is the invariant vector of the irreducible stochastic matrix G [7]. We denote K be the matrix satisfying the following equation

$$K = \sum_{k=0}^{\infty} C_k^* G^k,$$

and κ be the invariant vector of K such that $\kappa K = \kappa$, $\kappa e = 1$. In evaluating π_i, $i = 0, 1, \cdots$, there is a small advantage in denoting the vector $\tilde{\mathbf{u}}_1$ by

$$\tilde{\mathbf{u}}_1 = [I - C + (e - \beta)g]^{-1} e, \tag{3}$$

where $C \equiv \sum_{k=1}^{\infty} C_k = R_1 + R_2$ and $\beta \equiv \sum_{k=1}^{\infty} k C_k e$. To find π_0, we introduce two vectors ϕ_1 and ϕ_2 for simplicity, by using $\tilde{\mathbf{u}}_1$ in (3),

$$\phi_1 = \left[I - C_0 - \sum_{k=1}^{\infty} C_k G^{k-1} \right] \tilde{\mathbf{u}}_1 + (1 - \rho)^{-1} C_0 e,$$

$$\phi_2 = e + \left[\sum_{k=1}^{\infty} C_k^* - \sum_{k=1}^{\infty} C_k^* G^{k-1} \right] \tilde{\mathbf{u}}_1 + (1 - \rho)^{-1} \sum_{k=1}^{\infty} (k-1) C_k^* e. \tag{4}$$

We now define $\tilde{\kappa}_1$ by the same matrix as [7]. Then we can express $\tilde{\kappa}_1$ in terms of the vectors ϕ_1 and ϕ_2 in (4) as follows

$$\tilde{\kappa}_1 = \phi_2 + \sum_{k=1}^{\infty} C_k^* G^{k-1} \left[I - \sum_{k=1}^{\infty} C_k G^{k-1} \right]^{-1} \phi_1.$$

Thus the vector π_0 is given by

$$\pi_0 = (\kappa \tilde{\kappa}_1)^{-1} \kappa.$$

Once the vectors π_0 is known, it is possible to determine also the vectors π_i, $i \geq 1$, by using the following equation

$$\pi_i = \left[\pi_0 \bar{C}_i^* + \sum_{k=1}^{i-1} \pi_k \bar{C}_{i-k+1} \right] (I - \bar{C}_1)^{-1}, \quad i \geq 1,$$

where $\bar{C}_i^* = \sum_{k=i}^{\infty} C_k^* G^{k-i}$, and $\bar{C}_i = \sum_{k=i}^{\infty} C_k G^{k-i}$, $i \geq 0$.

4 Derivation of the Waiting Time Distribution

We assume that $(Q_t, J_t) \equiv (Q(t), J(t))$, $t \geq 0$, is stationary. We will find the stationary distribution of $(Q(t), J(t))$ at an arbitrary point. Let

$$x_{k,j} = P\{(Q(\tau), J(\tau)) = (k,j)\}, \quad k = 0, 1, 2, \cdots, \ 1 \leq j \leq T,$$

and $x = (x_0, x_1, x_2, \cdots)$ with $x_k = (x_{k,1}, x_{k,2}, \cdots, x_{k,T})$, $k = 0, 1, 2, \cdots$. By the same manner as in Yegani[8], we can find the stationary probability vector x. To find the stationary probability vector x, let the random variable η be the steady state of server, namely $\eta = 1$ if the server is busy and $\eta = 0$ if the server is idle. Then the probability of the system being busy is given by

$$\eta_1 \equiv P\{\eta = 1\} = \frac{S}{S + \pi_0[I - P\Lambda(0)]^{-1}e}.$$

In addition, we have the stationary distribution x_k with η_1 as follows

$$x_k = \frac{\eta_1}{S} \sum_{m=1}^{k} \sum_{t=0}^{S-1} (\pi_0 A_m + \pi_m) P(k-m, t), \quad k = 1, 2, \cdots,$$

where A_m is given in (2) and $P(k, 0) = \delta_{k0}I$. On the other hand, the vector x_0 satisfies

$$x_0 = \frac{\pi_0[I - P\Lambda(0)]^{-1}}{S + \pi_0[I - P\Lambda(0)]^{-1}e}.$$

By Little's low, the mean cell delay $E[W_C]$ for an arbitrary cell is given by

$$E[W_C] = \frac{\sum_{k=1}^{\infty} k x_k e}{B'(1) + d/T} = \frac{S}{\rho} \sum_{k=1}^{\infty} k x_k e.$$

To obtain the distribution of the waiting times W_{2C} and W_{2P} of a CBR cell and a CBR packet in the system, respectively, let consider a cell arriving at time τ and take the service interval $[t_l, t_{l+1})$ containing τ. We define the joint distributions by

$$\gamma_j(k, t) \equiv P\{Q(\tau) = k, J(\tau) = j, S_f = t \mid \eta = 1\},$$

where S_f is the remaining service time at time τ. Then we find the waiting time W_{2q} of the first cell of the CBR stream in the queue except service time as follows

$$P\{W_{2q} = w\} = \begin{cases} T\left(x_{0,T-1} + \eta_1 \gamma_{T-1}(1,1)\right), & w = 0, \\ \eta_1 T \gamma_{T-1}\left(\left[\frac{w}{S}\right] + 1, w - \left[\frac{w}{S}\right]S + 1\right), & w \geq 1. \end{cases} \quad (5)$$

where $[x]$ is the Gaussian integer. So we can find $P\{W_{2q} = w\}$ by $\gamma_j(k,t)$ in the same manner as [8]. The distributions of W_{2C} and W_{2P} are given by

$$P\{W_{2C} = w\} = \frac{1}{d} \sum_{i=1}^{d} P\{W_{2q} = w - iS\},$$

$$P\{W_{2P} = w\} = P\{W_{2q} = w - dS\}.$$

Then we can derive the mean cell delay $E[W_{1C}]$ for an arbitrary cell of the bursty stream from

$$E[W_{1C}] = \frac{1}{B'(1)}\left[\left(B'(1) + \frac{d}{T}\right)E[W_C] - \frac{d}{T}E[W_{2C}]\right].$$

5 Interdeparture Time Distribution of CBR Traffic

In this section, we derive the interdeparture time distribution of CBR stream. Let the time interval $[n-1, n)$ be referred to as the n-th slot. Consider the time epoch n as the right boundary of the n-th slot and let $U(n)$ be the unfinished work at time n. We let $J(n)$ have the same interpretation as in the previous section. Then $U(n+)$ and $U(n-)$ mean the unfinished works just after time n and just prior n, respectively. We note that $U(n-)$ is equal to the waiting time of CBR packet in the queue $W_{2q}(n)$ at time n. In this section, we expend the definition of the common distribution $b_k = P\{B_n = k\}$ as follows

$$b_{i/S} = \begin{cases} 0, & i/S \neq 0, 1, 2, \cdots, \\ b_k, & k = i/S = 0, 1, 2, \cdots. \end{cases}$$

From the definition of $U(n-)$ and $b_{i/S}$, we have, for $1 \leq m \leq T$,

$$P\{U(n-) = k \mid J(n) = T, U((n-m)+) = i\} = b_{(k-i+m)/S}^{(m-1)}, \quad i \geq m,$$
$$P\{U(n-) = k \mid J(n) = T, U((n-m)+) = i\}, \quad 0 \leq i \leq m-1.$$

At the first step, we have

$$P\{U(n-) = k \mid J(n) = T, U((n-1)+) = 0\} = \delta_{k0}.$$

For $m = 2, 3, \cdots, T$, $i = 1, 2, \cdots, m-1$, and $i = 0$ we have

$$P\{U(n-) = k \mid J(n) = T, U((n-m)+) = i\}$$
$$= \sum_{l=0}^{\infty} b_l^{(i)} P\{U(n-) = k \mid J(n) = T, U((n-m+i)+) = lS\},$$
$$P\{U(n-) = k \mid J(n) = T, U((n-m)+) = 0\}$$
$$= \sum_{i=1}^{m-1}\sum_{l=1}^{\infty} P\{U(n-) = k \mid J(n) = T, U((n-m+i)+) = lS\}b_0^{i-1}b_l$$
$$+ P\{U(n-) = k \mid J(n) = T, U((n-1)+) = 0\}b_0^{m-1}.$$

Thus we have obtained the conditional probability that the unfinished work just prior n equals to k, $k = 0, 1, 2, \cdots$, given that the unfinished work just after $n - m$ equals to i, $i = 0, 1, 2, \cdots$. To find the interdeparture time distribution of CBR stream, we must find the joint probability $P\{W_{2q}(n-T) = i, W_{2q}(n) = k\}$

that the waiting times of CBR packets in the queue are i at time $n-T$ and k at time n, respectively. To find the above joint probability, we have

$$P\{W_{2q}(n)=k\,|\,W_{2q}(n-T)=i\}$$
$$=P\{U(n-)=k\,|\,W_{2q}(n-T)=i\}$$
$$=\sum_{l=0}^{\infty}b_l P\{U(n-)=k\,|\,J(n)=T, U((n-T)+)=i+(d+l)S)\}. \quad (6)$$

We denote D_P as the interdeparture time distribution of CBR packet. Then by the definition of W_{2q} and the equations (5) and (6), we have

$$P\{D_P=v\}=\sum_{\substack{i=0\\i+v-T\geq 0}}^{\infty}P\{W_{2q}(n-T)=i, W_{2q}(n)=i+v-T\}.$$

Finally, we have the interdeparture time distribution D_C of CBR cells,

$$P\{D_C=v\}=\begin{cases}\frac{d-1}{d}+\frac{1}{d}P\{D_P=dS\}, & v=S,\\ \frac{1}{d}P\{D_P=v+(d-1)S\}, & v>S.\end{cases}$$

6 Numerical Results

In this section, we present some numerical results to show the mean delay in the basis of cell and packet. We take the CBR packet of $T=15$ cell times, the service time $S=3$ cell times for segmentation, and the batch size $d=2$ of a CBR packet(a packet is segmented by 2 cells), respectively. Therefore the arrival rate of CBR traffic is $\lambda_C=2/15$ and the service rate is $1/S=1/3$. In the case of the bursty background traffic, we take $b_k=P\{B_n=k\}\sim B(15,a)$ as the batch size distribution of bursty traffic, that is, $b_k=\binom{15}{k}a^k(1-a)^{15-k}$, then the mean arrival rate of the bursty traffic becomes $\lambda_B=15a$ in cell. After all, we have the traffic intensity of $\rho=(15a+d/T)S=3(15a+2/15)$.

In Figure 2, we show the effect of traffic load on the mean delay. We denote $E[W_{1C}]$, $E[W_{2C}]$ and $E[W_C]$ as the mean cell delay of bursty traffic, that of CBR traffic and an arbitrary cell, respectively. As the traffic(in precise the bursty traffic) load increases when CBR traffic load is fixed, the mean cell delay of CBR traffic increases. This fact shows that the mean waiting time of CBR traffic depends on the background bursty traffic. In this section, we assume that CBR traffic load is fixed. Since the CBR traffic is sensitive to delay, the heavy load of bursty traffic gives a negative effect on the QoS of CBR traffic.

In Figure 3, we give the effect of traffic load on the mean delay of CBR traffic. We denote $E[W_{2P}]$ as the mean packet delay of CBR traffic. The packet waiting time W_{2P} is the sojourn time in the system until the last cell of a packet departs the system and the cell waiting time W_{2C} is the sojourn time in the system until an arbitrary cell of a packet departs the system. Since the service

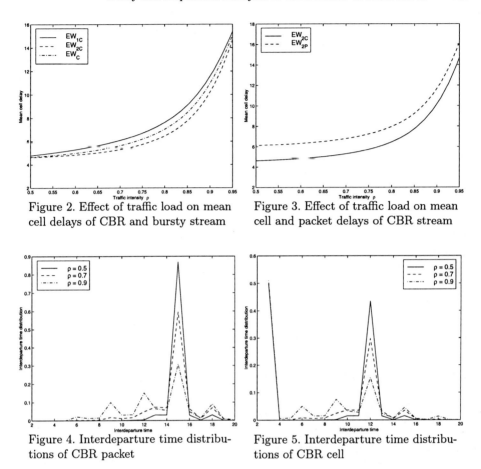

Figure 2. Effect of traffic load on mean cell delays of CBR and bursty stream

Figure 3. Effect of traffic load on mean cell and packet delays of CBR stream

Figure 4. Interdeparture time distributions of CBR packet

Figure 5. Interdeparture time distributions of CBR cell

time $S = 3$ and the batch size of a CBR packet is 2, the figure 3 shows the difference $[3 \times 2 - 3 \times (1+2)/2 = 1.5]$ between W_{2P} and W_{2C} curves.

In Figure 4, we give the interdeparture time distribution of CBR packets. We focus on the probability level, when the interdeparture time is 15. Now we remember that the mean interarrival time of CBR packets is 15. This figure shows that the probabilities decrease at $D_P = 15$, when the traffic intensity ρ increases from 0.5 to 0.9. From this fact, we can conceive that when ρ is lower, the system is more idle, and so the interarrival time holds better at the departure point.

In Figure 5, we present the interdeparture time distribution of CBR cells. We can see that the probabilities are high, when the interdeparture times are 3 and 12 in accordance with parameters $T = 15$, $d = 2$, and $S = 3$. We can also see that the probabilities decreases at the interdeparture time 12, when the traffic intensity varies from $\rho = 0.5$ to $\rho = 0.9$, and so the proportion of the idle time decreases in the system. Thus we have studied the delay parameters and interdeparture time distributions of CBR traffic.

7 Conclusion

In this paper, we model and evaluate the AAL MUX to analyze the AAL protocol and CDV of a delay sensitive traffic stream in ATM networks. We concentrate on the performance measures of both packet-level and cell-level. We model the AAL MUX with infinite buffer and two independent input streams as a $B^X + D/D/1$ queue and analyze this queueing system for illustrating the cell depart u re process of AAL MUX in order to analyze the CDV of CBR traffic. This system example is applied to the VoD service using a CBR transport stream with the background traffic stream such as IP and signaling data packets from the other connection. In this paper, we obtain cell delay time and cell interdeparture time distribution of delay sensitive CBR cell stream. Delay parameters consist of mean cell delay of an arbitrary cell and the waiting time of a deterministic CBR cell and the waiting time of a CBR packet in the system. In addition, delay parameters also contain the mean cell delay for a bursty stream. Moreover, the Cell Delay Variation tolerance value by a real time CBR stream can be seriously influenced by the AAL FIFO MUX. To illustrate this CDV of CBR stream, we obtain the cell interdeparture time distribution of a CBR stream.

References

1. R.O. Onvural, Asynchronous Transfer Mode Networks, Artech House (1995)
2. J. Roberts, U. Mocci and J. Virtao, Broadband Network Teletraffic: Final Report of Action COST 242, Springer (1996)
3. J. Roberts and F. Guillemin, Jitter in ATM networks and its impact on the peak rate enforcement, Performance Evaluation **16** (1992) 35–48
4. ATM Forum: Video on Demand Specification 1.1, SAA-AMS, March (1997)
5. G.I. Stassinopoulos, I.S. Venieris, K.P. Petropoulis and E.M. Protonotarios, Performance evaluation on adaptation functions in the ATM environment, IEEE Tans. on Comm. **42** (1994) 2335–2344
6. O. Hashida, Y.Takahashi and S. Shimogawa, Switched batch Bernoulli process SBBP. and the discrete-time SBBP/G/1 queue with application to statistical multiplexer performance, IEEE JSAC **9** (1991) 394–401
7. M.F. Neuts, Structured Stochastic Matrices of M/G/1 Type and Their Applications, Marcel Dekker, INC. (1989)
8. P. Yegani, Performance models for ATM switching of mixed continuous bit rate and bursty traffic with threshold based discarding, Proc. ICC'92 (1992) 354.3.1–3.7

Virtual Path Layout in ATM Path with Given Hop Count

Sébastien Choplin

Projet MASCOTTE
I3S-CNRS/INRIA/Université De Nice-Sophia Antipolis,
BP 93, F-06902 Sophia Antipolis Cedex, France.
Tel : (+33) (0) 4 92 38 78 80
Fax : (+33) (0) 4 92 38 79 71
Sebastien.Choplin@sophia.inria.fr

Abstract. Motivated by Asynchronous Transfer Mode (ATM) in telecommunication networks, we investigate the problem of designing a virtual topology on a physical topology, which consists of finding a set of *virtual paths (VPs)* satisfying some constraints in terms of *load* (the number of VPs sharing a physical link) and *hop count* (the number of VPs used to establish a connection). For particular network: paths, we give tight bounds on the *network capacity* (the maximum load of a physical link) as a function of the *virtual diameter* (the maximum hop count for each connection).

1 Introduction

The advent of fiber optic media has dramatically changed the classical views on the role and structure of digital communication networks. Specifically, the sharp distinction between telephone networks, cable television networks, and computer networks, has been replaced by a unified approach.

One of the most prevalent solutions for this new network challenge is called *Asynchronous Transfer Mode* (ATM for short), and is thoroughly described in the literature [KG98,Pr95]. The transfer of data in ATM is based on packets of fixed length, called *cells*. Each cell is routed independently, based on two routing fields at the cell header, called *virtual channel identifier* (VCI) and *virtual path identifier* (VPI). This method effectively creates two types of predetermined simple routes in the network, namely routes which are based on VPIs (called *virtual paths* or VPs) and routes based on VCIs and VPIs (called *virtual channels* or VCs). VCs are used for connecting network users (e.g., a telephone call); VPs are used for simplifying network management - routing of VCs in particular. Thus the route of a VC may be viewed as a concatenation of complete VPs. A major problem in this framework consists in defining the set of VPs in such a way that some good properties are achieved.

A capacity (or bandwidth) is assigned to each VP. The sum of the capacities of the VPs that share a physical link constitutes the *load* of this link. Naturally, this load must not exceed the link capacity, i.e., the amount of data it can carry.

The maximum load of the links (called the load of the physical network for this set of VPs) is a major component in the cost of the network, and should be kept as low as possible.

The maximum number of VPs in a virtual channel, called *hop count* in the literature, should also be kept as low as possible so as to guarantee low set up times for the virtual channels and high data transfer rates. In its most general formulation, the *Virtual Path Layout (VPL)* problem is an optimization problem in which, given a certain communication demand between pairs of nodes and constraints on the maximum load and hop count, it is first required to design a system of VPs satisfying the constraints and then minimizing some given function of the load and hop count.

We employ a restricted model similar to the one presented in [GZ94]. In particular, we assume that all VPs have equal capacities, normalized to 1. Hence the load of a physical link is simply the number of VPs sharing this link and we don't focus on the number of VCs contained in a VP. Although links based on optical fibers and cables are directed, traditional research uses an undirected model. Indeed, this model imposes the requirement that if there exists a VP from u to v then there exists also a VP from v to u. In fact, that is the way ATM networks are implemented at the present time. Therefore, we use an undirected model. The directed model has been studied in [BMPP98].

We focus on the *all-to-all problem* (all pairs of nodes are equally likely to communicate). Thus, the resulting maximum hop count can be viewed as the *diameter* of the graph induced by the set of VPs. More formally, given a communication network, the VPs form a virtual graph on the top of the physical one, with the same set of vertices but with a different set of edges. Specifically, a VP between u and v is represented by an edge between u and v in the virtual graph. This virtual graph provides a VPL for the physical graph. Each VC can be viewed as a simple path in the virtual graph. Therefore, a central problem is to find a tradeoff between the maximum load and the diameter of the virtual graph.

Here we consider the following restricted problem: the physical graph will be a path and given the diameter of the virtual graph we want to minimize the maximum load of a VPL on the path. The figure 1 is an example of VPL on the path with hop count 2 and maximum load 4. For example, the VP between the node 0 and the node 4 uses the physical links 0-1, 1-2, 2-3 and 3-4. The VC used for the request between the nodes 0 and 6 is formed by the VPs 0-4 and 4-6, with an hop count of 2. The maximum load reached on the physical links 3-4 and 4-5 is equal to 4.

2 Related Work

Many articles deal with the VPL problem defined above. Most of them consider the minimization of the diameter of the virtual graph given a load (or capacity) of the physical network, see [SV96,GZ94,GWZ95a,SV96,GCZ96a,KKP97,EFZ97] for the undirected case and [BMPP98] for the directed case. The dual problem

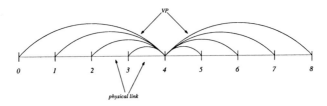

Fig. 1. Example of VPL on the path with hop count 2 and maximum load 4

we consider here has been less studied. Several bounds can be derived from the preceding problems but they are not tight ; a lower bound for planar graphs is given in [BG97]. The particular case of path with the one to many request has been studied in [FZ97,GWZ95a]. In that case, one wants to minimize the eccentricity of the sender in the virtual graph rather than the diameter. This work is sometimes used to give bound on the minimum diameter since it is at most twice the minimum eccentricity. The reader can find an excellent survey of results of the undirected model in [Zak97]. A related problem has been studied in [ABC+97].

3 The Model

We use the model described in [BMPP99,Cha98,CGZ96b]. The physical network is represented by an undirected graph $G = (V, E)$ with V the set of nodes of the network and E the set of physical links between them. Usually, we have a family of request $R \subset V \times V$. We are interested in a specific one: the *all-to-all* case (Gossipping). In this case a connection is required between all pairs of vertices; namely, R is formed by all $\binom{n}{2}$ couples of distinct elements of V. Therefore the maximal hop count corresponds to the diameter of the virtual graph. A VPL is a collection of simple paths in the network G, called VP.

Definition 1. *A VPL(H,P) is defined by a virtual graph $H = (V, E')$ with the same vertices as G and a routing function P which associates a path $P(e')$ in G to each edge e' in E' .*

Definition 2. *Given a VPL(H,P), the load $l(e)$ of an edge $e \in E$ of the physical graph is the number of VP's that include e; namely, $l(e) = |\{e' \in E' | e \in P(e')\}|$.*

Definition 3. *Given a VPL(H,P), the maximal edge load $\pi(G, H, P) = \max_{e \in E} l(e)$.*

Given a graph H, we are looking for a routing function P which minimizes $\pi(G, H, P)$.

Definition 4. *The minimal load of G for H is $\pi(G, H) = \min_P \pi(G, H, P)$.*

The problem stated in the introduction can be formulated as follow:

A.2 Upper Bound for $n_{\mathcal{AA}}(\pi, h)$

Remark 12 (interval). Given a virtual graph H of diameter h on P_n, there exists an integer k and k vertices $a_1 < a_2 < \ldots < a_k$ such that

- $\forall 0 \leq i \leq k$: all the vertices in $[a_i + 1, a_{i+1} - 1]$ are outward vertices.
- $\forall 0 \leq i \leq k-1$ a vertex in $[a_i + 1, a_{i+1}]$ is not an outward vertex.

(with $a_0 = -1$ and $a_{k+1} = n$). We will denote by J_i the interval $[a_i + 1, a_{i+1}]$ and I_i the outward interval $[a_i + 1, a_{i+1} - 1]$.

Now we call the intervals J_i *blocks of level 1* of the topology H. We note that the graph of the interval $\{J_i\}_{1 \leq i \leq k-1}$ (or blocks) of level 1 of H is also a virtual topology on the path but with k vertices; we note it H_1. The diameter of this topology is at most $h - 2$ because each of the interval has at least one non-outward vertex. In an iterative way, we define some blocks of blocks. The blocks from H_1 are called *blocks of level 2*. Formally:

- The *blocks of level l* from a virtual graph H of diameter h are the blocks of level 1 from the topology H_{l-1}.
- The topology H_l is the graph formed by the blocks from H_{l-1}.
- The vertices from P_n are the blocks of level 0.
- $H_0 = H$.

Proposition 13. *For $i \leq \frac{h}{2} - 1$ the diameter of the topology H_i is at most $h - 2i$.*

Proof. Directly by induction.

Definition 14. *The* available load *of an interval $I \in \{I_i\}_i$ ($\forall i, I_i$ is an interval and $\cup_i I_i = V$) is $\pi - \pi_I$ where π_I is the load induced by the edges between I and all the other intervals.*

Proposition 15. *The cardinal of an outward interval with an available load l is at most $2l$.*

Proof. If the interval has k vertices, k edges load the 2 extreme edges of the interval. Then $k \leq 2l$.

Definition 16. *Let denote by $B(l, c)$ the maximum number of vertices in a block of level l with an available load c.*

Proposition 17. *for $h = 2p + 1$ odd, $n_{\mathcal{AA}}(\pi, h) \leq \overline{n_{\mathcal{AA}}}(\pi, h)$ with*
$$\overline{n_{\mathcal{AA}}}(\pi, h) = \max_{k \leq 2\sqrt{\pi}+1} \left\{ \sum_{i=0}^{k-1} B(p, \pi - i(k - i - 1)) + B(p, \pi) \right\}$$

Proof. Let H be a virtual graph of diameter $h = 2p+1$ and load π on the path P_n. The graph H_p of the blocks of level p is the complete graph according to the Proposition 13. Let us fix $k = |V(H_p)|$. Now we consider the blocks of level p $\{J_i\}_{0 \leq i \leq k-1}$; they induce a complete graph with a load $i(k-i-1)^1$ on all the physical edges of the block J_i because of the lemma 8. Then the available load of the block J_i is at most $\pi - i(k-i-1)$. So for $0 \leq i \leq k-1$, $|J_i| \leq B(p, \pi - i(k-i-1))$. To finish we have to consider the interval J_k; this interval is an outward one and so its cardinal is lower than the one of the block of level p then $|J_k| \leq B(p, \pi)$. Then, $n_{AA}(\pi, 2) \leq \left(\sum_{i=0}^{k-1} B(p, \pi - i(k-i-1))\right) + B(p, \pi)$.

Remark 18. We can consider that $k \leq 2\sqrt{\pi} + 1$, because if $k > 2\sqrt{\pi} + 1$ then it exists i so that $\pi - i(k-i-1) < 0$, that's a nonsense.

Proposition 19. *for $h = 2p$ even, $n_{AA}(\pi, h) \leq \overline{n_{AA}}(\pi, h)$ with*
$$\overline{n_{AA}}(\pi, h) = \max_{k \leq 2\pi} \left\{ \left(2 \sum_{i=0}^{\lfloor \frac{k-1}{2} \rfloor} B(p-1, \pi - i)\right) + B(p-1, \pi) \right\}$$

Proof. Let H be a virtual graph of diameter $h = 2p$ and load π on the path P_n. The graph H_p of the blocks of level p is a graph of diameter 2 according to the Proposition 13. We fix $k = |V(H_{p-1})|$ and now we consider the blocks of level $p-1$ $\{J_i\}_{0 \leq i \leq k-1}$; they induce a graph of diameter 2 with a load $\min\{i, n-i-1\}^2$ on all the physical edges of the block J_i because of the lemma 10. Then the available load of the block J_i is at most $\pi - \min\{i, k-i-1\}$. So for $0 \leq i \leq k-1$, $|J_i| \leq B(p, \pi - \min\{i, k-i-1\})$. To finish we have to consider the interval J_k; this interval is an outward one and so its cardinal is lower than the one of the block of level p ; therefore $|J_k| \leq B(p, \pi)$. Then, we obtain the result because: $\min\{i, k-i-1\} = i$ if $i \leq \frac{k-1}{2}$ and $\min\{i, k-i-1\} = k-i-1$ otherwise.

A.3 Lower Bound for $n_{AA}(\pi, h)$

The techniques employed in the proof of the lower bound tell us that the good virtual topologies with diameter $2p+1$ may be constructed with blocks of level p linked together in a complete graph. We can note that the blocks of level p are sets of vertices which we can get out of in p jumps exactly. Then we can choose as blocks of level p some maximal trees with eccentricity p, the same as the one described by Gerstel, Wool and Zaks in [GWZ95b]. An equivalent proof can be given for the case h is even.

Proposition 20. *for all $h = 2p+1$ odd, for all k, $\left(\frac{k}{2}\right)^2 \leq \pi$ we have: $n_{AA}(h, \pi) \geq \underline{n_{AA}}(\pi, h)$ with $\underline{n_{AA}}(\pi, h) = \max_{k \leq 2\sqrt{\pi}} \left\{ \sum_{i=0}^{k-1} ball(p, \pi - i(k-i)) \right\}$ with $ball(p, c) = \sum_{i=0}^{\min\{p,c\}} 2^i \binom{l}{i}\binom{r}{i}$ the number of internal points in the p-dimensional Sphere of radius c.*

---
[1] $k-i-1$ because we don't count the load induced by J_i into the complete graph.
[2] as in the Proposition 17 we don't count the load induced by J_i.

Proof. We obtain the construction as follows: we choose k vertices $a_0, a_1, \ldots, a_{k-1}$ on the path and we link them with a complete graph, then we root 2 trees in each vertex (one on the right and one on the left) which use all of the available load and with eccentricity p as shown by the figure 3. We note $n(p, c)$ the maximal number of vertices that a tree of available load c and eccentricity p can contain. Then we have the following recurrence relation: $n(p, c) = n(p, c-1) + n(p-1, c-1) + n(p-1, c) = \frac{1}{2} \text{ball}(p, c) + \frac{1}{2}$, with $n(1, c) = c$ and $n(p, 0) = 1$ (There are the trees used in [GWZ95b] for the *One-To-All*-case).

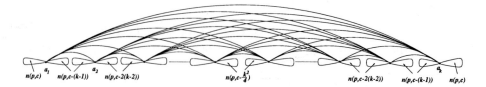

Fig. 3. Optimal construction on the path for h odd.

The available load: on the left of a_0 is π, between a_{i-1} and a_i is $\pi - i(k-i)$ for $0 \leq i \leq k$, on the right of a_k is π. Then we can root: on the left of a_0 a tree of eccentricity p with a load π which contains $n(p, \pi)$ vertices, for $0 \leq i \leq k$, between a_i and a_{i+1} 2 trees of eccentricity p with a load $\pi - i(k-i)$ which contain $2n(p, \pi - i(k-i))$ on the right of a_k: a tree of eccentricity p with a load π which contains $n(p, \pi)$ vertices. So the total number of vertices is $2 \sum_{i=0}^{k-1} n(p, \pi - i(k-i)) - k$.

Remark 21. A similar construction can be realized in the case h is even ($h = 2p$), then the graph of blocks of level $p - 1$ is isomorph to the graph of diameter 2 used in the case $h = 2$. Each vertex of this graph is taken over from an optimal tree of eccentricity $p - 1$ and a maximal available load.

A.4 Determination of $n_{\mathcal{A}\mathcal{A}}(\pi, h)$

$\overline{n_{\mathcal{A}\mathcal{A}}}(\pi, h)$: **Calculation of the Upper Bound of $n_{\mathcal{A}\mathcal{A}}(\pi, h)$.** We proceed as follows: the available load of a block of level l is c, then we determine the maximal number of vertices $B(l, c)$ that it can contain according to $B(l-1, i)$. This relation is given by the following recursion equation:

Lemma 22. $B(l, c) = 2(B(l-1, 1) + B(l-1, 2) + \cdots + B(l-1, c)) + B(l-1, c)$

Proof. A block of level l with an available c is formed with some blocks of level $l - 1$ and all (except one) have an outward vertex. We can suppose without lost of generality that all the edges don't cross each other (Gerstel, Wool and Zaks [GWZ95b]). Then, a block of level l with an available load c is formed with at most: 2 blocks of level $l - 1$ with an available load equal to 1, 2 blocks of level $l-1$ with an available load equal to 2, ..., 2 blocks of level $l-1$ with an available load equal to c. Now we have to add a block of level $l - 1$ with an available load equal to at most c (this is the block which is not outward), as we can see in the figure 4.

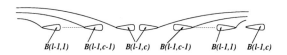

Fig. 4. Composition of a block of level l with an available load c.

In order to determine $B(l,c)$, we introduce the generating function: $\mathcal{B}(l,z) = \sum_{i=0}^{\infty} B(l,i)z^i$. The blocks of level 0 are reduced to vertices, $B(0,c) = 1$ and then $\mathcal{B}(0,z) = \frac{1}{1-z}$. Then we have, $B(l,c) = \frac{2^l c^l}{l!} + o(c^l)$.

Proposition 23. *for all* $h = 2p+1$ *odd*,
$$\overline{n_{\mathcal{A}\mathcal{A}}}(\pi,h) = 2^p \pi^{h/2} \sup_{0 \le \alpha \le 2} \left\{ \sum_{j=0}^{p} \frac{\alpha j!(-\alpha^2)^j}{(p-j)!(2j+1)!} \right\} + o(\pi^{h/2})$$

Proof. according to Lemma 22: $\overline{n_{\mathcal{A}\mathcal{A}}}(\pi,h) = \frac{2^p}{p!} \max_{k \le 2\sqrt{\pi}+1} \left\{ \sum_{i=0}^{k} (\pi - i(k-i))^p \right\}$
$+ o(\pi^{h/2})$ because of $\sum_{i=0}^{k}(\pi - i(k-i))^p = \sum_{j=0}^{p} \binom{p}{j} \pi^{p-j}(-1)^j \sum_{i=0}^{k} i^j(k-i)^j$
and we can approach $\sum_{i=0}^{k} i^j(k-i)^j$ using the *Euler-Mac Laurin*'s summation formula, in $\frac{(j!)^2}{(2j+1)!}k^{2j+1} + o(k^{2j+1})$. So we have:
$$\overline{n_{\mathcal{A}\mathcal{A}}}(\pi,h) = 2^p \pi^{p+1/2} \max_{0 \le \alpha \le 2+O(n^{1/h})} \alpha \sum_{j=0}^{p} \frac{j!}{(p-j)!(2j+1)!}(-\alpha^2)^j + o(\pi^{h/2})$$

Proposition 24. *for all* $h = 2p$ *even*, $\overline{n_{\mathcal{A}\mathcal{A}}}(\pi,h) = \frac{2^p \pi^p}{p!} + o(\pi^{h/2})$

Proof. $\overline{n_{\mathcal{A}\mathcal{A}}}(\pi,h) = \max_{k \le 2\pi} \left\{ \left(2 \sum_{i=0}^{\lfloor \frac{k-1}{2} \rfloor - 1} B(p-1, \pi-i) \right) + B(p-1,\pi) \right\}$
$\overline{n_{\mathcal{A}\mathcal{A}}}(\pi,h) = \frac{2^p}{(p-1)!} \max_{k \le 2\pi} \left\{ \sum_{i=0}^{\lfloor \frac{k-1}{2} \rfloor - 1} (\pi - i)^{p-1} \right\} + o(\pi^p)$

Clearly, this function of k reaches its maximum when k is maximum, i.e. $k = 2\pi$. Using the integral calculation, we obtain: $\overline{n_{\mathcal{A}\mathcal{A}}}(\pi,h) = \frac{2^p \pi^p}{p!} + o(\pi^p)$.

$\underline{n_{\mathcal{A}\mathcal{A}}}(\pi,h)$: Calculation of the Lower Bound of $n_{\mathcal{A}\mathcal{A}}(\pi,h)$.

Proposition 25. *for all* h, $\underline{n_{\mathcal{A}\mathcal{A}}}(\pi,h) = \overline{n_{\mathcal{A}\mathcal{A}}}(\pi,h) + o(\pi^{h/2})$.

Proof. Note that $ball(p,c) = \sum_{i=0}^{\min(p,c)} 2^i \binom{l}{i}\binom{c}{i} = \frac{2^p c^p}{p!} + o(c^p)$ if $p \ll c$.
In the odd case: $h = 2p+1$, with the Proposition 20, we can say that:
$\underline{n_{\mathcal{A}\mathcal{A}}}(\pi,h) = \frac{2^p}{p!} \max_{k \le 2\sqrt{\pi}} \left\{ \sum_{i=0}^{k}(\pi - i(k-i))^p \right\} + o(\pi^{h/2})$. Then $\underline{n_{\mathcal{A}\mathcal{A}}}(\pi,h) = \overline{n_{\mathcal{A}\mathcal{A}}}(\pi,h) + o(\pi^{h/2})$. The same is true in the even case.

Corollary 26. $n_{\mathcal{AA}}(\pi,h) = \frac{2^p \pi^p}{p!} + o(\pi^{h/2})$ for $h = 2p$

$n_{\mathcal{AA}}(\pi,h) = 2^p \pi^{(2p+1)/2} \sup\limits_{0 \leq \alpha \leq 2}\left\{ \alpha \sum_{j=0}^p \frac{j!(-\alpha^2)^j}{(p-j)!(2j+1)!} \right\} + o(\pi^{h/2})$ for $h = 2p+1$

Proof. $\underline{n_{\mathcal{AA}}}(\pi,h) \leq n_{\mathcal{AA}}(\pi,h) \leq \overline{n_{\mathcal{AA}}}(\pi,h)$ and $\underline{n_{\mathcal{AA}}}(\pi,h) = \overline{n_{\mathcal{AA}}}(\pi,h) + o(\pi^{h/2})$ then $n_{\mathcal{AA}}(\pi,h) = \underline{n_{\mathcal{AA}}}(\pi,h) + o(\pi^{h/2}) = \overline{n_{\mathcal{AA}}}(\pi,h) + o(\pi^{h/2})$.

Theorem 27. $\pi(P_n, h) = \frac{(p!)^{1/p}}{2} n^{2/h} + o(n^{2/h})$ for $h = 2p$

$\pi(P_n, h) = \frac{2^{1/h}}{2(M_p)^{2/h}} n^{2/h} + o(n^{2/h})$ for $h = 2p+1$ with

$M_p = \sup\limits_{0 \leq \alpha \leq 2}\left\{ \alpha \sum_{j=0}^p \frac{j!(-\alpha^2)^j}{(p-j)!(2j+1)!} \right\}$

References

[ABC+97] W. Aiello, S. N. Bhatt, F. R. K. Chung, A. L. Rosenber, and R. K. Sitaraman. Augmented ring networks. Technical Report UM-CS-1997-036, University of Massachusetts, Amherst, Computer Science, June, 1997.

[BG97] Becchetti and Gaibisso. Lower bounds for the virtual path layout problem in ATM networks. In *Theory and Practice of Informatics, Seminar on Current Trends in Theory and Practice of Informatics, LNCS*, volume 24. 1997.

[BMPP98] J-C. Bermond, N. Marlin, D. Peleg, and S. Perennes. Directed Virtual Path layout in ATM networks. In *Proc. of the 12th International Conference on Distributed Computing, Andros Greece*, 1998. LNCS 1499, pages 75-88.

[BMPP99] J-C. Bermond, N. Marlin, D. Peleg, and S. Perennes. Virtual Paths Layout with Low Congestion or Low Diameter in ATM Networks. In 1ère rencontre francophone... ALGOTEL '99, Mai 1999.

[Cha98] Pascal Chanas. *Dimensionnement de réseaux ATM*. PhD thesis, CNET Sophia, Sept. 1998.

[EFZ97] T. Eilam, M. Flammini, and S. Zaks. A complete characterization of the path layout construction problem for ATM networks with given hop count and load. In *24th International Colloquium on Automata, Languages and Programming (ICALP)*, volume 1256 of *Lecture Notes in Computer Science*, pages 527-537. Springer-Verlag, 1997.

[FZ97] M. Feighlstein and S. Zaks. Duality in chain ATM virtual path layouts. In *4th International Colloquium on Structural Information and Communication Complexity (SIROCCO)*, Monte Verita, Ascona, Switzerland, July 1997.

[Gau95] G. Gauyacq. *Routages uniformes dans les graphes sommet-transitifs*. PhD thesis, Université Bordeaux I, 1995.

[GCZ96a] O. Gerstel, I. Cidon, and S. Zaks. The layout of virtual paths in ATM networks. *IEEE/ACM Transactions on Networking*, 4(6):873-884, 1996.

[GCZ96b] Ornan Gerstel, Israel Cidon, and Shmuel Zaks. Optimal Virtual Path Layout in ATM Networks with Shared Routing Table Switches. *Chicago Journal of Theoretical Computer Science*, 1996(3), October 1996.

[GWZ95a] O. Gerstel, A. Wool, and S. Zaks. Optimal Layouts on a Chain ATM Network. In *3rd Annual European Symposium on Algorithms*, volume LNCS 979, pages 508–522. Springer Verlag, 1995.

[GWZ95b] O. Gerstel, A. Wool, and S. Zaks. Optimal layouts on a chain ATM network. In *3rd Annual European Symposium on Algorithms (ESA), (LNCS 979), Corfu, Greece*, pages 508–522, 1995. To appear in *Discrete Applied Mathematics*.

[GZ94] Ornan Gerstel and Shmuel Zaks. The virtual path layout problem in fast networks. In *Symposium on Principles of Distributed Computing (PODC '94)*, pages 235–243, New York, USA, August 1994. ACM Press.

[KG98] D. Kofman and M. Gagnaire. Réseaux haut débit, réseaux ATM, réseaux locaux et réseaux tout-optiques. *InterEditions-Masson*, 1998.

[KKP97] E. Kranakis, D. Krizanc, and A. Pelc. Hop-congestion trade-offs for high-speed networks. *International Journal of Foundations of Computer Science*, 8:117–126, 1997.

[MHS89] J-C. Meyer, M-C. Heydemann, and D. Sotteau. On Forwarding indices of networks. *Discrete Applied Mathematics 23:103-123*, 1989.

[Pri95] M. De Pricker. Asynchronous Transfer Mode, Solution for Broadband ISDN. *Prentice Hall*, August 1995.

[SV96] L. Stacho and I. Vrt'o. Virtual path layouts in atm networks. In *Structure, Information and Communication Complexity, 3rd Colloquium, SIROCCO*, pages 269–278. Carleton University Press, 1996.

[Zak97] S. Zaks. Path Layout in ATM Networks - A Survey. In *The DIMACS Workshop on Networks in Distributed Computing, DIMACS Center, Rutgers University*, Oct. 1997. manuscript.

Simulation-Based Stability of a Representative, Large-Scale ATM Network for a Distributed Call Processing Architecture

Ricardo Citro[1] and Sumit Ghosh[2]

[1] Intel Corporation, Chandler, Arizona
ricardo.citro@intel.com
[2] Secure Network Design Laboratory,
Department of Electrical & Computer Engineering
Stevens Institute of Technology, Hoboken, NJ 07030
sghosh2@attila.stevens-tech.edu

Abstract. In response to the lack of realistic performance guarantees in "store and forward" networks, the asynchronous transfer mode (ATM) network has resurrected the idea of connection oriented calls. Thus, for every user call, the ATM network first determines and verifies a route, where possible, that meets the user's quality of service (QoS) requirements, and then transports the traffic along the route. In classic telephony, a typical call processing event P_e requires 200 ms while an average call duration C_d lasts 90 s. The ratio $P_e/C_d = 450$, is large enough to render the call processing effort meaningful. While the link bandwidth in classic telephony is 64 Kb/s and each user call is allocated an identical 8 Kb/s bandwidth, links bandwidths in ATM equal or exceed 155 Mb/s and users may be allocated, in general, much higher, variable bandwidths depending on need and availability. This paper projects that the trend of short call durations, of ftp, http, image retrieval, and email messages type, will gain increasing processing prominence in the future, fueled by our impatience and the rapid growth in the backbone link bandwidths that is already outpacing the expected increase in the average message size transported across the networks. As backbone speeds increase and call durations decrease, networks will be perceived as more responsive and potentially attract a very high rate of users interacting with the networks. Under these circumstances, it is unlikely for the traditional uniprocessor based ATM node architecture to successfully address the increased call processing demand, resulting in overall performance degradation. Therefore, new call processing architectures for ATM node are necessary. An efficient distributed call processor architecture has been presented in [1] to accelerate the call setup process. Since the architecture is designed to address a highly stressed network, just short of driving it into "instability," it is imperative that while preserving the quality of service (QoS) guarantees, the distributed call processing architecture prevent perturbations on the network during peek traffic hours. This need translates into the stability of the network in the presence of user call requests and QoS metrics. Network stability is viewed as a condition that must be satisfied so that a large scale decentralized representative high speed network will consistently perform without degradation in the presence of unexpected perturbation. This paper presents the stability study for a representative, large-scale, Private Network Network Interface (PNNI) ATM network under the distributed call processing introduced in [1].

1 Introduction

The issue of stability in large-scale distributed systems was first introduced in [2] and subsequently detailed in [3] [4]. This paper is the first to report on the stability analysis of a large-scale PNNI network under distributed call processing architecture. Most previous published studies have relied on mathematical queueing theory to model a single processor node network [5], or a source-destination paradigm neglecting important information that may be inhibited on intermediate hops for a large real scale network. In this study, network stability is a condition that must be satisfied so that a large scale decentralized representative high speed network will consistently perform without degradation in the presence of unexpected perturbation. This perturbation may be characterized by a sudden or an abrupt increase in traffic intensity at any time. This intensity may be conditioned to burstiness caused by peak hours of network utilization or by increased congestion experienced in the intermediate paths of the network. Here, congestion is directly related to resource utilization which translates to long call setup times.

2 Modeling the Distributed Call Processing

The call processing architecture is modeled for a 15-node and 50-node network derived from the very high performance Backbone Network Service (vBNS) topology, which is the representative high performance network service implemented by MCI [6] and sponsored by the National Science Foundation (NSF). By using the vBNS high-level topology, a 15-node and a 50-node representative PNNI networks are constructed of which the 15-node network is shown in Figure 1. On the top of Figure 1, a high level hierarchy is constructed with higher-level nodes corresponding to major US cities, forming intergroups. In the middle of Figure 1, lower-level nodes are constructed and organized into peer groups. Each peer group constitutes a higher-level node and can consist of N nodes having one or more incoming and outgoing links. When looking at the 15-node network in Figure 1, it is clearly noticeable that each peer group has three nodes forming intragroups with the shaded node assigned as a peer group leader. In an intragroup, each node corresponds to a city district surrounding the major city with neighboring cities not distant more than 40 miles from each other. The 50-node low level hierarchy, not presented here, is formed by 9 peer groups with most of the peer groups containing 6 nodes, except for the two left edges that contain 4 nodes each. Here, intragroup nodes are also labeled to a city in the vicinity region with district cities apart from each other no more than 40 miles. Associated with every link is the value of the propagation delay and distance in miles. The propagation delay is computed from dividing the actual distance between the corresponding cities by the speed of the electromagnetic transmission in optical fibers. Also, associated with every link is its maximum capacity which is the basic 155.52 Mbits/sec. Hence, a node consists of two components: the processing system or call processor and the switching fabric. The call processor is slower but contains all the higher level information which in turn is more powerful and sophisticated. The switch fabric is a fast hardware but with a more limited functionality. Traffic cells which need to be multiplexed through the node are handled by the switch fabric without any involvement of the processing system. On the other hand, all signaling control packets associated with call setup are processed by the call processor. The signaling control involves call admission control, routing, QoS management, virtual path/virtual channel connection and permanent virtual path/virtual channel connection.

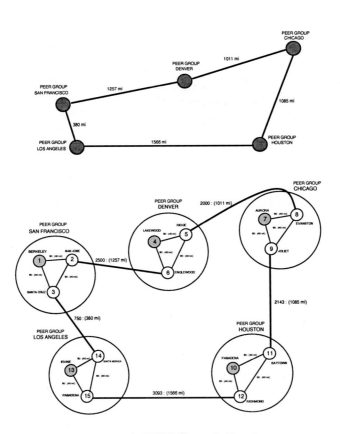

Fig. 1. 15-node PNNI Network Topology

2.1 Modeling Memory Contention and Processor Service Time

In the shared memory multiprocessor model, memory contention may occur when more than one processor has to wait in order to reference data in memory. In this model, memory contention is a function of three parameters: namely, (1) the lock operation (accessing the critical section), (2) the spin-lock time and (3) the memory access time. Hence, the model for memory contention can be expressed as

$$C_T(N) = \sum_N (M_A^i + L_A^i) + \sum_N P_L^i \qquad (1)$$

where C_T is the contention time for N number of processors; M_A is the memory access time for processor N; L_A is the memory access time to lock structure, and P_L is the processor spin-lock time.

The memory access time is calculated as follows. It is well known the difficulty in estimating the memory access time if no representative real software is available in source code in order to count the number of read/write accesses. For this reason, an accurate software PNNI-ATM network simulator is implemented which is closely consistent with the PNNI signaling protocol. Therefore, a more precise estimation on the number of memory accesses is quantified when building the network topology state information. This value is obtained by counting the number of accesses during an initiate call setup. Since the simulator executes on a Pentium 90 Mhz, the processor clock cycle time is 11 ns. The memory speed used in the Pentium workstation is 60 ns. The topology database for each call is built dynamically, creating routing map for intragroup and for intergroup. Intragroup is a collection of nodes within a peer group, and intergroup is a collection of border nodes connecting peer groups. The number of memory accesses for an intragroup topology is counted to be approximately equal to 280,000 and for intergroup to be approximately equal to 600,000. The number of memory access obtained for intragroup and intergroup are an overestimate. The actual number is probably much lower. Therefore, the actual results are likely to be much better than the data reported here. Now, it is possible to find the memory access time for one processor, which is given by

$$M_A = \eta_a * \epsilon \qquad (2)$$

where η_a is the number of memory access and ϵ is the memory speed that here is a constant value. Therefore, for intragroup and intergroup the time a processor takes to access the shared memory when creating the routing path is 16.8 ms and 36 ms respectively. Here, the time to access the memory is modeled by a uniform distribution between 16 ms - 36 ms.

In this research, the processor service time is derived from actual measurement obtained from simulation. Here, the processor service time is represented by an exponential distribution with mean $\lambda = 43$ ms. This value is found as follows.

more calls are now processed by a higher number of processors, the call setup will tend to worsen. That's why it is done all the stability study for 1-processor, 4-, 10-, 15-, and 20-processors. The behavior of the call setup time is not one of continual increase.

Table 1. Call Traffic Parameters - Scenario 1

Service	Transmission Rate (bps)	Session Duration (sec)
Voice and Audio	64K	90
File Transfer	1M-6M	1

Table 2. Call Traffic Parameters - Scenario 2

Service	Transmission Rate (bps)	Session Duration (sec)
Voice and Audio	64K	90
File Transfer	1M-6M	0.054

The distribution generated using the traffic parameters for a 15-node network indeed produces a stable network. Figures 2 (a), (b), and (c) show the call setup time as a function of simulation time for the peer group leader node 1, for the border node 8 and for the peer group leader node 10 with all the nodes utilizing 1 call processor model. These are representative nodes since they are involved in processing signaling cells at different points in the network and, therefore, offering insights on the nature of the traffic of calls that are propagated from nodes 1, 8, and 10 to nodes 2, 9, and 11 during simulation time. By analyzing Figure 2, it can be inferred that the network shows a stable rate of assertion of user call setup requests characterized by the uniform fluctuation of the points in the graph. Figure 3 (a)-(c), and Figure 4 (a)-(c) all show the call setup time as a function of simulation time relative to the same nodes but for a different number of call processors. The sharp peaks presented in the Figure are related to calls that are addressed to nodes that belong to other peer group boundaries, which then present a higher call setup time. The gaps between consecutive calls are due to the high percentage of calls asserted that are rejected. Therefore, not only all the plots show a uniform behavior that evidence the network stability, but also they show that the network stability improves with the increase in the number of processors for this particular traffic.

For a 50-node network, the traffic parameters are utilized and the distribution generated satisfies the criteria for network stability. The distribution generated produces a stable network as shown in Figure 5 (a)-(c). The Figure shows the call setup time, in timesteps, as a function of simulation time for nodes 5, 20, and 45, each utilizing 1 processor. The plots show all the calls that are originated from nodes 5, 20, and 45, which then are propagated to the adjacent nodes 4, 15, and 47. The plots also capture the stable rate of assertion of user call setup

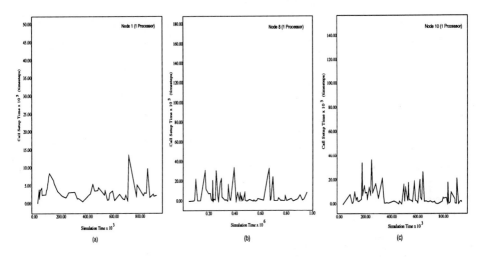

Fig. 2. Call Setup Time as a Function of Simulation Time (timesteps) for (a) source node 1 to destination node 2 (1 processor), (b) source node 8 to destination node 9 (1 processor) and (c) source node 10 to destination node 12 (1 processor) (relative to a 15-node network)

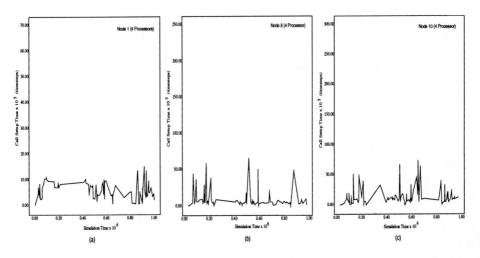

Fig. 3. Call Setup Time as a Function of Simulation Time (timesteps) for (a) source node 1 to destination node 2 (4 processor), (b) source node 8 to destination node 9 (4 processor) and (c) source node 10 to destination node 12 (4 processor) (relative to a 15-node network)

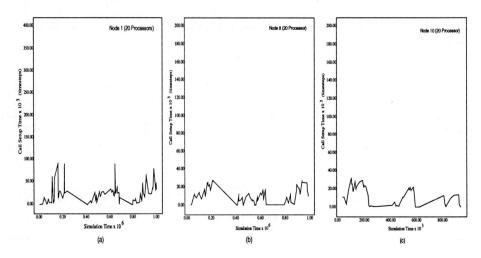

Fig. 4. Call Setup Time as a Function of Simulation Time (timesteps) for (a) source node 1 to destination node 2 (20 processor), (b) source node 8 to destination node 9 (20 processor) and (c) source node 10 to destination node 12 (20 processor) (relative to a 15-node network)

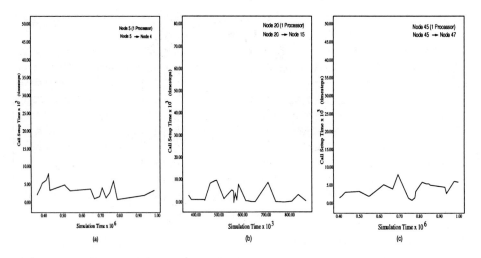

Fig. 5. Call Setup Time as a Function of Simulation Time for (a) source node 5 to destination node 4 (1 processor), (b) source node 20 to destination node 15 (1 processor) and (c) source node 45 to destination node 47 (1 processor) (relative to a 50-node network)

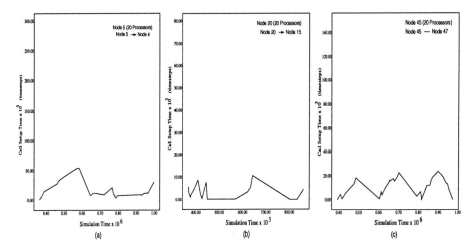

Fig. 6. Call Setup Time as a Function of Simulation Time for (a) source node 5 to a destination node 4 (20 processor), (b) source node 20 to a destination node 15 (20 processor) and (c) source node 45 to a destination node 47 (20 processor) (relative to 50-node network)

requests. It may be noticed that call setup time stays mostly uniform during simulation time which characterizes a stable network.

The network stability can be even further investigated by analyzing Figure Figure 6 (a)-(c) which shows the rate of assertions of user call setup with respect to the progress of the network operation for switches 5, 20, and 45, using 20-processors. The high peaks in a few of the graphs describe the longer call setup times that some of the 20% of intergroup calls experience. It may be observed that the behavior of the call setup time, for all involved calls, during simulation time are not one of the continual increase which characterizes a stable network for this traffic.

4 Conclusion

This paper presents the stability study for the distributed call processing described in [1] for a real high-speed high-scale Private-Network-Network-Interface (PNNI) ATM network [7]. The Stability study is performed for nodes using 1-, 4-, 10-, 15-, and 20-processors. the study reported here is unique in that no stability analysis has been presented for a large scale PNNI network with varied number of processors per node. The study has shown that the behavior of the call setup time, for all involved calls, during simulation time are not one of the continual increase which then characterizes a stable network for the particular traffic. In addition to uncovering the significance of distributed call processing in future ATM networks, this study offers a systematic and scientific approach, through behavior modeling and asynchronous distributed simulation, to examine network stability and validate proposed ATM call processing architecture designs under realistic conditions. Also, since more complex topological networks are expected to be introduced especially to support geographically disperse centers, network stability must be predicted before actual implementation.

References

1. Ricardo Citro, Distributed Call Processing and Reconfigurable Buffer Management Architectures in Asynchronous Transfer Mode Networks, *PhD Dissertation, Networking and Distributed Algorithms Lab., Department of Computer Science and Engineering, Arizona State University*, May 1999.
2. Tony Lee and Sumit Ghosh, A Distributed Approach to Real-Time Payements-Processing in a Partially-Connected Network of Banks, *Journal of the Society for Computer Simulation, San Diego, CA 92177*, Vol. 62(No. 3):180–201, March 1994.
3. Sumit Ghosh and Tony Lee, *Modeling and Asynchronous Distributed Simulation: Analyzing Complex Systems*. IEEE Press, June 2000
4. Tony Lee and Sumit Ghosh. On "Stability" in Asynchronous, Distributed, Decision-making Systems. *IEE Transactions on Systems, Man, and Cybernetics*, 30, Part B(4):549–561, August 2000.
5. Chengzhi Li, Raba, A and Wei Zhao, Stability in ATM Networks, *Proceedings of IEEE INFOCOM '97*, Vol. 1:160–167, 1997.
6. "MCI Systems Technical Documentation". (*www.mci.com*), 1998.
7. "ATM Foruim Technical Committee. Private Network-Network Interface Specification Version 1.0 (PNNI 1.0). Internet version af-pnni-0055.000", (*available online:www.atmforum/atmforum/specs/approved.html (March)*), 1996.

Proposed Architectures for the Integration of H.323 and QoS over IP Networks

Rafael Estepa, Juan Leal, Juan A. Ternero, and Juan M. Vozmediano

Área de Ingeniería Telemática, Universidad de Sevilla
Camino de los Descubrimientos s/n, E-41092 Sevilla
{rafa, jleal, jternero, jvt}@trajano.us.es

Abstract. The voice service over IP networks (VoIP), as a supplement, substitution or evolution of classic telephone service, is being developed due mainly to economical reasons. However, it faces the lack of quality of service (QoS) guarantees in IP networks. The joint use of the QoS scheme proposed by the IETF and the VoIP architecture proposed by the ITU-T (H.323) implies a number of interoperation issues that must be addressed. On one hand, it is necessary to maintain information on the users' authorization in each of the systems. On the other, both services must be billed. This article first describes the state of art of QoS in IP networks and H.323. Next, five architectures for the integration of the IETF QoS and the H.323 schemas are proposed. For each of these architectures, features, advantages and shortcomings are analyzed. Finally, three scenarios using the proposed architectures are studied.

1 Introduction

This section introduces a short description of VoIP, continues with QoS support on IP networks, and ends setting up the problems of user authorization[1].

1.1 VoIP Schemes

Early standards [1] on voice transport over packet based networks without QoS guarantees come from the International Telecommunications Union - Telecommunication Standardization Sector (ITU-T). Recommendation H.323 [2] is the main ITU-T standard for VoIP. It is an architectural standard which refers to other specific standards on signaling [3, 4] inter-gateway communication [5], supplementary services [6], etc. These standards define the necessary elements for voice transmission over this kind of networks, namely:

- Terminals: communication endpoints, equivalent to telephone handsets, which allow users to maintain audiovisual communications.
- Gatekeeper: server that performs centralized tasks, such as directory services, aliases and telephone numbers translation, call control, billing, etc.

---

[1] The work leading to this article has been partly supported by CICYT and the EU under contract number 1FD97-1003-C03-03.

- Gateway: it makes possible the communication between H.323 terminals and PSTN terminals like POTS, ISDN, GSM, etc.
- MCU (Multipoint Control Unit): it supports conferences among three or more endpoints.

The collection of all Terminals, Gateways, and Multipoint Control Units managed by a single gatekeeper is known as H.323 zone. Terminals can either set up calls on their own or using Gatekeepers, which are optional. When Gatekeepers receive set up requests from terminals, they are responsible for locating the destination, accepting or rejecting the call and, optionally, routing the signaling and/or media streams. The decision of whether to route the signaling and media or not corresponds to the Gatekeeper.

Alternatively, the Internet Engineering Task Force (IETF) has proposed a different VoIP model known as Session Initiation Protocol (SIP) [7] which is not considered in this work. At the moment, work is being done on the harmonization of both scenarios (H.323 and SIP). The European Telecommunication Standardization Institute (ETSI), in the Telephony and Internet Protocols Harmonization Over Networks (TIPHON) [8] program, is one of the main organizations working on this respect.

However, neither SIP nor H.323 overcome the lack of QoS supporting mechanisms, which is the main VoIP limitation. This fact is nowadays the most important difference between switched circuit and packet based networks, and is crucial for the success and universal deployment of VoIP [9].

1.2 QoS over IP Networks

In order to assure a certain QoS parameters (bandwidth, delay and packet loss) to data flows containing time sensitive information (such as coded voice), a number of methods are being developed [10].

The IETF developed the Integrated Services (IntServ) [11] model, which defines three classes of service based on application's delay requirements. These are the Guaranteed-Service (GS) [12] class, which provides for delay-bounded service agreements; the Controlled-Load (CL) [13] service class, which provides a form of statistical delay service agreements (nominal mean delay) that will not be violated more often than in an unloaded network and the well-known best-effort service. The Resource Reservation Protocol (RSVP) [14] is used for signaling in the former two service classes. Sender and receiver periodically send RSVP messages (reservations are maintained in a 'soft' state) and these messages make the routers to reserve resources. The support of per flow guarantees poses severe scalability problems. RSVP is useful for access networks, but demands high capacity in the core as the number of flows grows [10].

The Differentiated Services (DiffServ) [15] architecture was developed by the IETF to leverage that scalability problem. This architecture simplifies the operation of the intermediate nodes, moving the complexity to the edge nodes. DiffServ may be used in conjunction with IntServ applied in different parts of a network; the former in the core and the latter in the edge [16].

1.3 Authorization Issues

Independently of the architecture used, two conditions must be fulfilled for a QoS request to be granted. In the first place, enough resources must be available. Secondly, the administration system must authorize the request. Because of the complexity of the authorization problem, which depends on multiple factors, the IETF has defined the Policy Based Admission Control architecture [17]. This architecture consists of two fundamental elements: the Policy Decision Point (PDP), which acts as authorization server, and the Policy Enforcement Point (PEP), which uses the Common Open Policy Server (COPS) protocol [18] to communicate with PDP. PEPs must reside in the routers while the PDPs are generally separate servers.

Interdomain authorization is another aspect of the problem, but it's generally assumed that it will be addressed by Service Level Agreements (SLA).

Moreover, as mentioned above, both H.323 and QoS schemes rely on centralized servers (Gatekeepers and PDPs respectively): The PDP is the element that directly controls the QoS policies applied on routers, while Gatekeepers are necessary for the optimal control of Gateways, aliases and telephone numbers translation, signaling and media routing, call control, etc. Both types of servers complement each other for providing VoIP services with QoS support. Besides, Gatekeepers provide network based services, that is, those which need to be performed independently of the terminal or when the terminal is turned off (e.g. registration, admission, monitoring availability, call redirection, etc.).

Next section describes five alternative architectures for the integration between Gatekeepers and PDPs. The analysis of these architectures focuses on two administrative issues: authorization and accounting.

2 Proposed Integration Architectures

The elements involved in all the architectures are the same. Terminals, Gatekeepers and Gateways use H.323 signaling for communication set up. Once the communication is established, terminals, routers and policy servers negotiate the QoS using the RSVP and COPS protocols. Finally, terminals start transmitting the media streams. H.323 aliases to IP addresses translation and Gateways management are performed by Gatekeepers in all five architectures. These architectures are described below.

2.1 Architecture 1: No Integration

This architecture arises from the straight application of H.323 over a QoS enabled IP network. No interaction exists between H.323 servers (Gatekeepers) and QoS policy servers (PDPs), so H.323 and QoS domains work independently. Gatekeepers are responsible for terminals authorization based on their own criteria and billing for the use of gateways, while authorization and billing for the use of QoS are PDPs' responsibility. Therefore these tasks are performed autonomously.

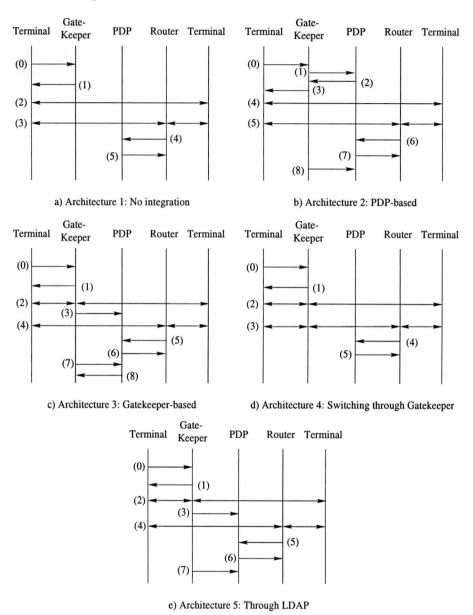

Fig. 1. Steps for call establishment in the different proposed architectures

Figure 1(a) illustrates the necessary steps to establish a call between two terminals with QoS support. The process starts when the calling party sends the Gatekeeper a call request containing the called party's alias address (0). The Gatekeeper translates the alias to the transport address and authorizes the call based on its own criteria. If succeeded, the Gatekeeper sends the called party's transport address back to the terminal (1), so that the terminals can establish the H.323 signaling channels (2). Once the terminals have agreed the media channels to be used, the terminals signal the QoS reservation sending RSVP messages (3). When the routers in the path receive these messages, they ask PDP for authorization (4), which responds based on its own policy, unrelated to the Gatekeeper's (5).

Attending to authorization and billing this architecture generates two different accounting records, one in the PDP for QoS and another in the Gatekeeper for gateway usage. Similarly, two separate authorization databases exist, and therefore consistency of stored data in both must be assured by administrative agreements. Consequently, this architecture is appropriate when the H.323 zone and the QoS domain belong to different organizations.

Major advantages of this architecture emerge from the separation between the H.323 and QoS systems, allowing independent evolution of both technologies, thanks to the absence of interaction between their elements. Besides, this solution is extensible to other kinds of services that need QoS guarantees such as video streaming. Finally, no additional development needs to be done.

Nevertheless, a terminal not registered in the Gatekeeper may obtain QoS guarantees for its calls, which constitutes an important shortcoming of this architecture.

2.2 Architecture 2: PDP Based Architecture

The PDP can be defined in short words as the QoS authorization and billing server. In this architecture the PDP works also as the VoIP server (for authorization and billing). Gatekeepers must include a PEP in order to ask the PDP for authorization, similarly to routers. Thus, QoS and VoIP authorizations are responsibility of the PDP, and the Gatekeeper's answers to the user are based on decisions made by the PDP. The Gatekeeper must also inform the PDP of Gateway usage. Hence, the whole billing information is centralized in the PDP.

Figure 1(b) shows the steps followed in this architecture. The call is initiated when the calling party sends the Gatekeeper a call request containing the called party's alias address (0). The Gatekeeper translates the alias to the transport address and asks the PDP for authorization using COPS protocol (1). The PDP responds to the Gatekeeper granting or denying the call (2). If succeeded, the PDP records the call request in order to later authorize the QoS for that call. This forces terminals to use the Gatekeeper to obtain QoS guarantees on their calls. Then, Gatekeeper sends the called party's transport address back to the terminal (3), so that the terminals can establish the H.323 signaling channels (4). Once the terminals have agreed the media channels to be used, the terminals signal the QoS reservation sending RSVP messages (5). When the routers in the path receive these messages, they ask the PDP for authorization (6). The response given by the PDP to routers is consistent with that given to the Gatekeeper (7). This procedure assures that only those calls that the Gatekeeper asked permission for may obtain QoS guarantees. Finally, the Gatekeeper informs the PDP about gateways usage for billing purposes (8).

The previous description shows that authorization tasks are dealt on the PDP, which simplifies administration but imposes a tight relationship between the PDP and Gatekeepers. Consequently, this architecture is suited when the H.323 zone and QoS domain belong to the same organization.

One of the main advantages of this schema is forbidding unregistered terminals to obtain QoS for their calls. This implies that the use of Gatekeepers is mandatory.

The main shortcoming is scalability, which derives from its centralized structure. For huge networks the PDP might become a bottleneck and its administration can be complex. Also, the request from the Gatekeeper to the PDP increases call set-up time. Besides, COPS usage for H.323 must be defined, which does not allow independent evolution of H.323 and QoS technologies.

2.3 Architecture 3: Gatekeeper Based Architecture

Gatekeepers are considered centralized servers in the H.323 standard. Their missions are, among others, authorization, billing and bandwidth management. From this point of view, it is reasonable that gatekeepers control QoS authorization and billing for H.323 terminals, that is, they perform some of the PDP's functionalities. In order to do that, the Gatekeepers ask the PDP to authorize QoS requests to be made by H.323 terminals and the PDP reports the Gatekeepers about resource usage necessary for billing. Besides, for the Gatekeeper to know the parameters of the media channels used for the call, it must route terminals' signaling.

The process to establish a typical call with QoS using this architecture is shown in figure 1(c). The process starts when the calling party sends a call request containing the called party's alias address to the Gatekeeper (0). This Gatekeeper translates the alias to the transport address and authorizes the call based on its own criteria. If succeeded, it sends the called party's transport address back to the terminal (1). The terminals establish the H.323 signaling channels using Gatekeeper-based routing (2). Once the terminals have agreed the media channels to be used, the Gatekeeper requires the PDP to grant the QoS using COPS protocol (3). After that, the terminals start sending RSVP messages (4). When the routers along the path receive these messages, they ask the PDP for permission (5), which responds based on the previous Gatekeeper requirement (6). When the communication is over, the Gatekeeper demands the PDP to revoke the authorization (7), and the PDP reports QoS usage to the Gatekeeper for billing purposes (8).

Authorization and billing in this architecture are centralized in the Gatekeeper, which simplifies administration and allows separate management of H.323 zones. However, the PDP must trust the Gatekeepers for VoIP QoS request authorization. Therefore, this architecture is appropriate in case H.323 zone and QoS domain belong to the same organization.

This architecture, as the previous one, forbids unregistered terminals to obtain QoS for their calls, so the use of Gatekeepers is mandatory.

One problem of this architecture is that it forces the dependent evolution of H.323 and QoS technologies. Besides, COPS usage for H.323 must be defined.

Another problem is that the signaling through the Gatekeeper and the communication between the GateKeeper and the PDP delays the call set-up completion.

2.4 Architecture 4: Call Switching through Gatekeeper

This architecture arises when considering that the Gatekeeper can route the whole calls, this is, signaling and media channels. This behavior resembles PABXs in circuit switched networks. Terminals establish their connections to their local Gatekeeper. Calls to and from the Gatekeeper are always granted in the PDP in a hard state, so no communication between those elements is needed. It is mandatory for Gatekeepers to support RSVP, so terminals without RSVP signaling can get QoS support in the path between the Gatekeeper and the remote terminal.

Figure 1(d) shows the steps followed this architecture. As usual, the call is initiated when the calling party sends the Gatekeeper a call request containing the called party's alias address (0). The Gatekeeper then translates the alias to the transport address and authorizes the call based on its own criteria. If succeeded, the Gatekeeper sends its own transport address back to the terminals (1), so that the terminals can establish the H.323 signaling with the Gatekeeper (2). Terminals and Gatekeepers reserve the QoS for the media streams which will be routed by the Gatekeepers (3). When the routers in the path receive the RSVP messages, they ask the PDP for authorization (4), which responds allowing the reservations, because all the calls to and from Gatekeepers are always authorized (5).

As described above, authorizations and billing are centralized in the Gatekeeper, easing the management. This architecture is, thus, suitable for independent administration of H.323 and QoS zones.

Athough there is no communication between the GateKeeper and the PDP, the signaling through the GateKeeper and the RSVP reservation in the path through the Gatekeeper increase call set-up time. Once the call is established, forcing the voice packets through the Gatekeeper will also increment the delay. Besides, this route may be non optimal. The Gatekeeper may also became a bottleneck when there are many terminals per H.323 zone.

2.5 Architecture 5: Interaction through LDAP

In this architecture, H.323 terminals' authorizations are centralized on the Gatekeeper and, therefore, VoIP users must only be provisioned in the Gatekeeper's database, but not in the PDP's. Since the PDP's database can be accessed through LDAP (Lightweight Directory Access Protocol), the Gatekeepers dynamically update the database of the PDP, so authorizations made by the PDP are consistent with those made by the Gatekeepers. The Gatekeepers must route terminals' signaling in order to know the parameters of the media channels used for the call, and appropriately update PDP's database.

The process for setting up a call is shown in figure 1(e). Upon the reception of a call request containing the called party's alias address (0), the Gatekeeper translates the alias to the transport address and authorizes the call based on its own criteria. If succeeded, it sends the called party's transport address back to the terminal (1). The terminals establish the H.323 signaling channels, using Gatekeeper-based routing (2). Once the terminals have agreed the media channels to be used, the Gatekeeper updates the LDAP database of the PDP (3). Then, the terminals signal the QoS reservation sending RSVP messages (4). When the routers in the path receive these messages, they ask the PDP for authorization (5), which responds based on the previous Gatekeeper indication (6). When the communication ends, the Gatekeeper updates the PDP's authorization LDAP database for that terminal, restoring its original state (7).

Authorization in this architecture is centralized in the Gatekeepers, which simplifies administration and allows separate management of H.323 zones. However, regarding accounting, this architecture generates two different records: one in the PDP for QoS and another in the Gatekeepers for gateway usage. This architecture is appropriate when the different H.323 zones are independently managed, but the QoS administration is unique.

A great advantage of this architecture, like those where the Gatekeeper centralizes the authorizations, is denying QoS authorization for the calls of non registered terminals, so in this architecture the use of Gatekeepers is mandatory.

The drawback of this architecture is that it forces the Gatekeeper to specify authorizations in the LDAP policy representation schema.

Call set-up delays are similar to those in the Gatekeeper-based architecture, where there is signaling through the GateKeeper and communication between the Gatekeeper and the PDP.

3 Applications

From the previous discussion of these architectures derives that no optimal architecture for every possible scenario exists.

Accounting integration is a requirement when VoIP services with QoS guarantee are provided and billed. However, accounting integration does not provide any advantage when one of the services is either not billed or included in the other. This may happen when providing VoIP without quality guarantees or when billing only QoS, be it utilized by VoIP traffic or by any other.

To show that there is not an optimal solution common to every case, three particular scenarios using H.323 and QoS over IP networks can be considered. For each of them, different integration architectures will be suitable.

The first scenario consists of a network formed by multiple LANs connected to a backbone whose resources must be optimized, and linked to the PSTN through a number of Gateways. The whole network is administered by a unique organization. This is the typical case in corporate networks or university networks. Let this network be QoS enabled.

To provide H.323-based VoIP services, Gateways and Gatekeepers should be placed in the network splitting it in zones which optimize network utilization. The administration of each H.323 zone could be the responsibility of every organizational unit, while the global QoS policy would normally be decided by the management center, unique to the whole organization. Therefore, the most appropriate architecture for this scenario would be the interaction through LDAP (architecture 5). Using this architecture, administrators of every H.323 zone do not need to register in the management center every new user in their zone. Instead, users would be automatically authorized in the QoS system when provisioned in the Gatekeeper. The fact that using this architecture Gatekeepers have access to the PDP's LDAP database is not a concern, as administrators of each H.323 zone and the QoS system belong to the same organization.

Let us consider a second scenario consisting of a local Internet Service Provider (ISP), QoS enabled network over which the ISP desires to offer H.323 based VoIP service to its subscribers. These networks typically consist of one or more access systems using ADSL, cablemodem or PSTN access, a number of servers for mail, web, etc. and a router that provides Internet access. These elements are connected by a private high speed network. In local ISP networks, there are few Gatekeepers and are located in the servers segment, while PDP is connected to the network management segment.

From the management point of view it is desirable that, on one hand, user provisioning is unique and, on the other hand, accounting for both services is obtained together so users can be billed for all the resources used in their VoIP calls (that is, QoS and Gateway usage) at the same time. Therefore, for this scenario the most appropriate architectures would be number 2 (PDP based) and number 3 (Gatekeeper based). These architectures are feasible because Gatekeepers and PDPs belong to the same organization, so the relationship between them can be very close.

Finally, a third scenario is a variation of the previous one, in which VoIP service is provided by an separate organization. The VoIP service provider uses the ISP's access network. VoIP subscribers do not need to sign up for the ISP service. Instead, the VoIP service provider bills its subscribers for the whole service (Gateway and QoS usage).

Because Gatekeepers and PDPs belong to different organizations, interaction between them is discouraged. Therefore architectures number 2, 3 and 5 are inappropriate.

Both of the remaining architectures are possible. Architecture number 1 requires administrative communication between ISP and VoIP service provider for user provisioning and QoS usage reporting. Application of architecture number 4 is straightforward because billing and authorization are centralized in the Gatekeeper.

4 Conclusions

In this article, five architectures for the integration of H.323 and QoS over IP networks have been proposed. These architectures achieve different degrees of interaction by means of various mechanisms for the communication between the H.323 Gatekeepers and the PDP. They have been analyzed regarding authorization and accounting, and the main advantages and shortcomings have been outlined. Finally, three application scenarios have been presented which lead to different optimal integration architectures. This fact shows that none of the proposed architectures is optimal for every possible scenario.

References

1. Toga, J., Ott J. ITU-T Standardization Activities for Interactive Multimedia Communications on Packet Networks: H.323 and Related Recommendations. Computer Networks, Vol. 31 (1999) 20–23.
2. ITU-T Recommendation H.323. Packet-based Multimedia communications systems. (11-2000)
3. ITU-T Recommendation H.225.0. Call signalling protocols and media stream packetization for packet based multimedia communications systems. (1999).
4. ITU-T Recommendation H.245. Control Protocol for multimedia communication. (2000).
5. ITU-T Recommendation H.248. Gateway Control Protocol. (2000).
6. ITU-T Recommendation H.450.1 Generic functional protocol for the support of supplementary services in H.323. (2-1998)
7. IETF RFC 2543. SIP: Session Initiation Protocol. (3-1999)
8. ETSI TR 101300 v2.1.1 Telecommunications and Internet Protocol Harmonization Over Network (TIPHON); Description of Technical Issues. (10-1999)
9. Li, J., Hamdi, M., Jiang, D., Cao, X. QoS enabled Voice Support in the Next-Generation Internet: Issues, Existing Approaches and Challenges. IEEE Communications Magazine. Vol 38-4 (2000) 54–61
10. Mathy,L., Edwards,C., Hutchison,D. The Internet: A global Telecommunications Solution?. IEEE Network Magazine, Vol 14-4 (2000) 46–57.
11. IETF RFC 1633. Integrated Services in the Internet Architecture: an Overview. (6-1994)
12. IETF RFC 2212. Specification of Guaranteed Quality of Service. (7-1997)
13. IETF RFC 2211. Specification of the Controlled-Load Network Element Service. (7-1997)
14. IETF RFC 2205. Resource ReSerVation Protocol (RSVP) – Version 1 Functional Specification. (7-1997)
15. IETF RFC 2475. An Architecture for Differentiated Services. (12-1998)
16. IETF Work in Progress, Bernet et al. A Framework For Integrated Services Operation Over Diffserv Networks . (5-2000)
17. IETF RFC 2753. A Framework for Policy-based Admission Control . (1-2000)
18. IETF RFC 2748. The COPS (Common Open Policy Service) Protocol . (1-2000)

Third-Party Call Control in H.323 Networks – A Case Study

A. Miloslavski, V. Antonov, L. Yegoshin, S. Shkrabov,
J. Boyle, G. Pogosyants, and N. Anisimov

Genesys Telecommunication Labs (an Alcatel company)
1155 Market St., San Francisco, CA 94103, USA
http://www.genesyslab.com

Abstract. This paper addresses the problem of building third-party call control in a VoIP network based on ITU-T Recommendation H.323. We consider a specific case that could be used in the call center environment. We will suggest and advocate a solution with H.323 gatekeepers connected to CTI Server via special protocol. The CTI Server contains all the logic for service execution (a call model) while a gatekeeper implements only elementary switching functions. Some basic features of the protocol are outlined. We also present the architecture of a real-world network where the approach is implemented.

1 Introduction

One of the most important applications of Computer Telephony Integration (CTI) is the call center (CC) [1]. The key element on which all the software of the call center relies is a CTI-link containing a communication protocol between a telephony switch and a computer that is often referred to as a CTI server. CTI protocol allows the CTI Server to monitor and control processing of telephony calls, i.e., it provides third-party call control. Modern call centers are comprised of complex and expensive software called call center applications that provide the processing of inbound and outbound calls from customers. For instance, a CC application may include preliminary call processing by Interactive Voice Response (IVR), routing the call to the most appropriate operator (agent), assisting the agent in call processing, managing customer relations, reporting, etc.

Using emerging Voice over IP (VoIP) technologies promises to substantially improve the performance of the call center. At the same time, in order to painlessly transition from PBX-based to VoIP networks in call centers one needs to solve certain technical problems, such as how to substitute these networks without changing other call center software. In other words, we need to build third-party call control in VoIP networks in order to provide PBX-based services, such as call transfer, conferencing, and automatic call distribution.

In this paper we will consider one of the solutions for the problem using the example of call center software developed at Genesys Telecommunication Labs, a wholly owned subsidiary of Alcatel. We will consider the case of when VoIP net is based on ITU-T Recommendation H.323 [2].

2 The Problem

The typical structure of a call center employing CTI-link is depicted in Fig.1. The key element of the call center is a CTI-link that connects PBX and CTI Server to provide visibility and control of the switching domain by the computing domain. The CTI link makes it possible to organize call processing in a flexible way. For example, the processing of an inbound call may involve the following steps. The CTI Server learns about the incoming call due to an event received via the CTI-link. Then it routes the call to an IVR that may collect additional information about the call and the customer. Then the Router routes the call to the most appropriate agent to be processed. The telephony call is transferred along with related data. During the call processing the agent may use information from the database (about the call and the customer), may consult with another agent, organize a conference or transfer the call to another agent. To initiate all these PBX-based services the agent usually uses a special desktop application that, in its turn, exploits the third-party call control functions. It should be mentioned that the Genesys CTI Server uses a call model that has much in common with the SCTA call model [3].

Now we want to substitute the PBX network with the VoIP network based on the H.323 Recommendation. In other words we want to build a third-party call control in the H.323 network which is exactly the same as that in the PBX network. We should stress the complexity of the problem that stems from the nature of the networks. Indeed, the CTI-link assumes centralization, while H.323 is built on a distributive principle. Third-party control may require additional complex protocol to connect endpoints and CTI Server as it suggested in [4].

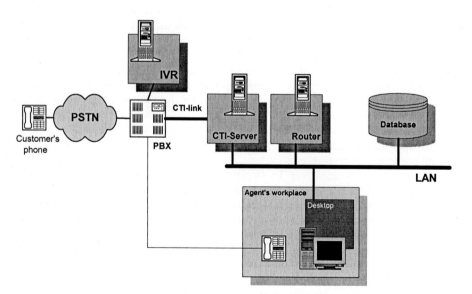

Fig. 1. Call Center Environment

One reasonable approach is to move CTI Server functions into a special server connected to the gatekeeper(s). The one example of this approach is the concept of "thin gatekeeper" developed within Dialogic [5]. According to this approach a gatekeeper is connected to a special server via CSTA protocol. This solution assumes that the gatekeeper understands and operates in accordance with CSTA protocol. However endowing a gatekeeper with CTI Server functionality is not a good solution because it will result in a complex and inflexible gatekeeper implementation. Moreover, it is not clear what to do when there are several gatekeepers in the network.

At the same time, notice that CSTA has been developed to connect switching and computing domains [3]. In fact, CSTA provides a copy of the switching call model in the CTI Server. Moreover, it keeps these call models consistent using a request/event mechanism. In our point of view this solution is still complicated because in the VoIP network there is no longer any boundary between the switching and computing domains. Everything is in the computing domain and it is reasonable to have only one call model rather than two.

3 Proposed Solution

In this paper we suggest a configuration where all service logic is placed in the CTI Server while gatekeepers perform more elementary operations.

3.1 Architectural Environment

Traditionally, CTI protocols were largely vendor specific, thus forcing CTI software vendors to develop separate software modules for each switch model. A number of protocols were proposed as potential standards for CTI links, such as CSTA [3], MGCP/Megaco/H.248 [7] and newer revisions of TAPI [8]. Unfortunately, such attempts at such CTI protocol standardization are not likely to produce compatible implementations. It is easy to see that any non-trivial CTI software suite has a need to maintain an accurate replica of the switch state, which in practice means that the CTI software has to replicate the call model of the particular switch, as shown in Fig.2(i). Any discrepancy between the actual call model implemented by the switch vendor and its reverse-engineered replica in the CTI software causes loss of coherency between the actual state and its image in the control software. Worse yet, practically all switch vendors introduce subtle changes to their call model in successive versions of switch software (this is unavoidable when new features are added and programming errors are corrected). Packet-switched telephony makes call models even more complicated by replacing centralized switches with a heterogeneous, distributed switching environment.

One approach to solving the problem of call model incompatibility would be to standardize the call model itself, providing a rigid specification for switch behavior. This, in practice, is a very challenging task because of the richness of possible call behavior and call control features. The current generation of telephone switches provides hundreds of these features to accommodate diverse customer requirements. Any single standard defining the switch behavior is bound to be overly restrictive;

therefore, vendors reacting to customer demand will be compelled to expand it, thus defeating the very purpose of standardization.

As a result, none of the proposed CTI standards even attempt to specify a standard call model (beyond some elementary features), thus leaving the actual behavior of switches to the discretion of the manufacturer. While helping to solve the easy problem of CTI message format encoding and decoding, they leave the complicated task of switch behavior modeling completely untouched.

In this paper we advocate another approach based on Simple Media Control Protocol (SMCP) [6] developed within Genesys Labs. The SMCP solves the behavioral compatibility problem by providing CTI software access to the basic switching functions, with no proprietary call model overlay. In SMCP architecture, a switch vendor not have to implement an elaborate call model, a thin SMCP protocol stack is sufficient. The PBX call control functionality can be external and developed separately; see Fig.2(ii). A switch vendor supporting SMCP could use third-party SMCP-based PBX call control software instead of doing costly in-house PBX design or integration. Note that in SMCP architecture, only one call model is used in CTI-controlled operation, thus eliminating the problem of CTI software and switch call model incoherence.

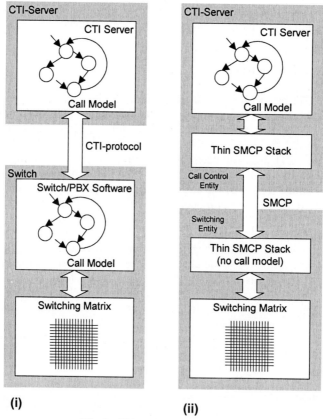

Fig.2. Old and new architectures

Many existing VoIP devices like phones, gateways, gatekeepers are separate devices that do not connected by common logic of services. The distributed nature of equipment and limitations on a network bandwidth result in a need to support services on a distributed basis. Thus, a CTI Server is a natural place for a centralized control of services execution.

The simplicity of the SMCP model allows it to incorporate functions not generally found in conventional voice-only call models. Indeed, the call commutation model defined in this document is in no way specific to voice, and can be used to control a very wide variety of call-like interactions, such as analog voice, synchronous or packetized digital voice, interactive video, chat, whiteboard or desktop sharing sessions.

3.2 Summary of SMCP

In this section we briefly consider SMCP. The full specification could be found in [6].

3.2.1 Basic Notions

SMCP is a protocol that regulates a communication between two objects – Call Control Entity (CCE) and Switching Entity (SWE), see Fig.2 (ii). The Switching Entity is a module that fulfills only switching functions and does not aware about services. The Call Control Entity contains service logic and implements it with the aid of SMCP instructing SWE what to do in order to provide the service. Thus SMCP is a protocol of a low level that provides only basic switching functions and does not contain service logic.

Note that CCE and SWE are abstract modules in the sense that their physical embodiment depends on the architecture of VoIP network. For example, if SMCP is used in an H.323 network, SWE may be implemented as a Gatekeeper. In an MGCP based network the role of SWE can be played by a Media Gateway Controller (MGC).

The SMCP assumes that the SWE implements an abstraction of a commutating device allowing separate control of call legs (call leg is a half of a connection). In other words, SWE must make sure that externally visible calls are kept established even when the commutating device tears down and reestablishes internal connections between them. Fig.3 shows how the SMCP commutator model maps to a typical enterprise VoIP configuration.

The call legs are terminated at endpoints and are associated with physical or virtual ports (there are situations when endpoints do not have associated ports). Some ports can be used for support of multiple simultaneous connections. The parts of call legs outside of a switch or a gateway are called exterior, connections between them inside switches are interior. More than two call legs can be connected together to form a multi party conversation.

Typically, endpoints can terminate only one call leg at a time. However, in some cases (for example, call treatment sources, such as music-on-hold) a single endpoint can be connected to multiple call legs. All endpoints have unique endpoint addresses represented as ASCII character strings; such strings can be telephone numbers, URLs, E-mails, etc; the exact naming scheme is application specific. Ports have unique vendor-specific names.

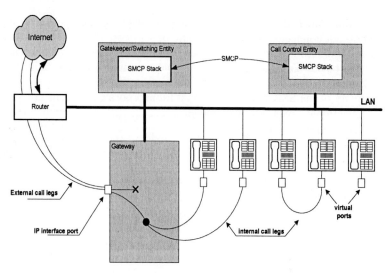

Fig.3. VoIP System as a SMCP Commutator

As in many other Internet protocols (e.g., HTTP, SIP, MGCP), SMCP messages are text based, i.e., all messages may contain only ASCII printable characters, spaces and tabs.

3.2.2 Exterior Call Control

The protocol consists of several procedures. However, in this paper we will consider only the main procedures related to call control. The other procedures such as configuration management, fault tolerance management, encryption, etc. are out of the scope of this paper and can be found in [6].

The exterior call control commands and events provide a mechanism for manipulating the endpoint sides of call legs.

Outbound Calls. For managing outbound calls SMCP uses the following protocol data units:

CALL	Make an outbound call;
BUSY	Port is busy;
CALLING	Outbound call request acknowledgment;
GOTCA	Got call alerting from endpoint;
ANSWERED	Call answered;
CONNECTED	Connection established.

Outbound calls generally progress through several stages, see Fig.4:
- CCE commands SWE to initiate the connection with a **CALL** request.
- If the request is not valid, SWE responds with an **ERROR** reply; if the selected port is busy, SWE responds with a **BUSY** reply; otherwise the request is considered valid and SWE allocates a call leg identifier and returns it to CCE with a **CALLING** reply.
- SWE initiates a transport connection to the endpoint, and sends a Call Setup message to the endpoint.

- The endpoint may reply with a Call Proceeding message to indicate that it accepted further responsibility for call delivery.
- The endpoint may then reply with a Call Alerting message to indicate that the call has been delivered to the end-user device, which started alerting the user.
- If the callee answers the call, the endpoint device replies with a Setup message, thus initiating the process of establishing a transport connection. SWE informs CCE about this event with an **ANSWERED** notification.
- When a transport connection is established the endpoint device replies with a Connected message. After that the call is completely established, and SWE sends a **CONNECTED** notification to the client.
- At any time after the **CALLING** notification, SWE may inform CCE that the connection cannot be completed with a **DROPPED** notification. If CCE did not send a **HANGUP** request for this call already, this request must be sent to release the connection after receipt of a **DROPPED** notification.

A diagram of CCE state transitions while performing an outbound call is shown in **Fig.** . Note that the connection from a switch to a directly attached phone set is also an outbound call in the SMCP call model. Making outbound calls generally requires specification of a switch or gateway port, and the additional target address (which can be for directly attached devices, can contain an E.164 phone number for PSTN calls, or an E-mail address or URL for VoIP calls). If SWE implements a numbering plan, the port specification may be omitted and selected automatically by the target address using the plan's target lookup table.

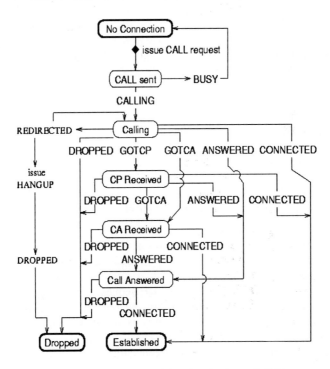

Fig. 4. CCE State Machine for Outbound Calls

Incoming Calls. For managing inbound calls SMCP uses the following protocol data units:

RING	Notify about incoming call;
SENDCP	Send Call Proceeding message;
SENDCA	Send Call Alerting message;
ANSWER	Answer the call.

A typical incoming call scenario is:
- SWE detects the incoming attempt to establish a transport connection, accepts the transport connection, and reads the Setup message.
- SWE alerts CCE about the incoming call with a RING notification.
- CCE may instruct SWE to send a Call Proceeding message to the calling party with a SENDCP request, and wait for an OK reply.
- CCE may then instruct SWE to send a Call Alerting message to the calling party with a SENDCA request, and wait for an OK reply.
- When CCE wishes to answer the call, it asks SWE to perform the transport protocol negotiation with the ANSWER request, and wait for an OK reply.
- SWE informs CCE about success or failure of the negotiation; if the negotiation was successful the connection is considered established and SWE sends a CONNECTED notification to the CCE.
- At any time after the original RING notification, SWE may inform CCE that the connection was abandoned with a DROPPED notification. If CCE did not send a HANGUP request for this call, this request must be sent to release the connection after receipt of a DROPPED notification.
- After receiving any reply or notification pertaining to the current call leg CCE may command SWE to drop the connection with a HANGUP request. After issuing the HANGUP request CCE should wait for the DROPPED notification from SWE before assuming that the port is available.

A diagram of CCE state transitions while answering an incoming call is shown in Fig.5.

Call Release. An established connection may be released by either the local or remote party. CCE may request releasing the call leg with a HANGUP request. When the call is dropped due to an endpoint's action or a transport disconnection, SWE sends a DROPPED notification to CCE. CCE then must release the call leg with a HANGUP request. SWE may reuse the call leg identifier only after it has sent a DROPPED notification and received a valid HANGUP request.

3.2.3 Interior Media Stream Commutation

The interior media stream commutation allows interconnecting existing exterior call legs in an arbitrary manner. The connections are established with a MAKE request and are torn down with BREAK or HANGUP requests. It is important to note that interior connections for different media streams related to the same call leg are different connections, and can be independently established or torn down.

MAKE - Make Connection between Call Legs
BREAK - Break interior connection between call legs

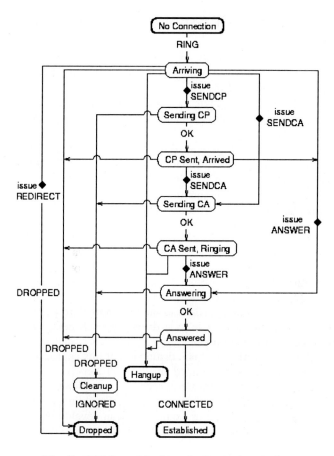

Fig. 5. CCE State Machine for Incoming Calls

3.3 Third-Party Service Creation

Exterior and interior leg control procedures allow one to develop different services whose logic is placed into the CTI Server as in Figure 2 (ii). In particular there may be standard PBX-based services like Plain Old Telephone Service (POTS) and complementary services like Call Transfer, Call Forwarding, Conferencing, Call Hold, etc. Moreover, we can design specific services that are not supported by standard PBXs.

For example, the implementation of Hold/Retrieve service is simply implemented by SMCP PDUs **BREAK** and **MAKE**, respectively. The two-step transfer consists in sending PDUs **BREAK**, **MAKE** and **HANGUP** for corresponding legs.

4 Architecture of SMCP Network

In Fig.6 the real-world architecture of VoIP network with SMCP is presented. It assumes using three additional modules: CTI-Server, Distributed Media Exchanger (DMX) and Stream Manager (SM).
- CTI Server contains a logic of a call model and complementary services, supports address resolution functions, and provides an interface with third-party applications;
- DMX talks with H.323 units (endpoints, gateways, MCUs) according to H.225 and H.245 recommendations, reports to CTI Server about events, executes commands from CTI Server, and controls SM;
- SM commutes media streams playing a role of a proxy, converts codecs if necessary and plays files to stream channels (e.g., playing a music file while the connection is on hold).

5 Conclusion

In this paper we suggested an approach to the design of third-party control in H.323 networks primarily intended for use in a distributed call center environment. This approach is based on a new protocol called SMCP that connects a CTI Server with elements of a H.323 network. The main distinction of this approach is that it contains only one call model that is maintained by the CTI Server. On the one hand this approach makes it possible to eliminate unnecessary complexity from the elements of H.323 networks. On the other hand, it gives more flexibility in creating third-party services for call center operation. For example we can develop more specific services for call centers than are allowed by approaches based on conventional CTI protocols like CSTA.

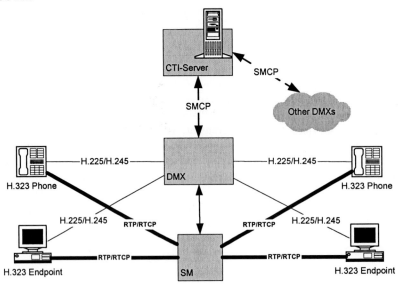

Fig.6. VoIP network with SMCP.

Acknowledgments

The authors are grateful to David Solnit and Steven Hamilton for technical help.

References

1. R.Walters. CTI in Action, Wiley, 1997.
2. ITU-T Recommendation H323. "Packet-Based Multimedia Communication Systems", Geneva, Switzerland, September 1999.
3. ECMA-269. Services for Computer Supported Telecommunications Applications (CSTA) Phase III, 3rd edition, December 1998. See http://www.ecma.ch
4. N.Oliver, T.Miller, J.D. Smith (Eds) Notes on CSTA in IP Telecommunications, Working paper by ECMA TC32-TG11, 15 May, 2000.
5. M. Robins. Pain-Free Internet Telephony-Powered CT-Apps, Internet Telephony Magazine, August 1999.
6. V.Antonov, L.Yegoshin. Simple Media Control Protocol, Genesys Telecommunication Labs. October 2000. See http://www.genesyslab.com
7. T.Taylor. Magaco/H.248: A New Standard for Media Gateway Control. IEEE Communication Magazine. October 2000, Vol.38, No.10, pp.124-132.
8. IP Telephony with TAPI 3.0. White Paper. Microsoft Corporation.

Measurement-Based MMPP Modeling of Voice Traffic in Computer Networks Using Moments of Packet Interarrival Times

N.S. Kambo [1], Dervis Z. Deniz [2], and Taswar Iqbal [2]

[1] Department of Industrial Engineering
[2] Department of Electrical and Electronic Engineering
Eastern Mediterranean University
Gazi Magusa, Mersin 10, Turkey.
Tel: +90-392-630-1300   Fax: +90-392-365-0240
dervis.deniz@emu.edu.tr

Abstract. In this paper we propose a novel technique for modeling packet voice traffic in computer networks based upon the measurement and estimation of the first three moments of packet interarrival times. The moments of an interarrival process are estimated using the approximation technique proposed in [1] which is applicable to any arbitrary interarrival time distribution. We then use a technique based on the pioneering method of Heffes and Lucantoni [2] for approximating the superposition of packetized voice streams from heterogeneous sources using a correlated Markov modulated Poisson process (MMPP). This is carried out in such a way that several important characteristics of the superposed traffic are matched with those of a two-state MMPP. The performance of a voice multiplexer with heterogeneous superposed packet voice traffic is then found by modeling the multiplexer as an MMPP/D/1 queuing system. A simulation model of the system is developed. Simulation results agree more closely with results obtained by the proposed technique than by the Heffes-Lucantoni technique. Practicality of the proposed technique has been demonstrated.

1 Introduction and the Proposed Technique

It is expected that the variable bit rate voice will make up a significant portion of the traffic load on broadband networks like Broadband Integrated Services Digital Networks and Asynchronous Transfer Mode networks in not too distant future. In order to design networks which meet the demands of this application, it is clear that we need a better understanding (or characterization) of the voice traffic and its behavior in queuing systems. To evaluate the performance of voice traffic multiplexed statistically in such queuing systems, there is also the need to characterize superposed traffic streams which have the properties like correlation and burstiness.

To approximate the superposed voice streams, Markov modulated Poisson process (MMPP) has been used because of its ability to capture the time varying behavior of superposed voice traffic, its versatility, as well as its analytical tractability in queuing performance evaluation. The matching of the characteristics of the MMPP and the

Measurement-Based MMPP Modeling of Voice Traffic in Computer Networks   571

superposed voice traffic is focused upon in such a way that MMPP satisfactorily approximates the important characteristics of the approximated process (superposed voice streams). Various matching techniques have been proposed in the literature [2-5]. In this paper the powerful pioneering technique of Heffes and Lucantoni [2] is modified by using distribution free approximations to the moments of a renewal process given in [1]. This results in a simplified technique and has the following advantages over the Heffes et. al. technique:

1. The new technique is based on the first three moments of the interarrival time between voice packets; estimated from the real traffic. While in [2], an *a priori* interarrival time distribution is assumed. Hence, the new technique is independent of the interarrival time distribution.
2. Proposed technique does not require numerical inversion of Laplace transforms as in [2].
3. Finally, in [2] steady state behavior of traffic is modeled whereas it is found that the new technique approximates both the transient (to some extend) and the steady state behavior of the arrival rate, the index of dispersion for the counts (variance-to-mean ratio), and third central moment of the number of arrivals of the superposed traffic streams.

The derivation of the technique is as follows. Consider a two-state MMPP. Let σ_1^{-1} and σ_2^{-1} be the mean sojourn times of the underlying Markov chain in states 1 and 2, respectively and λ_j ($j = 1, 2$) be the arrival rate of the Poisson process when the chain is in state j.

Heffes and Lucantoni [2] chose the four parameters σ_1, σ_2, λ_1 and λ_2 of the approximating two-state MMPP so that the following four characteristics of the superposition are matched with the corresponding characteristics of the MMPP:

1. The steady-state mean arrival rate $\lim_{t \to \infty} \dfrac{E[N^s(0,t)]}{t}$;

2. The variance-to-mean ratio $\dfrac{\text{var}[N^s(0,t_1)]}{E[N^s(0,t_1)]}$ of the number of arrivals in $(0, t_1)$;

3. The long term variance-to-mean ratio $\lim_{t \to \infty} \dfrac{\text{var}[N^s(0,t)]}{E[N^s(0,t)]}$ of the number of arrivals;

4. The third central moment $\mu_3^s(0, t_2)$ of the number of arrivals in $(0, t_2)$.

Here, $N^s(0,t)$ denotes the number of arrivals in (0,t) for the superposed traffic. Expressions for the mean, variance-to-mean ratio and third central moment of the MMPP with parameters, σ_1, σ_2, λ_1 and λ_2 are given in [2]. Using these expressions, the above characteristics 1-4 can be obtained. Further, we can use the following approximations [1] for the moments of the number of renewals, $N(0,t)$, in (0,t) to find the above four characteristics for the superposed traffic:

$$M_1(t) = E\left[N(0,t)\right] \approx \frac{t}{\mu} + \frac{(\mu_2' - 2\mu^2)}{2\mu^2}\left(1 - e^{s_1 t}\right). \tag{1}$$

$$M_2(t) - M_1^2(t) = Var[N(0,t)] \approx \frac{\sigma^2}{\mu^3}t + \left[\frac{2\sigma^2}{\mu^2} + \frac{3}{4} + \frac{5\sigma^4}{4\mu^4} - \frac{2\mu_3'}{3\mu^3}\right]$$

$$-\left[\frac{5\sigma^2}{2\mu^2} + \frac{1}{2} + \frac{\sigma^4}{\mu^4} - \frac{2\mu_3'}{3\mu^3}\right]e^{s_1 t} + 2tE_1\left(\frac{1}{\mu} + E_1 s_1\right)e^{s_1 t} - E_1^2 e^{2s_1 t}. \tag{2}$$

$$M_3(t) = E[N^3(0,t)] \approx E_1 + \frac{12E_1}{\mu s_1} + \frac{18E_1}{\mu^2 s_1^2} + \frac{36E_1^2}{\mu s_1} - 6E_1^2\left(\frac{1}{s_1} - E_1\right)$$

$$+ t\left[\frac{1}{\mu} + \frac{12E_1}{\mu} + \frac{18E_1}{\mu^2 s_1} + \frac{18E_1^2}{\mu}\right] + t^2\left(\frac{3}{\mu^2} + \frac{9E_1}{\mu^2}\right) + \frac{t^3}{\mu^3}$$

$$+ \left[-E_1 - \frac{12E_1}{\mu s_1} - \frac{18E_1}{\mu^2 s_1^2} - \frac{36E_1^2}{\mu s_1} + 6E_1^2\left(\frac{1}{s_1} - E_1\right)\right]e^{s_1 t}$$

$$+ t\left[\frac{18E_1^2}{\mu} - 6E_1^2(1 - E_1 s_1)\right]e^{s_1 t} + t^2\left[3E_1^2(s_1 - E_1 s_1^2)\right]e^{s_1 t}. \tag{3}$$

where $E_1 = (\mu_2' - 2\mu^2)/2\mu^2$, $s_1 = 6\mu(\mu_2' - 2\mu^2)/(3\mu_2'^2 - 2\mu\mu_3')$, $\mu_n' = E(X^n)$,

$\mu = \mu_1'$ and X is the interarrival time random variable. This matching procedure leads to the proposed method.

It may be remarked here that the approximation for $M_1(t)$ is exact when the interarrival time distribution is exponential, 2-Erlang or a mixture of two exponentials. The approximations are valid provided $(s_1 \leq 0)$. Kambo [1] has shown that s_1 is indeed non-positive when the interarrival time distribution is uniform, gamma, m-stage mixed exponential, lognormal or Weibull.

2 Numerical Results

In Heffes-Lucantoni [2] the packet arrival process from a single ON-OFF voice source is considered as a renewal process with interarrival cumulative distribution function (cdf) given by;

$$F_X(t) = [(1 - \alpha T) + \alpha T(1 - e^{-\beta(t-T)})] U(t-T), \quad t \geq 0. \tag{4}$$

where $U(t)$ is the unit step function, α^{-1} is the mean duration of talkspurt, β^{-1} is the mean duration of silent period and T is the interpacket time.

The results of the study using the proposed technique, the simulation model developed and the Heffes-Lucantoni technique are given in Figures 1-5. All simulation results are obtained are within 95% confidence intervals. In these graphs the parameter values of $\alpha^{-1} = 352ms$, $\beta^{-1} = 650ms$ and $T = 16ms$ are used. In Figures 1-3, plots for mean, variance-to-mean ratio and third central moments of the number of arrivals in the case of a single voice source are shown. The plots of the mean and third central moment are shown in Figures 4 and 5 for twenty superposed ON OFF sources. It is observed from these figures that in the transient part of the curves Heffes and Lucantoni technique underestimates the mean but overestimates the variance-to-mean ratio as well as the third central moment. The results of the proposed method agree with those obtained using the simulation results. In steady-state all the results agree.

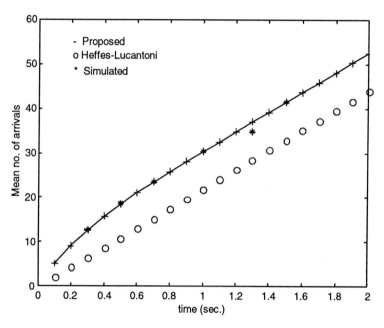

Fig. 1. First moment of the counting process for a single voice source

Next, the performance of the proposed matching technique for MMPP based modeling of superposed voice traffic is evaluated. We model the statistical multiplexer as an MMPP/D/1 queue. This means that the input consisting of superposition of voice streams to the multiplexer is approximated by a two-state MMPP and a fixed packet service time (transmission time on the line) is used. For calculation of the mean and variance of waiting time in steady state the method given in Heffes & Lucantoni [2] is followed. The results are also obtained by specializing the Lucantoni [9] (p. 40-42) results for the 2-state MMPP/G/1 queue to the

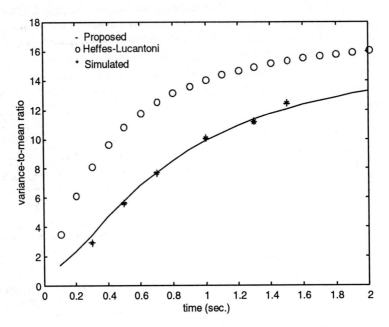

Fig. 2. Variance-to-mean ratio of the counting process for a single voice source

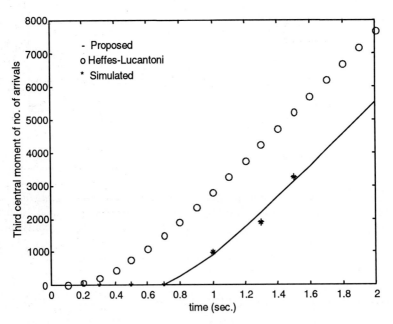

Fig. 3. Third central moment of the counting process for a single voice source

MMPP/D/1 queue case. The results obtained by both methods agreed. Table 1 gives these results for different number of superposed sources. Simulation results for the mean and the variance of the waiting times in the queue are found to be more close to the results obtained by the proposed technique than the results obtained by the Heffes-Lucantoni technique. All simulation results in Table 1 are obtained using 100,000 packet completions and thirty replications, except the one for 120 sources which is obtained using 1 million packet completions with thirty replications. Figure 6 depicts the comparison of variance-to-mean ratio of the actual generated traffic, and the matched MMPP process using the proposed and the Heffes-Lucantoni technique. It is observed that the variance-to-mean ratio curve for the traffic is quite close to the matched MMPP process curve using the proposed technique and Heffes-Lucantoni technique over estimates this curve.

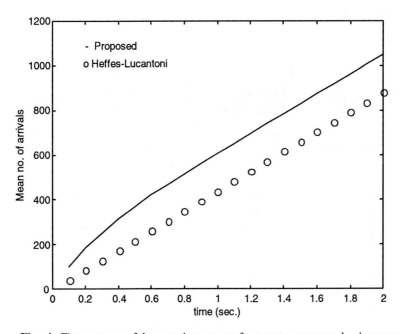

Fig. 4. First moment of the counting process for twenty superposed voice sources

It is also noticeable that the first moment approximation in the proposed matching technique starts characterizing the non-linear behavior of the first moment of the arrival process and it is in good agreement with the simulated results. This phenomenon is true even when the number of superposed sources is increased and at high traffic intensities.

In order to characterize the burstiness as well as correlation properties of the approximated process, the approximating two-state MMPP is fitted to the approximated process by matching the variance-to-mean ratio of the approximated and approximating processes [2]. Now in order to find the best approximation over a reasonably large range that can be considered as an entire range of variance-to-mean ratio the following technique suggested in [4] is used:

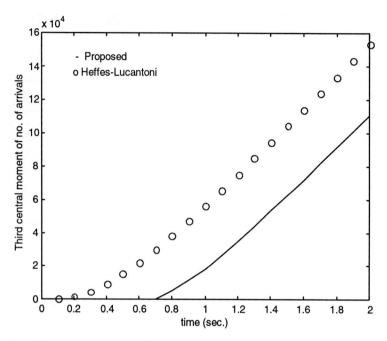

Fig. 5. Third central moment of the counting process for twenty superposed voice sources

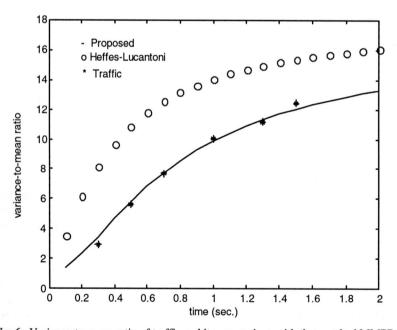

Fig. 6. Variance-to-mean ratio of traffic and its comparison with the matched MMPP process

i)   Start with any values of t_1, t_2 and calculate the parameters of 2-state MMPP.
ii) Using the parameters obtained in step (i) find the variance-to-mean ratio of the two-state MMPP using the given equation.
iii) Calculate the error which is the sum of the squared deviations of variance-to-mean ratio's, one obtained in step (ii) and the other one being approximated.
iv) Repeat steps (i) to (iii) until the error is less than a given tolerance.

In this work the duration of the matching range is chosen to be 20 seconds for the variance-to-mean ratio to match. The above selection seems reasonable because the approximated quantities reach their steady-state behavior reasonably well within this time duration. The accuracy used in this work is 10^{-6} and 200 points are considered to calculate the error. It is verified that the value of t_2 has a marginal effect on the fitting procedure. This is in accordance with the findings of [2]. The parameters of the two-state MMPP that result in minimum error are considered for the approximation purposes. It is also found that the fitting procedure is independent of the number of sources being approximated. The variance-to-mean ratio curves for the approximated and approximating processes for all the matching techniques mentioned above after implementing the fitting algorithm are obtained. The first two moments of waiting time for different matching techniques as well as the simulation are obtained and plotted.

Table 1. First two moments of waiting time in an MMPP/D/1 queue obtained using the *Heffes-Lucantoni* and the *proposed* techniques for different number of superposed sources, N ($\alpha^{-1} = 0.352$ s, $\beta^{-1} = 0.65$ s and $T = 0.016$ s)

$E_1(w)$ = Mean waiting time for the *Heffes-Lucantoni* matching technique
$E_2(w)$ = Mean waiting time for *proposed* matching technique
$E_3(w)$ = Mean waiting time obtained using the *simulator*
SD = Standard deviation

N	ρ	$E_1(w)$ m sec.	$E_2(w)$ m sec.	$E_3(w)$ m sec.	SD_1 m sec.	SD_2 m sec.	SD_3 m sec.
20	0.1464	0.02905	0.02876	0.0276	0.08538	0.08650	0.09717
40	0.2927	0.07063	0.06963	0.0706	0.14262	0.14653	0.20331
60	0.4391	0.1355	0.1324	0.1338	0.21939	0.22695	0.33184
80	0.5854	0.2524	0.2419	0.2354	0.34286	0.36610	0.49788
100	0.7318	0.5395	0.4857	0.4366	0.61945	0.72877	0.83339
120	0.8782	3.1870	1.6912	2.6110	2.23740	4.82882	8.81915

3 Conclusions

In this paper a novel parameter matching technique is proposed that uses the measurement based estimation of the moments of interarrival times presented in [1] and the general technique proposed in [2]. It has several advantages over the techniques it is based upon. One of the advantages is its practicability since on-the-fly measurements and estimation of first three moments of any traffic source can be carried out; the technique of [1] is distribution independent. This has been demonstrated through the cases presented in the results section. A matching range of 20 seconds is found to be satisfactory to adequately match the proposed parameters allowing real-time implementation of the technique. An additional advantage is the fact that the proposed technique does not require numerical inversion of Laplace transforms for estimating the moments of the arrival process to be used, hence improving practicability. Due to these features, transient behavior of voice traffic may also be reasonably modeled if sufficient time horizon is allowed.

The potential of the new technique is encouraging. Application of the technique to new cases and its extension for characterizing the multi-media traffic is an exciting research direction.

References

1. Kambo, N.S.: Distribution Free Approximations to Moments of a Renewal Process. Unpublished work.
2. Heffes, H., Lucantoni, D.M.: A Markov Modulated Characterization of Packetized Voice, Data Traffic and Related Statistical Multiplexer Performance. IEEE J. Select. Areas Commun., Vol. SAC-4, No. 6. (1986) 856-868
3. Fischer, W., Meier-Hellestern, K.: The MMPP Cookbook. Performance Evaluation. Vol.18, No. 2. (1992) 149-171
4. Gusella, R.: Characterizing the Variability of Arrival Processes with Indexes of Dispersions, IEEE J. Select. Areas Commun., Vol. SAC-9, No. 2. (1991) 203-212
5. Kang, S.H., Oh, C., Sung, D.K.: A Traffic Measurement-Based Modeling of Superposed ATM Cell Streams. IEICE Trans. Commun., Vol. E80-B, No.3, (1997) 434-441
6. Brady, P.T.: A Statistical Analysis of ON-OFF Patterns in 16 Conversations. BSTJ, Vol. 47 (1968) 73-91
7. S-Heydari, S., Le-Ngoc, T.: MMPP Modeling of Aggregated ATM Traffic. Proc. Canadian Conference on Electrical and Computer Engineering (CCECE'98), Waterloo, Canada (1998)
8. Lucantoni, D.M.: New Results on the Single Server Queue with a Batch Markovian Arrival Process. Commun. Statist.- Stochastic Models, Vol.7, No:1 (1991) 1-46

Evaluation of End-to-End QoS Mechanisms in IP Networks

Fayaz A. Shaikh[1] and Stan McClellan[1, 2]

[1] Center of Telecommunications, Education, and Research, Department of Electrical and Computer Engineering, University of Alabama at Birmingham, 1150 Tenth Avenue, South, BEC 253, Birmingham, AL 35294-4461, USA
Fayaz@ieee.org

[2] IP Architecture Group, Technology & Systems Engineering, Compaq Telecommunications, 1255 West Fifteenth Street, Suite 8000, Plano, TX 75075, USA
S.mcclellan@ieee.org

Abstract. The primary challenges in the deployment of Voice over Internet Protocol technology include Quality of Service (QoS) guarantees in the form of stringent bounds on end-to-end delay, jitter and loss, as well as rigorous validation of subjective and objective end-to-end quality. The performance of subjective tests to indirectly optimize transported voice quality requires test administrators to modify and compare the effect of network configurations for a large population of users. The impasse between control of network configuration and control of effective test design has prompted us to approach this joint optimization problem from the perspective of a programmable network testbed. The Video and Voice over IP Environment for Research (VVIPER) has been designed for this purpose. This paper describes design and implementation of this testbed, and presents samples of the performance results gathered for various popular IP-based QoS technologies.

Introduction

Voice over IP (VoIP) is one of the fastest growing technologies in the world today, and has tremendous potential for future deployment. The success or failure of VoIP deployments depend, in large measure, on the performance of the network elements that carry and route the real-time packets as well as the digital signal processing (DSP) techniques that are used in voice codecs. Additionally, there are many technical challenges that need to be overcome before deployment of converged networks is widespread. These challenges include guarantees in the form of stringent bounds on end-to-end packet delay, jitter and loss, and statistically rigorous measurements of subjective and objective end-to-end multimedia quality.

The measurement and validation of these network parameters is very difficult, particularly when phrased in the context of a highly subjective and variable phenomenon such as voice quality. Experienced network designers who have little experience in subjective testing methodologies and the particular nuances of voice quality tests tend to guide most of the technical implementations of VoIP technology. As a result, VoIP networks may be optimized according to ineffective parameters, or implementations may be grossly over-engineered in an attempt to ensure sufficient resources for multi-service real-time traffic. Unfortunately, many of the popular remedies for packet-based network issues simply aren't directly applicable to voice-grade transport. A different, more abstract or indirect form of "network optimization" is required in many of these cases.

A particularly troublesome dichotomy involved in optimizing voice transport is the apparent need for "third party control" of some intimate aspects of network configuration. This is particularly true in case of different per-hop behaviors (PHBs). To enable particular PHBs, the network administrator must perform a series of low-level configurations on IP routers and other network elements. Aside from being a fairly complex and high-security working environment, the potential is extremely high for slight misconfigurations to cause serious network outages or performance degradations. Unfortunately, in most cases, these configuration parameters are precisely the source of trouble in the end-to-end QoS of converged networks.

The use of subjective tests to indirectly optimize voice transport quality is an approach that may have substantial utility. However, the test administrator must be able to access, modify, and compare the effect of specific network parameters on end-to-end subjective voice quality for a large population of users. Additionally, the correlation between subjective results and objective measurements can yield significant insights into network performance.

The impasse between control of network configuration and control of effective test design has prompted us to approach the problem of network performance optimization from the perspective of a programmable network testbed. In the programmable network testbed, the designer/administrator of subjectively oriented tests must be able to invoke particular network configurations across a wide variety of network elements. To maintain precise, yet high-level control of the testing environment, many fundamental functions of network administration must be outsourced to the test designer. To be effective, this outsourcing of network parameters requires an easily configurable test environment, which is capable of controlling complex, heterogeneous network architectures from a very abstract perspective.

Environment Design

Conceptually, such a system requires the integration of subjective testing of voice quality and well-defined configuration of network parameters with support for sufficient objective measurements. Thus, the system must automate all activities involved in network configuration to enable the collection of subjective data from users as well as objective data from measurement equipment. All aspects of network configuration are essential parts of such an environment, including periodic variation in network parameters, call generation, and data collection, as well as a capacity for remote administration of the test environment.

The abstraction of such an environment is shown in Figure 1. The suggested environment provides a platform where various network scenarios can be deployed in a controlled fashion so that appropriate information regarding voice quality can be obtained. In this system, a network scenario corresponds to a particular QoS technology to be deployed in an IP-based network. Fundamental to the design of the system is the ability to collect large samples of data regarding transmitted voice quality using a web-based interface, and to have this corpus of data be associated with the network configuration under which it was collected. The subjective data, which is provided by untrained participants, is recorded in terms of perceived voice quality.

The objective data, which is collected from various measurement equipment, is recorded in terms of packet loss, delay, or other metrics. To create a more realistic multi-service environment, high-rate streaming video can also be injected into the network. In this fashion, QoS technologies can be analyzed for effectiveness in the simultaneous transport of several kinds of real-time traffic. Objective and subjective data collected by the system can be analyzed to determine several facets of the correlation between network QoS configuration and end-to-end quality.

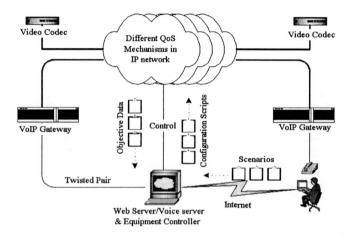

Fig. 1. Abstract Design

Web-based control over the sophisticated QoS parameters in various networking equipment is necessary for the system to provide synchronization, automation and controlled state transition during tests. A web-based call generation facility provides the additional required automation for the test participant, eliminating unnecessary human involvement in the test taking process. To simplify the construction of subjective tests, administrators can select between several configurations of network parameters, as well as separately generated non-speech "noise" which congests the transport path. This "noise" is used to challenge the techniques used in popular standardized QoS technologies as well as those specified by particular vendors.

During an iteration of a "test," participants are asked to rate the quality of voice for randomly selected audio clips, which are transmitted in real-time through several different, automatically configured network scenarios. These subjective rankings and objective measurements are collected, categorized by the network configuration under which they were collected, sorted according to the user who produced the data, and stored for later use in comparing the relative, overall subjective performance of the various network parameters. The Voice over IP Environment for Research (VIPER) was designed according to these requirements and specifications [1]. Initial design of VIPER did not include the video traffic and objective data collection features. Video and Voice over IP Environment for Research (VVIPER), a more robust and full-featured version of the older VIPER system has been implemented. This paper presents the features and implementation of the VVIPER system along with preliminary results and conclusions. The approach taken by VIPER and VVIPER is significant, because commercial implementations and configurations of various QoS technologies target the optimization of various network parameters. However, the relationship between perceived quality and "optimal" network parameters has not been approached from an integrated and automatic perspective.

QoS Mechanisms

The network QoS parameters related to the flow of multimedia traffic are bandwidth, delay, jitter and packet loss. Standardized approaches for IP-based QoS technologies fall into the categories of prioritization (Differentiated Services, or DiffServ) and reservation (Integrated Services, or IntServ). DiffServ classifies per-hop behaviors on the basis of a DiffServ Code Point (DSCP) [2] attached to the Type of Service (TOS) byte in each packet's IP header. This approach is a form of "soft" QoS, and is a rather coarse way of classifying services through packet marking. Depending on the actual queuing/forwarding implementation, the Expedited Forwarding (EF) class, typically DSCP value of 46, minimizes delay & jitter and provides the highest level of aggregate QoS. The effectiveness of transport for EF-marked packets is heavily dependent on the per-hop implementation. Another class of DiffServ called Assured Forwarding (AF), allows the administrator to assign the pre-set drop precedence to different traffic. IntServ on the other hand, emulating the resource allocation concept of circuit switching, apportions resources according to requests made by each application. Table 1 lists the QoS mechanisms whose behaviors have been tested using the VVIPER system, and following sections present brief descriptions of each technology.

Table 1. Different QoS technologies tested in VVIPER

Queuing	Other
• FIFO (Best Effort)	• RSVP (only)
• PQ	• RTP Priority (only)
• CQ	• RSVP with RTP Priority
• WFQ	• CAR
• WFQ with RTP priority	
• CBWFQ	
• LLQ with CBWFQ	

Queuing Strategies

Different queuing strategies determine the scope of congestion management. These strategies handle the incoming traffic in a particular manner by sorting and then prioritizing it into an output link. The following queuing strategies are supported by

Cisco Systems, Inc. Internetworking Operating System (IOS), and have been used in our test cases:

First In First Out (FIFO). Simply forwards the packets in the order of arrival.

Priority Queuing (PQ). Allows prioritization on some defined criteria, called policy. Four queues (high, medium, normal and low) are filled with arriving packets according to the policies defined. DSCP packets marking can be used to prioritize such traffic.

Custom Queuing (CQ). This allows allocating a specific amount of a queue to each class while leaving the rest of the queue to be filled in round-robin fashion. It essentially provides a way to allocate multiple classes priority in queuing.

Weighted Fair Queuing (WFQ). A proprietary queuing strategy developed by Cisco. WFQ schedules interactive traffic to the front of the queue to reduce response time, and fairly shares the remaining bandwidth among high-bandwidth flows.

Class Based Weighted Fair Queuing (CBWFQ). Combining custom and weighted fair queuing, class-based WFQ gives higher weight to higher priority traffic defined in classes using WFQ processing.

Low Latency Queuing (LLQ). This feature brings strict priority queuing to CBWFQ. LLQ gives delay-sensitive data, such as voice, preferential treatment over other traffic. Using LLQ, delay-sensitive packets are forwarded prior to the packets in other queues.

Reservation, Allocation, and Policing Strategies

Reservation, allocation, and policing strategies determine the commitment of network resources. These strategies handle the incoming traffic by signaling reservation requests, allocating bandwidth, and classifying streams. The following strategies are used in our test cases:

Resource Reservation Protocol (RSVP). RSVP is a signaling protocol that provides reservation setup and control to enable resource reservation prescribed by IntServ. Hosts and routers use RSVP to deliver QoS requests to the routers along the paths of the data stream and to maintain router and host state to provide the requested service, usually bandwidth and latency.

Real Time Protocol priority (RTP priority). This is another way of prioritizing voice traffic. Voice packets usually use User Datagram Protocol (UDP) with RTP headers. RTP priority takes a range of UDP ports and treats them with strict priority.

Committed Access Rate (CAR). Committed Access Rate (CAR) is traffic policing mechanism which provides the means to allocate bandwidth commitments and limitations to traffic sources and destinations while specifying policies for handling traffic that exceeds the bandwidth allocation. CAR policies may be utilized at either the ingress or egress of the network. CAR thresholds may be applied by access port, IP address or by application flow.

Environment Architecture

Figure 2 shows the network testbed set up in the CTER laboratory [3]. The main purpose of this network testbed is to provide a realistic simulation environment of an actual network which can be used as a basis for various test scenarios. This network is comprised of several pieces of state-of-the-art networking equipment, including Cisco 7200VXR routers at the core, and Cisco 3640 VoIP gateways at the edges. For core routing, we use the Interior Gateway Routing Protocol (IGRP). The edge-based VoIP gateways use the H.323 protocol for voice call signaling. For encoding and decoding of the real-time video traffic, VBrick MPEG-1 codecs are used.

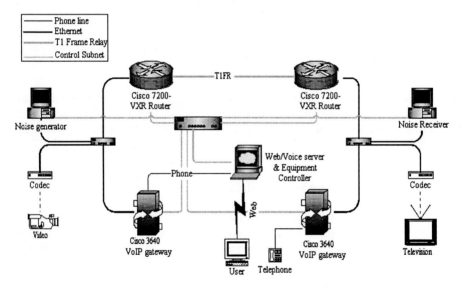

Fig. 2. Network testbed for VVIPER

Congestion and Control

An important part of the architecture is that the control subnet is separated from other network connections. This configuration allows the "control" traffic that we use to dynamically reconfigure the network to be isolated from the primary network connection which carries bearer (voice, video & signaling) traffic. For the results presented here, the core routers are connected via a T1 "trunk" with Frame Relay encapsulation. At the edge of the network, each of the Cisco 3640 VoIP gateways has a port that is directly connected to a Plain Old Telephone System (POTS) telephone handset or modem. Since congestion is a common (and significant) problem in IP networks, we use a configurable noise-generating tool to create "noise" which congests the common T1 trunk completely. All voice, video and noise traffic passes through the same T1 trunk link. Using QoS mechanisms implemented in the core routers, the multimedia traffic is prioritized or resources are reserved even though the link is completely congested.

Proxy Agents

There are six main functional components of the VVIPER system. Each component represents a proxy agent integrated into the system, except the web interface. Figure 3 shows the system components of the VVIPER software along with process sequence. In the figure, "user" represents the test. As part of the test process, the user enters session information, clicks on a dial link, listens to the phone, and submits an opinion regarding perceived voice quality. No other processes are apparent to the user. Each component in Figure 3 is explained briefly below.

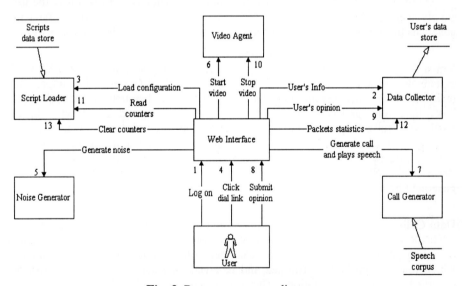

Fig. 3. Process sequence diagram

Web Interface. As shown in Figure 3, the system is controlled and integrated by the web interface, which executes each process in a controlled fashion. Using various web technologies, the system becomes remotely accessible and administrable in addition to being platform-independent on the client end. The web interface generates messages according to the sequence shown in Figure 3. Due to the fact that major functionalities are associated with the user interface, it must be actively dynamic. As a result, the VVIPER Web Interface component is implemented using PHP 4.0.

Script Loader. As soon as a participant starts a test, the Script Loader agent fetches a configuration from the script data store and executes it. Each script in a configuration scenario corresponds to a particular QoS mechanism, and is used to configure the core routers accordingly. These scripts are written using Expect 5.3.2, a command line scripting language.

Noise Generator. Noise, or congestion traffic, is generated using the Iperf 1.1.1 software tool. This tool is designed to generate the required traffic, which is then analyzed for measuring bandwidth and response time between two nodes. VVIPER uses two nodes as the noise generator and the noise receiver. UDP datagrams of equal size and equal interval are sent from the generator to the receiver to fill the trunk channel close to its physical capacity. As a result, voice and video traffic generated through calls between VoIP gateways and video codecs respectively, are challenged against the QoS mechanisms used by the routers.

Call Generator. The Call Generator agent is responsible for generating the telephone call to the user and playing a pre-recorded speech sample randomly chosen from the stored speech corpus. Call forwarding and routing is performed through the H.323 protocol used by the Cisco voice gateways. Vgetty - a Linux based tool - is used to generate the calls and pump the voice files into the channel. During a test, the user simply needs to click a hyperlink. In response, this agent of the system automatically generates the appropriate actions to connect a VoIP call through the voice gateways.

Video Agent. In contrast with the original VIPER voice-only testbed, the VVIPER system also includes real-time video traffic. The video MPEG-1 encoder compress the real-time video traffic coming from the attached camera, then transmits it to the same link where voice and noise traffic is routed. On the other end of the network, this traffic is received by an MPEG-1 decoder which, after decompressing the data, displays it on a video monitor. In the tests described here, video traffic has been given a lower priority than voice. The Video Agent performs the appropriate actions of starting or stopping transmission of the video stream according to the messages generated by the Web Interface.

Data Collector. To collect users' opinions and objective data, MySQL is used as the VVIPER database server. Each user is prompted to submit a subjective opinion regarding perceived voice quality for each network scenario. Additionally, counters which read the packets transmitted and received on each side of the network are observed for each test iteration. These data items are collected into the database, categorized by test setup and participant information, and available for remote access and administration by a test administrator.

Test Results

There are two sets of results collected from this environment. The first set of results is collected from earlier version of the system named VIPER, and is presented in [1].

The second set of results shown in Tables 2 & 3 are collected from VVIPER. The "ranking" approach used in [1] is related to the well-known Mean Opinion Score (MOS), but it has some significant differences. Since the VIPER testing attempts to obtain a relative ranking between specific network QoS mechanisms rather than an absolute rating of quality, we slightly redefine some components of the MOS approach. This results in the "pseudo" mean opinion score (pMOS). While the pMOS scores loosely represent an MOS measure of "quality", they are not directly comparable to MOS measurements obtained under other circumstances. From a comparison of the tabulated pMOS scores, the effectiveness of particular QoS techniques in a completely congested network can easily be observed. Since different QoS techniques incorporate different policies for routing and packet forwarding, the end-to-end nature of the pMOS scores tends to magnify those factors, which are most effective in preserving "high quality" voice transport via a congested network environment.

Tables 2 & 3 show results collected using VVIPER. This data consists of objective and subjective results for voice, and objective results for video. The major difference between the VVIPER results in the tables and the VIPER results in [1] is the sample size. The VIPER results presented in [1] were collected with a significantly higher numbers of users than the VVIPER results in Tables 2 & 3. According to the objective behavior of each QoS mechanism, the technologies can be divided into two categories. Table 2 shows the QoS mechanisms for which H.245 signaling call control could not establish a logical channel [4], resulting in failure of voice packet transmission. Therefore, the packet loss data for voice traffic (marked with EF) for this group of QoS technologies could not be collected. However, the average packet drops of video packets (marked with AF43) are presented in the table. This table presents the comparative performance of each QoS for video traffic transport in a congested link.

Table 2. Video packet drop in QoS using VVIPER (voice call signaling failure).

QoS Mechanism	Average Video Packets Dropped (%)
Best Effort only	72.95
RSVP only	56.15
CQ	16.26
PQ	0.36
IPRTP	73.29

Due to the failure of voice signaling in these scenarios, video packets have been treated with either highest priority or best effort service. For example, in the case of RSVP only, significant bandwidth was reserved for voice, whereas no reservation was made for video packets. As a result, the video has been treated as noise, resulting in higher average packet drop. In the case of IPRTP priority, the voice packets were the only packets treated with forwarding priority; therefore, video packets were transported with best effort forwarding. Note the similarity in drop percentage between these cases. In the case of PQ, the video packets are assigned to the second highest queue, which gets highest priority in this case since there are no voice packets. As a result, the average video packet drop is negligible. In the case of CQ, the voice is assigned the highest number of byte-counts, which means that the maximum number of voice packets will be de-queued before the next class of traffic (video) will be served. Due to the fact that video packets were assigned less byte-counts than voice, the results show that average video packet drop is slightly higher than it should have been in the case of no voice traffic.

In contrast, Table 3 shows another group of QoS technologies where voice signaling packets were successfully transported, the H.245 channel was established, and transmission of voice packets ensued. Shown in Table 3 are the pMOS ratings and average packet drops of voice and video traffic collected with VVIPER. These results were collected from a different population of users than the population that produced VIPER results in [1]. The QoS mechanisms presented in Table 3 are further divided into two groups. One group consists of mechanisms which produced subjective ratings consistent with those presented in [1], and the second group have subjective ratings which differ from previous results. We discuss these results separately.

Table 3. Subjective and Objective results using VVIPER (voice call signaling success).

QoS Mechanism	pMOS (95% Conf. Interval)	Average Voice Packet Drop (%)	Average Video Packet Drop (%)
WFQ	3.5 ± 0.30	0.00	27.79
CBWFQ	4.0 ± 0.24	0.00	14.17
RSVP + IPRTP	1.4 ± 0.43	13.66	50.49
IPRTP + WFQ	2.0 ± 0.24	56.37	28.58
CAR	3.3 ± 0.28	1.20	28.65
CBWFQ + LLQ	3.8 ± 0.22	0.00	23.62

In case of the WFQ, where packets containing interactive data are kept in front of the queue to be forwarded first, subjective quality is similar to results in [1]. Voice packets are not dropped in this case. This result shows that dropped packets are not the only factor affecting perceived voice quality. Video packets are dropped with relatively higher precedence than voice packets. In the case of CBWFQ, the situation remains similar except that CBWFQ performs relatively better objectively as well as subjectively for both voice and video. This is due to the fact that CBWFQ uses WFQ mechanisms within each class whose weight and specifications are defined by the test administrator.

The combined IPRTP and RSVP mechanism maintains a similar subjective rating regardless of the population of users. It improves the treatment of voice packets without significantly improving the subjective rating. This QoS approach (due to the limitation of the implementation) does not treat video preferably; therefore, video packet loss is significantly high on average. From Table 3, and in comparison with [1], we can see that the relative benefit of the combined IPRTP and WFQ approach is subjectively worse with this sample of users. This reduced rating is also directly reflected in the objective data of Table 3. This can be explained by considering the fact that WFQ does not recognize the DiffServ packet markings, and so eventually mixes up the real-time voice and video traffic. As a result, the presence of video seems to affect voice quality negatively.

With respect to the previous collection of VIPER test data in [1], the VVIPER subjective results of CAR and CBWFQ+LLQ listed in Table 3 have improved with the addition of video traffic and the new collection of test participants. However, from the objective results, the voice packet drop rate is clearly very low, while there is significant video packet drop. This shows that these QoS mechanisms are predictable, and are producing consistent qualitative and quantitative performance.

Conclusions

The VVIPER system explained in this paper was used to evaluate various QoS mechanisms for end-to-end transport of real-time, multimedia traffic in the presence of significant congestion. The subjective and objective results presented here indicate that joint optimization of network characteristics in the presence of general application-level traffic is an extremely complex issue. Queuing strategies, call admission controls, congestion avoidance mechanisms, and traffic shaping & policing technologies must each be optimized for adequate QoS. This is true especially when these technologies are deployed in a distributed system, where highly inter-related components can seriously affect overall performance. Different QoS technologies - each addressing different aspects of end-to-end performance and each implemented in particular ways by equipment vendors - need to be evaluated simultaneously to achieve adequate QoS for VoIP deployments. The VVIPER testbed, as presented here, is an effective approach to accomplishing these goals.

Acknowledgements

This work was supported, in part, by Cisco Systems, Inc. and Compaq Telecommunications.

References

1. Shaikh, Fayaz A., McClellan, Stan: An Integrated Environment for Voice Quality Evaluation in IP Networks. Proceedings for Advanced Simulation Technologies Conference. Seattle (2001)
2. Nicholas, K., et al.: Definition of the Differentiated Services Field (DS Field) in the IPv4 and IPv6 Headers. IETF RFC 2474 (1998)
3. Center of Telecommunications, Education and Research, University of Alabama at Birmingham
4. Collins, Daniel: Carrier Grade Voice over IP. 1st edn. McGraw-Hill, NewYork (2001)

Web-Enabled Voice over IP Call Center
An Open Source Based Implementation

Stefan Kuhlins[1] and Didier Gutacker[2]

[1] University of Mannheim
Department of Information Systems III
D–68131 Mannheim
Germany
stefan@kuhlins.de
http://www.wifo.uni-mannheim.de/~kuhlins/
[2] OSI mbH
Rudolf-Diesel-Straße 3
D–55286 Wörrstadt
Germany
gutacker@osi-online.de
http://www.osi-online.de/

Abstract. This paper describes the concept and open source based implementation for combining Internet commerce solutions and World-Wide Web access to information (e.g., Internet and telephone banking) with a voice call button that allows immediate access to a call center agent from any PC over a single network connection via *voice over IP*. A special strength of this integrated solution is that it is context sensitive. Hence, it allows a kind of collaborative browsing where customer and agent work synchronously with the same document. We call this a "Web-enabled voice over IP call center."

1 Motivation

Internet transactions cost about one-tenth as much as live telephone transactions, therefore the economic benefit of a self-service Web site is apparent. Unfortunately, self-service is not satisfying to all customers; there will always be those who prefer to speak with natural persons such as call center agents.

Imagine the following scenario: a bank customer uses Internet banking and looks for offers on his bank's Web site. He finds a very interesting offer, but now he needs further information. In order to get answers to his questions as soon as possible he makes a call to his bank's call center. There he must authenticate himself again and it is necessary to tell the call center agent about the offer in question.

In view of this scenario, the main benefit of a Web-enabled VoIP call center is that a customer only needs to click a button instead of making an ordinary call to his bank's call center. After clicking the button, a connection to the call center is established using *voice over Internet protocol* (VoIP). At the same time, the customer's data is extracted from a database on the server and the Web page loaded by the customer is determined. This information is automatically presented on the call center agent's terminal to help the agent assist the customer. Thus, questions can be answered quickly. As a consequence, customer and agent spend less time on things having nothing to do with bank services. Moreover, transactions may be done, which are not intended for the Web.

The most important benefits of a Web-enabled VoIP call center are:

- Customers get context sensitive support, e.g., for filling out forms or searching for specific content.
- Customers are not forced to repeat input, which they have already made in the Web, because it is automatically provided to the call center agent.
- Usually, calls by VoIP reduce telephone costs compared to phone calls via the public switched telephone network (PSTN). That applies not only to customers, but also to the call center, because it saves long distance toll charges that are incurred with an *800 number* [3]. Plus, there are no charges for hold time on the Internet.
- The combination of passive and active channels improves the quality of information altogether.
- If customers own only one line for surfing the Internet and making calls, they may do both at the same time by using VoIP. Therefore, VoIP removes the need for customers to disconnect their Internet connection to make a call [4].

The remainder of this paper is organized as follows: In Section 2 we specify the components of our solution. Section 3 details the implementation. Next, Section 4 gives directions for future work. In Section 5 we describe some limitations of VoIP in general. Finally, Section 6 concludes the paper with a summary.

2 Components

On principle, the required techniques for surfing and Internet telephony exist. Just an appropriate combination and integration is necessary to realize a Web-enabled VoIP call center.

The main goal of our project is to build a Web-enabled VoIP call center based on freely available software, preferably open source software. Such a solution should be suitable for small and mid-range companies, which are not able (or unwilling) to spend much money on a sophisticated system by one of the major players in the VoIP market.

On server's side, the following components are used (see Fig. 1):

- *Apache* Web server [1] for handling HTTP requests (Hypertext Transfer Protocol [19])
- *Tomcat* [2] for handling *JavaServer Pages* [17]
- *Java* [14] for implementing the application server
- *OpenH323* [10] as telephony software in the call center

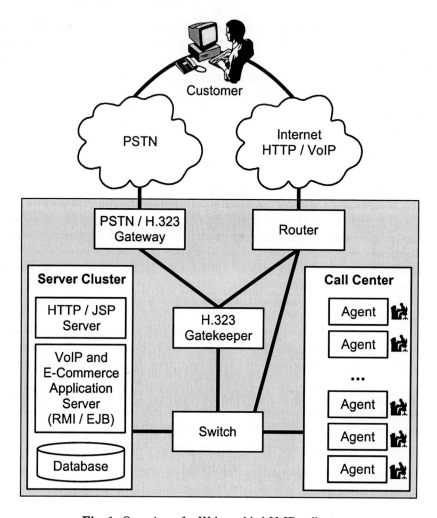

Fig. 1. Overview of a Web-enabled VoIP call center

Tomcat is the reference implementation for JSP [2]. JSP separates the user interface from content generation enabling designers to change the overall page layout without altering the underlying dynamic content [17]. The *OpenH323* project aims to create an Open Source implementation of the ITU H.323 teleconferencing protocol that can be used by personal developers and commercial users without charge [10]. H.323 is a standard for real-time multimedia communications and conferencing over IP networks [5].

On customer's side, the requirements should be as small as possible. Thus, customers only need a multimedia PC, Internet access, a Web browser, and a H.323 compliant telephony software such as *OpenH323* [10] or *Microsoft's NetMeeting* [7], which is already part of newer versions of Windows.

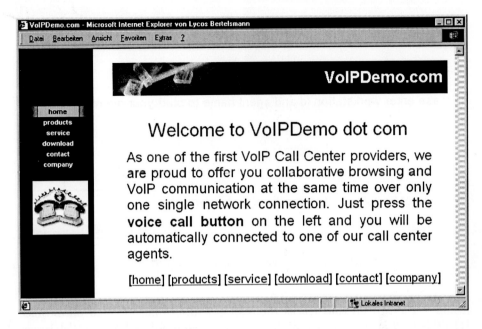

Fig. 2. HTML page with voice call button

While the communication with the customer's side has to deal with a myriad of possible software and hardware combinations, the equipment on the call center's side is well known. Hence, there are no obstacles to an installation of a comfortable environment. In addition to the requirements to the customer's side, the agents in the call center need a *Java Plug-in* [15] for the Web browser or the brand-new *Java Web Start* architecture [18], so applets or applications with a *Swing*-based graphical user interface [13] can be used.

A Web-enabled VoIP call center replaces a regular call center as well, which handles "normal" phone calls via the PSTN. Such calls are routed by a PSTN / H.323 gateway. A gateway translates between the traditional PSTN and packet-based data networks such as the Internet. A gatekeeper acts as the central point for all calls within its zone and provides call control services to registered endpoints. In the following, we will concentrate on VoIP calls and ignore this part concerning the PSTN.

3 Implementation

The primary purpose of our prototype is to illustrate the core technological concepts. In the following, we describe a typical scenario and the implementation techniques used behind the scenes.

First of all, a customer visits the Web site by loading on of its pages with a Web browser. What he sees is an HTML page (see Fig. 2), but in fact, the Web server generated this page based on a *JavaServer Page* (JSP). This page

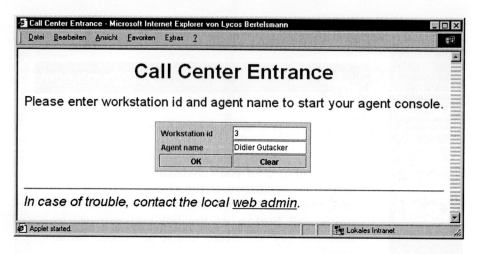

Fig. 3. Call center agent's login screen with applet

includes a *JavaBean* that tracks the session and customer related data. Since HTTP is by design a stateless protocol, sessions administered by the Web server are required to relate different requests to the same customer.

Besides session tracking, the *JavaBean* records every URL (uniform resource locator) that the customer visits. Moreover, it stores the customer's IP address, which is part of every HTTP request.

A connection via VoIP should be established as comfortable as possible. Therefore, a voice call button is placed on every page. If the customer pushes the button, a VoIP connection between the call center and the customer has to be established. This is realized by a link to a special JSP, which includes the aforementioned *JavaBean*. The *JavaBean* connects via Java's *Remote Method Invocation* (RMI [16]) to a central server object. The server communicates with a Java applet that is part of an HTML page, which is loaded by all call center agents, because it is the login for agents (see Fig. 3). With the help of *JSObject* [9], a Java wrapper around a *JavaScript* object [8], the applet is in a position to control the call center agent's browser. This is used to open a simple dialog box that asks a selected agent, if he accepts the voice call request. In case he does, the applet starts the telephony software with the customer's IP address through a system call, which is checked by the security manager.

On the other side, the customer's telephony software has to be started unless already running. The easiest way to do this is to start it manually by the user before the voice call button is pushed. Automatic activation can be realized using a signed Java applet that starts the telephony software through a system call similar to the agent's side.

Another solution, which only works with *Microsoft's Internet Explorer* in cooperation with *NetMeeting*, is the *callto* protocol [6]. This allows links to start *NetMeeting* automatically (e.g., <a href=callto:134.155.57.111+type=ip>). Aside from its proprietary nature, the call would go from the customer to the

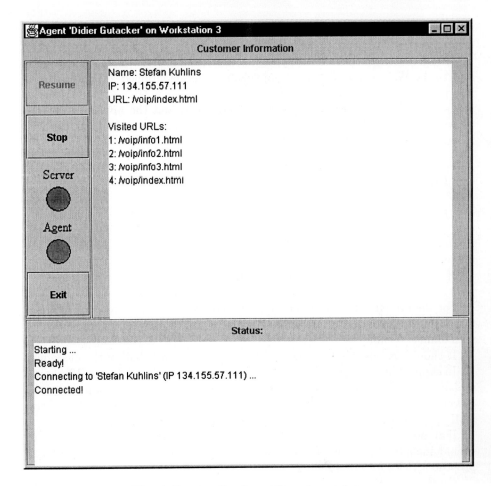

Fig. 4. Java application with customer data

call center, but we need it the other way round to avoid service malfunction and security problems. Hence, it is of no help in here, but nevertheless it is a nice feature.

If everything is fine so far, the customer is able to accept the call back of the agent and the telephony connection is established. In case no agent is available, this information is sent to the customer as an HTML page. The customer can retry later and "surf" the Internet in the meantime.

In parallel to set up the voice call connection, the server presents the customer's data to the agent (see Fig. 4). For that purpose, a Java application is started after login. If the customer has registered himself to the system his name is known. The customer's data includes visited URLs and the page that the customer is currently viewing (see Fig. 2). The voice call button of this page was pushed. This page is also the starting point for collaborative browsing. The agent sees it in a separate browser window (similar to Fig. 2), which was automatically opened by the Java application.

If the customer clicks on a hyperlink, a request for the underlying page is sent to the Web server. As a result of this, the aforementioned *JavaBean* comes into play again. It informs the RMI server about the new URL. The RMI server sends the call center agent's applet the new URL, and the applet causes the browser to load this page via *JavaScript*. Now, customer and agent are both looking at the new page.

4 Future Work

The goal of the first prototype is to provide clearer insights regarding the nature of VoIP and collaborative browsing. Hence, our initial "proof of concept" prototype works, but it is only a first step. There are many ways to improve the implementation.

If no agent accepts the call immediately, dynamic queue statistics such as position in queue and expected wait time should be provided.

For the present, the prototype supports only collaborative browsing in direction from the customer to the agent. The opposite direction can be implemented with similar techniques.

It should be possible that the agent fills in forms on behalf of the customer.

In support of a better scalability, an *Enterprise Java Beans Server* [12] should replace the RMI server. In result of this replacement, another weak point of the current prototype—the issue of persistence—would be eliminated; the central RMI server holds all data in memory, instead of using a database.

If an online shop uses the system, the customer's data presented to the agent should include the items of the "shopping cart." In this way, the agent can propose articles for cross selling.

On the customer's side, a seamless integration of telephony software into operating systems and browsers is desirable. In this case, customers are not forced to download and install telephony software themselves. Moreover, the telephony software should start automatically, similar to Microsoft's aforementioned *callto* protocol.

Provided that fast communications technologies will allow the addition of video to the connections, H.323 based video between agent and customer could be added.

5 Limitations

In the early days of IP telephony, most systems were half duplex, and all were incompatible with one another, let alone with conventional telephone systems. In the meantime, things have changed, but only when VoIP provides a level of quality that nearly equals the PSTN people will accept it.

At present, some restrictions on the customer's side concerning the technical equipment have to be considered, but in the near future, this should be no more obstacle. To achieve a suitable quality of service, customers must have an Internet connection with sufficient bandwidth. At least an ISDN line should be available. Additional equipment like a sound card, headset or speaker and microphone is needed. Today, especially headsets are not widely used.

6 Summary

Access to call center facilities via the Internet is a valuable adjunct to electronic commerce applications. Internet call center access enables a customer who has questions about a product being offered over the Internet to access customer service agents online. This combination of VoIP with point-of-service applications shows great promise for the longer term [11].

Our practical tests have shown that, apart from some minor restrictions concerning the comfort, solutions for making automated calls via VoIP and collaborative browsing are possible. Our implementation is only a first step towards a "Web-enabled voice over IP call center." More work has to be done for a portable solution with maximum comfort and speech quality.

References

1. Apache Software Foundation: Apache Web Server, http://httpd.apache.org/
2. Apache Software Foundation: The Jakarta Project – Tomcat, http://jakarta.apache.org/tomcat/index.html
3. Dialogic Corporation: IP Telephony in the Call Center, 1999, ftp://ftp.dialogic.com/www/pdf/5007.pdf
4. Genesys Telecommunications Laboratories, Inc.: Customer Service On The Internet – Redefining the Call Center, 1998, http://www.telemkt.com/whitepapers/genesys-redefining.html
5. International Telecommunication Union (ITU): Recommendation H.323, Packet-Based Multimedia Communications Systems, http://www.itu.int/itudoc/itu-t/rec/h/h323.html
6. Microsoft Corporation: CallTo URL Syntax, December 05, 2000, http://msdn.microsoft.com/library/psdk/netmeet/nm3_114o.htm
7. Microsoft Corporation: NetMeeting, http://www.microsoft.com/windows/netmeeting/
8. Mozilla Organization: JavaScript, http://www.mozilla.org/js/
9. Netscape Communications Corporation: Class netscape.javascript.JSObject, http://home.netscape.com/eng/mozilla/3.0/handbook/plugins/doc/netscape.javascript.JSObject.html
10. OpenH323 Project: http://www.openh323.org/
11. Ryan, J.: Voice over IP, The Technology Guide Series techguide.com, 1998, http://www.techguide.com/comm/voiceip.shtml
12. Sun Microsystems, Inc.: Enterprise JavaBeans Technology, http://java.sun.com/products/ejb/
13. Sun Microsystems, Inc.: Java Foundation Classes, JFC/Swing, http://java.sun.com/products/jfc/
14. Sun Microsystems, Inc.: Java Homepage, http://java.sun.com/

15. Sun Microsystems, Inc.: Java Plug-in, http://java.sun.com/products/plugin/
16. Sun Microsystems, Inc.: Java Remote Method Invocation Specification, Java 2 SDK, version 1.3.0, December 1999, ftp://ftp.java.sun.com/docs/j2se1.3/rmi-spec-1.3.pdf
17. Sun Microsystems, Inc.: JavaServer Pages, http://java.sun.com/products/jsp/
18. Sun Microsystems, Inc.: Java Web Start, http://java.sun.com/products/javawebstart/architecture.html
19. W3C: HTTP – Hypertext Transfer Protocol, March 27, 2001, http://www.w3.org/Protocols/

ANMP: Active Network Management Platform for Telecommunications Applications

Won-Kyu Hong and Mun-Jo Jung

Telecommunications Network Lab., R&D Group, Korea Telecom
463-1, Junmin-dong, Yusung-gu, Taejeon, 305-390 Korea
Phone: +82-42-870-8254, Fax: +82-42-870-8229
{wkhong,mjjung}@kt.co.kr

Abstract. The ability to rapidly create and deploy new services in response to the market demands will be the key factor in determining the success of future network service provider. There are several works to meet this requirement in the field of active network management [1,2,10–15]. However the traditional active network management is essentially based on the revision of network elements, which subsequently delays the provision of new service due to the hardware revision. This paper proposes an Active Network Management Platform (ANMP) for rapid provisioning of new telecommunications network management functions as we locate most of the necessary control functions of active network management not on the network elements but on the ANMP. To do this, this paper proposes the reference model of active network management composed of three hierarchical levels: the network element level, the active network management platform level based on CORBA, and the network management level. Based on the reference model, we propose the detailed system architecture for the active network management platform and programmable APIs designed by CORBA IDL [5] for provisioning of new service, modification of existing service, withdrawal of obsolete service and authentication. Lastly, this paper verifies the performance of the proposed ANMP with the implementation using the IONA Orbix [4] through the case study of the reconfiguration of the ATM VP tunnels connecting routers.

1 Introduction

Because the active networks play a significant role in the evolution of packet-switched networks, it leads to more general functionality supporting dynamic control and modification of network behavior from traditional packet-forwarding engines. The ability to rapidly create and deploy new services in response to market demands will be the key factor in determining the success of the future network service provider. As the high-speed switching and communication infrastructure is improved and transmission speed is increased, it is anticipated that the competition for product differentiation will increasingly depend on the degree of flexibility and the speed of deployment of new services that a future provider can offer. These factors in turn depend heavily on the flexibility of the software architecture of vendor-specific network elements.

A number of ways to provide the customer programmable network provisioning in terms of active network are as follows:

- Provisioning of software development facilities "on-board", together with a standard API. This would add considerably to the cost of network equipment, but, conceivably, it would be possible to design the switch or router of the future with a plug-in software development module as an optional feature, in the same way that today one can get an optional redundant controller card or PSU. Equipment without such a board could upload run-time code, but could not compile or debug programs [2].
- Use of a common logical model of the hardware at a node. The Java Virtual Machine is an obvious candidate here. Java is increasingly being promoted for use in embedded systems. It is to be presumed that the real-time performance required in networking applications will be supported by Java in the near future, as compiler technology is improved, and specialized hardware is rolled out. It is imperative that the technique adopted uses portable code, since otherwise we are adding complexity where we sought to remove it [2].

However these are major barriers to provide rapid provisioning of new services. It is most reasonable and cost effective for the distributed computing approach to rapidly meet the market demands because it need not directly control the network element itself and can solve the problems caused by the centralized network management. The paper proposes the distributed active network management platform for provisioning of new service, modification of existing service, withdrawal of obsolete service and authentication using the CORBA technologies such as IIOP, Naming Service, Event Service, etc [5, 6]. It provides the two major aspects of the elimination of network element revision and the utilization of the distributed nature of telecommunications network applications. To do this, this paper proposes the reference model of active network management composed of three hierarchical levels: the network element level, the active network management platform level based on CORBA, and the network management level.

The programmable APIs essentially needed in active network management are provided not at the level of network element but at the level of distributed middleware, *Active Network Management Platform (ANMP)*, taking into account the distributed nature of telecommunications network applications and the rapid deployment of new services without any modification of network elements. In this paper, we describe the programmable APIs using CORBA IDL [5, 6].

In addition, this paper verifies performance of the proposed ANMP with the implementation using the IONA Orbix [4] through the case study of the reconfiguration of the ATM VP tunnels connecting routers in case of performance degradation. It also describes the scenarios for ATM VP reconfiguration under the ANMP.

The rest of this paper is organized as follows: in the next section, we define a reference model of active network composed of three hierarchical functional levels. In section 3, we present the system architecture for the distributed active network management model, which describes the distributed active network management system components and their relationships for active network management. In addition, the distributed programmable APIs are described. In section 4, we show the performance of the ANMP in terms of ATM VP tunnel modification and creation when there are IP level performance degradation. Finally, we summarize our study and further studies.

2 A Reference Model of Active Network Management

For the rapid provisioning of new telecommunications network management functions, we revise only the distributed computing components running over Active Network Management Platform (ANMP) instead of the revision of the network element. It takes more than six months to add new service feature into the network elements because of persuading vendors to implement new service feature and reverting to legacy protocol, etc. To provision the new telecommunications services, this paper proposes an active network management model as shown in Fig. 1.

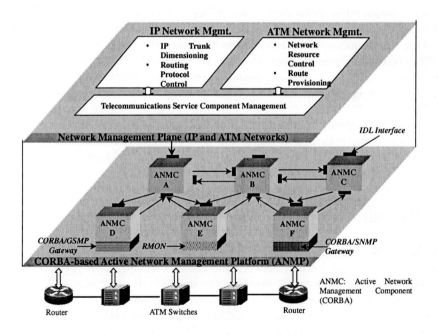

Fig. 1. Reference model for active network management

It consists of three hierarchical levels: the network element composed of ATM switches and routers, the active network management platform based on CORBA, and the network management plane. The *network element level* represents the network element itself such as ATM switch, router, etc. In our reference model, the network element is not modified or customized for supporting active network management. In case of the existing active network management, most of control functions for supporting active network management are embedded in network element [1, 2]. However we do not change any portion of network elements.

The *active network management platform* represents the distributed middleware for providing the active network management functions. Most of the active network management system components including distributed programmable APIs are resided in this level. All of the system components, Active Network Management Component (ANMC), use the CORBA services such as IIOP, Naming Service (NS), Event Service (ES), etc. It also includes the COBRA/SNMP [23, 24] gateway to interact with routers and CORBA/GSMP [14, 15] gateway to interact with ATM switches and CORBA/RMON gateway to monitor the network status.

The *network management plane level* represents the integrated IP/ATM network management functions. There are intrinsic management functions of IP and ATM such as IP trunk provisioning among routers, routing protocol designation, domain management, network resource control, VP dimensioning, route dimensioning, and so on. In addition to them, we newly define the telecommunications service component management functions taking the role of design and creation of new telecommunications services according to the policy of the network service provider to meet customer needs and face up with the future market demands.

Our active management model aims to design the necessary active network management functions to guarantee the IP QoS over ATM at the distributed middleware not at the network element. There are client/server network relationship between IP and ATM networks to support IP QoS over ATM based on ITU-T G.805 layering concept [3]. In order to guarantee IP QoS over ATM, the major IP network management functions can be Service Level Agreement (SLA) and performance monitoring and that of ATM can be ATM VP tunnel provision connecting routers and the reconfiguration of VP tunnel in the case of performance degradation and abnormalities.

3 System Architecture for ANMP

In this section, we define the system architecture for ANMP. It is the distributed middleware located between the network management plane level and the network element level. It provides the distributed programmable APIs to the telecommunications service management component of network management plane level for provisioning, modification, authentication, and withdrawal of telecommunications services. In addition, it takes the roles of the control of network elements in line with the service provisioning logic of the distributed

ANMP. The ANMP uses the CORBA [5, 6] as the distributed platform to deploy the active network management components in location transparency. Because CORBA gives the several transparencies of location, migration, interaction transparencies, etc., the system components composing ANMP can be dispersed into different systems. Fig. 2 shows the detail system components of the ANMP.

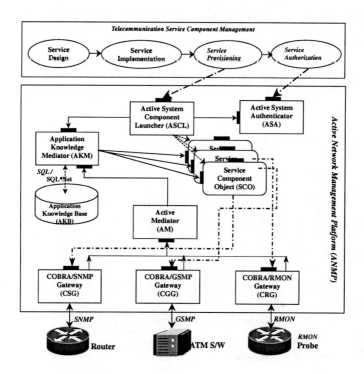

Fig. 2. System architecture for ANMP

To provide a new telecommunications service, the network service provider does following steps at the Network Management Plane:

- *Service Design* - At first, the network service provider designs the new service taking into account the modularity and interoperability with existing services.
- *Service Implementation* - With the results of service design, the network service provider implements the new service using the distributed implementation technologies such as CORBA.
- *Service Provisioning* - After implementing the new service, the network service provider uploads the implemented CORBA objects into the active network management platform via the FTP mechanism [20, 21]. The uploaded CORBA objects are composed of client and server stubs and implementation body.

– *Service Authorization* - After unloading the new service component into the active network management platform, the network service provider registers the authentication information such as the service invocation, modification or withdrawal.

The ANMP is composed of *Active System Component Launcher (ASCL), Active Mediator (AM), Application Knowledge Mediator (AKM), and Service Component Object (SCO)*. All of the objects are distributed objects running over CORBA with the location, interaction and migration transparencies. The interfaces provided by each distributed objects composing the ANMP are shown in Fig. 3 and the object interaction flow for provisioning a new service under the ANMP is shown in Fig. 4.

```
module ANMP {
    interface ActiveComponentLauncherIf {
        Result_t pushApp (in AppObject_t appObject, in AppType_t appType);
        Result_t popupApp (in AppType_t appType);
        Result_t activateApp (in AppType_t appType);
        Result_t deactivateApp(in AppType_t appType);
    };
    interface ApplicationKnowledgeMediatorIf {
        Result_t addAppKnowledge(in AppKnowledge_t appKnowledge,
                                  in AppKnowledgeId_t appKnowledgeId);
        Result_t removeAppKnowledge(in AppKnowledgeId_t appKnowledgeId);
        Result_t modifyAppKnowledge(in AppKnowledge_t appKnowledge,
                                     in AppKnowledgeId_t appKnowledgeId);
        void oneway activateAppService (in serviceInfo_t serviceInfo);
    };
    interface activeMediatorIf {
        Result_t setEFD(in EfdId_t efdId, in Action_t action);
        Result_t getEFD(in EfdId_t efdId, out Action_t action);
        void oneway notifyAlarm(in NotificationInfo_t notificationInfo);
    };
    interface activeSystemAuthenticatorIf {
        Result_t addAuthority(in AppType_t appType,
                              in Authority_t authority);
        Result_t removeAuthority(in AppType_t appType,
                                  in Authority_t authority);
        Result_t verifyAuthority(in AppType_t appType,
                                  in Authority_t authority);
    };
} // end of module
```

Fig. 3. CORBA IDL interface for ANMP

Active System Component Launcher (ASCL) takes the key role of the distributed ANMP. It provides the *pushApp()* interface that allows the network operators to deploy a newly implemented Service Component Object (SCO) into the ANMP, which has the software upload functions implemented by FTP [20, 21]. If the *pushApp()* operation is called, the Active System Component Launcher registers the provisioned Service Component Objects to CORBA Naming Service (NS) to provide the object location transparency and load balancing. It also provides the interfaces to activate and deactivate Service Component Objects such as *activeateApp()* and *deactiveApp()* as shown in Fig. 3 and 4 , respectively. After registering Service Component Objects to CORBA Naming Service (NS) in consequence of the *pushApp()* operation execution, the network provider calls the *activateApp()* operation of Active System Component Launcher to register

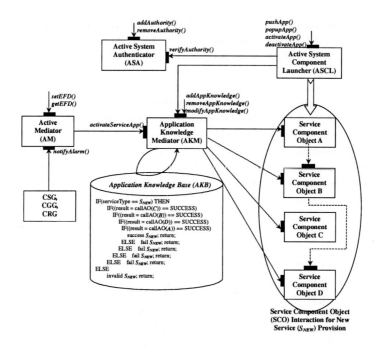

Fig. 4. Object interaction for new service provision

the necessary service provisioning scenario designated application knowledge to Application Knowledge Mediator (AKM) by calling the *addAppKnowledge()* operation. In addition, it provides *popupApp()* to network operator to remove the pre-provisioned Service Component Objects that is obsolete. If this operation is called, the Active System Component Launcher removes the named Service Component Objects from ANMP and deregisters them from CORBA NS. After removing obsolete Service Component Objects from ANMP, the network operator invokes the *deactiveteApp()* operation of Active System Component Launcher and the Active System Component Launcher subsequently invokes the *removeAppKnowledge()* operation of Application Knowledge Mediator (AKM) to remove the application knowledge corresponding to the removed Service Component Objects from the Application Knowledge Base (AKB). It provides the distributed CORBA IDL interfaces for the telecommunications application provisioning such as new service component registration, withdrawal, and service activation. In addition, it provides the CORBA IDL interface upload the implemented service components that are designed and implemented by network service provider.

Active System Authenticator (ASA) maintains the authentication information provisioned at the process of service authentication of network management plane and authenticates new service provisioning, modification of existing services, and the withdrawal of old services.

Application Knowledge Mediator (AKM) manages the Service Component Objects (SCOs) invocation scenario that is stored in database for persistency. Let's assume that there are four distributed application objects of A, B, C, and D. If one provides a new service of SNEW, the application objects of C, B, D, and A are sequentially activated. Such kinds of Service Component Object invocation rules are maintained and managed by the Application Knowledge Mediator as shown in Fig. 4. The object invocation scenario can be added, modified and removed with the *addAppKnowledge(), modifyAppKnowledge()* and *removeAppKnowledge()* operation of Application Knowledge Mediator, respectively. The object invocation scenario can be directly provisioned by the network operator or indirectly provisioned by the Active System Component Launcher (ASCL) in subsequence of Service Component Object (SCO) provisioning.

Service Component Object (SCO) represents the part of newly added telecommunications services, which is also implemented CORBA object. It has the client and server stubs for supporting IIOP communications under the CORBA distributed environment. There can be several distributed CORBA objects interacting with each other to provide a new service. The decomposition granularity of application objects to provide a new service depends tightly on the network provider's decision taking into account the object distribution overhead, reuse, etc. The provisioning of the application objects into ANMP is controlled by the *pushApp()* and *popupApp()* operations of Active System Component Launcher (ASCL).

Active Mediator (AM) provides the wicket IDL interface to CORBA/SNMP Gateway (CSG), CORBA/GSMP Gateway (CGG), and CORBA/RMON Gateway (CRG). On receiving any kinds of notifications from CSG, CGG, and CRG via *notifyAlarm()*, the active mediatator filters the alarm according to the "event forwarding descriminator (EFD)" that is managed by the interfaces of *setEFD()* and *getEFD()*. If an alarm is critical, the active mediator requests the activation of the telecommunications applications for coping with the alarm to Application Knowledge Mediator (AKM) by invoking the interface of *activateServiceApp()*.

4 A Case Study of ANMP

In case of IP over ATM environment, IP takes the roles of network management system function of Service Level Agreement (SLA), performance monitoring, etc. On the other hand, ATM takes the roles of the provisioning of the ATM VP PVCs for connecting routers and the reconfiguration of VP PVCs in the cases of performance degradation and abnormalities. In order to verify the proposed Active Network Management Platform (ANMP), we experimented the reconfiguration service of ATM Virtual Path (VP) tunnels connecting routers under environments as shown in Fig. 5. For the experimental network configuration, we use the four ASX-200 ATM switches [7] and three GRF1600 routers [8, 9].

The Connection Admission Control Object (CACO) takes the role of connection admission control on the selected routing path with the status of network resources. The ATM Routing Table Management Service Object (AMMSO) finds

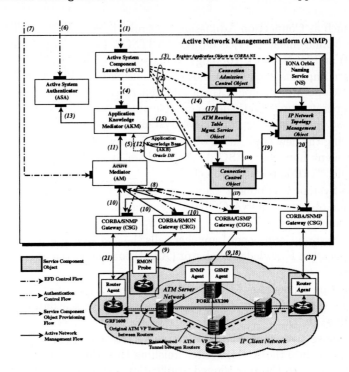

Fig. 5. ATM VP tunnelling service provisioning

the optimal route and alternative route based on the ATM network topology information. The Connection Control Object (CCO) maintains the ATM PVC information connecting routers and controls ATM switches to setup, release and modify the cross connections within ATM switches. In addition, it requests the optimal and alternative route provisioning to AMMSO in order to setup new ATM PVC or to reconfigure the existing ATM PVC suffered from congestion. The IP Network Topology Management Object (INTMO) maintains the IP network topology that is composed of routers and the IP links connecting routers. The IP links connecting routers are corresponding to the ATM VP PVC that is created by CCO. It also keeps the relationship information between IP link and ATM VP PVC, which represents the client/server relationship between IP link and ATM VP VPC.

For this case study of new service provisioning according to the proposed ANMP, we deploy the three GRF1600 routers connected with 10 Mbps ATM Virtual Path Connections (PVC). We assume that the router A detects the bottleneck at the output port that terminates the 10 Mbps ATM VPC between router A and B. The scenario to provide the new ATM VP tunnelling service into ANMP is as follows:

1. The network operator designs and implements the application objects to provide ATM VP tunnelling service that is composed of four application ob-

jects such as Connection Admission Control Object (CACO), ATM Routing Table Management Service Object (ARTSO), Connection Control Object (CCO) and IP Network Topology Management Object (INTMS). These objects are implemented as CORBA objects using the IONA Orbix3.0 [4]. The network operator complies them and generates executable codes. The network operator provisions the executable application objects to ANMP using the *pushApp()* operation of Active System Component Launcher (ASCL).

2. Receiving the new application objects from network operator, Active System Component Launcher (ASCL) locates them at the predefined location, for example, "\u1\applicationobject\".
3. Active System Component Launcher (ASCL) registers the provisioned application objects to IONA CORBA Naming Service (NS) for object location transparency.
4. Active System Component Launcher calls the *addAppKnowledge()* operation of Application Knowledge Mediator (AKM) to register the object interaction scenario.
5. Application Knowledge Mediator stores the received object interaction scenario for supporting VP tunnelling service into the Oracle database of Knowledge Base (KB).
6. With above five steps, the new active service for ATM VP tunnelling is provisioned into the ANMP. After finishing this, the network operator gives the authority to the provisioned application services using the interface of Active System Authenticator (ASA).
7. After furnishing the authority on the provisioned application services, the network operator controls the Event Forwarding Discriminator (EFD) by calling the interface, *setEFD()*, of Active Mediator (AM).
8. Because the EFD can be related with IP or ATM, Active Mediator (AM) controls the each Gateway according to the type of EFD.

With above eight steps, the new active service for ATM tunnelling is provisioned into the distributed ANMP. After finishing the application service provisioning, the provisioning scenario of the real VP tunnelling service is as follows:

9. The RMON Probe for IP segments and the GSMP Agents for ATM switches gather performance data [19] from routers and switches. If there is any symptom of performance degradation, they notify it to their corresponding gateways. If there is any fault, they notify an alarm to their corresponding gateways.
10. Each gateways notify the performance degradation or fault to Active Mediator (AM) via the *notifyAlarm()* interface.
11. On receiving the alarm caused by the performance degradation or fault, Active Mediator decides whether it is needed to activate the active network management function that is composed of several Service Control Objects (SCOs) or it is not needed to activate based on the EDF information.

12. On receiving the service activation request due to the congestion of the VP tunnel termination point, Application Knowledge Mediator refers to the Application Knowledge Base (AKB) to determine the object interaction scenario.
13. After identifying the service interaction scenario, Application Knowledge Mediator verifies whether it has the proper authority to activate the service of VP tunnel reconfiguration referring to the authority information that is maintained by the Active System Authenticator (ASA) using the *verifyAuthority()* interface.
14. With the object interaction scenario getting from Application Knowledge Base (AKB), the Application Knowledge Mediator (AKM) calls the IP Network Topology Management Object (INTMO) to get the corresponding ATM VP tunnel.
15. After getting the ATM VP tunnel causing the congestion from INTMO, Application Knowledge Mediator issues the reconfiguration request of ATM VP tunnel to the Connection Control Object (CCO).
16. The CCO requests the alternative route allocation to ATM Routing Table Management Service Object (ARTMSO).
17. On receiving the request of the alternative route provisioning from CCO, ARTMSO selects the optimal alternative route and requests the CAC to the Connection Admission Control Object (CACO). Subsequently, Connection Control Object (CCO) requests the ATM VP tunnel reconfiguration to CORBA/GSMP Gateway (SGG) [16–18, 22].
18. On receiving the VP tunnel reconfiguration request from CCO, CSG controls ATM switches (FORE ASX-200BX) to reconfigure the ATM VP tunnel according to the found alternative routes.
19. After finishing the ATM VP tunnel reconfiguration by CCO, CCO requests the reconfiguration of the VP tunnel termination points to IP Network Topology Management Object (INTMO).
20. INTMO requests the reconfiguration of ATM VP tunnel termination points to CORBA/SNMP Gateway (CSG) [23, 24]. In this case study, there are two ATM VP tunnel termination points: one is terminated at the router A and the other is terminated at the router B.
21. The CSGs managing the router A and B issues the reconfiguration request of ATM VP tunnel termination points to their corresponding router agents.

With above 13 steps from 9 to 21, the ATM VP tunnel connecting router A and B is reconfigured by the interaction of the active application objects in ANMP. If one wants to reconfigure the ATM VP tunnel with another routing policies, the network provider simply changes the application objects of ATM Routing Table Management Service Object (ARTMSO). If one wants to deploy another CAC algorithm, the network operator simply replaces the Connection Admission Control Object (CACO). Thus, this distributed active network management platform gives the high flexibility for new service provisioning at the level of middleware not at the level of network element. It takes 1.45 seconds to reconfigure the congested ATM VP PVC that traverses three ATM switches.

5 Concluding Remarks

This paper proposed the Distributed Active Network Management Platform (ANMP) that can provide the customized and rapid new telecommunications services. It could provide the rapid new service provisioning without the network element revision and maximize the efficiency for the deployment of new network management applications.

In order to design the efficient distributed active network management platform, this paper proposed the reference model of active network management composed of three hierarchical levels: the network element level, the active network management platform level based on CORBA, and the network management level. The programmable APIs essentially needed in active network management are provided not at the level of network element but at the level of distributed middleware of Active Network Management Platform (ANMP). In addition, this paper proposed the detailed system architecture for the active network management platform for provisioning of new service, modification of existing service, withdrawal of obsolete service and authentication. We also defined several distributed programmable APIs using CORBA IDL [5, 6] for the customized and rapid provisioning of the distributed telecommunications network applications.

We validated the proposed ANMP with the case study of the reconfiguration of ATM Virtual Path (VP) tunnel connecting two active nodes where there is congestion, which took 1.45 seconds to reconfigure the congested ATM VP PVC that traverses three ATM switches. With this validation, we concluded that the ANMP could be the highly applicable platform for telecommunications network management to provide the new network management functions without any modification of network elements. However, the mechanisms should be further refined to provide the programmable APIs to telecommunications service users as well as telecommunications network operators.

References

1. Kennrth L. Calvert, Samrat Bhattachharjee, Ellen Zegura, and Janmes Sterbenz: Directions in Active Networks, IEEE Communications Magazine, October 1999.
2. Martin Collier: Netlets: The Future of Networking?, First Conference on Open Architectures and Network Programming (OPENARCH), San Francisco, 3-4 April 1998.
3. ITU-T G.805: Generic Function Architecture Of Transport Networks, November 1995.
4. IONA Technologies Ltd.: Orbix Programming Guide - Release 3.2, November 1997.
5. OMG: CORBAServices: Common Object Service Specification, March 31, 1995.
6. OMG: The Common Object Request Broker: Architecture and Specification, July 1995.
7. http://www.marconi.com/products/communications/routing_switching/atm_core/asx-200/asx-200_hp.html.
8. http://www.lucent.com/ins/products/grf/grf1600.html.
9. http://www.lucent.com/ins/products/grf/grf_a.html.

10. Dan S. Decasper, Bernhard Plattner, Guru M. Parukar, Sumi Choi, John D. Dehart and Tilman Wolf: A Scalable High-Performance Active Network Node, IEEE Network, January/Febuary 1999.
11. D.Scott Alexander, Willam A. Arbaugh, Angelos D. Keromytis, and Jonathan M. Smith: A Secure Active Network Environment Architecture: Realization in SwitchWare, IEEE Network, May/June 1998.
12. Maraica Calderon, Marifeli Sedano, Arturo Azcorra, and Cristian Alonso: Active Network Support for Multicast Applications, IEEE Network, May/June 1998.
13. David L. Tennenhouse, Jonathan M. Simth, W.Savid Scincoskie, David J. Wetherall, and Gary J. Minden: A Survey of Active Network Research, IEEE Communications Magazine, January 1997.
14. Danny Raz and Yuval Shavitt: Active Networks for Efficient Distributed Network Management, IEEE Communications Magazine, March 2000.
15. D.Scott Alexander, Willam A. Arbaugh, Angelos D. Keromytis, and Jonathan M. Smith: The SwitchWare Active Network Architecture, IEEE Network, May/June 1998.
16. Balaji Srinivasan: Definitions of Managed Objects for the General Switch Management Protocol (GSMP), draft-ietf-sgmp-mib-02.txt, June 2000.
17. Tom Worster: General Switch Management Protocol V3, draft-ietf-gsmp-06.txt, January 2001.
18. Avri Doria: GSMP Packet Encapsulations for ATM, Ethernet and TCP, draft-ietf-gsmp-encaps-02.txt, July 2000.
19. Adny Bierman: Performance Measurement Capabilities MIB, draft-ietf-rmonmib-pmcaps-01.txt, 14 July 2000.
20. IETF RFC 2389: Future Negotiation Mechanism for the File Transfer Protocol, August 1999.
21. IETF RFC 2640: Internationalization of the File Transfer Protocol, July 1999.
22. IETF RFC 2515: Definitions of Managed Objects for ATM Management, February 1999.
23. IETF RFC 2571: An Architecture for Describing SNMP Management Frameworks, May 1999.
24. R. Presuhen: Management Information Base for the Simple Network Management Protocol, draft-ietf-snmpv3-update-mib-v5.txt, 9 August 2000.

An Active Network Architecture: Distributed Computer or Transport Medium

Eva Hladká and Zdeněk Salvet

[1] Faculty of Informatics
eva@fi.muni.cz
[2] Institute of Computer Science,
Masaryk University, 602 00 Brno,
and also CESNET, z.s.p.o., Prague
Czech Republic
salvet@ics.muni.cz

Abstract. Future computer networks must be more flexible and faster then today. Active network paradigm is the way how to add flexibility to networks. During the last five years, several active network architecture models were presented. A new one, based on model of active nodes is presented here. The key features of this architecture are the separation of session or connection management functions from the bulk data packet processing functions and associated session management protocol that facilitates user control over active network processing. This architecture is designed to be sufficiently general to accommodate and build on top of any packet-based networking technology. The description of the model is followed by brief of prototype implementation using PC-class computers with NetBSD operating system.

1 Introduction

Introduction of new innovative Internet applications sometimes demands special network services which are very difficult or impossible to provide in efficient manner using contemporary standard networks behaving as a passive transport medium. Some of such new complex services may be naturally supported by adding various levels of user programmability to the routing elements inside the network (routers, switches, optical cross-connect systems, etc.). In the last few years, several of such programmable network architectures are being developed, commonly called "active networks".

Active networks were introduced in order to provide more flexibility within the network itself — network with active routing and switching elements can be seen as a special type of distributed computing facility [1], [2]. Like other types of distributed computing facilities (or even more so), this may pose a lot of problems to solve, depending on type of services supplied to (end-)user. These problems range from need of high-level authentication, authorization, and accounting schemes and policies usable with multiple administrative domains through task scheduling and resource allocation, reservation, and management to lower-level performance, scalability, and device management issues. Also, in some cases confidentiality, non-repudiability or other special security services may be required. While not all these concerns are applicable to all active network architectures and service levels that may be supported, it is important to make sure they have been addressed before real-world deployment of active network.

Active networks provide excellent development environment for new network protocols. Our interest in active networks stems from this ability, combined with the interest in the real time transmissions of the audio and video streams and the associated requirements for quality of service (QoS). We find currently available technologies insufficient for use in situations where more sophisticated QoS management is needed. In typical IP networks, best-effort approach is used, and even best available implementations of RSVP and ATM protocols are limited to raw bandwidth and transmission delay management only. There is no support for more structure- or content-oriented features like, for example, multi-priority packet drop and hierarchical data streams. (Moreover, ATM is still and may remain too expensive for connecting ordinary end-user machines.) We try to construct AN architecture, which will help us not only to develop new network protocols tailored to new applications, but also gives us possibility to implement new QoS capabilities.

We have developed an active network architecture using the concept of active nodes and we are working on the implementation of this architecture using software routers based on the NetBSD (UNIX) open-source operating system. The main features of our active network model and its (prototype) implementation are influenced by the desire for support of novel QoS-oriented features that we plan to implement using the new active network model and its first implementation.

2 The New Active Network Model

The proposed active network architecture uses "active node" approach to active networking and concept of "sessions" similar to connections in connection-oriented networks or sessions in RSVP protocol.

Structure of an active node (router) plays a key role in our model. It is a network element which is able to accept user-supplied programs and to execute them. The processing of user code consists of two separate (communicating) processes. The first one controls the session establishment and management. It has the role of control plane in active router processing and includes (initial or later) load of user functions onto the routers along the path between source and destination address or addresses and execution of bookkeeping functions. The second part of user code, initiated by the first one, forms the central part of the data packet processing itself.

An *active session* consists of state information in active routers that controls handling of traffic among communicating endpoints. The most important part of an active session is the set of running instances of user code session programs. They manage other parts of state, i.e. spawning per-packet processing code, setting input packet filters and classifiers and configuration of output queues, using the active node programming interface (API). The session itself can be controlled (created, spread to new active nodes, modified by setting various parameters, partially or fully destroyed) in response to user requests but also from the inside — by session programs. An active session can be managed either remotely using *session management protocol* or locally using management API of active node management software. In the typical scenario, user application communicates with first-hop active router using the session management protocol and starts the first instance of appropriate session program with desired parameters. The session program then computes some additional (next-hop) active nodes that should take part in active processing, starts new instances of session program on them and programs per-packet processing functions. The whole process repeats with any new active nodes used. If any failure occurs during session setup, crankback technique can be used to find alternative paths in network. The location of first-hop active router (set of routers) may be either included in the end user configuration data (manual configuration, DHCP data, etc.) or determined by lower-layer specific active router discovery protocol. The software running on the active node consists of four sets of programs:

- *session programs* that are supposed to take care of establishing connections to other active nodes, computing routing information, negotiation of QoS parameters and other functions related to user-defined session management. They are supplied by user or other active node during session establishment along with paramcters and can be stopped or changed later during session lifetime.
- programs performing *per-packet processing* (forwarding, replication, policing, packet scheduling, etc.). These are supplied by user or other active node during session establishment as parameters or embedded parts of session programs. Session program can manipulate them and configure their relationships with standard packet processing modules (packet filters, queues, routing tables, output schedulers, device interface drivers).
- *basic system software* which implements session establishment and management protocol, basic execution environment for programs using native code, task management and basic resource management, security infrastructure (authentication and authorization policies), local session management API and facilities for device management by network administrator. This software can be changed only with administrative privileges.
- pluggable modules for optional *system software extensions*, e. g.
    - address/protocol families (the architecture is independent of lower networking layers)
    - authentication, authorization, and accounting protocols
    - program interpreters and runtime support (e. g. interpreters for interpreted languages, runtime linker, JIT compilers, module that downloads given URL for execution, etc.)

The model is designed as general as possible in order not to be restricted by the architecture in later development and deployment stages. All extensible information elements in session establishment and management protocol are designed with tagged polymorphic types. These modules can be downloaded by system software on demand from the code repository defined by the administrator or managed with device management facilities. IPv4 and IPv6 protocol families and C language binding will be supported by the prototype implementation.

The proposed active network architecture uses a connection-oriented (state keeping) approach. Obviously, this approach is more prone to scalability problems than capsule or "fixed function set" schemes but we believe that only small fraction of all data streams in typical network will require special QoS features or other algorithms running on active nodes, so the amount of state information which must be kept within the network elements will be almost always sufficiently low and not exceeding the processing and memory capacity of the active nodes. In overload situation, new session establishment request can be simply refused, currently running session programs can be asked to release some of already allocated resources, or alternative (possibly suboptimal in other situations) paths in active network can be searched. Second, our *active network* is meant as overlay network on more conventional network that may use "dumb" but fast elements for data transmission only (these are located especially in the core of the network) and the active elements will be placed only in key spots of the network where special features are needed (typically near bottleneck links, at the ingress and egress of the fast core, and on gateways between different networking technologies).

Using our session establishment and management protocol, it is possible to request change in parameters in atomic way (something that many signalling protocols cannot do without tearing down the running connection and trying to reestablishing it with new parameters) and negotiate permissible values of parameters (unnecessary trials and errors can be avoided in algorithms that are adaptable to various environmental conditions). The active nature of sessions have some other pleasant consequences:

- In case of failure or overload of some element (even not active element itself but some "dumb" router or physical link in between), connections may be re-established or rerouted automatically without any end user or network administrator intervention and without the need to implement these actions in performance-critical packet processing code.
- The session parameters can be changed dynamically by end users — such a change of parameters is processed in the similar way as the initial establishment of the session.

The active router model extends the "dumb" router functionality through a family of new protocols. The protocols can be divided to two groups relating to two phases of the active packet processing — protocols for network connectivity and program loading and those for processing loaded programs. Security of this active network model is based on authentication and authorization protocols not described in this paper and security properties of other used protocols.

One of our primary goals was to develop a general network protocol which would allow us to use existing network protocols or their modifications at the presentation and application layers. To achieve that, active node code may be authorized to process not only packets from active networking aware applications but also data sent by selected "legacy" applications. In such a case, separate application has to initiate the session with active node that resides on path used by application data and use appropriate input packet filter to capture the desired packet flow.

In session management protocol, the program is defined in generic way — only by code type, code length, and the program code itself. This allows different types of programs (e.g. compiled C or interpreted Java) to be run on the same active router. Due to split between session management and per-packet processing functions, it is possible to use different languages for these functions. Typical use of this feature would probably include running optimized native code in performance-critical per-packet program and implementing rest of code in convenient scripting language. Different performance characteristics of different types of code should be taken in account when calculating processing resource limits (CPU, memory, etc.), but this is an open issue in our architecture and subject to further research.

2.1 Data Plane

The actual processing of active data packets is done by program, which is (or whose identification is) supplied to the session program as part of its parameters. The session program is responsible for allocating necessary resources and setting parameters of per-packet tasks (either supplied by user or determined from the present active network state) and calling the active node API to start it up. The per-packet program is executed in a loop, this loop is permanent while the session is alive.

The active program registers active packets filter — e.g. the IP source and destination addresses, and the port number or precomputed flow label value — for this particular active connection and relies on the active router which sends all such packets to the active program loop. The packets are processed (e.g. modified, deleted, duplicated, ...) and than sent to appropriate active router port or ports (e.g. in case of multicast-like data). The fairness between different active programs running on the same active router is guaranteed by the active router core, which uses on one side the authorization data and on the other the statistics which are collected for each individual active program (CPU, memory consumption, etc.). The active programs are scheduled with respect to already consumed resources and to the actual resources they are authorized to use. In such a way the order of processing of data packets of different active connections is not strictly dependent on packet arrival order, but it is controlled by active program scheduler.

3 Implementation

We are working on implementation of the architecture described above using software routers based on the NetBSD (UNIX) operating system and off-the-shelf IBM PC-compatible hardware. Such routers are widely used in our metropolitan area network and they represent a general programmable platform suitable for implementation of novel network architectures and protocols. In addition to our previous experience, another reason to choose the NetBSD operating system was its well designed infrastructure for creating alternative OS personalities (emulations) which enables us to create (relatively easily) an execution environment where native binary (e.g. compiled C or C++) active programs can be securely executed.

While this operating system environment influences decisions about computer languages for implementation and the use of particular system functions, the active network model is not restricted in any way to this environment (i.e. the NetBSD based PC routers) and is implementable in any general enough computing environment.

Adapting a PC router to an active router means to introduce new features on the kernel and user levels. The most important components of our active router architecture are depicted in the Fig. 1. On the user level, the authentication and authorization module must be added, together with the full environment for the session and packet processing programs including parts of the resource management and statistics collection.

On the kernel level, it is necessary to create restricted execution interface for secure execution of native binary active programs, to add more general interface to packet classification code (standard part of PC router) which recognizes data packets of active connections, and also to add some changes to support more fine-grained scheduling for active programs.

Our implementation uses the C language to implement both the user and kernel level modifications and extensions of the PC router. While this means that the implementation will take more time then using high level approach with Java or similar programming environment (the implementation times compared in [4] are 3 years against 6 months), the result should be much more efficient. We believe that highly efficient implementation is necessary if we are to show the advantages of active networks.

4 Related Work

During the relatively short history of active networks research, several models or network architectures have been developed. These models can be divided to two basic groups based on means by which user code is supplied to programmable elements of the active network. The first group places the code of active program inside the data packet that will be processed by the program ("active packet" or "capsule"). The other approach ("active nodes") uses other ways of program distribution, e.g. preloading of fixed function sets active nodes, on-demand program downloading, distributed caching of programs, etc. For example, typical

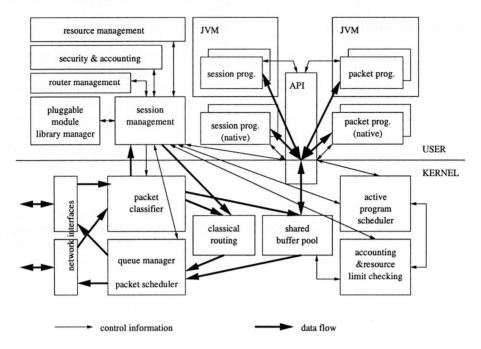

Fig. 1. Active Router Architecture

representatives of the first group are *Smart packets* project [7], *ANTS* [8], and *PAN* [9]; representatives of the second group are *SwitchWare* [5] or *Pronto* [6] systems. Both these two architectures have specific advantages and could be combined in one model.

Pronto is quite similar to our implementation of active node operating system support, in fact it could serve as alternative base for implementation of the architecture presented. The most important difference is lack of support for untrusted binary code execution in Pronto (user code interfaces with C++ class library).

The SwitchWare architecture also has some features in common with the model we proposed, e.g. support for multiple programming languages for user code. On the other hand, in SwitchWare, active packets always carry code in addition to data and this code cannot be completely preloaded on active node, unless it is defined as extension of active router.

5 Conclusion

The active network protocol design and its implementation within PC routers was motivated by the work in the area of quality of services for multicast, and voice and video applications in connectionless networks [3]. The active network technology looks promising for solving these and related problems and it also provides convenient environment for research, where different QoS and/or multicast protocols can be implemented and tested.

Acknowledgements

The authors would like to thank Luděk Matyska for support of our work, proof reading this text and stimulating discussion about the quality of services in computer networks now and in the future.

The work was supported by the CESNET Research intent MSM000000001, and one of the authors (EH) also acknowledge support from Universities development fund grant No:0430/2001.

References

1. D.L. Tennehouse, J.M. Smith, G.M.W. Sincoskie, D. J. Wetherall and G. J. Minden. *A survey of active network research.* IEEE Communications Magazine 35, 1997.
2. K. Psounis. *Active networks: Applications, security, safety and architectures.* IEEE Communication Surveys, 1999.
3. D. J. Wetherall et.al. *Introducing New Internet Services: Why and How.* IEEE Network, May/June 1998.
4. J. P. Gelas, L. Lefevre. *TAMANOIR: A High Performance Active Network Framework.* Workshop on Active Middleware Services 2000 at the Ninth IEEE International Symposium on High Performance Distributed Computing, Pittsburgh, Pennsylvania, USA, August 2000.
5. D. Scott Alexander, William A. Arbaugh, Michael W. Hicks, Pankaj Kakkar, Angelos D. Keromytis, Jonathan T. Moore, Carl A. Gunter, Scott M. Nettles, and Jonathan M. Smith. *The SwitchWare Active Network Architecture.* IEEE Network Special Issue on Active and Controllable Networks, vol. 12 no. 3, pp. 29–36.
6. G. Hjálmtýsson. *The Pronto Platform: A Flexible Toolkit for Programming Networks using a Commodity Operating System.* In IEEE OPENARCH 2000, March 2000.
7. Schwartz, Beverly I., W. Zhou, A. W. Jackson, W. T. Strayer, D. Rockwell, and C. Partridge. *Smart Packets for Active Networks.* Proceedings of InfoComm, New York, 1999.
8. D. J. Wetherall, J. V. Guttag and D. L. Tennehouse. *ANTS: A Toolkit for Building and Dynamically Deploying Network Protocols.* In IEEE OPENARCH'98, April 1998.
9. E.L. Nygren, S.J. Garland and M.F. Kaashoek. *PAN: A High–Performance Active Network Node Supporting Multiple Mobile Code Systems.* In IEEE OPENARCH'99, March 1999.

An Active Network for Improving Performance of Traffic Flow over Conventional ATM Service

Emad Rashid[1], Takashi Araki[1], and Tadao Nakamura[2]

[1] Dept. of Science and Technology,
Hirosaki University, Japan
`rashid@ieee.org`
[2] Graduate School of Information Sciences,
Tohoku University, Japan
`nakamura@archi.is.tohoku.ac.jp`

Abstract. In Unspecified Bit Rate (UBR) service provided by ATM networks, user sends cells into networks with no feedback, no guarantee cells may be dropped during congestion. This paper describes a congestion control scheme using active network technology to avoid congestion in a conventional ATM network. Ants routing algorithm for congestion control and associated with active network fashion is implemented. The routing algorithm uses information that is monitored and revised according to the congestion status of output ports and sends commands and executable bits stream to other switching elements. The Active switching elements share information about traffic levels of an interconnection network and reciprocate customized programs to rerouting cells in the presence of congestion at switching element's output ports. As a result, it is possible to build active switching network that can avoid congestion in a conventional ATM network.

1 Introduction

The Unspecified Bit Rate (UBR) [1] service category provided by ATM networks does not specify traffic related service guarantees and has no explicit congestion control mechanisms. In multimedia communication, the aggregate number of cells is characterized in a variable transmission rate, and the number of cells arriving at each link may be too large for a switch's buffer to handle. However, the cells will continue arriving at switching elements even if there is congestion in a network. As a result, the network may be prone to congestion. This phenomenon leads to a high cell loss probability inside the network, and to low network efficiency. There are several requirements for the implementation of future switching systems in a B-ISDN environment. The most important things to be emphasized in the switch designs are as follows: First, the internal fabric of a switch should be able to route cells from any of its input ports to any of its output ports without blocking. The blocking characteristic seriously limits the throughput of a switching element. Second, a switch should be intelligent and to be active on demand to avoid interconnection congestion and to gain high throughput and recognize from cells that carry active information or statistic (Passive) information. To satisfy these requirements, this paper proposes a new control scheme in switching element that has the ability to use active information to update its information, and has capability to share active packets with adjacent switching elements concerning the congestion levels in the interconnection fabric. The congestion is controlled by regulating the input traffic of cells in a switching element that has congestion at one of its output ports.

The UBR service, however, is intended for non-real-time applications, i.e., those not requiring tightly constrained delay and delay variation. Therefore, the cells that have specified as UBR cells can be deflected to reduce the high congestion level at that switch's output port. The cells that have been deflected will be distributed to arrive at other links. The control mechanism is achieved by antecedent rerouting some of the cells that are destined for a heavy-traffic link. This new routing technique, called *Ants routing* [2] is based on the information that has been collected in the switching elements concerning the traffic status of the output ports.

The performance of the *Routing* scheme is then evaluated in terms of the probability of cell loss. The analytical results demonstrate that the *Ants Routing* scheme provides significant advantages by relieving network link's congestion and reducing the consequences of the switching elements' buffer sizes. They also show how various parameters can be used to satisfy the requirements concerning cells loss in an interconnection network. The congestion control mechanism can lead to a lower cell loss probability and a higher utilization of interconnection network resources by deflecting some cells to a route that has light traffic.

The cell format is native ATM cell, which carries 48 Bytes data and 5 bytes header. There is no encapsulation issue on the cell traveling from switch to switch, but the different is the contents of the data carried by the cells whether the cells are carrying active data or passive data. However, at switch element, it has being business as usual for switching operation cells in and out, the switch will take action only if the cells arrived have active information and addressees to its destination.

2 An Active Network

In this section, we provide a detail of active packet format and how can a node distinguish between packet that has active information or programming code with packet that has passive information. The switching node also needs to have supplementary components to encounter a new format of active packets.

2.1 Active and Passive Packets Format

There are two types of the packets flow in the network at the physical layer; first types of the packets are user information packets or passive packets where switching nodes just route these packet from source to their destination. The second types of the packets are programming code information packets or active packets where the switching nodes take consideration to accumulate the codes and execute them to perform specific action such as active congestion control or change routing table. The switching node needs to know the type of the packet and to discern which packet is active or passive. This paper proposes a way that the switching node can discern the kind of the packet by modifying the unused bits in ATM header, i.e. Payload type PT, and from Virtual Path Identifier (VPI) and Virtual Channel Identifier (VCI) in the ATM header format.

ATM standard defines a fixed size cell with a length of 53 octets comprised of a 5-octet header and a 48-octet payload. Two headers format are also defined for user-network interface UNI and network-network interface NNI (as shown in Fig. 1).

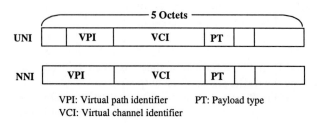

VPI: Virtual path identifier   PT: Payload type
VCI: Virtual channel identifier

Fig. 1. Cell header at the UNI and NNI

ATM is a connection-oriented protocol needs a connection to be established before user traffic can start flowing between communication entities. The connection setup is preformed via signaling processing to establish new VCI, VPI and virtual path connection identifier VPCI. The VPCI is introduced to discrepate the connection whether the connection between an end station and a network or between two networks. For active packets that directed to a specific node the value of VPCI and VPI are equal in the routing table at that node. Another active packet discerner is payload type field in the ATM header, this field is three bits size and used to carry information that network needs for its operation and maintenance as well as congestion indication bit. The payload type identifier has some values reserved for future functions, e.g. PT value (111), this value can be used for Active packets and the node can distinguish active packets from passive packets.

2.2 Active Switching Node Architecture

The switch element (shown in Fig. 2) has Active control unit that performs routing of passive packets and executing a program code in active packets. Beside the routing table that switching node has, there is a supplement memory called *Binary Link Status (BLS)* table for storing a congestion status of switching input/output ports. The information in the *BLS* table that shares with adjacent switching elements has three traffic levels (weights). These levels value represent congestion level at the input/output links that are used to reroute the packets in the interconnection network.

The control unit (shown in Fig. 3) is perform routing of passive packets in interconnection network by modify the VPI field which is corresponding to the VPCI value in the routing table. When the packet arrival at the node has VPI value equal to the VPCI in the routing table and also has PT value equal to (111) the node will consider this packet as active packet. The node will accumulate all active packets arrived in the accumulator part and loaded to the memory unit to be ready for execution. The effect of program execution is the production of processed data or results, which may implement modifying the routing table or updating the BLS table related to congestion level at the links.

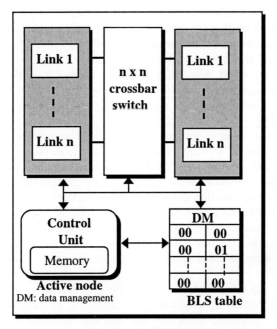

Fig. 2. Switching Architecture

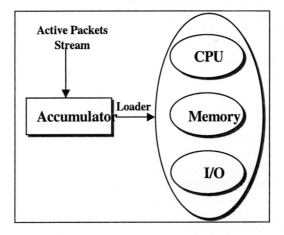

Fig. 3. Active node Control Unit

3 Congestion Control

A multi-stage non-blocking interconnection network is proposed as a means of constructing self-routing switch elements ATM interconnection networks. Neighboring switching elements are connected to each other through control links, to transfer information about the traffic status of links.

The traffic on a link may be at a light or normal level β a heavy level ζ or a congested level Ξ, where β, ζ and Ξ are represented in binary form as 00, 01, and 10, respectively. All this information on each link's status is stored in the *BLS* table (as shown in Fig. 1).

A state of congestion is detected by the buffer controller which is part of control unit, when the load at an output buffer exceeds the throughput of the link, that is, when the number of cells inside the buffer is larger than a certain threshold. Furthermore, the controller supervises the traffic passing through the output buffer. In our switch, we assume that traffic at an output buffer is normal when the number of cells in the buffer is less than 80% of the buffer size, heavy when the number of cells in the buffer exceeds 80% of the buffer size, and congestion when the number of cells in the queue reaches or exceeds 90% of the total buffer size. The buffer controller can monitor these states and set a value in the corresponding field of the *BLS* table.

Each switch has three kinds of control: First, in normal traffic, the switch's control unit routes the cell to a destination port according to each arriving cell's header. Second, in heavy traffic, when the link has few available resources, the switch's control unit starts deflecting cells from their original links to other (alternative) links. Third, when there is congestion at one of the switch's output ports, the link has no more available resources to accept arriving cells. Therefore, the control unit has to change the information in the *BLS* table that is related to the traffic status of the switch's input port. The *BLS* table's new values for the input ports will be transferred to neighboring switching elements. The status of each link based on the three traffic levels are represented in binary form (00, 01, 10) and stored in the *BLS* table.

3.1 Cell Sequence Integrity Assurance

In the Routing scheme, some cells that have the same virtual path may be deflected to alternative routes. Therefore, cells with the same virtual path will travel by using different routes. Several problems, however, may arise for deflecting packets: these problems are the arrival sequence of cells and delay of cells. The cells may arrive at the destination out of order. In other words, later cells may arrive before earlier cells.

For cell delay problem, the routing scheme is applied to UBR service, which is intended for non-real time applications, such as file transfer and email. Therefore, cells delay are not requiring tightly constrained.

In ATM networks, all cells that have the same virtual path must be received at the end of the connection in the same order as they were transmitted. It means that all cells must be in sequence. In normal traffic conditions, since cells with the same virtual path take the same path across the interconnection network, it is possible to keep track of the cells' sequence. When the traffic at the output port becomes heavy, this will be reflected in the value in the BLS table. The value of the corresponding output link in the BLS table will change from state β(00) to ζ(01) Therefore, cells addressed to this output port will immediately be affected by the change and will be routed to other links. As a result, cells that have the same virtual path may have different actual paths. The delivery of the cells in the correct sequence and with no errors should be guaranteed. In our network, the cell sequence correction process can be performed at the end of the connection, that is, at a destination node. The destination consists of the cell distributor CD, main buffer MBF, lateral buffer LBF, and selector (as shown in Fig. 4). The LBF, which is similar to the virtual channel described by Dally [3], consists of several independent lanes, each of which can hold one cell. In addition to the address field in the cell header, there is another field called the cell sequence number (SN) to ensure that the cells' sequence is maintained. The cells' sequence numbers were fixed when the cells are generated by a source node. The main function of the distributor is to check at the sequence number (SN) field. Cells that have the correct SN will be sent to the main buffer, and other cells will be delayed in the LBF until all previous cells in the sequence have passed through the MBF. The purpose of this scheme is to sequence cells so that the cells of the received message will be in the same order as those of the transmitted message.

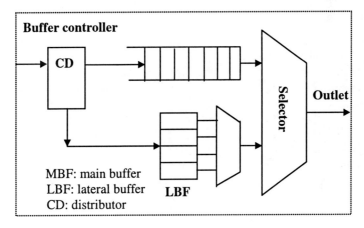

Fig. 4. Buffer controller at the destination node

3.2 Analytical Results

This subsection presents the results obtained through the analysis in the previous subsection. The purpose of the analysis is to compare the cell loss probability with the *Ants Routing* strategy and that without the *Ants Routing*. So far, the analysis of the performance of the switch elements under the *Ants Routing* scheme has been based on the input load deflection of the switching elements. The arrival rate of cells at a switch with congestion at one of its output ports will be reduced to a specific level. The deflected cells will be distributed among other switching elements that have low traffic congestion levels. The results show the probability of cell loss at one stage that has congestion in one of its switch elements. Our main concern is to minimize the size of the buffer required, while maintaining a low cell loss probability, i.e. around 10^{-9}.

As an example, we assume that the number of switch ports is 4 and that the number of switching elements at one stage is 4. We suppose that the traffic level at a certain output port reaches the congestion level ($\lambda_n^o = 0.9$), which means that the traffic arriving at a certain output port will occupy 90% of the output buffer. The cell loss probability is raised by increasing the load on one output port. One approach to maintaining a low cell loss rate is to increase the buffer size. However, increasing the buffer size is costly and contrary to our policy of minimizing the buffer size. Fig. 5 shows the cell loss probability as a function of the buffer size for various values of the offered load ρ on each switch element at a stage. The figure shows the cell loss probability without the *Ants Routing* scheme (dashed curves), and with the *Ants Routing* scheme (solid curves). We set the deflection rate at γ = 0.3. The figure that the required buffer size with *Routing* scheme must be more than 30 cells per output port in order to preserve a cell loss rate of less than 10^{-9}. The required buffer size (H) is very important for minimizing the switch element size and for reducing the switch cost.

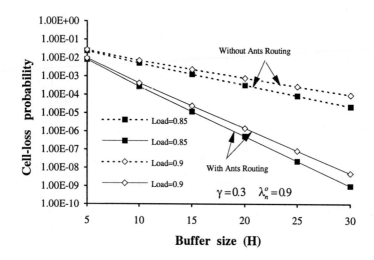

Fig. 5. Cell-loss probability relative to the load on a certain output port

Fig. 6 shows the effect of buffer size in the *Ants Routing* scheme on the cell loss probability for various load levels ($\lambda_n^o = 0.85, 0.9, 0.95$) at a congested output port λ_n^o. Note that dashed lines in the figure indicate a network without the *Ants Routing* control scheme. From the results (shown in Fig. 6), the required buffer size depends on the loads in the switch elements and the load at a certain switch's output ports. Therefore, if we set the deflection rate to γ= 0.4 and the offered load ρ on the switch element to 0.9, we can satisfy the requirement of a cell loss rate below 10^{-9}. Consequently, *Ants Routing* can be a powerful technique for minimizing buffer requirements, particularly for non-uniform traffic. It can be seen from the figure that the required buffer size without using the *Ants Routing* scheme has to be very large to obtain the same performance as with *Ants Routing*.

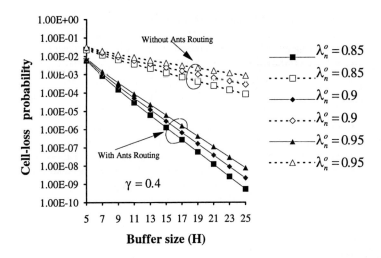

Fig. 6. Probability of cell loss relative to the buffer size (H) for various load levels at a certain output port

4 Concluding Remarks

In this paper, we propound a new self-routing switch algorithm called Ants Routing for congested networks. We asserted that switch elements should be intelligent and should activate switch control algorithms in response to traffic congestion. The switch network performance can be enhanced using active network technology to avoid congestion in a conventional ATM network. Analytical results show that our technique can significantly reduce the cell loss probability and relieve network congestion by deflecting cells to other paths. As a result, switch interconnection network resources can be used more efficiently.

References

1. ATM Forum Technical Committee: Traffic Management Specification Version 4.0, (1996) af-tm-0056.00.
2. E. Rashid, H. Kobayashi, T. Nakamura: Ants Routing: An adaptive Packets Flow Control Scheme in Multimedia Networks. Proc. of the IEEE ICUPC '93 (CANADA), (1993) 228-234.
3. W. J. Dally: Virtual-Channel Flow Control. IEEE Transactions on Parallel and Dist. Sys., vol. 3, no. 2, (1992) 194-205.

An Active Programmable Harness for Measurement of Composite Network States

Javed I. Khan and Asrar U. Haque

Internetworking and Media Communications Research Laboratories
Department of Math & Computer Science, Kent State University
233 MSB, Kent, OH 44242

Abstract. In this paper we present the active harness, a scalable and versatile means for deep probing of network local states. The approach makes a clear separation between the "communication" from the "information" part of the probing process. The composition of the "information" component is handled by means of network embedded harness plug-ins. It can facilitate not only a new generation of network aware applications but also network problems.

1 Introduction

Measurement and exchange of state information inside a network is a meta service. The ability to construct any global optimization service, whether in network layer or above, depends on how well the decision systems are aware of and up-to-date on the relevant local states in pertinent network elements. For example, local link state information is the building block for routing services and BGP, OSPF, or MPLS [2] all require sophisticated cost metric from local nodes following complex propagation patterns. There are also host of other advanced network services (such as dynamic QoS provisioning, mobile IP forwarding), and new generation of applications (such as distributed server, proxy prefetch) all requiring exchange of such network state information. A prerequisite to any QoS provisioning, whether by dynamic resource allocation or by reservation, is the knowledge about the resources. Distributed caching requires information such as current congestion and current delay to assign best mirror server to a request. Clearly, with the complexity of the Internet and the push for optimized services and applications it is becoming increasingly more important to find the means for monitoring and propagating various complex network state measurements. In this paper we present the concept of **active harness** designed for provisioning this crucial meta internet service. Before presenting the proposed system below a brief account of the current state of the Internet probing technologies is given.

1.1 Related Work

Simple Network Management Protocol (SNMP), designed in 1990 and (RFC 1157 and RFC 1155) latest updated in 1995 was the first to provide a simple but systematic way of monitoring the Internet states. It's current version SNMPv2 [2,11] now provides a simple point-to-point measurement framework. It defines a set of important variables into a database called management information base organized into 10 categories of information (Current version MIB-II) and allows nodes to send simple point-to-point UDP based queries. SNMP has been very successful in providing the required standardized organization, syntax and representation (ASN.1) for a base set of critical state information. However, its communication model is still point-to-point. Consequently, composite measurements cannot be made easily using SNMP. Because of the difficulty in its scalability, SNMP has been used more successfully in network management and monitoring than in dynamic run-time querying. Optimizing systems that need network wide information propagation therefore generally build their own custom mechanism. Routing information propagation in IGP, OSPF, BGP all are classical examples of this situation. Video multicast group management and synchronization require tracking of jitter and round trip delay. RTP and RTCP [12] have been proposed to collect custom time related statistics over a multicast tree. Similarly, *pathchar* [7], *cprobe, bprobe,* [1,9], *woodpecker* [3], and host distance estimation tool [5] have been proposed targeting various specific measurements. In this context, this research proposes a novel system that can address two emergent issues in network probing-- scalability and versatility, and these are illustrated below.

1.2 Scalability

Point-to-point mode of communication often severely limits scalability in a large network. For example, for measurement of path statistics the requesting node is required to send individual SNMP messages to all intermediate nodes. Consequently, redundant information flows inside the network increasing the overhead, severely reducing the transparency of the measurement process. Dissemination/aggregation of information with only a point-to-point communication means creates excessive traffic on the network severely limiting the scalability. It seems much of the limitations arises because SNMP cannot extract any intelligence from the intermediate nodes. Since there is no means for in network composition, all compositions must be done at the end-points, only after polling all state information there.

1.3 Versatility

On the other hand, the specific probing kits provide greater scalability but hardly can be reused for other measurements. Nevertheless, the trend suggests that versatility of the information is becoming equally important. Exactly, what measurement is useful depends on the optimization objective. For example, in a video server scenario, whether the jitter or the hard delay is more important is dependent on the specific video repair algorithm. In a different scenario, a server before sending data may want to poll information about the speed of only the last link to the home user's computers. In some other scenario the min/max of the path downstream delays and jitters from various junction nodes can help in strategically placing jitter-absorbing buffers in a multimedia streaming virtual private network tree. Emerging tele-interaction applications (such as tele-surgery, remote instrument control) will require handle on the delay incurred at the video frame level, which is exactly not the same as the packet delay. The trend suggests that as more advanced, and complex netcentric applications are being envisioned more versatile network state information would have to be exchanged.

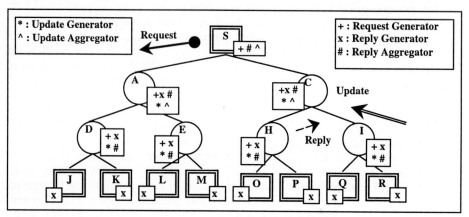

Fig-1 Components of Active Harness

1.4 Our Approach

Can scalability and versatility both be retained simultaneously? Apparently, there may not be any efficient answer in an end-to-end paradigm. In the general case, a network can choke with polynomial messaging at the end-points. However, the recent advent of **Active Network** technology seems to offer an innovative way out from this dichotomy [13,14,8]. Active network allows programmable modules to be embedded inside network junctions. In this research we propose an experimental dynamic mechanism for state information polling and propagation inside network with similar embedded information synthesizers, which seems to be both scalable and versatile. The approach first makes a clear separation between the "communication" from the "information" of the state exchange and propagation process. Communication is handled by the component called "harness". Harness propagates all information via coordinated messaging. On the other hand the "information" component of the process is controlled by a set of soft programmable plug-ins. These plug-ins decide the content of the messages propagated by the harness. In this paper in section 3 first presents the architecture of the proposed system. Section 4 then illustrates the operation of the harness for a typical application. Finally, in section 5 we share both analytical and empirical results depicting the performance of the proposed harness system.

2 Harness Architecture

The harness is responsible for initiating, propagating and responding to a series of well-coordinated messages between the nodes in a network. The harness once installed in network nodes, can act in three roles-- *initiator*, state *synthesizer*, and *terminals*. The initiator acts as the communication agent in the network layer for the application that actually requires the information. The synthesizer propagates the state requests and processes the returning states from the terminals.

The harness controls the communication pattern and thus deals with the efficiency of messaging. Harness system accepts a set of plug-ins, which determines the content of these messages, and how they are propagated and aggregated at the junction points.

2.1 Messaging

The harness system has been designed to operate with a novel request-reply update messaging scheme. It has three types of messages *request, reply* and *update*. A request message contains fields indicating what information it needs, and dictating how far down the network the probing session should propagate, i.e. level, and any information needed by the receiver to compute required data, e.g. to compute the jitter a receiver needs to know the time stamp of sending successive data. The *request initiator* decides how often a request is generated. The request messages are sent to the *terminals* if they are immediately connected, or to synthesizers for further downstream propagation. A synthesizer upon receiving a request propagates the query by generating a new request message to the down-stream nodes. However, at the same time it might also generate an immediate reply for the requestor. The replies from synthesizers may contain current local state and/or past remote states. The terminal nodes send replies to their respective requestors. The terminal reply contains locally retrieved current states. In the return trip of information, the synthesizer nodes aggregate the information and at each stage generate update messages for their requestors. Once a node receives all or specific number of update messages from its immediate down-stream nodes or on timeout, it updates the network local state variables and generates a new update message. The update message contains a synthesized summary of information calculated from all its immediate downstream nodes.

This three-part request-reply-update communication model, if needed, allows the information to be collected without working in lockstep. Even if a node downstream is delayed or silent, it does not hold the entire system; the estimation process can proceed for remaining nodes. The update phase is further equipped with optional and configurable timers to avoid update lockup. In essence, the request-reply phase allows collection of local immediate states. The reply mechanism allows immediate probing into current local states and past synthesized remote states, while the update message retrieves latest remote states.

2.2 State Composition

Harness system accepts a set of five plug-ins which are called *request generator, reply generator, reply aggregator, update generator,* and *update aggregator.* These modules together determine the content of these messages, and how they are aggregated at the junction points. They work via a virtual *slate*. A copy of which is maintained in each of the nodes. The slate works as the local abstract data structures. The slate is programmable and is defined at the session initiation phase. The *request generator* specifies the request message describing the fields it wants from the slate of its down-stream node. At individual nodes the model supports MIB-II and thus acts as a superset of SNMP. The terminal nodes can read/copy MIB variables (or their processed combination) existing in the local slate into variables marked for reply. The harness then invokes the reply messages with the designated slate variables. *Reply aggregator* (or *update aggregators*) in a similar fashion is invoked each time a reply (or update) is received by the harness. They perform domain specific processing of the reply message fields and similarly update their own slate variables. The *update generator* is invoked when a special trigger variable becomes true. The trigger variable is a set of conditions such as all, any, or a specified number of down-stream nodes have updated/replied, or a timer fires. The update generator sends the slate variables synthesized by the update aggregators to the upstream node. Fig-1 describes the architecture of the proposed harness system. It shows the roles, the typical locations of the plug-in modules and the direction of the messages.

At the heart of the composition ability is the transfer functions of the intermediate synthesizers. The request and update phase can be represented by equations:

$$S_t^j = \Phi(\vec{E}(Q_t^{i,j,-}), M_t^j, S_{t-1}^j), \text{ and } Q_t^{-,j,k} = \vec{F}(S_t^t) \quad ....(1)$$

$$S_t^j = \Psi(\vec{F}(P_t^{-,j,k}), M_t^j, S_{t-1}^j), \text{ and } P_t^{i,j,-} = \vec{E}(S_t^t) \quad ....(2)$$

Here S_t^j is the local *slate state* at event time t at node j, λ_n^o is the *request receiving filter* (RRF), M is the *local network state* (such as MIB variable), $\vec{F}$ is the *request forwarding* filter (RFF). $Q_t^{i,j,-}$ is the arrived request from parent i, to node j and $Q_t^{-,j,k}$ is the propagated requests to children k. $\vec{E}$ is the *update forwarding* filter (UFF), $\vec{F}$ is the *update receiving* filter (URF). $P_t^{-,j,k}$ is the arrived update from child k, and $P_t^{i,j,-}$ is the propagated update to parent i. While, the filters determined the information propagation rules, composition functions $\Phi()$ and $\Psi()$ together determine the message content. While, in principle each of these components for each of the individual harness sites can be programmed differently, however the associated management will be intractable. In this harness we divide the network nodes into subsets based on their role in the topology. Nodes in the topological subsets then inherit uniform programmed behavior. Thus, we need only five distinct programmed modules (plug-ins) to be supplied by the harness programmer.

3   Harness Execution Model

The harness operates through 8 states. Fig-2 shows the state transition diagram. The oval shaped boxes describe activities and the square boxes indicate plug-in modules used for those activities. The initiator have states 1-3-4-5-1 if it generates a request of level 1 and if reply is expected, otherwise if level is 1 and reply is not expected then it has states 1-3-1. If the initiator generates a request message of level more than one, and reply and update is expected, then it has states 1-3-4-5-6-7-1. The terminal nodes have states 1-2-1. Parents of terminal nodes (if both reply and update is expected) have states 1-2-3-4-5-8-1. Other synthesizers (if both reply and update is expected) have states 1-2-3-4-5-6-7-8-1. As fig-2 suggests, other combinations are possible depending on whether reply and/or update is not expected.

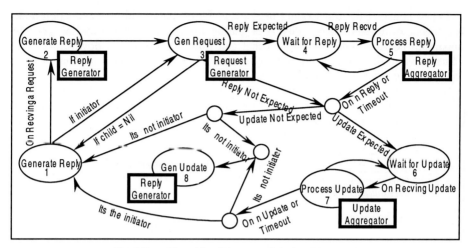

Fig.2 State Transition Diagram of Active Harness

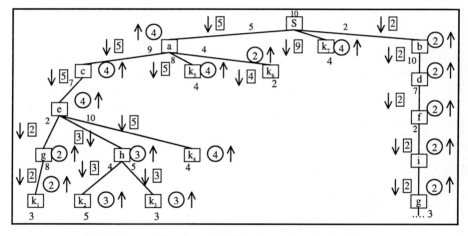

Fig-3 Transcoder Auto Stabilization using Active Harness

4 Example Probing: Transcoder Step Computation

Below we present an example of custom harness service that helps in configuring an embedded video transcoding tree network. In a multicast distribution tree a video transcoder [15] sits in the junction nodes and steps down the video rate to match the downstream path capacities. The rate of the individual links however cannot be determined just with node or link local information. The step-down parameters are quasi-global state dependent. The optimum rate assignment is a function of the upstream and downstream paths, requirements of the sinks as well as the delivery capacity of the source. Below we show how the harness system can be 'programmed' to configure the transcoding network in a two-phase request/update cycle.

Let us consider the transcoding network of Fig-3. Each transcoder junction node has to know the amount of video it is getting from its parent via up-link and the maximum possible amount of video that can flow through all its down-links. The values have to be determined optimally so that it does not receive more than what it can deliver or what is required down the stream.

```
    AppxChildBW  [ ]   :  float,
UsableChildBW [ ] : float,
    TreeLevel : int,
    MaxDownLinkTreeBW : float,
    UpLinkBW : float,
    SinkOrSourceCapacity : float
         (All are initialized to –1)
```

Fig-4(a) Slate

RequestMsg: MsgType:HarnessVariable, DeliverableRate: float,   Level: int
ReplyMsg: MsgType: HarnessVariable, Recvable: float
UpdateMsg: MsgType: HarnessVariable, MaxRecvable:float

Fig-4(b) Message Definitions

```
Request Generator Module {
if (it is a source)
     UpLinkBW= SinkOrSourceCapacity;
for each child I {
   RequestMsg.DeliverableRate
      = min (UpLinkBW, AppxChildBW[I]);
   >> Send RequestMsg to child [I]; <<
   }
}
Reply Generator Module {
UpLinkBW=RequestMsg.DeliverableRate;
if (Terminal Node)
     ReplyMsg.Recvable = min (UpLinkBW,
     SinkOrSourceCapacity);
  >> Send Reply Message; <<
}
Reply Aggregator Module {
if (it parent of terminal I){
     MaxDownLinkTreeBW = max
     (MaxDownLinkTreeBW, ReplyMsg.Recvable);
     UsableChildBW[I] = ReplyMsg.Recvable;
     }
}
Update Generator Module {
    UpdateMsg.MaxRecvable = MaxDownLinkTreeBW;
    >> Send Update Message; <<
}
Update Aggregator Module {
    MaxDownLinkTreeBW = max
    (UpdateMsg.MaxRecvable,MaxDownLinkTreeBW);
    UsableChildBW[I] = UpdateMsg.MaxRecvable;
}
(Note: The messages enclosed by << and >> are
send  by the harness)
```

Fig-4(c) Relevant Pseudocode of Modules

We map the initiator at the video source and the terminals at the video sinks and the synthesizers in each of the intermediate nodes. The slate, messages and the five plug-in modules are shown in Fig-4 (a)-(c). The process begins from the source. It sends a request, which contains its maximum delivery capacity. Each of the synthesizers regenerates a new downward request, which shows the *maximum deliverable video rate*, recomputed by the request generator plug-in modified by its local downward link capacity. When, the terminal receives the request it then compares the value with its sink capacity, and determines the *bandwidth receivable* based on the deliverable and its sink capacity. The synthesizers above the terminals collect the *bandwidth receivable* values and compute their own bandwidth receivable based on the maximum demand and past value of *bandwidth deliverable*. The information eventually propagates upward to the source. In this example (fig-3) there is a source – S, a, b, etc. are intermediate nodes, and k_1, k_2, etc are video sinks. The numbers beside the edges denote bandwidth of that link. For example, the bandwidth of link ac is 9 Mbps. The source, S, can generate video at a maximum rate of 10 Mbps and the capacity of sink k_1 is 3 Mbps. The numbers in rectangles indicate bandwidth deliverable that the parent sends to respective child in the request message. The numbers inside the circles indicate the bandwidth receivable that can be absorbed by down links and are carried by the update messages. The update messages propagate from the sink towards the source carrying the maximum units that can be absorbed. Before a request is generated, the concerned nodes approximate the bandwidth of all the children using SNMP.

Each node maintains two lists. The list AppxChildBW initially stores the approximated bandwidth of respective child and the other list UsableChildBW finally has the maximum bandwidth that respective child should receive. MaxDownLinkTreeBW stores maximum bandwidth up to level of TreeLevel. UpLinkBW stores the bandwidth of the link connecting it to its parent. SinkOrSourceCapacity stores the amount produced or absorbed by it if it is a source or a sink respectively.

Here we have demonstrated an instance how an application can benefit from the harness in deep probing. The harness system can also facilitate collection of other common forms of global information such as jitter, and end-to-end delay. It can also probe variety of precise and detail pseudo-global and local network states which are not easily accessible today, such as "end-points of the most constrained link in a path", "jitter across the m-th hop down-steam", "aggregate of outgoing bandwidth from a target node" etc. All it needs are different composition rules.

5 Performance

Below we provide estimates of the impact on the links and on the nodes due to the harness operation respectively for one execution/propagation wave. Here L and B stand for a probing depth level and branching factor of the context network tree respectively. The sizes of the Request, Reply, and Update are of r, p, and u bytes respectively and computational impact due to *request generator, reply generator, reply aggregator, update generator* and *update aggregator* are rg, pg, pa, ug, and ua respectively. Table-1 shows the network wide traffic, message per link and the byte density. The terminals do not generate updates hence terminal links do not carry update messages. Table-2 shows the computational impact on the network nodes due to the plug-in. A particular state probing session may be launched with a subset of capabilities (such as no reply, but update). The design objective is to provide the least impact communication for the given application scenario.

| Table-1 Link Traffic Impact ||||
Type	Byte	Max Msg	Msg/link	Bytes/link
Request	r	B^L	1	r
Reply	p	B^{L+1}	1	p
Update	u	B^L	1	u

| Table-2 Node Processing Impact ||||||
| Node | Request || Reply || Update ||
	Generator	Generator	Aggregator	Generator	Aggregator
Initiator	θ(B,rg)	0	θ(B,pa)	0	θ(B,ua)
Synthesizer	θ(B,rg)	θ(B,pg)	θ(B,pa)	θ(B,ug)	θ(B,ua)
Terminal	0	θ(B,pg)	0	0	0

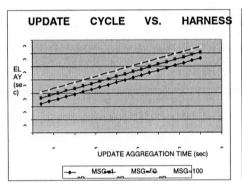

Fig-5(a)

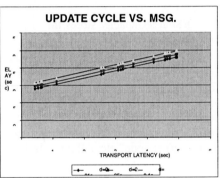

Fig-5(b)

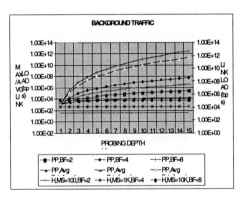

Fig. 5(c)

We have also performed statistical simulation to project the performance of the harness system under various constraints. The performance depends on the characteristics of the programmable components (complexity of the plug-ins, message size etc.) as well as on the network (such as bandwidth, topology, probing depth etc.) and platform characteristics (scheduling delay, messaging delay, etc). Fig-5(a) shows how one request/response cycle time varies for various plug-in execution time (x-axis) of the update for three orders of message sizes (1K, 100 bytes, and 10 bytes). In cyclic mode this also represents worst case bound on the information recency. Fig-5(b) shows the update delay with respect to the transport layer messaging delay. Another important metric from the network point of view is the background traffic that the harness operation creates for information collection. Fig-5(c) compares the scalability of the harness system as compared to a point-to-point mechanism. It plots the network "hum" (the background traffic generated by the probing process) for various depth and branching factor of the topology. It plots nine cases. The first three curves shows how the *hot-spot* "hum" (y-axis) near the probing root grows in the case of point-to-point mechanism with the increase in the depth of probing (x-axis) for three branching factors (BF=2,4,8). The second set of three curves shows the *average* hum for the same three cases, which is somewhat lesser but still grows rapidly with the probing depth. The average hum provides a measure that how much traffic on the average will be contributed by random spontaneous probing processes appearing at various network locations. On the other hand, the 'hot-spot' hum indicates the change that the application trying to probe, itself has, to be chocked with excessive surrounding probing traffic. The last three plots show the "hum" due to harness process (for three message sizes) (right y-axis). The dramatic scalability of the harness probing is quite apparent. As, can be seen that the hum in harness scheme is not only low but also flat. The result is not unexpected. The network embedded synthesis removes traffic information redundancy.

6 Conclusions

The key to the system's **scalability** and **versatility** are the embedded aggregators. Since local state dependent aggregation is performed inside a network, it reduces communication and thus enhances the system's scalability. Aggregators also provide the ability to compute network relative deep composite statistics, over the elementary MIB-II variables, thus enhancing the versatility of its ability to collect network states.

The scope of this paper does not permit discussion on implementation. It is non-trivial nevertheless can be realized at user space as *deamons*. Embedded implementation can cut down some overhead and will be critical for sub-second range probing cycles. Implementation on some form of active platform [8,13,14] can further facilitate matters such as remote deployment, and seamless secured execution of the plug-ins. The harness plug-ins require very limited form of programmability compared to general active net proposal. Also, the read-write suggestions are through local slate variables only. These characteristics assuage many of the security concerns. The proposed harness is perhaps one of those cases where provisioning even very low-grade programmability can be highly rewarding. The harness increases state visibility of network. In effect it facilitates high pay off smart optimizations for numerous applications, which are not possible today due to the black box nature of current network. Interestingly, such a network layer utility is not only crucial for building a new generation of network aware applications but it is also vital for many of the current problems internet is grappling with. Interestingly many of which are arguably artifacts of the opacity of current network design. Currently, we are exploring its active network based simulation. The work is being supported by the DARPA active network Research Grant F30602-99-1-0515.

7 References

1. Carter, Robert L., Mark E. Crovella. Measuring Bottleneck Link Speed in Packet-Switched Networks. Performance Evaluation 27 & 28 (1996), 297-318.
2. Comer D. E., Internetworking with TCP/IP, Principles, Protocols, and Architectures, 4th Ed, Pretice Hall, New Jersey, USA, ISBN- 0-13-018380-6, 2000
3. Dong, Yingfeng, Yiwei Thomas Hou, Zhi-Li Zhang, Tomohiko Taniguchi. A server-based non-intrusive measurement Infrastructure for Enterprise Networks. Performance Evaluation 36-37 (1-4), 1999, 233-247.
4. Downey Allen B., Using Pathchar to Estimate Internet Link Characteristics. http://ee.lbl.gov/nrg-talks.html, April 1997.
5. Francis, P., Sugih Jamin, Vern Paxson, Lixia Zhang, Daniel F. Gryniewicz, Yixin Jin. An Architecture for a Global Internet Host Distance Estimation Service. Proceedings IEEE INFOCOM New York, 1999.
6. Jacobson, V., Traceroute, [URL: ftp://ftp.ee.ibl.gov/traceroute.tar.Z,] 1989.
7. Jacobson, V, Pathchar- a tool to infer characteristics of Internet paths, [URL: http://ee.ibl.gov/nrg-talks.html], April 1997.
8. Javed I. Khan, S. S. Yang, Medianet Active Switch Architecture, Technical Report: 2000-01-02, Kent State University, [available at URL http://medianet.kent. edu/ technicalreports.html, also mirrored at http:// bristi.facnet.mcs.kent.edu/medianet]
9. Matt Mathis, Jamshid Mahdavi. Diagnosing Internet Congestion with a Transport Layer Performance Tool. Proceedings INET, 1996, Montreal Canada.
10. Paxson, V., Jamshid Mahdavi, Andrew Adams and Matt Mathis. An Architecture for Large-Scale Internet Measurement. IEEE Communication Magazine, August 1998, 48-54.
11. Rose, M.T., & McCloghrie, K. How to manage your Network Using SNMP, Englewood Cliffs, NJ, Pretice Hall, 1995.
12. Schulzrinne, H., S. Casner, R. Frederick, and V. Jacobson. RTP: A Trasport Protocol for Real Time Applications, RFC 1889, 1996.
13. Tennenhouse, D. L., J. Smith, D. Sincoskie, D. Wetherall & G. Minden.. A Survey of Active Network Research. IEEE Communications Magazine, Vol. 35, No. 1, Jan 97, pp 80-86
14. Wetherall, Guttag, Tennenhouse. ANTS: A Tool kit for Building and Dynamically Deploying Network Protocols. IEEE OPENARCH'98, San Francisco, April 1998. Available at: http://www.tns.lcs.mit.edu/publications/openarch98.html
15. Javed I. Khan & S. S. Yang. Resource Adaptive Nomadic Transcoding on Active Network, Applied Informatics. AI 2001, February 19-22, 2001, Insbruck, Austria, [available at URL http://medianet.kent.edu/, also mirrored at http:// bristi.facnet.mcs.kent.edu/medianet] (in press).

Protocol Design of MPEG-4 Media Delivery with Active Networks

Sunning Go[1], Johnny W. Wong[1], and Zhengda Wu[2]

[1] Department of Computer Science,
University of Waterloo, Waterloo, Ontario, Canada N2L 3G1
{sugo,jwwong}@bbcr.uwaterloo.ca
[2] School of Information Technology,
Bond University, Gold Coast, Queensland 4229, Australia
wz@bond.edu.au

Abstract. Traditionally, the Internet supports best effort service. The quality of service available to real time applications such as the delivery of video and audio is therefore difficult to predict and control. Two new technologies have emerged recently, namely the MPEG-4 standard and active network, for media data compression and data networking, respectively. MPEG-4 uses a content or object-based representation. It supports object-based scalability and, in particular, it allows the modification or editing of objects without having to first decode them. Active network allows program code to be loaded dynamically into network nodes. In this paper, we describe a system for real-time media delivery that exploits the strengths of MPEG-4 media encoding and active networks. It is based on a best-effort transport service. Our system performs congestion avoidance, error concealment, and media scalability.

1 Introduction

The delivery of various media, such as audio, video and still images, is growing explosively in the Internet today. However, the current Internet typically supports best-effort service. The quality of service (QoS) available to real time applications such as the delivery of video and audio is therefore difficult to predict and control. QoS support may be achieved by a transport service with resource reservation, e.g., RSVP [1] or differentiated services [2]. Media delivery control techniques can also be used at the end systems to overcome the deficiency of the best-effort service provided by the transport network. These include congestion and jitter control performed at the source or the receiver or both (see for example [3] [4]). Most techniques are designed for the MPEG-1 or MPEG-2 media data compression standards.

Two new technologies have emerged recently, namely the MPEG-4 standard [5] and active network [6], for media data compression and data networking, respectively. MPEG-4 uses a content or object-based representation rather than a pixel-based representation as in MPEG-1 and MPEG-2. It supports object-based scalability and, in particular, it allows the modification or editing of objects without having to first decode them. Active network allows program code to be loaded dynamically into network nodes. The code can perform tasks specific to a stream of packets or even individual packets. This makes application-specific processing possible. In the context of media delivery, an active network can provide the following benefits:

- the network can react to congestion using more sophisticated traffic editing than simply dropping packets at routers;
- rate-based congestion control schemes can be extended by re-pacing packets at routers closer to the site of congestion until the source re-adjusts its rate; and
- the congestion control scheme used can be changed according to the nature of the traffic flowing through the network.

In this paper, we present the design of a system for real-time media delivery that exploits the strengths of MPEG-4 and active networks. This system is based on a best-effort transport service. It performs congestion avoidance, error concealment, and media scalability. Our design is conceptual in nature. Using MPEG-4, it is possible to drop some of the less important objects in case of congestion, and using active networks, it is possible for the application to decide which objects are less important.

The rest of this paper is organized as follows. In Section 2, MPEG-4 media objects and their transmission are introduced. The features of the active network used in our system are described in Section 3. Our protocol for media delivery is presented in Section 4. Some related works are discussed in Section 5. Finally, Section 6 contains some concluding remarks.

2 Media Objects and Their Transmission

In MPEG-4, audiovisual information is organized into *media objects*. These objects can be of natural or synthetic origins. Video sequence and sound are examples of media of natural origins, while 2D and 3D graphics are examples of media objects of synthetic origins. An audiovisual *scene* in MPEG-4 consists of a hierarchy of objects, as illustrated in Figure 1 [5]. Media objects at the leaves of a hierarchy are called *primary media objects*. They can be classified as video objects, audio objects, still images and text.

MPEG-4 provides object and content based scalability. For example, some video objects can be suppressed during media transmission while other objects, such as audio, still image and text remain in the scene of a media delivery. In addition, individual objects can be further scaled. Each media object can be represented independently of its surroundings or background. MPEG-4 also provides facilities to compose a set of objects into a scene. The necessary composition information constitutes the *scene description* (SD), which is coded and transmitted together with the media objects. A scenario of original media encoded into a sequence of scenes and then packetized together with their SD's is shown in Figure 2. Note that

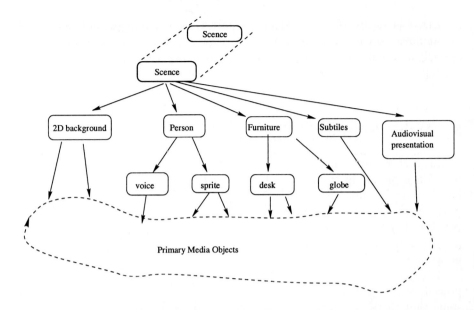

Fig. 1. The logical structure of a scene

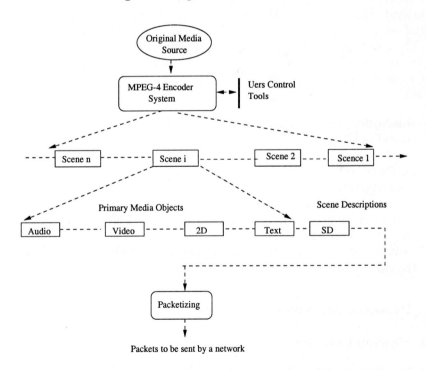

Fig. 2. Encoding, scenes, primary media objects and their packetizing

- SD objects are transmitted in a separate stream and provide pointers to audiovisual elementary streams;
- editing the scene can be performed by changing the SD stream; and
- full SD is transmitted periodically and SD updates are sent when there is a change.

3 Active Network

In this section, we describe the features of the active network used in our system. Each packet in the active network contains a program. The network allows any user to inject program-carrying packets. The language used to write a program has a primitive **remEval** for performing remote evaluation. The call **remEval(i,n)** causes the procedure invocation i to take place at node n. The node n does not have to be a neighboring node. An example of such an active network is PLAN [7].

The primitive **remEval** can be unreliable since the packet containing the procedure invocation i may be dropped before it reaches n. We therefore assume that there is another primitive **relRemEval** for remote evaluation that is reliable. Presumably, executing **relRemEval** is more costly than executing **remEval**. Thus, we use **relRemEval** only for important packets (such as the initial set-up packet - see subsection 4.3 below).

The router programming interface is similar to that in PLAN [7]. The key primitives are described below. In our description, we use *current node* to mean the node at which the packet resides (or the node at which the primitive is called).

- **thisNode**: returns the address of the current node.
- **routeLookUp(n)**: returns the neighboring node that one should go to in order to reach the node n.
- **getCurrentTime**: returns the current time.
- **transRate(n)**: returns the available bandwidth from the current node to the neighboring node n.
- **deliver(d,p)**: delivers the data d to the process p. This primitive is available only on end systems.
- **previousNode**: returns the previous node from which the packet containing the call arrived.

4 Protocol for Media Delivery

4.1 System Overview

We now describe our protocol for media delivery. The basic operation is the delivery of media data from a server to a client. The data stream is assumed to be routed over a fixed path. This is illustrated in Figure 3 where $N_1, N_2, ...N_n$ are nodes along the path from server to client. Our system is adaptive in the sense that the quality of the scenes transmitted is adjusted according to network conditions. Specifically, each node along the path estimates the capacity available at the outgoing channel. This information is reported to the server using the **relRemEval** primitive. The server therefore has a picture of the available bandwidth end-to-end, and the scene description can be modified according to the bandwidth available. Furthermore, during the transmission of scene data, if a node observes congestion at its outgoing channel, it may request the previous node to edit the scene description cached at that node such that the data rate is reduced. Finally, if a node observes an increase in the available capacity downstream, it can report this information to the server. The server then has a more up-to-date picture of the available bandwidth end-to-end, and may transmit a modified scene description to take advantage of the increased bandwidth.

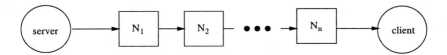

Fig. 3. Transmission of media objects

4.2 Packet Types

There are five types of packets in our system. They differ in the programs they carry. These programs implement our protocol for media delivery. Note that a scene description (SD) or a scene object may be transmitted in two or more packets. Also, we use "session" to mean a given delivery of media data from beginning to end. The packet types are as follows:

- **Set-up packet**: the first packet of a new session.
- **Non-terminal SD packet**: part of an SD, but not the last packet of that SD.
- **Terminal SD packet**: part of an SD and is also the last packet of that SD.
- **Scene packet**: part of a scene.
- **Reset packet**: indicates the end of session.

Each node along the path from the server to the client keeps track of a number of variables and arrays. These variables and arrays are allocated by the Set-up packet, read and written by subsequent packets, and released by the Reset packet. They are listed below.

- **nextNode**: records the next node in the path.
- **prevNode**: records the previous node in the path.
- **scenePcktNum**: records the number assigned to the Scene packet that has most recently arrived at the node.
- **nScenePckts**: keeps track of the number of packets of the current scene that have arrived at the node.

- **startTime**: records the time at which the first packet of a scene arrived at the node.
- **X**: keeps track of the currently available bandwidth to the next node for this session. It is updated by the last packet of every scene.
- **sdCache**: an array used to store the SD of the current scene. It is edited when there is a need to change the transmission rate. A scene packet reads the information in this array to determine whether it should drop itself.
- **sceneCache**: an array used to store the packets of a scene. It is used for error concealment.
- **sceneSize**: records the size of the current scene.

In addition, the server process has a *global data rate vector* $\mathbf{G} = [X_1, ...X_n]$ that keeps track of the bandwidth currently available to the session at each hop towards the client.

In the following subsections, we give a brief description of programs contained in each of the five types of packets.

4.3 Set-Up Packet

This is the first packet sent by the server. The program contained in this packet is responsible for initializing the path from the server to the client. More specifically, when the program is executed at a node, the variables and arrays described above are created and initialized. The **routeLookUp** primitive is next used to determine the next node along the path to the client. The **relRemEval** primitive is then used to report to the server the available bandwidth to this next node. This will initialize the corresponding element of $\mathbf{G}$. Finally, the **relRemEval** primitive is used to propagates the packet to the next node towards the client.

When the Set-up packet reaches the client, the appropriate variables and arrays will have been created and initialized at all nodes along the path, and the $\mathbf{G}$ at the server will have been initialized with the available bandwidth at every hop.

4.4 Non-terminal SD Packet

The Non-terminal SD packet carries scene description data. If the current node is the client, the **deliver** primitive is used to deliver the payload to the client, otherwise a copy of the payload is entered into a local cache **sdCache**. The **relRemEval** primitive is then used to propagates the packet to the next node towards the client.

4.5 Terminal SD Packet

The program contained in a Terminal SD packet is the same as that in a non-terminal SD packet, except that it also sets **sceneSize** to the size of the scene described by the SD.

4.6 Scene Packet

The Scene packet carries media object data. If the packet is at the client, the payload is simply delivered to the client using the **deliver** primitive.

On the other hand, if the packet is at an intermediate node, one must take into consideration that some packets in a scene may be missing. The missing packets can be explained as follows. As part of the adaptation scheme, a node may have edited the scene description stored in its local cache to reflect a reduced bandwidth downstream (this scheme is described below). Some packets may be dropped by this node according to the modified scene description. The next node may therefore find that some packets belong to a scene are missing.

When the program in the Scene packet is executed, a copy of the payload is entered into a local cache **sceneCache**. The packets that may need to be propagated to the next node are determined next. These include the current packet and any missing packets since the last packet propagated (the number assigned to this packet can be found in **scenePcktNum**). For each of these packets, the scene description in the local cache **sdCache** is checked to determine whether the packet should be dropped or not. If not, the **relRemEval** primitive is used to propagate the packet to the next node. For any missing packet, the copy stored at **sceneCache** is propagated. Note that this cached copy is a good approximation of the corresponding packet in the current scene, and substituting a missing packet by that from the previous scene is how we do error concealment.

The program contained in a Scene packet also implements our adaptation scheme. Specifically, if the Scene packet is the first of a scene, then the variable **startTime** is initialized to the current time as seen by this first packet. If the Scene packet is the last one of a scene, then a new value of **X**, the available bandwidth to the next node downstream, is calculated. This new value is given by

$$Xnew = \frac{K}{T - startTime} \quad (1)$$

where K is the number of bits in the current scene that were transmitted downstream and T is the current time as seen by the last packet of a scene. The **relRemEval** primitive is then used to report this new value to the server.

As part of the adaptation scheme, the state of the outgoing channel is checked to see whether overflow or underflow is about to occur. This can be based on parameters such as the length of the input queues, the length of the output queues, the difference between the new value and the old value of **X**, or some combination of these quantities. If it is determined that an overflow is about to occur, then a packet serving as an explicit congestion notification is evaluated at the previous node using the **relRemEval** primitive. This packet takes a parameter which is the rate at which the previous node should reduce its transmission rate to. The transmission rate is reduced by invoking a media editor which edits the scene description stored in **sdCache**. The details of the media editor is out of the scope of this paper.

On the other hand, if it is determined that underflow is about to occur, then a packet is sent to the server. This packet takes a parameter which is the new rate the outgoing channel is able to support. This rate may be different from $Xnew$. The server will then try to scale up the quality of the media since there is now more bandwidth available. More specifically, the server compares the new rate with the entries in the global data rate vector **G**. If the new rate is lower than every other entry in **G**, then the server will scale up the media to this new rate, otherwise the request to scale up is ignored.

4.7 Reset Packet

This is the last packet of a session. All variables and arrays at intermediate routers related to the session are released. Cached copies of the scene description and scene data are removed as well.

5 Related Work

Among the related works are those by Najafi and Leon-Garcia [8], and by Bhattacharjee, et al. [9]. In [8], the concept of "active video" is introduced. This concept is based on MPEG-4 and active network techniques. A number of advantages of the active video over existing solutions for video distribution over the Internet are discussed. In [9], simulations were performed to demonstrate that network-based adaptation could yield significant performance gains for multicasting MPEG-1 or MPEG-2 video distribution.

6 Conclusion

In this paper, we have described a protocol for delivering MPEG-4 real-time media. The protocol takes advantage of both the programmability of active networks and the content scalability of MPEG-4 encoding. The programmability of an active network gives more flexibility to the protocol designer/application developer. For example, different applications can use different criteria for determining whether congestion is about to occur. All these variants of the protocol can co-exist in the same network, and so there is no need to standardize on any one particular variant. On the other hand, the content scalability of MPEG-4 encoding makes it possible for an application to drop packets based on content, rather than just uniformly degrading the quality of, say the video component. This makes it possible for an application to drop entirely the objects that are deemed to be less important, without affecting the quality of the other objects. Since only an application can decide which objects are important, an active network is an attractive platform for an application developer to design a variant of the protocol that is specifically tailored to the application.

Acknowledgement

This work was supported by a grant from the Canadian Institute for Telecommunications Research under the National Centre of Excellence Program of the Government of Canada.

References

1. L. Zhang, S. Deering, D. Estrin, S. Shenker and D. Zappala, "RSVP: A New ReSerVation Protocol", *IEEE Network*, September 1993.
2. S. Blake, D. Black, Carlson, E. Davies, Z. Wang and W. Weiss, "An architecture for differentiated service", *Internet RFC 2475*, December 1998.
3. X. Li, M. Ammar and S. Paul, "Layered video multicast with retransmission (LVMR): Evaluation of hierarchical rate control", *Proc. IEEE INFOCOM*, March 1998.
4. W. Feng, M. Liu, B. Krishnaswami, and A. Prabhudev, "A priority based technique for the best-effort delivery of stored video", *Proc. Multimedia Computing and Networking*, January 1999.
5. ISO/IEC JTC1/SC29/WG11 N2725, Coding of Moving Pictures and Audio, "Overview of the MPEG-4 Standard", Seoul, South Korea, March 1999.
6. D.L. Tennenhouse, J.M. Smith, W.D. Sincoskie, D.J. Wethrall and G.J. Minden, "A Survey of active network research", *IEEE Communications Magazine*, vol. 35, no. 1, Jan. 1997.
7. M. Hicks, J.T. Moore, D.S. Alexander, C.A. Gunter and S.M. Nettles, "PLANet: An Active Internetwork", *Proc. IEEE INFOCOM*, 1999.
8. K. Najafi and A. Leon-Garcia, "Active Video: A Novel Approach to Video Distribution" *Proc. Second IFIP/IEEE int. Conference on Management of Multimedia Networks and Services*, Paris, France, November 1998.
9. S. Bhattacharjee, K. L. Calvert, and E. W. Zegura, "Network support for multimedia video distribution", *Technical Report CIT-CC-98-16*, College of Computing, Georgia Institute of Technology, 1998.

Prediction and Control of Short-Term Congestion in ATM Networks Using Artificial Intelligence Techniques

Guiomar Corral, Agustín Zaballos, Joan Camps, and Josep M. Garrell

Enginyeria i Arquitectura La Salle, Universitat Ramon Llull,
Passeig Bonanova 8, 08022 Barcelona, Spain
{guiomar, zaballos, joanc, josepmg}@salleURL.edu

Abstract. Nowadays high-speed transmissions and heterogeneous traffic are some of the most essential requirements that a communication network must satisfy. Therefore, the design and management of such networks must consider these requirements. Network congestion is a very important point that must be taken into consideration when a management system is designed. ATM networks support different types of services and this fact makes them less predictable networks. Congestion can be defined as a state of network elements in which the network cannot guarantee the established connections the negotiated QoS. This paper proposes a system to reduce short-term congestion in ATM networks. This system uses Artificial Intelligence techniques to predict future states of network congestion in order to take less drastic measures in advance.

1 Introduction

The design, configuration and management of real communication networks are phases that must be taken into account in order to achieve network performance objectives. These objectives include high-rate transmissions and heterogeneous traffic.

Asynchronous Transfer Mode (ATM) networks can fulfill these requirements with very high rates and different types of services. On the other hand, they are less predictable networks [1] than other networks that only support one type of service. The reason for this is that trusted statistics from traffic characteristics and connection establishment patterns aren't given. ATM networks also guarantee the engaged Quality of Service (QoS), so when a connection is accepted, the network must fulfill all the agreements reached with the user. But due to the statistical multiplexing of the traffic, an inherent feature of ATM, it is possible to have congestion even though all connections carry out their contracts.

ATM congestion can be defined as a state of network elements in which the network is not able to guarantee the negotiated Network Performance objectives for the already established connections [2]. Congestion can be caused by unpredictable statistical fluctuations of traffic flows or fault conditions within the network [2]. Since an ATM network supports a large number of bursty traffic sources, statistical multiplexing can be used to gain bandwidth efficiency, allowing more traffic sources to share the bandwidth. But if a large number of traffic sources become active simultaneously, severe network congestion can result [3].

Two different types of congestion, long-term congestion and short-term congestion, can be distinguished. The former takes place when network load overcomes network resources and can be avoided with an accurate network resource planning and preventive controls like Call Admission Control (CAC). The latter lasts less than the round-trip delay and it is difficult to foresee because of dynamic variations of cell interarrival times due to bursty traffic sources. However, contention techniques could minimize its consequences [4].

This paper proposes a system to reduce short-term congestion in ATM networks. This system uses Artificial Intelligence (AI) techniques to predict future states of network congestion. An existing buffer control algorithm [4] is used. Using historical data files and AI techniques, a synthetic-generated mathematical function (a model) is created. This model is used to generate several input parameters of the buffer control algorithm in order to adapt its behavior to the situation. Results show that the system improves network performance. [1]

This paper is divided into five sections as follows. The first section introduces and locates the problem that is going to be solved. The second section focuses on the congestion concept with congestion control schemes. The third section describes the implemented solution, which predicts short-term congestion using AI techniques. In the fourth and fifth sections, the simulation environment and results are given. Finally, results, conclusions and further work are presented.

2 Problem Analysis

Congestion occurs when network performance falls off dramatically [5]. Moreover, in ATM networks, congestion is defined as the network state in which the necessary resources to provide the QoS contracted by the users are not available [2]. This QoS, which must be assured, is negotiated in the connection establishment [2].

Monitoring ATM networks performance for traffic and congestion control purposes leads to difficulty because there are some parameters whose values must be tuned. These parameters are not only based on users' traffic descriptors but also on the different traffic control policies implemented [6].

Congestion control schemes can be classified depending on the congestion duration. The longer the congestion duration, the wider the perspective that will be used to control it. Some of the control mechanisms, arranged in order of congestion duration, are the following: network design, Connection Admission Control (CAC), routing, traffic shaping, end-to-end feedback, hop-by-hop control and buffering.

CAC is used during the call set-up phase to determine whether a connection can be accepted or rejected. A new connection request in an ATM network must contain traffic descriptor parameters that CAC uses to determine whether the connection may be established. By accepting the connection, the network forms a traffic contract with the user. Later on, Usage Parameter Control (UPC) will monitor the connection to determine whether the traffic conforms to the traffic contract [5] [6].

Priority control is useful when end-systems can generate different priority traffic flows using the Cell Loss Priority (CLP) bit of the ATM cell header [7]. Then, the

---

[1] This work is included in a research project partially supported by Spanish research program (CICYT TEL98-0408).

network may selectively discard cells with low priority, if necessary, in order to protect network performance for high priority cells. Traffic shaping is a mechanism that alters cell stream traffic characteristics to ensure its conformance, whereas feedback is used to regulate the traffic submitted according to the network's state.

Different performance parameters can be used to monitor ATM QoS, like Cell Loss Rate (CLR), Cell Misinsertion Rate, Cell Error Ratio, Cell Transfer delay (CTD), Cell Delay Variation and so on [2]. But the most important parameters related to congestion are CLR and CTD. However, since speeding up the ATM switching highway and the transmission link can satisfy CTD requirements, only CLR will be used [4].

[4] proposes a priority control for cell loss quality that attempts to tolerate short-term congestion situations at the expense of the source's low-priority traffic, which has the least restrictive traffic descriptors. In fact, this mechanism could be classified as a Partial Buffer Sharing policy with a priority control. Generally, a Partial Buffer Sharing policy, in contrast to Complete Buffer Partitioning policy, drops an incoming cell only if the total number of cells in the buffer exceeds a given threshold [8]. This means that a cell is only discarded if there is not enough free space in the buffer.

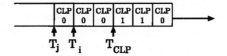

Fig. 1. A Partial Buffer Sharing Policy with three defined thresholds

It is very important to focus on the pursued goal, which consists of discarding the minimum cells whose CLP field is disabled by UPC. Congestion would cause the cells that had been previously accepted by UPC to be discarded, and this is what must be avoided. In the end, the cells tagged as nonconforming cells are those that must be dropped first if congestion occurs. If the system discards cells with enabled CLP field before a heavy congestion situation happens, the amount of discarded cells with enabled CLP could be minimized.

Therefore, accurate thresholds should be computed. Then, when the utilization thresholds are exceeded, low-priority cells are discarded. But these thresholds are static values, not dynamic. Thus, they cannot adapt to different network situations. Our proposal improves threshold selection by using AI techniques to predict a future short-term congestion based on the temporary evolution of the buffer utilization.

3 Short-Term Congestion Prediction Using AI Techniques

In order to predict short-term congestion in a network, AI techniques have been implemented. First of all, the prediction goal and input data must be explained. Then, the AI technique will be introduced, focusing on its application in our problem.

3.1 Prediction Goal and Input Data

The system should predict when a short-congestion situation is going to happen. If this moment is foreseen before it happens, some measures can be taken in advance to avoid this future and undesirable situation. The system can predict it, using information based on the temporary evolution of the buffer utilization.

AI techniques learn from past situations and apply their knowledge to predict new situations. The AI technique used is Genetic Programming (GP) [9]. GPs are a machine learning method useful for prediction model generation. In this case, it will predict the future buffer utilization. Like the majority of machine learning methods, GPs need a representative set of problem samples to be trained. Once the training phase is completed, the output (the model) is used to make the prediction.

The test bed used to train our GP is made up of a set of samples, where every set of samples is included in a file. These samples come from previous network simulations. Thus a file contains a time-ordered list of buffer utilization states. The GP system needs a data set that reflects not only the past and present situation, but also the future buffer utilization, in order to achieve a good training. Therefore, the data set will be fragmented in windows and every window will be processed as a new sample. Every window includes an account of buffer utilization in the previous instants as well as the future buffer utilization. The account is summarized in intervals and every interval is represented by the maximum, the minimum and the mean buffer utilization.

We must consider that there are different traffic sources and they can generate traffic of different priorities. So the intervals will contain the utilization percentage for every priority. On the other hand, the future interval buffer utilization will only be represented by the utilization mean. Future buffer utilization will be the predicted value. Consequently, a file containing buffer utilizations for every traffic priority in a period of time will be given to GP.

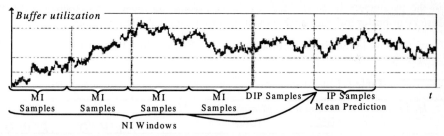

Fig. 2. Example of a window and how it is fragmented in different intervals. The graph illustrates the buffer utilization along simulation time. MI samples correspond to the samples of a measurement interval (MI). Some MI samples will be used to predict the future utilization of a prediction interval (IP). The distance between MI samples and IP is DIP (Prediction Interval Distance).

3.2 Genetic Programming Approach

As stated before, Genetic Programming [9] has been chosen to predict ATM node queue utilization. This is an AI technique related to evolutionary computation, which is inspired by Charles Darwin's theories on evolution and by Gregor Mendel's genetic inheritance, where the existence of a population is simulated and it evolves as time goes by.

The basic GP algorithm imitates nature simulating a population of individuals that clone, recombine with others or disappear. The better-adapted individuals to the environment are those that have more possibilities of survival and become progenitors of the new generations of individuals. The population spreads toward a group of better individuals, due to these transformations and to natural selection. The population's global tendency will maintain and replicate the important information, while it will lose the less representative one. In evolutionary computation, the environment is the outlined problem that wants to be solved. Likewise, each one of the population's individuals is a potential solution. The better the solution, the more possibilities it has to survive from one generation to another.

GP considers that each individual is a program that can solve the outlined problem. Thus the population is considered as a group of programs that will be possible solutions. From one generation to another, these programs will be cloned, crossed and mutated. Among all the programs, only the selected ones will be used. This selection depends on the fitness of each program, on the similitude between the desired solution and the obtained solution for each program. In the end, a group of sufficiently good programs is obtained and one of those is chosen, usually the best one.

Symbolic regression has been the strategy used to obtain a GP prediction model, so specific formulas that solve the proposed problem can be obtained. Then, a formula will be used to predict the mean utilization of an ATM node queue. With symbolic regression, the population's individuals are reduced to formulas. The fitness population formula is evaluated according to the prediction success. The formulas that achieve a more accurate prediction of the prospective utilization will have more probability of persisting.

The obtained formulas that predict mean buffer utilization after DIP contain some of the parameters listed in Table 1. These parameters must be calculated using the history of buffer utilization. One example of a formula is the following:

Future Mean Buffer Utilization = (P3 / P13) − P8 + P16 + P17   (1)

Table 1. List of parameters that can appear in the GP prediction model

Historical	Maximum value	Minimum value	Mean	%Low priority	%High priority
Window 1	P1	P2	P3	P4	P5
Window 2	P6	P7	P8	P9	P10
Window 3	P11	P12	P13	P14	P15
Window 4	P16	P17	P18	P19	P20

4 Simulation Test-Bed

The AI algorithm will predict congestion situations based on the buffer utilization; thus, a mechanism that works closely with the buffer should be chosen. [4] is a UPC mechanism that monitors the traffic behavior generated by the user's sources and follows a performance policy. To determine the performance of the new proposal a simulation environment in Mil3 OPNET has been programmed. The scenarios and test-bed implemented are explained in this section.

4.1 Simulation

The Mil3 OPNET simulator has been used to analyze the improvements introduced by the AI implementation. Firstly, a control congestion environment without AI was programmed. Any network node will detect the congestion situation monitoring its buffer utilization. Depending on previously defined thresholds, [4] always eliminates fewer high-priority cells than low-priority cells. Different simulation scenarios are implemented with different types of sources and different traffic loads. Thresholds, which have been calculated previously, will determine a CLR for each traffic type. Later, this environment was adapted using AI to control the congestion.

Opnet simulator uses nodes to carry out the different functions. Figure 3 shows the node design which implements [4].

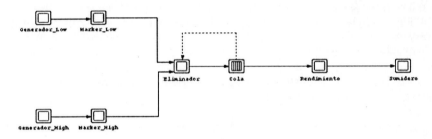

Fig. 3. Node design

An *Opnet* node is composed of several processes. The processes that form the control congestion node are the following:
- *Generador_High* and *Generador_Low*: these processes generate ATM traffic following almost any probabilistic or bursty distribution. The first one produces high priority traffic and the second one generates low priority traffic.
- *Marker_Low* and *Marker_High*: As the generated cells by OPNET sources do not tag any CLP field cell and the algorithm works closely with the UPC techniques, it was necessary to program this process. Its main function is the implementation of a Jumping Window algorithm [10].
- *Cola*: This is a programmable FIFO buffer queue of an ATM device.
- *Rendimiento*: This process is used to centralize link performance statistics.
- *Sumidero*: This is the cell sink that avoids the collapse of the simulation system.
- *Eliminador*: It implements [4]. The Finite State Machine is represented in figure 4.

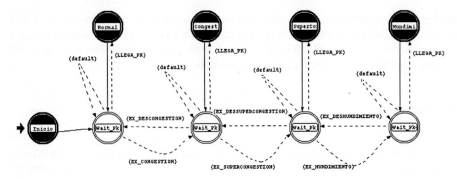

Fig. 4. State Machine of *'Eliminador'* Node

The *'Eliminador'* process had to be modified in order to adapt the system to the use of AI information. This calculates the estimated buffer utilization using the obtained prediction model and carries out an appropriate threshold policy. Depending on the prediction, thresholds will be more or less restrictive, that is to say, the node will drop cells sooner or later. The instant when the node starts dropping cells is fixed according to the buffer length and threshold policies. These thresholds are X_1, X_2 and X_3. When buffer length overcomes X_2, any cell with enabled CLP is discarded. If, far from improving, the congestion level increases, even overcoming X_3, the system will discard low-priority cells with disabled CLP. X_1 is used to implement a hysteresis cycle avoiding oscillation.

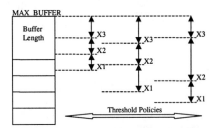

Fig. 5. Threshold policies applied to the buffer.

4.2 Traffic Sources

In order to evaluate the implemented model in an exhaustive way, the maximum number of possible scenarios should be simulated. So different source types are used in every simulation allowing us to study a high number of real cases.

Persistent, stochastic and bursty sources have been implemented in our simulations. A persistent source uses a constant distribution and data generation speed is constant. On the other hand, stochastic sources transmit depending on a certain probability distribution. In our case, exponential and normal distribution stochastic sources have been used. Bursty sources are characterized by an on-off behaviour because they alternate between active and idle periods. On-off models have been usually adopted in order to simulate real traffic sources [11]. In active periods, cells are generated at a peak bit rate, whereas in idle periods no cells are generated.

5 Results

After the implementation of both models in OPNET, a qualitative comparison can be performed. Thus, a measurement to evaluate both models is needed. The chosen measurement has been the number of high-priority and low-priority cells discarded by both algorithms. As in previous simulations, both algorithms fulfil initial specifications, that is to say, the number of low priority cells dropped are bigger than the number of high priority cells.

After the simulation of both algorithms using different traffic sources (exponential, normal and bursty ones) and varying the network load, obtained results are all similar in appearance to Figure 6:

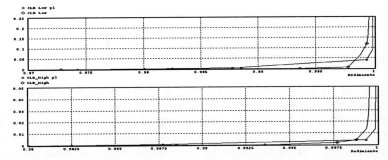

Fig. 6. The upper figure shows dropped low priority cells and the lower figure shows dropped high priority cells. One trace of each figure shows original algorithm results and the other one shows modified algorithm results.

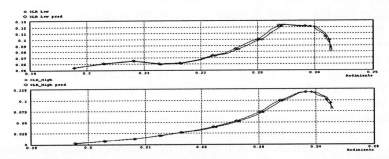

Fig. 7. The upper figure shows dropped low priority cells and the lower figure shows dropped high priority cells. One trace of each figure shows original algorithm results and the other one shows modified algorithm results.

Figure 6 shows that, in heavy load situations, our proposal eliminates fewer cells than the other one. However, in poor load situations, the original algorithm drops a barely noticeable number of cells than the modified ones.

Previous simulations analyzed all discarded cells. Now, if only CLP_0 discarded cells are taken into account, similar results are obtained. These results are shown in figure 7.

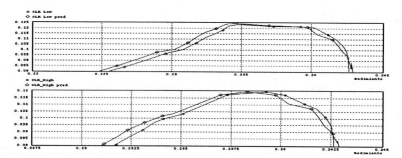

Fig. 8. This figure enlarges Figure 7. It focuses on the point where our algorithm improves the original one. From this point, the enhanced algorithm eliminates a smaller number of cells than the first one [4]

6 Conclusions and Further Work

To predict and control short-term congestion in ATM networks, a control method using Artificial Intelligence techniques has been introduced in this paper. After an exhaustive analysis over previous papers, a UPC algorithm that works with buffer utilization has been chosen. This algorithm has been adapted in order to improve the congestion control. The prediction of the future buffer utilization is the main goal of the new algorithm. This prediction will be obtained from the results of a formula given by AI. Using the account of buffer utilization and AI techniques, this mathematical function is created. The new algorithm uses this formula in order to foresee future states of network congestion. Then, it will apply this new knowledge with the purpose of minimizing the congestion effects.

Another goal has focused on obtaining a qualitative evaluation of AI application in ATM networks congestion control. Furthermore, it was important to determine its viability and effectiveness. Finally, it has been proved that AI contributes, in general, to the improvement of the mechanism capabilities in which it is applied. It is important to find out which of the processes used to control ATM congestion are likely to be improved by AI. In our concrete application case, the AI learning process could be improved using other techniques of evolutionary computation or increasing or reducing distance prediction. It could be interesting to analyse how different threshold policies could change algorithm performance. Then, it might be possible to obtain a concrete AI model that will improve the behavior of short-term congestion control whatever the real traffic pattern was.

To improve AI application advantages, several questions regarding this work still exist. These questions include the AI application formula, the methodology used to obtain this prediction formula, the utilization of different traffic sources and the threshold policy used. If, for example, threshold policy is modified to exclusively use more restrictive thresholds than the reference ones and never less restrictive thresholds, a new algorithm could be obtained. This new algorithm would eliminate a smaller number of cells than the original algorithm whatever the traffic load and whatever the traffic sources. All these points remain a subject to be studied.

Acknowledgements

We would like to thank the *Comisión Interministerial de Ciencia y Tecnología* for its support under grant number CICYT/Tel1998-0408-02. The results of this work were obtained using the equipment co-funded by the *Direcció General de Recerca de la Generalitat de Catalunya* (D.O.G.C. 30/12/1997). We would also like to thank *Enginyeria i Arquitectura La Salle* (EALS), *Universitat Ramon Llull* for their support to our research group.

References

1. M. de Prycker: Asynchronous Transfer Mode, solutions for Broadband ISDN, 2nd edn. Ellis Horwood, ISBN 0-13-178542-7
2. The ATM Forum: ATM User-Network Interface (UNI) Signalling Specification, Version 3.1, p. 69, September 1994.
3. Salim Hariri and Bei Lu: ATM-Based Parallel and Distributed Computing, 1996
4. S.Abe, T.Soumiya: A traffic Control Method for Service Quality Assurance in ATM Networks, IEEE Journal on Selected Areas in Communications Vol. 12, n.2, February 1994
5. William Stallings: Data & Computer Communitacions, 6th. ed., Prentice Hall, June 2000
6. D. Gaïti and G. Pujolle: Performance Management Issues in ATM Networks: Traffic and Congestion Control, IEEE Transactions on Networking, vol. 4, n.2, April 1996
7. Ramesh, Rosenberg and Kumar: Revenue Maximization in ATM Networks Using the CLP Capability and Buffer Priority Management, IEEE/ACM Transactions on Networking, Vo. 4, N.6, December 1996
8. Norio Matsufuru and Reiji Aibara: Flexible QoS Using Partial Buffer Sharing with UPC, IEICE Trans. Commun. Vol . E83-B, N.2, February 2000
9. John R. Koza: Genetic Programming, 6th Printing, Massachusetts Institute of Technology, ISBN 0-262-11170-5
10. Hilde Hemmer and Per Thomas Huth: Evaluation of Policing Functions in ATM Networks, Elsevier Science Publishers B.V., 1991
11. Tsern-Huei Lee, Kuen-Chu Lai: Design of a Real-Time Call Admission Controller for ATM Networks, IEEE/ACM Transactions on Networking, vol. 4, N.5, October 1996

Monitoring the Quality of Service on an ATM Network Carrying MMB Traffic Using Weighted Significance Data

A. Abdul Karim Mouharam and M.J. Tunnicliffe

School of Computer Science and Information Systems,
Kingston University, Kingston-Upon-Thames, KT1 2EE, U.K.,
Tel.0044 (208) 5472000 ext. 2674
Fax: 0044 (208)5477824
K929925@atlas.king.ac.uk

Abstract. Recent years have shown a great change in the telecommunication environment. Thus far, separated voice and data networks are being replaced by heterogeneous networks capable of supporting a diverse variety of traffic. Quality of Service (QoS) has become an important factor in the deployment of the next-generation of data networks. The continuing increase in the volumes of data to be carried by ATM networks has boosted the need to administrate the QoS. Although the Connection Admission Control (CAC) algorithm is not specified by the ITU-T, it is still widely used to moderate bandwidth allocation and User Parameter Control (UPC) algorithms can ensure that contractual stipulations are met. However, if an *accurate* QoS monitoring technique is implemented, both the CAC and UPC mechanisms will have a firmer foundation upon which to base their decisions. This will mean a firmer control on the bandwidth allocation. This paper focuses on the use of data interpretation to monitor the QoS of source bursty traffic based upon delay.

1 Introduction

Network bandwidth is being consumed by hungry applications like Internet telephony, video conferencing and video on demand. The proliferation of personal computers, together with the increased use of local area networks has caused an increase in traffic which had a severe impact upon the network QoS. This Explains the growth of the data communications market at an estimated 30% per annum [1], while the bandwidth representing data (as opposed to voice) traffic has expanded from 25% to over 61% in the years 1995-8. Figure 1 demonstrates the increase of data bandwidth relative to voice bandwidth. The fact is that many networking applications like the World Wide Web, Internet Telephony, Desktop Video Conferencing and Video-on-Demand have produced an increased demand for bandwidth, together with the support of heterogeneous services on a single network. Asynchronous Transfer Mode (ATM) is one of these broadband networks which has been deployed to cope with the increasing demand on bandwidth. Recommendations for its use in the implementation of a broadband digital service has been made by the International Telecommunication Union (ITU-T) and the American National Standard Institute (ANSI). Therefor it has been recognized as the future communication network. ATM is based on switching small, fixed size packets called cells (53 bytes each) using a connection oriented protocol. When an ATM contract is being established between the user and the network Administrator, the user than can establish a virtual connection (VC) for the user cells to travel from source to the destination. A number of these VCs created in parallel it is referred to as Virtual Path (VP). The small size cells will guarantee a faster transmission rate. It is capable of carrying video, voice, multimedia and other data services through a single network.

While the dominant Available Bit Rate (ABR) services can - if necessary - tolerate large delays, they are highly vulnerable to data loss. Their connection contracts therefore specify a minimum cell rate MCR [2]. However, for the purposes of real-time applications (voice, video etc.) cell delay and cell delay variation should also be kept to minimum.

These demands on the data network bandwidth have produced an increasing need to monitor the quality of service (QoS) provided. Data networks need to be monitored in order to allow network managers to keep up-to-date with new applications and their demand for bandwidth. As described in the following section, the monitoring of QoS will involve a continuous collection of data and interpretation of this data to detect intermittent error conditions, together with any trouble resulting from a gradual deterioration of the network. This allows network operators to take precise action if deterioration in the QoS is discerned.

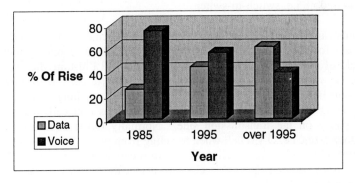

Fig. 1. Date traffic increase relative to voice traffic (after Atkins [1])

This paper reports on research aimed to establish a better method of monitoring the QoS of an ATM network without having to check every cell transmitted. This can be achieved by using additional monitor cells to detect the current state of the network. QoS in terms of delay is monitored since cell loss in a *real* ATM network is extremely low. (It is estimated that Cell Loss probability in a real ATM network is close to 10^{-9} [3]). In view of the inaccessibility of "real" network data, the integrity of a monitoring strategy must be verified by a computer simulation. For this purpose, a simulation program was written in C++. A multi-state Markov-modulated Bernoulli traffic streams has been monitored in the presence of heterogeneous MPEG cross-traffic.

2 ATM (QoS)

QoS monitoring in ATM networks will guarantee the user that there is no disturbance of the bandwidth/QoS agreed in the traffic contract. ATM covers important features of networking like QoS, statistical multiplexing and traffic loss prioritization. In ATM, QoS may be quantified by several parameters, most notably the cell-loss probability (CLP), the cell transfer delay and cell delay variation (CDV).

2.1 Operation and Maintenance Cell (OAM)

The ITU-T recommendation 1.610, define a cell with a special format called the Operation and Maintenance Cell (OAM). There are 2 types of OAM cells in the ATM layer and they are as follows:

1. Virtual Path Level, called F4 [4]. This provides an end-to-end flow from one end-point at the same level to another end-point.

2. Virtual Channel Level, called F5 [4]. This provides a segment flow, which is a concatenation of virtual path or virtual channels link from one point to another connection point, i.e. switch-to-switch.

OAM cells can be used for fault management or performance management. In this paper we are focusing on using OAM cells for performance management in the specific field of monitoring. The following scenario will illustrate the use of OAM cells: Suppose we have two end-points A and B. OAM cells will flow in each direction, A to B and B to A. In a real ATM network, launching an OAM cell will involve injecting the cell into the path, counting previous user cells and calculating the 16-bit parity on the transmission side. Normally OAM cell flow may have different block sizes, i.e. OAM cell when traveling from point A to B may be injected for every 100 cells, and when traveling from B to A can be injected for every 200 user cell. Performance Monitoring OAM cells will detect the following important ATM QoS parameters.

1. Cell Loss Ratio (CLR). A cell loss usually occurs when a cell arrives at a switch buffer which is already full to capacity. Since there is no available memory space to store the cell, it is discarded and lost. CLR in this case, can be represented in the simulation. (Cell loss may also be caused by system noise or by a component failure.)
2. Maximum Cell Transfer Delay (MaxCTD), This is the maximum elapsed time between the launching of a cell onto the network and its arrival at the destination node. It is therefore the sum of the delays of all the individual links in the path.
3. Cell Delay Variation (CDV), This is the difference between the maximum and minimum cell transfer delays. It is a particularly important parameter in real-time services such as video.

These parameters can then be used to analyze the QoS of the network.

2.2 Connection Admission Control (CAC)

When a user negotiates a traffic contract with the network management system to setup a virtual connection between a source and a destination, a number of traffic descriptors will need to be set (e.g. Peak Cell Rate, Sustainable Cell Rate). Every CAC of a node of the requested VP connection needs to agree with the contract, Otherwise it will reduce the values in the traffic descriptors. Therefore the duties of the CAC of the switch can be summarized as the intelligent determination of a suitable traffic model to match the traffic descriptors. Once a suitable traffic model is found then it is combined with a complex queuing model to determine whether the system has enough resources to abide by the contract. If the resources are found to be enough then bandwidth is allocated to this contract. ATM network efficiency depends on the CAC and when admitting a new connection. The CAC algorithm is still being researched in order to improve the bandwidth allocation efficiency [5]. It is worth mentioning that CAC algorithms are not specified by the ITU-T. Since the bandwidth allocated in the contract can be exceeded by the user (causing a decrease in the QoS of other users) a Usage Parameter Control (UPC) mechanism has been introduced. This mechanism carries out a policing role for the network, to make sure that users do not exceed what they have been granted in the contract.

3 Network Model

This section describes the network model that has been implemented. A random number generation scheme is used to introduce cell-streams into the switches. Figure 2 shows the simulated network configuration, which represents a cross-section of six ATM switches.

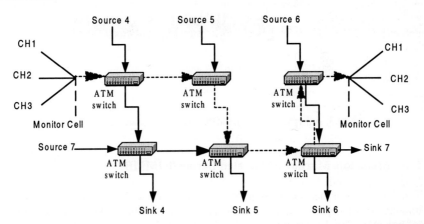

Fig. 2. The simulated Data Network. Dotted arrow indicates the direction of travel.

Three virtual channel connections labeled CH1, CH2 and CH3 are combined to a common virtual path connection. The network cross-traffic is represented by four traffic sources labeled Source 4, 5, 6 and 7.

A Multi-state Markov Modulated Bernoulli (MMB) traffic model was applied to the three virtual channel connections (CH1, CH2 and CH3). Similarly an MPEG traffic model was applied to the four cross-traffic sources. The MMB connection was monitored by an additional connection CH0, through which monitor cells were injected every 1000 ATM cycles. (Monitor cells could represent OAM cells in a real ATM network.)

3.1 Simulated MPEG Video Traffic

This represents the cross network traffic, which is the background load on the network. A high bandwidth is required to obtain an image from the receiving end, because of the large number of pixels involved. Image coding is therefore provided in order to reduce the pressure on the bandwidth. MPEG coding is based on a combination of a block based intra- and interframe techniques [6]. These techniques aim, as much as possible, to reduce the spatial correlation existing between individual pixels in the image. Bit rates can vary from zero to a maximum allowed rate, or a clear stepped variation across a range of quantified rate intensities. This source has been based on work carried out at Loughborough University [6].

It is certain that a relatively high proportion of future network traffic in local metropolitan and telecommunication networks will be of video traffic type. This video traffic would be generated by many sources like video conferencing and video telephony. Figure 3 shows a typical traffic profile generated by the simulation.

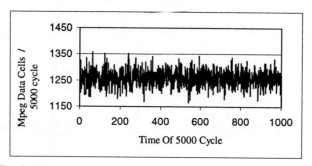

Fig. 3. A Typical output of MPEJ traffic generated by the program.

3.2 Mulitstate Markov-Modulated Bernoulli (MMB) Traffic

The VBR traffic has been of interst to many researchers [e.g. 9], and it is been implemeted using a muliti-state traffic model. This experiment introduced two refinements: Firstly, attention was given to the fact that cell loss in a *real* ATM network is extremely low (typically 10^{-9})[3]. Therefore QoS monitoring was focussed on data delay rather than data loss. Figure 4 shows typical delay profiles generated by the simulation. From the diagram it can be can seen that, the maximum tolerable delay has been set to 12 in the simulation. Therefore any cell arriving beyond the time limit of 12 cycles will considered a failure.

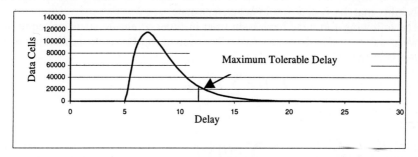

Fig. 4. Shows typical delay profiles generated by the simulation. The diagram represents the total number of cells dropped into the network by the monitored channel against the waiting time taken to transmit those cells

The Markov chain model is an established conceptual device for describing the way in which a stochastic system's behavior is related to its state transition probabilities. This is the traffic, which is being monitored by this program. There is no doubt that the Markov chain approach has received a wide acceptance for traffic modeling. Figure 5 shows the model used to generate MMB traffic used in these simulations. It shows that each state would have 6 possible destination states, each of which has its own cell-generation probability.

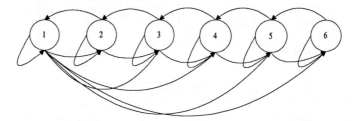

Fig. 5. Model used to generate the Markov modulated Bernoulli chain traffic.

$$Q(i,j) = \begin{bmatrix} q_{11} & q_{12} & q_{13} & q_{14} & q_{15} & q_{16} \\ q_{21} & q_{22} & q_{23} & q_{24} & q_{25} & q_{26} \\ \cdot & \cdot & \cdot & \cdot & \cdot & \cdot \\ \cdot & \cdot & \cdot & \cdot & \cdot & \cdot \\ \cdot & \cdot & \cdot & \cdot & \cdot & \cdot \\ q_{61} & q_{62} & q_{63} & q_{64} & q_{65} & q_{66} \end{bmatrix} \quad (1)$$

This model consists of 6 states, which is sufficient to represent quite a complex traffic source. The stochastic state transition matrix Q represents the transition probabilities from each state to every other state. This shown in Eqn. 1. For example, the element

q_{ij} from Q represents the state transition from state i to state j and q_{ii} represents probability that the system would not change state.

The following formulae have been used in order to ensure statistical equilibrium between states. If S is the state space of the system and P(i) is the probability that the system is in state i.

$$P(j) = \sum_{i \in s} P(i).qi,j \qquad (2)$$

Where

$$\sum_{i \in s} P(i) = 1 \qquad (3)$$

The traffic profile generated by this model has been plotted in the following figure. Which clearly show the jumping between states returning different rate of generation of cells into the network.

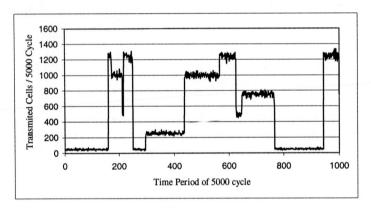

Fig. 6. Shows a typical traffic profile produced by this model and generated by the simulation

4 Experiment

The simulation was run for a period of 5,000,000 cycles (just over 13 seconds of real-time). All channels CH1, CH2 and CH3 in the virtual path connection were monitored to see whether the failed monitor-cell ratios were reflected in the corresponding data cells. Monitor cells were injected every 1000 cycles, and their observed failure rates were used to compute the 95% confidence interval for the failure probability [3].

The second experiment was designed to determine how many hits (i.e. failure rates within the 95% confidence interval) would be obtained when running twenty simulations at a time. In this case only one channel was monitored, and each time the simulation was run for a period of 5,000,000 cycle. Again monitor cells were injected every 1000 cycles, and their observed failure rates were used to compute the 95%

confidence interval for the failure probability. Figure 8 shows the upper and lower confidence limit along with loss probability when running the simulation for 20 run time.

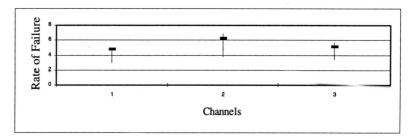

Fig. 7. shows the results generated by the simulation. It is clear that the *actual* failure percentages for all the three channels fall uniformly within this confidence interval. (Note that the vertical lines stretch from the upper to the lower confidence limits and the horizontal "dash" sign represents the "actual" failure probability.)

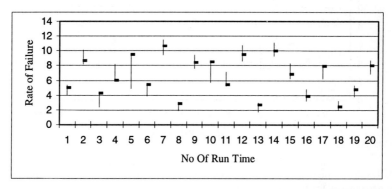

Fig. 8. Results of the experiment when running the simulation for 20 runs at a time. The *actual* failure percentages for the monitored channels fall most of the time within the predicted confidence intervals. (Note that the vertical lines stretch from the upper to the lower confidence limits and the horizontal "dash" sign represents the "actual" failure probability.) This result uses weighted significance algorithm.

5   Result Analyses

The Bayesian technique (which has been used to determine confidence intervals in this experiment) makes use of both subjective probability and measurement to determine the degree of belief about a value or unknown parameter. The subjective probability is called *prior* distribution. According to Harold [7], "…..the prior distribution of a parameter θ is a probability function or probability density function expressing our degree of belief about the value of θ, prior to observing a sample of a random variable X whose distribution function is indexed by θ.". The *posterior* distribution represents the way in which this belief is modified by subsequent observation, and would be used to build an estimator of the unknown parameter.

The simulation calculates the upper and lower confidence interval for the network performance parameters. The confidence interval is used to estimate that an unknown parameter (which is *the rate of failure*) is an interval so that:

$$\text{Upper limit} \leq \text{rate of failure} \leq \text{Lower Limit} \quad [8]$$

where the values of the *upper* and *lower* limits depend on numerical value of the sample statistics of the network.

The inaccuracy which happened in the past is believed to be a consequence of the additional load applied to the network over the high bit-rate excursions, during which a larger proportion of cells are transmitted than during quieter periods [9]. The monitor cells are injected uniformly across the time domain, and therefore form a poor representation of the source's characteristics. Figure 9 demonstrate the inaccuracy reading from the simulation which happened in the past.

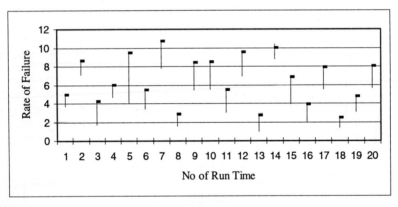

Fig. 9. Results of the experiment when running the simulation for 20 runs at a time. The *actual* failure percentages for the monitored channels fall most of the time above the predicted higher confidence limit. (Note that the vertical lines stretch from the upper to the lower confidence limits and the horizontal "dash" sign represents the "actual" failure probability.) Inaccuracy when not using weighted significance.

Therefore, a system of "weighted-significance" has been applied to the data prior to the computation of the confidence limits. The way is has been implemented is explained below:

Let us assume that each state *i* in the channel would generate N*i* cells of which J*i* are the successful cells received at the other end. Therefore the total number of transmitted cells would be $N=N_1+N_2+N_3+...+N_i$ and in the same way the total number of successfully received cells would be $J= J_1+J_2+J_3+...+J_i$. Lets assume a function δ (which we can call the *"significance factor"* between the bit rates for each state) thus:

$$\delta_n = \frac{N_n}{N_{\max\_bit\_rate}} \quad (4)$$

The values of N and J would therefore be implemented like the following formulas. The value of N and J would then be fed into the simulation to calculate the upper and lower confidence interval.

$$N = \sum_{r=1}^{i} \delta_r N_r \qquad (5)$$

$$J = \sum_{r=1}^{i} \delta_r N_r \qquad (6)$$

These formulas were used to modify the monitor cell data shown in Fig.8. It is clear that the rate of successful performance predictions is significantly improved.

6 Conclusions

Network bandwidth is being consumed by hungry applications like the Internet, Video conferencing and video on demand. The demand on bandwidth has produced a high need to monitor the QoS on networks. This paper focuses on establishing a better way to monitor the QoS of ATM network with out having to check every single cell transmitted. A "weighted significance" is proven to provide an accurate result and most of the time the *actual* failure percentages for the monitored channel fall within the predicted confidence intervals. Once the study is finalized, both CAC and UPC mechanism will have a firmer foundation upon which to base their decision.

References

1. Atkins, John.: Total area networking : ATM, IP, frame relay, and SMDS explained. 2nd edn. Chichester. John Wiley (1999)
2. George Kesidis.: ATM network Performance. 2nd edn. kluwer academic publishers (2000)
3. Mouharam, M. Tunnicliffe, A. J. Curley.: Monitoring Quality of Service on an ATM Network. Proc. IEE Colloquium on Control of Next Generation Networks, Savoy Place, London (1999)
4. McDysan, David E.: ATM theory and application. McGraw-Hill, New York, London(1998)
5. Natalie Giroux,Sudhakar Ganti.: Quality Of Service in ATM Networks. Prentice Hall (1998)
6. R. Marquez.: Statistical processing for telecommunication networks applied to ATM traffic monitoring. PhD thesis, Loughborough University (1997)
7. Larson, Harold J.: Introduction to probability theory and statistical inference. 2nd edn., Wiley (1974)
8. Lee, Peter M.: Bayesian statistics : an introduction. 2nd edn. Arnold, London 1997.
9. Mouharam, M. Tunnicliffe, A. J. Curley.: Monitoring VBR Traffic and markov chain model on ATM Network. Pgnet 2000, Liverpool (2000)

ATM Traffic Prediction Using Artificial Neural Networks and Wavelet Transforms

Priscilla Solís Barreto[1] and Rodrigo Pinto Lemos[2]

[1]Catholic University of Goiás-UCG, Department of Computer Science, Av. Universitária 1440 Setor Universitário, CEP 74605-010, Goiânia, Goiás, Brazil.
pris@zaz.com.br
[2]Federal University of Goiás –UFG, Department of Electrical Engineering, Praça Universitária s/n, Setor Universitário, CEP 74605-220, Goiânia, Goiás, Brazil.
lemos@eee.ufg.br

Abstract. This work proposes a method of combining wavelet transforms to feed forward artificial neural networks for ATM (Asynchronous Transfer Mode) traffic prediction. Wavelet transforms are used to preprocess the nonlinear time- series in order to provide a step-closer phase learning paradigm to the artificial neural network. The network uses a variable length time window on approximation coefficients over all scales. It was observed that this approach could improve the generalization ability as well as the accuracy of the artificial neural network for ATM traffic prediction.

1 Introduction

During the last years many researchers have observed the self-similarity of Asynchronous Transfer Mode (ATM) traffic. The certain modeling of the statistical multiplexing process, that promotes the traffic switching with variable bandwidth on different applications, is fundamental for the prediction process. Recently, several works modeled ATM traffic as a fractal phenomenon, with self-similarity and long dependence characteristics, especially in the case of Ethernet video traffic [1][2][3].
The multiplexing process of nodes in ATM networks uses a finite buffer, the long-time dependence produces delays and lost of cells, compromising the network quality service parameter and producing congestion situations. Then the traffic prediction task with some antecedence may provide the network controllers a multiplexing adjustment to avoid or minimize these effects.
Artificial neural networks (ANN) have demonstrated great potential for time-series prediction. Several works point their capabilities for nonlinear time-series forecasting that appear especially interesting for signal prediction models.
This correspondence presents a learning paradigm called phase learning that improves the generalization ability and reduces the computational cost of neural networks for ATM traffic prediction. To demonstrate the efficacy of this method, two comparative prediction experiments are driven on real ATM traffic.

1.1 Wavelet Transforms

In the signal processing analysis there are several mathematical tools for signal decomposition. The wavelet transforms functions make possible to frame a signal in regions of variable size that allows the analysis of the signal in different time scales, producing an automatic mapping of a signal in a time-frequency plane. The use of wavelets is especially interesting for signals that have non-stationary variations and self-similarity characteristics.

In the wavelet analysis, the signal low frequencies are identified as the approximation coefficients and the high frequencies are identified as the details. This can be seen as the original signal passing in a filtering process; with a lowpass filter producing the approximation coefficients and a highpass filter producing the details. Given a scale function ϕ_o and a basic wavelet ψ_o, the discrete wavelet transform (DWT) is defined as:

$$X(t) \rightarrow \{\{a_{j,k}, k \in Z\}, \{d_{jk}, j=1,2,....,j,k \in Z\}\} \quad (1)$$

The coefficients are defined as the internal product between X and two sets of functions.

$$a_{jk} = <X, \phi_o^{j,k}> \quad d_{jk} = <X, \varphi^0_{j,k}> \quad (2)$$

$\psi^0_{j,k}$ (respectively $\phi^0_{j,k}$) represents translations and dilations of ψ^0 (respectively ϕ^0), called basic wavelet function.

The multiresolution analysis (MRA) resides in the computation of a wavelet system for the decomposition and reconstruction of a signal $x(t)$ using and ortonormal base. The decomposition permits the approximation of a signal at different resolutions. There are two fundamental functions used to obtain a wavelet system: the scale function and the basic wavelet function, as seen in equation 5 and 6 respectively, where Z are the integer set and a_k are the wavelet coefficients. Both functions are prototypes of a class of ortonormal functions.

$$\phi(t) = \sum_{k \in Z} a_k \phi(2t - k) \quad (3)$$

$$\psi(t) = \sum_{k \in Z} (-1)^k a_{k+1} \phi(2t - k) \quad (4)$$

The phase learning paradigm introduced in this work decomposes the signal at different levels of detail using wavelet transforms. This decomposition produces different versions of the original signal, with coarsest versions in the higher scales and finest versions in the lower scales.

2 Models for ATM Traffic Prediction

Given a nonlinear time series with n samples, $\{y[i], i=1,2,...n\}$, representing the number of cells produced by an ATM source in n discrete time intervals, the goal is to predict the series value in the interval $n+k$, $k=1,2,...$, considering the series values until time interval n. In order to compare the quality prediction, the normalized mean square error was adopted as a performance parameter. The *NMSE* is computed as:

$$NMSE = \frac{1}{\sigma^2 p} \sum_{k=1}^{p} [(y(k) - \hat{y}(k)]^2 \tag{5}$$

where $y(k)$ is the observed value of the time series at time k, $\hat{y}(k)$ is de predicted value of $y(k)$ and σ^2 is the variance of the time series over the prediction duration. Thus, a NSME=1 corresponds to predicting the estimated mean of the series.

2.1 Model 1: ATM Traffic Prediction Using ANN

The model consists of a MLP network, with sigmoid activation functions on every layer, one output neuron, m input units and two hidden layers, as illustrated in figure 1.

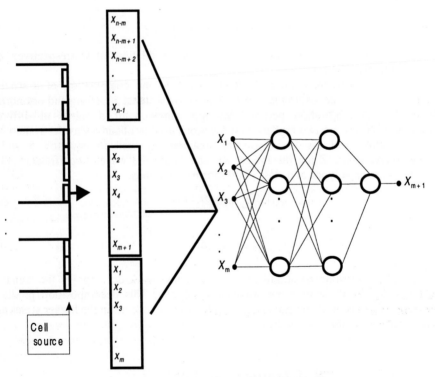

Fig. 1. MLP network used to predict ATM traffic.

The Levenberg-Marquadt algorithm was used in the learning process training, which uses numerical optimization techniques that appear to give a higher convergence speed for neural networks learning [4]. The initialization function of the weights is based on Nguyen and Widrow [5] technique. This function is implemented on the Neural Network Toolbox of Matlab 5.2.0.03084, January 1998 for Windows.

Several works have been done in predicting ATM traffic using models based on neural networks. In [6], they are used for predicting VBR traffic and in [7] three different models are used for the interpretation of aggregated traffic. In [8], a three layer MLP is used to predict ATM traffic in a network with military purposes. A major problem in neural learning is the overfitting that results when the neural network also memorizes from the training sets the statistical noise, making it difficult to establish statistical relations between the data. The result is generalization incapacity i.e., the neural network cannot generalize well to cases that were not included in the training set.

2.2 Model 2: ATM Traffic Prediction Using ANN and Wavelet Transforms

Several works seeking the ATM traffic prediction task using wavelet transforms have been done. In [9], Dantas proposes four analysis methods for ATM multiscale traffic prediction. The fact that the ATM traffic has a fractal behavior makes the prediction task more difficult, because the stochastic processes are not stationary or long dependence stationary. Also, Liang and Page [10] discuss about the learning capacity of an ANN and a multiresolution paradigm that breaks the original signal in several versions using wavelet transforms is proposed, searching the improvement of the ANN prediction task.

Using the above works as a reference, the proposal of this predictor consists to input the ANN a set of training sets preprocessed using wavelet transforms. The training sets are formed with vectors of different sizes, unlike to traditional network learning which employs a single signal representation for the entire training process. The zero scale corresponds to the original signal. The ANN will be trained with different scale versions of the original signal, trying to achieve a faster learning time and a less computational effort

The model consists of an ANN similar to the one described in model 1 but with a prior preprocessing of the signal using the Haar wavelet transform in the following manner: being X^s the original signal, $T_r(x^r)$ is the learning activity done in the x^r scale of the original signal X^s. The resulting x^r signal is composed only by the approximation coefficients. The details of the signal are discarded in each calculated scale. When r increases, the quantity of points of the original signal in the scale x^r diminishes in a proportion 2^r. Though, this can be interpreted as a visualization of the signal when fewer details and a lesser quantity of points. Interpreting the notation $T_j(x^j) \rightarrow T_i(x^i)$ as the training activity T_j with training set x^j preceding the training activity T_i with training set x^i, the training activities are given by the set $\{T_r(x^r), r=1,2,3,...n\}$, where $T_j(x^j) \rightarrow T_i(x^i)$ for $i=j-1$ e $n \geq j \geq 1$. Clearly appears that the set $T_0(x^0)$ denotes the training activity with the original signal. Figure 2 illustrates the model for the prediction task. The ANN weights are initialized just once during the first training activity (with the higher scale), afterwards they are just adjusted by the next learning activities with the lower scales.

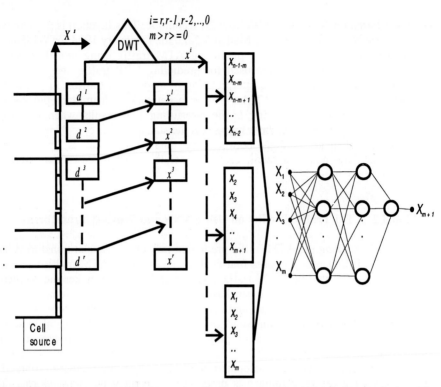

Fig. 2. MLP network with DWT Haar signal preprocessing for ATM traffic prediction. The ANN input consists of *m* neurons, with two hidden layers and one output neuron. The vectors x^i, resulting of the approximation coefficients of Haar wavelet transform are the different input training sets.

3. Experimental Results

3.1 The Time Series

The temporal series where obtained from real data from Bellcore of Ethernet network traffic. It was necessary to make a conversion of the data to characterize the ATM traffic. Two resulting series where obtained with the conversion: BC-OCTEXT and BC-OCTINT. The series BC-OCTINT consists of 1,759 seconds, in a time scale of 1 second, corresponding to 29 minutes and 31 seconds of internal traffic. The series BC-OCTEXT consists of 122,797.83 seconds, corresponding to 1,000,000 of packets of external traffic in a minute scale, resulting in 2046 points corresponding to 34 hours and 10 minutes of traffic. Figure 3(A) and 3(B) shows the series BC-OCTINT and BC-OCTEXT respectively.

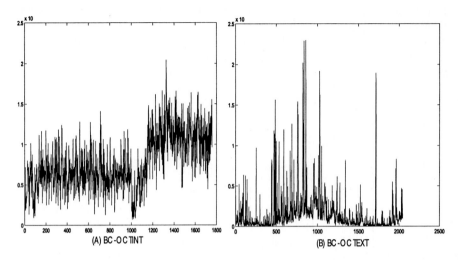

Fig. 3. (A) BC-OCTINT in second scale and BC-OCTEXT in minute scale testing series.

3.2 MLP ATM Traffic Prediction

In the case of the BC-OCTEXT time series were used three sets: a learning set, formed by elements 1 to 800, a validation set formed by elements from 801 to 1000 and a test set formed by elements 1001 to 2000. The prediction was made using a MLP network with $m=10$, two neurons in each hidden layer, resulting in a configuration of 10x2x2x1 and a learning rate of 0.01. In this case, the prediction quality was resulting NSME=0,42506. Figure 4(A) shows the prediction result.

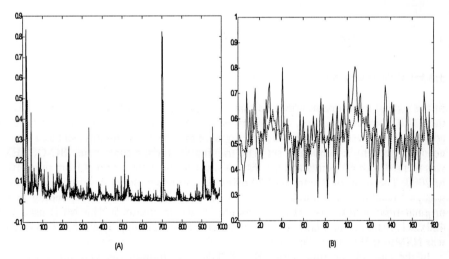

Fig. 4. (A) BC-OCTEXT prediction with NSME=0,42506. Real time series solid line, predicted time series dotted line. (B) BC-OCTINT prediction with NSME=0,92008. Real time series solid line, predicted time series dotted line.

In the case of the BC-OCTINT time series, the training set was formed with 600 points (800 to 1499), the validation set from points 100 to 599 and the test set with points from 1500 to 1699. A MLP 20x10x10x1 was used with a learning rate of 0.01 and the prediction result quality was an NSME=0,92008, that shows a low generalization capacity of the network, as seen in figure 4(B).

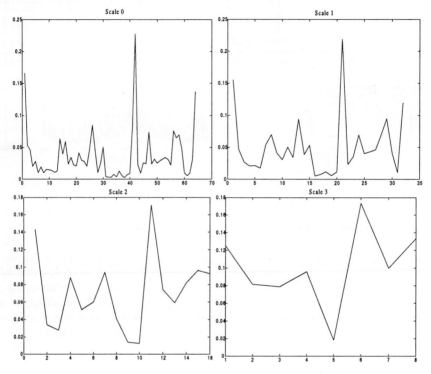

Fig. 5. BC-OCTEXT: training sets, four scales x^0, x^1, x^2, x^3.

3.3 MLP and Haar Wavelet Transform for ATM Traffic Prediction

In the second experiment, appeared that the training sets could be reduced as well as the MLP complexity. Also, there was observed that the early stopping procedure adopted in the first experiment did not improve the prediction quality, so there was no need for defining a validation set. The training set for the time series BC-OCTEXT was formed with 64 points and the test set was formed with elements from point 1001 to 2000. The scale of decomposition using the predictor model described in figure 2 was $j=3$ resulting in 4 training sets x^3, x^2, x^1 and x^0, the latter corresponding to the original time series. The training sets are shown in figure 5. The MLP used was 4x2x2x1, resulting in a less computational effort for training. The prediction quality was NSME=0,39951, as illustrated in figure 6.

In the case of the time series BC-OCTINT the training set was formed by the elements 1 to 256 and the test set with elements from point 1000 to 1699. It is important to note that in this second test set, there is an interval of abrupt variation with increases the prediction difficult. As with the time series BC-OCTEXT, the MLP used was a 4x2x2x1 and the prediction quality was NSME=0,46195, showing a significant improvement. The prediction result is illustrated in figure 7.

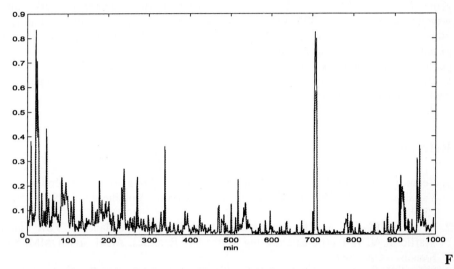

Fig. 6. (A) BC-OCTEXT MLP-wavelet transform prediction with NSME=0,39951, real time series solid line, predicted time series dotted line.

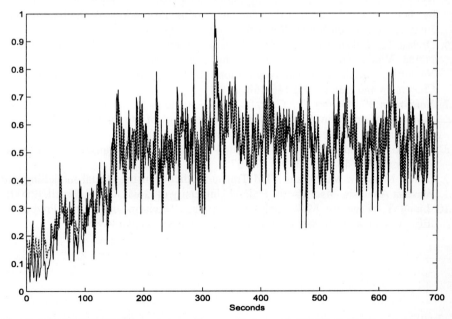

Fig. 7. BC-OCTINT MLP-wavelet transform prediction with NSME=0,46195, real time series solid line, predicted time series dotted line.

4 Conclusions

The conventional learning process in neural networks is sometimes inadequate for nonlinear, nonstationary signal prediction problems and often yields poor generalization performance. The combination of artificial neural networks and wavelet transforms for ATM traffic prediction permits a preprocessing of nonlinear time-series with fractal behavior. The neural network learns gradually from simpler to more complex training sets introducing a phase learning paradigm. This process permits a neural network complexity reduction and the pursuit of predictors that have a better generalization capacity with a low computational cost that appears to perform well enough to be of practical value.

References

1. Beran J., Sherman R., Taqqu M.S, Willenger W.: Long Range Dependence in Variable bit-rate Video Traffic. IEEE Trans. Comm, Volume 43, (1995), 1566-1579.
2. Ryu B.K., Elwalid A.: The importance of Long-Range Dependance of VBR Video Traffic in ATM Traffic Engineering Myths and Realities. ACM SIGCOMM, San Francisco CA, 1996.
3. Leland W.E. et al.: On the Self-Similar Nature of Ethernet Traffic (extended version). IEEE/ACM Trans. Net., Volume 2, (1994) 1-15.
4. Mathworks, Inc.: Neural Network Toolbox User´s Guide, 1997.
5. Nguyen D., Widrow B.: Improving the learning speed of 2-layer neural networks by choosing initial values of the adaptive weights. Proceedings of the International Joint Conference on Neural Networks, Volume 2. (1993) 357-363.
6. Drossu R., Lakshman T.V., Obradovic Z., Raghanvendra C.: Single Multiple Frame Video Traffic Prediction Using Neural Networks Models. Computer Networks, Architecture and Applications. S.V. Raghavanan, B.N. Jain Editors. Chapmann and Hall (1995) 146-158.
7. Nordstrom E., Galmo O., Gusraffson M., Asplund L. : Neural Networks for Admission Control in ATM Networks. Proceedings of the First Swedish National Conference on Connectionism. Skivde, Sweden (1992).
8. Lobejko W. Traffic Prediction by Neural Approach. IEEE Transactions on Signal Processing, Vol 45 N.11 (1996) 571-575.
9. Dantas Maria J.: Multiscale ATM Traffic Prediction using fractal modelling and Wavelet Analysis. Msc. Thesis. Federal University of Goiás. Goiás, Brazil. (2000)
10. Liang Y., Page E.: Multiresolution Learning Paradigm and Signal Prediction. IEEE Transactions on Signal Processing, Vol 45. (1997) 2858-2864.

Threshold-Based Connection Admission Control Scheme in ATM Networks: A Simulation Study

Xiaohong Yuan[1] and Mohammad Ilyas[2]

[1]Dept. of Computer Science, North Carolina A & T State University,
1601 East Market Street, Greensboro, NC 27411.
xhyuan@ncat.edu
[2]Dept. of Computer Science and Engineering, Florida Atlantic University,
777 Glades Road, Boca Raton, FL 33431.
mohammad@cse.fau.edu

Abstract. In this paper, a new threshold-based connection admission control scheme is proposed and analyzed. The scheme uses effective bandwidth for making decision about acceptance or rejection of a connection request. This threshold and effective bandwidth based method is simulated on an 8-node network model. The traffic used in the simulation is a multiplexed stream of cells from video, voice and data sources, which is typical to ATM environment. Simulations on the 8-node network model shows that, in a network that supports several service categories, having a threshold-based connection admission control affects the blocking probabilities of each type of traffic. In some environments, having a threshold is advantageous over the case without a threshold in terms of cell loss ratio, cell transfer delay and power (defined as throughput divided by cell transfer delay).

1 Introduction

Connection Admission Control (CAC) is one of the functions defined by ATM forum for managing and controlling traffic and congestion in ATM networks [1, 2, 3]. It is defined as the set of actions taken by the network during the call set-up phase in order to determine whether a connection request can be accepted or should be rejected. Based on the source of information used to make admission decision, the existing CAC schemes fall into two categories: model-based CAC and measurement-based CAC [4]. Model-based CAC schemes require each source characterize the offered traffic a priori in terms of the parameters of a deterministic or stochastic model, while measurement-based CAC schemes rely mainly on real-time traffic measurements to make admission decisions, although they may also rely on information provided by the user. Typical model-based CAC schemes include effective bandwidth based approaches [5, 6, 7] and cell loss probability estimation based approaches [8, 9, 10].

In this paper a new threshold-based CAC method is proposed. The basic idea is, when a new connection request arrives, instead of checking whether the available resource can accommodate the new connection or not, it is proposed to check whether a percentage of the available resource would accommodate this new connection. Let x represent the percentage, and assume the traffic request parameters required by the new connection are represented by the requested resource, then the if-then connection admission rule can be stated as follows:

If x × available resource ≥ requested resource,
Then accept the connection. Otherwise, reject the connection.

Adding a threshold x to available resource in making admission decision implies that, a certain amount of resource is always reserved for some later use. Through simulation it is found that with a threshold, some type of traffic is blocked in favor of other types of traffic. Having a threshold may also affect the network throughput, loss or delay performance. Given a network, by adjusting the threshold a better network throughput, or better loss or delay performance can be achieved.

To apply the threshold-based CAC, the effective bandwidth method due to Guérin *et al.* [5] is adopted to represent the available resource and the requested resource. Effective bandwidth is a measure of a connection's bandwidth requirement relative to the required QoS constraint, such as the delay and/or loss experienced by a connection's cells. The effective bandwidth concept is convenient to implement the threshold-based CAC, since one could just compare the effective bandwidth of the coming connection with certain percentage of available bandwidth (which is link capacity minus the sum of the effective bandwidth of all existing connections).

This paper presents a simulation study of the threshold-based CAC scheme. The threshold-based CAC scheme is simulated in an 8-node network model. The traffic used is a multiplexed stream of cells from data, voice and video sources typical to ATM environment. In what follows, the multiplexed traffic used in the simulation, the simulation model and algorithm, the verification of the simulation model are introduced. The simulation results are presented and summarized.

2 Generating Multiplexed Traffic and Measuring the Traffic Characteristics

In ATM environment, traffic is usually a multiplexed stream of cells from data, voice and video sources. When data, voice and video traffic are multiplexed over one output link, video traffic is given the highest priority, voice traffic is given the next highest priority, and data traffic is given the lowest priority when there is contention for the output link [11]. With video and data traffic modeled as Markov-modulated Bernoulli processes and voice traffic represented by ON-OFF Bernoulli process [3, 12, 13, 14], the video, voice and data sources being multiplexed can be characterized by parameters in Table 1 [15].

In Table 1, the loading factor specifies the percentage of time when a source actually generates traffic. Each source can be represented as a two-state process as shown in Fig. 1. For video traffic, the states are burst and non-burst, for data traffic, the states are high-rate and low-rate states. According to the source descriptions in Table 1, the data rate at each state, and the values of α, β can be obtained.

The discrete event simulation [16, 17] is used to generate the multiplexed traffic. According to the two-state process model, one could schedule the cell arrival process for a single video, voice or data source. The multiplexing of the three sources is implemented by selecting one cell from a voice, a video and a data cell that are scheduled to transmit according to the priority policy. This implementation assumes that the buffer at the multiplexer has infinite length.

Table 1. Source descriptions

Service Categories	Parameters
η_1: Data traffic (Markov-modulated Bernoulli process)	High bit rate = λ_{1h} Low bit rate = λ_{1l} Transition probability from high-to-low rate = P_{HL} Low-to-high rate = P_{LH} Loading factor: l_1^*
η_2: Voice traffic (CBR: On-Off Bernoulli process)	On-state bit rate = λ_2 On-to-on state probability = P_{onI} Off-to-off state probability = P_{offT} Loading factor: l_2^*
η_3: Video traffic (VBR: Markov-modulated Bernoulli process)	Non-bursty bit rate = λ_3 0Burstiness ratio = BP Transition probability from bursty-to-nonbursty state = P_{BN} Transition probability from nonbursty-to-bursty state = P_{NB} Loading factor: l_3^*

(Note: $l_1^* + l_2^* + l_3^* = 1$)

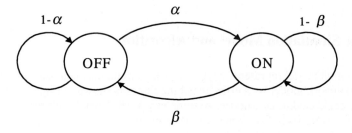

Fig. 1. Two-state process

The multiplexed traffic generated will be fed to the proposed network model to evaluate network performance with certain connection admission control schemes. When the user requests a connection, it is assumed that the user provides the source traffic descriptor parameters at the time of requesting a connection. What traffic descriptors the user needs to provide depends on the CAC scheme. The threshold-based CAC scheme is based on the effective bandwidth by Guérin et al. [5]. Assuming a single source feeding a finite capacity queue of size K with constant service rate. The source is an Interrupted Fluid Process (IFP) characterized by (R, r, b), where R is the peak rate, r is the fraction of time the source is active, and b is the mean duration of the active period, the equivalent capacity (c) of the source that corresponds to a cell loss ratio (ε) is given by:

$$c = \frac{a - K + \sqrt{(a-K)^2 + 4Kar}}{2a} R \qquad (1)$$

where $a = \ln(1/\varepsilon)b(1-r)R$.

For sources with generally distributed burst or idle period, the first and second moments of the burst/idle period are mapped into the first moment of an equivalent, exponentially distributed burst/idle period. The equivalent mean burst and idle periods b' and I' are given by $b' = \delta b$ and $I' = \delta I$, where b and I are the original mean burst and idle periods, and δ is selected as:

$$\delta = \frac{\overline{b^2}}{2b^2} \times \frac{\overline{I^2}}{2I^2}. \qquad (2)$$

The effective bandwidth is then calculated using this equivalent IFP source.

The multiplexed traffic generated is an ON-OFF fluid process but with nonexponential burst/idle period. Therefore to calculate the effective bandwidth, the peak cell rate (PCR), sustainable cell rate (SCR), b, I, $\overline{b^2}$ and $\overline{I^2}$ are measured on the multiplexed traffic generated and it is assumed that they are provided by the user when the user requests for a connection.

3 The Simulation Model and Algorithm

The simulation model represents a typical virtual circuit between users of a broadband ISDN. A network model with eight switching nodes is shown in Fig. 2. The switching nodes are connected to user nodes. Assume user A, B, F send multimedia data to user D, E and C respectively from time to time. From user A there are three routes to get to user D. They are: (1) from user A to node 1 to 2 to 3 to 4 and then to user D; (2) from user A to node 1 to 7 to 8 to 4 and then to user D; (3) from user A to node 1 to 6 to 5 to 4 and then to user D. From user B there are also three routes to get to user E. They are: (1) from user B to node 2 to 3 to 8 to 5 and then to user E; (2) from user B to node 2 to 7 to 8 to 5 and then to user E; (3) from user B to node 2 to 7 to 6 to 5 and then to user E. From user F there are three routes to get to user C also: (1) from user F to node 6 to 7 to 8 to 3 and then user C; (2) from user F to node 6 to 5 to 8 to 3 and then user C; (3) from user F to 6 to 7 to 2 to 3 and user C. Non-blocking switches are assumed in the model, and it is assumed that each output port has its own dedicated buffer.

It is assumed that this network supports three service categories, each of which has different traffic characteristics and requires different amount of resources. Each service category is a multiplexed stream of cells from video, voice and data sources. User A, B and F keep generating connection requests of three types to the network. The arrival process of the connection requests of each type from each user is assumed to follow Poisson process, and the holding time of a connection admitted to the network is assumed to have exponential distribution. If a connection is admitted to the network, then cells are generated for this connection and are transmitted through the network. If a connection is rejected, it is cleared. In the simulation, the traffic descriptor used includes: PCR, SCR, b, I, $\overline{b^2}$ and $\overline{I^2}$, and CLR is specified as the requested QoS parameter. The statistics collected in the simulation are: blocking probability, cell loss ratio (CLR), average cell transfer delay (CTD) and power (defined as throughput divided by cell transfer delay).

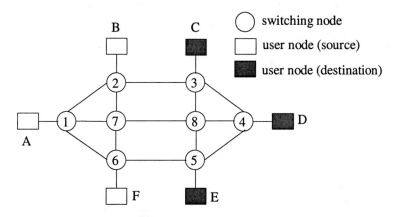

Fig. 2. An 8-node network model

In the simulation algorithm, the process of determining the route is embedded in the process of making admission decision. When a connection request arrives, the connection admission controller searches among all possible routes the one that has the maximum available bandwidth. Then the effective bandwidth of this connection is compared with the x percent of this maximum available bandwidth. If the former is less than or equal to the latter, then the connection request is accepted, otherwise it is rejected. If the connection is accepted, then it should take the route that has the maximum available bandwidth. This routing algorithm helps to balance the load at each route or link, so that the network will not have some links overly loaded but have some other links under loaded. The simulation program is written in C and the event scheduling approach has been used so as to minimize execution time.

4 Model Verification

The queueing theory results are used to verify the simulation program. Due to the complexity of the 8-node network model, the closed form solution for the blocking probability of each type of traffic is very difficult to obtain. The following techniques are then used to verify the simulation program: (1) Run the simulation for simpler cases. For example, we could assume only one user is generating traffic, and it is generating only one type of traffic. The closed form solution for the blocking probability for this situation could be obtained. (2) By letting some parameters in the 8-node network model take some special values, the 8-node network model could reduce into a simpler network model for which the closed form solution for the blocking probability could be solved.

When only user A is generating traffic, and it only generates one type of traffic, the blocking probability of the connection requests can be solved [18, 19, 20]. A program is written to calculate the blocking probability. The calculated theoretical values are compared with the simulation results to verify the correctness of the simulation program.

Let the links from node 1 to 2, from 1 to 6, from 7 to 6, and from 2 to 3 take value 0, and assume only user A and user B are generating traffic, the 8-node network model reduces to the model shown in Fig. 3 (the links that the traffic can never get to are not shown in the figure).

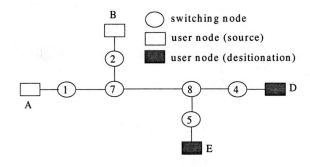

Fig. 3. The reduced network model

In this reduced network model, assume three types of connection requests are generated from user A to user D, and from user B to user E. The closed form solution for the blocking probability of each type of traffic can be solved [20]. A program is written to calculate the blocking probabilities of the three types of traffic and the theoretical values are compared with the simulation result to verify the correctness of the simulation.

Little's theorem is also used to verify the simulation model [21]. Applying Little's theorem to our network situation, we have: mean number of cells in the network = cell arrival rate × mean time a cell spends in the network. This relation is verified in our simulation.

5   Simulation Results

Assume the average arrival rates and holding times of the connections of the three types of traffic take the values in Table 2. Also assume the buffer size at each node is 50. The descriptions of the traffic sources and traffic contracts for the three service categories used in the simulation are given in [20]. Each service category is the multiplexed traffic of a video, a voice and a data source. The descriptions of the sources being multiplexed are used in generating the multiplexed traffic. The values of the parameters in the traffic contract are obtained through measurement on the generated multiplexed traffic. They are provided to the connection admission controller to make admission decision.

Table 2. Average arrival rate and holding time of the connections

Service Category	Arrival Rate	Holding Time
I	5×10^{-5}	1.0×10^{5}
II	5×10^{-5}	7.5×10^{4}
III	5×10^{-5}	5.0×10^{4}

The threshold-based CAC scheme is simulated in the 8-node network model under a variety of situations. Fig. 4 shows how the blocking probability, CLR, the CTD and the power change when the threshold x changes from 0.1 to 1.0. Fig. 5 shows how blocking probability, CLR, CTD and power change as the average arrival rate of type I calls changes, when $x = 1$ (i.e., without threshold) and when $x = 0.5$. Fig. 6 shows how blocking probability, CLR, CTD and power change as the average arrival rate of type II calls changes, when $x = 1$ and when the thresholds for type I, type II and type III calls are 0.5, 1 and 0.5 respectively.

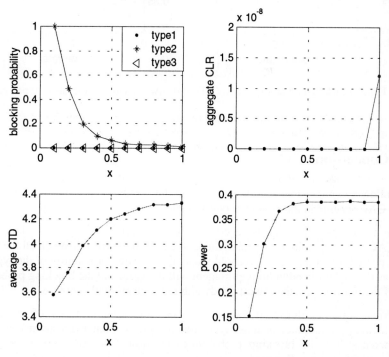

Fig. 4. Blocking probability, CLR, CTD, power versus threshold x

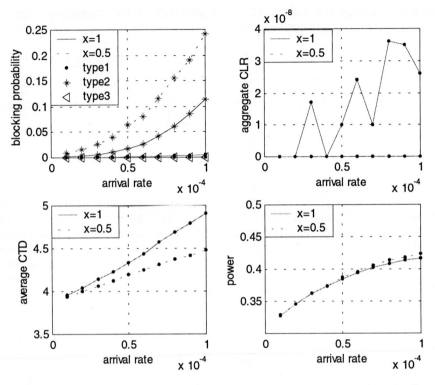

Fig. 5. Blocking probability, CLR, CTD, and power versus arrival rate of type I calls

6 Conclusion

The following conclusions can be drawn from the simulation results of the threshold-based CAC scheme in the 8-node network model:

- In a network that supports several service categories, having a threshold-based CAC affects the blocking probabilities of each type of traffic. When all the traffic types are given the same threshold value, the blocking probability of the type that has the highest bandwidth requirement will increase, and the blocking probability of the type that has the lowest bandwidth requirement will decrease.
- When different threshold values are used for different types of traffic, the blocking probabilities of those types that are given smaller threshold values will increase, and the blocking probabilities of those types that are given higher threshold values will decrease.
- The simulation results show that, with the threshold-based CAC, the aggregate cell-loss ratio and average cell transfer delay are smaller, and the power is higher compared with the case without a threshold.

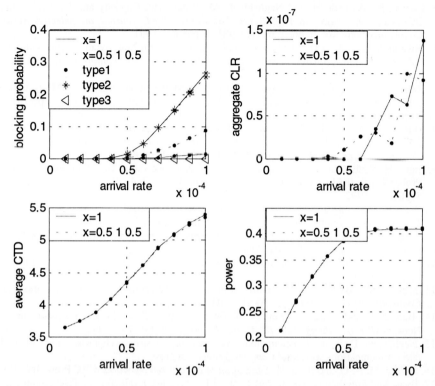

Fig. 6 Blocking probability, CLR, CTD, power versus arrival rate of type II calls with different thresholds for different types

- The power changes as threshold takes different values between 0.1 to 1. In some network environments, the power reaches its peak value when x is between 0.1 and 1.

Our future research will involve searching for mechanisms for changing x dynamically as the traffic condition changes in the network so as to improve network performance.

References

1. Adas, A., "Traffic Models in Broadband Networks," *IEEE Communications Magazine*, pp. 82-89, July 1997.
2. ATM Forum, "Traffic Management Specification," Version 4.1, March 1999.
3. Perros, H.G. and Elsayed, K. M. "Call Admission Control Schemes: A Review," *IEEE Communications Magazine*, vol. 34, no. 11, pp. 82-91, 1996.
4. Ren, Q. and Ramamurthy, G. "A Hybrid Model and Measurement Based Connection Admission Control and Bandwidth Allocation Scheme for ATM Networks," *GLOBECOM 98, IEEE Global Telecommunications Conference*, pp. 1016-1023, 1998.

5. Guérin, R., Ahmadi, H. and Naghshineh, M. "Equivalent Capacity and Its Application to Bandwidth Allocation in High-Speed Networks," *IEEE Journal on Selected Areas in Communications,* vol. 9, no. 7, pp. 968-981, September 1991.
6. Elwalid A. and Mitra D., "Effective Bandwidth of General Markovian Traffic Sources and Admission Control of High Speed Networks," *IEEE/ACM Transactions on Networking*, vol. 1, no. 3, pp. 329-343, June 1993.
7. Rege, K. M. "Equivalent bandwidth and related admission criteria for ATM stystems – a performance study," *International Journal of Communication Systems*, vol. 7, pp.181-197, 1994.
8. Murase, T., Suzuki, H., Sato, S. and Takeuchi, T. "A Call Admission Control Scheme for ATM Networks Using a Simple Quality Estimate," *IEEE Journal on Selected Areas In Communications*, vol. 9, no. 9, pp. 1461-1470, December 1991.
9. Saito, H. "Call Admission Control in an ATM Network Using Upper Bound of Cell Loss Probability," *IEEE Transactions on Communications*, vol. 40, no. 9, pp. 1512-1521, September 1992.
10. Lee, S. and Song J. "An Integrated Call Admission Control in ATM Networks," IEICE Transactions on Communications, vol. E82-B, no. 5, pp. 704-711, 1999.
11. Neelakanta, P. S. *A Textbook on ATM Telecommunications: Principles and Implementation.* CRC Press, Boca Raton, Florida, 2000.
12. Adas, A., "Traffic Models in Broadband Networks," *IEEE Communications Magazine*, pp. 82-89, July 1997.
13. Frost V. S. and Melamed B., "Traffic Modeling for Telecommunications Networks," *IEEE Communications Magazine*, pp. 70-81, 1994.
14. Bae, J. J. and Suda, T. "Survey of Traffic Control Schemes and Protocols in ATM Networks," *Proceedings of the IEEE*, vol. 79, no. 2, pp. 170-189, 1991.
15. Neelakanta, P. S. and Deecharoenkul, W. "A complex system characterization of modern telecommunications services," *Complex Systems*, in press.
16. Sadiku, M. N. O. and Ilyas, M. *Simulation of Local Area Networks*. CRC Press, Inc. 1995.
17. Ilyas, M. "Simulation: A Powerful Tool for Performance Evaluation of Telecommunication Networks," *Technical Report (TR-CSE-92-9)*. Department of Computer Science and Engineering, Florida Atlantic University, 1992.
18. Cooper, R. B. *Introduction to Queueing Theory (Third Edition)*. CEEPress Books, 1990.
19. Tilt, B. *Solution Manuals for Robert B. Cooper's Introduction to Queueing Theory, Second Edition*. Elsevier Science Publishing Co., Inc. 1981.
20. Yuan, X. *A Study on ATM Multiplexing and Threshold-Based Connection Admission Control in Connection-Oriented Packet Networks*. Dissertation, Dept. of Computer Science and Engineering, Florida Atlantic University, 2000.
21. Jain, R. *The Art of Computer Systems Performance Analysis: Techniques for Experimental Design, Measurement, Simulation, and Modeling*. John Wiley & Sons, Inc. 1991.

ABR Congestion Control in ATM Networks Using Neural Networks

K. Dimyati[1] and C.O. Chow[2]

Department of Electrical Engineering, Faculty of Engineering, University of Malaya
50603 Kuala Lumpur MALAYSIA
[1]kahar@fk.um.edu.my, [2]cochow@tm.net.my

Abstract. This paper presents the Neural Indicates Explicit Rate (NIER) scheme, in which an artificial neural network (ANN) is used to obtain suitable Explicit Rate (ER) for congestion control of Available Bit Rate (ABR) service in ATM networks. The neural network used in this scheme will monitor the queue status in switches' buffer and output suitable ER value to be carried by Resource Management (RM) cells to the sources for their rate adjustment. A comprehensive comparison study between the proposed algorithm and two well-known algorithms is done and simulation results prove that the proposed algorithm gives good performance in both LAN and WAN topologies.

1 Introduction

According to the ATM forum [1], there are currently six services for ATM networks but only Available Bit Rate (ABR) and Unspecified Bit Rate (UBR) are designed to handle bursty data applications. ABR service is believed to have better performance in handling data applications because it is associated with traffic management that allows the switches to monitor the network behavior and feedback related information by resource management (RM) cells to sources. The sources will adjust their own data rate based on the information obtained. This traffic management task is indispensable because it is the core that controls network efficiency and delivers the negotiated Quality of Service (QoS). Congestion control is one of the most important tasks of traffic management. Congestion control is a dynamic problem that cannot be solved by conventional static solution [2][3]. This can be proven by simple simulation showing that congestion still occurs in the network even with the increment of buffer size, bandwidth or processing speeds. Therefore, neural network is used to handle this difficult task in this paper because neural network perform powerful mapping and adaptation ability. Conventionally, only EFCI bit is used for congestion control purpose [4]. Further researches propose that ER indication could provide faster and better performance [2] and researchers have proposed different approaches in determining ER, such as NIST ER [5] and EPRCA [6]. ER is defined as the maximum allowable cell rate for the source in the next cycle.

The remainder of the paper is organized as follows: in section 2, the proposed algorithm is explained; Section 3 presents the simulations model and the choices of suitable parameters. The results obtained from the simulations are reported in section

4 and followed by related discussion. Finally, the properties of the proposed algorithm are explained.

2 NIER Scheme

The switching scheme presented in this paper is known as Neural Indicates Explicit Rate (NIER) scheme. The main motivation in designing this scheme is to implement neural network to provide simple but effective congestion control. Several researchers have suggested the use of neural network [7][8][9], but they focused on using neural network in predicting traffic pattern. Hence, this paper presents a complete algorithm using neural network to monitor network status and decide the amount of bandwidth that should be occupied.

The NIER scheme is a rate-based end-to-end closed-loop feedback scheme and uses RM cells to carry feedback information. It follows all the rules set by ATM Forum [1]. The operation of this switching scheme can best be represented with Fig. 1.

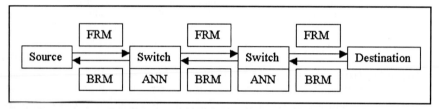

Fig. 1. Operation of the proposed NIER scheme

As mentioned before, the scheme follows the specification by the ATM Forum, therefore the basic operation is the same with other switching schemes. The major difference is how the ER is measured. The neural network monitors the queue status in switches' buffer for every fixed measurement interval (N). After every N ABR cells have been received, the neural network will calculate a suitable ER value based on the readings of queue length for current and previous intervals. When a BRM cell reaches the switch, the calculated value is compared with the ER value in the BRM cell, if the calculated value is smaller than the current ER value, the ER field in BRM cell will be updated. Anyway, if the calculated value is greater than current ER value, the ER field in BRM cell will be maintained.

In our algorithm, two inputs are needed. They are the current and previous queue lengths. These two inputs are essential in providing the neural network enough information about the status of the network. From the current queue length, the neural network able to know the queue occupancy and the level of congestion. Besides, the neural network will compare current queue length with the previous queue length to obtain information about the rate of change for the buffer. This is a good measurement of whether there is a possibility for the network to be congested or not. If the network is not congested, this is also a very good indication for the network to fully utilize the network resources.

The output of the neural network is the link fraction (LF) that should be used by the VC based on the current load condition. The value for LF is ranged from 0 to 1; a value closes to 0 indicates the network is facing congestion or there is a potential for the network to be congested, therefore the sources should send at lower or minimum cell rate. In contrast, a value of 1 indicates that all links should be fully utilized. The explicit rate (ER) for each VC is calculated by multiplying LF with the link speed of that VC:

$$ER = LF * \text{Link-Speed} \tag{1}$$

All the switches will maintain an ER value and update the value every N ABR cells are received. For every BRM cell travels through a switch, the switch will examine the ER field of that BRM cell. If the value in the BRM ER field is greater than the switch maintain ER value, the ER field in BRM cell is updated with the switch maintained ER value. If the value in BRM field in smaller, no action will be taken. After this stage, the BRM cell is forwarded to the next node until it reaches the appropriate source. All sources will adjust their data rate based on this ER value with the help of the CI and NI bit in the same BRM cell. The rules for source rate adjustment are given in table 1 [1].

Table 1. Source Rate Adjustments

NI	CI	Action
0	0	ACR ← Min(ER, ACR+RIF×PCR, PCR)
0	1	ACR ← Min(ER, ACR−RDF×ACR)
1	0	ACR ← Min(ER, ACR)
1	1	ACR ← Min(ER, ACR−RDF×ACR)

As a summary, the pseudocode for the proposed algorithm is as below:

```
while (online)

if (receive ABR cell)
   if (queue_length < output_queue_size)
      add cell to buffer
      count ++
      if (count % N == 0)
      link_fraction = NN(current_queue, previous_queue)
      ER =  link_fraction * link_speed
   else
      drop cell

if (schedule cell to link)
   if (cell is BRM cell)
      if(BRM.ER > ER)
         BRM.ER = ER
```

*Note: NN is the function of Neural Network operation

3 Simulation Model

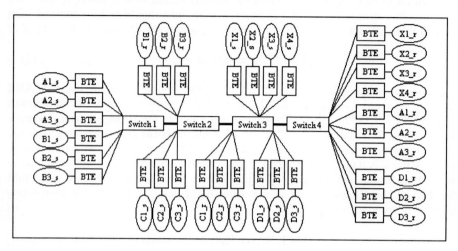

Fig. 2. Network Topology used for simulation and comparison study. It consists of five groups of VCs, where four groups are ABR connection and one group is CBR connection.

Fig. 2 is the topology used in this simulation. It consists of four groups of ABR VCs, named as group A, B, C and D, with three VCs each. Besides, it also contains four CBR VCs, which is named group X. There are four switches in the network and connected by three links, which are called Interlink 1, Interlink 2 and Interlink 3. All VCs are single hop connection, except VCs from group A, which are 3-hops connection. The ABR sources are set to 31Mbps, 93Mbps, 155Mbps, 217Mbps and 279Mbps respectively while the CBR is fixed at 31Mbps for all simulations. The main function of the CBR source is to occupy certain amount of bandwidth in Interlink 3 and observe its effects to this configuration. For the switch, the slot time is 155Mbps with buffer size of 1024 cells. There are two reasons for choosing small buffer size. First of all, it is the trend of switch's design to keep the buffer size to minimum. Besides, small buffer enable us to observe the congestion control scheme performance when minimum resources are available.

The link between BTE and switch is set to 2km each with link speed of 155Mbps. The links between switches have speed of 155Mbps as well, but the distance is 10km for LAN and 100km for WAN. Each simulation runs for 500msec and the cells received for each receiver and cell loss in each switch are recorded. In this simulation, only two performance metrics are observed, that is the throughput and cell loss.

Besides the proposed NIER algorithm, the above simulation is repeated by using two other well-known algorithms accepted by the ATM Forum, the NIST ER [3] and EPRCA schemes [5]. For the above algorithms, the default parameters are used in order to obtain their maximum performance [5][6]. For the proposed algorithm, the only parameter, i.e. the measurement interval (N), is set to 100 cells. This value is obtained through trial-and-error method in order to obtain maximum performance.

4 Results and Discussions

Fig. 3 and 4 are the plots for throughput and cell loss for various switching schemes under the ATM LAN environment. It is clear that the NIER scheme gives the most superior performance in the ATM LAN with highest throughput and lowest cell loss. The total network throughput for the NIER scheme is about 10Mbps above the NIST ER scheme and 50Mbps better than the EPRCA. For cell loss ratio, the NIER scheme is less than 0.2% while both NIST ER scheme and EPRCA have cell loss ratio of 3% and 17% respectively. Fig. 5, 6 and 7 show the throughput for a representative from each group, except the CBR connection, for NIER scheme, NIST ER scheme and EPRCA respectively. From the topology, the link between switch 3 and 4 is shared by group A, D and X. Since group A and D are ABR connection, therefore a fixed amount of bandwidth is allocated to group X and the leftover bandwidth should be divided equally between group A and D. From the plot, it is clear that A1 and D1 are having the same throughput. Therefore, it is fair to say that the NIER scheme achieves the same fairness as presented by other algorithms.

For the ATM WAN Environment, the plots for throughput and cell loss are presented in Fig. 8 and 9. The NIER scheme has a comparable performance with the NIST ER scheme. Both have throughput of about 300Mbps. In term of cell loss, the NIST ER scheme gives the best performance while the NIER scheme has cell loss of about 0.6%. The EPRCA still performs badly in the ATM WAN with throughput below 260Mbps and cell loss up to 18%. Same as the ATM LAN environment, all algorithms perform fairly.

It is believed that the proposed algorithm can give better performance if the fine-tuning is carried out in order to achieve more suitable N. This is because by changing this parameter, the sensitivity of the algorithm towards the buffer's change will increase and for sure will provide better control for the traffic flow. Anyway, this task is not carried out in the simulation to maintain the fairness in the comparison study. In conclusion, it has been proven that NIER is a better algorithm as compared with the switching schemes used as it gives good performance in both LAN and WAN environments.

5 Properties of NIER Scheme

The following properties of the proposed NIER scheme are the main attractive features for implementing this switching scheme:
- Efficiency: Simulations have shown that the proposed algorithm has high network throughput, high link utilization and low cell loss ratio. It has better performance than several existing switching schemes such as NIST ER and EPRCA.
- Robustness: Simulation shows that no collapse in throughput even under severe load (change of source rate from low to high). Besides, the same neural network can be used under different environments with minimum parameter changes.
- Simplicity: The implementation of the NIER scheme is simple because minimum parameter setting is involved and the step of calculation is straightforward. This greatly reduces the complexity in the switch design and construction.

- Scalability: The NIER scheme can operate in both ATM LAN and ATM WAN environment with modification only in one parameter that is measurement interval.
- Fairness: Fairness is achieved spontaneously without the need to keep track of the current individual cell rate. This is because the neural network gives output based on the current and previous queue length. Therefore, sources with the same conditions will be allocated the same resources.
- Response: Fast response is giving by the proposed NIER scheme. If there is no congestion, the network will achieve its PCR within very short time in both LAN and WAN topology. Also, when there is congestion in the network, it react faster than other schemes because it able to dramatically decrease the ACR.

Acknowledgements

The authors would like to acknowledge the National Institute of Standard and Technology (NIST) for the uses of NIST ATM/HFC Simulator Version 4.1 in this paper.

References

1. ATM Forum: Traffic Management Specification Version 4.1, AF-TM-0121.000, March 1999.
2. A. A. Tarraf, I. W. Habib and T. N. Saadawi: Intelligence Traffic Control for ATM Broadband Networks, *IEEE Communication Magazine*, Volume: 33, issue 10, Page(s): 76 - 82, October 1995.
3. J. L. Wang and L. T. Lee: Congestion Control Schemes for ATM Networks, *Proceeding of IEEE Singapore International Conference on Network and Information Engineering*, Page(s): 96 - 100, 1995.
4. A. Koyama, L. Barolli, S. Mirza and S. Yokoyama: An adaptive rate-based congestion control scheme for ATM Networks, *Proceeding of Information Networking, 1998. (ICOIN-12)*, 1998, Page(s): 14 -19
5. N. Golmie, Y. Saintillan and D. Su: ABR Switch Mechanisms: Design Issues and Performance Evaluation, *Computer Networks and ISDN Systems*, Volume 30, Page(s): 1749 - 1761, 1998.
6. L. Robert: Enhanced PRCA (Proportional Rate-Control Algorithm), ATM Forum 94-0735R1, August 1994.
7. N. J. H. Kotze and C. K. Pauw: The use of Neural Networks in ATM, *Proceeding of the Communications and Signal Processing, 1997. Proceedings of the 1997 South African Symposium on Communications and Signal Processing, COMSIG '97*, South African, Page(s): 115–119, 1987.
8. A. Hiramatsu: ATM Communications Network Control by Neural Network, *1989 International Joint Conference on Neural Networks, ICJNN'89*, Volume: 1, Page(s): 259-266, 1989.
9. A. Tarraf, I. HabiB and T. Saadawi: Congestion Control Mechanism for ATM Networks Using Neural Networks, *IEEE International Conference on Communications 1995, ICC '95 Seattle*, Volume: 1 Page(s): 206–210, 1995.

ABR Congestion Control in ATM Networks Using Neural Networks 693

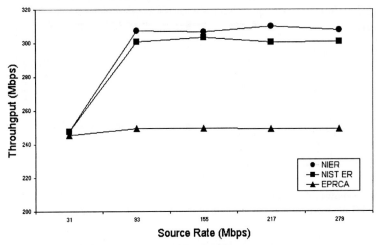

Fig. 3. Total Network Throughput for NIER, NIST ER and EPRCA schemes under ATM LAN environment. It can be observed that NIER gives the highest throughput. The NIST ER scheme is a bit lower while EPRCA performs badly.

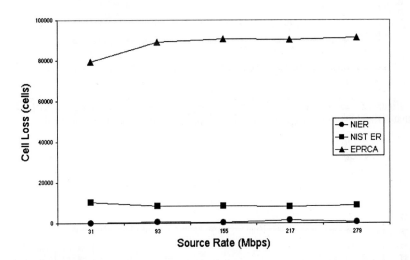

Fig. 4. Total Network Cell Loss for NIER, NIST ER and EPRCA schemes under ATM LAN environment. It can be observed that cell loss seldom occur in the network when NIER scheme is used. The NIST ER scheme also has low cell loss. The cell loss ratio is very high if the EPRCA scheme is used.

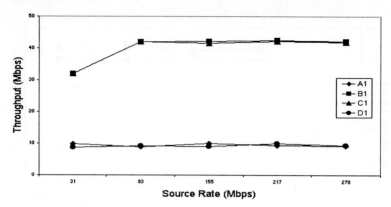

Fig. 5. Figure shows throughput for each VC using NIER scheme under ATM LAN. It shows that A1 and D1 are having the same throughput while B1 and C1 are identical.

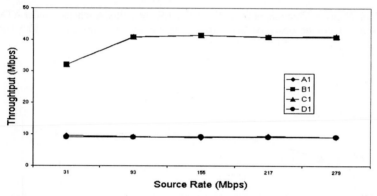

Fig. 6. Figure shows throughput for each VC using NIST ER scheme under ATM LAN. It shows that A1 and D1 are having the same throughput while B1 and C1 are identical.

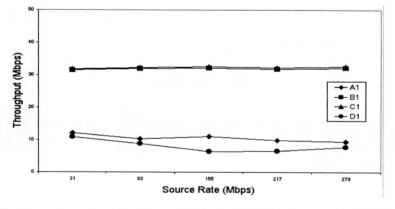

Fig. 7. Figure shows throughput for each VC using EPRCA scheme under ATM LAN. It shows that A1 and D1 are having the same throughput while B1 and C1 are identical.

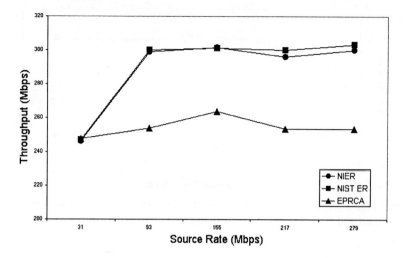

Fig. 8. Total Network Throughput for NIER, NIST ER and EPRCA schemes under ATM WAN environment. It can be observed that NIER and NIST ER are having very high throughput while EPRCA still performs badly.

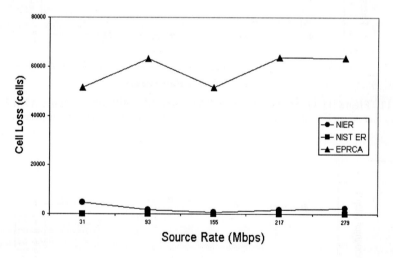

Fig. 9. Total Network Cell Loss for NIER, NIST ER and EPRCA schemes under ATM WAN environment. It can be observed that there is no cell loss when the NIST ER scheme is used. When the NIER scheme is used, there cell loss ratio is very low but it is very high when the EPRCA scheme is used.

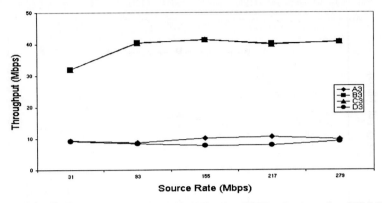

Fig. 10. Figure shows throughput for each VC using NIER scheme under ATM WAN. It shows that A1 and D1 are having the same throughput while B1 and C1 are identical.

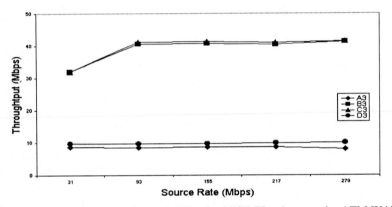

Fig. 11. Figure shows throughput for each VC using NIST ER scheme under ATM WAN. It shows that A1 and D1 are having the same throughput while B1 and C1 are identical.

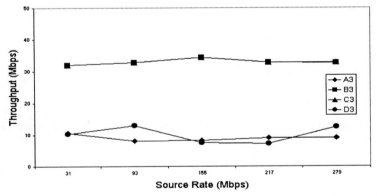

Fig. 12. Figure shows throughput for each VC using EPRCA scheme under ATM WAN. It shows that A1 and D1 are having the same throughput while B1 and C1 are identical.

An Architecture for the Transport of IP Telephony Services

Sidnei Guerra, Joan Vinyes, and David Fernández

Department of Telematic Systems Engineering, Technical University of Madrid, ETSI
Telecomunicación, Ciudad Universitaria s/n, 28040 Madrid, Spain.
Tel: +34 (91) 336.7939. Fax: +34 (91) 3367333.
{guerra, vinyes, david}@dit.upm.es

Abstract. The transport infrastructure to VoIP traffic is moving toward a multiple service platform of high-speed routers interconnected by optical network SDH/WDM. A multiple-service network allows operators to attend to clients through optimization of resources, differentiated quality of service and traffic engineering. This article describes a scalable architecture for the transport of IP telephony services based on MPLS and DiffServ as complementary solutions to provide QoS in access, collector portion, and core of the network. This work presents the state of the art of the protocols related with quality of service, routing and traffic engineering in the IP Telephony field. First we examine the network elements and their functionality inside each protocol, with a focus on the environment of the two workgroups, ITU-T and IETF. Finally we discuss the necessary changes for the introduction of MPLS (Multi Protocol Label Switching) as a significant strategy for the transport of IP telephony services, due to its capacity to emulate the orientation to connection provided by ATM technology. Moreover it offers scalability, low cost, and the possibility of automatization of several aspects of the traffic engineering function - such as the bearer capacity for VoIP, thereby providing more predictable, and even constrained QoS.

1 Introduction

In the evolution toward a network of multiple services there is a need to optimize and organize the traffic in order to respond to the congestion and changes in the traffic model. This is only possible through traffic engineering, which has ended up being an indispensable function in large IP based systems. The keys of the benefits associated with the use of traffic engineering techniques provide to network operators mechanisms to control and orientation of the traffics inside their networks, improving the efficiency in the use of network resources, as well as improving the quality of service offered to traffic flows. For example, in the case of VoIP traffics services, traffic engineering allows a better treatment of traffics inside the network, differentiating them from other traffics and offering them the required service levels in terms of packet looses, delays, throughputs, etc.

Traditional networks based on the use of classical dynamic routing protocols, such as OSPF or IS-IS, do not allow a tight control of traffics inside the network, mainly because they are based on shortest path algorithms. That makes difficult to exert an effective control on the network behavior, since the whole traffic has followed the best route. On the other hand, a model that leans on ATM technology as transport technology is a powerful solution that would allow a good control over network flows and applications, but it would present scalability problems and overhead for the support of IP based services.

Besides, traditional networks have serious problems when trying to satisfy all the quality of service requirements demanded by the real time applications, like IP telephony based on H.323 or SIP. They lacks mechanisms necessary to guarantee for each traffic the conditions of their demand, making the applications unusable in case of congestion problems. Typically, networks used nowadays to support real time traffics are overdimensioned to avoid congestion and satisfy the minimum QoS requirements of applications. But that solution is expensive in terms of the number of resources used, and does not scale well. Considering these conditions and the related above problems, the IETF - through their workgroups, presents some scalable solutions among which Differentiated Services (DiffServ) and Multi Protocol Label Switching (MPLS).

DiffServ [4] model is one of the main proposals to add QoS support to IP networks. It is based on the use of the formerly named Type of Service (ToS) field in the IP header (which has been renamed to Diffserv Code Point, DSCP), to allow different treatments or priorities among traffics going through the network. It is an scalable proposal that bases its scalability on the aggregation of individual flows to aggregated flows at the edge of DiffServ backbones. However, DiffServ does not offer traffic engineering capabilities, as it does not modify or interact with network routing - diverse network flows to the same destination can get different treatment by routers, but they all go through the same path.

MPLS [20] is an emerging standard that makes use of fixed size labels inserted before IP packets to classify and route differentiate traffics inside a network. MPLS uses mechanisms similar to the ones used in DiffServ to aggregate traffic at the edge of the network, but it adds new traffic engineering capabilities, making possible to reroute traffics to the same destination using different paths, allowing a more efficient use of network resources.

DiffServ and MPLS are not exclusive solutions. In fact, they can be properly combined to be a serious alternative for the transport of VoIP traffics in the public networks of Operators and ISPs. In our view, the combination DiffServ-MPLS has as main advantages the MPLS capacities for the traffic engineering and the flexibility of DiffServ for the user management with the different class of service. It is important to clarify the fact that the field that identifies the Quality of Service goes separated from the label which identifies the flow. This allows to apply different treatments at level of priorities to packets with oneself label (same FEC) but with different levels of quality. This has great utility when oneself segment of LSP transports flows of different qualities. In this case, it is not necessary to have identified the individual flows explicitly to offer them different services.

An IP network that uses MPLS for the encapsulation of the voice is given as Voice/RTP/UDP/IP/MPLS. The header compression techniques used increase those that are already defined in [7] and [5], providing extra capacity to compress multiple labels of the type MPLS. Also, the same ones can coexist with other compression standard that is executed on oneself connection (IP, IP/TCP and IP/UDP/RTP). This allows the creation of LSPs that can transport the voice in an efficient way, because these very LSPs support the multiplexing of multiple channels.

A reference model that uses MPLS for routing the voice over IP traffic includes a PSTN to IP interworking function and an IP to MPLS interworking function, which provides more predictability; that is to say, IP QoS bearer control through the use of LSP (Label-Switched Path). This interworking function may be implemented in the same physical device or as separate devices.

For the policy model this architecture need a management layer to define and control the available network resources to a particular session or groups of sessions. The PDPs (Policy Decision Points) perform a bandwidth management function to determine whether a session may be accepted and if so it's optimal route.

2 Background

Two divergent points of view exist over the definition of the IP Telephony, one as a technology and the other as a service concept. In the first case the public Internet is used to interconnect different users but without guarantees and quality of service control. Basically the user has an access best effort to connect PC-to-PC or PC-to-servers through programs of gratuitous distribution. In this case the quality of the communication depends on the user's connection.

In the second concept the voice over IP has more than enough it is viewed as a service of added value where the voice traffic travels over IP networks as well as over the public Internet. From that point of view it is necessary that the operators/ISPs define some priorities for the supply of IP Telephony service keeping in mind the following important factors [15]:

- The type of service or application that is being offered.
- Whether it is a new entrant or incumbent operator that is offering the service.
- The tariff structure of the incumbent operator, especially for access to the local loop.
- Whether or not proportionate return of traffic is a regulatory requirement on international routes.

For the operators and service providers it is already a reality that the growth of the data traffic is putting on above the voice traffic and that in the next future the routers and IP networks will find the necessary scalability to provide quality of service for the telecommunication networks. That phenomenon provides a true change in the current scenario of the telecommunications that during many years used the technology TDM (Time-Division Multiplexing) to build the networks, and that at the moment change to based network on the packets switch system. Besides, the fact that all that supposes a simplification of the network physical structure and the operational costs and equipment reduction. However, the complexity of the current networks and the problems related to the quality of service in the IP networks hinders the quantification of these potential benefits.

This paper contributes to the construction of news telecommunications networks based on packets, proposing a scalable architecture for the transport of IP Telephony services based on emergent technologies- such as MPLS and DiffServ, as complementary solutions to provide quality of service in the core as well as access network.

The main idea is centralized in introducing a multi-layer concept in the IP world [Figure 1] to differentiate the services and its quality levels. Thus, the IP network can

be represented as a set of layers of different priorities easing the task of planning and engineering. Taking as configuration an example give in [13] where a backbone supports three different classes of services:
1. Real time class
2. Committed bandwidth class
3. Best effort

Applying the multi-layer model to IP backbone, this can be divided in three different layers, where it is considered that each layer has a specified type of service and that its treatment will depend on priority when transporting over the network.

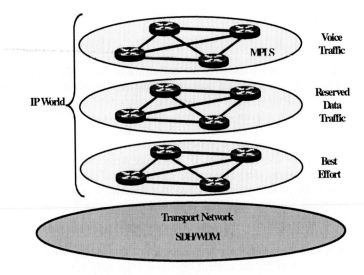

Fig. 1. Multilayer concept in the IP World

In this case, MPLS technology is used for voice treatment - which belongs to the real time class, thus allowing explicit routing of packets by putting labels on them, which can then be used to forward packets along specific Label Switched Paths (LSPs). A label switched path requires a type of set up, meaning that all the intermediate routers between the ingress and egress node are specified. The route selection refers to the method used for selecting the LSP for a particular FEC. For this case, it is considered an explicit routing, so the sequence of LSRs followed by an explicitly routed LSP may be chosen by a static model, or may be selected dynamically by a single node. For example, the egress node may make use of the topological information learned from a link state database in order to compute the entire path for the tree ending at that egress node [20].

In particular, the voice traffic can be seen as a multicommodity flow problem [1] where the routing can be determined by an off-line model taking for base estimated traffic of the traditional circuit network.

This model uses an external routing server that searches a feasible routing for the voice traffic compared to other types of traffic with smaller priority. The model consists in using mixed integer programming to solve, in a feasible manner, a multicommodity flow problem, where each commodity corresponds to an IP voice call. For it is necessary to consider a voice traffic behavior of the traditional circuit network to estimate an initial traffic matrix among the nodes of the packet network.

The result is a Virtual Voice Network (VVN) [Figure 2] with a level of quality that allows to offer service to an initial traffic profile. However, the network planning and traffic engineering for this case are generally complex tasks that require significant understanding about the underlying networking capability on the part of the network planner. A key attribute of a manageable networking environment is one where complexity is minimized. In that sense the VVN considers the Constraint-based Routing Label Distribution Protocol (CR-LDP), proposed by IETF, as online control protocol because the CR-LDP signaling allows the specification of traffic parameters to express the packet treatment at each node, also known as Per-Hop Behaviors (PHBs), for each differentiated service. It is this capability and the rules enforced at the edge that provide a powerful and flexible QoS support for end-to-end differentiated services.

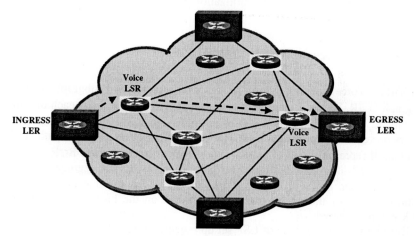

Fig. 2. Virtual Voice Network

These different services can be implemented in edge routers (traffic concentrators) using a field in the head of the IP packet to select the quality level that will be applied to the packets. The differentiated services enhancements to the Internet protocol are intended to enable scalable service discrimination in the Internet without the need for Per-flow State and signaling at every hop [4]. The differentiated service architecture is based on a simple model where traffic entering a network is classified and possibly conditioned at the boundaries of the network, and assigned to different behavior aggregates. Each behavior aggregate is identified by a single DS codepoint. Within the core of the network, packets are forwarded according to the per-hop behavior associated with the DS codepoint.

DiffServ uses a 6-bit header field, the DS-field. In IPv4 the TOS octet and in IPv6 the Traffic Class octet [6] acts as the DS-field. Even though the usage of TOS and Traffic Class octets are different from the usage of DS-field, the differences are such that deployment of DiffServ does not cause serious impacts on existing applications.

The requirements or rules of each service must be set through administrative policy mechanisms, for example COPS [10]. A differentiated services compliant network node includes a classifier that selects packets based on the value of the DS field, along with buffer management and packet scheduling mechanisms capable of delivering the specific packet forwarding treatment indicated by the DS field value.

A replacement header field, called the DS field, is defined, which is intended to supersede the existing definitions of the IPv4 TOS octet and the IPv6 Traffic Class octet. Six bits of the DS field are used as a codepoint (DSCP) to select the Per-Hop Behavior) a packet experience at each node. A two-bit currently unused (CU) field is reserved. An edge router, for example, may be configured so that it can be used to control how packets are scheduled onto an output interface [18].

There are several packet queuing mechanisms that can be utilized in determined nodes of the network as implementation of PHBs. What they all have in common is that the incoming packets are written into a queue, then the packets are read from the queue in some order and placed in the egress interfaces, possibly discarding excess packets. Some manufactures are using Weighted Fair Queuing (WFQ) as a scheduling algorithms that exhibit selective prioritization properties.

3   Virtual Voice Network Model

The Virtual Voice Network (VVN) Model is represented as a subgraph $G_v \subset G$, where the graph $G=(R, E)$ has R as a set of routers and E as a set of links, each of which has capacity C_{ij} (Kbits/sec). V is a subset of routers that is assumed as ingress-egress routers or voice routers of the VVN, between which labels switched path (LSP) can be set up. B represents a set of border routers (edge routers), where $B \subset R$.

Let D be a traffic demand in that an entry d_{ij} represents the aggregated bandwidth request for voice traffic from i to j, $\forall (i, j) \in B$, we assume that D considers voice traffic behavior of the circuit traditional network on the busy hour traffic that the trunk group is offered.

Given an offline sequence of LSP set up requests, it is NP-Hard to determine what is the maximum number of voice call requests that can be simultaneously routed in the network [14]. Due to the complex nature of the model, an optimization problem formulation is used as a multi-commodity flow problem that is solved by an offline server.

The general model of the routing problem can be formulated as the following linear program:

$$Min \ \beta = \sum_{ij}^{d} X_{ij}, \forall (i,j) \in E, \forall d \in D \qquad (1)$$

Subject to

$$\sum_{d} f_{ij}^{d} \leq (C_{ij} - \alpha_{ij}) X_{ij}, \forall (i,j) \in E \qquad (2)$$

$$\sum_{(i,j) \in E} f_{(j,i)}^{d} - \sum_{(i,j) \in E} f_{(i,j)}^{d} = \begin{cases} D_{(d,i)}, \forall i \in B \\ 0, \forall i \in V \end{cases} \qquad (3)$$

$$X_{ij} \in \{0,1\}, \forall (i,j) \in E \qquad (4)$$

$$f_{ij}^d \geq 0 \qquad (5)$$

The result of that model is a virtual network dedicated to the voice packet traffic over the network core (MPLS). The VVN allows solving the traffic-engineering problem through an offline model of optimization that simplifies the task of the online routing algorithms. After the optimization, with a feasible result it is possible to apply shortest path algorithm in the virtual network to reroute the labels, utilizing CR-LDP like protocol of control. Constraint-based Routing Label Distribution Protocol (CR-LDP) is a simple, scalable, open, non-proprietary, traffic engineering signaling protocol for MPLS IP networks. It is based in a generic label distribution protocol that uses TCP (Transmission Control Protocol) to maintain signaling sessions between the LSRs and exchange messages.

Due to the dynamic characteristics of the problem, the model adds a reserve coefficient α_{ij} in the capacity restriction of the connection between i and j, in this way, it is possible to absorb variations of the traffic allowing a slack capacity in the virtual network. With this coefficient the virtual network is dimensioned considering only one part of total link capacity. For example, an operator can to reserve only 80% of link capacity to the aggregated voice traffic.

The virtual voice network model can also be expanded to interconnect other secondary networks of the same operator, different operators or ISPs, allowing policy of interconnection among different networks for the negotiation of minimum quality levels for voice traffic routing [Figure 3].

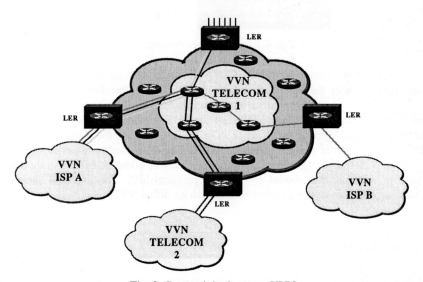

Fig. 3. Connectivity between VVNs

4 Study Methodology

The study network consists of a routing server meant to resolve all network complexity with respect to the traffic engineering. The communication between the routing server and network elements (routers) for configuration of the routing tables utilizes the protocol COPS-PR (COPS Usage for Policy Provisioning) proposed by IETF. The control policy specification determines the access to network resources and is based on the approaches established by the operator. The policy architecture used establishes that each domain has a routing server. This server solves the problem of optimization, and then the result of the model is stored in a database with the explicit routes that determine the VVN. In this architecture the management system is free to change the LSPs determined by the model or to add backup tunnels for the case of network failure.

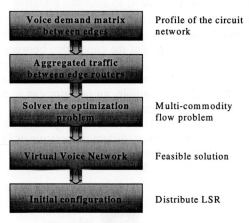

Fig. 4. Routing Server Procedure

Using the provisioning model of COPS-PR and separate data model allow to isolate the details of the label distribution mechanism in the data model. The LSR notifies the PDP of the label distribution mechanism it supports, for example by advertising its capabilities to the PDP, and the PDP provides decisions including the parameters needed by that mechanism. The LSR then uses the data to initiate the tunnel with whichever mechanism it supports. This also makes it easier to expand support for new traffic engineering features that may be added to MPLS in the future, by updating just the data model and not the protocol [19]. The Figure 4 shows the methodology used by routing server to establish an initial VVN.

5 Conclusion

This article presents an infrastructure that makes possible to solve the optimization problem through a static model that reduces complexity of the dynamic traffic management, allowing that set of procedure and policies to ensure that network congestion is minimized and controlled.

The connectivity model with other networks may be further expanded to support multiple core MPLS domains. In this case, the access domains may be provided by the ISDN, requiring a TDM to packet interworking function at the gateway to the core MPLS.

Due to the fact that the optimal solution to the optimization problem is computationally expensive, heuristics can be used to obtain suboptimal solutions for the virtual voice network.

Finally, this solution can be extended for use in optical network. These extensions include information pertaining to the support of multiple layers, diverse routing support, and resource utilization. In this case, other features must be introduced, such as support for bidirectional and backup paths, and fault tolerance.

Acknowledgments

We would like to thank PISCIS (2FD97-1003-C03-01 FEDER-CICYT Program) Project for its support to this work.

References

1. Ahuja, R., Magnanti, T., Orlin J.: Network Flows: Theory, Algorithms, and Applications. Prentice Hall, 1993.
2. Awduche, D., et al.: Requirements for Traffic Engineering Over MPLS. RFC 2702, September 1999.
3. Bhaniramka, P., Sun, W., Jain, R.: Quality of Service using Traffic Engineering over MPLS: An Analysis. Internet Draft, draft -bhani-mpls-te-anal-00.txt, March 1999.
4. Blake, S., et al.: An Architecture for Differentiated Services. RFC 2475, December 1998.
5. Casner, S. et al.: Compressing IP/UDP/RTP Headers for Low-Speed Serial Links. RFC 2508, February 1999.
6. Deering, S. et al.: Internet Protocol, Version 6 (IPv6) Specification. RFC 2460, December 1998.
7. Degermark, M. et al.: IP Header Compression. RFC 2507, February 1999.
8. Doverspike, R., Phillips, S., Westbrook, J.: Future Transport Network Architecture. IEEE Communications Magazine, 1999.
9. Doverspike, R., Phillips, S., Westbrook, J.: Transport Network Architecture in an IP World. IEEE INFOCOM 2000, Tel Aviv, Israel. March 26-30, 2000.
10. Durham, D. et al.: The COPS (Common Open Policy Service) Protocol. RFC 2748, January 2000.
11. Faucheur, F. et al.: MPLS Support of Differentiated Services. Internet Draft, Internet Engineering Task Force, February 2001, Work in progress.
12. Gleeson, B. et al.: A Framework for IP Based Virtual Private Networks. RFC 2764, February 2000.
13. Hersent, O., Gurle, D., Petit, J.: IP Telephony: Packet-based multimedia communications systems. Addison Wesley, 2000.
14. Horowitz, E., Sahni, S.: Fundamentals of Computer Algorithms. Computer Science Press, 1978.
15. International Telecommunications Union. IP Telephony Workshop, Geneva, 14-16 June 2000.
16. Kankkuene, A. et al.: VoIP over MPLS Framework. Internet Draft, Internet Engineering Task Force, July 2000.

17. Kodialam, M., Lakshman, T.: Minimum Interference Routing with Applications to MPLS Traffic Engineering. IEEE INFOCOM 2000, Tel Aviv, Israel. March 26-30, 2000.
18. Nichols, K. et al.: Definition of the Differentiated Services Field (DS Field) in the IPv4 and IPv6 Headers. RFC 2474, December 1998.
19. Reichmeyer, F. et al.: COPS Usage for MPLS/Traffic Engineering. Internet Draft, Internet Engineering Task Force, July 2000.
20. Rosen, E. et al.: Multiprotocol Label Switching Architecture. RFC 3031, January 2000.

Architectural Framework for Using Java Servlets in a SIP Environment

Roch Glitho, Riad Hamadi, and Robert Huie

Ericsson Research Canada
8400 Decarie Blvd, Mount Royal, H4P 2N2, QC, Canada
{lmcrogl, lmcriha, lmcrohu}@lmc.ericsson.se

Abstract. Internet telephony services enable a wealth of services. One of the important questions in engineering services is how to create the services. Another question is where to execute the services. This paper proposes a java-oriented solution to create services and also a new software component, which is able to control their execution.

1 Introduction

Internet telephony enables a wealth of services. It is critical to provide a means for creating, deploying and executing these services. It should not be necessary to add a new network element for each service. In addition, it should be easy for a third party to create new services. This paper proposes an approach based on Java servlets for creating and executing services. It introduces new components that can be added to a SIP model to control services. In addition it proposes an Application Programming Interface (API) that offers the possibility to create services with a high level of abstraction of SIP.

In this paper, we propose a new approach based on Java servlets for creating and executing services. The first section gives background information and examines the related work. The section after that gives a system view of our architecture. The last section presents the servlet API we propose for service creation and also an example of implementation of a service using API.

2 Background and Related Work

There is a strong connection between service engineering and the signaling protocols that are used in the network. In this paper we focus on services engineering in SIP environment. SIP is based on a request-response model. It uses simple messages composed by headers to describe the session and by a body to describe the media. Proxy, redirect servers, registrar servers, and user agent (a client and a server) are the key components used in a SIP environment [5].

In order to engineer services in a SIP environment, a logic [4] is added to the protocol to control services. When a proxy, a redirect server or a user agent receives a request, the stack parses it, and then it sends the information to the service logic who

decides which service to execute. Answers to the questions below will allow us to better understand the process in the specific context of Java servlets. It will also allow us to examine the related work.
- How to create the services?
- Where do the services reside?
- Where are the services executed?

2.1 Service Creation

Several approaches have been proposed to create services in a SIP environment using Java servlet APIs. Some of them extend the generic Java servlet API [1], [6] while others propose the use of Java servlets APIs developed from scratch [3].

Extended Generic Servlet API
This approach is similar to the one of HTTP servlet API [6]. It defines a *SipServlet* class from the generic class *GeneriServlet* of Java API, Fig 1. The API proposes a *SipServletRequest* class, which is an extension of the *ServletRequest*. *SipServletRequest* defines access to headers of a SIP request. The API defines too a *SipServletResponse* class which is an extension of the *ServletResponse* class of the generic API. *SipServletResponse* defines access to the headers of a SIP response.

If the proposed API has the advantage to use the Java generic servlet classes, it presents the following disadvantages:
- The API presents a class redundancy. We know that requests and responses in SIP are defined by different type of headers [2]. Request headers, response headers. Entity and general header that can defined in a request or in a SIP response. General and entity headers access is redundant because, *SipServletRequest* and *SipServletResponse* classes duplicate the implementation of there methods.
- Another disadvantage of the API is the difficulty to be used in a proxy server for the reason that the API that not offer the possibility to fork or to forward SIP messages.

IETF Servlet API
The proposed servlet API [3] is designed from scratch. The advantage of this API is that can be used by a user agent, in addition to redirect server, and a proxy server. It defines architecture of classes based on a good understanding and a good use of SIP model (i.e. *SipRequest*, *SipUrlAddress*). But, it does not offer a high level of abstraction for creating services, for example by offering access to databases (i.e. *ContactDataBase*).

2.2 Service Location and Execution

Services can reside either on the servers themselves or in other hosts. When services and servers co-reside, a simple API allows service execution. When services and servers are not in the same host, other protocol interfaces allow the execution of services, such as RPC, CORBA, or DCOM.

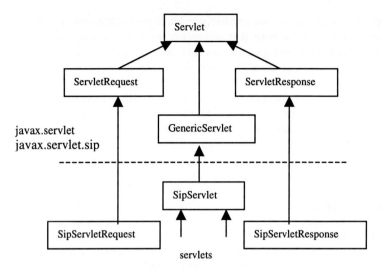

Fig. 1. SipServlet API

When the server receives requests such as an INVITE message, it invokes the service to be executed. The service can be active for the duration of the call or it can be controlled by the server who has some means to specify the points when a service starts its execution and when it is destroyed.

The server loads the service from a repository on its processor to run it locally. This approach gives the server, the functionality to control services and run them in addition to call processing.

3 System View of the Novel Architecture

The proposed architecture uses a SIP server/client, a policy server, a user profile, a servlet engine, and servlet repositories, Fig 2. The approach isolates the call processing to a SIP server and the service execution to a servlet engine. In addition this architecture allows a flexibility of creating services by offering a Java API. Services can be located outside the SIP servers.

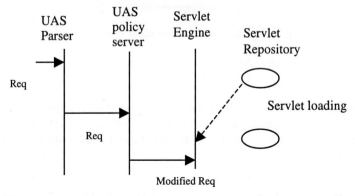

Fig. 2. The scenario used by the architecture to process a received request and loading a service

Before presenting in details our architecture, let us give some definitions.
- The User profile is a database that represents a set of rules defined by users in order to establish relationships between services and SIP messages. A relation is defined between services identifiers and fields' values of SIP messages. If no relationship is defined between fields of the received request and services, then the policy server cannot invoke services for the received SIP message.
- A policy server is the software component that uses rules of the user profile to take the decision to execute services or not when receiving SIP messages.
- A servlet repository is the host that contains SIP servlets.
- Servlet engine is the software component that executes services.

3.1 Policy Server

When a SIP Server receives a SIP request, the parser checks the correctness of the message. It processes the header fields of the request that describe the session, and the body of the request that describes the individual media. To take the decision of what to do with the received request (invoking services or not), we propose to integrate a policy server to SIP servers, Fig 3.

The policy server analyses the received message and decides if services can be executed for the received SIP message. To take the decision, the policy server refers to a user profile, Fig 4.

The policy server tries to instantiate rules of the user profile by fields' values of the received message. If an instantiation appears, the policy server identifies the services to execute for the received SIP message.

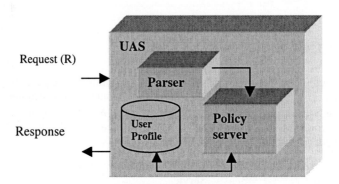

Fig. 3. Architecture of the user agent server

After the policy server identifies the service to execute, it modifies the received SIP request. It adds the identifier(s) of invoked service(s) in the received SIP message.

An extension of SIP is proposed to add the service identifier. It consists of creating a new header *ServiceRequest* in SIP messages: *ServiceRequest* (**service$_1$,..., service$_N$**). Parameters service$_1$...service$_N$ of *ServiceRequest* header are identifiers of invoked services.

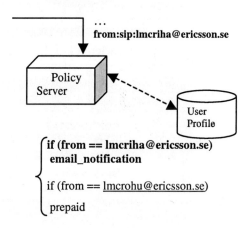

Fig. 4. Policy server and user profile communication

After this, the server forwards the modified SIP message to the servlet engine that downloads and executes invoked services.

3.2 Servlet Engine

A servlet engine is the software component composed by a **sip-stack** and by a **servlet processor,** The sip-stack sends, receives and processes SIP messages, and the servlet processor loads and controls the execution of services, Fig 5.

When the servlet engine receives a SIP message, the sip-stack parses it to check the correctness of the message and gets the identifier(s) of the service(s) to execute from the ServiceRequest header of the message. Identifier(s) is (are) sent via a **sip-stack API** to the servlet processor, Fig 5. The servlet processor finds the location of the service to execute, loads it from its servlet repository using an HTTP protocol [6], and then executes the service. The execution of a service is done in three steps:
- Initialization of the service.
- Execution of service.
- Destruction of the service.

To control the execution of services, the servlet engine uses a **Servlet API**, Fig 5. For example, to initialize a SIP servlet it invokes **init()** method of the servlet. The execution of the service depends on the method of the received message. To each SIP method corresponds a method in servlet API. For example, if an INVITE request is received, doInvite() method of the service is executed.

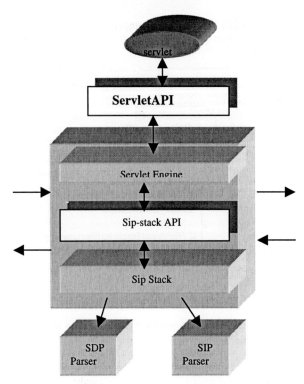

Fig. 5. Servlet engine architecture

Two architectures are proposed for servlet engine. In the first architecture the servlet engine is located in the same host as the server. This architecture is used especially to execute services created on a UAS.

In the second architecture, the servlet engine is an independent component (node) in the network and can be used by different servers such as proxy servers. This architecture is used to execute services created for proxy servers. If this architecture requires us to work on the communication security between the servlet engine and the server, and between servlet engine and servlet repository, it provides us with the following advantages:
1. The same servlet engine can be connected to many servers and many servlet repositories.
2. This approach centralizes service execution. Services can be executed on a same servlet engine.

4 Servlet API

This section presents a general description of the servlet API used in our architecture to built and to control the execution of services. The API offers a high level of abstraction and offers flexibility to SIP servlet developers to create services on proxy, redirect or UA.

The API is a specialization (extension) of the generic servlet package **javax.servlet**. We define a *SipServlet* class that implements the generic servlet interface. Each created service must specialize *SipServlet* class, Fig 6. *SipServlet* handles a received message via the generic method ***service(sipRequest,sipResponse)***. This method dispatches received messages to the correct method *doInvite(), doBye(), doCancel(),* ... of the servlet depending on the information received in the request message.

The API offers the common interface *SipMessage* that provides access to entity and general headers of a SIP messages [2]. This interface is specialized to a SIP request *SipRequest,* and to a SIP response *SipResponse,* Fig 6.

SipRequest class provides access to request headers such as the contact header. In addition This call allows the possibility to create and to send a requests to a specific server at a specific address. For example to send or forwards a message to a SIP server, the servlet uses the method **send(sipUrlAddress)**.

SipResponse provides access to response headers and information about the response, such the description of the received response **getResponseDescription()** or the status of the response; **getStatus()**. In addition *SipResponse* interface allows sending responses to the caller.

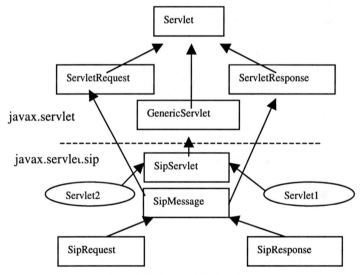

Fig. 6. SIP servlet API package

Each developed servlet using the API has to implement a set of methods. At first each servlet extends the *SipServlet* class. A set of the methods to implement is proposed:
- doInvite() in response to received INVITE message.
- doBye() in responses to received BYE message.
- doCancel() in response to received CANCEL message.

In addition, implementation of these methods allows the use of other methods of the API implemented by the servlet engine. For example the implementation of the *doInvite()* method can need the use of the *getFrom()* or *getTo()* methods that gets the values of from and to headers.

5 Prototype

We show here an example of a SIP servlet created using the proposed API. The personalized voice service plays specific message to each caller. For example if Robert who is calling, a message msg1 is played, but if it is Joe who is calling, a message msg2 is played, in addition to an e-mail that it is to Joe. Bellow is the implementation of doInvite() method of the service.

```
public void doInvite(SipServletRequest req,
                            SipServletResponse res)
{
  . . .
  String fromget = req.getFrom().getAddress();
  System.out.println("THE GET ADDRESS:"+fromget);
  String receivedfrom =
                fromget.substring(fromget.lastIndexOf('/')+1,from
        get.lastIndexOf(':'));
  if (req == null)
  String sender = "Who: "+req.getFrom().getAddress()+'\n';
  String calltime = "When: "+getTimeStamp2();
  . . .
  try {
  FileWriter fw = new FileWriter(filePrefix + ".txt");
  fw.write(sender, 0, sender.length());
  fw.write(calltime,0,calltime.length());
  fw.flush();      fw.close();
  } catch (IOException e) {. . .}
  if (receivedfile != null){
    try {
          DataInputStream dataInput = new DataInputStream(
          new BufferedInputStream( new
               FileInputStream("C:/Media/mailconfig.txt")));
     try { while (dataInput.available() != 0)
         { String s= dataInput.readLine();
           registeredCaller=s.substring(0,s.lastIndexOf('='));
              if ((registeredCaller.equals(receivedfrom)) &&
                           (receivedfile != null)) {
              mailTo = s.substring(s.lastIndexOf('=')+1); }
         }
     } catch (IOException e) {}
  } catch (FileNotFoundException e) {}
       sendfile(mailTo,mailFrom,SMTPsvr,subject,msgText,
        receivedfile, true); }
     else  sendfile(mailTo,mailFrom,SMTPsvr,subject,msgText,
         filePrefix+".txt", true); }
```

The servlet engine that controls its execution loads this servlet. The servlet engine (ServletEngine class) invokes doInvite() method of the service in response to the INVITE received message. Before it starts the execution of services, the servlet engine initialises its SIP stack, then the listener of the stack processes the request and loads the service, (processRequest method).

Architectural Framework for Using Java Servlets in a SIP Environment 715

```java
public ServletEngine(String _user, String _hostname,
                    int _port1, int _port2) {
    if (_port1 < 0)
       _port1 = 6060;
    m_sipStack = new JainSipStackImpl();
    m_sipStack.setStackName("Servlet Engine Stack");
    try{ m_listeningPoint = new
     ListeningPoint(InetAddress.getByName(_hostName),_port1);
    } catch (UnknownHostException e) { . . . }
    try {
      m_provider =
           m_sipStack.createProvider(m_listeningPoint);
    } catch (TPPeerUnavailableException e) {}
    catch (ListeningPointUnavailableException c) {. .  }
    m_listener = new JainSipListenerImpl(m_sipStack, m_provider,
    this);
    try {m_provider.addJainSipListener(m_listener);
    } catch (TooManyListenersException e) {. . . }
       catch (IPListenerAlreadyRegisteredException e) {. . . }
}

public void processRequest(RequestMessage _request, int
                   _transactionId) {
    SipParseInfo message = null;
    SessionDescription body = null;
    InputStream btmp;
    String servletIdf=null;
    SipServletRequest aSipRequest=null;
    String classname=null;
    String method = _request.getMethod();
    byte[] tmpBody=null;
// checks if receives an INVITE SIP message.
    if (method.equals(InviteMessage.method)) {
      try { tmpBody = _request.getBody();
      } catch (BodyNotSetException e) {}
      btmp = new ByteArrayInputStream(tmpBody);
      sdpParser.ReInit(btmp);
      body = new SessionDescription();
      try { sdpParser.session_description(body);
      } catch (ParseException e) {
            body = null; } catch (TokenMgrError e) {
            body = null; }
// gets the idf of the invoked service.
      servletIdf = body.sessionInfo;
      classname =
servletIdf.substring(0,servletIdf.lastIndexOf(".class"));}
    try {
       ServletLoader aLoader = new ServletLoader();
// loads the servlet
       Class ServletClass = aLoader.myload(classname,aLoader);
       SipServlet svl;
       svl = (SipServlet) ServletClass.newInstance();
       svl.init();   // initialize the servlet
        SipServletRequest aSipServletRequest=null;
       aSipServletRequest = new SipRequestImpl(this,_request,
                              m_servletEngine);
// invokes doInvite method because receives INVITE message
       svl.doInvite(aSipServletRequest,null);
         } catch (Exception e) { . . }
  }
```

6 Conclusion

The architecture offers the following advantages:
- Separates the service control and session processing
- Isolates the execution of services on servlet engine host.
- Offers the possibility to service providers to create services on different hosts.

However, we are still looking in the communication mechanisms between servers and servlet engine. Some of the issues that need to be addressed are:
- The number of servers that can be connected to a same servlet engines.
- The possibility to execute services simultaneously by the same servlet engine.

References

1. Ajay P. Deo, Kelvin R. Porter, Mark X. Johnson: Java SIP Servlet API Specification. MCI Worldcom, April 27, 2000.
2. M. Handley, H. Schulzinne, E. Schooler, J. Rosenberg: Session Initiation Protocol. RFC 2534, IETF 1999.
3. A. Kristensen, A. Byttner: The Sip Servlet API, Internal Draft, IETF 1999 (work in progress)
4. J. Rosenberg, J. Lennox, H. Schulzrinne: Programming Internet Telephony Services. IEEE Network, May/June 1999, Vol. 13 No. 13, 42- 48.
5. H. Schulzrinne, J. Rosenberg: The session Initiation Protocol: Internet-Centric Signalling. IEEE Communication Magazine, October 2000, 134-140.
6. Werner Van Leekwijck, Dirk Brouns:Siplets: Java-based Service Programming for IP telephony, ICIN 2000, 22-27.

A Practical Solution for Delivering Voice over IP

Stanislav Milanovic[1] and Zoran Petrovic[2]

[1] Serco Group plc, Via Sciadonna 24/26, 00044 Frascati (RM), Italy
`Stanislav.Milanovic@esa.int`
[2] University of Belgrade, Faculty of Electrical Engineering,
Bulevar Kralja Aleksandra 73, 11000 Belgrade, Yugoslavia
`zrpetrov@ubbg.etf.bg.ac.yu`

Abstract. This paper describes a practical deployment of the VoIP (Voice over Internet Protocol) in the aim of extending the portfolio of value-added services for Italian customers of a prominent European travel services organisation. This includes unified messaging, post paid and prepaid calling card services along with related validation, billing and payment systems.

1 Introduction

Data traffic has traditionally been forced to fit onto the voice network (using modems, for example). The Internet has created an opportunity to reverse this integration strategy — voice and facsimile can now be carried over IP networks, with the integration of video and other multimedia applications close behind. VoIP's appeal is based on its capability to facilitate voice and data convergence at an application layer. Increasingly, VoIP is being seen as the ideal last-mile solution for cable, DSL, and wireless networks because it allows service providers to bundle their offerings [1]. VoIP also offers service providers the ability to provision standalone local loop bypass and long distance arbitrage services. Additional bandwidth will be needed to support distributed, multimedia voice over IP applications building to multigigabit network backbones based on Gigabit Ethernet or Asynchronous Transfer Mode [2].

Customer demand is driving service providers to round out their service portfolios with a mix of voice, data, fax, and video services, and this trend is fuelling the convergence of voice and data network architectures. Merging voice traffic onto an existing data network represents significant cost savings and revenue opportunity for service providers. The popularity of the Internet is also driving the emergence of multimedia applications, many of which are ripe for the integration of voice and data traffic over a single network. These applications include Web-enabled call centres, unified messaging, real-time multimedia video/audio conferencing, distance learning, and the embedding of voice links into electronic documents. In fact, the full business potential of such applications is only beginning to be discovered [3]. But one thing is clear: these integrated voice-and-data applications will require a converged IP network.

Managing company's communications media requires speedy, reliable, and cost-effective messaging solutions. With the emergence of Internet-based applications, enhanced messaging systems are fast becoming tools that can dramatically increase employee productivity, simplify an IT management and reduce operating costs enterprise-wide. A unified messaging solution provides a single location for voice, fax, and email messages; users can have access to all of their messages from their desktop, cellular or wireless phone, or over the Internet and respond, forward, save, or delete messages. Messaging systems with conference calling capabilities let us automatically call the message sender, or click on a screen icon from a desktop and link several parties to the same discussion.

With unified messaging a single application can be used to store and retrieve an entire suite of message types. Voice-mail messages stored as WAV files can be downloaded as e-mail attachments while travelling, a response recorded and returned to the sender, all recipients, or an expanded list. e-mail can be retrieved via a telephony user interface, converted from text to speech, and reviewed from an airport lobby phone or cell phone. Infrastructure is decreased as now a single application can provide voice, e-mail, and fax. Productivity is increased because what were once disparate message types can be retrieved via the most convenient, or the user's preferred, interface.

The Internet and the corporate Intranet must soon be voice-enabled if they are to make the vision of "one-stop networking" a reality.

2 VoIP Overview

VoIP is an emerging technology that allows the systems and wires that connect computer networks to act as an alternative to phone lines — delivering real-time voice to both standard telephones and PCs. VoIP converts standard telephone voice signals into compressed data packets that can be sent locally over Ethernet or globally via an ISP's data networks rather than traditional phone lines [4]. Most serious voice traffic will not use the public Internet but will run on private IP-based global networks that can deliver voice data with minimal congestion. VoIP is vital, according to some analysts, to the continued existence and profitability of Internet Service Providers (ISPs), Application Service Providers ASPs), Local Exchange Carriers (LECs) and Inter Exchange Carriers (IXC). Enhanced services such as unified messaging, "follow-me", Web-based services, calling cards, and billing are revenue-generating services carriers can offer with very low overhead and a high payback [5]. Figure 1 illustrates the IP network protocols that are currently being used to implement VoIP.

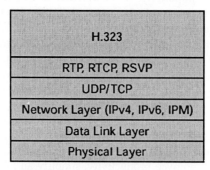

Fig. 1. VoIP protocol structure

H.323 is an ITU (International Telecommunication Union) standard that enables audio, video, and data communications across IP-based networks. H.323 standard has been defined to describe terminals (i.e. client end points), equipment and services for multimedia communication over networks (such as LANs or the Internet) that do not provide a guaranteed QoS [6]. H.323 is a family of software-based standards that define various options for compression and call control. RTP (Real-time Transport Protocol) is an IETF real-time end-to-end protocol utilising existing transport layers for data that has real-time properties. RTCP (RTP Control Protocol) is an IETF protocol to monitor the QoS and to convey information about the participants in an ongoing session; provides feedback on total performance and quality so that modifications can be made. RSVP (Resource Reservation Protocol) is an IETF general purpose signalling protocol allowing network resources to be reserved for a connectionless data stream, based on receiver-controlled requests.

The most important consideration at the network level is to minimise unnecessary data transfer delays. Providing sufficient node and link capacity and using congestion avoidance mechanisms (such as prioritisation, congestion control, and access controls) can help to reduce overall delay.

3 Objective

An acknowledged travel services organization planned to deliver telephone portal services with phone and Web access to customer's e-mail, fax, address book, calendar, file storage service and Web News — anywhere, anytime, from any phone. To this extent, an IP based network was envisaged to transform "best-effort" communications into functionality that can support both continuous, streaming voice conversations as well as bursty data transmissions. A converged network infrastructure should be deployed at Italian corporate headquarter and integrated with London based network of a global enabler of enhanced Internet services. A leading supplier of global enhanced telecommunications and information services would provide travel services organization's subscribers with telephone portal services (VOGO) [7], including telephone access to e-mail, unified messaging and other personal communications services.

In addition to this general capability, specific features supporting voice transmission must also be implemented in the aimed network platform. These features include [8]:
- Compression — Low bit-rate voice compression significantly reduces the amount of bandwidth used by a voice conversation while maintaining its high quality.
- Silence suppression — The ability to recover bandwidth during periods of silence in a conversation makes that bandwidth available for data transmission and other network activities.
- Quality of Service (QoS) functionality — Assuring priority for voice transmission is essential. On the Internet and other IP networks, QoS functionality is provided by the Resource Reservation Protocol (RSVP), which reserves resources across the network for a voice call.
- Signalling for voice traffic — Data network equipment can provide more sophisticated services (such as least-cost routing and virtual private networks)

than simple voice transmission, by recognising and responding to voice signalling.
- Voice switching — Data network equipment can not only perform sophisticated voice transmission between company locations, but can provide private branch exchange (PBX) functionality by performing call processing and voice switching capabilities either within a campus or over the WAN.

The challenge is to design such an information network that could be quickly and easily extended according to the projected business growth. This means that no large, up front investments for the future years should be required. The whole information system has to be manageable via user friendly and easy-to-use graphical user interface, which would reduce training and hand-over period at minimum. It is important to deploy management software that would include all standard options relating to control, configuration and voice over IP traffic analysis possibilities. This management software has to be modular and easy to upgrade in case of future increasing of the value-added services.

4 Requirements

Deployment of VoIP application for public use involves much more than simply adding compression functions to an IP network. Anyone must be able to call anyone else, regardless of location and form of network attachment (telephone, wireless phone or PC). Everyone must believe the service is as good as the traditional telephone network. Long-term costs (as opposed to simply avoiding regulatory costs) must make the investments in the infrastructure worthwhile. Any new approach to telephony will naturally be compared to the incumbent and must be seen as being no worse (i.e. the telephone still has to work if the power goes off), implying that all necessary management, security, and reliability functions are included.

Some of the functions that are required for practical VoIP solution include:
- Fault Management: One of the most critical tasks of any telecommunications management system is to assist with the identification and resolution of problems and failures. Full SNMP management capabilities using MIBs should be provided for enterprise-level equipment.
- Accounting/Billing: VoIP gateway must keep track of successful and unsuccessful calls. Call detail records that include such information as call start/stop times, dialled number, source/destination IP address, packets sent and received, etc. should be produced. This information would preferably be processed by the external accounting packages that are also used for the PSTN calls. The end user should not need to receive multiple bills.
- Configuration: An easy-to-use management interface is needed to configure the equipment (even while the service is running). A variety of parameters and options are involved. Examples include: telephony protocols, compression algorithm selection, dialling plans, access controls, PSTN fallback features, port arrangements and Internet timers.
- Authentication/Encryption: VoIP offers the potential for secure telephony by making use of the security services available in TCP/IP environments. Access

controls can be implemented using authentication and calls can be made private using encryption of the links.

5 The VoIP Solution

With reference to Figure 2, the following networking hardware had been deployed to deliver Voice over IP to subscribers of a rapidly evolving travel services organisation:
- Nokia's DX 220 IP Access integrates the PSTN and IP worlds to offer a high capacity solution for dial-up traffic, enabling a PSTN subscriber to access IP networks: company Intranets or ISP networks [9]. It offers telecom reliability and security for IP traffic, all in a compact package. DX 220 IP Access also provides the Always on Net service, which gives customers a constant connection to the Internet, and opens the window to next generation web-based businesses and services.
- Nuera's Open Reliable Communications Architecture (ORCA) GX-21 trunk gateway is the cost-effective solution for Digital Circuit Multiplication Equipment (DCME) applications. ORCA uses digital compression techniques to increase the capacity of cable links carrying voice, fax and voice-frequency modem traffic [10]. The ORCA GX-21 provides excellent line capacity gain and high quality transmission at all times, even under heavy traffic conditions in the facsimile demodulation mode. DCME applications transport many more voice channels over a transmission circuit than can normally be accommodated. ORCA-based DCME applications provide user-configured compression ratios from 4:1 to 20:1 with up to 17 compressed T1/E1 DCME trunks per chassis. Therefore, an E1 circuit that accommodates 32 voice channels can provide 640 voice channels when a DCME is deployed at each end. As a result, utilisation of the long-distance transmission circuits is dramatically increased.

This large difference can turn a financially-threatened business case into a profitable venture. DCMEs provide more than just voice service. They also provide fax and modem services. The method used to multiply the throughput of the transmission circuit varies depending on what type of service is being provided. The multiplication ratio varies depending on the type of traffic:
- Voice traffic multiplication is achieved by using voice compression technology to lower the bit rate. Over the years, lower and lower bit-rate vocoders have been developed. Traditional vocoder rates are 64Kb/s while currently 8Kb/s vocoders, with nearly the same voice quality, are common. Nuera offers the highest performance low-bit-rate vocoders in the industry, as proven in numerous independent lab tests. Generally, two people don't speak at the same time while having a conversation. In addition, there are intervals of time when neither party is speaking. Therefore, when using newer generation packet technology, these silence intervals (which add up to more than 50% of the two-way session) can be used to carry more voice traffic from other conversations.

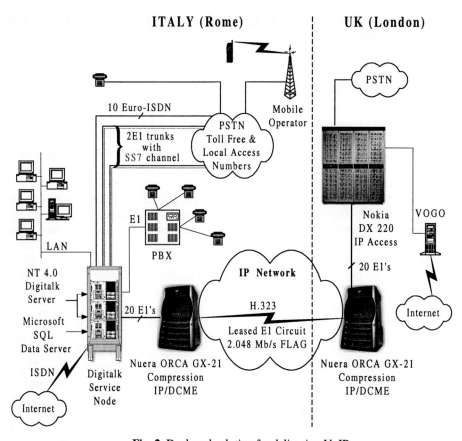

Fig. 2. Deployed solution for delivering VoIP

- Fax traffic multiplication is achieved by demodulating the fax modem (being carried in the 64Kb/s duplex PCM channel) at the near-end DCME to recover the data being transmitted. The base band data (which is typically 9.6Kb/s) is then transmitted to the destination where it is remodulated and delivered on a PCM (Pulse Code Modulation) channel. In the reverse direction, little information is transmitted. Therefore, the 64Kb/s of bandwidth in the reverse direction is available for carrying traffic from other channels.
- Modem traffic multiplication is achieved by transporting the modem waveforms in a lower-bit-rate vocoder when compared to 64Kb/s PCM. Typically, the DCME network bandwidth used in each direction is three times the modem rate and supports modems up to 19.2Kb/s. Therefore, a 9.6Kb/s modem uses approximately 30Kb/s in each direction as compared to 64Kb/s in each direction on a traditional circuit switched network.
- The Digitalk service node is an open standards-based switch, utilising industry-standard computer telephony hardware with support for tandem switching and prepaid interactive voice response [11]. Employed Digitalk service node includes

a signalling Gateway Server (consisting of two Compaq ProLiant 1600R machines), an SQL Data Server (Compaq Proliant 3000R) dedicated as a Data Warehouse, the DataKinetics Septel ISA E1/T1 PC Line Card for SS7 (providing 2E1 trunks with SS7 channel) [12] and an Internal ISDN modem for Remote Access Service (RAS). Each Compaq Proliant 1600R handles 16 E1's (4 quadspan E1 boards/4slots). The system features of deployed Digitalk service node include:

- Full Windows NT hosted administration and system configuration tools;
- SS7-based Call Control providing a seamless interoperability to the PSTN and 99.999% reliability;
- Fully featured least-cost routing based upon dialled digits or time of day/day of week;
- Alternate routing on primary route congestion or port failure, automatic re-routing programmable on call failure codes;
- Dynamic call editing allowing number modification for routing via alternate carriers;
- Sophisticated SQL based call record logging and system load profiling tools;
- Built-in protocol analyser tools and diagnostic test utilities;
- Fully integrated remote management tools.

Digitalk platform can be easily extended to a Tandem (redundant) switching architecture (see Figure 3) cost effectively by saving the original information platform. In this scenario, new technologies such as ATM and SQL server are integrated to allow massively scaleable solutions to be deployed.

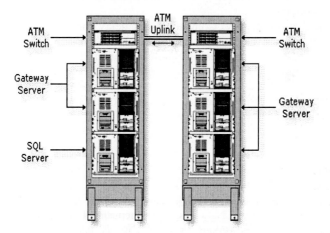

Fig. 3. Tandem switching architecture

Digitalk service node can be linked to other nodes via 155Mb/s ATM fibre links connected into a 4.2Gb/s backbone (up to 24 switch nodes). Incoming calls on any node in the system can be routed transparently to any other node via the ATM switching backbone. It is possible to host a combination of applications on the same switch that would allow new trial services to be launched on existing hardware without the need for a fixed allocation of hardware and software resources.

6 Conclusion

Rather than building separate networks to offer multiple services, an international travel services organisation integrated various traffic flows onto efficient, packet switching infrastructures based on statistical multiplexing, which maximises the use of network capacity. Using the same infrastructure for all traffic also holds the promise of significant savings because the equipment handles multiple functions and the operational costs are reduced.

Deployed solution for delivering Voice over IP provides the following features:
- Mediation or gateway services between the PSTN and the IP network for transparent internetworking.
- A high degree of scalability — the ability to affordably add processing and switching power as network expands.
- The deployment of new and enhanced services.
- Multiple services over a common infrastructure.
- Automated and integrated management functions.
- The ability to mix and match best-of-breed equipment and software from different vendors.

While enabling new value-added services for Italian customers, it was in foresight for the near future to deploy similar information systems in other countries, providing a global delivery of Voice over IP to the international subscribers of a well known travel services organization.

Convergence is on its way. Companies that delay for too long the adoption of first-wave convergence technologies, most notably VoIP, will suffer a substantial competitive disadvantage as these technologies enter the mainstream. There are, however, clear risks that need to be avoided at this early stage of converging technologies. To avoid these risks while still practising due diligence in advancing the corporate technology portfolio, decision-makers should carefully consider the use of intelligent multi-path gateway switching.

References

1. Stanislav Milanovic, Alessandro Maglianella, "ATM over ADSL Probe in Telecom Italia Environment", Computer Networks, the International Journal of Computer and Telecommunications Networking, Vol. 34, No. 6, pp. 965-980, November 2000, http://www.elsevier.com/inca/publications/store/5/0/5/6/0/6/index.htt. Proceedings of TERENA Networking Conference 2000, the Trans-European Research and Education Networking Association Conference: "Pioneering Tomorrow's Internet", May 2000, Lisbon, Portugal, http://www.terena.nl/tnc2000/proceedings/10A/10a3.pdf
2. Stanislav Milanovic, "At the Front End in Migrating to Gigabit Ethernet", Proceedings of SoftCOM 2000, the IEEE Conference on Software,

Telecommunications and Computer Networks, pp.369-378, October 2000, http://www.fesb.hr/SoftCOM/2000/IE/Network_Architectures.htm
3. "Migrating Corporate Voice Traffic to the Data Network - Strategies, Risks and Rewards", The Technology Guide Series, The Applied Technologies Group Inc., 2000.
4. "Migrating to Multiservice Networks — A Planning Guide", White Paper, Cisco Systems Inc., 1999.
5. "nuVOICE –Next Generation VoIP Solutions", White Paper, Nuera Communications Inc., 2000.
6. "Voice over IP (VoIP)", The Technology Guide Series, The Applied Technologies Group Inc., 1998.
7. http://www.vogo.net/
8. "Calling on the Network for Voice Communications", Packet Magazine Archives, Cisco Systems Inc., 1998.
9. "Nokia DX 220 IP Access Solution", White Paper, Nokia Networks, 2000.
10. "ORCA Brochure", White Paper, Nuera Communications Inc., 2000.
11. "Digitalk Gateway", White Paper, Digitalk Ltd, 2000.
12. "Septel ISA E1/T1 PC Line Card for SS7", White Paper, DataKinetics Ltd, 2000.

QoS Guaranteed Voice Traffic Multiplexing Scheme over VoIP Network Using DiffServ

Eun-Ju Ha, Joon-Heup Kwon, and Jong-Tae Park

Department of Electronic Engineering, Kyungpook National University,
1370 KanKyug-Dong, Buk-Gu, Taegu, Korea
ejha@ain.knu.ac.kr, joonheup@hotmail.com, park@ee.knu.ac.kr

Abstract. The current VoIP transfer methods using low bit rate codes such as G.723.1 and G.729 are still very inefficient due to its small payload size in comparison with its large overhead size. Besides, the traffic load on the access routers tends to geometrically increase when the incoming traffic flows of short packets increase. These factors cause problems such as delay, jitter, and packet loss, which seriously deteriorate the voice quality. We propose a new efficient voice traffic multiplexing scheme with guaranteed end-to-end QoS between VoIP access routers using differentiated services (DiffServ). The newly defined RTP/UDP/IP packets, namely, L_packet (long packet) using DiffServ are multiplexed at ingress routers for offering real-time communication services in very high-speed backbone network. We analyze the performance of proposed scheme for various bit rate type traffics and simulate the call blocking probability. It shows that the proposed scheme quite satisfactorily guarantees the end-to-end QoS requirements.

1 Introduction

The VoIP systems are expected to be very extensively used throughout the world because of their low price, better user interfaces, and potentiality for providing integrated multimedia applications over IP networks. Internet can't satisfy various quality of service (QoS) requirements mainly due to its transport mechanism based on best effort services. In order to transmit the voice over packet network, VoIP access routers are required. VoIP access router provides an interface between the existing circuit switched telephony system such as PSTN/PLMN and the packet switched IP networks.

To guarantee a quality of service for real-time audio and video applications, the IP based protocol stack of the real-time transport protocol (RTP) over user datagram protocol (UDP) is required. RTP is used as means of achieving interoperability among different implementations of network audio and video applications, and integrated into the H.323 protocol stack, which was standardized by the ITU-T. RTP is often used to unicast and multicast real-time traffic transmission of audio, video, and interactive simulation over Internet, but it can't support end-to-end QoS. So, we need an appropriate RTP/UDP/IP packet encapsulation procedure for the incoming voice traffic over access networks to guarantee QoS.

In a traditional VoIP application, telephone calls between PSTN/PLMN users interconnected by a pair of IP-ARs are carried by separate IP/UDP/RTP connections. At present, the low bit rate codes such as G.723.1 [1] and G.729 [2] are used to the baseline codec of VoIP. They compress incoming speech samples, and generate packets with sizes ranging from 5 to 20 bytes per speech sample. For example, G.723.1 compresses voice traffic to 5.3 kbps/6.3 kbps and the payload size is defined to be 20 bytes. That is, only one-third of all the traffic is payload and other two-thirds (i. e., RTP (12 bytes) + UDP (8 bytes) + IP (20 bytes)) is overhead. If voice traffic is sent in a RTP packet, this means that only 33 % of the total size of the packet is used to the payload. In addition, the traffic load on the routers would increase geometrically due to the traffic flows of many short packets. This causes the delay, jitter and packet loss problems. Thus, in order to guarantee end-to-end QoS requirements for real time voice service over IP network, we must take proper actions of decreasing this large overhead of the increased IP/UDP/RTP packet headers between VoIP access routers in the first place.

It is expected that the future backbone network will be constructed by using very high-speed WDM network with more than 10Giga bps. In this kind of very high-speed backbone network, the main cause of QoS deterioration is due to poor bandwidth control. To guarantee end-to-end QoS in high-speed backbone network, integrated services (IntServ) and resource reservation protocol (RSVP) has been proposed. However, these approaches are based on the per-flow management, which may cause a scalability problem. It is also known that these approaches cost very high for the construction. As another solution, the IETF has proposed a new architecture, called differentiated services (DiffServ). This architecture allows flow aggregation, that is, a number of traffic flows being handled as one unit. Except for fixed bandwidth allocation, it is shown that the flow aggregation based on DiffServ is efficient to guarantee end-to-end QoS in a high-speed backbone network [3]. Secondly, in order to guarantee QoS in very high-speed backbone network, the bandwidth control mechanism is required to incoming RTP/UDP/IP packets of ingress routers.

There are some related works dealing with these voice streams multiplexing issues [1-3] and DiffServ [4-5]. But, previous related researches have usually focused on access network, or backbone network, not both in an integrated way. Their solutions can't be applied for the satisfaction of the true end-to-end QoS requirements. In this paper, we propose a new end-to-end QoS guaranteeing mechanism by combining RTP/UDP/IP packet multiplexing scheme with DiffServ QoS architecture.

We adapt RTP/UDP/IP packet multiplexing scheme in Ref [6] to our scheme with some modification. At ingress router, the appropriate control mechanism is required to support end-to-end QoS for input traffic. We solve this problem by using DSCP (DiffServe Code Point) of DiffServ. We analyze the performance of proposed scheme using M/G/1 with HOL-NPR (head-of-line non-preemptive) queueing model. This scheme not only significantly reduces the short traffic flow management cost, but also guarantees the end-to-end QoS requirements. The remainder of this paper is organized as follows: Section 2, we present the overall architecture and multiplexing packet formats of voice traffic using DiffServ. In Section 3, we present the analysis results and Section 4 shows the numerical results. Our conclusion follows in Section 5.

2   VoIP System Architecture Using DiffServ

In this section, we present overall VoIP architecture and its multiplexing packet format using DiffServ. Figure 1 shows the overall architecture. The number of access networks, which are connected to one ingress router depends on the real capacity of ingress router. We use the RTP/UDP/IP packet multiplexing concept based on Ref[5] to reduce the packet overhead to 67%. However, the major problem of Ref[5] is that it does not take into account any QoS requirements. Thus, we modify voice traffic multiplexing scheme in Ref[5] in order to support QoS requirements. We propose the new multiplexing scheme using DSCP of DiffServ [6]. The key idea is as follows; *Add voice stream multiplexing scheme* **with identical destination IP address and DSCP (DiffServ Code Point)**

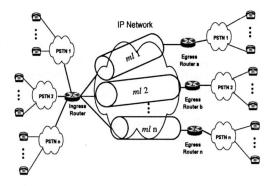

Fig. 1. Overall Architecture

Figure 2 presents the transmission link between ingress router and egress router in more detail. The same model can be applied to the other links. We define the several parameters shown in Figure 2. *ml* is defined as the multiplexing link. We assume that the capacity of link is infinite. *t* is defined as a trunk. In our model, trunk is defined to be a group of voice traffic, which has identical destination and DSCP. *vt* is defined as voice traffic. We assume that the capacity of *vt* is all the same.

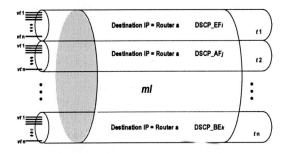

Fig. 2. Multiplexing link between Ingress router and Egress router

In order to guarantee the end-to-end QoS, we propose new RTP/UDP/IP packet format at ingress router. We define L_packet (Long Packet), which adapt DiffServ DSCP to RTP/UDP/IP packet format. Figure 3 shows the L_packet format, which uses the ingress, intermediate, and egress router. At ingress router, the L_packet, which has the same destination and DSCP, is classified and destined to egress router through several intermediate routers. At intermediate router, it only transits the L_packet without any modification based on DSCP value. At egress router, the arrived L_packets are separated and processed according to the priority of DSCP value.

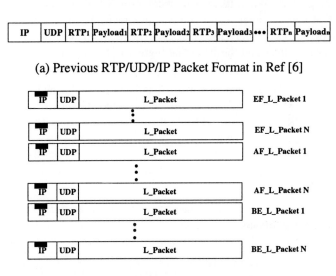

Fig. 3. Multiplexing RTP/UDP/IP Packet Format

3 Performance Evaluation

3.1 Mathematical Model

In this section, we first describe the mathematical model. In the near future, it is expected that the backbone network is to be installed with very high-speed WDM network with more than 10Giga bps. In this very high-speed backbone network, the main problem for the QoS deterioration is caused by poor bandwidth control. Thus, we describe the end-to-end voice traffic blocking probability based on M/G/1 with HOL-NPR (head-of-line non-preemptive) queueing model in Figure 2. We assume that at least one trunk can be established between ingress router and corresponding egress router. The voice traffic blocking occurs by two things. One is due to the contention among voice traffics sharing the same trunk, with the same destination of egress router. The other is caused by contention among trunks sharing the link. We first derive the mathematical model of the first part. In this case, we assume that the incoming voice traffics are classified into their priorities and complete the multiplexing procedures at ingress router..

Link capacity of an ingress router is denoted by $C(ml)$. The number of voice traffics accommodated at each trunk depends upon their traffic characteristics. Each trunk has different voice traffic arriving rate and mean service time. We assume that there are P types of queues. In a queue of type k ($k = 1, 2, ..., P$), there are m_k servers and no waiting room. When there is no trunk established between ingress router and corresponding egress router and a setup requirement of voice traffic between them newly arrives, ingress router checks whether there is still bandwidth available for a new trunk establishment on the link. The ingress router will reserve some amount of the bandwidth required by m_k voice traffic for a trunk of type k on the link if possible.

Furthermore, if the corresponding trunk is already established and it can further accommodate more flow, the flow will be accepted. Otherwise, the flow will be blocked. The trunk will be released when it has no voice traffic. Arrivals of customers at a queue of type k are characterized by an independent Poisson process with an arrival rate λ_k, and with an arbitrary or general service time distribution of form $B(x)$ with a finite mean h_k.

3.2 Performance Analysis

In this section, we analyze the mathematical model described in the previous section. It follows the Ref [2,3]. Let $\rho_k = \lambda_k h_k$ denote the traffic intensity in a trunk of type k. We denote by $P_k(i)$ the steady state probability of i ($i = 1, 2, ..., m_k$) customers at a trunk of type k using the Ref [7].

$$P_k(i) = P_k(0)\frac{\rho_k^i}{i!} \quad \text{Where,} \quad P_k(0) = [\sum_{i=0}^{m_k}\frac{\rho_k^i}{i!}]^{-1} \quad (1)$$

Let $B(m,\rho)$ denotes the blocking probability with traffic intensity ρ. It is well known that $B(m,\rho)$ can be computed by the recursion [8];

$$B(m,\rho) = \frac{\rho B(m-1,\rho)}{m + \rho B(m-1,\rho)}, \quad m = 1, 2, \ldots \quad (2)$$

with $B(0,\rho) = 1$. Thus, the conditional mean number M_k is given by

$$M_k = \frac{\sum_{i=1}^{m_k} i P_k(i)}{1 - P_k(0)} = \frac{\rho_k(1 - B(m_k, \rho_k))}{1 - P_k(0)} \quad (3)$$

Next, we describe the mean busy period H_k of a trunk of type k. Since the sequence of idle and busy period is regarded as an alternating renewal process, we have

$$H_k = \frac{1 - P_k(0)}{\lambda_k P_k(0)} \quad (4)$$

Customer loss occurs in two ways. One is that a customer arrives when his trunk is full, and the other is that the trunk is empty but the reservation at the second step is failed on arrival. We denote the blocking probability of customers arriving in trunks of type k in the former and the latter cases by $P_{loss}^{(1)}(k)$ and $P_{loss}^{(2)}$, respectively. Apparently, the overall blocking probability $P_{loss}(k)$ of customers arriving in a trunk of type k is given by $P_{loss}(k) = P_{loss}^{(1)}(k) + P_{loss}^{(2)}(k)$. We first consider $P_{loss}^{(1)}(k)$. We define that the probability $r(k)$ that a trunk of type k is reserving m_k servers using the Ref [7].

$$r(k) = \frac{U_k(N, C(ml))}{N_k} \quad (6)$$

Therefore, we have

$$P_{loss}^{(1)}(k) = r(k)\frac{B(m_k, \rho_k)}{1 - P_k(0)} \quad (7)$$

Next we consider the overall blocking probability $P_{loss}(k)$. Since the blocking probabiity is a fraction of lost customers to customers arriving in a unit time, we have

$$P_{loss}(k) = 1 - \frac{M_k U_k(N, C(ml))}{N_k \rho_k} \quad (8)$$

However, the above call blocking probility are effective in the case of the total waiting time for customers from type k in class N can't exceed the end-to-end delay bounds. The total waiting time (T_p) in our queueing model is defined as follows.

$$T_p = \frac{h_k(1-\sigma_k) + \sum_{i=1}^{m_k} \lambda_k h_k^2/2}{(1-\sigma_k)(1-\sigma_{k+1})} \quad \text{Where,} \quad \sigma_k = \sum_{i=1}^{m_k} \rho_k \quad (9)$$

4 Numerical Results

In this section, we evaluate the blocking probability using our proposed scheme. We treat the heterogeneous trunk case. We defined related parameters as follows;

Defined Parameters
- Link capacity: $C(ml)$ = 150 Mbps
- Allocated bandwidth for each voice traffic: $N(vt)$ = 10 kbps
- Number of traffic types: P = 2 (high and low)
- Number of low trunks (i. e. DSCP_AF and DSCP_BE): $N(L)$ = 100
- Number of high trunks (i. e. DSCP_EF): $N(H)$ = 10
- Mean traffic intensity for each low trunk: ρ_L = 2 voice traffic
- Mean holding time: h_L = 1 [min]
- Mean interarrival time: $1/\lambda_L$ = 30 [sec]
- Mean traffic intensity for each high trunk: ρ_H = 1250 voice traffic
- Mean holding time: h_H = 1 [min]
- Mean interarrival time : $1/\lambda_H$ = 48 [msec]
- Capacity of each of DSCP_AF and DSCP_BE trunks : m_L flows
- Capacity of each of DSCP_EF trunks : m_H flows

We assume two types of traffic such as high and low. The high traffic flows are preferentially assigned to the DSCP_EF trunks. The high trunks have larger capacity to meet their higher demand than the low trunks. The low traffic flows are assigned to the DSCP_AF and DSCP_BE trunks according to the trunk conditions. In our evaluation, the total amount of traffic is fixed to 12700 flows. So, the mean utilization becomes 84.7% if there is no blocking. Table 1 shows the voice traffic blocking probability according to the varying capacity of each of low class trunks. In this case, the amount of DSCP_EF is fixed to 1400 flows. From Table 1, the optimal value to minimize the voice traffic blocking probability for DSCP_AF and DSCP_BE trunks exists in which both DSCP_EF, DSCP_AF, and DSCP_BE trunk blocking probability becomes less then 10^{-4}. It demonstrates that the QoS requirements are well satisfied by using our proposed scheme. Also, the same results are obtained in case where the capacity of each of high-class trunks is varied as time goes by.

Table 1. Voice Traffic Blocking Probability

M_L	$P_{loss}(H)$	$P_{loss}(L)$
2	1e-06	0.798
4	1e-06	0.1
6	1e-06	0.098
8	1e-06	0.001
10	1e-06	0.0000429
12	1e-06	0.096
14	1e-06	0.1098
16	1e-06	0.689

5 Conclusion

We present the end-to-end QoS guaranteed voice traffic multiplexing scheme between VoIP access routers using differentiated services (DiffServ). At ingress router, the newly defined RTP/UDP/IP packets, namely, L_packet (long packet) are multiplexed according to same destination egress router and same DSCP. To prove the validity of our scheme, we have described the network model developed here to analyze the blocking performance. Also, we analyzed the model using M/G/1 with HOL-NPR queueing system. Finally, we have presented numerical results using the analysis. Through the results, the proposed scheme is shown to be very efficient to guarantee the QoS requirements over VoIP.

References

1. Floyd S, "Notes on CBQ and Guaranteed Service", ftp://ftp.ee.lbl.gov/papers/guarantedd.ps, July 1995.
2. Katsuyoshi IIDA, Tetsuya TAKINE, Hideki SUNAHARA, and Yuji OIE, "Delay Analysis for CBR traffic in static-priority scheduling: single-node and homogeneous CBR traffic case", IEEE SPIE'97.
3. Katsuyoshi IIDA, Kenji KAWAHARA, Tetsuya TAKINE, and Yuji OIE, "Performance Evaluation of the Architecture for End-to-End Quality-of-Service Provisioning", IEEE Communication Magazine, April 2000.

4. Giuseppe Biawnchi, Antonio Capone, and Chiara Petrioli, "Throughput Analysis of End-to-End Measurement-Based Admission Control in IP", IEEE INFOCOM'2000.
5. Tohru HOSHI, Keiko TANIWAQA, ad Koji TSUKADA, "Voice Stream Multiplexing between IP Telephony Gateways", IEICE Trans. INF. & SYST., Vol, E82-D, No. 4, April 1999.
6. Hassan Naser, and Alberto Leon-Garcia, "Voice over Differentiated Services", IEEE draft, draft-naser-voice-diffserv-eval-00.txt, December 1998.
7. Ross, K.W., "Multiservice Loss Models for Broadband Telecommunication Networks", Sprnger, Berlin, 1995.
8. Cooper, *Introduction to Queueing Theory*, CEEPress, Washington, D.C., 3rd edition, 1990.

VoIP over MPLS Networking Requirements

Jong-Moon Chung[1,*], Elie Marroun[2], Harman Sandhu[3], and Sang-Chul Kim[1]

[1] School of Electrical and Computer Engineering, Oklahoma State University,
Stillwater, OK 74078, U.S.A
{jchung, sangchu}@okstate.edu
[2] Cisco Systems, 1450 N. McDowell Blvd., Petaluma, California, 94954, U.S.A.
[3] Cisco Systems, 2200 E. Pres. George Bush Trnpk., Richardson, Texas 75082, U.S.A.

Abstract. Multiprotocol label switching (MPLS) is currently emerging as the protocol of the future due to several key features that MPLS networking provides through traffic engineering. Some of these features are possible due to its multiprotocol architecture where it utilizes a simple label switching procedure to provide various quality of service (QoS) features, explicit source routing, differentiated services, architectural scalability, integrated service features, and high quality end-to-end service features that are necessary for virtual private networks (VPNs). Currently, the constraint-based routing label distribution protocol (CR-LDP) and the extensions to the resource reservation protocol (E-RSVP) are the signaling algorithms under development that will be used for MPLS traffic engineering. In this paper, we investigate the requirements of MPLS networking for voice over IP (VoIP) applications, and propose extensions to the MPLS protocol, MPLS routers, CR-LDP, and E-RSVP for improved compatibility and system performance of VoIP over MPLS (VoMPLS) network services.

1 Introduction

Voice applications over IP routing utilizing MPLS networks require several changes in the signaling algorithm and voice data transfer mechanism in order to provide the enhanced QoS features that voice and real time data require. MPLS networking is under development to provide scalable integrated and differentiated service features that current IP networking cannot provide. The classification, queue, and scheduling (CQS) traffic engineering topologies of MPLS network routers enable the network to

---

This research was funded by Williams Communications, Inc., One Williams Center, Tulsa, Oklahoma 74172, U.S.A.

[*] J.-M. Chung is the Director of the Advanced Communication Systems Engineering Laboratory (ACSEL) and the Oklahoma Communication Laboratory for Networking and Bioengineering (OCLNB) at the Oklahoma State University.

provide controllable quality of service (QoS) and grade of service (GoS) features. MPLS networking also provides a solution to scalability and enables significant flexibility in routing [6]. MPLS was intended to solve the differentiated service (DS) problems of IP connectionless datagram networking. Hence MPLS routers (i.e., label edge routers (LERs) and label switching routers (LSRs)) need to incorporate the CQS parameters required by the application (voice, video, data, control data, etc.). This will require the LSRs and LERs to support and interact with the transport-layer, network access layer, and the physical layer up to some degree to provide enhanced service features.

The current MPLS networks are not fully equipped to support voice data applications properly. Some of these examples can be seen in the signaling protocol mechanisms. The label switching router (LSR) that conducts DS is required to conduct a three-step procedure to enable traffic engineering. These three basic steps are classification, queue, and scheduling (CQS). As label attached packets arrive at the input ports, the input label is used to identify the forwarding equivalent class (FEC) and the corresponding output label. The output label replaces the input label of the packet. Then, based on the output label and FEC, the packet is sent to the corresponding output queue where the scheduling multiplexer decides on the output order, timing, and the output port for the packet to be sent out. The setup of the LSR is done by the signaling protocols (LDP, CR-LDP, RSVP, E-RSVP).

The Real Time Protocol (RTP) is the Internet-standard protocol for the transport of real-time data, including audio and video. It can be used for media-on-demand as well as interactive services such as Internet telephony. RTP consists of a data and a control part. The latter is called Real Time Control Protocol (RTCP).

While UDP/IP is its initial target-networking environment, efforts have been made to make RTP transport-independent so that it could be used, say, over CLNP (Connection-Less Network Protocol), IPX (Internetwork Packet Exchange) or other protocols such as MPLS. RTP is currently also in experimental use directly over AAL5/ATM. RTP does not address the issue of resource reservation or quality of service control; instead, it relies on resource reservation protocols such as RSVP.

No end-to-end protocol, including RTP, can ensure in-time delivery. This always requires the support of lower layers that actually have control over resources in switches and routers. Therefore, the idea of making RTP run over CR-LDP in an MPLS network will provide certain reliability as well as add some constraints and QoS features.

The current RSVP and the E-RSVP are IP based soft-state protocols [3]. E-RSVP was developed to enable RSVP signaling for MPLS networks. RSVP and E-RSVP need to be able to signal RTP/UDP transport layer setup and prepare the MPLS network routers (i.e., LERs and LSRs) with the required service features to guarantee QoS and GoS for real-time constrained voice data packets.

On the other hand, LDP and the proposed CR-LDP relies on the user datagram protocol (UDP) for peer-to-peer and TCP for end-to-end data transport-layer operation [1,2]. Although TCP is the most widely used transport layer protocol, there are several weaknesses that TCP has which makes TCP difficult to be used for voice applications. First TCP is a point-to-point connection oriented protocol that makes its application unsuitable for multicasting distribution. Second, TCP applies an algorithm for retransmission of lost segments, which is not suitable for most real time applications. Third, TCP contains no mechanism for associating timing information with segments, e.g., timestamp, fraction loss, packet loss, inter-arrival jitter, time delay, etc. In addition, UDP is not equipped to support general-purpose tools for real-time applications of voice [18].

Several of these requirements for QoS and GoS in VoIP over MPLS (VoMPLS) networks are well described in [12]. Although, [12] does not provide methodologies of how to extend these features to the MPLS architecture, LER and LSR functionalities, E-RSVP, and CR-LDP signaling protocols. This paper therefore proposes several extensions to E-RSVP and CR-LDP that must be made to enable RTP over UDP protocols to work with E-RSVP and CR-LDP. Additionally, the MPLS routers should be able to operate with the required RTP parameters that are requested in the LSP setup through E-RSVP or CR-LDP for guaranteed QoS and GoS for VoIP over MPLS networks.

Based on the RTP standards [23] RFC 1890, 128 kinds of real-time payload types for standard audio and video encoding are each assigned a 7 bit code. This code can be embedded into the signaling protocol along with the indication of the RTP/UDP to the MPLS LSRs in the LSP setup mode. Currently, there are no Internet drafts or standards that specify the service level agreement (SLA) for the signaling protocols (E-RSVP or CR-LDP) at the edge routers [1]. Only the local behavior specifications of the LSRs are indicated. Therefore, the ingress LER which is operating the source explicit-routing procedures must be able to trigger the desired service requirement by indicating its type of connection requirements [1]. Current and past protocols were not able to provide guaranteed services. In order to enable MPLS to provide guaranteed services additional service payload recognition is required. In this paper, we propose novel methods to make this indication possible such that at the setup mode of the LSP, E-RSVP and CR-LDP can inform the LSRs more specifically about the traffic characteristics and correspondingly reserve the required resources for real-time high QoS or guaranteed services.

2 Extensions to E-RSVP for VoMPLS

Extensions to RSVP (E-RSVP) have been made and proposed to support explicit route LSPs as well as provide additional features to RSVP. Since the RSVP protocol was proposed to support MPLS LSP setups, a considerable amount of modifications and extensions have been made to the original protocol to cope up with the traffic engineering requirements. In addition to the extensions already made, more extensions need to be added to the RSVP protocol in order to support voice traffic over the MPLS architecture.

```
┌─────────────────────────────┐
│      Common Header          │
├─────────────────────────────┤
│  INTEGRITY (optional)       │
├─────────────────────────────┤
│         SESSION             │
├─────────────────────────────┤
│        RSVP_HOP             │
├─────────────────────────────┤
│       TIME_VALUES           │
├─────────────────────────────┤
│     EXPLICIT_ROUTE          │
│       (optional)            │
├─────────────────────────────┤
│     LABEL_REQUEST           │
├─────────────────────────────┤
│    SESSION_ATTRIBUTE        │
│       (optional)            │
├─────────────────────────────┤
│   POLICY_DATA (optional)    │
├─────────────────────────────┤
│   Other Optional Fields     │
├─────────────────────────────┤
│     Sender Descriptor       │
└─────────────────────────────┘
       (a) PATH message
```

```
┌─────────────────────────────┐
│     SENDER_TEMPLATE         │
├─────────────────────────────┤
│      SENDER_TSPEC           │
├─────────────────────────────┤
│     ADSPEC (optional)       │
├─────────────────────────────┤
│     RECORD_ROUTE            │
│       (optional)            │
└─────────────────────────────┘
       (b) Sender Descriptor
```

```
┌─────────────────────────────┐
│      Common Header          │
├─────────────────────────────┤
│  INTEGRITY (optional)       │
├─────────────────────────────┤
│         SESSION             │
├─────────────────────────────┤
│        RSVP_HOP             │
├─────────────────────────────┤
│       TIME_VALUES           │
├─────────────────────────────┤
│   RES_CONFIRM (optional)    │
├─────────────────────────────┤
│      SCOPE (optional)       │
├─────────────────────────────┤
│   POLICY_DATA (optional)    │
├─────────────────────────────┤
│   Other Optional Fields     │
├─────────────────────────────┤
│          STYLE              │
├─────────────────────────────┤
│    Flow descriptor list     │
└─────────────────────────────┘
       (c) RESV message
```

Fig. 1. Format of the (a) PATH message, (b) Sender Descriptor, and (c) RESV message. Not drawn to scale. Each field may be variable length based on protocol subfields applied.

Voice traffic is best handled by RTP/UDP at the transport and higher application layers. Hence, whenever voice is sent, the protocol ID in the IP datagram would be set to UDP. Since, RSVP uses raw IP datagram forwarding, RSVP signaling should be able to give enough information to indicate the kind of traffic that would be sent over the established path. We propose modifications to the LABEL_REQUEST and SESSION objects. Additional traffic specs can be added by using the SENDER_TSPEC field of the Sender Descriptor field.

The PATH message and the RESV message have the following format as shown in Fig. 1. The shadowed areas are the fields that will have the RTP/UDP protocol indication to trigger the desired operational features.

| Reserved (1 byte) | Protocol (1 byte) | L3PID (2 bytes) |

Fig. 2. Label Request object without Label Range (Class = 19, C_Type = 1).

Reserved (1 byte)		Protocol (1 byte)	L3PID (2 bytes)
M (1bit)	Res. (3bits)	Minimum VPI (12 bits)	Minimum VCI (2 bytes)
Res. (4 bits)		Maximum VPI (12 bits)	Maximum VCI (2 bytes)

Fig. 3. Label Request object with ATM Label Range (Class = 19, C_Type = 2).

The Label Request object in the PATH message was modified to include an 8 bit protocol field to indicate the RTP/UDP transport layer operation as well as the real-time payload type operation and corresponding requirements. Fig. 2-4 show the fields where the Protocol field was assigned. Each of these fields was taken from the Reserved fields which were reserved for extensions to be made.

Reserved (1 byte)		Protocol (1 byte)	L3PID (2 bytes)
Reserved (6 bits)	DLI (2 bits)	Minimum DLCI	
Reserved (1 bytes)		Maximum DLCI	

Fig. 4. Label Request object with Frame Relay Label Range (Class = 19, C_Type = 3).

IPv4 tunnel end point address (4 bytes)		
Reserved (1 byte)	Protocol (1 byte)	Tunnel ID (2 bytes)
Extended Tunnel ID (4 bytes)		

Fig. 5. LSP_TUNNEL_IPv4 Session Object (Class = SESSION, LSP_TUNNEL_IPv4, C_Type = 7).

The Session object in the RESV message was modified the same way as the Label Request message to include an 8-bit protocol field to indicate the RTP/UDP transport layer operation as well as the real-time payload type operation and corresponding requirements. Fig. 5 and 6 show the fields where the Protocol field was assigned.

	IPv6 tunnel end point address (16 bytes)		
Reserved (1 byte)	Protocol (1 byte)		Tunnel ID (2 bytes)
	Extended Tunnel ID (16 bytes)		

Fig. 6. LSP_TUNNEL_Ipv6 Session Object (Class = SESSION, LSP_TUNNEL_IPv6, C_Type = 8).

3 Extensions to CR-LDP for VoMPLS

Extensions to CR-LDP are still under research. Many parameters are being included in the optional field of the CR-LDP Label Request Message and the CR-LDP Label Mapping Message to support certain services. RTP does not define any mechanisms for recovery of packet loss. Such mechanisms are likely to be highly dependent on the packet content or the underlying technology used. The idea of including RTP in the CR-LDP messages, as an optional field, will help having a reliable real time application like Voice over IP over MPLS (VoMPLS) since CR-LDP could be used as a bearer control for voice traffic. This application will have an end-to-end connection including all the MPLS features such as QoS, traffic engineering and protection. The CR-LDP messages will enable the MPLS LSRs to detect voice traffic traveling through the network and initialize an RTP/UDP/IP/MPLS encapsulation type.

Since the optional parameters may appear in any order, RTP could be placed in different ways in the CR-LDP messages, for example in the traffic TLV parameters. However, the message could be composed of variable length, ranging from 0 to 31 bits. It should contain bits to initialize the RTP session over MPLS network. Besides, important voice issues must be included in the RTP parameter about Packetization/De-packetization, compression and multiplexing. Addressing these issues in the RTP message will help other applications like H.323 to work efficiently in the voice transport mechanism.

Since RTP does not support any mechanisms for recovery of packet loss, future research could be done on addressing this problem. One solution for that is to include some features in the RTP parameter in the CR-LDP message notifying the sending user of the lost packets and about the status of the network, i.e., congestion and queuing.

The encoding for the CR-LDP Label Mapping Message is extended with an Aux LSPID TLV, Route Record TLV, and Stacked Label TLV as shown in Fig. 8-9.

0	Label Request (0x0401) (15 bits)	Message Length (2 bytes)
Message ID (4 bytes)		
FEC TLV		
LSPID TLV (CR-LDP, mandatory)		
ER-TLV (CR-LDP, optional)		
Traffic TLV (CR-LDP, optional)		
Pinning TLV (CR-LDP, optional)		
Resource Class TLV (CR-LDP, optional)		
Pre-emption TLV (CR-LDP, optional)		
L3PID TLV (optional)		
Diff-Serv TLV (optional)		
Aux LSPID TLV (optional)		
Route Record TLV (optional)		

Fig. 7. Extensions to CR-LDP Label Request Message.

0	Label Mapping (0x0400) (15 bits)	Message Length (2 bytes)
Message ID (4 bytes)		
FEC TLV		
Label TLV		
Label Request Message ID TLV		
LSPID TLV (CR-LDP, optional)		
Traffic TLV (CR-LDP, optional)		
Diff-Serv TLV (optional)		
Aux LSPID TLV (CR-LDP, optional)		
Route Record TLV (optional)		
Stacked Label TLV (optional)		

Fig. 8. Extensions to CR-LDP Label Mapping Message.

0	0	Type = 0x0810 (14 bits)	Length = 24 (2 bytes)	
Flags (1 byte)		Frequency (1 byte)	Protocol (1 byte)	Weight (1 byte)
Peak Data Rate (PDR)				
Peak Burst Size (PBS)				
Committed Data Rate (CDR)				
Committed Burst Size (CBS)				
Excess Burst Size (EBS)				

Fig. 9. Traffic Parameters TLV protocol format.

4 RTP Relay Functions

A RTP relay operating at a given protocol layer is an intermediate system that acts as both a destination and a source in a data transfer. RTP allows the use of two kinds of RTP relay systems, namely translators and mixers [18]. A mixer is a system that receives stream of RTP data packets from one or more sources, where the mixer may possibly change the data format and/or combine the data streams and then forward the combined stream. In general, the timing among multiple input sources will not be synchronized, therefore, the mixer will make timing adjustments among the streams and generate its own timing for the combined stream, thus being the synchronization source (SSRC) [22, 23]. On the other hand, a translator forwards RTP packets with their SSRC identifiers, which makes it possible for receivers to identify individual sources even though packets from all the sources pass through the same translator and carry the translator's network source address. Based on the service requirements, translators may change the encoding of the data and thus the RTP data payload type and timestamp. If multiple data packets are combined into one packet, or vice versa, a translator will assign new sequence numbers to the outgoing packets [22, 23]. The true multiprotocol architecture of the MPLS protocol becomes more appropriate for the relay operations. Unlike ATM cell types that depend on link connection type as well as the bit rate format (e.g., constant or variable bit rates), MPLS LSRs will operate on a more flexible traffic engineering concept through the FECs of selection through the consideration of the CoS and application requirements.

5 RTCP over MPLS Functions

The real time control protocol (RTCP) provides network congestion control services for real-time data transportation. Some of the basic functions of RTCP include analyzing sender and receiver reports and maintaining the measurement and calculation of the packet loss and jitter. Packet loss is measured to provide persistent congestion control. Packet loss requires the cumulative number of packets lost and fractions lost to be measured. The cumulative number of packets lost is used in the calculation of network throughput. The fraction lost is used when shortage of reception station information and interval between reports becomes long [22, 23]. Interarrival jitter is used for transient congestion control. The interarrival jitter field provides a second short-term measure of the network congestion. Packet loss tracks persistent congestion while the jitter measures the transient congestion. The jitter measure may indicate congestion before it leads to packet loss [22, 23]. The CQS architecture of MPLS routers enables the network to provide adaptive traffic engineering QoS and GoS features. The MPLS network will need to incorporate the RTP and RTCP service fields and map them into MPLS CQS parameters corresponding to application (voice, video, data, control data, etc.). This will require the LSRs and LERs to support and interact with the RTP/RTCP transport-layer, network access layer, and the physical layer up to some degree to provide enhanced service features.

6 Conclusion

Currently, based on the knowledge of the authors, there is little provision within IETF drafts or articles that connect the real-time QoS issues to the MPLS protocol and networking architecture. The real-time transport functions are well prepared in RTP and RTCP. Hence, this paper proposes methods of how to enable MPLS LSPs to acknowledge the RTP/RTCP information and correspondingly provide appropriate actions for real-time quality services in MPLS networks. Interfacing MPLS with RTP is necessary since RTP does not address the issue of resource reservation or QoS management; instead, it relies on resource reservation protocols such as RSVP, extensions to RSVP, or CR-LDP for this functionality. In addition, it is impossible for a transport layer end-to-end protocol alone to ensure in-time delivery. Real-time high quality transport services require the support of the lower layers, which are actually in control over the resources in the switches and routers. In order to accomplish guaranteed real-time data services, the lower layers need to interact with the RTP and RTCP fields and take certain information (e.g., payload type and/or interarrival jitter parameters) into account in the service delivery mechanism.

References

[1] O. Aboul-Magd, L. Andersson, P. Ashwood-Smith, F. Hellstrand, K. Sundell, R. Callon, R. Dantu, P. Doolan, T. Worster, N. Feldman, A. Fredette, M. Girish, E. Gray, J. Halpern, J. Heinanen,T. Kilty, A. Malis, P. Vaananen, and L. Wu, "Constraint-Based LSP Setup using LDP," IETF Draft, http://www.ietf.org/internet-drafts/draft-ietf-mpls-cr-ldp-04.txt, July 2000.
[2] L. Andersson, P. Doolan, N. Feldman, A. Fredette, and B. Thomas, "LDP Specification," IETF Draft, http://www.ietf.org/internet-drafts/draft-ietf-mpls-ldp-11.txt, Aug. 2000.
[3] D. O. Auduche, L. Berger, D. H. Gan, T. Li, V. Srinivasan, and G. Swallow, "RSVP-TE: Extensions to RSVP for LSP Tunnels," IETF Drafts, http://www.ietf.org/internet-drafts/draft-ietf-mpls-rsvp-lsp-tunnel-06.txt, July 2000.
[4] D. Awduche, A. Chiu, A. Elwalid, I. Widjaja and X. Xiao, "A Framework for Internet Traffic Engineering," IETF Draft, http://www.ietf.org/internet-drafts/draft-ietf-tewg-framework-02.txt, July 2000.
[5] C. Casetti, J. C. De Martin, and M. Meo, "A Framework for the Analysis of Adaptive Voice over IP," former IETF Draft, (currently removed from IETF links.)
[6] J.-M. Chung, (Invited Paper) "Analysis of MPLS Traffic Engineering," *Proceedings of the IEEE Midwest Symposium on Circuits and Systems 2000 Conference* (IEEE MWSCAS'00), East Lansing, MI, USA, Aug. 8-11, 2000.

[7] N. Demizu, K. Nagami, H. Esaki, Y. Katsube, and P. Doolan, "VCID Notification over ATM link for LDP," IETF Draft, http://www.ietf.org/internet-drafts/draft-ietf-mpls-vcid-atm-05.txt, Aug. 2000.

[8] D. De Vleeschauwer, B. Steyaert, and G. H. Petit, "Performance analysis of voice over packet based networks." submitted to ITC 16, Edingburgh, U.K., June 7-10, 1999.

[9] F. Le Faucheur, "IETF Multiprotocol Label Switching (MPLS) Architecture," *Proceedings of the 1st IEEE International Conference on ATM (ICATM-98)*, pp. 6-15, June 22-24, 1998.

[10] F. Le Faucheur, L. Wu, B. Davie, S. Davari, P. Vaananen, R. Krishnan, P. Cheval and J. Heinanen, "MPLS Support of Differentiated Services," IEFT Draft, http://www.ieft.org/internet-drafts/Draft-ietf-mpls-diff-ext-07.txt, Aug. 2000.

[11] O. Hagsand, K. Hanson, I. Marsh, "Measuring Internet Telephony Quality: Where are we today?" *Proceedings of the IEEE Global Telecommunications Conference,* vol. 3, pp 1838-1842, 1999.

[12] A. Kankkunen, G. Ash, A. Chiu, J. Hopkins, J. Jeffords, F. Le Faucheur, B. Rosen, D. Stacey, A. Yelundur, L. Berger "VoIP over MPLS Framework," IETF Draft, http://www.ietf.org/internet-drafts/draft-kankkunen-vompls-fw-01.txt, Jan. 2001.

[13] H. Liu and P. Mouchtaris, "Voice over IP Signaling: H.323 and Beyond," RFC 2474, NIC, 1998.

[14] P. Mishra and H. Saran, "Capacity Management and Routing Policies For Voice Over IP Traffic," *IEEE Commun. Mag.*, vol. 14, pp. 24-33 ,Mar.-Apr. 2000.

[15] Y. Rekhter and E.C. Rosen, "Carrying Label Information in BGP-4," IETF Draft, http://www.ietf.org/internet-drafts/draft-ietf-mpls-bgp4-mpls-04.txt, Jan. 2000.

[16] T. Senevirathne and P. Billinghurst, "Use of CR-LDP or RSVP-TE to Extend 802.1Q Virtual LANs across MPLS Networks," IETF Draft, http://www.ietf.org/internet-drafts/draft-tsenevir-8021qmpls-01.txt, Oct. 2000.

[17] P. A. Smith, B. Jamoussi, D. Fedyk and D. Skalecki, "Improving Topology Data Base Accuracy with LSP Feedback," IETF Draft, http://www.ietf.org/internet-drafts/draft-ietf-mpls-te-feed-01.txt, July 2000.

[18] W. Stallings, *High-Speed Networks, TCP/IP and ATM design principles*. Upper Saddle River, N.J.: Prentice Hall, 1998.

[19] G. Swallow, "MPLS Advantages for Traffic Engineering," *IEEE Communications Magazine*, vol. 37, no. 12, pp. 54 -57, Dec. 1999.

[20] P. Van Mieghem, "A lower bound for the end-to-end delay in networks: Application to Voice over IP," *Proceedings of Globecom'98*, pp. 2508 –2513, 1998.

[21] RFC 1771, *A Border Gateway Protocol (BGP-4)*. Internet Architecture Board (IAB), Mar. 1995.

[22] RFC 1889, *RTP: A Transport Protocol for Real-Time Applications*. Internet Architecture Board (IAB), Jan. 1996.

[23] RFC 1890, *RTP Profile for Audio and Video Conferences with Minimal Control*. Internet Architecture Board (IAB), Jan. 1996.

A System Level Framework for Streaming 3-D Meshes over Packet Networks

Ghassan Al-Regib and Yucel Altunbasak

School of Electrical and Computer Engineering
Georgia Institute of Technology
Atlanta, GA 30332-0250, USA
{gregib,yucel}@ece.gatech.edu

Abstract. In this paper, a system-level framework is proposed for 3-D graphics streaming. The proposed architecture is scalable with respect to the variations in both bandwidth and channel error characteristics. We consider the end-to-end system, and jointly optimize pre-processing and post-processing solutions. In particular, forward error correction (FEC) and multiple description (MD) codes are applied to the base-layer mesh; FEC codes are applied to enhancement layer connectivity data; and unequal error protection (UEP) and error-concealment methods are applied to the geometry data. Re-transmission is utilized wherever it is applicable as determined by the delay requirements. Furthermore, the performance analysis of the system is also provided.

1 Introduction

The internet evolves from being a place for users to communicate through media objects into a place where users meet and interact with each other within a virtual world. 3-D graphics applications are the driving force for this "virtual reality over IP" concept. These applications utilize highly detailed 3-D models, giving rise to huge amount of data to be stored, transmitted, and rendered. To alleviate these limitations, several single-layer compression techniques have been developed to minimize the storage space for a given model [1,2]. However, the resulting compressed mesh still requires significant time to be fully downloaded before being displayed on the client's screen. To reduce this time, progressive compression techniques have been designed so that a coarse representation of the model is downloaded first and then the enhancement layers are streamed out to the client [3–5].

In a typical network, there are several factors that considerably affect the quality of the received 3-D mesh. These factors are:

1. The bandwidth between the server and the client
2. The clients' graphics card capabilities
3. The channel/link packet bit-error characteristics
4. The channel/link packet loss characteristics

The existing 3-D graphics compression techniques are designed to reduce the bandwidth and storage requirements, and do not deal with transmission aspects, hence they only provide a solution for the first and the second constraint. In this paper, we propose a system-level framework that provides a solution considering all these aspects.

The research efforts in 3-D graphics are mainly limited to compression and simplification. Not many research efforts have been undertaken to stream out 3-D graphics except MPEG-4 related research. However, even MPEG-4 3-D mesh research efforts concentrate on compression and integration with video, but they do not provide error handling methods specific to 3-D graphics streaming and/or transmission. In this paper, we provide a robust 3-D graphics streaming framework that borrows several ideas from existing robust video streaming research, but adapts them for our particular purpose. In particular, we consider server-side pre-processing, network processing and client-side post-processing at the same time, and jointly optimize them.

2 Progressive Compression

Progressive compression schemes can be categorized into two main classes: Wavelet-based [5] and vertex-split-based [3, 4] schemes. We adopted the latter class due to its popularity, simplicity and its good performance for a variety of 3-D models. A progressive compression scheme has been proposed by Hoppe [3] for the first time. In his work, Hoppe simplifies a mesh by applying a sequence of edge-collapse operations to arrive at the coarsest mesh M^0. In practice, this base mesh, M^0, is first transmitted to the client, and then a sequence of vertex split operations are streamed out until the fully refined mesh, M^n, is constructed at the client. These two operations are illustrated in Figure 1. Each vertex split command specifies: i) the new vertex, ii) two new faces, and iii) the attributes of the new vertices and the faces.

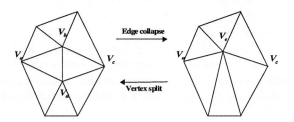

Fig. 1. Edge-collapse and vertex-split commands.

Pajarola and Rossignac [4] improved Hoppe's work by grouping vertex split operations into batches, where each batch splits half of the previously decoded vertices using edge collapse operations. The encoder works as follows:

1. The base mesh, M^0, is compressed using a single-level 3-D mesh compression scheme, such as [2].
2. At each level i, the vertex spanning tree is traversed. If the vertex is split, a bit "1" is inserted; otherwise a bit "0" is inserted into the bit-stream.
3. For the vertex v to be split, the edges are stored clockwise starting from the edge connecting v to its parent in the spanning tree. The pair of indices corresponding to the two edges affected by the vertex-split operation is encoded using $\lceil \log_2(\frac{d}{2}) \rceil$ bits, where d is the degree of vertex v.
4. A prediction algorithm is applied to predict the position of the vertices from the already encoded vertices. The prediction error is then quantized and entropy-encoded.

Accordingly, the decoder works as follows:

1. The base mesh is received and decoded.
2. Then, the stream containing the batches (*e.g.*, enhancement layers) is received. The bit-stream is parsed one bit at a time. If the current bit is "0", then the next bit is read. However, if it is "1", then the vertex is split, and the next $\lceil \log_2(\frac{d}{2}) \rceil$ bits are decoded to identify which pair of edges are affected by the split. Then, the prediction error is decoded completing the vertex split operation for this particular vertex.
3. Go to step 2 until the bit rate is completely utilized or the whole bit stream is parsed.

The time to display the mesh is the time needed to download the base mesh. As it will be shown in the next section, this accounts for less than 10% of the total download time. However, although progressive compression methods produce a bandwidth scalable bit-stream, they are not robust against packet bit-errors and/or packet losses. The next section discusses the proposed error-handling mechanism for streaming 3-D meshes over packet networks.

3 Error-Handling Framework

3.1 The Bit-Stream Content

The bit-stream is composed of three main parts as follows:

1. *The base mesh data:* Typically, simplifying mesh M^{i+1} to mesh M^i reduces the number of vertices by 30% [4]. Hence, applying n-level batches produces a base mesh, M^0, with $((\frac{2}{3})^n \times V)$ vertices, where V is the number of vertices in the original 3-D mesh. We used the Touma-Gosta (TG) single-level compression algorithm due to its superior performance upon other algorithms. On the average, this method achieves a compression ratio of 10.4 bits per vertex for both the mesh connectivity and the geometry [2]. As a result, the output stream for the base mesh contains $(10.4 \times (\frac{2}{3})^n \times V)$ bits.
2. *Enhancement layer connectivity data:* The batch b_i that produces M^{i+1} from M^i contains $(\frac{2}{3} \times V)$ bits that specify either a vertex is being split or not.

In addition, this batch contains ($\frac{2}{3} \times V \times \lceil \log_2(\frac{d_j}{2}) \rceil$) bits for each vertex split since the two collapsed edges for a vertex can be encoded by $\lceil \log_2(\frac{d_j}{2}) \rceil$ bits, where d_j is the degree of vertex j. Hence, this batch contains ($\frac{1}{2} \times \frac{2}{3} \times V \times \log_2(\frac{d_j}{2})$) bits, where the $\frac{1}{2}$ factor accounts for the fact that half of the vertices in M^i are split to produce M^{i+1}. Thus, the total number of bits to represent the mesh connectivity in all batches is

$$\sum_{i=1}^{n} \left(\frac{2}{3}\right)^{n-(i-1)} \times V + \sum_{i=1}^{n} \frac{1}{2} \times \left(\frac{2}{3}\right)^{n-(i-1)} \times \lceil \log_2(\frac{d}{2}) \rceil \quad (1)$$

In practice, 5 bits are adequate to specify the two edges being collapsed per vertex split. In other words, $\lceil \log_2(\frac{d_j}{2}) \rceil$ in the above equation can be replaced by (5) [4].

3. *Enhancement layer geometry data:* Each batch contains bits for the prediction error. The size of this segment is variable due to entropy encoding. Assuming each batch, b_i, contains E_i bits describing the geometry information, then the total number of bits in the enhancement bit stream describing geometry information is $\sum_{i=1}^{n} E_i$ bits.

Table 1 lists these statistics for different 3-D mesh models. The data in the table will be used in performance analysis of the proposed system in the following subsections.

3.2 Error-Handling and Packet Format

According to the bit-stream content-analysis in section 3.1, the base mesh constitute approximately 1-3% of the total mesh data (See Table 1). This portion of the bit stream is of utmost importance, hence, it should be strongly protected against channel degradations and delivered error-free to the client within a certain period of time as specified by the synchronization protocol. We protect the base mesh bit-stream by two mechanisms: i) Reed-Solomon (RS) code against the bit-errors, and ii) multiple description codes (MDC) against the packet losses [6]. MDC is only utilized in cases where packet re-transmission is prohibited due to the stringent time-delay contraints.

The enhancement layer mesh connectivity information also needs to be strongly protected against channel errors. Any bit-error and/or packet-loss (in this part of the bit-stream) at a certain level causes catastrophic errors in all subsequent levels. For example, if a single bit describing a vertex split command is received erroneously, then it will cause a considerable distortion in the mesh in all subsequent levels. Therefore, we apply strong channel encoding to the enhancement layer data by utilizing the RS code. However, we do not employ MDC in this case, since the connectivity data constitute a substantial percentage (30%) of the total rate. The packet-errors are handled through re-transmission at the expense of increased delay. However, since the base-layer mesh is already

Table 1. The bit stream content for different models. The numbers in the parenthesis are percentage of the bits in that category to the total number of bits in the bit stream.

Model (♯ Vertices)	♯ Levels (n)	M^0 (bits)	Batch-0 (bits)	Batch-n (bits)	Total Batches (bits)	Geometry (bits)
horse (10811)	8	4048 (1.7)	1362 (0.6)	24970 (11.0)	70810 (31.2)	153190 (67.4)
skull (10952)	8	4101 (1.7)	1380 (0.58)	25300 (10.6)	71730 (29.9)	163940 (68.4)
fohe (4005)	7	2272 (2.2)	764 (0.75)	9250 (9.1)	25730 (25.4)	80713 (79.4)
fandisk (6475)	7	3674 (2.5)	1236 (0.85)	14960 (10.3)	41590 (28.7)	99411 (68.6)
triceratops (2832)	7	1607 (3.1)	540 (1.03)	6540 (25.5)	18190 (34.6)	32718 (62.1)
shape (2562)	7	1454 (2.7)	489 (0.92)	5920 (11.1)	16460 (30.9)	35236 (66.2)
bunny (34834)	9	8609 (1.1)	2897 (0.36)	53110 (6.6)	231040 (28.5)	570570 (70.4)
phone (83044)	10	13545 (0.8)	4558 (0.27)	83560 (4.9)	55360 (33.2)	1144900 (66.8)
happy (49794)	9	12306 (1.0)	4141 (0.34)	75920 (6.2)	330270 (27.1)	875350 (71.8)

available at the client, the delay induced by the re-transmission of the enhancement layer connectivity data packets is more tolerable to the end-user in most applications.

The remaining portion of the bit stream is the geometry information in the batches. Such information constitutes more than 60% of the total bit stream. Here, we propose to utilize unequal error protection (UEP). We label each vertex either as "critical vertex" or "non-critical-vertex" depending on whether its position can be predicted from already decoded vertex locations within a predetermined threshold. Since the non-critical vertex locations can be interpolated, no error protection is applied. On the other hand, "critical vertices" are protected through RS since there is no way of recovering/concealing these vertices. If the synchronization and delay requirements permits, re-transmission is used in all cases.

The bit stream is organized as shown in Figure 2. In this figure, RSR, C_{b_i} and G_{b_i} stand for Reed-Solomon Redundancy, enhancement layer mesh connectivity and geometry information in the batch b_i, respectively. To prevent error propagation, re-synchronizing codewords are inserted after every batch as well as the base mesh.

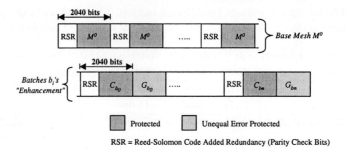

Fig. 2. Bit stream format using $RS(255, 223)$ code.

3.3 Performance Analysis

The proposed pre-processing methods, such as FEC and MD codes, provide error-resilience at the expense of increased redundancy. This added redundancy causes an increase in i) the bandwidth requirements, and in ii) the time-delay. Particularly, such delay may considerably affect synchronization among the clients. However, we observe that if the link suffers from packet errors/losses considerably, then the delay induced by the proposed error-resilience methods is much smaller than the delay caused by solely re-transmission based solutions. Furthermore, application of error concealment at the client side to recover the geometry information reduces the re-transmission rate.

Table 2 shows the redundancy introduced by employing $RS(255, 223)$ code for a number of 3-D meshes. The last two columns in this table refer to the amount of time consumed by the redundancy added by the FEC for two different links, $56Kbps$ and $1Mbps$, respectively.

As depicted in Table 2, on a LAN, the redundancy added by the FEC code is in the order of milliseconds while on a $56K$-link, the delay is in the range of $0.06 \sim 1.4$ seconds. The latter delay is significant enough to affect the synchronization in some interactive applications. Fortunately, this simple analysis is not an accurate assessment of the proposed system since the bandwidth scalability prevents streaming the whole bit-stream over such a low bandwidth link, but rather the base mesh with the first few enhancement layers are streamed out to the client. This observation leads to the conclusion that the amount of data being streamed out is determined mainly by two factors: i) the allowable delay from the synchronization algorithm, α (in seconds), and ii) the link bandwidth, β (in bps).

The number of transmitted bits should be less than $(\alpha \times \beta)$ in our method using the $RS(255, 223)$ code. That is,

$$(M^0 + 256 \times \left\lceil \frac{M^0}{1784} \right\rceil + \sum_{i=0}^{n} (b_i + 256 \times \left\lceil \frac{b_i}{1784} \right\rceil)) \leq \alpha \times \beta \qquad (2)$$

The server needs to satisfy the constraint in Equation 2 for each user and the corresponding synchronization and bandwidth requirements. However, the

Table 2. The delay introduced by applying $RS(255, 223)$ on two different links for various 3-D meshes.

Model	Added Redundancy (bits)	Delay on 56Kbps (sec.)	Delay on 1Mbps (sec.)
horse	12032	0.21	0.012
skull	12288	0.21	0.012
fohe	5120	0.09	0.005
fandisk	7680	0.13	0.007
triceratops	3840	0.07	0.004
shape	3584	0.06	0.003
bunny	35328	0.62	0.034
phone	82432	1.44	0.079
happy	50688	0.88	0.048

following inequality has to be satisfied to stream out the base mesh:

$$M^0 + 256 \times \left\lceil \frac{M^0}{1784} \right\rceil \leq \alpha \times \beta \quad (3)$$

Equation 3 describes both the synchronization and the bandwidth scalability requirements.

Note that we do not limit the server to use the $RS(255, 223)$ code. In general, it can choose among several available FEC codes that will satisfy the conditions in Equation 4 and Equation 5:

$$(M^0 + (n-k) \times m \times \left\lceil \frac{M^0}{k \times m} \right\rceil + \sum_{i=0}^{n}(b_i + (n-k) \times m \times \left\lceil \frac{b_i}{k \times m} \right\rceil)) \leq \alpha \times \beta, \quad (4)$$

$$M^0 + (n-k) \times m \times \left\lceil \frac{M^0}{k \times m} \right\rceil \leq \alpha \times \beta \quad (5)$$

where the $RS(n, k, m)$ code is used. Such a selection guarantees that the client will at least receive a coarse representation of the mesh in the allowable amount of time.

Equations 4 and 5 can be re-written in terms of $t = \frac{n-k}{2}$ to result in Equations 6 and 7, respectively.

$$(M^0 + 16 \times t \times \left\lceil \frac{M^0}{8 \times (255 - 2 \times t)} \right\rceil + \sum_{i=0}^{n}(b_i + 16 \times t \times \left\lceil \frac{b_i}{8 \times (255 - 2 \times t)} \right\rceil)) \leq \alpha \times \beta \quad (6)$$

$$M^0 + 16 \times t \times \left\lceil \frac{M^0}{8 \times (255 - 2 \times t)} \right\rceil \leq \alpha \times \beta \quad (7)$$

Therefore, the server's task is to: i) calculate the value of t that satisfies Equation 7 and maximizes the left hand side of the inequality; and ii) continue streaming out as long as Equation 6 is satisfied.

Equation 7 is a constrained optimization problems. The optimum t can be computed for a given mesh, bandwidth, and delay parameter by solving this equation. Figure 3 illustrates t versus α for different bandwidths for the 3-D mesh *fohe*. As shown, for a specific value of α, t is scaled according to the bandwidth so that Equation 7 is satisfied and hence the client receives the base

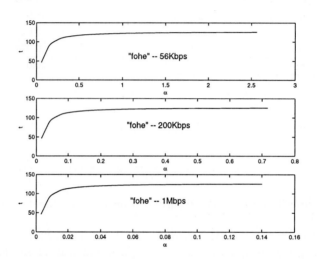

Fig. 3. Three plots of t vs. α for three different links for the mesh *fohe*, which has 4005 vertices.

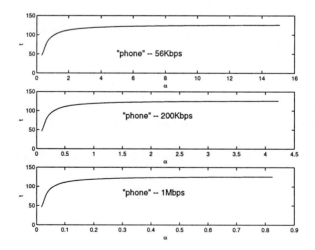

Fig. 4. Three plots of t vs. α for three different links for the mesh *phone*, which has 83044 vertices.

mesh during the allowable window of time. Similarly, Figure 4 plots the same curves for a more complicated model, the *phone* mesh. This increase in the number of vertices has an impact of lowering t for the same delay, α, over the same link, β.

4 Conclusion

In this paper, a system-level framework is proposed for 3-D graphics streaming. The proposed architecture is scalable with respect to variations in both bandwidth and channel error characteristics. We consider end-to-end system, and jointly optimize pre-processing and post-processing solutions. In particular, FEC and MD codes are applied at the client-side to the base-layer mesh; FEC codes are applied to enhancement layer connectivity data; and UEP and error-concealment methods are applied to the geometry data. Re-transmission is utilized wherever it is applicable as determined by the delay requirements.

References

1. G. Taubin and J. Rossignac, "Geometric compression through toplogical surgery," *ACM Transactions on Graphics*, vol. 17, no. 2, pp. 84–115, April 1998.
2. C. Touma and C. Gotsman, "Triangle mesh compression," in *Proceedings of Graphics Interface*, Vancouver, Canada, June 1998.
3. H. Hoppe, "Progressive meshes," in *Proceedings ACM SIGGRAPH'96*, 1996, pp. 99–108.
4. R. Pajarola and J. Rossignac, "Compressed progressive meshes," *IEEE Transactions on Visualization and Computer Graphics*, vol. 6, no. 1, pp. 79–93, January-March 2000.
5. A. Khodakovsky, P. Schroder, and W. Sweldens, "Progressive geometry compression," in *Proceedings of the on computer graphics*, 2000, pp. 271–278.
6. Y. Wang and Q.-F. Zhu, "Error control and concealment for video communication: A review," in *Proceedings of the IEEE*, May 1998, pp. 974–997.

Techniques to Improve Quality-of-Service in Video Communications via Best Effort Networks

B.E. Wolfinger and M. Zaddach

Dept. of Computer Science, Telecommunications and Computer Networks Division
Hamburg University, Vogt-Kölln-Straße 30, D-22527 Hamburg, Germany
Phone: ++49(40)42883 2424, Fax: ++49(40)42883 2345
{wolfinger, zaddach}@informatik.uni-hamburg.de

Abstract. Provisioning of real-time video communication services with sufficiently high quality is getting increasingly important in nowadays computer networks. This task is particularly challenging in the case of the presently typical communication networks which offer their data transport services in a best effort manner, i.e. without any guarantees regarding service quality. This paper classifies the principal techniques for quality improvement in video communications, which can be embedded in a communication network or in dedicated, network-external middleware components. With these techniques we distinguish, on one hand, mechanisms achieving quality improvement by means of local decisions within the sending or receiving endsystem and, on the other hand, mechanisms requiring communication and cooperation between middleware components on both, sending and receiving side. We also emphasize the combined usage of different techniques for quality-of-service (QoS) improvement. In order to judge the current service quality on different layers within the protocol hierarchy we introduce a set of QoS measures for each of the basic service interfaces identified. Finally, we solve the mapping problems between these QoS measures by way of example.

1 Introduction

Transmission of traffic of different type via so-called *best effort* networks is becoming an increasingly relevant task. Thereby, a complex mix of traffic may result of new kinds of applications, which, besides requiring traditional data communication services (e.g. file transfer), tend to demand transmission of new types of information, such as, speech, audio or video sequences, etc. The typical characteristic of best effort networks is, that they offer their communication and cooperation services without any guarantee regarding the quality-of-service (QoS) being assured a priori by the network. Thanks to their significantly lower complexity as compared to networks providing QoS guarantees (such as ATM networks [20]), presently, best effort communication networks are dominating by far within the class of Local Area Networks with Ethernet and TCP/IP based Intranets as the outstanding representatives. Moreover, also in the class of Metropolitan, Wide and Global Area Networks best effort networks are again dominating because the traditional Internet architecture leads to best effort service provisioning. Furthermore, the increasingly important class of mobile communication systems based on wireless transmissions is typically unable, too, to provide QoS guarantees.

In best effort networks we need mechanisms to improve the quality of basic communication services in order to thus satisfy the requirements of (human) endusers regarding the quality of application-oriented services with sufficiently high probability. The judgement of quality of a video sequence by a human observer after its transmission in real-time via a communication network is evidently quite subjective, not only dependent on human perception [21] but also on the chosen application domain [5]. Independent of the applications it is possible, however, to identify the following two dimensions which are relevant when judging the quality of media communications at an interface close to the application [17]:

- media quality, i.e. in video communications e.g. colour fidelity, sufficiently high video resolution, adequate picture format, no noticeable freezing or corruptions of video sequence (e.g. in case of video frame losses);
- media relations, i.e. timing properties within a single stream as well as the timing relations between video and audio streams (e.g. lip synchronization [21]). These relations are not in the focus of this contribution.

Therefore, the demand for good quality of a video communication typically implies the following QoS requirements as observed at the communication interfaces close to the enduser:

- transfer at a high data rate (yielding to advantages: large picture format, high resolution and frame rate as well as very good color fidelity thus achievable);
- strongly limited delay and possibly low delay jitter between originated and displayed video sequence (despite of the real-time requirements for the video communication, low transfer delay may allow retransmissions in case of data packet losses or corruptions respectively);
- sufficiently infrequent corruptions of the data units transmitted representing the video stream;
- possibly security requirements, which are not subject of this contribution.

This paper is concerned with QoS provisioning within real-time video communication systems. At first, fundamental methods for QoS improvement in video communications will be classified by means of an elementary architectural model and most important techniques will be illustrated by way of example (Sect. 2). Besides QoS improvements which result from mechanisms being embedded into the communication network proper, we distinguish improvements resulting from local mechanisms in a communicating endsystem (Sect. 3) and such ones which rely on communication and cooperation between corresponding transformation processes in the sending as well as the receiving endsystem (Sect. 4). We observe that complex interdependencies typically exist between different methods for QoS improvement [2]. In Sect. 5 we introduce QoS measures at various service interfaces and we illustrate how to map these measures onto each other by way of example. This enables an adaptive and dynamic decision support for QoS management components which have to choose, to adequately combine and parametrize a set of appropriate methods for QoS improvement.

2 Basic Methods for QoS Improvements and Their Embedding in a Video Communication System

In order to support a more in-depth discussion of techniques for QoS improvements in video communications, we now want to introduce an elementary architectural model of a video communication system, including the following main components/constituents:

- *Human users*, interacting with a distributed video application.
- A *distributed video application*, which comprises ≥ 1 video source(s) and ≥ 1 video sink(s); without loss of generality, in the following we will assume exactly 1 source and 1 sink, as multipoint connections just represent some straight-forward generalization.
- A *communication network*, which handles the video communications between communicating endsystems.
- *Middleware*, which improves and/or supplements the functionality and performance of network services in such a way that the requirements (related to functionality and to QoS) of the distributed video application and their users can be satisfied. As a refinement of the middleware component we want to distinguish in particular
  - *application-related middleware* (a_MW), comprising e.g. functions such as video encoding/compression,
  - *communication-related middleware* (c_MW), comprising e.g. functions to directly improve the quality of the offered network services in an application-independent manner.
- *Management component*, which receives, interprets and possibly forwards state information of other main components and coordinates decisions for QoS improvement.

We also assume for our architectural model that the (main) components of a video communication system as identified by us are interacting via well-defined interfaces. Thus, we obtain an architectural model as depicted by Fig. 1 including interfaces $IF_{\cdot,\alpha}$ on sending and $IF_{\cdot,\beta}$ on receiving site. Evidently, we can evaluate the QoS at all of the service interfaces introduced by Fig. 1. This leads to a view which corresponds to the same kind of QoS interfaces as assumed by ZITTERBART [25] in her suggested reference model for a QoS classification, if we just identify:

- $IF_{0,\cdot}$ with *(end)user related QoS*, for short: U_QoS,
- $IF_{1,\cdot}$ with *application-oriented QoS*, for short: A_QoS,
- $IF_{2,\cdot}$ with *communication specific QoS*, for short: C_QoS, and finally,
- $IF_{3,\cdot}$ with *network dependent QoS*, for short: N_QoS.

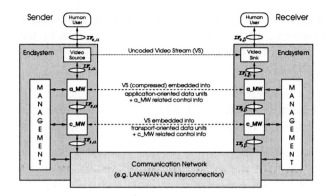

Fig. 1. Main components and typical service interfaces of a distributed video communication system

A closer look at the architectural model of Fig. 1 indicates that a QoS improvement may be achieved, in principle, by means of embedding adequate mechanisms into very different parts of the architectural model (components or interfaces), in particular:

- *at the service interfaces* (e.g. at $IF_{2,\alpha}$ or $IF_{3,\alpha}$ by manipulating the stream of data units to be transferred, e.g. "traffic smoothing" [3]);
- *by local decisions* within the main components involved in the process of either sending or receiving the video stream, in particular the middleware (a_MW, c_MW), e.g. mechanisms to mask the negative impact of data losses during the displaying of the video sequence at the receiver;
- *by decisions* within the middleware components (a_MW or c_MW) *taking into account partner feedback*, i.e. status information of the communication partner of the same architectural layer; example for such decisions: forward-error-correction (FEC) and dynamically increased amount of redundancy on sending side in case of large loss rates as signalled by the receiver [4];
- by some dedicated *QoS improvement within the communication network*.

The emphasis of this paper is placed on those QoS improvements which are achievable by means of adaptive middleware components [13] (cf. details in Sect. 3 and 4). Therefore, the first and last class of mechanisms as introduced above, namely QoS improvement at network-external service interfaces as well as improvements within the communication network proper will only be treated quite shortly in our contribution (mainly in the remainder of this section). Services offered by a communication network to support video communication typically become more valuable from applications point of view, if

- the achievable data throughput (e.g. as observed at interface $IF_{3,\beta}$ in Fig. 1) is increased → increased throughput may be used, e.g., for FEC;
- the number of packet corruptions or losses is reduced → this, in general, reduces directly the number of losses of video frames;

- the maximum overall delays in the communication network is reduced → this may be used, e.g., for a smoothing of the video stream by the sender;
- the delay jitter in the communication network is reduced → this, in general, reduces buffer requirements in the endsystems.

These goals in the desire to increase the service quality of a communication network in the context of video communication can be achieved, among others, by taking the following network-internal actions and measures (cf., e.g., [20]):

- use of better technology and of parallel processing (to reduce the service times of data transfer requests or to create gracefully degrading components),
- improvement of communication functions (such as routing, error-, flow- or congestion control),
- transition from a dynamic resource allocation (e.g. to single packets) to some longer-term resource reservation (e.g. during connection establishment),
- control of the behaviour of the network environment, i.e. its service users, at the service access points of the communication network (e.g. giving priorities to data units or to connections,...).

Amelioration of the quality of the network services per se can be complemented by taking, e.g., measures having already an impact at network-external interfaces, such as:

- reduction of the required throughput related to transfer of user data e.g. by reduction of number or size of data units (corresponding, e.g., to video frames) to be transferred or by splitting the overall video stream to be transferred onto different connections (thus reducing the throughput requirement per connection);
- giving application-oriented priorities to data units (e.g. according to their importance for the receiver);
- smoothing of the arrival process of data units to be transferred (e.g. avoidance of bursts in the time domain as well as those caused by strongly varying lengths of data units);
- reduction of time constraints for the execution of data transfers (this may allow to accept a larger limit for maximum delay or delay jitter).

3 QoS Improvement by Means of Local Decisions of Middleware Components

In the remainder of this paper we will restrict ourselves to QoS improvements by using appropriate middleware components. So, what are the kind of network-external measures which can lead, in principle, to some QoS improvement in video communications ? We can identify three different ways to achieve ameliorations in service quality:

a) By means of network-external decisions, we can try to improve the quality in which the communication network is able to offer its services, e.g.

- the throughput requirements and the load of the network by user data to be transmitted can be reduced as a consequence of external decisions,
- the communication network can be informed about different degrees of importance of the data units to be transmitted (e.g. packets, ATM cells).

b) By means of network-external measures and actions, we can try to improve (from applications' point of view) the quality in which the network offered its services by implementing additional service functionality outside the network, e.g. data losses in the network can be masked by using network-external FEC.

c) We could tolerate insufficient QoS even at an application-oriented interface by taking application-specific reactions, which could hide as much quality degradation as is possible from the human enduser, e.g. a video stream could be temporarily frozen in case of video frame losses.

The approaches b) and c) do not try to improve the quality of the network per se, but they try to present an acceptable video quality to the human enduser by dedicated middleware and application-specific reactions. Therefore, following [24], we can denote approaches b) and c) as QoS improvement by means of so-called "smart applications". If measures of type a) to c) are taken by a middleware component in one and only one of the communicating endsystems, we call this: "QoS improvement by means of local decisions" (as opposed to "decisions with feedback", cf. Sect. 4).

The following important classes of local decisions for QoS improvement, which can be taken at the sending side in the middleware components a_MW and c_MW (cf. Fig. 1), can be identified :

- *Information dispersal* [1], i.e. splitting of the user data to be transmitted onto several separate connections of the network (the goal being e.g. reduction of the packet loss probability to be expected) → typically implemented as part of c_MW.
- *Giving priorities to data units to be transmitted* → e.g. priorities could be associated to video frames to be transmitted dependent on the frame type (by a_MW) or priorities could be associated to packets dependent on the relevance of the frame information they contain (by c_MW).
- *Advantageous influencing of load*, i.e. the video data to be transmitted → e.g. reduction of load at interface $IF_2,.$, (cf. Fig. 1) by means of varying the factor of quantization, by skipping video frames, by an adequate choice of the Group of Pictures (GOP) patterns in algorithms such as MPEG [21] or favourable partitioning of video frames to be transmitted ("Application Level Framing", ALF [6]); load reduction at interface $IF_3,.$ (e.g., by means of smoothing the video stream to be transmitted).

At the receiving side we see the following possibilities to improve QoS by local decisions in middleware components a_MW and c_MW:

- *Masking of errors* → e.g. in component a_MW missing (portions of) video frames could be replaced by means of interpolation using other correctly

received frames or the receiver could freeze the last frame [7, 9] as long as (partially) corrupted frames are received.
- *Usage of playout buffer* in a_MW in order to eliminate delay jitter at the receiver [8]; basic idea: intentional additional waiting time before starting to display the total video stream (limitations: real-time requirements, size of receiver's buffer).

Evidently, it is possible to further refine the local decisions within the communicating endsystems by taking into account whether the decisions include

- only the local state of the middleware component which takes the decision,
- the local state and the state of the underlying service interface,
- additional information concerning the present state of the communication network, including its present level of service quality offered.

It is possible to complement the above-mentioned state information by establishing some feedback to the communication partner of the remote endsystem. Of course, this kind of additional state information provides far-reaching possibilities to improve QoS, which will be discussed in the following section.

4   QoS Improvement by Means of Decisions of Communicating Middleware Components

Middleware components which can coordinate their actions and decisions regarding QoS improvement (in particular by mutually exchanging control information) are thus able to significantly ameliorate quality of a video communication. The possible approaches comprise :

- *Forward error correction* (FEC), both on the level of communicating a_MW or c_MW components; under appropriate boundary conditions, FEC algorithms allow considerable quality improvements [24].
- *Error correction by retransmission* [20], again on the level of a_MW or c_MW components, can be used under the assumption that the application's real-time requirements and the delays in the communication network still allow the retransmission of data units.
- *Limitation of error propagation at the receiver* (supported by the sender) in particular on the level of communicating a_MW components, e.g. by setting checkpoints or taking other actions to limit error propagation such as
  - appropriate choice of GOP patterns [18] in algorithms as MPEG (cf. above),
  - horizontal or vertical interleaving of picture data corresponding to subsequent video frames,
  - synchronization of block groups in algorithms as H.263 [7, 9],
  - adding the feature to decode macro-blocks in reverse order (reverse decoding) [14].

For many of the techniques to achieve QoS improvement in video communications there exist some published studies regarding their inherent potential for quality improvements. However, up to now, only very few studies are available which allow one to judge in detail the potential for QoS improvements of adequately combined mechanisms. Therefore, now we will take a closer look on relationships between QoS measures. These relationships are fundamental when initiating, combining and managing mechanisms to provide fault-tolerance.

5 QoS-Mapping for Adaptive Decision-Support: Some Examples

In the previous sections we have reviewed and classified a variety of quite well-understood techniques to improve video transmissions. Yet, if we want to control these mechanisms in order to guarantee some specified video quality, we must be able to evaluate QoS measures on different layers of the protocol/service hierarchy. Furthermore, there exists the necessity to develop methods and calculation algorithms to solve the mutual mapping between QoS measures of neighbouring layers [16].

In this section we are going to suggest a set of QoS measures allowing to express, in a compact way, the service quality at all of the service interfaces as introduced by Fig. 1. We then illustrate by way of example a potential mapping between these QoS measures, in particular from N_QoS onto C_QoS as well as from C_QoS onto A_ QoS. We will assume that A_QoS related measures directly reflect the percepted (subjective) video quality at user level (U_QoS), i.e. we can neglect the corresponding mapping problem.

5.1 Mapping of N_QoS onto C_QoS

Quality of a network, i.e. QoS as observed at $IF_{3,\cdot}$ (cf. Fig. 1), typically refers to delay, delay jitter and loss of data units (e.g. packets, ATM cells) within the network. On transport system level real-time protocols (such as RTP) [20] support and define isochronous media time instances, $T_S := \{t_{S,i} = t_0 + \frac{i}{\nu} : i \in \mathbb{N}\}$ at the interface $IF_{2,\alpha}$ (sender), and $T_R := \{t_{R,i} = t_0 + \frac{i}{\nu} + \Delta_T : i \in \mathbb{N}\}$ at the interface $IF_{2,\beta}$ (receiver), where ν denotes the framing frequency, t_0 the start of the stream, and $\Delta_T \in \mathbb{R}_+$ the real time relationship between sender and receiver. Typically, a video frame is segmented according to an ALF strategy, perhaps protected by FEC and maybe reverse decoding is used. Based on these strategies the loss of single packets directly determines the relative proportion r_i of the frame structure being still correctly decodable, $r_i \in [0,1]$. Thus, QoS at $IF_{2,\cdot}$ can be expressed by $(\Delta_T, R = (r_i)_{i \in \mathbb{N}})$. We observe that R is quite hard to handle because one would have to keep a lot of "history". Therefore, R should be replaced by a new and more compact measure. A good idea seems to be to calculate $\varepsilon_i := a\, r_{i-1} + (1-a)\,\varepsilon_{i-1}$, where ε_i represents a transient estimate for that proportion of the actual video frame which is still correct and ε_i is based on some geometric weighting with parameter $a \in (0,1)$. So we can consider $(\Delta_T, \varepsilon_i)$ as an estimate for the current state of the real-time channel at a given time instance $t_{R,i}$. Comprehensive case studies showed this estimate to be very useful and applicable in a direct and efficient way [15]. In particular, ε_i can be effectively computed for given packet loss traces corresponding to N_QoS measurements. Evidently, definition of ε_i could be modified to get estimates which are more or less dynamic (e.g. by changing the weighting function). For a particular system we would suggest to choose a definition of estimate ε_i dependent on network characteristics and on the application context used. In order to limit the expenditure for measurements it would be desirable to measure directly C_QoS (rather than N_QoS), i.e. to measure r_i and then to calculate ε_i. Estimates $(\varepsilon_i)_{i \in \mathbb{N}}$ can then be used to support QoS management decisions both to control behaviour of a_MW as well as of c_MW.

5.2 Mapping of C_QoS onto A_QoS

QoS measures regarding A_QoS should reflect not only the aspects of video encoding by the application but also should take into account subjective video quality as percepted by the human user. So, it is primarily important to consider in detail the way in which errors propagate within a video sequence. Unless they are corrected by some refreshing of corresponding erroneous portions of a video frame errors remain persistent in future frames and may even be overlayed by later errors. Frame refresh typically is achieved by intra-encoded frames (I-Frames), cf. H.26x and MPEG. We will assume that at a distance of n frames such an I-Frame refresh is enforced and, thus, a given frame is affected by the interference at the current media time instance as well as by all the interferences since the last refresh. Therefore, we want to propose a QoS measure ζ_i which estimates the percepted accumulated interference between two I-Frames I_1 and I_2 in such a way that the expected interferences estimated at time $t_{R,i}$, namely ε_i, is summed up for all the frames being displayed between I_1 and I_2, i.e. $\zeta_i := \sum_{j=1}^{n} j \varepsilon_i = \varepsilon_i \frac{1}{2} n(n+1)$. In [22] it is shown that, with a fixed quantization, the measure ζ_i has a behaviour which is proportional to the mean sum of squared errors (MSE) on picture level. Moreover, the peak signal to noise ratio (PSNR) is just the logarithmic representation of MSE. So, ζ_i allows to estimate PSNR. This is good news because long term investigations have shown that the PSNR is the best known measure for judging the visual perceptions of human beings [23]. If Δ_{Enc} denotes the overall sum of encoding, decoding and frame processing time on application layer, we can define $\Delta_E := \Delta_T + \Delta_{\text{Enc}}$. Therefore, an adequate measure for A_QoS is given by (Δ_E, ζ_i), if the quantization level used is kept constant.

So, we have completed our proposals for QoS measures regarding all the service interfaces we wanted to cover. Thanks to the solutions provided to solve the mapping problems between these measures collection of measurements can be strongly simplified. If a management component is assumed with a global view onto all interfaces it will be sufficient to characterize the state of only one or two interfaces by dedicated measurements - the state of the remaining interfaces can be approximated by calculations referring to the mapping of QoS measures. QoS management could be performed, e.g., as follows:

- An application could specify a certain minimum video quality still acceptable in terms of an A_QoS related measure; then, management (MG) could determine the implications for the required quality at C_QoS and, finally, the control of a_MW and c_MW layers could be executed by MG.
- As an alternative or just as a supplement, the actual service quality could be measured on N_QoS or C_QoS interface and, if QoS turns out to be insufficient, MG could directly stimulate adequate reactions by c_MW.

6 Conclusion and Outlook

Management decisions to achieve high quality services in video communications via best effort networks should be supported by a variety of techniques for fault-tolerance and by other mechanisms used to be able to tolerate an insufficient quality of network services. We have identified a large variety of techniques which allow one to increase the quality of video communications as percepted by human endusers. We also showed that these techniques can be embedded into different layers of a video communication system and we have argued to use not only one but a combination of appropriate techniques.

The initiation and usage of adaptive fault-tolerance mechanisms should be based on QoS measures and on models which allow to map these measures between neighbouring layers. We advocated for usage of mechanisms for adaptive fault-tolerance controlled by dedicated QoS management components. For this purpose we introduced a set of QoS measures for various service interfaces including a proposal of how to solve the QoS mapping. Our suggestions for QoS management in video communications, e.g., allow the adaptive video encoding taking into account the estimated (measured and/or calculated) state of the communication network. However, QoS management should not be restricted to reduce or to limit the negative impact of packet losses or intolerably high packet delays on the video stream as displayed by the receiver. In order to be fair to other users of the communication network, QoS management decisions should also try to adapt the load, as it is generated by the controlled application, to the presently observed behaviour of the network. Combining load and error control facilities will be an adequate foundation to keep the video quality at a level acceptable to the enduser, even if best effort networks, such as the Internet, are used.

References

1. Albanese, A., S. Siemsglüss, B. E. Wolfinger "Information Dispersal to Improve Quality-of-Service in the Internet" *Proc. of SPIE Intern. Symp. on Voice, Video and Data Communications*, Vol. 3529, Boston (1998)
2. Aurrecoechea, C., A. Campbell, L. Hauw "A Survey of QoS Architectures" *Multimedia Systems*, Vol. 6, No. 2 (1998)

3. Bai, G., B. E. Wolfinger "Possibilities and Limitations in Smoothing MPEG-coded Video Streams: A Measurement-based Investigation" *Proc. of MMB'97*, VDE-Verlag, Freiberg (1997)
4. Carle, G. "Error Control for Real-Time Audio-Visual Services" *Seminar on High Performance Networks for Multimedia Applications*, Dagstuhl, Germany, 1997.
5. Chen, Z., S. M. Tan, R. H. Campbell, Y. Li "Real Time Video and Audio in the World Wide Web" *World Wide Web Journal*, Vol. 1 (1996)
6. Clark, D. D., D. L. Tennenhouse "Architectural Considerations for a New Generation of Protocols" *Proc. of SIGCOMM'90*, Philadelphia (1990)
7. Färber, N., B. Girod, J. Villasenor "Extensions of ITU-T Recommendation H.324 for Error-Resilent Video Transmission", *IEEE Communications Magazine*, Vol. 36, No. 6 (1998)
8. Ferrari, D., A. Gupta, M. Moran, B. E. Wolfinger "A Continuous Media Communication Service and its Implementation" *Globecom '92 Conf.*, Orlando (1992)
9. Girod, B., N. Färber, E. Steinbach "Error-Resilient Coding for H.263" D. Bull, N. Canagarajah, A. Nix (eds.) *Insights into Multimedia Communications* Academic Press (1999)
10. Girod, B., H. Stuhlmüller, M. Link, U. Horn "Packet Loss Resilent Internet Video Streaming" *SPIE Visual Communications and Image Processing '99*, San Jose, Invited Paper (1999)
11. Hafid, A., P. Dini "A Quality of Service Degradation Model for Multimedia Applications" *1st IEEE International World Conference on Systems Simulation (WCSS'97)*, Singapore (1997)
12. Heidtmann K., J. Kerse, T. Suchanek, B. E. Wolfinger, M. Zaddach, "Fehlertolerante Videokommunikation über verlustbehaftete Paketvermittlungsnetze" *Kommunikation in verteilten Systemen (KIVS'01)*, Hamburg (2001)
13. Jacobs, S., A. Eleftheraidis "Adaptive video applications for non-QoS networks" *Proc. of the 5th Intern. Workshop on Quality of Service (IWQOS'97)*, New York (1997)
14. Kamosa, G. "Video Coding and Robust Transmission in Error-Prone Environments", Technical Report, Univ. of California, Los Angeles (1998)
15. Kerse, J. "Realisierung von Stabilisierungsmechanismen für H.261/H.263 codierte Videoströme in IP-basierten Rechnernetzen", Diploma thesis, Dept. of. Comp. Sc., Hamburg Univ. (2000)
16. Knoche, H., H. de Meer "Quantitative QoS Mapping: A Unifying Approach" *Proc. of the 5th Intern. Workshop on Quality of Service (IWQOS'97)*, New York (1997)
17. Nahrstedt, K. "Quality of Service in networked Multimedia Systems" in B. Fuhrt (ed.) *Handbook of Internet and Multimedia Systems and Applications*, CRC Press/IEEE Press (1999)
18. Rhee, I. "Error Control Techniques for Interactive Low-bit Rate Video Transmission over the Internet" *ACM SIGCOMM'98*, Vancouver (1998)
19. Schneier, B. *Applied Cryptography* 2nd ed., John Wiley (1996)
20. Stallings, W. *High Speed Networks - TCP/IP and ATM Design Principles* Prentice Hall (1998)
21. Steinmetz, R., K. Nahrstedt *Multimedia Computing, Communications & Applications* Prentice Hall (1995)
22. Suchanek, T. "Untersuchung der Bildqualität übertragener Videos in Abhängigkeit von Übertragungsverlusten und der Qualitätsverbesserung mit Hilfe von Fehlertoleranzverfahren im Empfängerendsystem", Diploma thesis, Dept. of. Comp. Sc., Hamburg Univ. (2001)

23. VQEG "Final Report from the Video Quality Experts Group on the Validation of Objective Models of Video Quality Assessment" *Video Quality Experts Group* (2000)
24. Wolfinger, B. E. "On the Potential of FEC Algorithms in Building Fault-tolerant Distributed Applications to Support High QoS Video Communication" *ACM Symp. on Principles of Distrib. Comp., PODC'97*, Santa Barbara (1997)
25. Zitterbart, M. "High Speed Transport Components" *IEEE Network*, Vol. 5, No. 1 (1991)

Simulation of a Video Surveillance Network Using Remote Intelligent Security Cameras

J.R. Renno[1], M.J. Tunnicliffe[1], G.A. Jones[1], D.J. Parish[2]

[1] School of Computing and Information Systems, Kingston University, Penhryn Road, Kingston-on-Thames, Surrey, KT1 2EE, U.K. Tel. +44 (0)208 2000,
{J.Renno, M.J.Tunnicliffe, G.Jones}@king.ac.uk

[2] Department of Electronic and Electrical Engineering, Loughborough University, Ashby Road, Loughborough, Leicestershire, U.K. Tel. +44 (0)1509 227078,
D.J.Parish@lboro.ac.uk

Abstract. The high continuous bit-rates carried by digital fiber-based video surveillance networks have prompted demands for intelligent sensor devices to reduce bandwidth requirements. These devices detect and report only significant events, thus optimizing the use of recording and transmission hardware. The Remote Intelligent Security Camera (R.I.S.C.) concept devolves local autonomy to geographically distant cameras, enabling them to switch between tasks in response to external events and produce output streams of varying bandwidth and priority. This paper presents an investigation of the behavior of such a network in a simulation environment, with a view to designing suitable algorithms and network topologies to achieve maximum quality of service and efficient bandwidth utilization.

1 Introduction

The video surveillance market has experienced tremendous growth throughout the 1990s, the bulk of expenditure being in the area of video hardware (cameras, recording equipment, control/monitoring stations). The traditional arrangement of analogue sensors, transmitting via bundled cables to central control monitors under human supervision, is being increasingly superseded by digital multiplexing and transmission of high bandwidth video signals over wireless, ISDN and LAN networks. The fact that a continuous high bandwidth is required, even where the bulk of transmission contains nothing of interest, has led to demands for intelligent algorithms to ease the network load.

Intelligent algorithms for video surveillance are now capable of identifying and tracking the trajectories of objects [1], including personnel [2] and vehicles [3,4] against non-stationary backgrounds. The deployment of such algorithms allows the automatic tracking of objects of interest, leaving human operators free to perform other support tasks (such as co-ordinating an appropriate response to any perceived threat). Classes of objects (people, cars, etc.) can be differentiated by their shape and motion characteristics [5] and "suspicious" behaviour can be detected against a baseline of benign activity [6]. Specialised behavioural analysis algorithms [7,8] require a substantial degree of computational power, and are best employed analyzing those video streams containing events of interest.

The Remote Intelligent Security Camera (R.I.S.C.) takes this concept a step further, devolving intelligent algorithms to the cameras themselves and giving them a degree of local autonomy. Cameras are therefore enabled to switch between different tasks in response to external events, generating output streams of varying bandwidth and priority. For example, while tracking objects exhibiting "normal" (non-suspicious) behaviour, the camera may produce a semantic textual data stream describing trajectories and behaviour. Once a threat has been detected, a medium-quality high priority video report will be generated for the immediate attention of human operators. In addition to this video report, a high-quality forensic standard video record may be generated and stored locally.

The aim of this project[1] is to investigate the behaviour of such a network in a simulation environment, using statistical algorithms to mimic the behavior of the cameras, network and security-operators. The results of these simulations should produce a clearer understanding of the traffic/network interaction, thus aiding the development of the algorithms and topologies required to achieve maximum quality of service (QoS) and bandwidth utilization efficiency.

2 Experimental Methodology

The simulation software used in this study was written in Microsoft Visual C++, and runs upon an 850MHz Pentium III PC. The program contains modules representing the network architecture (which was assumed to be a broadband fibre-based network), the intelligent cameras and the central control/monitoring station. The latter is usually referred to as the "in-station" of the network, while the video cameras are the "out-stations". The network nodes are described as "mid-stations"; each services a cluster of out-stations (cameras) and reports to/receives commands from the central in-station (see Figure 1).

2.1 Network Architectures and Protocols

2.1.1 Asynchronous Transfer Mode (ATM)

The Asynchronous Transfer Mode (ATM) network protocol typically runs at 155Mbit/s, and carries data in 53 octet "cells" (48-octet data field plus a 5-octet header). Data is transferred on a connection-based scheme in which all cells in a particular stream follow a common "virtual channel connection" (VCC) through the network. Each node (or "switch") in an ATM network has a series of input and output ports: Cells arriving at a particular input port are interrogated for the "virtual channel identifier" (VCI) in the header field, which is used to select the appropriate output port for that cell. (The mapping between VCI's and port numbers is specified in a switching table, which is initialized when the connection is established.) The switch also contains storage buffers to hold cells until the output ports are ready for them.

---

[1] The project is funded by the Engineering and Physical Sciences Research Council.

These buffers are FIFO (first-in-first-out) may be either on the input or the output ports of the switch.

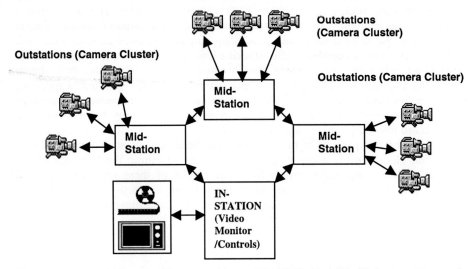

Fig. 1. Typical video surveillance architecture, with "out-stations" (camera sensors), "mid-stations" (network nodes/switches) and in-stations (central monitoring/control node (after Nche et al. [9]).

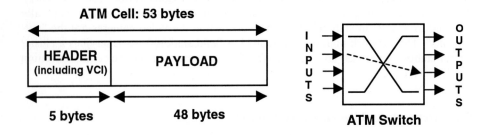

Fig. 2. ATM cell *(left)* and switch architecture *(right)*. The broken line indicates a virtual channel connection (VCC) between an input port and an output port of the switch, along which all cells in a particular stream travel. Many interconnected switches form an ATM network.

2.1.2 ATM "Active Bus" Architecture

Although ATM provides an ideal platform for video surveillance, its complete implementation would be an unnecessary expense. For this reason, workers at Loughborough University have developed a functional subset of ATM specifically for surveillance applications [9]. Figure 3 shows the basic architecture: Data cells flow both ways along the bus, camera output data in one direction and camera control data in the other. Each cluster of cameras (out-stations) is connected to a mid-station which contains an "M" (multiplex) box and a "D" (de-multiplex) box. The M-box allows the cell-streams generated by the cameras to be merged together with those already on the network. It also contains FIFO buffers to store cells temporarily during periods of congestion. (When this buffer becomes full, any further cells entering the M-box are dropped from the network.) The D-box allows control cells intended for a particular camera to be de-multiplexed out of the bus and to reach their required destinations.

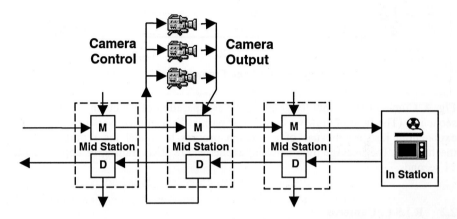

Fig. 3. Simplified illustration of the "active bus" video surveillance network architecture developed at Loughborough University [9], upon which the simulation experiments in this paper are based.

This architecture is clearly an example of an ATM network. The video, text and control streams of each camera form VCC's and the M- and D-boxes may be considered ATM switches. While the M-box has many input ports and one output port, the D-box has multiple outputs and one input.

The network simulator used in this study operates at "cell-level": The progress of each cell is tracked by the computer and cell transfer delay distributions are recorded. Most of the experiments assumed input-port buffering, though output buffering was also tested. The simultaneous arrival of multiple cells in an M-box necessitates a priority assignment system, and two such schemes were tested: (i) All incoming cells were given equal priority and were selected at random for forwarding. (ii) Cells arriving from the upstream mid-station were given automatic priority over those arriving from local cameras, in order to compensate for their longer journey to the in-station. Furthermore, each control cell generated by the in-station is destined for a specific camera; there are no broadcast/multicast cells.

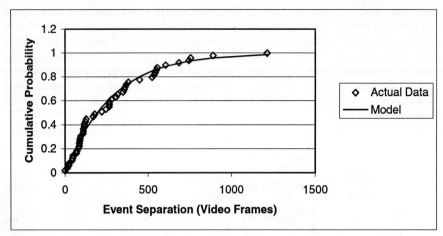

Fig. 4. Cumulative probability graph for time-separation of events in a video camera sequence. (The field of view was a staff car park at Kingston University.) The near-fit exponential justifies the use of a Poisson process to model the report-generation mechanism. The average time between events is 278 PAL frames, or approximately 11 seconds.

2.2 R.I.S.C. Cameras

In order to mimic the operation of a R.I.S.C. camera, some preliminary experiments were performed using a 20 minute video sequence of a car park. The sequence was examined frame-by-frame for "significant events" of the sort which might be detected by an intelligent algorithm (e.g. "car enters car park", "cyclist leaves car park"). The times between the end of each event and the onset of the next was recorded and the results were plotted as a cumulative probability graph. Figure 4 shows that this data approximately follows a curve of the form

$$F(t) = 1 - e^{-\lambda t} \qquad (1)$$

where λ is the reciprocal of the mean inter-event time. This indicates that the probability of event-occurrence is governed by a Poisson distribution. Hence a Poisson process was implemented in the simulation program to mimic the event-report mechanism.

At this stage in the project, several simplifying assumptions have been made concerning the camera output characteristics. When no video report is generated, each camera is assumed to send a low bandwidth, constant bit-rate (CBR) text stream to the in-station. When a video report is generated, the camera sends a higher bit-rate video stream, together with a message informing the in-station that a video report has arrived.

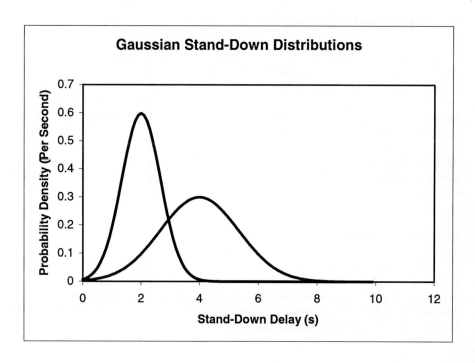

Fig. 5. Gaussian distributions used to model stand-down delays in reported simulations using means are 2 and 4 seconds (standard deviations are one third of these values). These may be seen as representing the responses of alert and sluggish operators respectively.

2.3 The In-station

Upon receiving a video report, the in-station waits for a randomly selected period of time before responding. This is the "stand-down delay" and represents the time required by the security operator to analyze and identify the reported event. Once the stand-down delay has elapsed, the in-station sends a control cell to the reporting camera instructing it to stand-down (i.e. to terminate its video report). When the camera receives this message, it resumes the low bit-rate CBR transmission prior to the video report.

The process of identifying the reported object/event is likely to require a period of cogitation, the Poisson distribution is an unsuitable representation of the stand-down process. Stand-down delays are likely to be clustered symmetrically around a central mean value, and for this reason a Gaussian distribution was chosen. While the mean of this distribution was a user-adjusted parameter, the standard deviation was set to exactly one third of the value. Figure 5 shows Gaussian curves for mean stand-down delays of 2 and 4 seconds (representing more alert and more sluggish security operators respectively.)

3 Experimental Results

Throughout these experiments, the physical layer speed was assumed to be 155Mbit/s, which is typical for fiber-based ATM. The video-report bit-rate was set to 15Mbit/s and the standby "test" data rate was 80kbit/s. The network buffers were dimensioned such that no cells were dropped due to buffer-overflow. (Typical loss rate on a well-dimensioned ATM network is of the order of 1 cell in 10^9 transmissions.) Network performance was quantified in terms of the cell transfer delay distributions, which could be determined for the network as a whole and for individual camera streams. All simulations were run for 500 seconds (simulated time) which was found sufficient to ensure statistically significant results. This was verified by performing multiple simulations with identical parameters and comparing the respective outcomes.

3.1 Network Topology

The first set of experiments investigated the effects of network topology upon the quality of service. Figures 6 and 7 shows the effects of adding more cameras and mid-stations to an existing network; the larger the network becomes, the greater the traffic load and the longer the average cell delay. (In these experiments, the inputs to each M-box were assumed to have equal priority.)

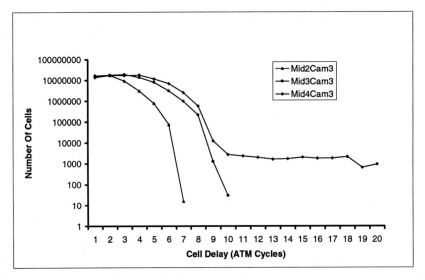

Fig. 6. Cell delay distributions for entire data on networks of 2, 3 and 4 mid-stations (Mid2Cam3, Mid3Cam3 and Mid4Cam3 respectively). Each mid-station services three cameras. Mean inter-event time was 11 seconds and mean stand-down time was 6 secs.

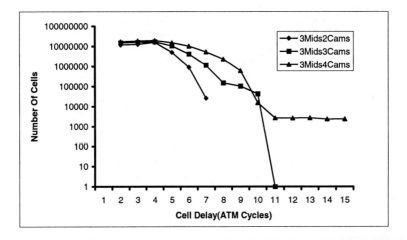

Fig. 7. Cell delay distributions for entire data on networks of 2, 3 and 4 cameras per mid-station (3Mids2Cams, 3Mids3Cams and 3Mids4Cams respectively). All three networks contained 3 mid-stations. Mean inter-event time was 11 seconds and mean stand-down time was 6 secs.

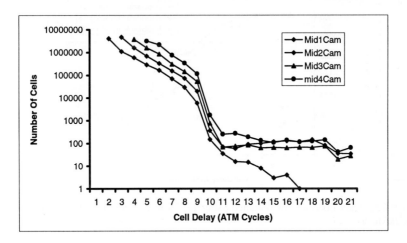

Fig. 8. Cell delay statistics for video-stream cells from four individual cameras on a network of four mid-stations. Mid1Cam and Mid4Cam are cameras on the midstations closest to and furthest away from the in-station respectively.

Figure 8 shows cell-delay distributions for individual cameras at different points in a network. It is noticeable that the cameras further away from the in-station experience longer delays, since their cells have further to travel to the in-station. This experiment was repeated, giving automatic priority to cells from upstream mid-stations (Fig.9).

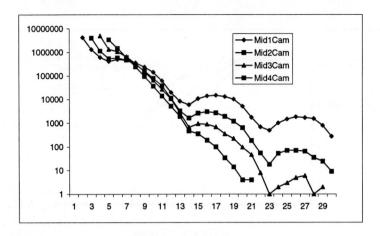

Fig. 9. The experiment of Figure 8 repeated, giving automatic priority to all cells from "upstream" mid-stations to compensate for their longer journey.

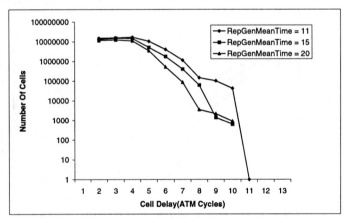

Fig. 10. Cell-delay distributions obtained using 11, 15 and 20 seconds mean inter-event time (RepGenMeanTime).

3.2 Network Activity

Figure 10 shows the results of another simulation experiment, in which the effect of changing the level of network activity was observed. In all cases, the network consisted of three mid-stations, each of which served a cluster of three cameras. The mean time between events (RepGenMeanTime) was set to 11, 15 and 20 seconds and the effect upon the cell delay distributons was observed. The results show that the greater network activity associated with a shorter inter-event time increases the overall load on the network and increases the average cell delay.

4 Conclusions and Future Work

This paper has outlined some of the recent trends in the development of video surveillance networks, and shown the benefits of building intelligence into individual camera sensors. A simple cell-level simulator of a network of intelligent cameras, based upon an ATM "active bus" architecture has been established and tested. This simulator has then been used to explore the effects upon QoS of different network topologies and video-report generation frequencies.

A number of assumptions have been made in the formulation of the model; firstly the arrival of events observed by each camera has been assumed to be random and uncorrelated, and therefore governed by a Poisson process. (This assumption was partially justified by observation of a limited data set.) A further assumption is that no correlation between events detected by *different* cameras; each is governed by an independent statistical process. This is likely to be untrue in practice, since a single suspicious vehicle or individual would almost certainly be detected by more than one camera; hence the arrival of an event on one camera would increase the likelihood of an event on neighbouring cameras.

These results provide a foundation for a more realistic simulation study, which should ultimately allow recommendations to be made about optimum network configurations.

References

1. Owell,J, Remagino,P, Jones,G.A.: From Connected Components to Object Sequences. Proc. 1st. IEEE International Workshop on Performance Evaluation of Tracking and Surveillance. Grenoble, France, 31 March (2000) 72-79
2. Iketani,A, Kuno,Y, Shimada,N, Shirai,Y.: Real-Time Surveillance System Detecting Persons in Complex Scenes. Proc. 10th. International Conference on Image Analysis and Processing. Venice, Italy, 27-29 Septtember (1999) 1112-1117
3. Harrison, I, Lupton,D.: Automatic Road Traffic Event Monitoring Information System ARTEMIS. IEE Seminar on CCTV and Road Surveillance (1999/126), London, U.K., 21 May (1999) 6/1-4
4. Collinson,P.A.: The Application of Camera Based Traffic Monitoring Systems. IEE Seminar on CCTV and Road Surveillance (1999/126), London, U.K., 21 May (1999) 8/1-6
5. Haritaoglu,I, Harwood,D, Davis,L.S.: Active Outdoor Surveillance. Proc. 10th. International Conference on Image Analysis and Processing. Italy, 27-29 September (1999) 1096-1099
6. Thiel,G, "Automatic CCTV Surveillance - towards the VIRTUAL GUARD", Proc. 33rd. International Carnahan Conference on Security Technology, Madrid, Spain, 5-7 Oct., pp.42-8, 1999.
7. Remagnino, P., Orwell, J., Jones, G.A., "Visual Interpretation of People and Vehicle Behaviours using a Society of Agents", Congress of the Italian Association on Artificial Intelligence, Bologna, pp. 333-342, 1999.
8. Orwell, J., Massey, S., Remagnino, P., Greenhill, D., Jones, G.A., "Multi-Agent Framework for Visual Surveillance", IAPR International Conference on Image Analysis and Processing, Venice, pp. 1104-1107. 1999.
9. Nche,C.F., Parish,D.J., Phillips,I.W., Powell,W.H., "A New Architecture for Surveillance Video Networks", International Journal of Communication Systems, 9, pp.133-42, 1996.

Cooperative Video Caching for Interactive and Scalable VoD Systems

Edison Ishikawa[1,2] and Claudio Amorim[1]

[1] COPPE - Systems Engineering and Computer Science Program,
Federal University of Rio de Janeiro,
Centro de Tecnologia H-318 Ilha do Fundao, Rio de Janeiro, RJ, Brazil,
CEP 21945-970
{edsoni,amorim}@cos.ufrj.br
[2] Military Institute of Engineering, IME
Rio de Janeiro, RJ, Brazil
http://www.ime.eb.br
ishikawa@ime.eb.br

Abstract. In this work, we introduce a novel technique called Cooperative Video Caching (CVC) that enables scalable and interactive Video-on-Demand (SI-VoD) servers to be developed. The key insight of CVC is to exploit the client buffers as a large global distributed memory that can be randomly accessed over the communication network. CVC uses patching and chaining but in such a way that minimizes communication traffic, eliminates the chaining interruption problem, and most important, enables VCR-like interaction. Through detailed CVC simulations our results show that CVC-based servers are potentially scalable. In addition, we show that interactive CVC servers significantly outperform conventional interactive VoD servers. These preliminary results suggest that CVC is a promising technique to develop SI-VoD servers

1 Introduction

Video-on-demand (VoD) systems offer a promising solution for delivering video-based applications in important areas such as Distance Learning and Home Entertainment. To date, current VoD implementations support only a limited amount of clients, which restrict the widespread use of VoD applications. Therefore, multicast-based Near-VoD systems (N-VoD) [4,6,1] have been proposed, as a scalable solution. However, N-VoD systems faces the problems that client requests may have to wait for the next appropriate video multicast, and offer restricted interaction rather than VCR-like operations.

Alternatively, the VoD scalability problem can be addressed by exploiting the extra communication bandwidth at the client (machine) side, using techniques such as patching [7] and chaining [9]. The key idea behind such techniques is to allow the vast amount of video that is stored in the client buffers throughout the system to be reused to generate new video streams in such a way that reduces server occupancy and improves significantly the system throughput. However, such schemes have two main drawbacks: (1) The amount of traffic generated over the network can grow explosively; and (2) VCR-like interaction is not supported.

In this work, we introduce a new technique called Cooperative Video Caching (CVC) that exploits also the client buffers but in a novel way. First, CVC organizes all the client buffers as a large global distributed memory [5,2,11,10] that can be randomly accessed over the communication network. Second, CVC uses patching and chaining but in a different and efficient manner. More specifically, CVC implementation introduces three distinct features: (1) Clients instead of servers can send patches between them; (2) Chaining is implemented in such a way that minimizes communication traffic, and (3) CVC eliminates the chaining interruption problem [9]. In addition, CVC enables VCR-like interaction whereas current schemes do not.

CVC requires the use of digital broadband communication networks [1] where each client can receive and transmit video streams. In this way, CVC-based VoD servers are ideal to small-to-large size audiences that can be found in academic and enterprise environments where desktop computers are typically interconnected by high-speed networks with symmetrical access.

The remainder is organized as follows. The next section describes CVC and its mechanisms that support scalable and interactive VoD servers. Section 3, presents our experimental methodology and discuss simulated performance results. Finally, section 4, presents our conclusions and outline future works.

2 Cooperative Video Caching

In this section, we describe the CVC technique that manages all the client buffers as a single global distributed memory. For this purpose, CVC uses the group of frames (GoF) as the memory access unit, organizes the client buffers as a single distributed memory over the network, and introduces a new scalable and interactive VoD protocol that implements the cooperative video caching, as described next.

2.1 CVC Access Unit

A GoF [8] is a self-contained video unit in which all the required information to decode the GoF is stored within the GoF structure. In addition, each GoF has a timestamp associated with, which CVC uses to randomly address GoFs within a video stream. In this way, GoF timestamps can be used within video streams like word addresses in a RAM.

2.2 The Client Buffer Organization

In a VoD system, the client buffers store GoFs to deal with system jitter. CVC extends the control of client buffers so that other clients can share all the buffer

---

[1] This is one of the networks described in VoD specification [3]

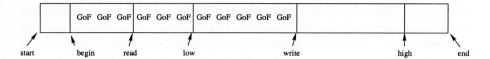

Fig. 1. CVC organization of a client buffer

contents. Figure 1 shows the client buffer organization in CVC. As can be seen, seven control pointers implement the buffer control structure. The *start* and *end* pointers define the first and last locations in the buffer. The *write* and *read* pointers indicate the buffer positions to store and show the next GoF, respectively. The *begin* pointer defines the location of the earliest received GoF and is updated whenever a GoF is discarded. The *low* and *high* pointers determine the minimum and maximum buffer levels at which warnings are sent to the CVC manager that will take appropriate actions to prevent either buffer underflow or overflow, accordingly.

2.3 CVC Basic Protocol

Given the client buffer organization and access modes, let us describe the basic CVC implementation. Also, consider the case of a single-video server, for the sake of simplicity. On arriving the first request the server will start transmitting a new video stream to the client. As soon as the client buffer reaches its minimum level it will start showing the movie (figure 2(a)). So far, CVC behaves like an ordinary server.

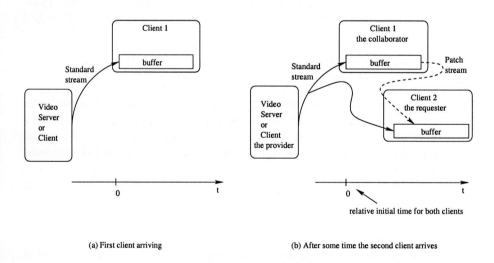

Fig. 2. The use of Patching in CVC

When a second request arrives, the CVC manager looks up its table for a client, called the provider, that holds the video initial part (VIP) in its local buffer (figure 2(b)). To determine whether or not a client holds the VIP the server needs to estimate how much time of video stream is left within a client buffer. Let t_b be the amount of video stored in a client buffer in seconds[2]. A provider is a client that received the first video GoF at most t_b seconds before the second request arrival time. If there is more than one provider, CVC selects the one that its relative initial time is $t_b/2$ seconds smaller than the current time. In this case, the provider will insert the requester into its multicast group. Also, as the requester missed the VIP, another client of the multicast group, called the collaborator, will provide the VIP to the requester. In case of no such a provider be available, then the last client that received the VIP at less than t_b seconds becomes the provider and implements the chaining that links its buffer to the requester buffer (figure 3). Otherwise, the server will start a new video stream that is transmitted to the requester.

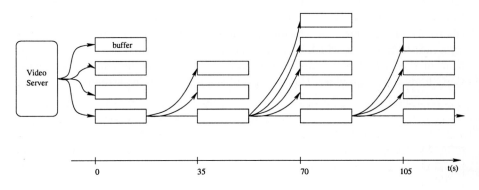

Fig. 3. Optimized Chaining in CVC

Whenever a provider fails silently, its client will consume GoFs until reaching the buffer's minimum level at which the CVC manager either determines another provider or starts a new video stream (figure 4). As a result of earlier detection of video underflow, the client will have GoFs to show while CVC searches for a new provider, thus solving the interrupt chaining problem in a simple and efficient way.

The pause operation is also simple to implement in CVC. If the requester that issues a pause is in a tree leaf then the solution is trivial. If the requester is a provider (figure 4), the CVC manager will walk towards the root until it finds the next client, which is the provider of the current provider. If there are other clients in the provider's multicast group, CVC chooses one from the group as the new provider.

---

[2] Note that the size t_b and $t_b/2$ are a trafeoff between using the multicast stream in full against making long pathces or limiting the multicast to transmit shorter patches

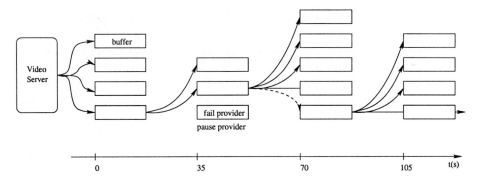

Fig. 4. Client's failure or pause operation in CVC

When a client sends a resume command to the CVC manager, it starts showing the video using the GoFs stored in the client's buffer. At the same time, CVC traverses the chain towards the same stream direction until it finds a provider that has approximately the same elapsed time from the beginning of the movie (figure 5), and defines it as the new provider. Note that the client has from now on a new initial time, which corresponds to the current time less the timestamp time of the last received GoF. Most important, only if the chain ends before CVC finds a new provider, the server will start a new video stream.

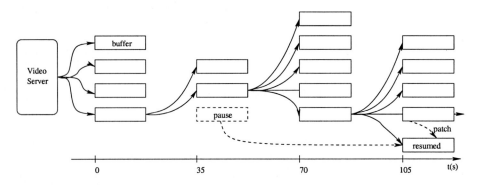

Fig. 5. Resume operation after 70 seconds

2.4 The CVC Interactive Protocol

CVC implements fast forward (FF) and rewind (RW) operations either in a discrete or continuous mode. In the discrete mode, CVC uses the average MPEG1 throughput to determine how many GoFs it needs to either forward or backward depending on whether the operation is FF or RW. For a small number of GoFs the buffer size suffices. Otherwise, CVC must find a new provider that holds the video stream at the selected point (figure 6).

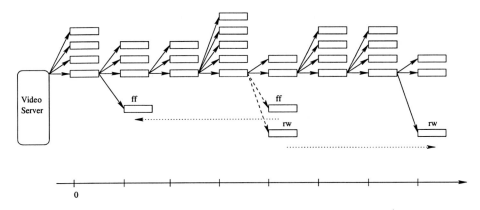

Fig. 6. Discrete ff/rw operation in CVC

The implementation of continuous FF/RW operations is similar to the discrete ones. The only difference is that while traversing the chain CVC will request and show buffer contents along the way to the client (figure 7).

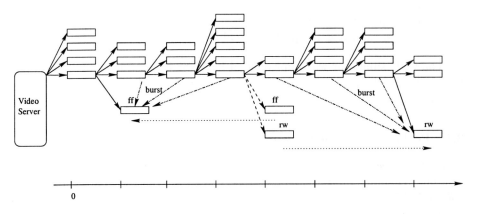

Fig. 7. Continuous ff/rw operation in CVC

3 Experimental Methodology

In this paper we present preliminary performance results based on CVC simulations that assume the following system configuration: (1) Servers and client stations are interconnected by an ATM network; (2) The network topology is fully connected; and (3) The network implements a deterministic service.

Table 1 summarizes system workload and simulation parameters, with associate standard values and/or value range when appropriate. It is important to notice that the parameter values we chose allow us to examine the scalability of CVC at a relatively low transmission speed, using low-cost servers and basic client resources.

$(\alpha_o, \beta_o, \gamma_o)$ produces a bit-stream at a rate of $\mathcal{B}_o$. Assuming that the target bit-rate is $\mathcal{B}_t$, we can pose the problem as estimating the resolution levels $(\alpha_t, \beta_t, \gamma_t)$ that will optimize the re-coded video quality (either in terms of subjective quality or in Peak Signal to Noise Ratio, PSNR). In general, optimal $(\alpha_t, \beta_t, \gamma_t)$ is a function of $\mathcal{B}_o$, $\mathcal{B}_t$, $(\alpha_o, \beta_o, \gamma_o)$, and the video sequence v itself. That is,

$$(\alpha_t, \beta_t, \gamma_t) = \mathcal{M}(\alpha_o, \beta_o, \gamma_o; \mathcal{B}_o; \mathcal{B}_t; v) \tag{1}$$

Although this mapping $\mathcal{M}$ may be automatically determined by examining transform coefficients and/or motion vectors, in this paper we will limit ourselves to a supervised classification approach, where the mapping is pre-estimated and stored for various video sequences. Figure 1 depicts the rate shaping module comprised of a bit-stream extraction module and scaling information database. A feedback channel providing information about the network state and the client capabilities is assumed. Our main objective is concentrated on building the scaling information database in Fig. 1. However, rather than estimating this labor-intensive multi-dimensional mapping for each individual video sequences, we classify the video sequences into a few classes and estimate a single mapping for each class of video. This is further motivated by our observation that the mappings for the video sequences within the same class are very similar.

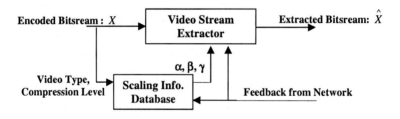

Fig. 1. Optimum Rate Shaping Module

It should also be noted that the display size and display frame rate are kept the same as determined by the terminal screen size and frame refresh rate in all the scalability options (through temporal frame repetition and spatial bi-linear-interpolation).

4 Simulation Results and Discussion

Two different sets of experiments are carried out in order to demonstrate the efficacy of the proposed rate shaping method. The first sets of experiments use MPEG coded video whereas the video sequences in the second sets of experiments are coded with a 3-D subband coder [8]. This particular subband coder is comprised of 3-D subband coding, progressive quantization, and arithmetic coding providing fine scalability spanning over spatial, temporal, and SNR resolutions.

4.1 MPEG Simulations

In the first sets of experiments, three different video sequences, $V1$, $V2$, and $V3$, with different amount of motion content, are utilized. Each video is originally encoded at $\mathcal{B}_o = 2$ Mbps, and scaled down to four different bit rates. The results are tabulated in Table 1, where subjective quality of the scaled-down video is measured by a star rating system. A four star corresponds to the best quality whereas a single star indicates the worst quality. The ratings are relative among the same target output rate. A group of 8 Ph.D. students are asked to relatively rate the output video sequences with the star rating system. The median rating is then reported in Table 1. The first raws in each sub-table (in Table 1) denote the target output rate, $\mathcal{B}_t$. The first columns in each sub-table indicate the scaling performed in each video. Two spatial resolutions $\alpha_t =$ CIF (352x240) and $\alpha_t =$ QCIF (176x120); and two temporal resolutions $\beta_t = 30$ fps, and $\beta_t = 15$ fps, resulting in 4 different scalability combinations, are taken into account. SNR scaling is automatically adjusted to meet the target bit rate. Thus, 30fps CIF resolution video is implicitly the most coarsely quantized video with lowest value of γ_t (low SNR resolution) among the same bit rate video streams.

Table 1. Quality Report from Scaling Using MPEG

V1: Head and Shoulder Video Sequence				
	200 Kbps	500 Kbps	1 Mbps	1.5 Mbps
30 fps, CIF	*	*	***	****
30 fps, QCIF	**	**	**	**
15 fps, CIF	****	****	****	****
15 fps, QCIF	***	***	**	**

V2: Video Sequence with Moderate Motion				
	200 Kbps	500 Kbps	1 Mbps	1.5 Mbps
30 fps, CIF	*	**	****	****
30 fps, QCIF	***	**	**	**
15 fps, CIF	**	***	***	***
15 fps, QCIF	***	**	**	**

V3: Video Sequence with High Motion				
	200 Kbps	500 Kbps	1 Mbps	1.5 Mbps
30 fps, CIF	*	*	**	****
30 fps, QCIF	**	***	***	**
15 fps, CIF	**	**	**	**
15 fps, QCIF	***	****	****	**

Relative Rating with * = worst and **** = best

Illustrated in all three sub-tables (in Table 1), the relative rating for CIF resolution video sequences at 30 fps is very poor at low bit rates ($\mathcal{B}_t = 200\text{Kbps}$ and $\mathcal{B}_t = 500\text{Kbps}$). Because no data reduction from neither the spatial resolution nor the temporal resolution is made, video sequences have to be very coarsely quantized to meet the target bit rate. Thus, they fail to produce good quality video at the receiver side. This coarse quantization is also in agreement with the traditional rate shaping methods where only SNR scalability is exploited. It is clearly observed that SNR scalability alone can not yield the best rate shaping. The temporally scaled down video sequence, CIF at 15fps, is evaluated the best in all available target bit rate for the video sequence $V1$. In the video sequence with high motion, $V3$, the output sequence with a reduction in both temporal and spatial resolution (QCIF at 15fps) is evaluated the best for first three lowest target bit rates. Only when the target bit rate reaches 1.5 Mbps, 30fps CIF resolution video become the highest quality. Compared to the video sequences with distinct characteristics (low motion in $V1$ and high motion in $V3$), the video sequence with moderate motion, $V2$, does not exhibit any unique pattern in its rating.

4.2 Three-Dimensional Subband Coding Simulation

In the second sets of experiments, three video sequences ("head and shoulder" video sequence, the "football" sequence, and the "remote lecture" sequence) are considered. The motion contents of these video sequences are quite different. All three scalability options (e.g., spatial, temporal and SNR scalability) are considered. Three different spatial resolutions corresponding to $\alpha_t = 352\text{x}240$, $\alpha_t = 176\text{x}120$, and $\alpha_t = 88\text{x}60$ are considered. Three temporal resolution corresponding to $\beta_t = 30\text{fps}$, $\beta_t = 15\text{fps}$, and $\beta_t = 7.5\text{fps}$ are considered. Finally, six SNR resolution scaling, $\gamma = 0$ representing highest SNR resolution though 5 representing lowest quality resolution, are considered. The simulation is performed over the video samples coded with an average bit-rate, $\mathcal{B}_o$, ranging from 500Kbps to 2Mbps. The video quality degradation is measured using the Mean-Square-Error (MSE) metric.

Effect of Different Temporal, Spatial and SNR Resolutions. Table 2 compares the video quality with different scaling options for the high motion football sequence originally coded at $\mathcal{B}_o = 1000$ Kbps. The second, third and fourth columns show the spatial, temporal and SNR resolutions, respectively. The resulting MSE is tabulated in the fifth column. The performance data confirms that the different combinations of scalability options (e.g., resolutions) yielding the same target output bit-rate exhibit substantial differences in decoded video quality.

Under the same target bit-rate $\mathcal{B}_t$, the decoded video quality with lower temporal frame rate, indicating more temporal scaling, always performs worse (for this video sequence). Due to the fast motion in this video sequence, considerable dissimilarity between consecutive frames exists and temporal interpolation does

Table 2. 3-D subband coding results

Bit rate	α	β	γ	MSE	Bit rate	α	β	γ	MSE
0.9 Mbps	176x120	30 fps	4	263	0.5 Mbps	88x60	30 fps	5	328
	352x240	15 fps	3	557		352x240	15 fps	4	584
0.4 Mbps	176x120	7.5 fps	3	1245	0.25 Mbps	88x60	15 fps	5	740
	88x60	15 fps	4	735		176x120	7.5 fps	4	1181

Notation: α Spatial resolution, β Temporal resolution, γ SNR resoution
(for SNR resolution: 0 = finest, 6 = most coarse)

not improve the decoded video quality. This loss in temporal resolution cause the significant information loss. With fast motion video sequence, temporal resolution scaling should be avoided if good quality video is desired at the receiver end.

Effect of Different Video Classes. Classifying video streams according to their characteristics has a major impact on the video quality when different scalability combinations of (α, β, γ) are employed. Figure 2, 3, and 4 depict the reduction in spatial, temporal and SNR resolutions that provides the best video quality for the "head and shoulder", "football" and "remote lecture" video sequences, respectively. In all cases, the initial video sequences are encoded at 1Mbps. The horizontal axis measures overall bit rate reduction ratio from the original encoded bit rate ($\frac{B_t}{B_o}$), whereas the vertical axis represents reduction ratio from each scalability level. All three curves together express the optimal scalability combination. For example, in Fig. 2, 85% reduction in overall bit rate ($x = 85\%$) obtained by 45% reduction in SNR resolution, 75% reduction in temporal resolution, and no reduction in spatial resolution provides the best video quality for the "head and shoulder" video sequence.

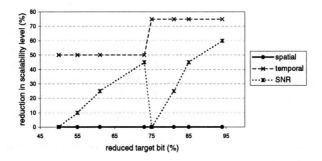

Fig. 2. Optimum Scaling for Head and Shoulder Sequence

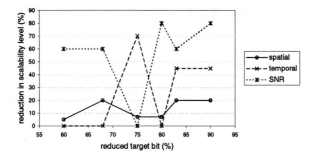

Fig. 3. Optimum Scaling for Football Sequence

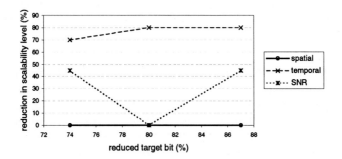

Fig. 4. Optimum Scaling for Remote Lecture Sequence

The "head and shoulder" video sequence has a very low motion content. The major percentage of the bit-rate reduction is provided by temporal scaling, as indicated by the long-dashed curve in Fig. 2. It jumps from 50% to 75% at $x = 75\%$. However, temporal scaling does not account for all the reduction in the bit-rate. In fact, the appropriate target bit-rate is satisfied using the fine grain SNR scalability. It is represented by a short-dashed curve gradually increasing from 0% upto 45% in Fig. 2. The particular 3-D subband coder [8] used in our experiments only allow both spatial and temporal resolutions to be reduced by a factor of two at a time. Once SNR scalability reduction reaches its maximum[2], then the temporal resolution reduction is needed to further cut down the rate. There is no reduction in spatial resolution throughout all output bit-rates. Temporal resolution scaling is favorable in "head and shoulder" video sequence. Since temporal interpolation (though frame repeats) at the receiver produce a reconstructed video comparable to the original video, the decrease in temporal resolution is not critical here. Bit- allocation to higher SNR and spatial resolutions is preferred.

In contrast, for "football" sequence in Fig. 3, keeping the highest temporal resolution is recommended until no data reduction can further be achieved with

---

[2] It is only maximum under the given target bit rate. It does not refer to the maximum achievable scaling level.

spatial and SNR scaling alone to meet the target bit-rate. Once the spatial and SNR resolution reduction reaches its maximum (located at $x = 68\%$), only then, the temporal scaling is performed for further bit rate reduction (temporal scaling at $x = 75\%$). At next available target bit rate ($x = 80\%$), temporal reduction level resumes to zero again.

5 Conclusions and Future Research

In this paper, we proposed a novel rate shaping approach applicable to both MPEG and subband coded video streaming. The proposed method considers the full spectrum of scalability options: spatial, temporal, and SNR scalabilities. Thus, it ensures an optimum video quality under various channel conditions.

From the results, it is evident that rate shaping with spatial, temporal and SNR scalability options provides significant gains in video quality over the existing approaches where only SNR scalability has been exploited. Furthermore, once the database is built, which is an off-line process, the complexity of our approach is no more than SNR scalability based approaches. The experimental results also show that optimal scalability options for different video classes even under the same initial and target bit rate is quite different.

Our current experimentation classifies each video in its entirety observing its content. However, since a single video stream may contain different characteristics during its duration, video classification should be performed at the Group-Of-Pictures (GOP) level. Automatic estimation of the mapping $\mathcal{M}$ is also underway.

References

[1] R.J. Safranek, C.R.Kalmanek Jr., and R. Garg, "Methods for matching compressed video to ATM networks," *Proceedings of the International Conference on Image Processing*, vol. 1, pp. 13–16, October 1995.

[2] A. Eleftheriadis and D. Anastassiou, "Constrained and general dynamic rate shaping of compressed digital video," *International Conference on Image Processing*, vol. 3, pp. 396–399, 1995.

[3] N.Celandroni, E. Ferro, F. Potorti, A. Chimienti, and M. Lucenteforte, "DRS compression applied to MPEG-2 video data transmitted over a satellite channel," *Proceedings of the Fifth IEEE Symposium on Computers and Communications*, pp. 259–266, 2000.

[4] S. Jacobs and A. Eleftheriadis, "Streaming video using dyanmic rate shaping and TCP congestion control," *Journal of Visual Communication and Image Representation*, vol. 9, pp. 211–222, 1998.

[5] Z. L. Zhang, S. Nelakuditi, R. Aggarwal, and R. P. Tsang, "Efficient selective frame discard algorithms for stored video delivery across resource constrained networks," *Proceedings of IEEE Infocom*, vol. 2, pp. 472–479, March 1999.

[6] W. Zeng and Bede Liu, "Rate shaping by block dropping for transmission of MPEG-precoded video over channels of dynamic bandwidth," *Proceedings of the fourth ACM International Conference on Multimedia*, pp. 385–393, 1996.

[7] R. Rejaie, M. Handley, and D. Estrin, "Quality adaptation for congestion controlled video playback over the Internet," *Proceedings of ACM SIGCOMM*, pp. 189–200, August 1999.

[8] D. Taubman and A. Zakhor, "Multirate 3-D subband coding of video," *IEEE Transaction On Image Processing*, vol. 1, no. 5, pp. 572–588, September 1994.

IP Stack Emulation over ATM

Ir. Geert Goossens and Ir. Marnix Goossens

VUB INFO/TW, Pleinlaan 2, 1050 Brussels, Belgium
geert@info.vub.ac.be

Abstract. This paper describes the ideas behind a new way of running Internet applications on an ATM network. ATM already contains some functionality of the network layer and therefore there is no need to duplicate them in the IP stack. Instead of adapting the TCP/IP stack to ATM at the bottom, we replace the TCP/IP stack with something else that is capable of supporting the QoS needs of applications. The intention is to have Internet applications working over ATM with the same performance as a native ATM application could obtain.

Introduction

Internet applications are traditionally only involved in the transfer of non real-time data. These applications do not need anything better than a best effort service from the network. There are no conditions on delay, delay-variation or bandwidth. In the latest years, however, there has been an increasing interest in (mostly high-bandwidth) real-time applications, like multimedia. The industry has therefore been going more and more towards a convergence of the Internet and telecommunications. This requires an enhanced network that is able to support both types of traffic. ATM (Asynchronous Transfer Mode) was developed with this idea in mind. ATM has many features that come from the telecom world (broadband ISDN). It is a connection-oriented protocol, a clear distinction from the connection-less Internet, and it is able to support strict Quality-of-Service (QoS) on each connection. ATM also supports a best effort service. This can be used by traditional Internet applications. ATM therefore would seem to be an ideal choice to support this convergence. The problem with ATM has been that it has been marketed as a fast LAN technology (CLIP [3], LANE 1.0 [1], LANE 2.0 [2], and MPOA [14]), or as a backbone or access technology. In the first case, QoS in CLIP and LANE is almost non-existent. LANE 2.0 does make limited use of QoS. MPOA can make better use of QoS, but is a rather complicated system. In the latter case, some QoS features are used (reserved bandwidth) but not dynamically (administratively configured PVCs) and not end-to-end (not over the entire path from the sending host to the receiving host).

In this paper, we present a new method for running Internet applications over ATM. It is the intention to fully support all the available QoS features of the ATM network. ATM contains a number of functions of the network layer. It provides routing, error checking, fragmentation and reassembly of transport-layer packets. Because of this, we want to investigate if it is not possible to eliminate the entire IP layer. Of course, it will not be possible to eliminate IP addresses, because these are used by the applications. We will assume that we have a way to convert an IP address to ATM addresses and we assume for this discussion that this method can work in a global ATM network. We will briefly show a possible method in this paper. Since our solution consists in eliminating the IP stack and replacing it with something else, we call it IP Stack Emulation (IPSE). The applications still work as if there was a normal TCP/IP stack beneath them. In this paper we will look especially at how unicast UDP applications can be made to run over ATM.

This paper is organized as follows. In Section 1 we look at the differences and similarities between ATM and IP. In Section 2 we discuss how to map connection-less (UDP) data onto VCs. In Section 3 we see how we want to add QoS support to applications. In Section 4 we look at how we want to use VCs and how we can change the QoS for an application dynamically. In Section 5 we discuss briefly some problems that we encounter when trying to run TCP over ATM. In Section 6 we give a short description of a system that can translate IP addresses to ATM addresses in a large ATM network. Finally, Section 7 concludes the paper.

Comparison between ATM and IP

The functionality of ATM overlaps in a number of ways with that of IP [20]. Although ATM is often represented as a layer 2 protocol, it also works as a layer 3 protocol. If we use AAL5 as adaptation layer, it is possible to transport UDP datagrams [18] with the same service as a native UDP implementation.

AAL5 can transport packets up to 64K, whereas the maximum payload size for UDP is 65468 byte. From this follows that our header is limited to a maximum of 68 byte. Since this is more than enough for the header, we do not need to provide additional fragmentation or reassembly in IPSE.

AAL5 uses a redundancy check to detect errors in the packet (bit errors or lost cells). Hence, we do not need a separate UDP checksum in the packet.

Because ATM can do end-to-end routing just like IP, we can setup a connection directly to the destination (or to an edge device if the destination is not connected to the ATM network). In any case, there is no need for routers in this design. We do need a way to translate an IP address into the corresponding ATM address (see later).

Since AAL5 cannot do multiplexing like UDP, we have to prepend a header to the payload containing the source and destination port numbers. The IP addresses are fixed for a particular VC, but it can be difficult to signal these at VC setup time (see later). So, we put the source and destination IP addresses in the header as well. This results in a header of 12 byte, which is well below the limit of 68 byte.

Mapping Connection-Less Data over VCs

Flows and FlowPaths

We want to support Internet applications in the best way possible. For this it is necessary to understand how such applications communicate with TCP/IP. We take the popular Berkeley socket interface as starting point. In this API an application first needs to create a socket (an application end-point). This can then be used to send and receive data. We will focus here mainly on UDP applications (connection-less). Datagrams that are sent on the same local socket to the same remote socket constitute a flow. We need to introduce the concept of flows if we want to match application-level packets to VCs. This means that we only need to check once (when the flow is created) what QoS the flow needs and find or setup a VC accordingly. Because a connection-less application does not mark the end of live of a flow explicitly, we delete the flow either when the application destroys its end-point or when the flow has been idle (unused) for a while. Remark that a flow in this definition is the logical representation of the connection between two application end-points.

In addition to an application end-point, we also associate the flow with something we call a flowpath. A flowpath groups together all the flows that have the same two hosts as end-points. The flowpath object is created when the first flow is created between a pair of hosts. The flowpath is only destroyed if there have not been any flows on the flowpath for some time. The flowpath will be used to store the common VC that is used for flows that do not need a higher QoS than best effort.

Mapping Flows to VCs

We make a coarse distinction between two groups of flows: best effort flows and flows with a required QoS. By default, a flow is assumed to be best effort. We will see later how a QoS can be attached to a flow. We also need to take into account that VCs are a scarce resource in an ATM network of a given size. Every VC consumes resources in every ATM switch along the path. Also, the time that it takes to establish a VC through the ATM network is not negligible. It would therefore be best to keep the number of VCs as small as is necessary to provide the required service.

For the first category of flows - best effort - it is not necessary to provide a separate VC for each flow between the same two hosts. We create only one VC - called the *common* or *multiplexed* VC - to transport all best effort flows. Because all the best effort flows belonging to the same flow path will use this VC, we associate this VC with the flowpath. The common VC is established from the moment that an application sends data over a best effort flow. The common VC is closed only when the flowpath is destroyed or after it has been idle for a preconfigured time, whichever comes first.

For the second category, we need to provide a specific QoS. Because we want to use the full capabilities of ATM, we need to setup a separate VC for each flow, called a *private* or *non-multiplexed* VC. We establish this VC with the required QoS when the first packet is sent over the associated flow. We tear down the VC when the flow is destroyed or after a preconfigured idle time. In case the VC cannot be established because the network is unable to guarantee the QoS, we may fallback on trying to establish a VC with a lower QoS or simply use the common VC of the associated flowpath.

Quality of Service Support

We will focus here on the support of applications that provide streaming video and audio. There are two strategies to support this class of applications. Either we change the applications to work around the limitations of a best effort service (as far as this is possible) or we let the network provide QoS to the applications. Obviously, we are interested in doing the latter. This allows users to view a video stream at a constant quality. If it is the intention to provide the same experience as with the current television system (the idea of convergence), then this is the approach that we will have to follow. This means that there must exist a way to associate a QoS with a flow. We distinguish three methods.

The most obvious way is to modify the application to provide information about the QoS that it requires. Of course, the application is the best authority in knowing what is required for the particular data that it transports. Additions to the socket interface (like Winsock 2.0 [4]) make this possible.

It is not practical or even possible to modify all applications in this way. Therefore, we provide a way for the administrator of the host to specify a number of rules. These rules associate the flow parameters - source and destination IP address, source and destination UDP (or TCP) port - with specific QoS parameters. The flow parameters can also be wildcards. So, it is possible to say that all flows with destination port 5000 have to use a VC with a specific QoS. In addition it is also possible to restrict a rule to a specific application. When a flow is created, IPSE checks if the flow falls under one of these rules. If it does, a private VC will be created for this flow.

However, in some cases it still is difficult to know what QoS an application will need. A user can select different codecs or set different quality parameters for a codec, each having its own QoS requirements. Also, it is not always possible to know what ports an application will use. To overcome these problems, we provide a tool that visualizes the flows. The user can use this tool to look at the flows that each application has created and he can inspect the statistics of the flow (number of bytes sent/received, idle time, ...). In addition the user can change the QoS dynamically, without stopping or interrupting the application. In this way, each flow will normally start off as best effort. When the user notices that the quality is less than what he would expect, then he can set a higher QoS. Instead of filling out all the QoS parameters, the user can also use QoS templates. This simplifies the task for the user of figuring out what type of stream (MPEG, H.261, ...) needs what kind of QoS. It also allows the user to rapidly change a QoS. The user can specify the QoS in the way described in [6], or he can specify the ATM service [7] (UBR, ABR, nrt-VBR, rt-VBR or CBR) directly and give appropriate values for the parameters of the service. In the first case, a translation must be done to the corresponding ATM values. RFC 2381 [8] describes how to do this for a guaranteed and controlled-load service. How it is done exactly may depend on administrative settings and on what services are available. For end users it is probably better to specify the QoS in this general way, rather than working with ATM concepts.

One of these methods sets the 'ideal' QoS that the flow should have. It will not always be possible for the ATM network to deliver this, of course.

Using VCs

Establishing VCs

In ATM every VC needs to connect to a SAP (Service Access Point) on the remote host. We create a SAP per IP address on each host. This is the SAP that common VCs connect to. For every application end-point (socket), we also create a SAP. This is the SAP that a private VC must connect to. The name of the SAP can be derived from the UDP port number. The Broadband High Layer Information (BHLI) element of the UNI specification [15] [16] is used for this purpose. This element must be present in the connection setup and it must be set as follows. The High Layer Information Type is set to 1 (User Specific) and the High Layer Info field is set to an 8 character identifier, starting with 'UDPE' followed by the hexadecimal ASCII representation of the port number. This has the advantage that the call is cleared immediately if the destination end-point does not exist. If the VC had a QoS, the reserved bandwidth will become again available to the network as fast as possible.

A problem poses itself for the remote host when it accepts a VC. Since neither the IP address nor the UDP port is contained in the signaling message, the remote host does not know what flow or flowpath this VC is for. Since this information is fixed per VC, we would ideally like to put this information in the initial signaling messages. RFC 3033 [5] addresses this issue. However, it does not always seem possible to do so, since this depends on the configuration of the ATM switches along the path. Probably not all switches even have the possibility to allow these information elements in the signaling messages. Therefore, we give this information in the header of each packet. This means that the peer's IP address and/or UDP port is only known when the first packet has been successfully received over the VC. We will see later that this presents a problem when we want to change the QoS dynamically.

Another problem that we may face is that both hosts establish a VC to each other for the same flow or flowpath at approximately the same moment. This can be easily detected as soon as packets are being received over both VCs. Because VCs take up resources in the network, we would like to get rid of the superfluous VC. If the QoS for the associated flow has been negotiated (see later) then we take the VC that has a QoS that is closest to the negotiated QoS. Otherwise, we take the VC with the largest QoS. If both VCs have equal QoS, the host with the largest ATM address will start using the VC of the other host as soon as contention is detected. It will then close its own VC. The other host simply continues to use its own VC. It will, of course, continue to accept packets on the other VC until this one is closed.

Sending Data

In a connection-less service the application provides the destination address for each send operation. Normally we would have to check for each packet being sent what QoS to use and to search for the matching VC. Instead, we look in the list of flows that start from that application end-point. If the flow does not yet exist, we create a new flow descriptor and find out which VC it must use and establish the VC with the correct QoS if necessary. Once we have the flow information, we can direct the packet immediately over the correct VC.

Receiving Data

A VC either has a flowpath or a flow associated with it. In the first case, when we receive data on such a VC, we look in the list of flows that belong to the flowpath. In the second case, we have the flow immediately. Every flow is associated with an application end-point, so we can deliver the packet to the application immediately. If the VC is not yet associated with any flow or flowpath, we must lookup and if necessary create these objects. Here it is possible that we find out that the flow or flowpath is already assigned to another VC. In that case, one of them will be dropped according to the rules described above.

Changing the QoS

It is possible that the QoS requirements of a flow change during the lifetime of a flow, either because the application wants a different QoS or because of a user action. At any time one of the two hosts can start a procedure to negotiate new QoS parameters. The messages are sent as control packets (all header fields set to zero). They contain the flow specification (IP addresses, port numbers) and the desired QoS for both directions. These messages can be exchanged over the existing VC between the two applications or over the common VC between the two hosts (if this VC exists). If negotiation was successful, the host that started the negotiation establishes a new VC with the negotiated QoS. Now, there are two possibilities. Either the network supports including extra information elements in the SETUP message. In that case, the receiving host knows the flow for which this VC is intended at the same time as the VC is established. In case it is not possible to signal this information at VC establishment time, we have a problem. The problem occurs when the host that established the connection does not have anything to send or if the message that he did send was lost. Because we provide an unreliable service (like UDP) there is no guarantee that a message will arrive. To solve this problem, the accepting host starts a VC Identification Protocol. The receiving host does this if it does not know by any other means what flow the VC represents. The receiving host sends a control message over the VC (all header fields set to zero) announcing its own flow parameters (IP address and destination port). The peer must respond with its flow parameters. When the first host sees the response, it knows what flow has to be connected to the new VC and the old one can be dropped. The host must retry transmission of the VC Identification Packet when it does not get a response in time. When the host receives a normal packet that identifies the flow completely, it may stop the identification protocol.

TCP Emulation

We talked mainly about UDP. Instead of the TCP protocol [19], we can use any other protocol as long as it provides the same service to the application as TCP does. At the moment, however, we use a normal implementation of the TCP protocol and run it over ATM in a similar way as UDP. However, we do not need to introduce a flow concept here, since TCP already is a connection-oriented protocol. When the application opens a TCP connection, we figure out if a QoS needs to be used or not and we establish a VC if necessary. If the application sends data, we forward it immediately to the correct VC, because this VC is associated with the TCP connection. If the application closes the TCP connection and it used a private VC, then the private VC is closed immediately. The connection-oriented TCP protocol now runs directly over the connection-oriented ATM protocol. There is no connection-less IP layer between them as in CLIP [3].

To improve TCP performance over ATM, it is advantageous to set a large MTU (Maximum Transmit Unit) to decrease the header overhead and the overhead involved in processing packets. We could easily set the MTU to 16K or more. This has a side effect on the Nagle algorithm [17]. This algorithm prevents TCP from sending any more packets if a maximum size packet has been sent, but has not been acknowledged yet. This is to prevent TCP from sending a lot of small packets on the network. If the MTU is large, then it's possible that the application is unable to provide the full MTU to TCP. This would downgrade performance, since transmission would be halted until the ACK is received. To solve this, we set a lower watermark (let's say 4K) that we use for the Nagle algorithm instead of the real MTU. From the moment that there is this much data available, TCP is allowed to send a packet.

A large MTU also influences congestion control [21], since the congestion window typically opens with one MTU per acknowledgement (during slow start). For large MTU's this could open the window very fast. This can cause congestion in the network because all packets (and cells) are sent in burst. Even with smaller MTU's TCP can exhibit a bursty behavior, which causes a lot of cells to arrive all at once in an ATM switch. If the switch lacks sufficient buffer capacity at that moment to store all these cells, cells and, consequently packets, will be dropped. It should be possible in ATM to pace the TCP stream [11] [12] at the cell level, so that the buffer utilization of ATM switches is minimized. In that case, the ATM card receives instructions from TCP at which cell rate a packet must be transmitted on the network.

Other improvements can make use of the fact that all packets are automatically delivered in order over an ATM VC. This allows the receiver to quickly detect a lost packet in a stream and to request retransmission [13].

Address Resolution

Since Internet applications only work with IP addresses, we need to have a way to convert an IP address into the corresponding ATM address. The ATMARP system used in CLIP [3] is inadequate because of scaling problems in large networks. Potentially the number of hosts in our network could be very high, since the ATM cloud is not divided into subnets by routers. We want a solution that scales and that is stable with respect to changes in the network topology. We will present here a short overview of a proposed solution. A complete description of the system will be presented later in a separate paper.

Small Networks

For compatibility reasons we want to keep the ATMARP protocol described in [3], since such a server is often conveniently implemented in a switch. This would require no extra server if the system were used in a small network.

Medium-Sized Networks

To provide scalability we create a DNS-like hierarchy, comparable with the in-addr.arpa domain [9]. This domain is used to translate an IP address into the corresponding host name. Here, we will create a similar domain to translate an IP address into the ATM address of the same machine. The IP address is subdivided in four bytes, which in reverse order each represent a subdomain. So, the highest level server could be responsible for the 134 domain, the next level server handles the 184.134 domain, the last one the 50.184.134 domain. The ATMARP servers are configured with the ATM address(es) of one or more DNS servers. When an ATMARP server has a host that belongs to a certain domain as specified above, it finds and contacts the DNS server that is responsible for this domain. It uses dynamic updating to modify the contents of the DNS server [10]. For relatively small networks we can use a caching timeout of 15 to 20 minutes, similar to the timeout in ATMARP. For the implementation on the hosts nothing changes. They continue to communicate with ATMARP servers using the classical ATMARP protocol.

Problem in Large Networks

In a global ATM network, the small caching time of 15 minutes could present problems. In a normal DNS system cache timeouts are in the order of days. We would therefore prefer to cache addresses for longer times. It should be noted that the binding between an ATM and an IP address is not so stable as between an IP address and a domain name. Indeed, simply changing a host from one ATM switch to another (or even to a different port on the same switch) makes it difficult to keep the ATM address the same. We need a more stable solution.

Virtual ATM Addresses

We divide the ATM cloud into smaller networks. We assume that each network is managed separately and interconnected in a well-defined way (like networks on the current Internet). We give each network an identification number. This number is completely arbitrary, but must be unique in the entire network. Each one of our DNS servers is configured with the number of the network that it belongs to. In addition to the real ATM address, a fixed virtual ATM address is associated with every IP host.

This ATM address consists of a network dependent part and a part (host identifier) that contains the IP address. In the DNS-like requests we provide a way to specify the network identifier of the requestor. If the field of the requestor matches up with the authoritative server's network identifier, the real ATM address is returned. Otherwise, the requestor receives the virtual ATM address. For this virtual address the cache time may be set very large.

Border Switches

We still need to guarantee that any host from outside the destination network can successfully connect to a virtual ATM address inside the network. Let's take two networks with different identifiers. We assume that these networks are connected to each other by a special kind of switch that we will call the border switch. The first part of the virtual ATM address refers to this border switch. In this manner, the connection setup is routed to this switch. The last part of the virtual ATM address contains the IP address of the host. Because the switch is part of the destination network, it can retrieve the real ATM address via this IP address. The switch can now proxy the signaling protocol to the real destination so that a VC is established between the two end hosts. The border switch then behaves like a normal switch for this VC. Alternatively, instead of a switch a gateway can be deployed that terminates the first VC and starts a new VC to the destination. The gateway then functions as a packet-level router, which can also be configured as a firewall.

Conclusion

We have presented here a method for running Internet applications over ATM by replacing the TCP/IP stack with something that takes the special features of ATM into account by design. The system is very radical in the sense that it tries to eliminate as much as possible from the TCP/IP stack and replace it directly with native ATM functionality. Applications that use QoS are supported. In addition the user can also indicate what QoS to use for a particular flow in case the application is not QoS-aware.

More work still needs to be done on adding better support for TCP, multicast UDP and QoS. Also in the future support for IPv6 will need to be added. Because we do not have packet compatibility with normal IP packets, we do have a problem when we want to connect this system to the rest of the Internet. This problem will need to be addressed as well.

References

1. ATM Forum Technical Committee: LAN Emulation Over ATM Version 1.0, January 1995, ftp://ftp.atmforum.com/pub/approved-specs/af-lane-0021.000.pdf
2. ATM Forum Technical Committee: LAN Emulation Over ATM Version 2 - LUNI Specification, July 1997, ftp://ftp.atmforum.com/pub/approved-specs/af-lane-0084.000.pdf
3. M. Laubach, J. Halpern: RFC 2225 Classical IP and ARP over ATM, April 1998
4. M. Hall e.a.: Windows Sockets 2 Application Programming Interface, May 1996, http://www.stardust.com/winsock
5. M. Suzuki: RFC 3033 The Assignment of the Information Field and Protocol Identifier in the Q.2941 Generic Identifier and Q.2957 User-to-user Signaling for the Internet Protocol, January 2001
6. C. Partridge: RFC 1363 A Proposed Flow Specification, September 1992
7. The ATM Forum Technical Committee: Traffic Management Specification Version 4.0, April 1996, ftp://ftp.atmforum.com/pub/approved-specs/af-tm-0056.000.pdf
8. M. Garrett, M. Borden: RFC 2381 Interoperation of Controlled-Load Service and Guaranteed Service with ATM, August 1998
9. P. Mockapetris, RFC 1034 Domain Names - Concepts and Facilities, November 1987
10. P. Vixie, RFC 2136 Dynamic Updates in the Domain Name System (DNS UPDATE), April 1997
11. J. Kulik, R. Coulter, D. Rockwell, C. Partridge: Paced TCP for High Delay-Bandwidth Networks, Proceedings of IEEE Globecom, December 1999
12. A. Aggarwal, S. Savage, T. Anderson: Understanding the Performance of TCP Pacing, Proceedings of IEEE Infocom, March 2000
13. F. Ansari: Adapting TCP/IP over ATM, Master of Science Thesis, University of Kansas, 1996
14. The ATM Forum Technical Committee: Multi-Protocol over ATM Version 1.1, May 1999, ftp://ftp.atmforum.com/pub/approved-specs/af-mpoa-0114.000.pdf
15. The ATM Forum Technical Committee: ATM User-Network Interface (UNI) Signalling Specification Version 4.0, July 1996, ftp://ftp.atmforum.com/pub/approved-specs/af-sig-0061.000.pdf
16. The ATM Forum Technical Committee: ATM User-Network Interface Specification Version 3.1, September 1994
17. J. Nagle: RFC 896 Congestion Control in IP/TCP internetworks, January 1984
18. J. Postel: RFC 768 User Datagram Protocol, August 1980
19. J. Postel: RFC 793 Transmission Control Protocol, September 1981
20. J. Postel: RFC 791 Internet Protocol, September 1981
21. M. Allman, V. Paxson, W. Stevens: RFC 2581 TCP Congestion Control, April 1999

ATM Network Restoration Using a Multiple Backup VPs Based Self-Healing Protocol

S.N. Ashraf[1] and C. Lac[2]

[1] Institut national des télécommunications
9, rue Charles Fourier, 91011 Evry Cedex, France
Tel: +33 1 60 76 47 29, Fax: +33 1 60 76 42 91
shahid.naeem-ashraf@int-evry.fr

[2] France Télécom R&D, RTA/ECD
2, avenue Pierre Marzin, 22307 Lannion Cedex, France
Tel: +33 2 96 05 27 36, Fax: +33 2 96 05 34 27
chidung.lac@francetelecom.com

Abstract: Physical route assignment and capacity allocation functions of a Virtual Path (VP) are decoupled in Asynchronous Transfer Mode (ATM) networks. This means a zero-bandwidth VP can be preconfigured as backup of a working VP. This feature makes self-healing ATM networks particularly interesting as faster and efficient restoration can be realized by capturing the necessary bandwidth on the backup VP after a failure. With an aim to further increase the restoration speed, we developed a new Multiple Backup VPs (MBV) based protocol that permits one or more backup VPs to protect a single working VP. This approach helps holding the spread of restoration information in reduced span leading to faster recovery. The multiple backup VPs can be setup either in overlapped or non-overlapped way. This paper describes the operation of the protocol and studies how it recovers from unidirectional and bidirectional single link failures using both non-overlapped and overlapped scenarios. Performances are evaluated in terms of restoration ratio for a meshed-topology.

1 Introduction

Asynchronous Transfer Mode (ATM) is the switching and multiplexing method adopted for implementing Broadband Integrated Services Digital Networks (B-ISDN). An infrastructure based upon fiber optics and high speed digital electronics allows ATM to reach high transfer rates. However, use of fiber optics implies more traffic being concentrated on fewer routes, any transmission link/system failure tends to cause extremely serious problems due to huge loss of bandwidth, loss of service to user and loss of revenue to the operating companies. Therefore, in B-ISDN networks, demands for network reliability have highly increased. In this context, restoration techniques that minimize damage by automatically rerouting the traffic away from the failure point play a key role in network reliability enhancement.

Originally proposed by Grover in 1987 [1], self-healing is a distributed control restoration scheme for networks with no topological restriction. Since 1992, Virtual Path (VP) layer based self-healing networks have been studied, and many schemes using ATM's characteristics have been proposed [2, 3]. VP restoration has several fundamental advantages made possible by the VP path layer concept of ATM network as standardized in [4]. VP restoration realizes simple and resource-efficient restoration architecture. One of the most striking characteristics is the ability to pre-establish backup paths using zero bandwidth VPs. Another advantage useful for self-healing is the OAM (Operation, Administration and Maintenance) cell mechanism possible with VPs.

End-to-end (ETE) backup VP based self-healing protocol is the most widely studied VP layer based protocol till recently [5]. In this restoration scheme, the main VP and corresponding backup VP are established between source-destination on link/node-disjoint routes to avoid simultaneous failures in both VPs. However, such use of backup VPs is not efficient from restoration speed's viewpoint in huge networks, as the restoration messages have to pass through all nodes on the failed and backup VP between the source and destination. In this context, we developed a protocol called Multiple Backup VPs (MBV) based self-healing protocol that allows a working VP to be split into one or more sections depending upon the length. A smaller VP that can be restored faster protects each part as presented in the following sections.

This article studies operation of the MBV protocol for unidirectional and bidirectional single link failure scenarios and evaluates its performances from restoration speed viewpoint.

Article is organized as follows. In section II, basic principle MBV self-healing protocol is presented. Restoration operation of the protocol for single link failure scenario is described in sections III and IV. Results are presented in section V, followed by the conclusion in section VI.

2 MBV Protocol

Services carried by an ATM transport network can be restored by using an ETE backup VP [5]. Initially, the network searches for several paths between source-destination pair. The best such path is designated as main VP, and the second best path is reserved as ETE backup path as depicted in Fig.1. VPI numbers and the required amount of bandwidth are assigned to the main VP. For efficient use of network resources, no bandwidth but only VPI numbers are assigned to the ETE

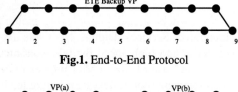

Fig.1. End-to-End Protocol

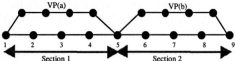

Fig.1. Multiple Backup VPs Protecting a main VP

backup VP. In case of a failure, next downstream node of the failure detects the fault. The failure is recovered by shifting transmission from the main VP to the preplanned alternative backup VP after proper bandwidth has been assigned to it.

Since ETE backup VP is established between source-destination on link/node disjoint paths, for longer VPs failure notifications and subsequent bandwidth capturing on the backup VP may take considerable time. However, if the restoration process can be localized, then the network can act and respond faster to failures and recover more rapidly than it is possible with ETE protocol.

In MBV protocol, the main path is divided into one or more restoration sections. For instance, in Fig.2, the normal VP of example shown in Fig.1 is split into two sections. Each section is protected by one backup VP. Depending upon the way backup VPs are organized, the protocol can be divided into two following types:

A.   Non-overlapped backup VPs,
B.   Overlapped backup VPs.

A. NON-OVERLAPPED BACKUP VPs

This case is shown in Fig.2 where two sequential sections have the same node as terminal node. Such a node is not protected by any of the backup VPs and therefore, its failure will be unrecoverable. Let us suppose that the bidirectional link 3-4 fails. Then restoration procedure can then be described by following phases:

- Phase 1: Failure Detection

OAM continuity check cells as described in ITU-T I.610 standard are used for failure detection. In case of supposed failure, nodes 3 and 4 detect the failure, thereby starting the restoration from both sides.

- Phase 2: Fault Notification

As soon as an intermediate node detects a fault, OAM Alarm Indication Signal (AIS) cells are generated and transmitted periodically towards end node through all active VPs that pass through the failed link or node. All the intermediate nodes can non-intrusively monitor these cells and declare VP AIS-State. Therefore, nodes along the failed VP can know about the failure. Node 3 sends the AIS message towards node 1 for notification whereas node 4 sends a similar message to node 9.

- Phase 3: Bandwidth Capture on Backup VP

On declaration of VP AIS-State, end nodes of the section send restoration messages from the respective side on the backup VP *a* protecting the section. Normally, bandwidth is captured step-by-step on the intermediate nodes in a way similar to the call-setup process shown in Fig. 6. However, in case of backup VPs with zero bandwidth assigned, due to the fact that the routing tables are preconfigured, a quasi-parallel way of bandwidth capturing as shown in Fig. 7 is proposed. In quasi-parallel way; the node of the failed VP connected to the backup VP sends in sequence two messages towards next node of backup VP as follows:

Bandwidth Capture Message: Each node on backup VP keeps a copy of the message and then simply switches it forward to the next node until it reaches the destination. This message helps intermediate nodes to know about the failure as well as the bandwidth to be captured.

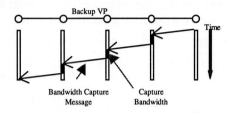

Fig.3. Step-by-Step Bandwidth Capture on Backup VP

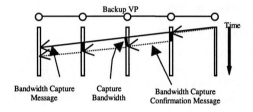

Fig. 4. Quasi-Parallel way of Bandwidth Capture on Backup VP

Bandwidth Capture Confirmation Message: This is a step-by-step confirmation message and is forwarded to the next node if required bandwidth is available.

Phase 4: Restoration Messages Collision
Arrival of a bandwidth capture message will cause collision at a node that have already captured the bandwidth due to a similar message received earlier from other side of the backup VP. Upon collision, the node transmits bandwidth capture acknowledgement towards both end nodes of the section, i.e., nodes 1 and 5. Multiple collisions may occur in two adjacent nodes as it is quite unlikely that bandwidth capture message arrive simultaneously at a node. Multiple collisions result multiple acknowledgements as shown in Fig.5.

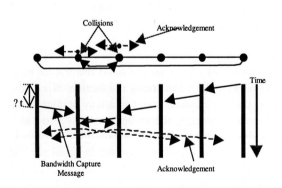

Fig.5. Multiple Collisions of Bandwidth Capture Messages on the Backup VP

- Phase 5: Switching to Backup VP

On reception of the acknowledgement, nodes 1 and 5 switch the traffic from the failed VP to the backup VP, thereby completing failure restoration for this VP.

The scheme can recover from all type of failures, either unidirectional or bidirectional of intermediate links or nodes on the main VP, except the node where adjacent backup VPs terminate (i.e., node 5). This shortcoming can be avoided by overlapping the backup VPs as explained in the following part.

B. OVERLAPPED BACKUP VPs

As adjacent backup VPs terminate on the same node, this single node failure can not be recovered by protecting the normal path with multiple backup VPs in non-overlapped way. In overlapped case, however, no single node affect more than one backup VPs; therefore, this scheme can recover from all type of failures of intermediate links/nodes of main path. Overlap of sections may consist of two or more nodes.

Fig.6 depicts a main VP consisting of nine nodes protected by two overlapped backup VPs. Section 1 extends from node 1 to node 6, whereas section 2 extends from node 4 to node 9. In case of node 4's failure, backup VP a is used for the recovery and likewise node 6's failure can be recovered by backup VP b.

Overlapping of backup VPs makes the protocol relatively complex as compared to the non-overlapped case. From complexity's viewpoint, the possible failure scenarios can be divided into two following groups:

    a. Non-overlapped zone failures,
    b. Overlapped zone failures.

Any failure in the non-overlapped zone is restored almost in similar fashion as in case of non-overlapped backup VPs and can be considered as a simple failure compared to the overlapped zone since there is always one backup VP to recover the failure. However, in the overlapped zone, each network element is protected by more than one VP. Therefore, a failure in this zone (e.g., node 5's failure) may start the restoration procedure independently from more than one backup VPs leading to a possibility of conflicts.

To resolve such a conflict overlapped zone is put under the responsibility of one of the two backup VPs as shown in Fig. 4. In this way, section 1 becomes 1–2–3–4–5–6 and section 2 becomes 6–7–8–9. This means that VP b does not have the same end nodes as those of the corresponding section 2. VP a will recover the failures detected by any node on section 1 of main VP and, likewise, VP b will do the same for section 2. Node 6 is a part of section 1 when the failure is on upstream side and of section 2 if the failure is on downstream side of the main VP.

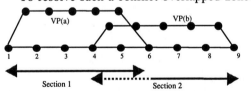

Fig.6. Overlapped Multiple Backup VPs Protecting the Main VP

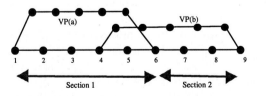

Fig.7. Overlapped Multiple Backup VPs With Non-Overlapped Sections

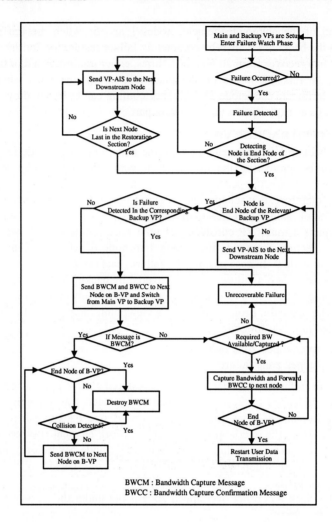

Fig.8. Flow Diagram for Restoration using MBV Protocol

Flow diagram of the restoration procedure using MBV protocol employing quasi-parallel bandwidth capturing method is shown in Fig.8.

3 Unidirectional Link Failure

Fig.9 illustrates how the network performs restoration from unidirectional single link (between nodes 3 and 4) failure. Fault-detecting node 4 generates and forwards AIS cells towards the end node 9. All the intermediate nodes 5, 6, 7 and 8 copy and then forward theses AIS cells. When the end node 9 receives AIS message about the failure, it sends a RDI message towards the upstream end node 1 through main VP so that the nodes upstream to the failure may know about the failure as well. Node 1 on

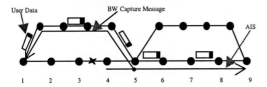

Fig. 9. Restoration from Unidirectional single Link Failure Using Two Backup VPs

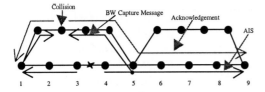

Fig. 10. Restoration from Bi-directional single Link Failure

arrival of such a message declares RDI-State and stops sending data until the arrival of a proper restoration message from other side via backup VP a. Node 5 on reading about the fault location from the AIS message takes up responsibility to perform restoration. Node 5 sends restoration messages towards node 1 via backup VP to capture the bandwidth for carrying the affected traffic. Node 1, on receiving the restoration message, switches from RDI to normal-state and starts sending data towards node 9 through new route. When data cells reach at the node 9, it switches from AIS to normal state as well. This completes the restoration.

4 Bidirectional Link Failure

In case of bidirectional failure node 3 and 4 detect the failure, thereby starting the restoration from both sides as depicted in the Fig.10. AIS message are sent towards respective end nodes for notification. Nodes 1 and 5 (which are in fact section's end nodes where failure has occurred) send restoration messages to the backup VP to capture the necessary bandwidth. Upon detection of the collision, acknowledgement messages are sent via section end nodes towards the VP end nodes to indicate successful bandwidth reservation on the backup VP so that data transmission can be resumed.

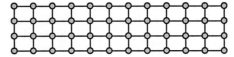

Fig. 11. Sample Network

Item	Value
Propagation Delay	0.25 msec/link
Failure Detection	20 msec
VP-AIS Processing	1 msec/node
VP-RDI Processing	1 msec/node
Bandwidth Capture	10 msec/node
Traffic Redirection	10 msec

Tab.1. Evaluation Parameters

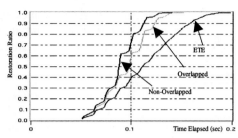

Fig.12. Comparison of Restoration Ratio Between ETE and MBV Protocols for Unidirectional Link Failure

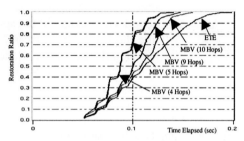

Fig.13. Restoration Ratio where VPs Longer than Certain Hop Length are Split into Two Sections

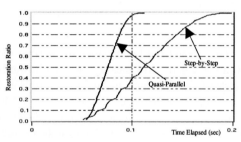

Fig.14. Restoration Ratio While Employing Step-by-Step and Quasi-Parallel Bandwidth Capture Method with ETE Protocol

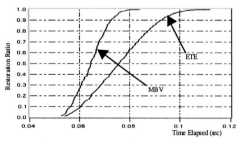

Fig.15. Comparison of ETE and MBV Protocols While Employing Quasi-Parallel Bandwidth Capture Method

5 Restoration Ratio

To quantify the performances of MBV protocol speed of restoration is used as criterion. Results show restoration ratio (capacity of restored VPs to capacity of failed VPs) as a function of time elapsed after failure. The restoration ratio results from all possible simulation cases of single link failure restored by ETE/MBV protocol for meshed topology shown in Fig.11 and using the parameters of Tab 1.

Unidirectional single link failure: Fig.12 shows gain in restoration speed compared to the ETE protocol. Here it is supposed that a VP longer than 6 hops is split into two sections. Fig.13 shows the effect on restoration ratio of variation in hop-length limit beyond which the VP is split into two sections. The results shown in Fig.12 and Fig.13 are observed while employing step-by-step bandwidth capturing approach on the backup VP. The gain in speed by using quasi-parallel bandwidth capturing approach is shown in Fig.14 for ETE protocol where as a comparison between ETE and MBV protocols is shown in Fig.15.

Bidirectional single link failure: As explained earlier bidirectional link failure is restored by restoration procedure triggered from both sides of the failed link and is likely to be restored faster as compared to the unidirectional link failure. Results similar to those for unidirectional link failure are presented for bidirectional link failure scenario

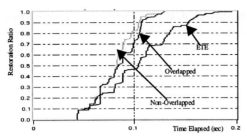

Fig.16. ETE vs. MBV Protocols for Restoration from a Bidirectional Link Failure

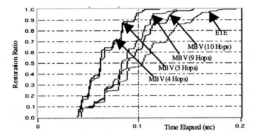

Fig.17. VPs Longer than Certain Hop Length are Split into Two Sections for Bidirectional Link Failure Scenario

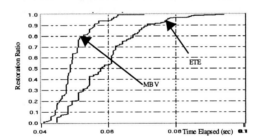

Fig.18. ETE vs. MBV Protocols While Employing Quasi-Parallel Bandwidth Capture Method

in self-explanatory Fig.16, Fig.17 and Fig.18. Hop-length limit for VP splitting is 9 for the results shown in Fig.16.

6 Conclusions

We have presented operation of a new self-healing protocol for ATM networks called multiple backup VPs based (MBV) protocol. This protocol is optimized for the restoration speed from a failure. Faster restoration is achieved by splitting the longer VPs into more than one section, thereby, keeping the restoration operation in reduced span around the failed element. The protocol can recover from all single link/node failures.

process. *Consolidation delay* arises when a BRM cell incurs additional delay at a branch-point because of consolidation. Usually a branch-point waits for a FRM cell from the upstream node or BRM cell(s) from downstream branch(es) before the consolidated BRM cell can be passed. *Consolidation loss* occurs from the excessive loss of BRM cells that are discarded at a branch-point in the consolidation process and thus resulting in the loss of information. It is desirable that the ratio of BRM cells to FRM cells at source should be one.

In this paper, we introduce a new approach for consolidation of BRM cells at branch-points, which utilizes the information contents of a BRM cell to decide whether to pass it or not and thus it avoids delay as well as loss of critical information.

2 Related Research-Work

A number of algorithms have been proposed in the literature for extending point-to-point ABR flow control to p-mp connections [2-7]. Various design options and implementation alternatives of existing algorithms can be found in [4]. In [2], consolidation algorithms have been classified based on (1) *feedback information storing method* for consolidation and, (2) *BRM cell returning condition*.

Most of the existing algorithms use per-VC accounting for storing feedback information. In per-VC accounting, a single set of variables is maintained at each branch-point of a VC to store the consolidated information. Whenever, a BRM cell returning condition occurs, a BRM cell is sent to the upstream node and these variables are refreshed to their default values. Depending upon what the BRM cell returning condition is, per-VC accounting may avoid the consolidation noise of type 1 but generally not of type 2. The reason is, in-between the time the variables are refreshed and the next BRM cell returning condition occurs, only the most constraining values are stored at a branch-point, and thus the consolidation does not immediately take note of the bandwidth constraints being relaxed at an earlier congested branch. Consequently, in [2], per-branch accounting for each VC has been proposed, where the consolidated information of the BRM cells is stored on a per-branch basis at each branch point of a connection. Using per-branch accounting, it is possible to keep up-to-date information of all branches at all the time and thus it can avoid consolidation noise of both types.

Now, we turn our focus on the BRM cell returning conditions, which can best be described by citing the consolidation algorithms. Roberts [5] proposed a consolidation algorithm a branch-point passes the consolidated information to its upstream node by generating a BRM cell immediately after receiving a FRM cell. Despite of its fast transient response, Roberts' algorithm suffers from high consolidation noise [4], because the sent BRM cell may not contain information even of a single branch. In order to reduce such noise, Tzeng [6] improved this algorithm by ensuring that a branch-point will return a BRM to its upstream node only when it receives at least one BRM cell from its branches after the last feedback was sent.

In [7], Ren et al have claimed that generating BRM cells leads to higher complexity at the switches and proposed two alternative algorithms. In the first algorithm, the first BRM cell, received after receiving a FRM cell, is passed to the upstream node. However, this algorithm exhibits similar consolidation noise as that shown by the scheme in [6]. In their second algorithm, which is known as 'wait-for-all' algorithm, Ren et al have proposed that a BRM cell should be passed only when a branch-point receives BRM cells from all of its branches. Since a branch-point has to wait for the BRM cells from all the branches, the waiting time may be long, resulting in very slow transient response. To alleviate the slow transient response problem, fast overload indication technique has been proposed in [4]. Although, the fast overload indication technique improves the transient response, it is unable to respond to an under-load condition where the source is entitled to increase its rate.

3 Proposed Algorithm

3.1 Responsibility of a Branch-Point

The fundamental responsibility of a branch-point is to notify the minimum cell-rate that can be supported by its all branches at any instant of time. To do so, it should send a BRM cell, whenever, any of the following two events takes place:
- The branch-point receives a BRM cell from one of its branches with lower ER value than the last ER value passed to the upstream node by this branch-point. This event implies that there is a new bandwidth constraint at the branch-point that must be made known to the upstream node, which may ultimately lead to reduction in the ACR of the source. Whenever, this event takes place, the branch-point will tag that branch as the bottleneck branch.
- The branch-point receives a BRM cell from the bottleneck branch with a higher ER value than that passed to the upstream node by this branch-point. This event implies that now the branch-point can support a higher throughput. In this case, the source may increase its ACR if all other branches of the multicast tree can support that.

Apart from these two cases, the BRM cells carry no additional information for a source to compute precise ACR value, and therefore, such BRM cells should be discarded. Thus, in our proposed algorithm, we focus on the above two simple events.

3.2 Proposed Algorithm

The key ideas in our basic algorithm are as follows:
1. Each branch-point stores the feedback information on a per-branch basis for each VC, and
2. A branch-point passes the BRM cell whenever any of the two conditions, as explained in Section 3.1, takes place.

Following variables are used in the implementation of the algorithm. Each branch-point maintains array of variables $MER[i]$, $MNI[i]$, and $MCI[i]$ to store the values of explicit rate, no increase indication and congestion indication, respectively, carried by the BRM cell of branch i. At each branch-point, a register $MER1$ stores the bottleneck ER value, and a pointer $PTR1$ points to the bottleneck branch, i.e., the branch that supports the minimum ER. Now we discuss the normal operation of the algorithm.

Whenever, a BRM cell is received at a branch-point from branch i, corresponding $MER[i]$, $MNI[i]$, and $MCI[i]$ values are updated, and the value of ER received, ER_{BRM}, is compared with that of the *MER1*, the current bottleneck value. If the ER_{BRM} is less than *MER1*, the branch-point discovers a new bottleneck ER, and thus that BRM cell is passed with the bottleneck ER value. The values of *MER1* and *PTR1* are assigned to ER_{BRM} and i, respectively.

In our algorithm, as the bottleneck branch is always tagged by the *PTR1* pointer, a BRM cell from branch *PTR1* with higher ER value signifies congestion relieve at the bottleneck branch, and thus the BRM cell is passed to its upstream node with min $MER[i]$. Consequently, the new bottleneck ER and the corresponding branch are assigned to the *MER1* and *PTR1*, respectively. In all other cases, the BRM cells are discarded at the branch-point. The Congestion Indication (CI) flag and the No Increase (NI) flag of a BRM cell also affect the computation of ACR at the source. So, our algorithm incorporates these two flags as well. The pseudo code of the algorithm is given below.

Upon receiving an FRM (ER, CI, NI) cell:
```
    Multicast the FRM cell to all participating
    branches at the branch-point.
```
Upon receiving a BRM (ER, CI, NI) cell from branch i:
```
    Let MER[i] = ER, MCI[i] = CI, and MNI[i] = NI;
    If (ER < MER1)
        MER1 = ER, PTR1 = i;
        Pass the BRM cell to the upstream node;
    Else if (( PTR1 == i) and (ER > MER1))
        Let MER1 = ER = min(MER[j]) , CI = or(MCI[j]),
        and NI = or(MNI[j]) for all participating
        branches j;
        Let PTR1 = j for which MER1 = = MER[j];
        Pass the BRM cell to the upstream node;
    Else Discard this RM cell.
```

At a branch-point in this algorithm, MER1 and *MER[i]* for all branches are initialized to PCR and the pointer *PTR1* is initialized to 0. A timer can be easily incorporated at each branch-point to detect whether or not the branch *PTR1* is responsive. If the timer finds the bottleneck branch *PTR1* non-responsive, that branch is cut off from the multicast tree and the first available BRM cell is passed with new bottleneck ER, and *MER1* and *PTR1* are updated accordingly.

4 Performance Evaluation

We compare the performance of our proposed algorithm with that of the wait-for-all algorithm since it has been incorporated in the ATM Forum recommendation [1].

4.1 Parameter Values

In simulation, the link length between the ABR end stations and their access switches is 10 Km, and each link has a bandwidth of 150 Mbps. All the sources are assumed to be persistent. The minimum cell rate (MCR), initial cell rate (ICR) and peak cell rate (PCR) for each ABR VC are 0.1, 3.0, and 150 Mbps, respectively. Other ABR parameters are identical for each VC and are selected as recommended in [1]. The switches are assumed to be adopting Explicit Rate Indication for Congestion Avoidance (ERICA) algorithm [9] with a target link utilization of 90%. Whenever a BRM cell is received at a branch-point, the received ER value is compared with the ER permitted by ERICA, and then the BRM cell is disposed or passed as decided by the consolidation algorithm.

4.2 Network Model I

Our first network model is selected to investigate the scalability of the consolidation algorithms in terms of number of hops and level of branching. Fig. 3.1 illustrates the Network Model I. The link distance between the switches S_1 and S_N is maintained at 1000 Km regardless of the value of N, which represents the levels of branching. In this model, p-mp ABR source is assumed to have 10 independent VCs with identical parameters so that the fairness among the VCs can also be observed. A point-to-point variable bit rate (VBR) VC is used to serve as the background traffic on the link connecting S_N and S_{N+1}, and thus it is the bottleneck link. The VBR source is periodically high (130 Mbps) and low (20 Mbps), each for 100 ms duration.

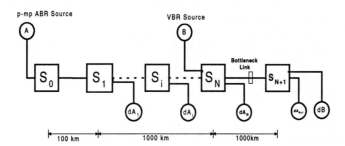

Fig. 1. Network Model I

In Fig. 2 and Fig. 3, we study the behavior of a single VC, which is randomly selected out of the 10 p-mp VCs. In Fig. 2, the value of N is 2, and in Fig. 3, it is 10. In the figures, the changes in the ACR of the selected VC are shown with respect to the changes in the VBR background load. Both the algorithms show smooth changes in the ACR with respect to the background load. However, the proposed algorithm shows faster response in comparison to the wait-for-all algorithm, which is more prominent when the value of N increases from 2 to 10. The reason is, the wait-for-all algorithm suffers from consolidation delay, which further increases with the increase in the number of hops. The slow response, in turn, limits the average throughput of the VC severely. Therefore, the wait-for-all algorithm is not scalable. On the other

cells to FRM cells close to unity at the source, we modify our algorithm so that the branch-point can pass the BRM cell with the current minimum ER value (*MER1*) if (*numBRMpassed/numFRMreceived*) < 0.90, even if the BRM cell does not contain any critical information. To guard against the possibility of feedback implosion, we incorporate *RMratio* (*numBRMpassed/numFRMreceived*) to determine the BRM cell returning condition in the following way. Whenever the ratio is greater than unity, ER < (*MER1 / RMratio*) is considered as the occurrence of new bottleneck and ER_{PTRi} > (*MER1 * RMratio*) signifies the congestion relieve at bottleneck link *i*, and a BRM cell is passed accordingly.

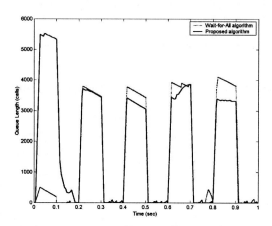

Fig. 5: Queue Length at Bottleneck Link in Network I, N = 10

Now we will focus on how to minimize the number of local variables to store the feedback information at a branch-point. In our algorithm, ER for each branch is stored so that whenever congestion is relieved on the bottleneck branch, the most constraining ER can be found from the stored variables. To reduce the number of variables at a branch-point, we can store just few lowest ER values with their corresponding branch pointers.

To illustrate the performance after modifications of our basic algorithm to cope with the first two problems and with a reduced number of variables at a branch-point, we consider Network Model II shown in Fig. 6. In the figure, each branch-point has 25 branches. Switch S_2 has 20 ABR destinations and it connects to 5 downstream switches, S_3 to S_7. The switches S_3 to S_7 have 25 ABR destinations each. Source A is a p-mp ABR source, and sources B to E provide on-off background traffic for each link from switch S_2 to other five switches (S_3 - S_7). The VBR sources are periodically high (130 Mbps) and low (20 Mbps) with 100 ms duration for each. The phase difference between two consecutive sources (B and C, C and D, and so on) is 10 ms.

In Fig. 7, the changes in the ACR of the p-mp ABR VC are compared of the per-branch accounting method with that of using reduced numbers of variables at branch-points. Although, the background load is not shown, the VBR source B starts transmission at time zero at high rate followed by the transmission at high rate by the source C after 10 ms, and so on. Therefore, the congestion at the link between switches S_2 and S_3 is relieved after 100 ms, the ACR does not increase before 150 ms due to congestion at the other links. So the changes in the ACR of the point-to-multipont VC is fully in accordance with the changes in the background traffic at different links. The ratio of the BRM cells to FRM cells at the source is 0.89. Instead of using 25 variables to store the ER values received from BRM cells at each branch-point, we used only 6 variables to store the lowest six ERs, which are extracted from the BRM cells received from the downstream nodes. In addition to this, six pointers are used to point to those six corresponding branches at each branch-point. The response to the changes in the background load using reduced numbers of variables is as fast as that with the per-branch accounting, though some glitches are introduced. The glitches occur at some of the instants when background load is changing. However, that may be tolerable compared to the benefits achieved using less number of variables at a branch-point, especially when there are a very large number of branches.

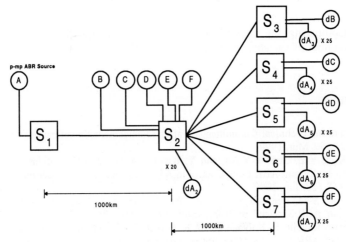

Fig. 6. Network Model II

5 Conclusions

There are two main contributions of this paper: (i) the fundamental responsibilities of a branch-point in a p-mp ABR connection have been clearly identified, and (ii) an efficient consolidation algorithm has been proposed.

In our basic algorithm, we adopted per-branch accounting to avoid consolidation noise of any types. Moreover, we used the information content of the BRM cells to decide whether to pass a BRM cell from a branch-point or not, and thus it is free from consolidation delay. We compared the performance of our basic algorithm with that of an existing wait-for-all algorithm. We found our algorithm is scalable in terms of number of levels of branching, has fast response, and it provides accurate feedback. The queue length at a bottleneck link is similar to the one achieved under the wait-for-all algorithm but a little bit higher initially due to the poor transient performance. The throughput of the bottleneck link is excellent, and the fairness among the VCs (point-to-point and p-mp) is superb.

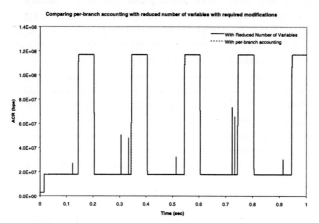

Fig. 7. Comparing the changes in ACR using per-branch accounting and using reduced number of variables with Network Model II

References

1. ATM Forum Technical Committee, "Traffic Management Specification Version 4.1", *af-tm-0121.000,* Mar. 1999.
2. D-H. Kim, Y-Z. Cho, Y-Y.An, and Y. Kwon, "A Scalable Consolidation Algorithm for Point-to-Multipoint ABR Flow Control in ATM Networks," in *Proceedings of ICC'99, vol. 1, pp. 118-123,* Jun. 1999.
3. Y-Z. Cho, S-M. Lee and M-Y. Lee, "An Efficient Rate-Based Algorithm for Point-to-Multipoint ABR Service," in *Proceedings of IEEE GLOBECOM'97,* Dec. 1997.
4. S. Fahmy, R. Jain, S. Kalyanaraman, R. Goyal, B. Vandalore, and X. Cai, "Performance analysis of ABR point-to-multipoint connections for bursty and nonbursty traffic with and without VBR background," *ATM Forum/97-0422,* Apr. 1997.
5. L. Roberts, " Rate based algorithm for point to multipoint ABR service," *ATM Forum/94-0772R1,* Nov. 1994.
6. H.-Y. Tzeng and K.-Y. Siu, "On Max-Min Fair Congestion Control for Multicast ABR Service in ATM," *IEEE Journal on Selected Areas in Communications, vol. 15, no. 3, pp. 545-555,* Apr. 1997.
7. W. Ren, K.-Y. Siu, and H. Suzuki, "On the performance of congestion control algorithms for multicast ABR service in ATM," in *Proceedings of IEEE ATM'96,* Aug. 1996.
8. R. Jain, S. Kalyanaraman, R. Goyal, S. Fahmy, and R. Viswanathan, "ERICA Switch Algorithm: A Complete Description," *ATM Forum/96-1172,* Aug 1996.

Experimental TCP Performance Evaluation on Diffserv Assured Forwarding over ATM SBR Service

Shigehiro Ano[1], Nicolas Decre[2,*], and Toru Hasegawa[1]

[1] KDDI R&D Laboratories Inc,
2-1-15 Ohara Kamifukuoka shi, Saitama 356-8502, Japan,
{ano,decre,hasegawa}hsc.kddilabs.co.jp
[2] Ecole Nationale Superieure des Telecommunications,
Paris, France,
decre@enst.fr

Abstract. In these days, the deployment of Diffserv (Differentiated Services) that enables the QoS guarantee is urgently required by the IP network customers. However, AF (Assured Forwarding) PHB (Per Hop Behavior) in Diffserv still has not been provided by conventional routers. It is a realistic solution that SBR3 (Statistical Bit Rate 3) of ATM emulates AF PHB, but it is not clear whether TCP traffic over AF PHB emulated by ATM is differentiated from the best effort TCP traffic over DF (Default Forwarding) PHB. To confirm the differentiation, we have experimentally studied TCP performance through the link into which TCP connections over AF PHB and DF PHB is aggregated. This paper describes the experimental results and discusses the possibility of the TCP performance differentiation between AF PHB and DF PHB over ATM.

1 Introduction

According to the growing demand for QoS guaranteed services over IP networks, the deployment of Differentiated Service (DiffServ)[1] is urgently required. Among the Diffserv PHBs (Per Hop Behaviors), EF (Expedited Forwarding) PHB[2] has been already supported by commercial routers. However, AF (Assured Forwarding) PHB[3] has been evaluated only by simulation studies[4][5]. Since AF PHB was not supported by commercial routers, it is an alternative solution to deploy AF PHB over ATM SBR3 (Statistical Bit Rate 3) [6] service with SCD (Selective Cell Discard) [7][8]. In order to evaluate the feasibility, we have experimentally evaluated the performances of AF PHB and DF (Default Forwarding) PHB over ATM services. The various experiments were performed to evaluate the differentiation of PHBs from the viewpoints of TCP throughput and fairness. The objective is to know whether TCP flow control mechanisms

---

* The author was engaged in this work through the internship at KDDI R&D Laboratories Inc.

can guarantee TCP throughput equal to the bandwidth provided by Diffserv AF PHB. The results show that Diffserv over ATM service is a practical solution for the early deployment of Diffserv services.

2 Diffserv AF PHB over ATM

2.1 Diffserv AF PHB

AF PHB is used to build an assured bandwidth end-to-end service without jitter/latency guarantee. Four classes are defined in terms of allocated network resources such as buffer space and bandwidth. Within each class, IP packets are marked with one of the three drop precedence values. The drop precedence values are changed by the policer that watches whether the traffic is conforming to the subscribed rate or not. Depending on the drop precedence values, the packets are scheduled to drop or queue in the congestion periods. The traffic contract of service implemented by AF PHB consists of assured packet rate and maximum burst size.

2.2 Mapping of PHBs and ATM Services

(1) AF PHB

It is a natural way to map AF PHB to SBR3 with SCD in ATM, as discussed by the ATM Forum[9]. In SBR3, a traffic contract consists of PCR for CLP (Cell Loss Priority) = 0+1 [PCR01], SCR (Sustainable Cell Rate) and MBS (Maximum Burst Size) values for CLP=0 [SCR0, MBS0].

The traffic contracts of AF PHB and ATM SBR3 are mapped in the following way: SCR0 and MBS0 correspond to "Assured Packet Rate" and "Maximum Burst Size", respectively. The CLP of SBR3 is mapped to AF drop precedence values. CLP=0 and CLP=1 correspond to high drop precedence value and medium / low drop precedence values, respectively.

The marking and dropping of AF PHB is emulated by ATM switches in the following way: When the queuing traffic exceeds the upper boundary ratio (SCD threshold) of ATM switch buffer size, SCD function starts the discard of cells with CLP=1 in advance to the cell discard by the buffer overflow. In case of no congestion, violation cells are only tagged to CLP=1.

(2) DF PHB

DF PHB is a best effort forwarding; so, it is mapped to ATM UBR (Unspecified Bit Rate) service.

3 Overview of Experiments

3.1 Testing Methods

1. To confirm the differentiation of TCP throughput level, we compare between AF PHB and DF PHB during the congestion periods.

2. PVP (Permanent Virtual Path) of SBR3 is used to transfer an aggregate of the traffic forwarded by AF PHB. A PVP of UBR is used to transfer an aggregate of DF PHB.
3. The above PVPs are concentrated on a trunk line between ATM WAN switches, and the congestion will occur at the output port to the trunk line where AF PHB is carried out.
4. Each PVP accommodates multiple TCP/IP connection.
5. The ATM traffic contract values of PVP for AF PHB are determined as follows:

PVP emulating AF PHB (SBR3 with SCD):

$$PCR01 = \text{``Trunk line speed''} \tag{1}$$

$$SCR0 = \frac{\text{``Trunk line speed''}}{\text{``Number of PVPs''}} \tag{2}$$

$$MBS0 \geq \sum_{i=1}^{\gamma} \lambda_i \tag{3}$$

Here,

$\gamma :=$ Number of accommodated TCP connections into this PVP connection

$\lambda_i :=$ Number of cells when the data corresponding to each TCP send/receive socket buffer size is consecutively transferred by ATM.

Equation (3) comes from the experimental results in reference [10].

3.2 Experimental Configuration

Figure 1 shows the configuration of the experiments. Eight PCs (Personal Computers: Pentium III 500MHz and Solaris 7) with an ATM NIC (Fore PCA-200) are used. These are connected to an ATM LAN switch (Fore ASX-200BX) and an ATM WAN switch (Fore ASX-200BX) via eight OC3-c lines. Each PC establishes TCP connections with different PCs. The number of TCP connections established by one PC is six and each PC pair has two TCP connections with the different TCP port numbers. Therefore, each of eight VPs which correspond to eight OC3-c lines has six TCP connections and the total number of TCP connections for this testing is 48. Each TCP connection is mapped to one VC. At the ATM LAN switch, the VCs with the same destination are switched into the same output OC3-c line. It should be noted that cell loss due to the buffer overflow does not occur at ATM LAN switch. At the ATM WAN switch, eight input lines are multiplexed into one OC3-c output line handled as the ATM

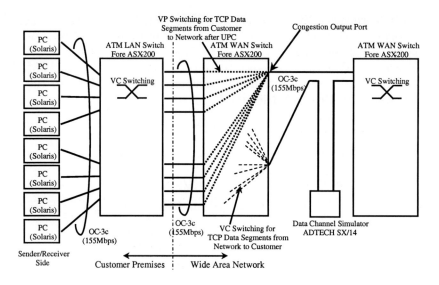

Fig. 1. Exprimental Configuration

WAN trunk line. The output buffer size for each PHB in the ATM WAN switch is set to 10000 cells. The data channel simulator (ADTECH SX/14) for insertion of propagation delay is connected. The VPs are maintained between the output ports of the ATM LAN switches and the input port of the second ATM WAN switch. VP level UPCs are performed at the first ATM WAN switch. We need to say that the second ATM WAN switch is introduced just because of the limitation of number of VPs supported by Fore ASX-200BX.

A free software module, ttcp, for TCP throughput measurement is used in the TCP/IP communication between PCs. It can calculate TCP throughput in the case that a greedy transmitter like ftp is used, by varying the values of various TCP parameters, such as TCP window size and the user data size. The TCP window size is set to 48 kbyte. The user data size and MSS (Maximum Segment Size) is fixed to 8192byte. Based on the principle described in section 2.1, ATM traffic contract values are set as follows:

$$PCR01 = 149.76 Mbit/s$$

$$SCR0 = 18.72 Mbit/s$$

$$MBS0 = 8256 cell$$

SCD threshold is fixed to 90% through the testing. During the ttcp execution, we also measure the packet queuing delay using 2048byte ICMP (Internet Control Message Protocol) packet by ping command over the route of TCP connections.

As for PCR shaping at the PCs, 35 Mbit/s including the cell header is adopted. The duration of each TCP throughput measurement is fixed to 180 seconds and the RTT (Round Trip Time) value is set to 20ms, 80ms or 160ms.

4 Results of TCP Performance Measurement

4.1 Differentiation between AF PHB and DF PHB

Under the configuration of Fig.1, we measured the throughput of each TCP connection under SBR3 with SCD for AF PHB and UBR for DF PHB. Two PVPs are devoted to AF PHB and other PVPs to DF PHB. Figure 2 shows each TCP throughput in the case of RTT = 80ms. TCP connections whose identifier are from #1 to #12 use PVPs for AF PHB and TCP connections whose identifier are from #13 to #48 use PVPs for DF PHB.

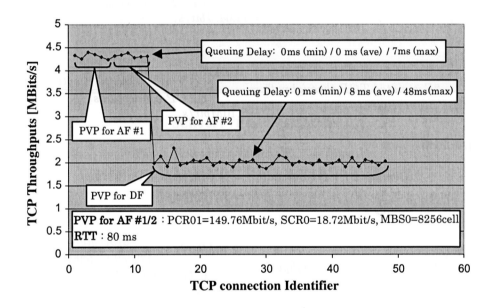

Fig. 2. Each TCP Throughput under PVPs for AF and DF PHBs

As shown in the figure, the AF and DF are differentiated from the viewpoint of TCP throughput. To analyze the throughput values quantitatively, estimated TCP throughput values for AF PHB and DF PHB are calculated based on the following assumption:

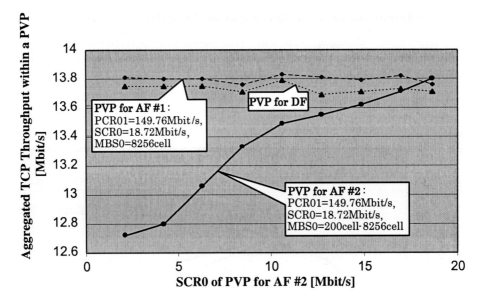

Fig. 4. Effect of SCR0 with RTT = 160ms

Table 1. TCP Throughput Aggregate in the Case of Various SCR0 Values

RTT	SCR0	18.72$Mbit/s$	14.84$Mbit/s$	10.60$Mbit/s$	6.32$Mbit/s$	2.12$Mbit/s$
80ms	Estimated	29.44$Mbit/s$	26.39$Mbit/s$	23.05$Mbit/s$	19.69$Mbit/s$	16.39$Mbit/s$
	Measured	25.72$Mbit/s$	25.52$Mbit/s$	24.01$Mbit/s$	21.23$Mbit/s$	14.33$Mbit/s$
160ms	Estimated	14.48$Mbit/s$	14.48$Mbit/s$	14.48$Mbit/s$	14.48$Mbit/s$	14.48$Mbit/s$
	Measured	13.80$Mbit/s$	13.62$Mbit/s$	13.49$Mbit/s$	13.06$Mbit/s$	12.72$Mbit/s$

Here,

$\mu :=$ "TCP send/receive socket buffer size set in PVP for AF#2"

$\delta :=$ "Data sending out time to ATM line"

$\alpha :=$ "Number of the accommodated TCP connections"

Table 1 shows the estimated values along with the measured values. In the case of RTT = 80ms, the measured values are almost the same as the estimated values below SCR0=14.84Mbit/s. This means that the TCP aggregate of AF #2 is allocated the fair share of the residual bandwidth. However, if the SCR0 becomes larger value than 14.84Mbit/s, the TCP aggregate of AF #2 cannot get the bandwidth equal to the fair share. The upper bound of SCR0 value becomes smaller value ($\simeq$10Mbit/s) in the case of RTT = 20ms. This is because the increase of SCR0 triggers the increase of the ATM switch buffer overflow that interferes to differentiate AF PHB from DF PHB by SCD function. In that sense, the upper bound of SCR0 value must be lowered in accordance with the increase of congestion level.

In the case of RTT = 160ms, the total bandwidth delay product of TCP connections is not always enough large to fulfill TCP data in the network. In other words, the link is not always utilized due to the small window sizes of TCP. In this case, the bandwidth of PVP whose SCR0 is smaller cannot get not only the subscribed rate but the residual bandwidth. This is because the number of discarded cells by SCR function increases according to the decrease of SCR0 value. Therefore, the lower bound of SCR0 value also be considered in the light congestion.

Based on the above discussions, it can be summarized that the differentiation to assured packet rates is realized within the range of certain SCR0 value. The upper/lower bound of SCR0 value is determined by the congestion level as described above.

4.2.2 Effect of MBS0

As for influence of MBS, no influences can be observed independent of the congestion level (RTT=20ms, 80ms, 160ms). Figure 5 shows the result. We can say that changing MBS0 values have no effect on the differentiation between AF PHB classes.

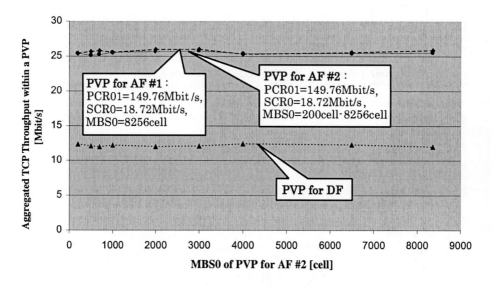

Fig. 5. Effect of MBS0 with RTT = 80ms

5 Conclusions

The goal of the above experiments was to study the differentiation between AF PHB and DF PHB over the ATM PVP services. We mapped AF PHB to SBR3 with SCD and DF PHB to UBR respectively. Using this mapping strategy, many experiments by altering the traffic contract values in the ATM WAN switches are carried out. The following results are obtained:

1. The TCP throughput differentiation between AF PHB and DF PHB over the ATM networks is realized during the congestion. ATM SBR3 with SCD can assign each TCP connection belonging to the PVPs for AF PHB the estimated TCP throughput. On the other hand, the average queuing delay of PVP for AF PHB is largely smaller than that for DF PHB.
2. The differentiation to assured packet rates can be realized under the following conditions. First, the upper bound of the setting rate for the fair share becomes smaller when the congestion level becomes heavier. On the other hand, the lower bound becomes larger when the congestion level becomes lighter. In any case, accommodated TCP aggregate must use window sizes large enough for the bandwidth delay product of the network.
3. The differentiation to maximum burst size cannot be realized independent of RTT value. It can be concluded that maximum burst size set for TCP aggregate has no effect on TCP throughput aggregate.

Consequently, ATM SBR3 with SCD is a practical way to realize the Diffserv Assured Forwarding using the commercial products.

References

1. S. Blake, D. Black, M. Carlson, E. Davies, Z. Wang, and W. Weiss, "An Architecture for Differentiated Services," IETF RFC2475, December 1998
2. V. Jacobson, K. Nichols, and K. Poduri, "An Expedited Forwarding PHB," IETF RFC2598, June 1999
3. J. Heinanen, F. Baker, W. Weiss, and J. Wroclawski, "Assured Forwarding PHB Group," IETF RFC2597, June 1999
4. D. Clark and W. Fang, "Explicit Allocation of Best Effort Packet Delivery Service," IEEE/ACM Transactions on Networking, Volume 1, Number 4, pp.397-413, August 1998
5. J. Ibanez and K. Nichols, "Preliminary Simulation Evaluation of an Assured Service," Internet Draft, ¡draft-ibanez-diffserv-assured-eval-00.txt¿, August 1998
6. ITU-T recommendation I.371, "Traffic Control and Congestion Control in B-ISDN," Geneva 1996
7. S. Ano, T. Hasegawa, and T. Kato, "A Study on Accommodation of TCP/IP Best Effort Traffic to Wide Area ATM Network with VBR Service Category Using Selective Cell Discard," IEEE ATM'99 Workshop, pp.535-540, May 1999
8. S. Ano, T. Hasegawa, and T. Kato, "An Experimental Study on Performance during Congestion for TCP/IP Traffic over Wide Area ATM Network Using VBR with Selective Cell Discard," IEICE Transactions on Communications, Volume E83-B, Number 2, pp.155-164, February 2000

9. O. Aboul-Magd et al., "Mapping of Diff-Serv to ATM Categories," The ATM Forum / 99-0093, 1999
10. S. Ano, T. Hasegawa, T. Kato, K. Narita, and K. Hokamura, "Performance Evaluation of TCP Traffic over Wide Area ATM Network," IEEE ATM'97 Workshop, pp.73-82, May 1997

PMS: A PVC Management System for ATM Networks

Chunsheng Yang and Sieu Phan

National Research Council of Canada, Canada
{Chunsheng.Yang, Sieu.Phan}@nrc.ca

Abstract. Reported in this paper is the developed PMS, a PVC management system for ATM networks. PMS provides a scalable, end-to-end path management solution required for managing today's complex ATM networks. It aims to assist the network operators to perform PVC operations with simplified procedures and automatic optimum route selection. It also aims to provide effective decision-making support for PVC fault identification and prevention to the network operators.

1 Introduction

ATM communication network is playing more and more important role in today's telecommunication networks. It has been widely used in backbone networks, transmission networks, access networks, and even enterprise networks. Such emerging large heterogeneous ATM networks have raised many new challenges for researchers and developers in the area of network management. In the management of ATM communication networks that have increased dramatically in size and complexity, the PVC (Permanent Virtual Circuit) management [8][9] is considered as one of the most important tasks. This task mainly consists of PVC operation, which includes path creation, path upgrade and path deletion; PVC fault identification; PVC fault correction; PVC fault prevention; and PVC QoS guarantee and service management. Existing COTS (Commercial Off-the-Shelf) software components provided by different ATM switch vendors can only provide partial support for operators to perform such task, and the procedure of PVC operation is disparate. It is difficulty for operator to perform automatic PVC operation and management. To assist the operator to perform automatic PVC operation and management in complex heterogeneous ATM networks, we developed PMS, a PVC management system for ATM networks based on our DPSAM (Distributed Proactive Self-Adjusting Management) framework [1]. DPSAM framework is generic framework that facilitates the incorporation of AI, distributed-computing, and web-based technologies in the development of network management system. The developing environment is Jess4.0, JDK1.1.6, OrbixWeb3.1, Apach1.3.0 Web-server for Solaris platform and Microsoft Peer Web-server for NT platform. PMS offers a simple mechanism for setting up Permanent Virtual Path Connections (PVPC) and Permanent Virtual Channel Connections (PVCC) with a point-and-click user interface. The PVC can be established automatically or manually according to user-specified requirement and QoS parameters such as throughput and delay. It also monitors and maintains the managed PVCs by performing real-time traffic data analysis, fundamental alarm correlation, and incident association. When PMS detects some problems or foresees some future anomalies, it will perform the necessary correction or prevention automatically whenever feasible. On the situations that automatic actions are not possible, PMS will notify network operator with detailed information such as the nature of the problems, the location where they occur, the reasons why they happen, and the procedures to correct or prevent them. The developed PMS is an ATM network management tool for service providers and enterprise network operators to effectively manage network resource, to provide good quality service, to improve network performance and to reduce downtime loss. During the research and development of PMS, we have focused on two main issues: PVC operation support with automatic optimum route selection and PVC fault identification and prevention. In this paper, the PMS system architecture and the implementation of PVC operation and management will be presented. This paper is organised as follows: Section 2 discusses the CORBA-based system architecture; Section 3 describes the PVC operations and management; Section 4 is on an experiment environment; and the final section concludes the paper.

2 CORBA-Based System Architecture

PMS is developed based on CORBA-based and three-tiered architecture. The architecture accommodates existing management protocol standards such as SNMPv1, SNMPv2, and CMIP, and uses CORBA as the underlying distributed middleware. The key characteristic of a three-tiered architecture is the separation of distributed computing environment into three layers: presentation, functionality, and data components. This is needed for building flexible, scalable, reusable, maintainable application. CORBA was chosen as the distributed middleware because it is a stable standard with mature products available. The Object Request Broker (ORB) provides a way to invoke methods on remote objects without necessarily knowing the location of those objects, or even their exact functionality. Thus, CORBA clients can manage distributed devices without explicit knowledge of the composition, size, or topology of the network. As shown in Fig.1, the top tier is the view layer. The middle tier is the service layer. The bottom tier is the data layer. For the developed PMS, these three tiers are implemented as follows.

View Tier. This tier is made up of user interface applications. These applications invoke the methods of the objects in the middle tier, which may be located in different locations and platforms. PMS View Tier provides web-based user graphical interface for operator to access and control ATM networks from remote locations by using Java-enable browser. These downloadable applications consist of the following user interfaces:
- PVC operation support;
- network representation and network element view;
- PVC status monitoring;
- decision-making support for fault management; and
- on-line knowledge base updates.

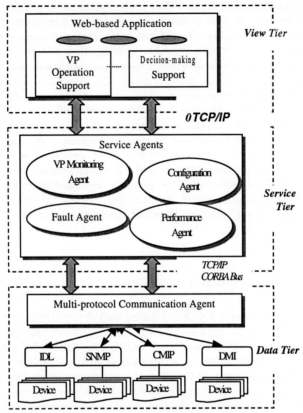

Fig. 1 A CORBA-Based Three-Tiered Architecture

Service Tier. It contains the service agents. The main service agents are configuration agent, fault agent, performance agent and PVC monitoring agent. They perform the PVC monitoring and management tasks that include the fundamental functionality recommended by the OSI and TMN. They provide service in response to the requests from the View Tier applications. These agents could be distributed on different locations and platforms, because they are designed to support CORBA IDL communication. These agents mainly perform the following PVC operation and management tasks:
- simplified PVC operations with automatic optimum path selection;
- fault identification and correction;
- fault prediction and prevention; and
- automatic fault correction and prevention.

Data Tier. This tier is usually made up of objects that interact with database management systems. Because PMS is developed to be able to manage ATM networks that contain multi-vendor equipment [10], this tier is made up of multi-protocol communication agents, which comprise different protocol objects such as SNMP, CMIP, and DMI. Multi-protocol communication agents perform all the interacting operations with device agents and cope with the requests from service agents to the device agents. From the viewpoint of devices, they map different protocol data objects to IDL data objects and delivers them to service agents. The details how to map the data between IDL and different protocols will be reported in a different paper.

3 PVC Operation and Management

The goals of PMS are to assist the network operator to perform PVC operation with simplified procedures and automatic path routing and selection, and to provide them with effective decision-making support for fault identification and prevention. In this section, we describe briefly the PVC operations, PVC fault identification, and PVC fault prevention.

3.1 PVC Operations

PVC operations and configuration are the principal PVC management tasks in ATM networks. The disparate and proprietary management procedure provided by individual ATM switch vendors has created an environment where it is very difficult to automate the process. The situation will become increasingly untenable in the future as networks continue to develop in size, intricacy and volatility. To solve this problem, Boyer et al [11] proposed to configure the PVC using intelligent mobile agents. Such approach requires that ATM switch can provide an environment to run mobile agents from elsewhere in the network. However, existing ATM switches almost provide SNMP or CMIP support for remote access. Considering such a situation, we are using distributed agents to access ATM switches sequentially by sending requests to each switch. Requests from CORBA clients are IDL data package. Service agent, configuration agent in PMS, will determine the PVC operation sequence (scenario), pack the operation scenarios and pass them to multi-protocol agent. According to operation scenario, multi-protocol agent will determine the action commands to perform PVC operation corresponding to different switches, which support different communication protocols such as SNMPv1, SNMPv3, or CMIP. This methodology is shown in Fig.2. In PMS, the system can perform PVC creation, PVC upgrade, and PVC deletion. PVC creation is to set up a path across a set of ATM switches within an ATM network; PVC upgrade is to route path or negotiate the parameter for existing path; and PVC deletion is to release a path across a set of ATM switches within an ATM network.

In this study, we have two objectives for providing PVC operations to operators. The first goal is to provide the simplified procedures to assist operators to perform PVC operations. Using such procedures, operators do not need to worry about the switch specification, operation environments, locations on network, and so on. What they need to do is to specify the node name, the name of group, VPI and VCI for each connected switch. All procedures for PVC operation will be hidden in the system.

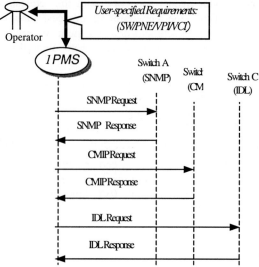

Fig. 2. The Methodology of PVC Operations

The second goal is to assist operators to perform PVC operation with the support of automatic path routing and path selection. In a way, operators do not need to determine the path route by themselves from the available resource, which contains thousands of VPI and VCI parameters and network topology. To this end, constraint-based reasoning is used to determine path route and to select an optimal path for the requested PVC path operations. The details how to automatically route and select path will be reported in a different paper due to space limitation.

3.2 PVC Fault Identification

PVC fault identification and prevention are the main PVC management tasks in PMS. To reduce the human error and misunderstanding in the process of fault identification, a knowledge-based approach is considered as one of the most effective approach [2][5][6][7]. There have been a number of achievements in applying knowledge-based approach to alarm correlation. However, these knowledge-based alarm correlation systems can only reduce the amount of alarms, it cannot detect the problem from the viewpoint of PVC management requirement, and it also cannot help operators to make decision for PVC fault identification and fault prevention. Therefore, the operator still needs to make decision for PVC fault correction and prevention by himself. To assist the operators to effectively manage the PVC, we developed a knowledge-based system to identify the PVC problems and make decision on fault correction. This knowledge-based system is implemented by using Jess4.0. The knowledge base consists of DPSAM knowledge, PMS system knowledge, Generic PVC management knowledge, and vendor's ATM specification knowledge. As shown in Fig. 3, PHB, PSB, PC and PO are defined as PVC fault problems [1]. They stand for path hardware break, path software break, path congestion, and path overload, respectively. IS_{r_b}, $IS_{r_{sb}}$, IS_{r_c} and IS_{r_o} are incident sets defined for each problem. PVC fault identification comprises four main inference procedures: data collection, fault detection, fault isolation, and fault correction. They are described as follows.

3.2.1 Data Collection

Fault is a disorder occurring in the hardware or software of the managed ATM switches, or is caused by network traffic density and path routing. Alarm events are external manifestations of the faults. Alarm events are defined by ATM vendors and generated by ATM equipment and they are observable to network operators. In PMS, alarm events are the most important data that must be collected. Alarm events are very useful for detecting PHB and PSH problems. In order to effectively detect path congestion and path overload, it is necessary to collect traffic data for the monitored PVCs. Another reason why we need traffic data is that alarm events might be lost and not real time because they are sent to management system via notification. Consequently, we collect the alarm events from all the managed ATM switches and the traffic data for the monitored PVCs.

3.2.2 Fault Detection

The task of fault detection is to find out the symptoms for PVC problem identification from the collected alarm events and traffic data. According to the definition of the incident, fault detection is to generate the corresponding incidents by using fundamental alarm correlation and traffic data analysis; then to associate the incidents and open an incident set for the PVC problems. The following is the description of alarm correlation, traffic data analysis and incident association.

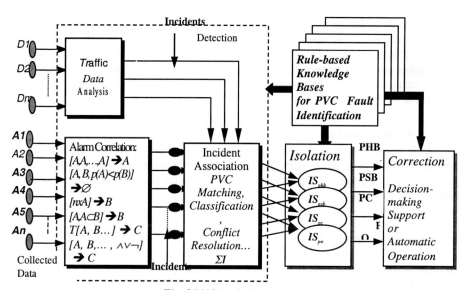

Fig. 3 PVC Fault Identification

3.2.2.1 Alarm Correlation

In the network management area, alarm correlation is often used to aid the operator to diagnose network faults by reducing the amount of alarms. In PMS, besides reducing

π_{loss} can be obtained as

$$\pi_{loss} = 1 - \frac{1 - q\pi_0}{pD}. \tag{23}$$

By solving for π_{loss} from Eqs. (21) and (23), the formula for loss probability (π_{loss}) can be found as

$$\pi_{loss} = \frac{1 - pD}{\frac{q}{\sum_{i=1}^{D-1} \sigma_i} - pD}. \tag{24}$$

From Eqs. (21) and (24), π_0 can be derived

$$\pi_0 = \frac{\pi_{loss}}{\sum_{i=1}^{D-1} \sigma_i}$$

$$= \frac{1 - pD}{q - pD \sum_{i=1}^{D-1} \sigma_i}. \tag{25}$$

The value $\sum_{i=1}^{D-1} \sigma_i$ that appeared in the Eqs.(24) and (25) will be derived in the next section.

3 Buffer-Size Approximation

We first find the value $\sum_{i=1}^{D-1} \sigma_i$ in terms of K. By Cramer's rule, we can obtain the values σ_i from Eq. (18) in the form

$$\sigma_j = \frac{M(j)}{M} \tag{26}$$

where

$$M(j) = \begin{bmatrix} 1 & z_1 & \cdots & z_1^{j-2} & qz_1^{-r} & z_1^j & \cdots & z_1^{D-2} \\ 1 & z_2 & \cdots & z_2^{j-2} & qz_2^{-r} & z_2^j & \cdots & z_2^{D-2} \\ 1 & z_3 & \cdots & z_3^{j-2} & qz_3^{-r} & z_3^j & \cdots & z_3^{D-2} \\ \vdots & \vdots & & \vdots & \vdots & \vdots & & \vdots \\ 1 & z_{D-2} & \cdots & z_{D-2}^{j-2} & qz_{D-2}^{-r} & z_{D-2}^j & \cdots & z_{D-2}^{D-2} \\ 1 & z_{D-1} & \cdots & z_{D-1}^{j-2} & qz_{D-1}^{-r} & z_{D-1}^j & \cdots & z_{D-1}^{D-2} \end{bmatrix},$$

and

$$M = \begin{bmatrix} 1 & z_1 & z_1^2 & \cdots & z_1^{D-2} \\ 1 & z_2 & z_2^2 & \cdots & z_2^{D-2} \\ 1 & z_3 & z_3^2 & \cdots & z_3^{D-2} \\ \vdots & \vdots & \vdots & & \vdots \\ 1 & z_{D-2} & z_{D-2}^2 & \cdots & z_{D-2}^{D-2} \\ 1 & z_{D-1} & z_{D-1}^2 & \cdots & z_{D-1}^{D-2} \end{bmatrix},$$

and $r = KD + 1$.

If we let $M_{ik}(j)$ be the ik^{th} major of matrix $M(j)$, and M_{ik} be the ik^{th} major of matrix M, then σ_j can be written as

$$\sigma_j = \frac{\sum_{i=1}^{D-1} M_{ij}(j)\beta_i}{\sum_{j=1}^{D-1} M_{ij} z_i^{j-1}} \tag{27}$$

where $\beta_i = q/z_i^r$.

The sum of σ_j can immediately be derived from Eq. (27) as

$$\sum_{j=1}^{D-1} \sigma_j = q \sum_{j=1}^{D-1} \frac{\sum_{i=1}^{D-1} M_{ij}(j) z_i^{-(KD+1)}}{\sum_{l=1}^{D-1} M_{kl} z_k^{l-1}} \tag{28}$$

$$= q \sum_{i=1}^{D-1} \frac{(\sum_{j=1}^{D-1} M_{ij}(j)) z_i^{-(KD+1)}}{\sum_{l=1}^{D-1} M_{kl} z_k^{l-1}} \tag{29}$$

$$= q \sum_{i=1}^{D-1} \frac{(\sum_{j=1}^{D-1} M_{ij}) z_i^{-(KD+1)}}{\sum_{l=1}^{D-1} M_{kl} z_k^{l-1}}, \tag{30}$$

where Eq. (30) comes from the fact that $M_{ij}(j) = M_{ij}$. Note that $\sum_{l=1}^{D-1} M_{kl} z_k^{l-1}$ is the determinant of the Vandermonde matrix M in which its determinant does not depend on k or l.

Since M is the Vandermonde matrix, we can rewrite Eq. (30) as

$$\sum_{i=1}^{D-1} \sigma_i = q \sum_{i=1}^{D-1} \frac{(\prod_{k \neq i}(1 - z_k)) z_i^{-(KD+1)}}{\prod_{k \neq i}(z_i - z_k)}$$

$$= -q \sum_{i=1}^{D-1} \frac{F'(1) z_i^{-(KD+1)}}{F'(z_i)}$$

$$= q \sum_{i=1}^{D-1} \frac{(1 - pD) z_i^{-(KD+1)}}{pD z_i^{(D-1)} - 1}. \tag{31}$$

Representing the $\{\sigma_i\}$ parts in loss probability expression by Eq.(31) therefore gives

$$\pi_{loss} = \frac{1 - pD}{\frac{1}{(1-pD)\sum_{i=1}^{D-1} \frac{z_i^{-(KD+1)}}{pDz_i^{D-1}-1}} - pD} \tag{32}$$

The buffer size can be derived by expressing the variable K in Eq. (32) in terms of other variables. However, it is impossible to derive the expression for buffer size (K) directly from Eq. (32). Therefore we simplify this expression by approximating the loss probability from the obtained loss probability expression. The approximation is done by leaving only the dominant root in the sum of the denominator part of Eq. (32).

$$\pi_{loss} \simeq \frac{1-pD}{\frac{z_{D-1}^{KD+1}(pDz_{D-1}^{D-1}-1)}{1-pD} - pD},\qquad(33)$$

where z_{D-1} is the root other than 1 of Eq. (16) with the least absolute value. Since pD is less than 1, z_{D-1} is therefore real and positive.

We thus compare the loss probability expression obtained from the approximation formula with that from the exact formula as shown in table 1.

Table 1. Loss Probability Approximation

D	K	p	loss	loss(approx)
5	3	0.19	0.08501909	0.08501909
5	21	0.19	0.00355178	0.00355178
5	210	0.19	1.3430416E-13	1.3430416E-13
5	2100	0.19	1.5421715E-117	1.5421715E-117
5	9100	0.19	1.6800793E-502	1.6800793E-502

As it is shown in table 1, the loss probability approaches zero as K tends to infinity.

We can see from the above table that approximation by using Eq. (33) was very good.

From Eq. (33), we can therefore approximate the value K in the form of Buffer size (D) and arrival probability (p) by using Eq. (34),

$$K = \lfloor \frac{1}{D} \log_{z_{D-1}} \frac{(1-pD)[pD + \frac{1-pD}{\pi_{loss}}]}{[pDz_{D-1}^{D-1} - 1]z_{D-1}} \rfloor,\qquad(34)$$

where $\lfloor x \rfloor$ is the function that gives the maximum integer value less than or equal to x.

4 Numerical Results

In this section, we compute the loss probability first from the specified K and D and then approximate the buffer size by the formula in Eq.(34). The results are as shown below.

As we can see from tables 2 and 3, the approximation for K was very good even when loss probability is infinitesimally small.

Table 2. Approximation of Buffer Size for D = 11

D	K	p	π_{loss}	K(approx)
11	3	0.08	0.0733298	2
11	23	0.08	1.975355E-4	23
11	230	0.08	4.240009E-29	230
11	2300	0.08	8.930368E-276	2300
11	5000	0.08	1.580943E-597	4999
11	7500	0.08	1.907397E-896	7499
11	10000	0.08	2.301261E-1192	9999
11	15000	0.08	3.349775E-1789	14999

Table 3. Approximation of Buffer Size for D = 12

D	K	p	π_{loss}	K(approx)
12	3	0.08	0.103686	2
12	23	0.08	0.0057816	22
12	230	0.08	5.780803E-11	229
12	2300	0.08	2.118674E-90	2300
12	5000	0.08	5.175661E-194	5000
12	7500	0.08	5.982012E-290	7500
12	10000	0.08	6.9139884E-386	9999
12	20000	0.08	1.2338269E-769	19999

5 Conclusion

We proposed the formula for approximating the buffer size of the Geo/D/1/K queueing system when loss probability is given. From the buffer-size approximation formula, our results showed that it gave a very good approximation As we further our study, we will apply our method to the discrete-time queueing systems with more complicated arrival processes such as MMBP (Markov Modulated Bernoulli Process).

References

1. G. Pujolle and H.G. Perros, "Queueing Systems for ATM networks modelling," Proc. Conf. on the Performance of Distributed Systems and Integrated Communication Networks, Tokyo, Japan, 1991.
2. G.L. Wu and J.W. Mark, "Discrete time analysis of leaky-bucket congestion control," Computer Networks and ISDN systems 26 (1993), pp. 79-94.
3. A.Gravey, J.R. Louvion and P. Boyer, "On the Geo/D/1 and Geo/D/1/n Queues," Performance Eval. 11, pp. 117-125, 1990.
4. H. Takagi, Queueing Analysis. North-Holland, 1993.
5. I. Khan and V.O.K. Li, "Traffic control in ATM networks," Computer Network and ISDN Systems 27, pp. 85-100, 1994.

Client-Server Design Alternatives:
Back to Pipes but with Threads

Boris Roussev[1] and Jie Wu[2]

[1] Information Systems Department, Susquehanna University, 514 University Avenue,
Selinsgrove, PA 17870, USA
roussev@roo.susqu.edu

[2] Department of Computer Science and Engineering, Florida Atlantic University,
Boca Raton, Fl 33431, USA
jie@cse.fau.edu

Abstract. In this paper we set out to theoretically explore and experimentally compare different client-server design alternatives implemented in Java. We introduce a new concurrent data structure, called concurrent hash table, for solving the synchronization problem in the classical producer/consumer model. The structure allows multiple reads and a single write to proceed concurrently. We look at the following TCP server designs: concurrent server–new thread per client; pre-threaded servers: locking around accept; socket passing through a shared buffer; socket passing through a concurrent queue; socket passing through a concurrent hash table; socket passing through pipes. The servers have been tested on a network of 35 workstations. The experimental results have shown that the server using pipes to pass tasks to the workers outperforms every other one. For all servers, better performance is achieved by using a number of worker threads in the range of one hundred rather than fifteen as commonly recommended.

1 Introduction

The third industrial revolution is all about what George Herbert Wells envisaged at the dawn of last century as "The World Brain." Today, we see the prophecy come true. The neurons of this brain constitute the interconnection of networked computers called the Internet where the processing elements interact with each other using the client-server pattern. In this paper we set out to explore and experimentally compare different client-server design alternatives implemented in Java. The application programming interface used is "Berkeley sockets" [9]. Threads, an abstract compute model, are used to structure the concurrent activities in the server programs.

The performance and correctness of the server programs is of crucial importance to the success of many network applications and distributed systems. With the explosion of the WWW, busy Web servers measure the number of connections per hour in the hundreds of thousands. Furthermore, many of these servers interact with backend database servers, which in turn have to process an even greater workload. In other words, if we are to build robust Internet applications materializing Wells's metaphor, more than ever before the network applications should be based on efficient client-server architectures. Today, many new technologies, like the Java Virtual Machine (JVM), multiprocessor and multithreaded kernels have matured. In JVM the time required to spawn a new thread or to obtain an object's lock in most implementations is negligible. The use of multiprocessor computing machines is a norm. Mapping of threads to processors has been optimized tremendously. As a result, designs that have been impossible until recently become viable propositions.

Java [4], an object-oriented language, has become popular because of its platform independence and safety. It has greatly simplified network programming [5] by providing elegant TCP/IP API, object serialization, network class loading (code mobility), remote method invocation, Servlets and built-in concurrent constructs. This along with its phenomenally growing popularity entails a rapidly expanding body of projects: Atlas, Charlotte, Javelin, JPVM, Globus, IceT, JavaSpaces, MPIJ, Bayanihan to mention but a few, that use Java as a language for high performance computing on networks of workstations [11]. The Syracuse workshop [3] discussed Java's possible role as the basic programming language for science and engineering–taking the role now played by Fortran 90 and C++–and concluded that Java could become dominant by adding the necessary functionality to the basic Web loosely coupled distributed model.

In this paper we compare the following TCP server designs: (1) iterative server; (2) concurrent server–new thread per client; (3) pre-threaded server with locking (mutex) around accept; (4) pre-threaded server–connected socket passing through a shared buffer; (5) pre-threaded server–connected socket passing through a concurrent queue [8]; (6) pre-threaded server–connected socket passing through a concurrent hash table; and (7) pre-threaded server–connected socket passing through pipes.

For all the server architectures, we evaluate experimentally parameters taken for granted for quite a long time, for example, the number of worker threads a server should spawn, the buffer capacity of the shared buffer and techniques for synchronizing the work of the thread accepting the connections and the threads carrying out the service requests. To evaluate the servers, we run the same client on 35 hosts on the same subnet on which the server being evaluated is running. Each client spawns between 4 and 20 child clients to create multiple simultaneous connections to the server, for a maximum of 700 simultaneous connections. We consider the effect of having too many/few threads.

The remainder of the paper is structured as follows. In Section 2, we review the TCP/IP protocol stack and Java concurrent and multithreading constructs. In Section 3, we describe the architecture and the implementation of the server designs. Next, in Section 4 we give theoretical analysis of the performance of the concurrent hash table and synchronization using pipes. Then, in Section 5 we present experimental results. In the final section we outline plans for future work and conclude.

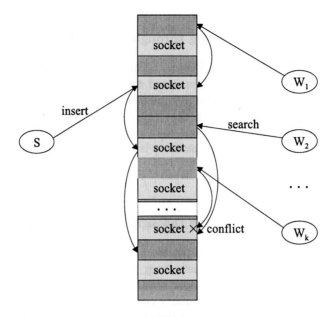

Fig. 5. Concurrent hash table. The main server thread inserts connected sockets in the table. The prethreaded worker threads retrieve these sockets

thread can access any slot without possessing a synchronization lock as a precondition. We have two different cases to consider corresponding to the methods insert and search: (i) The main server thread calls accept() and then inserts the returned connected socket in the table using the insert method of the concurrent hash table object. The insert method uses the hashCode() method of the Object class to compute the key of the connected socket and double hashing [1] to compute the slot in the table. (ii) Worker threads retrieve connected sockets from the table using the search method of the concurrent hash table object. They pass as an argument to the search method a pseudo randomly generated number. This number is used to calculate the first slot of the table to be checked for a connected socket. If the slot is not null, the connected socket is retrieved and the slot is set to null. Otherwise, the next slot in the probe sequence is calculated and checked. If a collision occurs, i.e. two or more workers access the same slot, because the table slots are objects of type SynchronizedRef only one of the workers will retrieve the connected socket and set the slot to null. The rest will see the null value and try to retrieve a connected socket from other slots.

3.6 Pre-threaded Server, Connected Socket Passing Through Pipes

The final modification, shown in Figure 6, of the pre-threaded server gets around the need for using a common buffer and synchronization between the main thread and the worker threads. The main server thread calls accept(). It keeps track

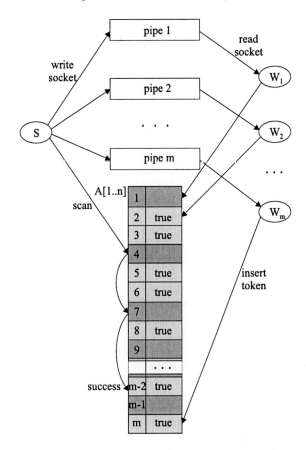

Fig. 6. TCP pre-threaded server, connected socket passing through pipes

of the worker threads being free to pass a new connected socket to a free worker through a pipe. We had to write our own pipe class since class `Socket` does not implement the `Serializable` interface and therefore objects of this class cannot be passed through Java communication pipes. When a new request arrives, the main server thread finds the first available free worker by scanning the array of `WorkerStatus` elements denoted as $A[1..n]$ in Figure 6 and passes the socket to that worker through its own pipe. Being finished with a client, the worker thread changes the status of its pipe back to ready state by writing `true` in its `WorkerStatus` element.

4 Theoretical Analysis

4.1 Analysis of Concurrent Hashing

Given a concurrent hash table T with m slots that stores n elements, we define the load factor α for T as n/m, $\alpha \leq 1$. We make the assumption of *uniform hashing*: each key considered is equally likely to have any of the $m!$ permutations of $\{0, 1, \ldots, m-1\}$ as its probe sequence. In our implementation we use double hashing which is a suitable approximation to uniform hashing. Double hashing uses a hash function of the form

$$h(k, i) = (h_1(k) + ih_2(k)) \bmod m$$

where h_1 and h_2 are auxiliary hash functions. The initial position probed is $T[h_1(k)]$; successive probe positions are offset from previous positions by the amount $h_2(k)$ modulo m. The value of $h_2(k)$ must be relatively prime to the concurrent hash table size m for the entire concurrent hash table to be searched. Otherwise, if m and $h_2(k)$ have greatest common divisor $d > 1$ for some key k, then a search for key k would examine only $1/d$th of the table.

Theorem 1. *Inserting an element into a concurrent hash table with load factor $\alpha < 1$ requires at most $1/(1-\alpha)$ probes on average, assuming uniform hashing.*

Proof Inserting an element requires an unsuccessful search followed by the placement of the element in the first empty slot found. In an unsuccessful search, every probe but the last accesses an occupied slot, and the last slot probed is empty. Let $p_i = \Pr\{\text{exactly } i \text{ probes access occupied slots}\}$ for $i = 0, 1, 2, \ldots$. For $i > n$, we have $p_i = 0$, since we can find at most n slots already occupied. Thus the expected number of probes is

$$1 + \sum_{i=0}^{\infty} i p_i \tag{1}$$

To evaluate (1) we define $q_i = \Pr\{\text{at least } i \text{ probes access occupied slots}\}$ for $i = 0, 1, 2, \ldots$. Since i takes on values from the natural numbers

$$\sum_{i=0}^{\infty} i p_i = \sum_{i=1}^{\infty} q_i$$

The probability that the first probe accesses an occupied slot is n/m. Thus

$$q_1 = \frac{n}{m}.$$

A second probe, if necessary, is to one of the remaining $m - 1$ unprobed slots, $n - 1$ of which are occupied, thus,

$$q_2 = \left(\frac{n}{m}\right)\left(\frac{n-1}{m-1}\right).$$

The ith probe is made only if the first $i - 1$ probes access occupied slots. Thus,

$$q_i = \left(\frac{n}{m}\right)\left(\frac{n-1}{m-1}\right)\cdots\left(\frac{n-(i-1)}{m-(i-1)}\right) \le \left(\frac{n}{m}\right)^i = \alpha^i,$$

since $(m-n-j)/(m-j) \leq (m-n)/m$ when $m-n \leq m$ and $j \geq 0$. Now, we can evaluate (1).

$$1 + \sum_{i=0}^{\infty} ip_i = 1 + \sum_{i=1}^{\infty} q_i \leq 1 + \alpha + \alpha^2 + \alpha^3 + \ldots = \frac{1}{1-\alpha}$$ ∎

If α is a constant, Theorem 1 predicts that inserting an element runs in $O(1)$ time.

Theorem 2. *In a concurrent hash table with load factor $\alpha = n/m < 1$, the expected number of probes in a successful search is at most $1/\alpha$.*

The proof is similar to that of Theorem 1. If α is a constant, Theorem 2 predicts that searching an element runs in $O(1)$ time. For example, if we have a table with $\alpha = 0.95$, then the average search will take 1.05 probes, meaning that when the server is overloaded, a worker thread can quickly retrieve a connected socket and service a pending request.

4.2 Analysis of Pipe Synchronization

Here, the roles of producer and consumer are reversed w.r.t. the concurrent hash table. The worker threads insert synchronization tokens (**true** values) while the main server thread searches for an empty slot to find a free worker. Similarly to hash table, given an array $A[1..m]$ with m elements n of which are set to **true**, we define the load factor β for A as n/m, $\beta \leq 1$.

Proposition 1. *Insertion of **true** by a worker thread takes $O(1)$ time.*

Theorem 3. *Assuming uniform hashing, searching for an element into array $A[1..m]$ with load factor $\beta < 1$ requires at most $1/\beta$ probes on average by the server thread S.*

The proof is similar to that of Theorem 1. If β is a constant, Theorem 3 predicts that searching an element runs in $O(1)$ time.

5 Experimental Results

To evaluate the server design alternatives, we run the same client on 35 hosts running Windows NT against each server, measuring the server wall-clock time required to process a fixed number of requests. We summarize all our CPU timings in Table 1. The readings recorded in column 3 (column 4) correspond to 250 milliseconds (500 milliseconds) server delay before sending back the response. Each client spawns 4 child clients to create 4 simultaneous connections to the server, for a maximum of 140 simultaneous connections at the server at any time. Further, each client makes 20 connections to the server amounting to 2800 connections altogether. For the tests involving multithreaded servers, the server

Table 1. Timing comparisons of the various server designs

#	Server description	CPU time	CPU time
1	Iterative server	700000	-
2	Concurrent server, one thread per child	12552	21499
3	Locking around accept with 50 threads	21153	33602
4	Shared buffer with 50 threads, 101 capacity	16562	28960
5	Concurrent queue with 70 threads	13048	-
6	Concurrent hash table with 50 threads, 101 capacity	14562	24265
7	Pipes with 150 threads	13195	19253

creates up to 200 worker threads when it starts. We consider also the effect of having too many/few threads. The clients send 1 byte to the server, which responds with 4000 bytes after waiting for a predefined interval of time specified in milliseconds. The number of worker threads and the buffer capacities recorded in the second column of Table 1 were found experimentally to give the best results for the corresponding server design. Table 2 shows how we have arrived at the figures for the server using a concurrent queue.

The small CPU time obtained for the concurrent server indicates that spawning a new thread may be less expensive than synchronizing a great number of pre-spawned threads. The smallest CPU time is for the server using pipes. In this design, not only is there no synchronization among the worker threads, but there is no additional cost for spawning worker threads. We were interested to find out the threshold after which the server using pipes denies service, i.e., the effect of having too many threads to synchronize. We ran experiments with up to 700 simultaneous clients, and the server was still providing a satisfactory service.

Table 2. Effect of threads number on the performance of the pre-threaded concurrent server using concurrent queue as a shared buffer

# threads	10	30	40	50	60	70	80
CPU time	70312	23781	20828	17549	17047	13048	18203

6 Conclusion

In this paper we compared the performance of seven server designs by running them against the same client. The experimental results show that only one pre-threaded server, the one using pipes, outperforms the classical concurrent server where the server spawns a new thread to handle the client connection. This leads us to the conclusion that spawning a new thread is less expensive than synchronizing a great number of threads. We introduced a new concurrent data structure called concurrent hash table. Although theoretically sound, in practice the performance of the concurrent hash table is not so good as the performance of the server using pipes. We also considered the number of worker threads the server should spawn in order to get maximum performance. We found out that better performance is achieved by using a greater number of worker threads, in the range of 100 rather than 15 as is commonly recommended.

We plan to run more experiments to fine-tune the behavior of the concurrent hash table. We are going to test this data structure on a multiprocessor and to gather data about the number of collisions. Similarly to hash tables this concurrent data structure is highly sensitive to the choice of the hash function and the capacity of the table. Theoretically, this server design should give good performance when the load factor α approaches 1.

References

1. T. Cormen et al., *Introduction to Algorithms*, The MIT Press, 1994.
2. S. Deering and R. Hinden, "Internet Protocol, Version 6 (IPv6) Specification," RFC 2460, 1998.
3. G.C. Fox and W. Furmanski, "Java for parallel computing and as a general language for scientific and engineering simulation and modeling," *Concurrency: Theory and Practice*, Vol 9(6), pp.415-425, 1997.
4. J. Gosling, B. Joy, and G. Steele, *The Java Language Specification*, Sun Microsystems, Inc., Palo Alto, CA, 1996.
5. E.R. Harold, *Java Network Programming*. O'Reilly & Associates, 1997.
6. C.A.R. Hoare, "Monitors, An Operating System Structuring Concept," *Communication of the ACM*, Vol.17, pp.549-557, Oct. 1974; Erratum in *Communication of the ACM*, Vol.18, p.95, Feb. 1975.
7. D. Lea, *Concurrent Programming in Java*, 2nd ed., Addison-Wesley, 1999.
8. M. Michael and M. Scott, "Simple, Fast, and Practical Non-Blocking and Blocking Concurrent Queue Algorithms," *In Proc. of the 15th ACM Symposium on Principles of Distributed Computing*, Philadelphia, Pennsylvania, pp. 267-276, May 1996.
9. J. Postel, ed., "Transmission Control Protocol," RFC 793, 1981.
10. J. Postel, "Internet Protocol," RFC 760, 1980.
11. B. Roussev and J. Wu, "Lottery-based scheduling of multithreaded Java applications on NOWs," *Annual Review of Scalable Computing*, Vol.3, 2001.
12. W.R. Stevens, *Unix Network Programming*, Vol.1, 2nd ed., Prentice Hall, 1998.

Appendix

```
class HashBuffer {
   protected SynchronizedRef[] table ;
   protected int capacity ;
   public HashBuffer( int capacity ) {
      this.capacity = capacity ;
      table = new SynchronizedRef[capacity] ;
      for ( int i = 0; i < capacity ; i++ )
         table[i] = new SynchronizedRef( null ) ;
   }
```

```
protected void insert( Object obj ) {
   int pos = 0 ;
   int key = obj.hashCode() ;
   if ( key < 0 ) key = -key ;
   int hash1 = key % capacity ;
   int hash2 = 1 + key % (capacity - 1) ;
   while ( true ) {
      pos = hash1 ;
      if ( table[pos].get() == null ) {
         table[pos].set( obj ) ;
         return ;
      }
      for ( int i = 0; i < capacity ; i++ ) {
         pos = (pos + hash2) % capacity ;
         if ( table[pos].get() == null ) {
            table[pos].set( obj ) ;
            return ;
         }
      }
      synchronized( obj ) {
         try { obj.wait( 500 ) ; }
         catch ( InterruptedException e ) {}
      }
   }
}
protected Object search( int key ) {
   Object temp = null ;
   int pos = 0 ;
   if ( key < 0 ) key = -key ;
   int hash1 = key % capacity ;
   int hash2 = 1 + key % (capacity - 1) ;
   while ( true ) {
      pos = hash1 ;
      if ( ( temp = table[pos].set( null ) ) != null )
         return temp ;
      for ( int i = 1; i < capacity ; i++ ) {
         pos = (pos + hash2) % capacity ;
         if ( ( temp = table[pos].set( null ) ) != null )
            return temp ;
      }
      synchronized( obj ) {
         try { obj.wait( 500 ) ; } catch(InterruptedException e){}
      }
   }
}
```

Towards a Descriptive Approach to Model Adaptable Communication Environments

Antônio Tadeu A. Gomes, Sérgio Colcher, and Luiz Fernando G. Soares

[1] Laboratório TeleMídia – Departamento de Informática – PUC-Rio
R. Marquês de São Vicente, 225 – Gávea
22453-900 – Rio de Janeiro – Brasil
{atagomes, colcher, lfgs}@inf.puc-rio.br

Abstract. One of the main challenges in the telecommunication sector has been to devise communication environments that allows: (i) the integration of a multitude of different services in a single and efficient communication system, and (ii) the rapid and easy creation, modification and continuous adaptation to new demands and conditions. In this paper, we present a recursive-structuring model that gives adequate support for defining both communication environments and their adaptation mechanisms. Second, we propose the use of frameworks as powerful artifacts that can help a service designer delineate common abstractions that appear within any communication environment, allowing design reuse. Finally, we discuss the role of software architecture, and more precisely of ADLs, as an appropriate way to describe communication environments according to the model and frameworks mentioned above.

1 Introduction

The new telecommunication marketplace has been characterized by an increasing interest in network technologies that can support different quality-of-service (QoS) requirements. The development of innovative applications and new media codification techniques has imposed a continuous research in the telecommunications sector regarding the definition of new services with diverse quality demands, which in general leads to continuous changes in switching systems. However, these changes were only possible, until recently, through slow operational procedures such as hardware or firmware updates.

The current trend in this sector has been the development of more versatile switching systems that could deal with the creation or modification of services by means of switching systems software explicitly adapted or programmed during network operation, possibly without any service disruption. In a scenario where telecommunications services are offered by adaptable or programmable communication environments, the complex tasks of service design, implementation and deployment have justified the creation of a *"Telecommunications Service Engineering"* [21]. Most concepts, principles and rules of this young discipline have been borrowed from software engineering, relying heavily on the object-oriented paradigm and on open distributed processing models. The aim of the present work is to propose a set of adequate tools for representing adaptations on generic communication environments by following the service engineering approach. Despite having some similarities with other existing proposals, our work proposes some additional contributions.

First, we present the *Service Composition Model* (SCM), a kind of abstract diagram-based model that allows service designers to describe a particular communication environment as well as the mechanisms in charge of adapting it. *Recursion* is an outstanding characteristic of SCM in the sense that various parts of a communication environment can be modeled similarly, thus improving the design reuse. However, our model has a purposely relaxed semantics so as to be sufficiently generic, which in turn hinders the identification of the points in a communication environment that can be adapted or programmed to support new services. Thus, we propose the use of object-oriented frameworks as artifacts that feed service designers with information about these "flexible" points. In fact, we have already illustrated on previous works [8] various QoS-specific functions represented with the help of frameworks. Moreover, because of the recursive nature of SCM we could also show how these functions are recurrent on the various parts of a communication environment and how they can cooperate to implement what we call *QoS orchestration*.

Nevertheless, even with the support provided by frameworks, the SCM "box-and-line" diagrams remain vague with regard to several aspects that are particularly important during service implementation. We can mention, as examples of these aspects, the interfaces of the components that make part of a communication environment and the maintenance of the overall integrity of the environment in the presence of adaptations. Thus, this paper also discusses the use of formal notations (i.e., *Architecture Description Languages* – ADLs) to describe these aspects according to SCM concepts. ADLs describe software architectures by means of coarse-grained elements, thus abstracting away from some implementation details while allowing the formal validation of others (e.g. the interactions among components). We make an investigation of several ADLs proposed in the literature and focus on some features that we consider highly desirable to represent component associations in the presence of adaptable QoS constraints.

The paper is structured as follows. Section 2 comments on some related works. Section 3 introduces the basic elements that comprise SCM. This section is a summary version of the work published in [4]. Following this, the role of frameworks in the context of SCM is presented in Section 4. Section 5 focuses on the description of the SCM meta service abstractions. Then, Section 6 goes into a insight on some ADLs proposed in the literature with respect to SCM concepts. Finally, some concluding remarks are given in Section 7.

2 Related Work

A multitude of works in the telecommunication sector has taken into consideration the service engineering approach as an adequate way for developing, operating and maintaining services. Most of these works hardly propose effective changes on the traditional service lifecycle, solely focusing on techniques for speeding the service development process. Such techniques include the use of design patterns applied to the domain of service construction [15], languages for modeling services [18] and programming libraries together with extensions in operating systems [9][20]. Despite their relevance, all these techniques do not take into account adaptations accomplished during service operation.

Other proposals rely on the concept of *generic communication environments*. In these environments, object models offer the necessary abstraction for the creation and adaptation of services by means of basic operations on system components, such as factoring, replacements and associations. Programming support offered by environments defined over CORBA and ODP, such as those of the TINA [2] and Binding [11] models, are important examples of this strategy. Extending this approach, the Open-ORB architecture [5] proposes the use of *procedural reflection* and *open implementations* [10] as the main concepts behind the design of adaptable communication environments. That is, not only components as a whole but also their internal implementation (which includes the environment itself) can be manipulated in order to adapt a service. All these works have as their main weakness the absence of mechanisms for constraining adaptations, which leads to the possibility of widespread changes that can compromise the integrity of the overall communication environment. SCM provides a conceptual basis that both unifies the ideas of the aforementioned works in a single abstract object model and has an inherent nested organization that regulates the definition of adaptation mechanisms in a homogenized way. We argue, however, that further semantics must be aggregated to SCM so that it can really help service designers model adaptations and its constraints.

The Aster Project [3] discusses the problem of adaptation constraining and proposes some add-ons to the aforementioned models, focusing on Open-ORB. The approach adopted is to apply formal descriptions on the definition of environment configurations. More specifically, it aims at supporting the systematic synthesis of these configurations from ADL notations. With the help of proper tools, ADLs can assist service designers in formally verifying an environment configuration, thus guaranteeing the correctness of the configuration before service implementation. As already mentioned, the present paper is also interested in applying the strength of ADL representations in the context of SCM, although we envision the use of ADLs in a different way, as we will show later.

3 Basic Elements of the Service Composition Model

Figure 1 illustrates the two basic elements that are used to represent a communication environment according to SCM: (i) *user components*, which correspond to entities that make direct use of services, and (ii) *service providers*, which are responsible for offering services. In SCM, the service provider abstraction is defined without any concern with the dispersion of its user components. For instance, components can be as close as objects contained in a single process or as far as objects located on internetworked machines. A special abstraction called *MediaPipe* is defined in SCM to represent the use of a service by a group of two (point-to-point) or more (multipoint) user components that wish to communicate. MediaPipes also capture the idea of virtual resources over which QoS requirements can be defined and handled.

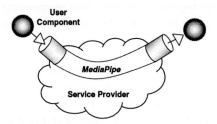

Fig. 1. Representation of a communication environment according to SCM.

MediaPipes are defined in terms of associations among service components, infrastructure providers and access providers. *Service components* accomplish the internal implementation of a provider. They may communicate with each other using more primitive providers, called *infrastructure providers*. Besides the infrastructure providers, there must also be *access providers* that permit user components to make use of MediaPipes through service components. Figure 2(a) shows the relationship among all these SCM elements.

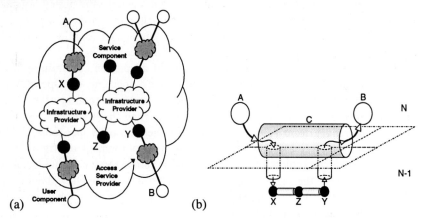

Fig. 2. Nested organization of SCM. (a) shows the internal structure of a hypothetical service provider, whereas (b) shows the decomposition of a MediaPipe in this service provider.

The infrastructure and access providers can also be structured so that the service components pictured in Figure 2(a) act as user components. Within these providers, other service components, infrastructure and access providers (and, consequently, MediaPipes) show up. This nested organization has some similarities with the layering principle defined by OSI-RM. Figure 2(b) suggests this analogy by illustrating the abstraction of a MediaPipe C between user components A and B, as well as its internal implementation by means of more primitive MediaPipes.

Despite its similarity with OSI-RM, SCM introduces other characteristics absent in the former model. As a little, yet very important example, the local system environments are outside the scope of OSI-RM, whereas in SCM they can be represented through access providers. Hence, SCM presents to service designers a most natural way to model QoS provision in an end-to-end scale. Since SCM suggests that the abstraction of a provider may represent all types of communication, it also has a strong potential for design and implementation *reuse*. For instance, we can define general interfaces and data structures to model some aspects that appear in any kind of communication environment. For example, when considering QoS-specific aspects, this homogenization lessens the work of introducing new QoS requirements and changing existing ones. In addition, as we will show in the following sections, it can facilitate the implementation of algorithms to offer end-to-end QoS.

4 Representing Service Aspects with Frameworks

Although SCM allows service designers to describe particular communication system architectures through composition of service providers, we envision that the model alone is not sufficient to help service designers build real communication systems. We argue that additional structures are desired to better delineate the recurrent interfaces and data structures that appear in such architectures. These structures would help service designers not only improve reuse but also regulate the adaptations.

In [8], we propose the use of object-oriented frameworks as a strategy to delineate such structures, focusing on QoS aspects. Frameworks capture design decisions that are common to a particular domain. Usually, various parts of a framework cannot be previewed, so they should be left incomplete or "subject to variations". These "*hot spots*" [19] allow the definition of adaptable structures and interfaces regardless of the idiosyncrasies of the domain.

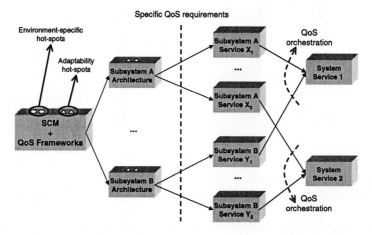

Fig. 3. The different types of hot spots in the service engineering process.

When concerning frameworks in the context of SCM, further considerations must be taken with regard to the points of adaptation. There are hot spots that should be fulfilled (adaptations should be done) before operation. They are usually only related to the environment dispersion level, that is, the specific subsystem (operating system, internetwork, and so on) in which the frameworks are being applied. There are other hot spots, however, that should be maintained incomplete to be fulfilled in execution time, as again taken QoS frameworks as examples, the points in the subsystems that are programmed to deal with different QoS requirements. In order to adapt the service provider during operation, some kind of adaptation service will be needed, as discussed in the next section. Figure 3 sketches the different types of QoS hot spots during service development.

5 Meta Services

According to SCM, in order to be adaptable during operation, a service provider must be the target of special *meta services* that act upon its elements, performing adaptations. Meta services can be structured, as any service, by means of components and providers. The same provider can be the target of several meta services and the same meta service can act upon multiple providers, regardless of their levels of dispersion. Moreover, meta services can be the target of other meta services, thus defining the concept of *meta service towers*.

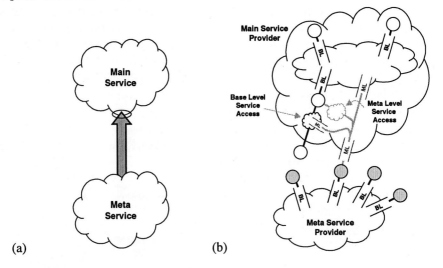

Fig. 4. Services and meta services. (a) shows a compact notation of the meta service relationship illustrated in (b).

The adaptation of components and providers of the target service provider is based on the concept of open implementations. Besides the *base-level* (BL) interfaces that allow "normal" interactions (access to the services), components and providers of the target service provider must also have *meta-level* (ML) interfaces that reveal some of their internal aspects, allowing adaptations. Figure 4 illustrates the concept of SCM meta services. As an example of the expressiveness of this concept, when the main service provider also the functions as a meta service provider, we open the perspective of components from both levels directly communicating among each other, which allows the representation of reflective architectures.

As an example, Figure 5 illustrates the distribution of the main functions modeled by the QoS frameworks presented in [8] among a typical meta service tower. Because of space restrictions, we omit more details about SCM meta services in this paper. Further information can be found in [4].

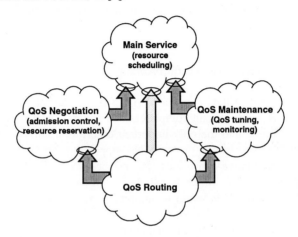

Fig. 5. QoS meta service tower.

As mentioned before, an important characteristic of SCM is its suitability for modeling QoS orchestration. Orchestration relies on concatenation heuristics that decide quantitatively how portions of QoS-provision responsibility will be divided, before *and* during operation, among the internal MediaPipes that compose a main provider MediaPipe. With the support provided by frameworks, SCM can provide similar QoS interfaces and data structures regardless of the dispersion levels of the QoS functions. Indeed, Figure 3 illustrates that all subsystem services, which collectively build a main system service, derive from a unique general set of interfaces and structures. The definition of uniform and well-known interfaces facilitates the implementation of concatenation heuristics in a seamless way.

6 The Need for a Descriptive Language for SCM

The use of SCM diagrams does raise the level of abstraction in which service designers reason about adaptable communication systems architectures. Despite its simplicity SCM encompasses nearly all paradigms of adaptability and programmability found in the literature (in [4] we show a classification of the main adaptation strategies according to SCM concepts). However, the SCM diagrams show various limitations, such as the absence of more precise information concerning the interfaces of the components and the nature of interaction among them, only to mention a few. These limitations reduce the utility of the SCM diagrams when the service designer faces the service implementation.

We believe that the use of ADLs is an appropriate way to describe communication systems and can also eases the development process with the help of proper tools. We have worked on investigating the applicability of several ADLs [1][12][13][14][17] on the context of SCM for the description of adaptable communication environments. The remaining of this section focuses on some issues pertaining to these ADLs, such as the interaction among service components, the support for adaptations and the role of architectural styles.

Describing the Quality of Interaction Among Components

One of the limitations of the investigated ADLs is the lack of support for specifying the *quality of interaction* among components. In general, ADLs define *connector* as the unit of interaction, and the only important description allowed in a connector is related with the communication protocol. No support is offered for the description of qualitative and quantitative constraints (for instance, throughput and delay) imposed by the type of interaction, which is mandatory when taking into consideration the issue of QoS provision. It is worth mentioning a new work in this area [6], which proposes an ADL that gives support to both QoS specification and dynamic QoS management (in terms of monitoring and control components).

Identifying the Points of Adaptation

Another important feature that an ADL should provide is the support for representing adaptations. Some investigated ADLs [13][14][17] represent adaptations by means of additions and replacements of components and connectors as well as whole architecture reconfiguration. These ADLs do not explicitly offer specific adaptation interfaces (the meta interfaces of SCM) or reflective abstractions that can reveal internal aspects of the components. When a service designer wishes to describe adaptable communication system architectures with the help of these ADLs, it may be hard to realize which parts of the architecture correspond to points of adaptation.

The Role of Architectural Styles

As it happens with the conceptual basis provided by SCM, an ADL alone may not be useful enough to service designers because the description of all structures of an architecture must be started from scratch. As seen before, frameworks can accomplish the task of describing such structures. However, the description of the QoS frameworks presented in [8] were made with the help of an extended version of the *Unified*

Modeling Language (UML) proposed in [7]. This notation is notoriously limited with regard to distinguishing among descriptions of families of architectures and descriptions of specific architecture configurations, being more adequate to the later. However, the SCM-related frameworks focus exactly on the description of aspects pertaining to the family of adaptable communication architectures. The work presented in [1] introduces the concept of *architectural style* to exploit commonalties across families of architectures. An architectural style is defined as a set of properties that are shared by various architectural configurations. We envision the use of architectural styles as a very useful tool for applying design decisions captured by the SCM-related frameworks while taking advantage of ADL features.

Another important characteristic of architectural styles is their support for constraining architectures. As proposed in [1], a style description may have *predicates* that must be obeyed by any architecture configuration that makes use of the style. Among other things, the constraints imposed by architectural styles would permit a service designer to explicitly define adaptation constraints, hence helping him to maintain the overall integrity of the environment.

7 Conclusion

This paper proposed a conceptual structuring model that permits the design of adaptable communication system architectures recursively, allowing improvements on design reuse. In addition to this main abstraction, frameworks are used to delineate important aspects that appear recurrently in a communication system, such as the mechanisms in charge of QoS provision. We also showed that, in spite of the strength of SCM representation, a service designer needs more precise information regarding finer-grained implementation details. Hence, we propose the use of ADLs to extend the SCM functionality. Some extensions have been discussed in recent works [3][6], but we envision that the generality of our model can lead to the definition of ADLs with larger scope.

At present, we have been working on validating the model by means of two main implementations. First, we modeled the int-serv and diff-serv architectures [16] according to SCM concepts and with the help of the QoS frameworks presented in [8]. The most important result of this research was the definition of a conceptual boundary among service definitions, signalling protocols and QoS functions performed in switched routers. This separation of concerns eases the introduction of new services when programmable-switched routers are available. The second experiment aims at defining appropriate QoS orchestration abstractions among multimedia client applications, multimedia servers and the communication infrastructure, which encompasses spatial and temporal inter-media synchronization, prefetch of multimedia objects and elastic adjustment of multimedia presentations. In all these experiments we wished for a tool that could ease not only the development process but also its documentation, in particular the hot-spot definition. Again, the use of an ADL with the extensions proposed in this paper would be of great help.

References

1. Allen, R. J.: A formal approach to software architecture. Ph.D. thesis, School of Computer Science, Carnegie Mellon University, May 1997.
2. Berndt, H., et al.: Service specification concepts in TINA-C. International Conference of Intelligence in Broadband Services and Networks, 1994.
3. Blair, G. S., Blair, L., Issarny, V., Tuma, P., Zarras, A.: The role of software architecture in constraining adaptation in component-based middleware platforms. Proceedings of the 2nd International Conference on Distributed Systems Platforms and Open Distributed Processing (Middleware'2000), IBM Palisades, New York, April 2000.
4. Colcher, S., Gomes, A. T. A., Rodrigues, M. A. A., Soares, L. F. G.: Modeling service aspects in generic communication environments. Tech Report, PUC-Rio, Brazil, 2000.
5. Costa, F. M., Duran-Limon, H. A., Parlavantzas, N., Saikoski, K. B., Blair, G., Coulson, G.: The role of reflective middleware in supporting the engineering of dynamic applications. Reflection and Software Engineering, June 2000.
6. Duran-Limon, H. A., Blair, G. S.: Specifying real-time behaviour in distributed software architectures. 3rd Australasian Workshop on Software and Systems Architectures, Sydney, Australia, November, 2000
7. Fontoura, M. F. M. C.: A systematic approach to framework development. D.Sc. thesis, PUC-Rio, Brazil, July 1999.
8. Gomes, A. T. A., Colcher, S., Soares, L. F. G.: Modeling QoS provision on adaptable communication environments. International Conference on Communications (ICC2001), Helsinki, Finland (to appear).
9. Hutchison, N. C., Peterson, L. L.: The x-Kernel: an architecture for implementing network protocols. IEEE Transactions of Software Engineering, January 1991.
10. Kiczales, G.: Towards a new model of abstraction in the engineering of software. Proceedings of IMSA'92, Tokio, Japan, November 1992.
11. Lazar, A. A., Lim, K. S., Marconcini, F.: Binding model: motivation and description. Tech. Report CTR 411-95-17, Columbia University, 1995.
12. Luckham, D. C., et al.: Specification and analysis of system architecture using Rapide. IEEE Transactions on Software Engineering, 1995.
13. Magee, J., Kramer, J.: Dynamic structure in software architectures. Proceedings of the 4th ACM Symposium on the Foundations of Software Engineering, 1996.
14. Medvidovic, N.: ADLs and dynamic architecture changes. Proceedings of the 2nd ACM SIGSOFT International Software Architecture Workshop, October 1996.
15. Meszaros, G.: Design patterns in telecommunications systems architecture. IEEE Communications Magazine, April 1999.
16. Mota, O. T. J., Gomes, A. T. A., Colcher, S., Soares, L. F. G.: An adaptable QoS architecture for the Internet. Submitted for Brazillian Symposium on Computer Networks (portuguese version).
17. Paula, V. C. C.: ZCL: A formal framework for specifying dynamic distributed software architectures. D.Sc. thesis, Federal University of Pernambuco, Brazil, May 1999.
18. Perumalla, K. S., Fujimoto, R. M., Ogielski, A. T.: MetaTeD – a meta language for modeling telecommunications networks. Tech. Report GIT-CC-96-32, Georgia Institute of Technology, Atlanta, 1996.
19. Pree, W.: Framework patterns. SIGS Books & Multimedia, 1996.
20. Schmidt, D. C.: The ADAPTATIVE communication environment. Proceedings of the 12th Annual Sun Users Group Conference, San Jose, CA, December 1993.
21. Znaty, S., Hubaux, J.: Telecommunications services engineering: principles, architectures and tools. Proceedings of the ECOOP'97 Workshops, Jyväskylä, Finland, June 1997.

All-to-All Personalized Communication Algorithms in Chordal Ring Networks

Hiroshi Masuyama[1], Hiroshi Taniguchi[2], and Tsutomu Miyoshi[3]

[1] Information and Knowledge Engineering, Tottori University, Tottori, 680-8552, Japan,
masuyama@ike.tottori-u.ac.jp
[2] Graduate School, Tottori University, Tottori, 680-8552, Japan,
itanigut@ike.tottori-u.ac.jp
[3] Information and Knowledge Engineering, Tottori University, Tottori, 680-8552, Japan,
mijosxi@ike.tottori-u.ac.jp

Abstract. Chordal ring networks are ring structured and they become unreliable when nodes or arcs in the ring break down. Such a shortcoming can be circumvented by adding duplicate rings. Though this inherent topology of preparing many loops in the network structure is advantageous to broadcasting, no such broadcasting algorithms have been reported up until now.

This paper discusses broadcasting algorithms for Chordal ring networks under the last one of four different communication primitives: One-to-all broadcasting, one-to-all personalized communication, all-to-all broadcasting, all-to-all personalized communication. In this paper, Chordal ring networks are assumed to be not under critical faults, that is, the maximum number of faults we consider is 2. The number of time units required to the proposed algorithm which can tolerate under to the critical faults is made clear.

1. Introduction

Parallel computer networks consisting of a large number of identical processing elements are currently in various stages of development. These networks enable the processing elements to intercommunicate.

Many reports related to the parallel computer networks such as Mesh connected [1], Cube connected [2], Ring connected [3], and Asynchronous Transfer Mode switching [4] are known. One of recent important issues related to such parallel computer networks is how to construct a high-performing broadcasting algorithm. The above networks have different topological properties that broadcasting is a high performance.

Single-loop networks present a better solution to local networking than non-loop networks do, since the network interface and the control software are simple. Single-loop networks, however, tend to become unreliable when nodes or arcs break down. This shortcoming can be circumvented by adding redundant links to single-loop networks. There have been several network architectures proposed for fault-tolerant communications in a distributed system [5]. A Chordal ring network [6] is one of such networks, as is a ring structured network. Chordal ring networks are in a family of regular graphs of degree 3. The diameter or maximum length message path for a properly constructed network was investigated. That the symmetry of the graphs makes it possible to determine massage routing by using a simple distributed algorithm was made clear. From the view point of finding the shortest bypass route in faulty Chordal rings, the reliability analysis of Chordal rings also has been done. In addition, the sorting ability of Chordal ring networks [7], and the ability to perform BPC permutations [8] for nonfaulty Chordal ring networks and [9] for faulty ones have been investigated. The diagnosis approaches for Chordal ring networks also have been presented [10].

In distributed memory system, communication among the processors is performed mainly via message passing. Since the communication time may be quite expensive compared to the computation time, efficient communication schemes are extremely important to achieve high performance in the system. Johnsson and Ho [11] introduced four different communication primitives:

1) one-to-all broadcasting (or single node broadcasting) in which a single node distributes common data to all other nodes,
2) one-to-all personalized communication (or scattering) in which a single node sends unique data to all other nodes,
3) all-to-all broadcasting (or multimode broadcasting) in which all nodes broadcast concurrently to all other nodes, and
4) all-to-all personalized communication (or total exchange) where each and every node sends unique data to every other node.

Many researches have proposed various communication algorithms for hypercube multicomputers [12], however, no report has been published for Chordal ring networks. This paper tries to propose a communication algorithm based on 4) of the above four different communication primitives for Chordal ring networks.

Communication algorithms can be implemented in either a *one-port* or an *n-port* model. In this paper, only an n-port model will be considered, since n-port models can be simulated by a one-port model without losing any efficiency. Hence, all incident links of a node can be used simultaneously for packet transmission and reception.

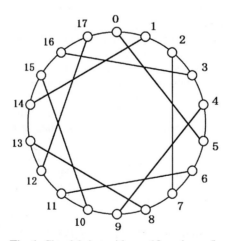

Fig. 1. Chordal ring with $n = 18$ and $\omega = 5$.

2. Preliminaries

Chordal ring consists of n nodes and two types of links; n arcs and some chords. The number n of nodes in a Chordal ring is assumed to be even and nodes are indexed $0, 1, 2, \ldots, n-1$ around the ring. Arcs are links of the type $(i, i+1 \bmod n)$, $i = 0, 1, 2, \ldots, n-1$. In this paper, for another type of link (chords), we restrict our attention to the case in which each odd-numbered node i ($i = 1, 3, \ldots, n-1$) is connected to a node $(i - \omega) \bmod n$. Accordingly, each even-numbered node j ($j = 0, 2, \ldots, n-2$) is connected to a node $(j + \omega) \bmod n$. Then ω is called chord length and is assumed to be positive odd number. Without loss of generality, we assume that $\omega \leq n/2$. For a given number n of nodes, a number of Chordal rings can be obtained for different values of chord length ω. An example of a Chordal ring for $n = 18$ and $\omega = 5$ is shown in Fig.1. The links can be either unidirectional or bidirectional, but in this paper only the bidirectional links are considered.

Note that Chordal ring structure is incrementally extensible by adding multiples of two nodes to the original configuration. However, as the number of nodes increases, the optimal chord length giving the minimum diameter changes. Since it is known that the optimal chord length is nearly equal to $\sqrt{n}$ when the diameter becomes nearly equal to $\sqrt{n}$, then we will consider such Chordal ring in the followings.

Communication algorithms for these Chordal ring networks can be implemented in either a one-port or a 3-port model. In a one-port model, a node can transmit a packet along at most one incident link and can simultaneously receive a packet along at most one incident link. On the other hand, in a 3-port model, all incident links or a node can be used simultaneously for packet transmission and reception. In this paper, only a 3-port model will be considered, since 3-port models can be generally simulated by a one-port model without losing any efficiency, as mentioned in [12]. An algorithm designed for a one-port model is simpler than one designed for a 3-port model and is not considered here. We assume that it takes one time unit to transmit a message from a node to an adjacent node.

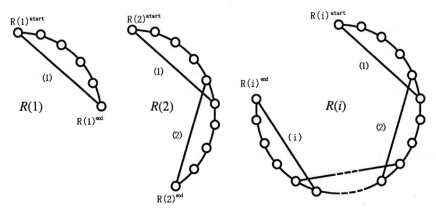

Fig. 2. $R(1)$, $R(2)$, and $R(i)$ in the case of $\omega = 5$.

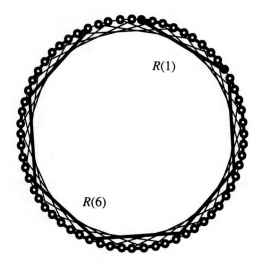

Fig. 3. Chordal ring with $n = 64$, $\omega = 9$ and rings $R(1)$ and $R(6)$.

The faults in Chordal ring networks we consider is one in which only the nodes fail because a faulty communication link can be accommodated by treating as a faulty node. In addition, we don't consider critical faults by which communications among nonfaulty nodes are destroyed, then in this paper we will treat single or double faults.

[Definition 1] The distance (or length) of the loop is the number of nodes on the loop.
(1) Ring $R(1)$ is a loop which is composed of a chord (labeled by (1) in Fig.2), ω arcs bypassed by the chord, and $(\omega + 1)$ nodes, in other words, the distance of $R(1)$ is $(\omega + 1)$.
(2) Ring $R(2)$ is a loop which is composed of 2 Chords which have only one common arc bypassed by 2 consecutive chords (labeled by (1) and (2)), and 2ω nodes, but not includes the common bypassed arc. The distance of $R(2)$ is 2ω.
(3) Ring $R(i)$ is a loop which is composed of ring $R(i-1)$ exclusive of an arc bypassed by the i-th additional new chord (labeled by (i)), the additional new chord, and $(\omega - 1)$ arcs and nodes bypassed by the additional new chord. The length $|R(i)|$ (that is, the number of nodes) of $R(i)$ is $(\omega - 1) + 2$. There exist $n/2$ $R(i)$ rings in a Chordal ring. The longest $R(i)$ is $R(\lceil (n-2)/(\omega - 1) \rceil)$. $R(i)$ is shown in Fig.2.

Two node labels $R(i)^{start}$ and $R(i)^{end}$ on each loop shown in Fig.2 are special marks to find the loop as mentioned in the followings. Let the distance between two nodes i and j be $d(i, j)$. We will call a loop which takes arcs unless it bypasses faults a single loop.

[Property 1] A Chordal ring has two isolated rings $R(1)$ and $R(i)$ on condition that $n \geq 2(\omega+1)$, where $i = \lfloor (n-\omega-3)/(\omega-1) \rfloor \geq 1$.

Proof: Obvious. A Chordal ring can't have two separated rings $R(1)$ and $R(i+1)$. When $R(i)^{start}$ is node $(\omega+2)$, any node j on ring $R(i)$ has two adjacent nodes defined as follows:
 i) Node $j\,(=\omega+2)$ has adjacent nodes $j+1$ and $j+\omega$,
 ii) Node $j\,(=\omega+2+\alpha(\omega-1))$ has adjacent nodes $j-1$ and $j+\omega$,
 iii) Node $j\,(=\omega+3+\alpha(\omega-1))$ has adjacent nodes $j+1$ and $j-\omega$,
 iv) Node $j\,(=\omega+3+i(\omega-1))$ has adjacent nodes $j-1$ and $j-\omega$,
 v) Node j (otherwise) has adjacent nodes $j-1$ and $j+1$,
where $\alpha = 1, 2, ..., i-1$. See Fig.3.

[Property 2] A Chordal ring has three isolated rings such as two consecutive rings $R(1)$ and the other ring $R(i)$ on condition that $n \geq 3(\omega+1)$, where $i = \lfloor (n-2\omega-4)/(\omega-1) \rfloor \geq 1$.

Proof: Obvious. The Chordal ring can't have ring $R(i+1)$ instead of $R(i)$. When $R(i)^{start}$ is node $2(\omega+1)+1$, any node j on ring $R(i)$ has adjacent nodes defined as follows:
 i) Node $j\,(=2(\omega+1)+1)$ has adjacent nodes $j+1$ and $j+\omega$,
 ii) Node $j\,(=2(\omega+1)+1+\alpha(\omega-1))$ has adjacent nodes $j-1$ and $j+\omega$,
 iii) Node $j\,(=2(\omega+1)+2+\alpha(\omega-1))$ has adjacent nodes $j+1$ and $j-\omega$,
 iv) Node $j\,(=2(\omega+1)+2+i(\omega-1))$ has adjacent nodes $j-1$ and $j-\omega$,
 v) Node j (otherwise) has adjacent nodes $j-1$ and $j+1$,
where $\alpha = 1, 2, ..., i-1$. See Fig.4.

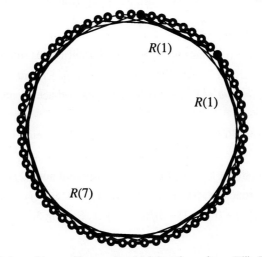

Fig.4. Chordal ring with $n=64$, $\omega=7$ which has three rings $R(1), R(1)$ and $R(7)$.

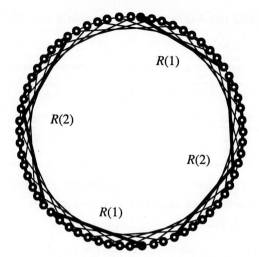

Fig.5. Chordal ring with $n = 64$, $\omega = 9$ which has four rings $R(1), R(2), R(1)$ and $R(2)$.

[Property 3] A Chordal ring has 4 isolated rings such as $R(1)$, $R(i_1)$, $R(1)$, and $R(i_2)$ clockwise on condition that $n \geq 4(\omega + 1)$, where $1 \leq i_1 \leq i_2$.

Proof: Obvious. When node $R(i_1)^{start}$ is node $(\omega + 2)$, any node j_1 on ring $R(i_1)$ has two adjacent nodes defined as follows:

i) Node $j_1 (= \omega + 2)$ has adjacent nodes $j_1 + 1$ and $j_1 + \omega$,

ii) Node $j_1 (= \omega + 2 + \alpha_1(\omega - 1))$ has adjacent nodes $j_1 - 1$ and $j_1 + \omega$,

iii) Node $j_1 (= \omega + 3 + \alpha_1(\omega - 1))$ has adjacent nodes $j_1 + 1$ and $j_1 - \omega$,

iv) Node $j_1 (= \omega + 3 + i_1(\omega - 1))$ has adjacent nodes $j_1 - 1$ and $j_1 - \omega$,

v) Node j_1 (otherwise) has adjacent nodes $j_1 - 1$ and $j_1 + 1$,

where $\alpha_1 = 1, 2, ..., \lfloor ((\beta - 1)(\omega + 1) - 2)/(\omega - 1) \rfloor - 1$ and β is the largest value given by $(\beta - 1)(\omega + 1) \leq d(R(i_1)^{start}, R(i_2)^{start})$.

Any node j_2 on ring $R(i_2)$ has two adjacent nodes defined as follows:

i) Node $j_2 (= \beta(\omega + 1) + 1)$ has adjacent nodes $j_2 + 1$ and $j_2 + \omega$,

ii) Node $j_2 (= \beta(\omega + 1) + 1 + \alpha_2(\omega - 1))$ has adjacent nodes $j_2 - 1$ and $j_2 + \omega$,

iii) Node $j_2 (= \beta(\omega + 1) + 2 + \alpha_2(\omega - 1))$ has adjacent nodes $j_2 + 1$ and $j + 1$,

iv) Node $j_2 (= \beta(\omega + 1) + 2 + i_2(\omega - 1))$ has adjacent nodes $j_2 - 1$ and $j_2 - \omega$,

v) Node j_2 (otherwise) has adjacent nodes $j_2 - 1$ and $j_2 + 1$,

where $\alpha_2 = 1, 2, ..., \lfloor (n - \beta(\omega + 1) - 2)/(\omega - 1) \rfloor - 1$. See Fig.5.

3. All-to-All personalized communication

3.1 No Fault

Send, for every node, its two personal packets to the farthest two nodes in each different directions, next send to the second farthest two nodes in each different directions, ..., and finally send to its both side adjacent nodes connected with an arc as shown in Fig.6. During these operations, for every node, send a packet arrived from its adjacent node to the opposite-side adjacent node connected with an arc. Since we use a Chordal ring as a single loop whose diameter is $n/2$, the all-to-all personalized communication can be performed at $\sum_{i=1}^{n/2} i$ time units.

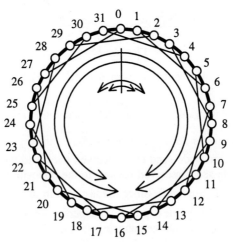

Fig.6. Packet transmissions from node 0.

3.2 Single Fault

Find a ring $R(1)$ with a faulty node f which is in the middle of $R(1)$, and let it be $R(1)^f$. Make the longest nonfaulty single loop which is composed of $R(i)^{start}$, $R(i)^{end}$, and all nodes except nodes on $R(1)^f$, and let it be $\overline{R(1)}^f$, as shown by bold lines in Fig.7.

All-to-all personalized communication can be performed at $\sum_{i=1}^{\alpha} i + \sum_{i=\alpha+1}^{\alpha+\beta} i$ time units where $\alpha = \lceil (n-(\omega-1))/2 \rceil$ and $\beta = \lceil (\omega-1)/2 \rceil$, as follows: First, For every personal packet on $\overline{R(1)}^f$, by sending to all nonfaulty nodes perform all-to-all personalized communication on condition that every node sends received packets to the opposite-side adjacent node (it takes $\sum_{i=1}^{\alpha} i + \sum_{i=\alpha+1}^{\alpha+\beta} i$ time units as shown in

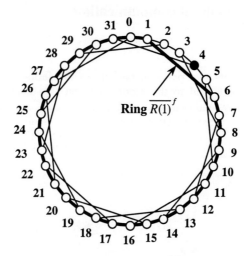

Fig. 7. Broadcasting in Chordal ring with $n = 32$, $\omega = 5$, and faulty node 4.

Fig.8). Next, for every nonfaulty node on $R(1)^f$ except $R(i)^{start}$ and $R(i)^{end}$, send every personal packet to its adjacent node on $\overline{R(1)}^f$ in larger order of destination distance, and the adjacent node send each received packet to the destination node in order, while every node sends received packets to the opposite-side adjacent node (it takes $\sum_{i=1}^{\alpha} i + \sum_{i=\alpha+1}^{\alpha+\beta} i + 1$ time unit). Since the last step in the first process mentioned above and the first step in the next process are performed simultaneously, then the required time unit is the one mentioned above.

3.3 Double Fault

Let the distance between two faults f_1 and f_2 (which also mean node indices, $f_1 \leq f_2$ for simplicity) be $d(f_1, f_2)$. We will discuss broadcasting algorithms and the required time units in the following divided four cases.

Case of $1 \leq d(f_1, f_2) \leq \omega$:

1) When $f_1 + 1 = f_2$ as shown in Fig.9 (a); find a ring $R(1)$ with f_1 and f_2 which are in the middle of $R(1)$. These two faults can be treated as a single fault. The broadcasting algorithm is the almost same as the one in the case of single fault, then it requires $\sum_{j=1}^{\alpha} j + \sum_{j=\alpha+1}^{\alpha+\beta} j$ time units.

2) When $f_1 + 2 = f_2$ as shown in Fig.9 (b); first, perform all-to-all personalized communication on fault-free ring $\overline{R(1)}^f$ in the same manner as in 3.1. Next, perform all-to-all personalized communication between nodes on $\overline{R(1)}^f$ and

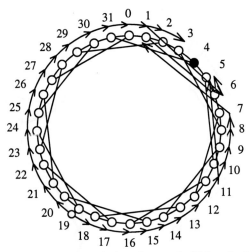

Fig. 8. Route of packet transmissions from node 19 (required time units = total distance of arrowheads from node 19).

fault-free nodes on $R(1)^f$ as follows: See an example shown Fig.9 (b). Send two packets arrived at node a to node b in two steps (that is, two time units) while send two packets from node b to ring $\overline{R(1)}^f$ in different directions. Then, we can estimate the required time units $\sum_{j=1}^{\alpha} j + \sum_{j=1}^{\alpha} (j+2)$.

3) Otherwise; all-to-all personalized communication can be performed at most at $\sum_{j=1}^{\gamma} j + \sum_{j=1}^{\gamma}(j+2)$ where $\gamma = \{(\omega-1)i + 2\}/2$, and $\sum_{j=1}^{\gamma} j + \sum_{j=1}^{\gamma}(j+2) + \sum_{j=1}^{\gamma}(j+3)$ time units when $(n-\omega-3)/(\omega-1)$ is an integer i and not integer, respectively, as follows:

(When $(n-\omega-3)/(\omega-1)$ is an integer i); Chordal ring can be modified as shown in Fig.9 (c). By using ring $R(i)$ instead of single loop in Fig.8, all-to-all personalized communication can be performed in the same manner as the one in the case of 2). Since the length of ring $R(i)$ is 2γ, then the above amount of time units is obtained.

(When $(n-\omega-3)/(\omega-1)$ is not integer); Fig.9(d) shows an example of this case. Every nonfaulty node not on ring $R(i)$ has a path of distance of at most 2 to the nearest node on $R(i)$, then the communication algorithm is as follows: First, perform all-to-all personalized communication on fault-free ring $R(i)$ in the same manner as in 3.1. Next perform all-to-all personalized communication between nodes on $\overline{R(1)}^f$ and fault-free nodes on $R(1)^f$ whose distance is 1 from ring $\overline{R(1)}^f$ (it takes $\sum_{j=1}^{\gamma}(j+2)$ time units), and finally perform all-to-all personalized

Formal and Practical Approach to Load Conditions in High Speed Networks

Armin Pollak

University of the Armed Forces Munich,
Department of Computer Science 3.3*,
Werner-Heisenberg-Weg 39, 85577 Neubiberg, Germany,
pollak@informatik.unibw-muenchen.de

Abstract. There has been little interest in research of high speed networks under heavy load. Only some special areas, like media access has been intensely studied. Within this paper we present an approach to a more formal view of load conditions in high speed networks. We clearly show what load conditions are and combine this work with a practical approach for generating load. We introduce an architecture for generating defined load conditions in high speed networks and show how to use them for performance evaluation.

1 Introduction

The evolution in networking has brought high bandwidth to the desktop. This changed the bottleneck in data transmission: from the (high speed) network itself to the network access points ([4][7]). Due to this tradeoff almost none research is done on heavy load in networks. Only some special areas has been intensly studied under heavy load or overload, with main focus on media access and fairness aspects.

We introduce within this paper a formal approach to load conditions in high speed networks and the notation of near-limit load conditions. Our aim is the extension of the simple bandwidth view, used when talking about performance of networks. We want to cover all aspects within high speed networks belonging to performance and want to detect all possible limitations. This requires a new formal view of a high speed network.

In the next section we present this formal view of such a network and derive load conditions. This leads towards a notation for load, heavy load and near limit load and shows the impact for performance studies of such networks.

Based on these definitions we introduce our framework for generating load conditions in high speed networks. Section 3 describes the implemented prototype for generating traffic and load conditions.

Finally we summarize in Sect. 4 our work and give an outlook for future research.

---

* I want to thank Prof. Dr. Helmut Rzehak, who made this work possible and supported its development with helpful hints and discussions.

2   Formal Approach

A common view of networks is based on the OSI reference model and its layered architecture. This approach defines a function oriented view where each layer offers services to the upper layer. Within this model it is not applicable to derive load conditions.

Our view is based on a model of interconnected nodes with a container[1] oriented data transmission. Each node offers and uses resources for data transmissions and is characterized by its qualities. Resources are scalable elements like buffer, datarate, packets or cells. Scalable means there exists a dimension like number of bits, time, ... Qualities define the characteristic for the usage of resources. These qualities describe the interfaces of the nodes. We do not distinguish between interface to the user and connections between nodes.

Definition 1 (node, nodes, network). *Let $\mathcal{R}$ be the set of all resources, and $\mathcal{Z}$ the set of all network access points with given qualities, then we define node $K = (R, Z)$ with $R \subseteq \mathcal{R}, Z \subseteq \mathcal{Z}$,
set of nodes $\mathcal{K} = \{(R_i, Z_i), i = 1 \ldots n, R_i \subseteq \mathcal{R}, Z_i \subseteq \mathcal{Z}\}$ and
network $N = \{K_i, i = 1 \ldots m, K_i \in \{\mathcal{K}\}\}$*

According to Def. 1 a network is a set of nodes. The connections are given through the qualities of the network access points. Each connection is a match of two different interfaces with the same qualities[2].

Now we define load condition as utilization of a resource or a set of resources. With utilization we mean the usage of a resource, which could be expressed as a percentage of utilization like "90% of available bandwidth" or with a formal description like "save transmission" or "cheap transmission", which allows to include non countable parameters like security or cost.

Definition 2 (utilization,load). *For all resources R we define utilization as*
$$B : \mathcal{R} \mapsto \{0, 1\} : B(R) = \begin{cases} 0 \; free \\ 1 \; used \end{cases}$$
and load as
$\exists i : B(R_i), R_i \subseteq \mathcal{R} \; with \; B(R) = 1 \; \forall R \in R_i$

Definition 2 gives us the possibility to specify various load conditions as function of resources to the dimension or the range of values of this resources. Near-limit load condition is then defined as the state where **one** resource is within the defined near-limit condition, which must be predetermined. This flexible approach to high speed networks allows to model different load conditions.

These definitions lead to a formal overall view of load conditions in high speed networks. The mentioned studies of media access or fairness only use a subset of the ressources belonging to a node or a network and therefore restrict performance issues to this isolated view. With our approach it is possible to expand performance issues to an overall view of high speed networks.

---
[1] A container is a transport entity (cell, packet, ...) which carries unspecified data.
[2] It is not necessary to distinguish between different connections between the nodes, because we are only interested *how* the connection is used.

3 Practical Approach

For the generation of these load conditions a special traffic generator is needed. The basic idea behind our traffic generator is the generation of a specific load in an existing, packet oriented network. The generation is based on sending a packet stream to a network access point, where the length of the packets, the used address information, the inter gap time between the packets and the content of the packets is variable. We think relative time stamps are adequate for our aims and there is no need for global watches. This concept allows to extend former load or traffic generators (for example Strazdas [6] or Yuksel [8]) and to integrate new traffic profiles like those mentioned in [1].

The innovative aspect of this traffic generator is the combination of various traffic sources for different network access points. This is necessary due to the limitations of the network access points. Special load conditions in high speed networks like near-limit load conditions could not be realized with **one** single traffic source. The traffic sources itself are coordinated by a central instance. We call a run of the traffic generator a test, and the autonomous traffic sources included in such a run a parttest. Each test consists of at least one parttest. The parttest is hardware dependent, whereas the test is hardware independent.

Our design concepts are a physical separation of sender and receiver, to exclude any dependencies between these components. Further we use a test management (central instance) and distinguish between two different operation modes: offline and online ([5]). During the offline mode the test and its included parttests are set up. In online mode the execution of the test is done and only some test handling like abort or runtime control are allowed. After a run of the generator the test results are analyzed during another offline phase.

In online mode the test management controls the parttests, which are sending back status information to the test management. The architecture of the traffic generator is shown in Fig. 1. The central part is the Load Control (LC) which controls the sender and receiver. LC uses the table management (TM) and the file system (FS). The LC interacts with the GUI, which allows a remote control. Each pair of Load Control Interface (LCI) and Load Processing Entity (LPE) is a sender (generating traffic) or a receiver (collecting data) and is bound to a specific hardware architecture.

4 Summary

In summary load conditions in high speed networks are more than restriction to available bandwidth. This often used simple view does not cover near-limit load conditions. We presented a new approach to this problem, which is very flexible and covers all aspects of high speed networks. So the definition of various load conditions including those with non countable quantities like security is possible.

With our traffic generator we showed that the necessary high effort for generating load conditions in high speed networks could be reduced. The remaining restrictions are based on the architecture of the available network access points.

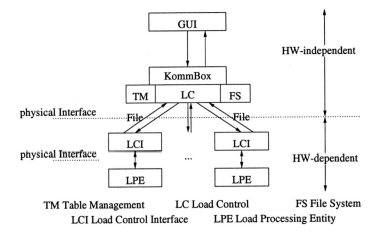

Fig. 1. Architecture of the traffic generator

Near-limit conditions are therefor the result of a combination of multiple traffic sources and could not be generated from one single source. So some restriction used in former investigations of heavy load like load generation from a single point or node distances are no longer valid ([2][3]). It should be proved in which way the practical aspects could be integrated in these theoretical investigations.

Our work excluded so far some important aspects like flow control, load control, routing or quality of service. It should be proved, how these aspects influences load conditions. Further the implemented traffic generator is far away from completeness and should be adapted to other network architectures.

References

[1] A. Adas. Traffic Models in Broadband Networks. *IEEE Communications Magazine*, 35(7):82–89, July 1997.
[2] M. Conti, E. Gregori, and L. Lenzini. Metropolitan Area Networks (MANs): Protocols, Modeling and Performance Evaluation. In *Proceedings Performance Evaluation of Computer an Communication Systems*, 1993.
[3] M. Kabatepe and K. Vastola. A delay analysis of FDQ and DQDB networks. *Computer Networks and ISDN Systems*, 30:1013–1027, 1998.
[4] L. Kleinrock. The Latency/Bandwidth Tradeoff in Gigabit Networks. *IEEE Communications Magazine*, 4:36–40, 1992.
[5] A. Pollak. Lastgenerator für ein Hochgeschwindigkeitsnetz. In *Proceedings ITG/GI-Fachtagung Messung, Modellierung und Bewertung*, pages 60–64, 1997.
[6] R. Strazdas. A Network Traffic Generator For DECNET. MIT/LCS/TM-127, March 1979.
[7] H. van As. Media Access Techniques: The Evolution towards terrabit/s LANs and MANs. *Computer Networks and ISDN Systems*, 26(4):603–656, 1994.
[8] M. Yuksel. Traffic Generator for an on-line simulator. Master's thesis, Rensselaer Polytechnic Institute, 1999.

Author Index

Abolhasan, M. II-186
Ahn, G. II-319
Akimaru, H. I-65
Al-Ibrahim, M. II-239
Al-Regib, G. II-745
Ali, M.M. I-429
Altman, E. II-97
Altunbasak, Y. II-745, II-786
Amorim, C. II-776
Anagnostopoulos, C. I-459
Anagnostopoulos, J. I-459
Anisimov, N. II-559
Ano, S. II-825
Antonakopoulos, T. II-497
Antonov, V. II-559
Araki, T. II-620
Asfour, T. I-589
Ashraf, S.N. II-805
Atmaca, T. I-107, I-348

Balbinot, L.F. I-204
Banerjee, G. II-302
Barakat, C. II-97
Bartoš, R. I-388
Bartz Ceccon, M. II-277
Becker Westphall, C. II-438
Beghdad, R. I-520
Benaboud, H. I-780
Bester, J. II-348
Bettahar, H. I-630
Blanpain, Y. II-259
Boavida, F. II-458
Bonnet, C. II-31
Bordes, J. I-124
Bosquiroli Almeida, M.J. II-277
Bouabdallah, A. I-630
Bouzida, Y. I-520
Boyle, J. II-559
Braun, T. II-206
Brogle, M. II-206
Budkowski, S. I-649

Cáceres Alvarez, L.M. II-438
Caminero, M.B. II-358

Camps, J. II-648
Capelle, A. Van de I-790
Carrapatoso, E. I-27
Carrión, C. II-358
Čavrak, I. II-399
Chang, S.W. I-709
Chao, H.-C. I-307, I-709
Chatterjee, M. I-157
Che, H. I-368
Chen, H.-C. II-225
Chen, J.-L. I-307, I-709
Chen, Y.C. II-82
Cheng, H.-C. I-307
Cherkasova, L. II-71
Choel, B.-Y. I-184
Choi, H. I-698
Choi, I. I-698
Choi, M.C. II-137
Choi, S. II-61
Choi, Y. I-771
Choi, Y.-I. II-380
Choplin, S. II-527
Chow, C.E. I-753
Chow, C.O. II-687
Chujo, T. I-753
Chun, W. II-319
Chung, J.-M. II-735
Čičić, T. I-488
Citro, R. II-538
Colcher, S. II-867
Corral, G. II-648
Czachórski, T. I-107

Das, S.K. I-157
Dastangoo, S. I-336
Davis, C. I-124
Davis IV, N.J. II-370
Decre, N. II-825
Demestichas, P. I-286
Deniz, D.Z. II-570
Denz, P.R. I-449
Dey, D. I-326
Di, Z. I-245
Dimyati, K. II-687

Droz, P. I-218
Duato, J. II-358
Dutkiewicz, E. I-720, II-186

Ehrensberger, J. I-378
Ehsan, N. II-51
Eissa, H.S. II-268
El-Sayed, A. I-610
Elbiaze, H. I-348
Estepa, R. II-549

Fabbri, A. I-620
Fan, L. I-147
Fan, Z. I-55
Fazekas, P. I-296
Fernández, D. II-697
Fornés, J.M. II-449
Freire, M.M. I-358
Fujii, A. II-845
Fulp, E.W. I-409

Gardiner, J.G. I-147
Garrell, J.M. II-648
Gaspary, L.P. I-204
Ghosh, S. II-538
Gjessing, S. I-488
Glitho, R. II-707
Go, S. II-639
Gomes, A.T.A. II-867
Goossens, I.G. II-795
Goossens, I.M. II-795
Graja, H. I-803
Günter, M. II-206
Guerra, S. II-697
Gupta, A.K. II-815
Gutacker, D. II-590

Ha, E.-J. II-726
Hämäläinen, T. II-250
Hagino, T. II-177
Hamadi, R. II-707
Han, D.H. II-517
Haque, A.U. II-628
Harle, D. I-829
Hasegawa, T. II-825
Hashmani, M. I-398
Hava Muntean, C. I-821
Hayashi, M. II-31

Hayes, J.F. I-429
He, J. I-753
He, X. I-368
Heemstra de Groot, S. I-326
Hegde, M.V. I-124
Hei, X. I-45
Hendler, M. I-478
Hladká, E. II-612
Hoang, D.-H. I-599
Hoang, T.T.M. I-743
Homan, P. II-348
Hommel, G. I-478
Hong, W.-K. II-599
Hsieh, H.-Y. I-569, II-259
Huang, C.-M. I-559
Huie, R. II-707

Ikenaga, T. I-398
Ilyas, M. II-677
Imre, S. I-296, II-11, II-468
Iqbal, T. II-570
Ishikawa, E. II-776

Jäger, R. I-419
Jakobs, K. I-194
Jedruś, S. I-107
Jones, G.A. II-766
Jun, K.-P. II-380
Jung, D.-S. I-1
Jung, M.-J. II-599

Kaario, K. II-250
Kabatnik, M. II-20
Kambo, N.S. II-570
Kamel, T. II-268
Kandel, O. I-657
Kanter, T. I-12
Kesidis, G. II-339
Khan, J.I. II-628
Khotimsky, D.A. I-137
Kim, C. II-61
Kim, K. I-117
Kim, M. II-786
Kim, S.-C. II-735
Kim, Y.-C. II-1
Kinicki, R. I-98
Klöcking, J.-U. I-498
Ko, J.-W. II-1

Kobayashi, T.   I-167
Konstantinopoulou, C.   I-286
Koutsopoulos, K.   I-286
Kovacevic, V.   I-75
Kovačić, D.   II-399
Kuhlins, S.   II-590
Kung, H.-Y.   I-559
Kuo, S.-Y.   I-317
Kure, Ø.   I-488
Kwon, J. H.   II-409, II-726

Laalaoua, R.   I-107
Lac, C.   II-805
Lau, C.-T.   II-168
Lavery, B.   I-37
Leal, J.   II-449, II-549
Lee, B.-J.   II-1
Lee, B.-S.   II-168, II-380, II-815
Lee, C.L.   II-82
Lee, D.-E.   II-1
Lee, S.-H.   I-1
Lee, Y.   I-771
Leiss, E.L.   I-549
Li, X.   I-235
Liao, H.Y.   II-82
Liao, W.-H.   II-158
Libeskind-Hadas, R.   I-508
Lin, F.Y-.S.   II-148
Linwong, P.   II-845
Liu, M.   II-51
Lorenz, P.   II-422

Magoni, D.   I-762
Maher, M.   I-124
Maihöfer, C.   I-498
Makios, V.   II-497
Mammeri, Z.   II-422
Mandyam, G.D.   I-157
Maniatis, S.   II-108
Manjanatha, S.   I-388
Marbukh, V.   II-309
Marias, G.F.   I-439
Marques, P.   II-458
Marroun, E.   II-735
Masuyama, H.   II-877
Matsikoudis, E.   I-286
McClellan, S.   II-579
McManis, J.   I-803, I-821

Melhem, R.   I-508
Merabti, M.   II-507
Merakos, L.   I-439
Michalareas, T.   I-687
Mikou, N.   I-780
Milanovic, S.   II-717
Miller, L.K.   I-549
Miloslavski, A.   II-559
Mirchandani, V.   I-720
Miyoshi, T.   II-877
Mokhtar, H.M.   II-507
Mouchtaris, P.   I-117
Mouftah, H.T.   I-245
Mouharam, A.A.K.   II-658
Müller, F.   I-276
Muntean, G.-M.   I-540
Murphy, J.   I-821
Murphy, L.   I-540
Mutka, M.W.   I-640, II-71

Nakamura, T.   II-620
Naraghi-Poor, M.   I-124
Nemoto, Y.   II-845
Nguyen, H.N.   I-37
Nguyen, H.-T.   I-256
Nikolouzou, E.G.   I-266
Nilsson, A.A.   I-449

Ogino, N.   I-673
Oie, Y.   I-398
Oktug, S.F.   II-390
Oliveira, J.   I-27
Onozato, Y.   I-579
Owen, H.L.   II-137

Pagani, E.   I-468
Palola, M.   II-478
Pansiot, J.-J.   I-762
Papadakis, A.E.   II-488
Papadopoulos, A.   II-497
Parish, D.J.   II-766
Park, C.G.   II-517
Park, H.-J.   II-380
Park, J.S.   II-370
Park, J.-T.   II-409, II-726
Park, S.-W.   I-1
Perdikeas, M.K.   II-488
Pereira, R.   II-507

Petrovic, Z.   II-717
Pham, H.   I-37
Phan, S.   II-836
Pieprzyk, J.   II-239
Pils, C.   I-194
Pinto Lemos, R.   II-668
Pletka, R.   I-218
Pogosyants, G.   II-559
Pollak, A.   II-890
Popovic, M.   I-75
Potemans, J.   I-790
Protonotarios, E.N.   I-459
Protopsaltis, N.G.   I-459
Pyrovolakis, O.I.   II-488

Quiles, F.J.   II-358

Radhakrishnan, S.   I-184
Räisänen, V.   II-127
Rao, N.S.V.   I-184
Rapp, J.   I-727
Rashid, E.   II-620
Reeves, D.S.   I-409
Renno, J.R.   II-766
Reschke, D.   I-599
Rhee, S.H.   I-698
Rindborg, T.   I-12
Risso, F.   I-85
Roca, V.   I-610
Rockenbach Tarouco, L.M.   I-204, II-277
Rodriguez, J.   II-422
Roque, R.   I-27
Rossi, G.P.   I-468
Rothermel, K.   I-498
Roussev, B.   II-854
Russell, L.   II-71
Rzehak, H.   I-256

Sacks, L.   I-687
Sahlin, D.   I-12
Saleh, H.   I-811
Salvador, M.R.   I-326
Salvet, Z.   II-612
Sampatakos, P.D.   I-266
Samtani, S.   I-117
Sandhu, H.   II-735
Saydam, T.   II-329
Scharinger, J.   II-196

Schmid, A.   I-657
Schmid, O.   I-124
Sedillot, S.   I-27
Seet, B.-C.   II-168
Sekkaki, A.   II-438
Serhrouchni, A.   I-589
Shaikh, F.A.   II-579
Shamsuzzaman, M.   II-815
Shao, H.-R.   I-235
Sheu, J.-P.   II-158
Shinagawa, N.   I-167
Shiraishi, Y.   I-579
Shkrabov, S.   II-559
Sidhu, D.   II-302
Silva, H.J.A. da   I-358
Silva, J.   II-458
Silva, L.   II-458
Simões, P.   II-458
Singh, A.K.   II-217
Sivakumar, R.   I-569, II-259
Smith, H.   I-640
Soares, L.F.G.   II-867
Sokol, J.   II-137
Solís Barreto, P.   II-668
Steigner, C.   I-657
Stiller, B.   I-218
Storch, R.   I-204
Stranjak, A.   II-399
Stuckmann, P.   I-276
Sue, C.-C.   I-317
Sugar, R.   II-468
Sung, K.-Y.   I-530
Suzuki, M.   I-673
Szabo, S.   II-11

Takahashi, Y.   I-167
Talpade, R.   I-117
Tanaka, Y.   I-65
Tang, W.   II-71
Taniguchi, H.   II-877
Tassiulas, L.   II-339
Tatsuya Watanabe, W.   II-438
Templemore-Finlayson, J.   I-649
Ternero, J.A.   II-549
Teughels, M.   I-790
Theologou, M.E.   I-286, I-459
Theunis, J.   I-790
Tizraoui, A.   I-811
Touvet, F.   I-829

Tsang, D.H.K. I-45
Tsaur, W.-J. I-174
Tseng, Y.-C. II-158
Tsetsekas, C. II-108
Tsykin, M. II-291
Tunnicliffe, M.J. II-658, II-766
Tutsch, D. I-478

Umemoto, M. I-737
Urien, P. I-811

Van Lil, E. I-790
Venieris, I.S. I-266, II-108, II-488
Vergados, D.D. I-459
Vinyes, J. II-697
Vozmediano, J.M. II-549

Wallbaum, M. I-194
Wang, S.-L. II-158
Wehrle, K. II-117
Wendt, F. I-204
Wolfinger, B.E. II-754
Wong, J.W. II-639
Wong, L. I-117

Woo, M. II-41
Woodward, M.E. I-147
Wu, H. I-235
Wu, J. II-854
Wu, Z. II-639

Wuu, L.-C. II-225
Wysocki, T. II-186

Yalamanchili, S. II-358
Yamamoto, U. I-579
Yamori, K. I-65
Yang, C. II-836
Yang, J. I-753
Yang, L. I-640
Yasuda, K. II-177
Yegoshin, L. II-559
Yen, H.-H. II-148
Yoon, J. I-698
Yoshida, M. I-398
Yuan, X. II-677
Yucel, S. II-329

Zaballos, A. II-648
Zaddach, M. II-754
Žagar, M. II-399
Zambenedetti Granville, L. II-277
Zappala, D. I-620
Zhang, J. II-250
Zhang, X. I-429
Zheng, Z. I-98
Zhou, J. I-579
Zhuklinets, I.A. I-137
Zorn, W. I-743
Zugenmaier, A. II-20

Lecture Notes in Computer Science

For information about Vols. 1–2015
please contact your bookseller or Springer-Verlag

Vol. 2016: S. Murugesan, Y. Deshpande (Eds.), WebEngineering. IX, 357 pages. 2001.

Vol. 2018: M. Pollefeys, L. Van Gool, A. Zisserman, A. Fitzgibbon (Eds.), 3D Structure from Images – SMILE 2000. Proceedings, 2000. X, 243 pages. 2001.

Vol. 2019: P. Stone, T. Balch, G. Kraetzschmar (Eds.), RoboCup 2000: Robot Soccer World Cup IV. XVII, 658 pages. 2001. (Subseries LNAI).

Vol. 2020: D. Naccache (Ed.), Topics in Cryptology – CT-RSA 2001. Proceedings, 2001. XII, 473 pages. 2001

Vol. 2021: J. N. Oliveira, P. Zave (Eds.), FME 2001: Formal Methods for Increasing Software Productivity. Proceedings, 2001. XIII, 629 pages. 2001.

Vol. 2022: A. Romanovsky, C. Dony, J. Lindskov Knudsen, A. Tripathi (Eds.), Advances in Exception Handling Techniques. XII, 289 pages. 2001

Vol. 2024: H. Kuchen, K. Ueda (Eds.), Functional and Logic Programming. Proceedings, 2001. X, 391 pages. 2001.

Vol. 2025: M. Kaufmann, D. Wagner (Eds.), Drawing Graphs. XIV, 312 pages. 2001.

Vol. 2026: F. Müller (Ed.), High-Level Parallel Programming Models and Supportive Environments. Proceedings, 2001. IX, 137 pages. 2001.

Vol. 2027: R. Wilhelm (Ed.), Compiler Construction. Proceedings, 2001. XI, 371 pages. 2001.

Vol. 2028: D. Sands (Ed.), Programming Languages and Systems. Proceedings, 2001. XIII, 433 pages. 2001.

Vol. 2029: H. Hussmann (Ed.), Fundamental Approaches to Software Engineering. Proceedings, 2001. XIII, 349 pages. 2001.

Vol. 2030: F. Honsell, M. Miculan (Eds.), Foundations of Software Science and Computation Structures. Proceedings, 2001. XII, 413 pages. 2001.

Vol. 2031: T. Margaria, W. Yi (Eds.), Tools and Algorithms for the Construction and Analysis of Systems. Proceedings, 2001. XIV, 588 pages. 2001.

Vol. 2032: R. Klette, T. Huang, G. Gimel'farb (Eds.), Multi-Image Analysis. Proceedings, 2000. VIII, 289 pages. 2001.

Vol. 2033: J. Liu, Y. Ye (Eds.), E-Commerce Agents. VI, 347 pages. 2001. (Subseries LNAI).

Vol. 2034: M.D. Di Benedetto, A. Sangiovanni-Vincentelli (Eds.), Hybrid Systems: Computation and Control. Proceedings, 2001. XIV, 516 pages. 2001.

Vol. 2035: D. Cheung, G.J. Williams, Q. Li (Eds.), Advances in Knowledge Discovery and Data Mining – PAKDD 2001. Proceedings, 2001. XVIII, 596 pages. 2001. (Subseries LNAI).

Vol. 2037: E.J.W. Boers et al. (Eds.), Applications of Evolutionary Computing. Proceedings, 2001. XIII, 516 pages. 2001.

Vol. 2038: J. Miller, M. Tomassini, P.L. Lanzi, C. Ryan, A.G.B. Tettamanzi, W.B. Langdon (Eds.), Genetic Programming. Proceedings, 2001. XI, 384 pages. 2001.

Vol. 2039: M. Schumacher, Objective Coordination in Multi-Agent System Engineering. XIV, 149 pages. 2001. (Subseries LNAI).

Vol. 2040: W. Kou, Y. Yesha, C.J. Tan (Eds.), Electronic Commerce Technologies. Proceedings, 2001. X, 187 pages. 2001.

Vol. 2041: I. Attali, T. Jensen (Eds.), Java on Smart Cards: Programming and Security. Proceedings, 2000. X, 163 pages. 2001.

Vol. 2042: K.-K. Lau (Ed.), Logic Based Program Synthesis and Transformation. Proceedings, 2000. VIII, 183 pages. 2001.

Vol. 2043: D. Craeynest, A. Strohmeier (Eds.), Reliable Software Technologies – Ada-Europe 2001. Proceedings, 2001. XV, 405 pages. 2001.

Vol. 2044: S. Abramsky (Ed.), Typed Lambda Calculi and Applications. Proceedings, 2001. XI, 431 pages. 2001.

Vol. 2045: B. Pfitzmann (Ed.), Advances in Cryptology – EUROCRYPT 2001. Proceedings, 2001. XII, 545 pages. 2001.

Vol. 2047: R. Dumke, C. Rautenstrauch, A. Schmietendorf, A. Scholz (Eds.), Performance Engineering. XIV, 349 pages. 2001.

Vol. 2048: J. Pauli, Learning Based Robot Vision. IX, 288 pages. 2001.

Vol. 2051: A. Middeldorp (Ed.), Rewriting Techniques and Applications. Proceedings, 2001. XII, 363 pages. 2001.

Vol. 2052: V.I. Gorodetski, V.A. Skormin, L.J. Popyack (Eds.), Information Assurance in Computer Networks. Proceedings, 2001. XIII, 313 pages. 2001.

Vol. 2053: O. Danvy, A. Filinski (Eds.), Programs as Data Objects. Proceedings, 2001. VIII, 279 pages. 2001.

Vol. 2054: A. Condon, G. Rozenberg (Eds.), DNA Computing. Proceedings, 2000. X, 271 pages. 2001.

Vol. 2055: M. Margenstern, Y. Rogozhin (Eds.), Machines, Computations, and Universality. Proceedings, 2001. VIII, 321 pages. 2001.

Vol. 2056: E. Stroulia, S. Matwin (Eds.), Advances in Artificial Intelligence. Proceedings, 2001. XII, 366 pages. 2001. (Subseries LNAI).

Vol. 2057: M. Dwyer (Ed.), Model Checking Software. Proceedings, 2001. X, 313 pages. 2001.

Vol. 2059: C. Arcelli, L.P. Cordella, G. Sanniti di Baja (Eds.), Visual Form 2001. Proceedings, 2001. XIV, 799 pages. 2001.

Vol. 2060: T. Böhme, H. Unger (Eds.), Innovative Internet Computing Systems. Proceedings, 2001. VIII, 183 pages. 2001.

Vol. 2062: A. Nareyek, Constraint-Based Agents. XIV, 178 pages. 2001. (Subseries LNAI).

Vol. 2064: J. Blanck, V. Brattka, P. Hertling (Eds.), Computability and Complexity in Analysis. Proceedings, 2000. VIII, 395 pages. 2001.

Vol. 2065: H. Balster, B. de Brock, S. Conrad (Eds.), Database Schema Evolution and Meta-Modeling. Proceedings, 2000. X, 245 pages. 2001.

Vol. 2066: O. Gascuel, M.-F. Sagot (Eds.), Computational Biology. Proceedings, 2000. X, 165 pages. 2001.

Vol. 2068: K.R. Dittrich, A. Geppert, M.C. Norrie (Eds.), Advanced Information Systems Engineering. Proceedings, 2001. XII, 484 pages. 2001.

Vol. 2070: L. Monostori, J. Váncza, M. Ali (Eds.), Engineering of Intelligent Systems. Proceedings, 2001. XVIII, 951 pages. 2001. (Subseries LNAI).

Vol. 2071: R. Harper (Ed.), Types in Compilation. Proceedings, 2000. IX, 207 pages. 2001.

Vol. 2072: J. Lindskov Knudsen (Ed.), ECOOP 2001 – Object-Oriented Programming. Proceedings, 2001. XIII, 429 pages. 2001.

Vol. 2073: V.N. Alexandrov, J.J. Dongarra, B.A. Juliano, R.S. Renner, C.J.K. Tan (Eds.), Computational Science – ICCS 2001. Part I. Proceedings, 2001. XXVIII, 1306 pages. 2001.

Vol. 2074: V.N. Alexandrov, J.J. Dongarra, B.A. Juliano, R.S. Renner, C.J.K. Tan (Eds.), Computational Science – ICCS 2001. Part II. Proceedings, 2001. XXVIII, 1076 pages. 2001.

Vol. 2075: J.-M. Colom, M. Koutny (Eds.), Applications and Theory of Petri Nets 2001. Proceedings, 2001. XII, 403 pages. 2001.

Vol. 2076: F. Orejas, P.G. Spirakis, J. van Leeuwen (Eds.), Automata, Languages and Programming. Proceedings, 2001. XIV, 1083 pages. 2001.

Vol. 2077: V. Ambriola (Ed.), Software Process Technology. Proceedings, 2001. VIII, 247 pages. 2001.

Vol. 2078: R. Reed, J. Reed (Eds.), SDL 2001: Meeting UML. Proceedings, 2001. XI, 439 pages. 2001.

Vol. 2081: K. Aardal, B. Gerards (Eds.), Integer Programming and Combinatorial Optimization. Proceedings, 2001. XI, 423 pages. 2001.

Vol. 2082: M.F. Insana, R.M. Leahy (Eds.), Information Processing in Medical Imaging. Proceedings, 2001. XVI, 537 pages. 2001.

Vol. 2083: R. Goré, A. Leitsch, T. Nipkow (Eds.), Automated Reasoning. Proceedings, 2001. XV, 708 pages. 2001. (Subseries LNAI).

Vol. 2084: J. Mira, A. Prieto (Eds.), Connectionist Models of Neurons, Learning Processes, and Artificial Intelligence. Proceedings, 2001. Part I. XXVII, 836 pages. 2001.

Vol. 2085: J. Mira, A. Prieto (Eds.), Bio-Inspired Applications of Connectionism. Proceedings, 2001. Part II. XXVII, 848 pages. 2001.

Vol. 2086: M. Luck, V. Mařík, O. Štěpánková, R. Trappl (Eds.), Multi-Agent Systems and Applications. Proceedings, 2001. X, 437 pages. 2001. (Subseries LNAI).

Vol. 2089: A. Amir, G.M. Landau (Eds.), Combinatorial Pattern Matching. Proceedings, 2001. VIII, 273 pages. 2001.

Vol. 2091: J. Bigun, F. Smeraldi (Eds.), Audio- and Video-Based Biometric Person Authentication. Proceedings, 2001. XIII, 374 pages. 2001.

Vol. 2092: L. Wolf, D. Hutchison, R. Steinmetz (Eds.), Quality of Service – IWQoS 2001. Proceedings, 2001. XII, 435 pages. 2001.

Vol. 2093: P. Lorenz (Ed.), Networking – ICN 2001. Proceedings, 2001. Part I. XXV, 843 pages. 2001.

Vol. 2094: P. Lorenz (Ed.), Networking – ICN 2001. Proceedings, 2001. Part II. XXV, 899 pages. 2001.

Vol. 2095: B. Schiele, G. Sagerer (Eds.), Computer Vision Systems. Proceedings, 2001. X, 313 pages. 2001.

Vol. 2096: J. Kittler, F. Roli (Eds.), Multiple Classifier Systems. Proceedings, 2001. XII, 456 pages. 2001.

Vol. 2097: B. Read (Ed.), Advances in Databases. Proceedings, 2001. X, 219 pages. 2001.

Vol. 2098: J. Akiyama, M. Kano, M. Urabe (Eds.), Discrete and Computational Geometry. Proceedings, 2000. XI, 381 pages. 2001.

Vol. 2099: P. de Groote, G. Morrill, C. Retoré (Eds.), Logical Aspects of Computational Linguistics. Proceedings, 2001. VIII, 311 pages. 2001. (Subseries LNAI).

Vol. 2101: S. Quaglini, P. Barahona, S. Andreassen (Eds.), Artificial Intelligence in Medicine. Proceedings, 2001. XIV, 469 pages. 2001. (Subseries LNAI).

Vol. 2105: W. Kim, T.-W. Ling, Y-J. Lee, S.-S. Park (Eds.), The Human Society and the Internet. Proceedings, 2001. XVI, 470 pages. 2001.

Vol. 2106: M. Kerckhove (Ed.), Scale-Space and Morphology in Computer Vision. Proceedings, 2001. XI, 435 pages. 2001.

Vol. 2109: M. Bauer, P.J. Gymtrasiewicz, J. Vassileva (Eds.), User Modelind 2001. Proceedings, 2001. XIII, 318 pages. 2001. (Subseries LNAI).

Vol. 2110: B. Hertzberger, A. Hoekstra, R. Williams (Eds.), High-Performance Computing and Networking. Proceedings, 2001. XVII, 733 pages. 2001.

Vol. 2118: X.S. Wang, G. Yu, H. Lu (Eds.), Advances in Web-Age Information Management. Proceedings, 2001. XV, 418 pages. 2001.

Vol. 2119: V. Varadharajan, Y. Mu (Eds.), Information Security and Privacy. Proceedings, 2001. XI, 522 pages. 2001.

Vol. 2121: C.S. Jensen, M. Schneider, B. Seeger, V.J. Tsotras (Eds.), Advances in Spatial and Temporal Databases. Proceedings, 2001. XI, 543 pages. 2001.

Vol. 2126: P. Cousot (Ed.), Static Analysis. Proceedings, 2001. XI, 439 pages. 2001.

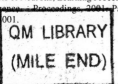